MULTICOLOUR EDITION

A Textbook of Engineering Mechanics

(SI Units)

Textbook for students of B.E., B.Tech., B.Sc. Engg., B.Arch., AMIE, Diploma Polytechnic, UPSC (IES)

As per the Latest Guidelines of AICTE (All India Council of Technical Education)

R.S. KHURMI
N. KHURMI

AF543515

S Chand And Company Limited
(ISO 9001 Certified Company)

S Chand And Company Limited

(ISO 9001 Certified Company)

Head Office: D-92, Sector–2, Noida – 201301, U.P. (India), Ph. 91-120-4682700

Registered Office: A-27, 2nd Floor, Mohan Co-operative Industrial Estate, New Delhi – 110 044, Phone: 011-49731800

www.**schandpublishing.com;** e-mail: **info@schandpublishing.com**

Marketing Offices:

Chennai : Ph: 23632120; chennai@schandpublishing.com
Guwahati : Ph: 2738811, 2735640; guwahati@schandpublishing.com
Hyderabad : Ph: 40186018; hyderabad@schandpublishing.com
Jalandhar : Ph: 4645630; jalandhar@schandpublishing.com
Kolkata : Ph: 23357458, 23353914; kolkata@schandpublishing.com
Lucknow : Ph: 4003633; lucknow@schandpublishing.com
Mumbai : Ph: 25000297; mumbai@schandpublishing.com
Patna : Ph: 2260011; patna@schandpublishing.com

© S Chand And Company Limited, 1967

All rights reserved. No part of this publication may be reproduced or copied in any material form (including photocopying or storing it in any medium in form of graphics, electronic or mechanical means and whether or not transient or incidental to some other use of this publication) without written permission of the copyright owner. Any breach of this will entail legal action and prosecution without further notice.

Jurisdiction: *All disputes with respect to this publication shall be subject to the jurisdiction of the Courts, Tribunals and Forums of New Delhi, India only.*

S. CHAND'S Seal of Trust

In our endeavour to protect you against counterfeit/fake books, we have pasted a hologram over the cover of this book. The hologram displays full visible effect, emboss effect, relief effect, mirror lens effect, pearl effect, motion effect, animated text, kinetic effect, concealed effect, micro structure, multicolour small text 'S.CHAND', nanotext '50 micron' 'ORIGINAL' 'S.CHAND', mirror strip '6.5 mm', mirror lens 3 mm with text 'SC', microtext 'OK', scratch strip '7 mm', color sparkling effect under scratch and QR code size 10 mm x 10 mm, etc.

A fake hologram does not display ALL these effects.

First Edition 1967

Subsequent Editions and Reprints of the Main Textbook 1968, 70, 71, 72, 73, 74, 75, 76 (Twice), 77 (Twice), 78, 79, 80, 81 (Twice) 82, 83 (Twice), 84 (Twice), 85 (Twice), 86 (Twice) 87 (Twice) 88, 89, 90, 91 (Twice) 93, 94, 95, 96, 97, 98, (Twice), 99, 2000, 2001, 2002, 2003, 2004, 2005, 2006, 2007, 2008, 2009 , 2010, 2011, 2012, 2013, 2014 (Twice), 2015

Concise Edition 2016

Revised Edition 2019; Reprints 2019, 2020 (Twice), 2021 (Thrice), 2022 (Twice)

Reprint 2023 (Twice)

ISBN: 978-93-528-3396-2 **Product Code:** H3EME62EMEC10ENAV19O

PRINTED IN INDIA

By Vikas Publishing House Private Limited, Plot 20/4, Site-IV, Industrial Area Sahibabad, Ghaziabad – 201 010 and Published by S Chand And Company Limited, A-27, 2nd Floor, Mohan Co-operative Industrial Estate, New Delhi – 110 044

To
My Revered Guru and Guide
Shree **B.L.Theraja**
A well-known author, among Engineering
students, both at home and abroad,
to whom I am ever indebted for
inspiration and guidance

Preface to The Twenty Second Edition

It gives a great pleasure in presenting the new version of twenty second edition of this popular book to innumerable students and academic staff of the Universities in India and Abroad. This book is **Recommended and Followed as Text-book cum Reference in many Syllabi**.

The favourable and warm reception, which our books have enjoyed all over India and Abroad, has been a matter of great satisfaction. The object of this book is to present the subject matter in most concise, compact and simple manner. To make the book useful at all levels the subject matter has been given in following manner.

1. This edition is in S.I. Units and adheres to latest trend of examinations.
2. Theory Articles are given in Logical Sequence to provide Conceptual Understanding of the subject.
3. Lucid presentation of Theorems, Hypotheses, Formulae and Working Equations.
4. Solved Numerical Examples collected from Examination Papers of various Indian and International Universities as well as Other Examining Bodies, have been Included after being Graded in Systematic Manner.
5. Unsolved Numerical Examples with Answers have been given for Practice.
6. Diagrams are given to Support the Theory and Quick Understanding.
7. Photographs are given to Correlate Subject Theory with Practical Reality.

Following New Chapters have been added:

1. **Forces in Space (In Vector Form)**
2. **Work Energy Method**
3. **Pappus—Guldinus Theorem**
4. **Product of Inertia**
5. **Rectangular and Cylindrical Coordinate System in Dynamics**
6. **Vibrations**

Sign of "* or **" is marked on Important and Very Important Articles and Examples inside the book. Correlation of Important Chapters & Articles with Syllabi of Various Exams is also given.

Any suggestions for the improvement of this volume can be sent at below mentioned email. The same shall be thankfully acknowledged and incorporated in the next edition.

R.S. KHURMI
N. KHURMI

E-mail : khurmieducation@yahoo.com
Website : www.khurmis.com

Preface to The First Edition

I take an opportunity to present this standard treatise entitled as A TEXTBOOK of ENGINEERIG MECHANICS to the Students of Degree, Diploma and A.M.I.E. (I) classes. This object of this book is to present the subject matter in a most concise, compact, to-the-point and lucid manner.

While writing this book, I have constantly kept in mind the requirements of all the students regarding the latest as well as the changing trends of their examination. To make it more useful, at all levels, the book has been written in an easy style. All along the approach to the subject matter, every care has been taken to arrange matter from simpler to harder, known to unknown with full details and illustrations. A large number of worked examples, mostly examination questions of Indian as well as foreign universities and professional examining bodies, have been given and graded in a systematic manner and logical sequence, to assist the students to understand the text of the subject. At the end of each chapter, a few exercises have been added, for the students, to solve them independently. Answers to these problems have been provided, but it is too much to hope that these are entirely free from errors. In short, it is expected that the book will embrace the requirements of the students, for which it has been designed.

Although every care has been taken to check mistakes and misprints, yet it is difficult to claim perfection. Any error, omission and suggestion for the improvement of this volume, brought to my notice, will be thankfully acknowledged and incorporated in the next edition.

Feb. 24, 1967 **R.S. KHURMI**

Disclaimer : While the authors of this book have made every effort to avoid any mistakes or omission and have used their skill, expertise and knowledge to the best of their capacity to provide accurate and updated information. The author and S. Chand does not give any representation or warranty with respect to the accuracy or completeness of the contents of this publication and are selling this publication on the condition and understanding that they shall not be made liable in any manner whatsoever. S.Chand and the author expressly disclaim all and any liability/responsibility to any person, whether a purchaser or reader of this publication or not, in respect of anything and everything forming part of the contents of this publication. S. Chand shall not be responsible for any errors, omissions or damages arising out of the use of the information contained in this publication.
Further, the appearance of the personal name, location, place and incidence, if any; in the illustrations used herein is purely coincidental and work of imagination. Thus the same should in no manner be termed as defamatory to any individual.

Acknowledgements

We are thankful to the following Indian and International Universities as well as Other Examining Bodies, whose syllabi have also been kept in view while writing this treatise. Moreover their examination questions on the subject have been included.

National Institutes of Technology (NIT):

Agartala, Allahabad, Bhopal, Calicut, Durgapur, Goa, Hamirpur, Jalandhar, Jaipur, Jammu, Jamshedpur, Kurukshetra, Nagpur, Patna, Raipur, Rourkela, Silchar, Srinagar, Surat, Surathkal, Trichy, Warangal.

Indian Institutes of Technology (IIT):

Delhi, Mumbai, Kanpur, Kharagpur, Roorkee, Guwahati, Chennai.

Technical Universities in India:

Guru Gobind Singh Indra Prastha University (GGSIPU) - New Delhi; Delhi Technological University (DTU) - Delhi; Jawaharlal Nehru Technological University (JNTU) - Anantapur, Hyderabad, Kakinada - Andhra Pradesh; Rajiv Gandhi University of Knowledge Technologies - Hyderabad - Andhara Pradesh; Mahamaya Technical University (MTU) - Noida - Uttar Pradesh; Gautam Buddh Technical University (GBTU) - Lucknow - Uttar Pradesh; Punjab Technical University (PTU) - Jalandhar - Punjab; Aryabhatta Knowledge University - Patna - Bihar; Uttarakhand Technical University (UTU) - Dehradun - Uttarakhand; Visveswariah Technological University (VTU) - Bengaluru - Karnataka; Swami Vivekanand Technical University (SVTU) - Bhillai - Chhattisgarh; Biju Patnayak University of Technology (BPUT) - Rourkela - Orissa; Veer Surendra Sai University of Technology (VSSUT) - Burla - Orissa; Mahatama Gandhi University (MGU) - Kerala; Cochin University of Science and Technology (CUST) - Kerala; Anna University of Technology (AUT) - Chennai, Coimbatore, Madurai, Tiruchirappalli, Tirunelveli - Tamil Nadu; MDU, KU and GJU - Haryana; YMCA University - Faridabad - Haryana; Himachal Pradesh Technical University (HPTU) - Hamirpur - Himachal Pradesh; Rajiv Gandhi Proudyogiki Vishwavidyalaya (RGPV) - Bhopal - Madhya Pradesh; Rajasthan Technical University (RTU) - Rajasthan; Gujarat Technical University (GTU) - Gujarat; West Bengal University of Technology (WBUT) - Kolkata - West Bengal; Dr. Babasaheb Ambedkar Technological University - Lonere - Maharashtra; University of Pune (PU) - Maharashtra; Islamic University of Science and Technology - Pulwama - J&K; Shri Mata Vaishno Devi University - Katra - J&K;

Other Universities in India:

Delhi University; Punjab University - Chandigarh; Dr. Bhim Rao Ambedkar University - Agra - Uttar Pradesh; Allahabad University; Annamalai University; Banaras University; Baroda University; Babasaheb Bhimrao Ambedkar Bihar University, Central University of Bihar; Gauhati University; Gorakhpur University; Indore University; Jabalpur University; Jadavpur University; Jiwaji University; Jodhpur University; Kanpur University; Kolkota University; Madurai University; Mysore University; University of Mumbai; University of Pune; Nagpur University - Maharashtra; Mahatma Jyotiba Phule Rohilkhand University (MJPRU) - Bareilly - Uttar Pradesh; Osmania University; Patna University - Bihar; Pune University; Ranchi University - Jharkhand; Roorkee University;

Deemed and Private Universities in India:

Amity University; Bharati Vidyapeeth University - Pune; Manipal University; Lingaya's University; Manav Rachna International University; Birla Institute of Technology; Homi Bhabha National Institute; Thapar University; Chitkara University; Lovely University; Sharda University;

Institutions in India:

Indian Institute of Science - Bangalore; Birla Institute of Technology and Science - Pilani and Ranchi; College of Engineering - Anna University - Chennai; Dayalbagh Engineering Institute - Agra; Delhi Institute of Technology - Delhi; Delhi College of Engineering - Delhi; Faculty of Engineering and Technology - Jadavpur - Kolkata; Government Engineering College - Ujjain; Guindy Engineering College - Chennai; Harcourt Butler Technological Institute - Kanpur; Institute of Technology - BHU - Varanasi;

Jamia Millia Islamia - Jamia Nagar - New Delhi; M.S. University of Baroda - Vadodra; Madhav Institute of Science and Technology - Gwalior; Netaji Subhas Institute of Technology - Delhi; North Eastern Regional Institute of Science and Technology - Itanagar; Osmania University - Hyderabad; PSG College of Technology - Coimbatore; Punjab Engineering College - Chandigarh; S.V. Univ. College of Engineering - Tirupati; Thapar Institute of Engg. and Technology - Patiala; Vellore Institute of Technology - Vellore; Victoria Jubilee Technical Institute - Mumbai; Z.H. College of Engg. and Technology - AMU - Aligarh; Institute of Engineering and Technology (IET) - Khandari - Agra; Thangal Kunju Musaliar College of Engineering - Quilon;

Universities and Institutions Abroad:

USA - Massachusetts Institute of Technology - Boston; California Institute of Technology - California; Princeton University; Stanford University; University of Texas - Austin; University of Michigan; Ohio State University; Pennsylvania State University; Cornell University; University of Illinois at Urbana Champaign; Georgia Institute of Technology; Canada: University of Toronto; University of Waterloo; UK - Oxford University; Cambridge University; London University - London; Birmingham University; Imperial College of Science and Technology - London; Netherland - Delft University of Technology; Australia - University of Melbourne; University of Sydney; University of Queensland Australia; Japan - University of Tokyo; Korea - Korea Advanced Institute of Science and Technology; Pohang University of Science and Technology; China - Peking University; Singapore - National University of Singapore (NUS); Nanyang Technological University of Singapore (NTU); Hong Kong - Hong Kong University; Thailand - Asian Institute of Technology - Bangkok; Iraq - Baghdad University; University of Technology – Baghdad; University of Basrah; Foundation of Technical Institute; Mousl; Iran - Arya Mehar University of Technology - Tehran; Azarabadegan - Tabriz; Pakistan - Mehran University of Engineering and Technology - Nawabshah; Eritea - University of Asmara; Botswana - Gaborone; Egypt - University of Cairo; Tanzania - Dar-E-Salaam; Libya - Garyounis - Benghazi; Nigeria - Ife-Ife; UAE - University of Dubai; Kuwait - University of Kuwait; Saudi Arabia - University of Riyadh; Sri Lanka - University of Colombo/Katubedda;

Examining Bodies:

Institution of Engineers - India (IEI) - AMIE Section A and B; Board of Technical Education of States conducting Diploma Engineering Courses (BTE); Institution of Electronics and Telecommunication Engineers - India (IETE) - AM-IETE and DIP-IETE.

Competitive and Recruitment Exams:

Graduate Aptitude Test in Engineering (GATE); U.P.S.C. (IES-Engineering Services and IAS-Administrative Services); Recruitment Boards;

R.S. KHURMI
N. KHURMI

CONTENTS

List of Symbols

Quantity	Symbol	Symbol Name	SI Units

Engineering Mechanics

Part 1

Quantity	Symbol	Symbol Name	SI Units
Length	L, l		m
Width	B, b		
Depth	D, d		
Height	H, h		m
Radius	R, r		
Diameter	D, d		
Area of cross section	A		
*Volume	V		m^3
Centre	O		
Centre of Gravity	G		
Centroid of Plane Area or Lamina	G		
Moment of Inertia	I		mm^4
Polar Moment of Inertia	J		mm^4
Radius of Gyration	k		
*Force, Pull, Push,	F, P		N
Resultant Force	R		
Component of Force along *XX* axis	P_X		
Component of Force along *YY* axis	P_Y		
Reaction or Reacting Force	R		
Vertical Reaction	R_A, R_B, R_C		
Horizontal Reaction	R_H		
Cartesian co-ordinates	$x, y\ z$		
Polar Co-ordinates	r, θ	(theta)	
Angles	$\theta, \alpha, \beta, \gamma$	(theta, alpha, beta, gamma)	rad
Section Modulus	Z		
Radius of Gyration	k		
Sum of Horizontal Forces	$\sum H$		
Sum of Vertical Forces	$\sum V$		
Sum of Moments	$\sum M$		
Co-efficient of Friction	μ	(mu)	
Efficiency	η	(neta)	
Mechanical Advantage	M.A.		
Velocity Ratio	V.R.		

Part 2

Quantity	Symbol	Symbol Name	SI Units
Distance	s, x		m
Time	T, t		s
*Velocity	v		m/s

Acceleration	a		
Horizontal Range of Projectile	R		N-m
Mass	M, m		kg
Impulse	I		
Momentum	M		
Speed	N		r.p.m.
Number	n		
Angular Velocity	ω	(omega)	rad/s
Angular Displacement	θ	(theta)	rad
Angular Acceleration	α	(alpha)	rad/s^2
Frequency	ν	(nu)	
Gauge	G		
Super-elevation	S		
Work	W		
*Power	P		
Energy	E		
Kinetic Energy	K.E., E_K		
Potential Enegy	P.E., E_P		
Stiffness of Spring	s		N/mm
Pitch	p		

CHAPTER

1 Introduction

Contents

1. SCIENCE

In this modern age, the word 'science' has got different meanings for different people. An ordinary man takes it as 'something' beyond his understanding, whereas others may take it as 'mysteries of research' which are understood only by a few persons working amidst complicated apparatus in a laboratory. A non-scientist feels that it is a 'subject' whose endeavour is aimed to improve the man's life on the earth. A business executive has the idea that it is 'something' which solves our day to day manufacturing and quality control problems, so that the nation's economic prosperity keeps on improving.

In fact, 'science' may be defined as the growth of ideas through observation and experimentation. In this sense, the subject of science does not, necessarily, has to contribute something to the welfare of the human life, although the man has received many benefits from the scientific investigations.

2. APPLIED SCIENCE

Strictly speaking, the world of science is so vast that the present day scientists and technologists have to group the various spheres of scientific activities according to some common characteristics to facilitate their training and research programmes. All these branches of science, still have the common principle of employing observation and experimentation. The branch of science, which co-ordinates the research work, for practical utility and services of the mankind, is known as Applied Science.

3. ENGINEERING MECHANICS

The subject of Engineering Mechanics is that branch of Applied Science, which deals with the laws and principles of Mechanics, alongwith their applications to engineering problems. As a matter of fact, knowledge of Engineering Mechanics is very essential for an engineer in planning, designing and construction of his various types of structures and machines. In order to take up his job more skilfully, an engineer must persue the study of Engineering Mechanics in a most systematic and scientific manner.

4. BEGINNING AND DEVELOPMENT OF ENGINEERING MECHANICS

It will be interesting to know, as to how the early man had been curious to know about the different processes going on the earth. In fact, he used to content himself, by holding gods responsible for all the processes. For a long time, the man had been trying to improve his ways of working. The first step, in this direction, was the discovery of a circular wheel, which led to the use of animal driven carts. The study of ancient civilization of Babylonians, Egyptians, Greeks and Roman reveal the use of water wheels and wind mills even during the pre-historic days.

It is believed that the word 'Mechanics' was coined by a Greek philosopher **Aristotle** (384–322 BC). He used this word for the problems of lever and the concept of centre of gravity. At that time, it included a few ideas, which were odd, unsystematic and based mostly on observations containing incomplete information. The first mathematical concept of this subject was developed by Archimedes (287–212 BC). The story, for the discovery of First Law of Hydrostatics, is very popular even today in the history of the development of Engineering Mechanics. In the normal course, Hieron king of Syracuse got a golden crown made for his use. He suspected that the crown has been made with an adultrated gold. The king liked the design of the crown so much that he did not want it to be melted,in order to check its purity. It is said that the king announced a huge reward for a person, who can check the purity of the crown gold without melting it. The legend goes that Archimedes, a pure mathematician, one day sitting in his bath room tub realised that if a body is immersed in water, its apparent weight is reduced. He thought that the apparent loss of weight of the immersed body is equal to the weight of the liquid displaced. It is believed that without further

Sir Issac Newton (1643–1727)

thought, **Archimedes** jumeped out of the bath tub and ran naked down the street shouting 'Eureka, eureka !' *i.e.* I have found it, I have found it !'

The subject did not receive any concrete contribution for nearly 1600 years. In 1325, Jean Buridan of Paris University proposed an idea that a body in motion possessed a certain impetus *i.e.* motion. In the period 1325–1350, a group of scientists led by the Thomas Bradwardene of Oxford University did lot of work on plane motion of bodies. Leonarodo Da Vinci (1452–1519), a great engineer and painter, gave many ideas in the study of mechanism, friction and motion of bodies on inclined planes. Galileo (1564–1642) established the theory of projectiles and gave a rudimentary idea of inertia. Huyghens (1629–1695) developed the analysis of motion of a pendulum.

As a matter of fact, scientific history of Engineering Mechanics starts with **Sir Issac Newton** (1643–1727). He introduced the concept of force and mass, and gave Laws of Motion in 1686. James Watt introduced the term horse power for comparing performance of his engines. John Bernoulli (1667–1748) enunciated the priciple of virtual work. In eighteenth century, the subject of Mechanics was termed as Newtonian Mechanics. A further development of the subject led to a controversy between those scientists who felt that the proper measure of force should be change in kinetic energy produced by it and those who preferred the change in momentum. In the nineteenth century, many scientists worked tirelessly and gave a no. of priciples, which enriched the scientific history of the subject.

In the early twentieth century, a new technique of research was pumped in all activities of science. It was based on the fact that progress in one branch of science, enriched most of the bordering branches of the same science or other sciences. Similarly with the passage of time, the concept of Engineering Mechanics aided by Mathematics and other physical sciences, started contributing and development of this subject gained new momentum in the second half of this century. Today, knowledge of Engineering Mechanics, coupled with the knowledge of other specialised subjects *e.g.* Calculus, Vector Algebra, Strength of Materials, Theory of Machines etc. has touched its present height. The knowledge of this subject is very essential for an engineer to enable him in designing his all types of structures and machines.

5. DIVISIONS OF ENGINEERING MECHANICS

The subject of Engineering Mechanics may be divided into the following two main groups:

1. **Statics** 2. **Dynamics.**

6. STATICS

It is that branch of Engineering Mechanics, which deals with the forces and their effects, while acting upon the bodies at rest.

7. DYNAMICS

It is that branch of Engineering Mechanics, which deals with the forces and their effects, while acting upon the bodies in motion. The subject of Dynamics may be further sub-divided into the following two branches :

1. **Kinematics.** 2. **Kinetics**

8. KINEMATICS

It is that branch of Dynamics, which deals with the bodies in motion, without any reference to the forces which are responsible for the motion.

9. KINETICS

It is the branch of Dynamics, which deals with the bodies in motion due to the application of forces.

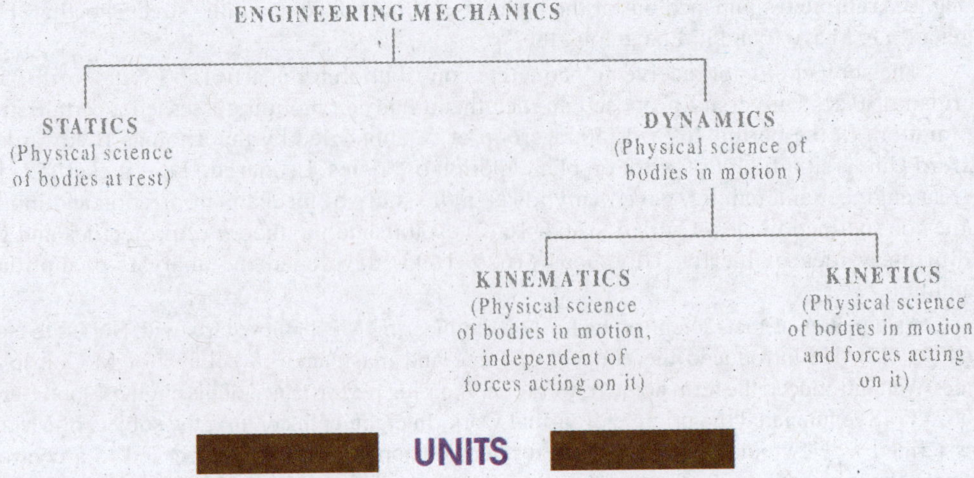

UNITS

10. FUNDAMENTAL UNITS

The measurement of physical quantities is one of the most important operations in engineering. Every quantity is measured in terms of some arbitrary, but internationally accepted units, called *fundamental units.*

All the physical quantities, met with in Engineering Mechanics, are expressed in terms of three fundamental quantities, *i.e.*

1. Length **2. Mass** **3. Time.**

11. DERIVED UNITS

Sometimes, the units are also expressed in other units (which are derived from fundamental units) known as *derived units e.g.* units of area, velocity, acceleration, pressure etc.

12. SYSTEMS OF UNITS

There are only four systems of units, which are commonly used and universally recognised. These are known as :

1. C.G.S. units **2. F.P.S. units** **3. M.K.S. units** **4. S.I. units.**

In this book, we shall use only the S.I. system of units, as the future courses of studies are conduced in this system of units only.

13. S.I. UNITS (INTERNATIONAL SYSTEM OF UNITS)

The eleventh *General Conference of Weights and Measures has recommended a unified and systematically constituted system of fundamental and derived units for international use. This system of units is now being used in many countries.

In India, the Standards of Weights and Measures Act of 1956 (vide which we switched over to M.K.S. units) has been revised to recognise all the S.I. units in industry and commerce.

In this system of units, the †fundamental units are metre (m), kilogram (kg) and second (s) respectively. But there is a slight variation in their derived units. The following derived units will be used in this book :

Density (or Mass density)	kg/m^3
Force (in Newtons)	N (= $kg\text{-}m/s^2$)
Pressure (in Pascals)	Pa (= $N/m^2 = 10^{-6}\ N/mm^2$)
Stress (in Pascals)	Pa (=$N/m^2 = 10^{-6}\ N/mm^2$)

Work done (in Joules) J (= N-m)
Power (in Watts) W (= J/s)

Note:

* General Conference of Weights and Measures (G.C.W.M.) is an international organisation of which most of the advanced and developing countries (including India) are members. This conference has been ensured the task of prescribing definitions of various units of weights and measures, which are the very basis of science and technology today.

† The other fundamental units are electric current, ampere (A), thermodynamic temperature, kelvin (K) and luminous intensity, candela (cd). These three units will not be used in this book.

14. METRE

The international metre may be defined as the shortest distance (at 0°C) between two parallel lines engraved upon the polished surface of the Platinum-Iridium bar, kept at the International Bureau of Weights and Measures at Sevres near Paris.

A bar of platinum - iridium metre kept at a temperature of 0º C.

The standard platinum - kilogram is kept at the International Bureau of Weights and Measures at Serves in France.

15. KILOGRAM

The international kilogram may be defined as the mass of the Platinum-Iridium cylinder, which is also kept at the International Bureau of Weights and Measures at Sevres near Paris.

16. SECOND

The fundamental unit of time for all the four systems is second, which is $1/(24 \times 60 \times 60) = 1/86\ 400$th of the mean solar day. A solar day may be defined as the interval of time between the instants at which the sun crosses the meridian on two consecutive days. This value varies throughout the year. The average of all the solar days, of one year, is called the mean solar day.

17. PRESENTATION OF UNITS AND THEIR VALUES

The frequent changes in the present day life are facililtated by an international body known as International Standard Organisation (ISO). The main function of this body is to make recommendations regarding international procedures. The implementation of ISO recommendations in a country is assisted by an organisation appointed for the purpose. In India, Bureau of Indian Standard formerly known as Indian Standards Institution (ISI) has been created for this purpose.

We have already discussed in the previous articles the units of length, mass and time. It is always necessary to express all lengths in metres, all masses in kilograms and all time in seconds. According to convenience, we also use larger multiples or smaller fractions of these units. As a typical example, although metre is the unit of length; yet a smaller length equal to one-thousandth of a metre proves to be more convenient unit especially in the dimensioning of drawings. Such convenient units are formed by using a prefix in front of the basic units to indicate the multiplier.

The full list of these prefixes is given in Table 1.1.

Table 1.1

Factor by which the unit is multiplied	*Standard form*	*Prefix*	*Abbreviation*
1000 000 000 000	10^{12}	Tera	T
1 000 000 000	10^{9}	giga	G
1 000 000	10^{6}	mega	M
1 000	10^{3}	kilo	k
100	10^{2}	hecto*	h
10	10^{1}	deca*	da
0.1	10^{-1}	deci*	d
0.01	10^{-2}	centi*	c
0.001	10^{-3}	milli	m
0.000 001	10^{-6}	micro	μ
0.000 000 001	10^{-9}	nano	n
0.000 000 000 001	10^{-12}	pico	p

Note : These prefixes are generally becoming obsolete probably due to possible confusion. Moreover, it is becoming a conventional practice to use only those powers of ten, which conform to $(10)^{3n}$ (where n is a positive or negative whole number).

18. RULES FOR S.I. UNITS

The Eleventh General Conference of Weights and Measures recommended only the fundamental and derived units of S.I. system. But it did not elaborate the rules for the usage of these units. Later on, many scientists and engineers held a no. of meetings for the style and usage of S.I. units. Some of the decisions of these meetings are :

1. A dash is to be used to separate units, which are multiplied together. For example, a newton-meter is written as N-m. It should no be confused with mN, which stands for millinewton.
2. For numbers having 5 or more digits, the digits should be placed in groups of three separated by spaces (instead of *commas) counting both to the left and right of the decimal point.
3. In a †four digit number, the space is not required unless the four digit number is used in a column of numbers with 5 or more digits.

At the time of revising this book, the author sought the advice of various international authorities regarding the use of units and their values, keeping in view the global reputation of the author as well as his books. It was then decided to ††present the units and their values as per the recommendations of ISO and ISI. It was decided to use :

4500	not	4 500	or	4,500
7 589 000	not	7589000	or	7,589,000
0.012 55	not	0.01255	or	.01255
30×10^6	not	3,00,00,000	or	3×10^7

Note:

* In certain countries, comma is still used as the decimal marker.

† In certain countries, space is used even in a four digit number.

†† In some question papers, standard values are not used. The author has tried to avoid such questions in the text of the book, in order to avoid possible confusion. But at certain places, such questions have been included keeping in view the importance of question from the reader's angle.

The above mentioned figures are meant for numerical values only. Now we shall discuss about the units. We know that the fundamental units in S.I. system for length, mass and time are metre, kilogram and second respectively. While expressing these quantities, we find it time-consuming to write these units such as metres, kilograms and seconds, in full, every time we use them. As a result of this, we find it quite convenient to use the following standard abberviations, which are internationally recognised. We shall use :

m	for metre or metres
km	for kilometre or kilometres
kg	for kilogram or kilograms
t	for tonne or tonnes
s	for second or seconds
min	for minute or minutes
N	for newton or newtons
N-m	for newton × metres (*i.e.*, work done)
kN-m	for kilonewton × metres
rad	for radian or radians
rev	for revolution or revolutions

MATHEMATICAL REVIEW

19. USEFUL DATA

The following data summarises the previous memory and formulae, the knowledge of which is very essential at this stage.

20. ALGEBRA

1. $a^0 = 1$ and $x^0 = 1$
(*i.e., Anything raised to the power zero is one.*)
2. $x^m \times x^n = x^{m+n}$
(*i.e., If the bases are same, in multiplication, the powers are added.*)
3. $\dfrac{x^m}{x^n} = x^{m-n}$
(*i.e., If the bases are same in division, the powers are subtracted.*)
4. If $ax^2 + bx + c = 0$
then $x = \dfrac{-b \pm \sqrt{b^2 - 4ac}}{2a}$
where a is the coefficient of x^2, b is the coefficient of x and c is the constant term.

21. TRIGONOMETRY

In a right-angled triangle *ABC* as shown in Fig. 1

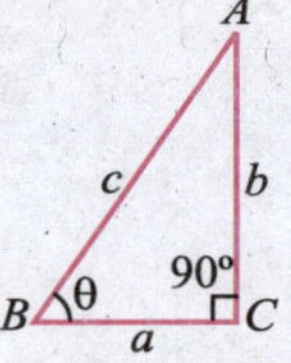

Fig. 1.

1. $\dfrac{b}{c} = \sin\theta$
2. $\dfrac{a}{c} = \cos\theta$
3. $\dfrac{b}{a} = \dfrac{\sin\theta}{\cos\theta} = \tan\theta$
4. $\dfrac{c}{b} = \dfrac{1}{\sin\theta} = \text{cosec}\,\theta$
5. $\dfrac{c}{a} = \dfrac{1}{\cos\theta} = \sec\theta$
6. $\dfrac{a}{b} = \dfrac{\cos\theta}{\sin\theta} = \dfrac{1}{\tan\theta} = \cot\theta$
7. The following table shows values of trigonometrical functions for some typical angles:

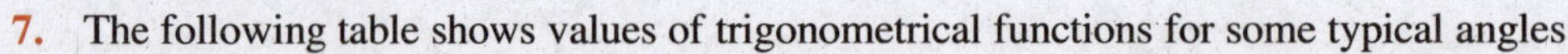

angle	0°	30°	45°	60°	90°
sin	0	$\frac{1}{2}$	$\frac{1}{\sqrt{2}}$	$\frac{\sqrt{3}}{2}$	1
cos	1	$\frac{\sqrt{3}}{2}$	$\frac{1}{\sqrt{2}}$	$\frac{1}{2}$	0
tan	0	$\frac{1}{\sqrt{3}}$	1	$\sqrt{3}$	∞

or in other words, for sin write

0°	30°	45°	60°	90°
$\frac{\sqrt{0}}{2}$	$\frac{\sqrt{1}}{2}$	$\frac{\sqrt{2}}{2}$	$\frac{\sqrt{3}}{2}$	$\frac{\sqrt{4}}{2}$
0	$\frac{1}{2}$	$\frac{1}{\sqrt{2}}$	$\frac{\sqrt{3}}{2}$	1

for cos write the values in reverse order ; for tan divide the value of sin by cos for the respective angle.

8. In the first quadrant (*i.e.*, 0° to 90°) all the trigonometrical ratios are positive.
9. In the second quadrant (*i.e.*, 90° to 180°) only sin θ and cosec θ are positive.
10. In the third quadrant (*i.e.*, 180° to 270°) only tan θ and cot θ are positive.
11. In the fourth quadrant (*i.e.*, 270° to 360°) only cos θ and sec θ are positive.
12. In any triangle *ABC*,

$$\frac{a}{\sin A} = \frac{b}{\sin B} = \frac{c}{\sin C}$$

where *a*, *b* and *c* are the lengths of the three sides of a triangle. *A*, *B* and *C* are opposite angles of the sides *a*, *b* and *c* respectively.

13. $\sin (A + B) = \sin A \cos B + \cos A \sin B$
14. $\sin (A - B) = \sin A \cos B - \cos A \sin B$
15. $\cos (A + B) = \cos A \cos B - \sin A \sin B$
16. $\cos (A - B) = \cos A \cos B + \sin A \sin B$
17. $\tan (A+B) = \dfrac{\tan A + \tan B}{1 - \tan A . \tan B}$
18. $\tan (A-B) = \dfrac{\tan A - \tan B}{1 + \tan A . \tan B}$
19. $\sin 2A = 2 \sin A \cos A$
20. $\sin^2 \theta + \cos^2 \theta = 1$
21. $1 + \tan^2 \theta = \sec^2 \theta$
22. $1 + \cot^2 \theta = \text{cosec}^2 \theta$
23. $\sin^2 A = \dfrac{1 - \cos 2A}{2}$
24. $\cos^2 A = \dfrac{1 + \cos 2A}{2}$
25. $2 \cos A \sin B = \sin (A + B) - \sin (A - B)$
26. Rules for the change of trigonometrical ratios:

$$(A) \begin{cases} \sin (-\theta) & = -\sin \theta \\ \cos (-\theta) & = \cos \theta \\ \tan (-\theta) & = -\tan \theta \\ \cot (-\theta) & = -\cot \theta \\ \sec (-\theta) & = \sec \theta \\ \text{cosec} (-\theta) & = -\text{cosec}\, \theta \end{cases}$$

$$(B)\begin{cases}\sin(90^\circ-\theta) &= \cos\theta\\ \cos(90^\circ-\theta) &= \sin\theta\\ \tan(90^\circ-\theta) &= \cot\theta\\ \cot(90^\circ-\theta) &= \tan\theta\\ \sec(90^\circ-\theta) &= \operatorname{cosec}\theta\\ \operatorname{cosec}(90^\circ-\theta) &= \sec\theta\end{cases}$$

$$(C)\begin{cases}\sin(90^\circ+\theta) &= \cos\theta\\ \cos(90^\circ+\theta) &= -\sin\theta\\ \tan(90^\circ+\theta) &= -\cot\theta\\ \cot(90^\circ+\theta) &= -\tan\theta\\ \sec(90^\circ+\theta) &= -\operatorname{cosec}\theta\\ \operatorname{cosec}(90^\circ+\theta) &= \sec\theta\end{cases}$$

$$(D)\begin{cases}\sin(180^\circ-\theta) &= \sin\theta\\ \cos(180^\circ-\theta) &= -\cos\theta\\ \tan(180^\circ-\theta) &= -\tan\theta\\ \cot(180^\circ-\theta) &= -\cot\theta\\ \sec(180^\circ-\theta) &= -\sec\theta\\ \operatorname{cosec}(180^\circ-\theta) &= \operatorname{cosec}\theta\end{cases}$$

$$(E)\begin{cases}\sin(180^\circ+\theta) &= -\sin\theta\\ \cos(180^\circ+\theta) &= -\cos\theta\\ \tan(180^\circ+\theta) &= \tan\theta\\ \cot(180^\circ+\theta) &= \cot\theta\\ \sec(180^\circ+\theta) &= -\sec\theta\\ \operatorname{cosec}(180^\circ+\theta) &= -\operatorname{cosec}\theta\end{cases}$$

Following are the rules to remember the above 30 formulae :

Rule 1. Trigonometrical ratio changes only when the angle is $(90^\circ - \theta)$or $(90^\circ + \theta)$. In all other cases, trigonometrical ratio remains the same.
Following is the law of change :
sin changes into cos and cos changes into sin,
tan changes into cot and cot changes into tan,
sec changes into cosec and cosec changes into sec.

Rule 2. Consider the angle θ to be a small angle and write the proper sign as per formulae 8 to 11 above.

22. DIFFERENTIAL CALCULUS

1. $\dfrac{d}{dx}$ ... is the sign of differentiation.

2. $\dfrac{d}{dx}(x)^n = nx^{n-1}$ ...$\left(e.g.\ \dfrac{d}{dx}(x)^8 = 8x^7 \text{ and } \dfrac{d}{dx}(x) = 1\right)$

(i.e., to differentiate any power of x, write the power before x and subtract on from the power).

3. $\frac{d}{dx}(C) = 0 \quad ...\left(e.g. \quad \frac{d}{dx}(7) = 0 \right)$

(i.e., differential coefficient of a constant is zero).

4. $\frac{d}{dx}(u.v) = u.\frac{dv}{dx} + v.\frac{du}{dx}$

$$\begin{bmatrix} \textit{i.e., Differential coefficient} \\ \textit{of product of any} \\ \textit{two functions} \end{bmatrix} = \begin{bmatrix} \textit{(1st function} \times \textit{Differential coefficient of second function)} \\ + \textit{(2nd function} \times \textit{Differential coefficient of first function)} \end{bmatrix}$$

5. $\frac{d}{dx}\left(\frac{u}{v}\right) = \frac{v.\frac{du}{dx} - u.\frac{dv}{dx}}{v^2}$

$$\begin{bmatrix} \textit{i.e., Differential coefficient of} \\ \textit{two functions when one is} \\ \textit{divided by the other} \end{bmatrix} = \begin{bmatrix} \dfrac{\textit{(Denominator} \times \textit{Differential coefficient of numerator)} - \textit{(Numerator} \times \textit{Differential coefficient of denominator)}}{\textit{Square of denominator}} \end{bmatrix}$$

6. Differential coefficient of trigonometrical functions

$\frac{d}{dx}(\sin x) = \cos x$ and $\frac{d}{dx}(\cos x) = -\sin x$

$\frac{d}{dx}(\tan x) = \sec^2 x$ and $\frac{d}{dx}(\cot x) = -\text{cosec}^2 x$

$\frac{d}{dx}(\sec x) = \sec x \,.\, \tan x$ and $\frac{d}{dx}(\text{cosec}\, x) = -\text{cosec}\, x \,.\, \cot x$

(i.e., The differential coefficient, whose trigonometrical function begins with co, is negative).

7. If the differential coefficient of a function is zero, the function is either maximum or minimum. *Conversely*, if the maximum or minimum value of a function is required, then differentiate the function and equate it to zero.

23. INTEGRAL CALCULUS

1. $\int dx$ ... is the sign of integration.

2. $\int x^n \, dx = \frac{x^{n+1}}{n+1} \quad ...\left(e.g. \quad \int x^6 \, dx = \frac{x^7}{7} \right)$

(i.e., to integration any power of x, add one to the power and divide by the new power).

3. $\int C \, dx = Cx \quad ...\left(e.g. \quad \int 7dx = 7x \right)$

(i.e., to integrate any constant, multiply the constant by x).

4. $\int (ax+b)^n \, dx = \frac{(ax+b)^{n+1}}{(n+1) \times a}$.

(i.e., to integrate any bracket with power, add one to the power and divide by the new power and also divide by the coefficient of x within the bracket).

VECTORS

24. SCALAR QUANTITIES

The scalar quantities (or sometimes known as scalars) are those quantities which have magnitude only such as length, mass, time, distance, volume, density, temperature, speed etc.

25. VECTOR QUANTITIES

The vector quantities (or sometimes known as vectors) are those quantities which have both magnitude and direction such as force, displacement, velocity, acceleration, momentum etc.

Following are the important features of vector quantities :

1. ***Representation of a vector.*** A vector is represented by a directed line as shown in Fig. 2. It may be noted that the length OA represents the magnitude of the vector $\overrightarrow{OA}$. The direction of the vector is $\overrightarrow{OA}$ is from O (*i.e.*, starting point) to A (*i.e.*, end point). It is also known as vector P.

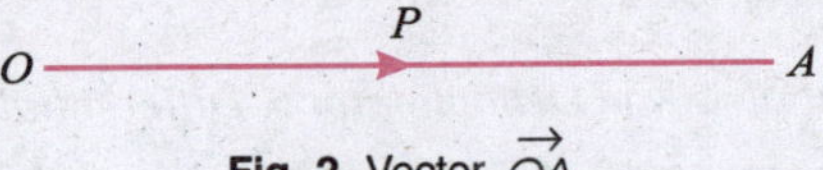

Fig. 2. Vector $\overrightarrow{OA}$

2. ***Unit vector.*** A vector, whose magnitude is unity, is known as unit vector.
3. ***Equal vectors.*** The vectors, which are parallel to each other and have same direction (*i.e.*, same sense) and equal magnitude are known as equal vectors.
4. ***Like vectors.*** The vectors, which are parallel to each other and have same sense but unequal magnitude, are known as like vectors.
5. ***Addition of vectors.*** Consider two vectors PQ and RS, which are required to be added as shown in Fig. 3. (a).

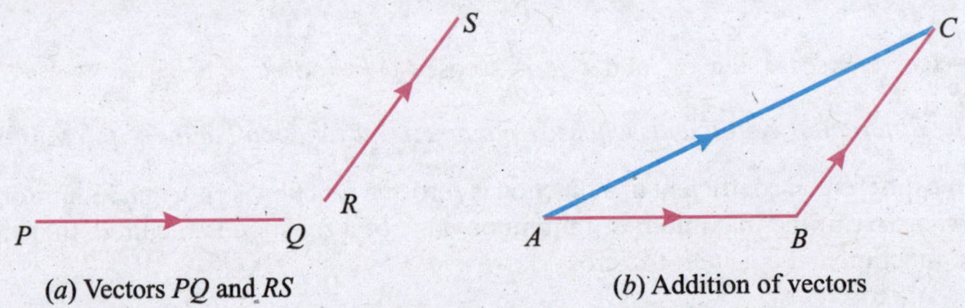

(a) Vectors PQ and RS (b) Addition of vectors

Fig. 3.

Take a point A, and draw line AB parallel and equal in magnitude to the vector PQ to some convenient scale. Through B, draw BC parallel and equal to vector RS to the same scale. Join AC which will give the required sum of vectors PQ and RS as shown in Fig. 3. (b).

This method of adding the two vectors is called the Triangle Law of Addition of Vectors. Similarly, if more than two vectors are to be added, the same may be done first by adding the two vectors, and then by adding the third vector to the resultant of the first two and so on. This method of adding more than two vectors is called Polygon Law of Addition of Vectors.

6. ***Subtraction of vectors.*** Consider two vectors PQ and RS in which the vector RS is required to be subtracted as shown in Fig. 4 (a)

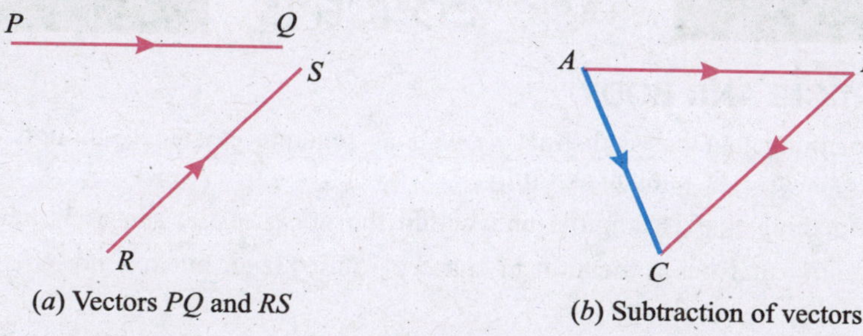

(*a*) Vectors *PQ* and *RS*

(*b*) Subtraction of vectors

Fig. 4.

Take a point *A*, and draw line *AB* parallel and equal in magnitude to the vector *PQ* to some convenient scale. Through *B*, draw *BC* parallel and equal to the vector *RS*, but in *opposite direction*, to that of the vector *RS* to the same scale. Join *AC*, which will give the resultant when the vector *PQ* is subtracted from vector *RS* as shown in Fig. 4 (*b*).

BASIC CONCEPTS

MATTER, PARTICLE AND BODY

Entity. Something that has a distinct as well as separate existence, though it need not be a material existence is known as *entity*.

Object. Something that is tangible and within the grasp of the senses is known as *object*.

Point. An entity that has a location in space or on a plane, but has no extent is known as a *point*.

Matter. Anything that a occupies space, possesses mass and offers resistance to any external force is known as *matter*.

Example: Iron, stone, wood salt, water and air *etc.*, represent matter.

Particle. An object that has infinitely small volume (occupies negligible space) but has a mass which can be considered to be concentrated at a point, is known as *particle*.

A body whose dimensions are practically negligible is called a *particle*. In any problem of mechanics, when the applied forces have no tendency to rotate the body on which they act, the body may be considered as a particle. Forces acting on the particle are concurrent, the point through which they pass being the point representing the particle.

Body. Anything that has a definite shape and consists of numbers of particles is known as *body*.

Rigid Body. Body which does not change its shape and size under the effect of forces acting over it is known as *rigid body*. A body in which the distance between any two particles remains constant, *i.e.*, the size and shape of the body do not change, is called a *rigid body*. A rigid body exhibits no relative deformation between two parts when it is acted upon by a force system.

Deformable Body. A body which undergoes deformation under the effect of forces acting on, is known an *deformable body*.

Elastic Body. A body which undergoes deformation under the effect of forces acting on it and returns to its original shape and size on removal of the forces acting on the body, is known an *elastic body*.

The rigidity of a body depends upon the fact that how far it undergoes deformaion under the effect of forces acting on it.

In real sense no solid body is perfectly rigid because every body changes its size and shape under the effect of forces acting on it. In actual practice, the deformation (i.e. change in shape and size of a body under the effect of forces acting on it) is very small and therefore it may be considered as a rigid body.

SPACE, MOTION AND TRAJECTORY

Space. A region which extends in all directions and contains every thing in it, is known as *space* Example: sun, moon, stars etc. In space position of a body is located with respect to a reference system. The position of an aircraft in space is found with respect to the earth. On earth, position of a body is determined with respect to a permanent station on earth by taking measurements.

Motion. When a body changes its position with respect to other bodies, then body is said to be in *motion.* In other words, relative change in position is *motion.* It involves both space and time.

Trajectory. The path followed by a body during its motion is known as *trajectory.* It may be a straight line or a curve. Straight line path is observed in *rectilinear motion,* whereas curved path is observed in *curvilinear motion.*

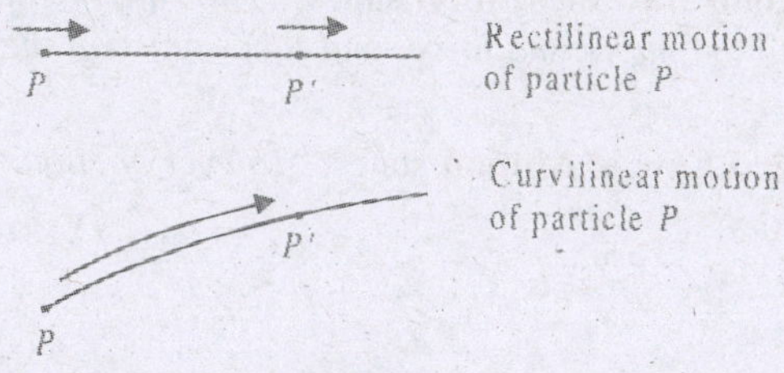

Fig. 5.

MASS AND WEIGHT

Mass. The amount of matter contained in a body is known as its *mass.* For most problems in mechanics for a given body, the mass may be considered constant.

The SI unit for mass is the kg.

Mass is a scalar quantity.

Weight. The force with which a body is attracted towards the centre of earth by the gravitational pull is known as its *weight.*

Mathematically,

The weight (w) of a body is given as,

$$w = mg$$

where (m) is the mass of the same body and (g) is the acceleration due to gravity.

The value of g is taken as 9.81 (usually) or 9.80 (to make the calculation work easier) in M.K.S. as well as in S.I. units.

The SI unit of weight is the N (newton).

Difference between Mass and Weight

Mass	*Weight*
1. It is the quantity of matter contained in a body.	**1.** It is the force with which the body is attracted towards the centre of earth.
2. It is constant at all places.	**2.** It is different at differentplaces.
3. It resists motion in the body.	**3.** It produces motion in the body.
4. It is a scalar quanity since it has magnitude only.	**4.** It is a vector quantity since it has magnitude as well as direction.
5. It can be measured by an mass balance. (consisting of a beam, fulcrum and two pans.)	**5.** It is measured by a spring balance. (consisting of a spring and a hook.)
6. It is never zero.	**6.** It is zero at the centre of earth.
7. It is measured is kilogram (kg) in M.K.S. units as well as in S.I. units.	**7.** It is measured in kilogram-weight (kg-wt) in M.K.S. units & in newton (N) in S.I.units.

NEWTON'S LAWS OF MOTION (STATEMENT)

Following are the three laws of motion, which were enunciated by Newton, who is regarded as father of the Science.

1. **Newton's First Law of Motion** states, "*Everybody continues in its state of rest or of uniform motion, in a straight line, unless it is acted upon by some external force.*"
2. **Newton's Second Law of Motion** states, "*The rate of change of momentum is directly proportional to the impressed force, and takes place in the same direction, in which the force acts.*"
3. **Newton's Third Law of Motion** states, "*To every action, there is always an equal and opposite reaction.*"

CHAPTER

2 Forces

Contents

1. Force

The **force** is defined as an agent which produces or tends to produce, destroys or tends to destroy motion.

Note: It may be noted that the force may have either of the two functions *i.e.*, produces or tends to produce motion. The second part of the definition is an application of the first part. In statics, we consider the second function of the force only *i.e.*, '*tends to produce motion*'.

Examples:

1. A horse applies force to pull a cart and to set it in motion.

2. The force is supplied by the muscular power

of our arms and shoulders to work on a air pump to fill air in tyre.

3. Sometimes, the applied force may not be sufficient to move a body, *e.g.*, if we try to lift a stone weighing 2 or 3 quintals, we fail to do so. In this case we exert a force, no doubt, but no motion is produced. This shows that a force may not necessarily produce a motion in a body ; but it may, simply, tend to do so.

4. In a tug-of-war the two parties, when balanced, neutralize each other's force. But the moment one party gets weaker, the other party pulls off, in spite of first party's best effort to destroy motion.

2. TYPES AND CLASSIFICATION OF FORCES

Forces can be classified into following types based on different criteria. Some of the important classifications are given as under:

1. According to the effect produced by the force

A force may produce the following **effects on the body, on which it acts** :

(*i*) It may give rise to the reactions at point of supports on the body, on which it acts. It may develop tension or compression in the strings attached to the body. We shall study these effects in chapter of "Equlibrium of Forces" of this book.

(*ii*) It may give rise to the internal stresses in the body, on which it acts. We shall study this effect in chapter of "Analysis of Perfect Frames" of this book.

(*iii*) It may change the motion of a body, on which it acts.
If a body is at rest, the force may set it in motion.
If the body is already in motion, the force may accelerate it or retard it or bring it to rest.

The above mentioned classifications are important from the scope of the subject.

But for extra information you may also read the classifications given below.

2. According to the origin of the force

(*i*) **External force.** The force which is *applied externally* to a body, is known as an *external force.*

(*ii*) **Internal force.** The force *developed inside* the body material to the resistance to deformation or change of shape, is known as an *internal force.*

(*iii*) **Active force.** A force which *causes* the motion of a body or change of shape of a body, is known as *active force.*

(*iv*) **Passive force.** A force which *prevents* the motion of a body or deformation of a body, is known as *passive force.*

3. According to nature of the force

(*i*) **Action and Reaction.** Whenever there are two bodies in contact, each exerts a force on the other. Out of these forces one is called action and other is called *reaction. Action and reaction are equal and opposite.*

(*ii*) **Attraction and Repulsion.** These are actually non-contacting forces exerted by one body on another body without any visible medium of transmission. Example: magnetic forces.

(*iii*) **Tension and Thrust.** When a body is dragged with a string the forces communicated to the body by the string is known as *tension*. If we push the body with a rod, the force exerted on the body is known as *thrust.*

4. According to whether the force acts at a point or is distributed over a large area

(*i*) **Concentrated force.** The force whose point of application is so small that it may be considered as a point is known a *concentrated force.*

(*ii*) **Distributed force.** The force whose place of application is area, is known as *distributed force*.

5. **According to whether the force acts at a distance or by contact**

(*i*) **Non-contacting forces or forces at a distance.** Example: Magnetic, electrical and gravitational forces.

(*ii*) **Contacting forces or forces by contact.** Example: The pressure of steam in a cylinder and pressure of the wheels of a locomotive on the rails.

3. CHARACTERISTICS AND REPRESENTATION OF A FORCE

CHARACTERISTICS OF A FORCE

[DTU, GTU, BPUT]

In order to determine the effects of a force, acting on a body, we must know the following characteristics of a force :

1. Magnitude of the force (*i.e.*, 100 N, 50 N, 20 kN, 5 kN, etc.)
2. The direction of the line, along which the force acts (*i.e.*, along *OX*, *OY*, at 30° North of East etc.). It is also known as line of action of the force.
3. Nature of the force (*i.e.*, whether the force is push or pull). This is denoted by placing an arrow head on the line of action of the force.
4. The point at which (or through which) the force acts on the body.

Characteristics of a force are shown in the Fig. 1.

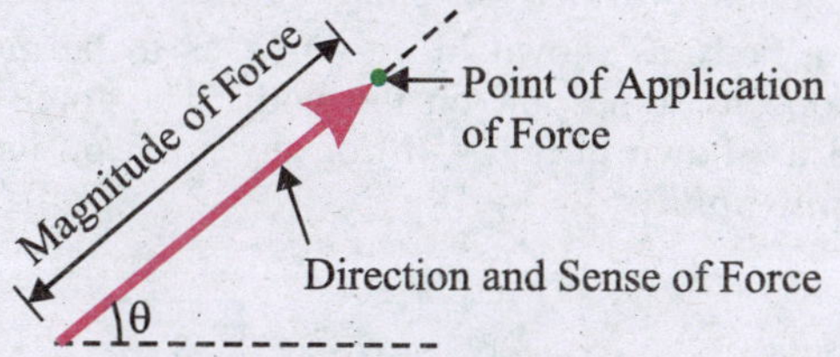

Fig. 1. Characteristics of a Force

REPRESENTATION OF A FORCE

[RGPV]

Forces may be representated in the following two ways:

1. Vector representation 2. Bow's notation.

Vector representation. It is a method of representing a force graphically by a vector. In Fig. 2, the Force *P* (6 N) is represented by vector $\overrightarrow{AB}$.

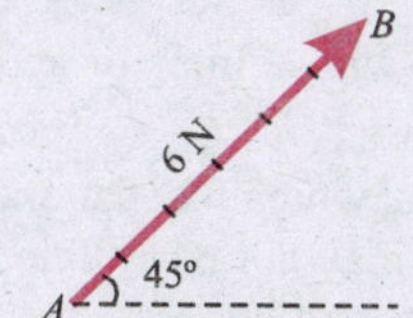

Fig. 2. Vector Representation of a Force

Bow's notation. It is a method of representing a force by writing two capital letters one on either side of the force. In Fig. 3, the force P_1 (20 kN) is represented by *AB* and the force P_2 (10 kN) by *CD*.

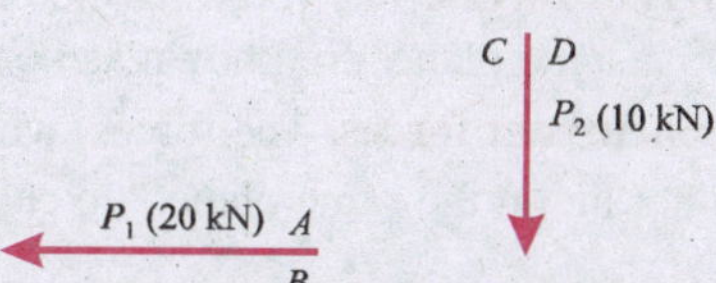

Fig. 3. Bow's Notation of Force

3.1. PARTICLE AND RIGID BODY

[WBUT]

It is important to take a review of the following two terms before we proceed.

* **Particle.** A particle may be defined as a body of infinitely small volume and is considered to be concentrated point.

** **Rigid body.** A rigid body may be defined as a body which can retain its shape and size, even if subjected to some external forces. Therefore in a rigid body there is no relative movement between the various points of the body and distance between any two points remains same at all times.

In actual practice, no body is perfectly rigid. But for the sake of simplicity, we take all the bodies as rigid bodies.

4. *PRINCIPLE OF PHYSICAL INDEPENDENCE OF FORCES

[UPTU, JNTU, AUT]

It states, "*If a number of forces are simultaneously acting on a *particle, then the resultant of these forces will have the same effect as produced by all the forces.*"

5. *PRINCIPLE OF TRANSMISSIBILITY OF FORCES

It states, "*If a force acts at any point on a **rigid body, it may also be considered to act at any other point on its line of action, provided this point is rigidly connected with the body.*"

Example. Suppose a body as shown in Fig. 4-(a) is to be moved by a horizontal force P applied by hooking a rope to some pint on the body. The force P will have the same effect if it is applied at 1, 2, 3 as shown in Fig. 4-(b) or any point on its line of action. This property of force is called *transmissibility*.

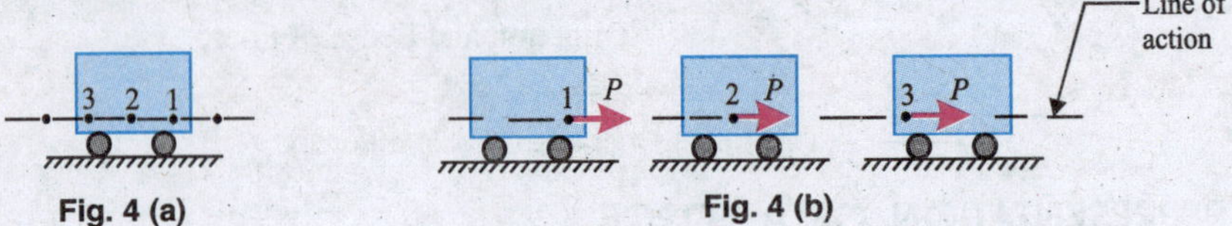

Fig. 4 (a) Fig. 4 (b)

6. *SYSTEMS OF FORCES

When two or more forces act on a body, they are called to form a *system of forces*. Following systems of forces are important from the subject point of view :

1. **Coplanar forces.** The forces, whose lines of action lie on the same plane, are known as coplanar forces.
2. **Collinear forces.** The forces, whose lines of action lie on the same line, are known as collinear forces.
3. **Concurrent forces.** The forces, which meet at one point, are known as concurrent forces. The concurrent forces may or may not be collinear.
4. **Coplanar Concurrent forces.** The forces, which meet at one point and their lines of action also lie on the same plane, are known as coplanar concurrent forces.
5. **Coplanar Non-Concurrent forces.** The forces, which do not meet at one point, but their lines of action lie on the same plane, are known as coplanar non-concurrent forces.
6. **Non-Coplanar Concurrent forces.** The forces, which meet at one point, but their lines of action do not lie on the same plane, are known as non-coplanar concurrent forces.
7. **Non-Coplanar Non-Concurrent forces.** The forces, which do not meet at one point and their lines of action do not lie on the same plane, are called non-coplanar non-concurrent forces.

The diagramatic representation of the different systems of forces is shown in Fig. 5.

Systems of Forces

According to Positions of Lines of Actions of Forces, the Forces may be classified into as follows:

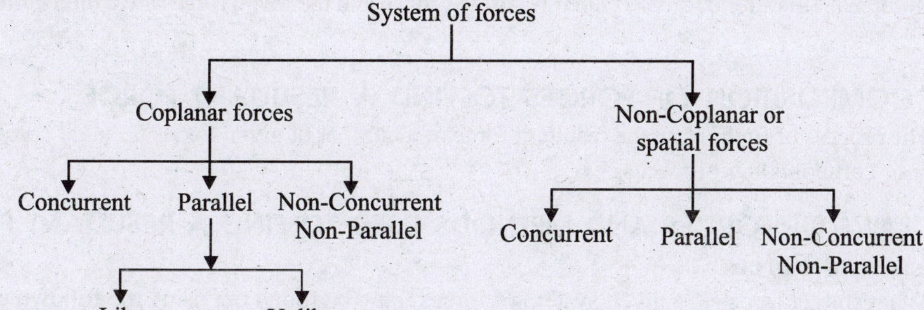

Collinear force system

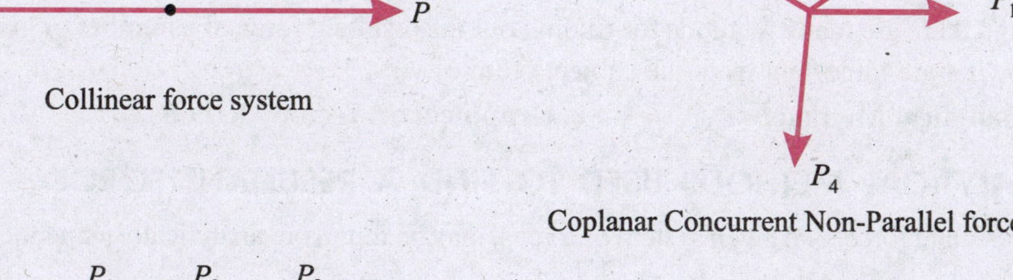

Coplanar Concurrent Non-Parallel force system

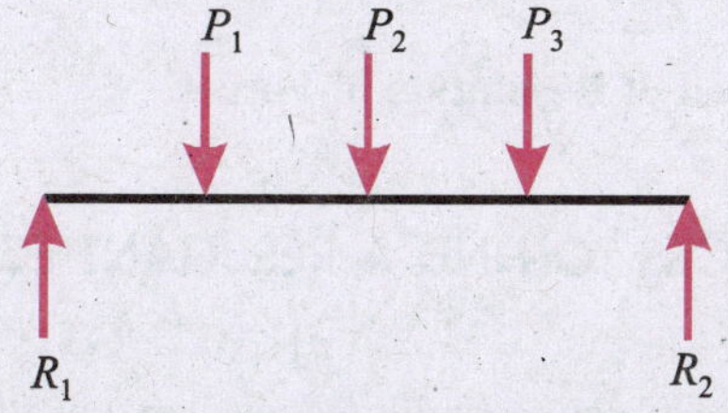

Coplanar Non-Concurrent Parallel force system

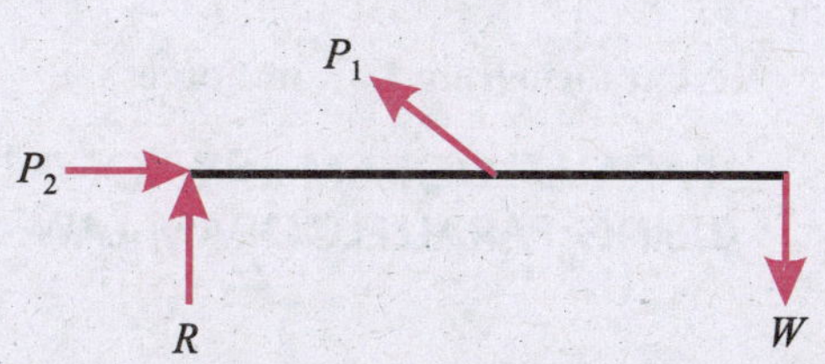

Coplanar Non-Concurrent Non-Parallel force system

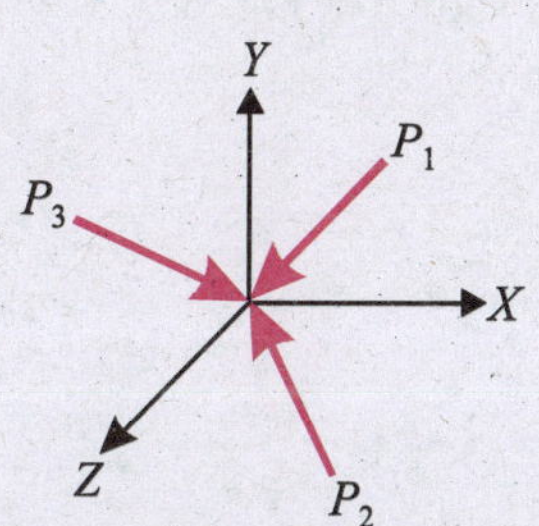

Non-Coplanar Concurrent forces system

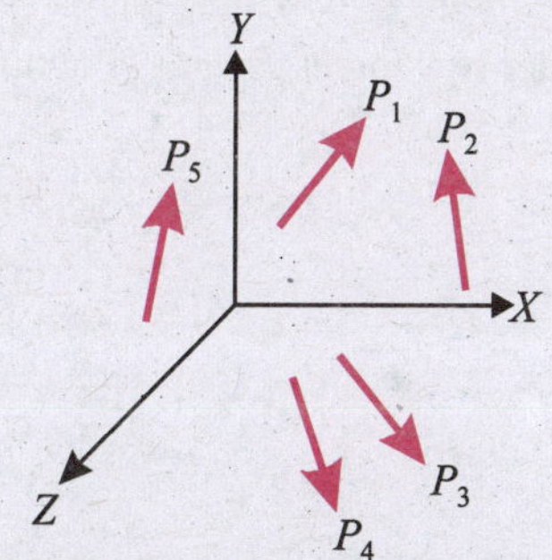

Non-Coplanar Non-Concurrent force system

Fig. 5.

7. *RESULTANT FORCE

If a number of forces are acting simultaneously on a particle, then it is possible to find out a single force which could replace them *i.e.*, which would produce the same effect as produced by all the given forces. This single force is called **resultant force** and the given forces are called **component forces**.

8. *COMPOSITION OF FORCES TO FIND A RESULTANT FORCE

The process of finding out the resultant force of a number of given forces is called *composition of forces* or compounding of forces.

9. *LAWS OF FORCES AND METHODS USED TO FIND A RESULTANT FORCE

Laws of Forces

The resultant force, of a given system of forces, may be found out using the following laws.

1. Parallelogram Law of Forces.
2. Triangle Law of Forces.
3. Polygon Law of Forces.

We shall discuss these laws in detail at respective places of their use in this chapter.

Methods Used To Find the Resultant Force

Though there are many methods for finding out the resultant force of a number of given forces, yet the following are important from the subject point of view :

1. **Analytical Methods.**
2. **Graphical or Vector Methods.**

10. ANALYTICAL METHODS USED TO FIND A RESULTANT FORCE

The resultant force, of a given system of forces, may be found out analytically using the following methods :

1. **Parallelogram Law of Forces.**
2. **Method of Resolution of Forces.**

11. *PARALLELOGRAM LAW OF FORCES (USING PARALLELOGRAM LAW OF FORCES TO FIND A RESULTANT FORCE)

[UPTU, GTU, AUT]

Parallelogram Law of Forces states that, "*If two forces, acting simultaneously on a particle, be represented in magnitude and direction by the two adjacent sides of a parallelogram, then their resultant may be represented in magnitude and direction by the diagonal of the parallelogram, which passes through their point of intersection.*"

Consider two forces F_1 and F_2 acting at same point O as shown in Fig. 6.

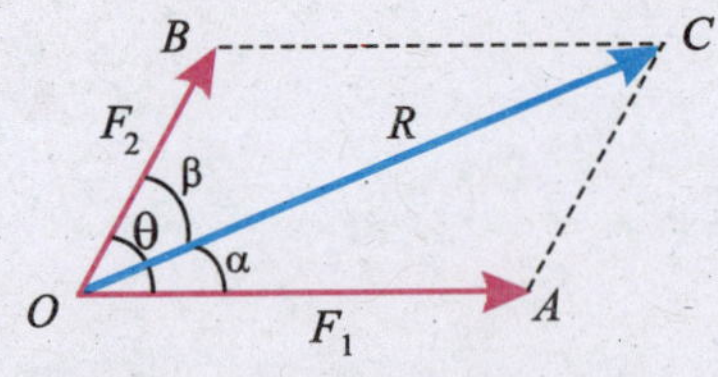

Fig. 6.

Let F_1 and F_2 = Forces whose resultant is required to be found out

R = Resultant of forces F_1 and F_2

θ = Angle between the forces F_1 and F_2

α = Angle which the resultant force R makes with one force F_1

β = Angle which the resultant force R makes with the other force F_2.

Now according to the Parallelogram Law of Forces, the Resulant (R) is given as,

$$R = \sqrt{F_1^2 + F_2^2 + 2F_1F_2 \cos\theta}$$

The Angles α and β can be determined from the relations,

$$\tan\alpha = \frac{F_2 \sin\theta}{F_1 + F_2 \cos\theta}$$

and
$$\tan\beta = \frac{F_1 \sin\theta}{F_2 + F_1 \cos\theta}$$

Important Results:

1. If $\theta = 0$ *i.e.*, when the forces act along the same line, then

$$R = F_1 + F_2 \qquad \text{...(Since } \cos 0° = 1)$$

2. If $\theta = 90°$ *i.e.*, when the forces act at right angle, then

$$R = \sqrt{F_1^2 + F_2^2} \qquad \text{...(Since } \cos 90° = 0)$$

3. If $\theta = 180°$ *i.e.*, when the forces act along the same straight line but in opposite directions, then $R = F_1 - F_2$ ...(Since $\cos 180° = -1$)

   In this case, the resultant force will act in the direction of the greater force.

4. If the two forces are equal *i.e.*, when $F_1 = F_2$

   then
$$R = \sqrt{F^2 + F^2 + 2F^2 \cos\theta} = \sqrt{2F^2(1+\cos\theta)}$$
$$= \sqrt{2F^2 \times 2\cos^2\left(\frac{\theta}{2}\right)} \qquad \text{...}\left[\because\ 1+\cos\theta = 2\cos^2\left(\frac{\theta}{2}\right)\right]$$
$$= \sqrt{4F^2 \cos^2\left(\frac{\theta}{2}\right)} = 2F\cos\left(\frac{\theta}{2}\right)$$

Example 1. *Two forces of 100 N and 150 N are acting simultaneously at a point. What is the resultant of these two forces, if the angle between them is 45°?*

Solution. Given : First force (F_1) = 100 N; Second force (F_2) = 150 N and angle between F_1 and F_2 (θ) = 45°.

We know that, the Resultant Force,

$$R = \sqrt{F_1^2 + F_2^2 + 2F_1F_2\cos\theta}$$
$$= \sqrt{(100)^2 + (150)^2 + 2 \times 100 \times 150 \cos 45°} \text{ N}$$
$$= \sqrt{10\,000 + 22\,500 + (30\,000 \times 0.707)} \text{ N}$$
$$= 232 \text{ N} \quad \textbf{Ans.}$$

Example 2. *Two forces act at an angle of 120°. The bigger force is of 40 N and the resultant is perpendicular to the smaller one. Find the smaller force.*

Solution. Given : Angle between the forces $\angle AOC = 120°$, Bigger force $(F_1) = 40$ N and angle between the resultant and F_2 ($\angle BOC$) $= 90°$;

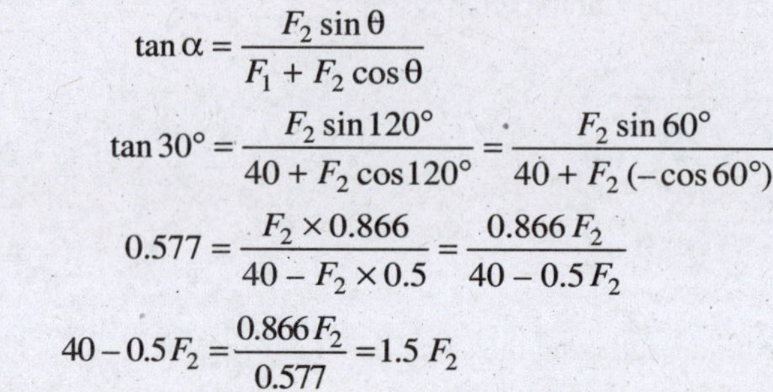

Fig. 7.

Let F_2 = Smaller force in N

From the geometry of the figure, we find that $\angle AOB$,

$$\alpha = 120° - 90° = 30°$$

We know that

$$\tan\alpha = \frac{F_2 \sin\theta}{F_1 + F_2\cos\theta}$$

or
$$\tan 30° = \frac{F_2 \sin 120°}{40 + F_2 \cos 120°} = \frac{F_2 \sin 60°}{40 + F_2(-\cos 60°)}$$

or
$$0.577 = \frac{F_2 \times 0.866}{40 - F_2 \times 0.5} = \frac{0.866 F_2}{40 - 0.5 F_2}$$

or
$$40 - 0.5F_2 = \frac{0.866 F_2}{0.577} = 1.5\, F_2$$

or
$$2F_2 = 40 \qquad \text{or} \qquad F_2 = 20 \quad \textbf{Ans.}$$

Example 3. *Find the magnitude of the two forces, such that if they act at right angles, their resultant is $\sqrt{10}$ N. But if they Act at 60°, their resultant is $\sqrt{13}$ N.*

Solution. Given : Two forces = F_1 and F_2.

First of all, consider the two forces acting at right angles. We know that when the angle between the two given forces is 90°, then the resultant force (*R*)

$$\sqrt{10} = \sqrt{F_1^2 + F_2^2}$$

or
$$10 = F_1^2 + F_2^2 \qquad \text{...(Squaring both sides)}$$

Similarly, when the angle between the two forces is 60°, then the resultant force (*R*)

$$\sqrt{13} = \sqrt{F_1^2 + F_2^2 + 2F_1F_2\cos 60°}$$

or
$$13 = F_1^2 + F_2^2 + 2F_1F_2 \times 0.5 \qquad \text{...(Squaring both sides)}$$

or
$$F_1F_2 = 13 - 10 = 3 \qquad \text{...(Substituting } F_1^2 + F_2^2 = 10)$$

We know that, $(F_1 + F_{2]})^2 = F_1^2 + F_2^2 + 2F_1F_2 = 10 + 6 = 16$

or
$$F_1 + F_2 = \sqrt{16} = 4 \qquad ...(i)$$

Similarly, $(F_1 - F_2)^2 = F_1^2 + F_2^2 - 2F_1F_2 = 10 - 6 = 4$

or
$$F_1 - F_2 = \sqrt{4} = 2 \qquad ...(ii)$$

Solving equations (*i*) and (*ii*),

$$F_1 = 3 \text{ N} \quad \text{and} \quad F_2 = 1 \text{ N} \quad \textbf{Ans.}$$

12. RESOLUTION OF A FORCE INTO COMPONENT FORCES

The process of splitting up the given force into two components, without changing its effect on the body is called **resolution of a force**.

The force which is split into two parts is called the **resolved force** and the parts are called **component forces** or **resolutes**.

In fact, the resolution of a force is the reverse action of the addition of the components. Generally a force is resolved into the following two types of components:

1. Multually perpendicular components **2.** Non-perpendicular components.

1. Multually perpendicular components.

Let the force F (to be resolved) be represented in magnitude and direction by OC as shown in Fig. 8. Let F_X be the component of force F in direction OA making an angle θ with the direction OC of the force. Complete the rectangle $OACB$. Then the other component F_Y at right angle to F_X will be represented by OB which is also equal to AC.

From the right-angled triangle OAC

$$F_X = OA = P\cos\theta$$
$$F_Y = AC = OB = P\sin\theta$$

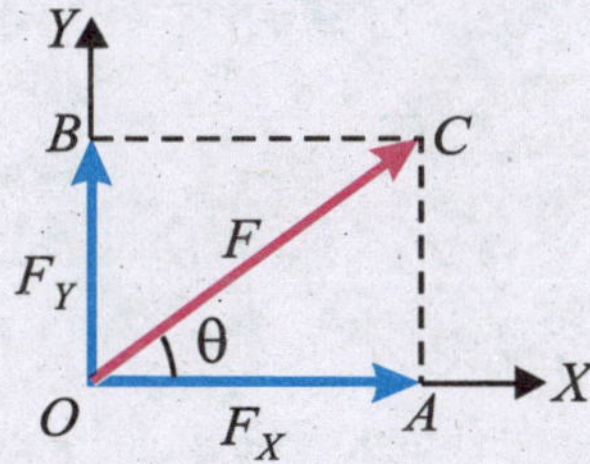

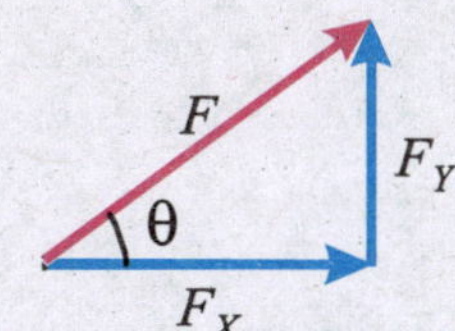

Fig. 8.

Note: The force F can be represented in vector form as $\vec{F} = \vec{F_X} \cdot \vec{i} + \vec{F_Y} \cdot \vec{j}$

... (where $\vec{i}$ and $\vec{j}$ are unit vectors.)

2. Non-perpendicular components.

Let the force F (to be resolved) be represented in magnitude and direction by OC as shown in Fig. 9. Draw OA and OB making angle α and β with OC. Through C draw CA parallel to OB and CB parallel to OA to complete the parallelogram $OACB$. Then the vector OA and OB represent in magnitude and direction (to the same scale) the components F_1 and F_2 respectively.

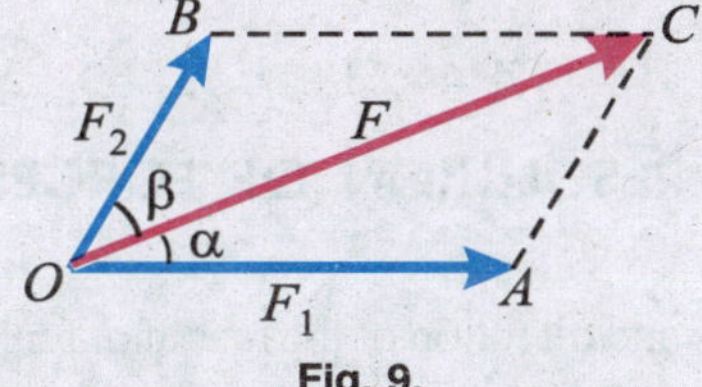

Fig. 9.

Now, from the triangle OAC, by applying sine rule,

$$\frac{OA}{\sin\beta} = \frac{OC}{\sin[180 - (\alpha + \beta)]} = \frac{AC}{\sin\alpha}$$

or $$\frac{F_1}{\sin\beta} = \frac{F}{\sin(\alpha+\beta)} = \frac{F_2}{\sin\alpha}$$

$\therefore$ $$F_1 = F \times \frac{\sin\beta}{\sin(\alpha+\beta)}$$

and $$F_2 = F \times \frac{\sin\alpha}{\sin(\alpha+\beta)}$$

13. PRINCIPLE OF RESOLUTION OF FORCES (OR PRINCIPLE OF RESOLVED PARTS)

It states, "*The algebraic sum of the resolved parts of a number of forces, in a given direction, is equal to the resolved parts of their resultant in the same direction.*"

Note : In general, the forces are resolved in the vertical and horizontal directions.

Example 4. *A machine component 1.5 m long and weight 1000 N is supported by two ropes AB and CD as shown in Fig. 10 given below. Calculate the tensions T_1 and T_2 in the ropes AB and CD.*

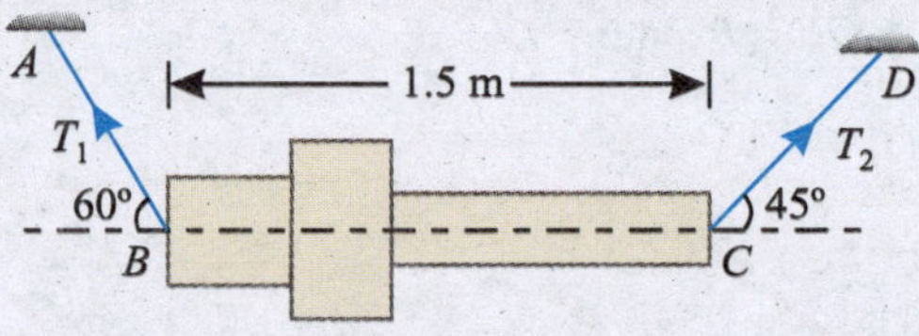

Fig. 10.

Solution. Given : Weight of the component = 1000 N

Resolving the forces horizontally (*i.e.*, along *BC*) and equating the forces acting in left direction with forces acting in right direction,

$$T_1 \cos 60° = T_2 \cos 45°$$

or $$T_1 = \frac{\cos 45°}{\cos 60°} \times T_2 = \frac{0.707}{0.5} \times T_2 = 1.414\, T_2$$

Resolving the forces vertically and equating the forces acting in upward direction with forces acting in downward direction,

$$T_1 \sin 60° + T_2 \sin 45° = 1000$$

or $$(1.414\, T_2)0.866 + (T_2 \times 0.707) = 1000$$

or $$1.93\, T_2 = 1000$$

or $$T_2 = \frac{1000}{1.93} = 518.1 \text{N} \quad \textbf{Ans.}$$

and $$T_1 = 1.414 \times 518.1 = 732.6 \text{N} \quad \textbf{Ans.}$$

14. USING METHOD OF RESOLUTION OF FORCES TO FIND A RESULTANT FORCE

Let us find out the magnitude and direction of the resultant force using the Method of Resolution of Forces. It is done as discussed below :

1. Resolve all the forces horizontally and find the algebraic sum of all the horizontal components (*i.e.*, ΣH).
2. Resolve all the forces vertically and find the algebraic sum of all the vertical

components (*i.e.*, ΣV).

3. The resultant R of the given forces will be given by the equation :

$$R = \sqrt{(\Sigma H)^2 + (\Sigma V)^2}$$

4. The resultant force will be inclined at an angle θ, with the horizontal, such that

$$\tan\theta = \frac{\Sigma V}{\Sigma H}$$

Notes : The value of the angle θ will vary depending upon the values of ΣV and ΣH as discussed below :

1. When ΣV is +ve, the resultant makes an angle between 0° and 180°. But when ΣV is –ve, the resultant makes an angle between 180° and 360°.
2. When ΣH is +ve, the resultant makes an angle between 0° to 90° or 270° to 360°. But when ΣH is –ve, the resultant makes an angle between 90° to 270°.

Example 5. *A triangle ABC has its side AB = 40 mm along positive x-axis and side BC = 30 mm along positive y-axis. Three forces of 40 N, 50 N and 30 N act along the sides AB, BC and CA respectively. Determine magnitude of the resultant of such a system of forces.*

Solution. The system of given forces is shown in Fig. 11.

In triangle ABC the side AB is along x-axis, and the side BC is along y-axis, therefore it is a right-angled triangle.

Now, $$AC = \sqrt{AB^2 + BC^2} = \sqrt{(40)^2 + (30)^2} = 50 \text{ mm}$$

Therefore $$\sin\theta = \frac{30}{50} = 0.6$$

and $$\cos\theta = \frac{40}{50} = 0.8$$

Resolving all the forces horizontally (*i.e.*, along AB), and find the algebraic sum of all the horizontal components,

$$\Sigma H = 40 - 30\cos\theta$$
$$= 40 - (30 \times 0.8) = 16 \text{ N}$$

Fig. 11.

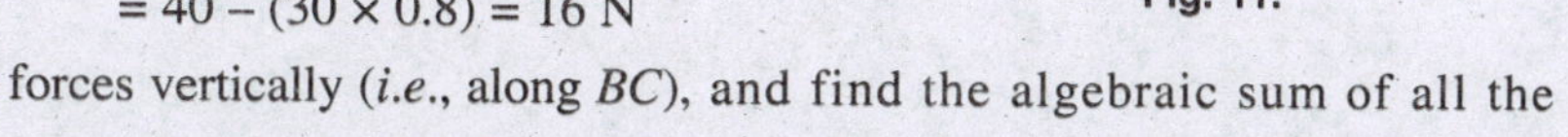

Resolving all the forces vertically (*i.e.*, along BC), and find the algebraic sum of all the vertical components,

$$\Sigma V = 50 - 30\sin\theta$$
$$= 50 - (30 \times 0.6) = 32 \text{ N}$$

We know that magnitude of the resultant force,

$$R = \sqrt{(\Sigma H)^2 + (\Sigma V)^2} = \sqrt{(16)^2 + (32)^2} = 35.8 \text{ N} \quad \textbf{Ans.}$$

Example 6. *A system of forces are acting at the corners of a rectangular block as shown in Fig. 12. Determine the magnitude and direction of the resultant force.*

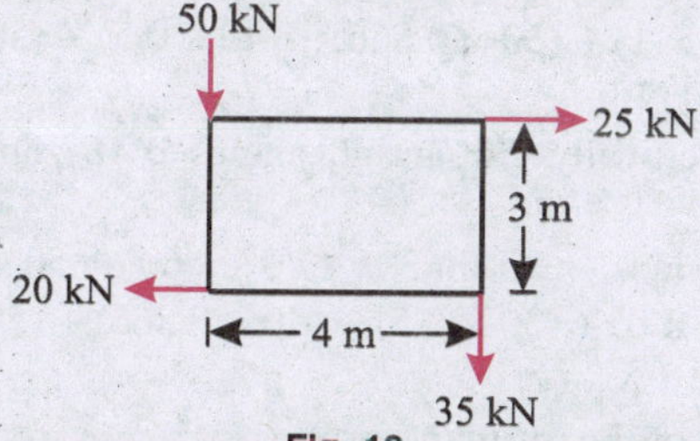

Fig. 12.

Solution. Given : System of forces

Magnitude of the resultant force

Resolving forces horizontally, and find the algebraic sum of all the horizontal components,

$$\Sigma H = 25 - 20 = 5 \text{ kN}$$

Resolving the forces vertically, and find the algebraic sum of all the vertical components,

$$\Sigma V = (-50) + (-35) = -85 \text{ kN}$$

∴ Magnitude of the resultant force,

$$R = \sqrt{(\Sigma H)^2 + (\Sigma V)^2} = \sqrt{(5)^2 + (-85)^2} = 85.15 \text{ kN} \quad \textbf{Ans.}$$

Direction of the resultant force

Let θ = Angle which the resultant force makes with the horizontal.

We know that

$$\tan\theta = \frac{\Sigma V}{\Sigma H} = \frac{-85}{5} = -17 \quad \text{or} \quad \theta = 86.6°$$

Since ΣH is positive and ΣV is negative, therefore resultant lies between 270° and 360°. Thus actual angle of the resultant force

$$= 360° - 86.6° = 273.4° \quad \textbf{Ans.}$$

Example 7. *The forces 20 N, 30 N, 40 N, 50 N and 60 N are acting at one of the angular points of a regular hexagon, towards the other five angular points, taken in order. Find the magnitude and direction of the resultant force.*

Solution. The system of given forces is shown in Fig. 13

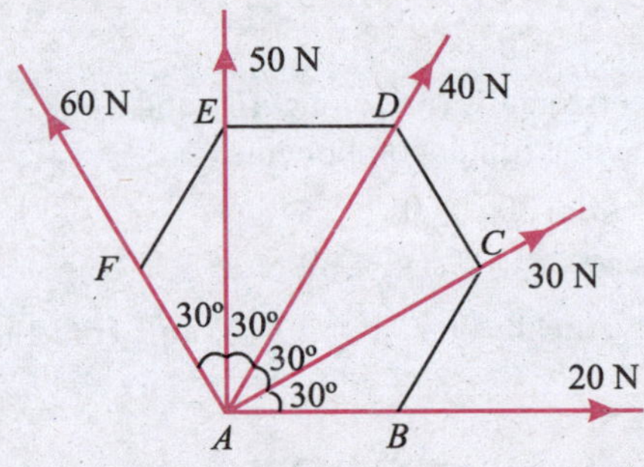

Fig. 13.

Magnitude of the resultant force

Resolving all the forces horizontally (*i.e.*, along *AB*), and find the algebraic sum of all the horizontal components,

$$\begin{aligned}\Sigma H &= 20\cos 0° + 30\cos 30° + 40\cos 60° + 50\cos 90° + 60\cos 120° \text{ N}\\ &= (20\times 1) + (30\times 0.866) + (40\times 0.5) + (50\times 0) + 60\,(-0.5) \text{ N}\\ &= 36.0 \text{ N} \qquad \text{...}(i)\end{aligned}$$

Resolving all the forces vertically (*i.e.*, at right angles to *AB*), and find the algebraic sum of all the vertical components,

$$\begin{aligned}\Sigma V &= 20\sin 0° + 30\sin 30° + 40\sin 60° + 50\sin 90° + 60\sin 120° \text{ N}\\ &= (20\times 0) + (30\times 0.5) + (40\times 0.866) + (50\times 1) + (60\times 0.866) \text{ N}\\ &= 151.6 \text{ N} \qquad \text{...}(ii)\end{aligned}$$

We know that magnitude of the resultant force,

$$R = \sqrt{(\Sigma H)^2 + (\Sigma V)^2} = \sqrt{(36.0)^2 + (151.6)^2} = 155.8 \text{ N} \quad \textbf{Ans.}$$

Direction of the resultant force

Let θ = Angle, which the resultant force makes with the horizontal (*i.e.*, *AB*).

We know that

$$\tan\theta = \frac{\Sigma V}{\Sigma H} = \frac{151.6}{36.0} = 4.211 \quad \text{or} \quad \theta = 76.6° \quad \textbf{Ans.}$$

Note. Since both the values of ΣH and ΣV are positive, therefore actual angle of resultant force lies between 0° and 90°.

***Example 8.** *The following forces act at a point :*

(*i*) *20 N inclined at 30° towards North of East,*

(*ii*) *25 N towards North,*

(*iii*) *30 N towards North West, and*

(*iv*) *35 N inclined at 40° towards South of West.*

Find the magnitude and direction of the resultant force. **[UPTU, GTU, PU, MU]**

Solution. The system of given forces is shown in Fig. 14.

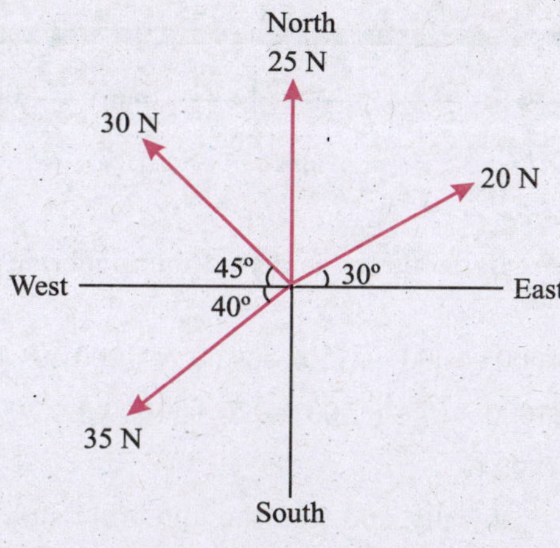

Fig. 14.

Magnitude of the resultant force

Resolving all the forces horizontally *i.e.*, along East-West line, and find the algebraic sum of all the horizontal components,

$$\begin{aligned}\Sigma H &= 20\cos 30° + 25\cos 90° + 30\cos 135° + 35\cos 220° \text{ N}\\ &= (20 \times 0.866) + (25 \times 0) + 30\,(-0.707) + 35\,(-0.766) \text{ N}\\ &= -30.7 \text{ N} \qquad ...(i)\end{aligned}$$

Resolving all the forces vertically *i.e.*, along North-South line, and find the algebraic sum of all the vertical components,

$$\begin{aligned}\Sigma V &= 20\sin 30° + 25\sin 90° + 30\sin 135° + 35\sin 220° \text{ N}\\ &= (20 \times 0.5) + (25 \times 1.0) + (30 \times 0.707) + 35\,(-0.6428) \text{ N}\\ &= 33.7 \text{ N} \qquad ...(ii)\end{aligned}$$

We know that magnitude of the resultant force,

$$R = \sqrt{(\Sigma H)^2 + (\Sigma V)^2} = \sqrt{(-30.7)^2 + (33.7)^2} = 45.6 \text{ N} \quad \textbf{Ans.}$$

Direction of the resultant force

Let θ = Angle, which the resultant force makes with the East.

We know that

$$\tan\theta = \frac{\Sigma V}{\Sigma H} = \frac{33.7}{-30.7} = -1.098 \quad \text{or} \quad \theta = 47.7°$$

Since ΣH is negative and ΣV is positive, therefore resultant lies between 90° and 180°. Thus actual angle of the resultant = 180° – 47.7° = 132.3° **Ans.**

Example 9. *A horizontal line PQRS is 12 m long, where PQ = QR = RS = 4 m. Forces of 1000 N, 1500 N, 1000 N and 500 N act at P, Q, R and S respectively with downward direction. The lines of action of these forces make angles of 90°, 60°, 45° and 30° respectively with PS. Find the magnitude, direction and position of the resultant force.*

Solution. The system of the given forces is shown in Fig. 15

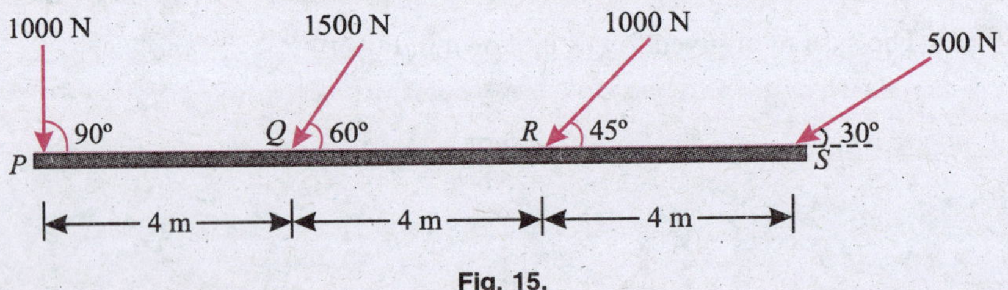

Fig. 15.

Magnitude of the resultant force

Resolving all the forces horizontally, and find the algebraic sum of all the horizontal components,

$$\Sigma H = 1000 \cos 90° + 1500 \cos 60° + 1000 \cos 45° + 500 \cos 30° \text{ N}$$
$$= (1000 \times 0) + (1500 \times 0.5) + (1000 \times 0.707) + (500 \times 0.866) \text{ N}$$
$$= 1890 \text{ N} \quad ...(i)$$

Resolving all the forces vertically, and find the algebraic sum of all the vertical components,

$$\Sigma V = 1000 \sin 90° + 1500 \sin 60° + 1000 \sin 45° + 500 \sin 30° \text{ N}$$
$$= (1000 \times 1.0) + (1500 \times 0.866) + (1000 \times 0.707) + (500 \times 0.5) \text{ N}$$
$$= 3256 \text{ N} \quad ...(ii)$$

We know that magnitude of the resultant force,

$$R = \sqrt{(\Sigma H)^2 + (\Sigma V)^2} = \sqrt{(1890)^2 + (3256)^2} = 3765 \text{ N} \quad \textbf{Ans.}$$

Direction of the resultant force

Let θ = Angle, which the resultant force makes with *PS*.

$$\therefore \quad \tan\theta = \frac{\Sigma V}{\Sigma H} = \frac{3256}{1890} = 1.722 \quad \text{or} \quad \theta = 59.8° \quad \textbf{Ans.}$$

Note. Since both the values of ΣH and ΣV are +ve. therefore resultant lies between 0° and 90°.

Position of the resultant force

Let x = Distance between P and the line of action of the resultant force.

Now taking **moments of the resultant force and the vertical components of the applied forces about *P*, and equating the same,**

$$3765\, x = (1000 \times 0) + (1500 \times 0.866)\, 4 + (1000 \times 0.707)\, 8 + (500 \times 0.5)\, 12$$
$$= 13\,852$$

or $$x = \frac{13\,852}{3765} = 3.68 \text{ m} \quad \textbf{Ans.}$$

EXERCISE 1

1. Find the resultant of two forces equal to 50 N and 30 N acting at an angle of 60°.
[Ans. 70 N ; 21.8°]

2. Two forces of 80 N and 70 N act simultaneously at a point. Find the resultant force, if the angle between them is 150°. **[Ans.** 106.3 N ; 61°]

3. Find the resultant of two forces 130 N and 110 N respectively, acting at an angle whose tangent is 12/5. **[Ans.** 185.7 N ; 30.5°]

***4.** A push of 180 N and pull of 350 N act simultaneously at a point. Find the resultant of the forces, if the angle between them be 135°. **[Ans.** 494 N ; 30°]

***5.** Find the angle between two equal forces P, when their resultant is equal to (*i*) P and (*ii*) $P/2$.
[AM-IETE] [Ans. 120° N ; 151°]

Hint. When resultant is equal to P, then

$$P = \sqrt{P^2 + P^2 + 2P.P \cos\theta} = P\sqrt{2 + 2\cos\theta}$$

$\therefore$ $2\cos\theta = -1$ or $\cos\theta = -0.5$ or $\theta = 120°$ **Ans.**

When resultant is equal to $P/2$, then

$$0.5P = \sqrt{P^2 + P^2 + 2P.P \cos\theta} = P\sqrt{2 + 2\cos\theta}$$

$\therefore$ $2\cos\theta = -1.75$ or $\cos\theta = -0.875$ or $\theta = 151°$ **Ans.**

***6.** The resultant of two forces P and Q is R. If Q is doubled, the new resultant is perpendicular to P. Prove that $Q = R$.

Hint. In first case, $R = \sqrt{P^2 + Q^2 + 2PQ\cos\theta}$

In second case, $\tan 90° = \dfrac{(2Q)\sin\theta}{P + (2Q)\cos\theta}$

Since $\tan 90° = \infty$, therefore $P + 2Q\cos\theta = 0$

15. **TRIANGLE LAW OF FORCES

Triangle Law of Forces states that, ***"If two forces acting simultaneously on a particle, be represented in magnitude and direction by the two sides of a triangle, taken in order ; their resultant may be represented in magnitude and direction by the third side of the triangle, taken in opposite order."***

16. **POLYGON LAW OF FORCES

It is an extension of Triangle Law of Forces for more than two forces. Polygon Law of Forces states that, ***"If a number of forces acting simultaneously on a particle, be represented in magnitude and direction, by the sides of a polygon taken in order ; then the resultant of all these forces may be represented, in magnitude and direction, by the closing side of the polygon, taken in opposite order."***

17. GRAPHICAL (VECTOR) METHOD TO FIND A RESULTANT FORCE. (USING POLYGON LAW OF FORCES TO FIND A RESULTANT FORCE)

Let us find out the magnitude and direction of the resultant force using the polygon law of forces. It is done as discussed below :

1. *Construction of space diagram (position diagram).* It means the construction of a diagram showing the various forces (or loads) alongwith their magnitude and lines of action.
2. *Use of Bow's notations.* All the forces in the space diagram are named by using the Bow's notations. It is a convenient method in which every force (or load) is named by two capital letters, placed on its either side in the space diagram.
3. *Construction of vector diagram (force diagram).* It means the construction of a diagram starting from a convenient point and then go on adding all the forces vectorially one by one (keeping in veiw the directions of the forces) to some suitable scale.

   Now the closing side of the polygon, taken in opposite order, will give the magnitude of the resultant force (to the scale) and its direction.

Example 10. *A particle is acted upon by three forces equal to 50 N, 100 N and 130 N, along the three sides of an equilateral triangle, taken in order. Find graphically the magnitude and direction of the resultant force.*

Solution. The system of given forces is shown in Fig. 16 (*a*)

First of all, name the forces according to Bow's notations as shown in Fig. 16 (*a*). The 50 N force is named as *AD*, 100 N force as *BD* and 130 N force as *CD*.

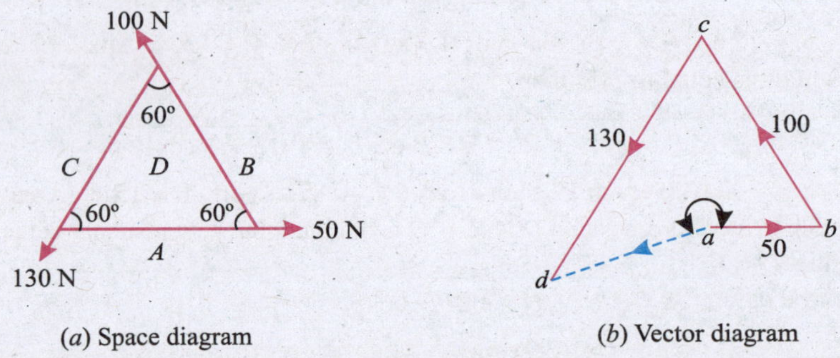

(*a*) Space diagram

(*b*) Vector diagram

Fig. 16.

Now draw the vector diagram for the given system of forces as shown in Fig. 16 (*b*) and as discussed below :

1. Select some suitable point *a* and draw *ab* equal to 50 N to some suitable scale and parallel to the 50 N force of the space diagram.
2. Through *b*, draw *bc* equal to 100 N to the scale and parallel to the 100 N force of the space diagram.
3. Similarly through *c*, draw *cd* equal to 130 N to the scale and parallel to the 130 N force of the space diagram.
4. Join *ad*, which gives the magnitude as well as direction of the resultant force.
5. By measurement, we find the magnitude of the resultant force is equal to 70 N and acting at an angle of 200° with *ab*. **Ans.**

Example 11 *The following forces act at a point :*

(*i*) *20 N inclined at 30° towards North of East.*

(*ii*) *25 N towards North.*

(*iii*) *30 N towards North West and*

(*iv*) *35 N inclined at 40° towards South of West.*

Find the magnitude and direction of the resultant froce.

***Solution.** The system of given forces is shown in Fig. 17 (*a*).

First of all, name the forces according to Bow's notations as shown in Fig. 17 (*a*). The 20 N force is named as *PQ*, 25 N force as *QR*, 30 N force as *RS* and 35 N force as *ST*.

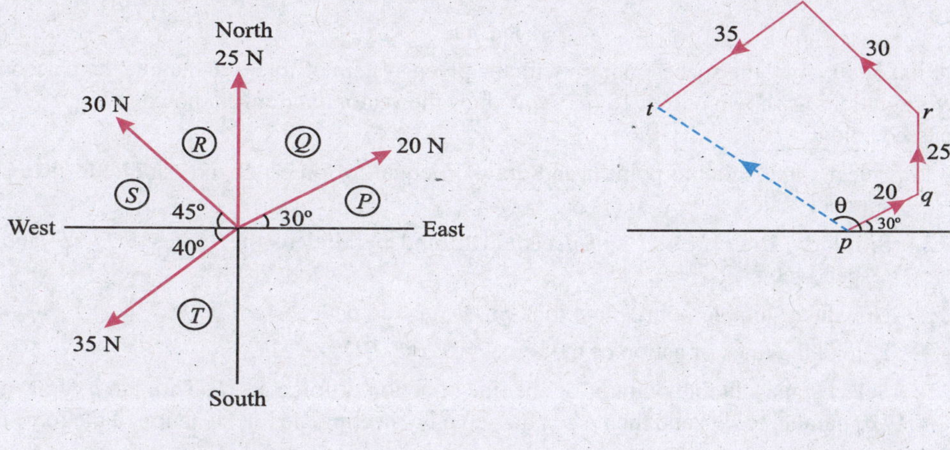

Fig. 17.

Now draw the vector diagram for the given system of forces as shown in Fig. 17 (*b*) and as discussed below :

1. Select some suitable point *p* and draw *pq* equal to 20 N to some suitable scale and parallel to the force *PQ*.
2. Through *q*, draw *qr* equal to 25 N to the scale and parallel to the force *QR* of the space diagram.
3. Now through *r*, draw *rs* equal to 30 N to the scale and parallel to the force *RS* of the space diagram.
4. Similarly, through *s*, draw *st* equal to 35 N to the scale and parallel to the force *ST* of the space diagram.
5. Joint *pt*, which gives the magnitude as well as direction of the resultant force.
6. By measurement, we find that the magnitude of the resultant force is equal to 45.6 N and acting at an angle of 132° with the horizontal *i.e.* East–West line. **Ans.**

* **Note:** We have already solved the above example analytically as example 7 in this chapter.

Example 12. *A horizontal line PQRS is 12 m long, where PQ = QR = RS = 4 m. Forces of 1000 N, 1500 N, 1000 N and 500 N act at P, Q, R and S respectively with downward direction. The lines of action of these forces make angles of 90°, 60°, 45° and 30° respectively with PS. Find the magnitude, direction and position of the resultant force.*

Solution. The system of the forces is shown Fig. 18.

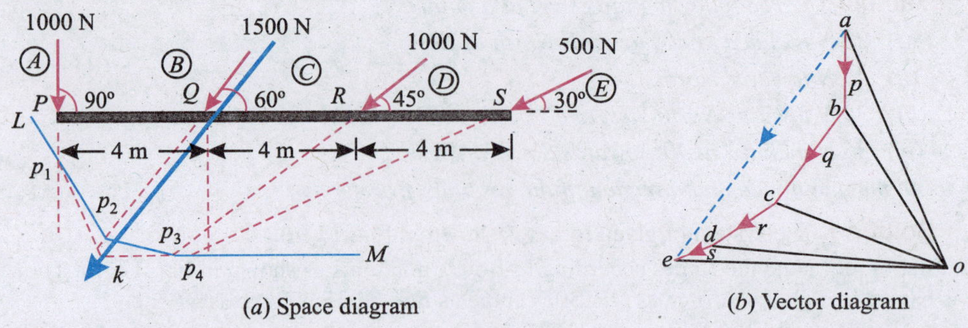

(*a*) Space diagram (*b*) Vector diagram

Fig. 18.

First of all, draw the space diagram with the given system of forces and name them according to Bow's notations as shown in Fig. 18 (*a*). Now draw the vector diagram as shown in Fig. 18 (*b*) and as discussed below :

1. Select some suitable point *a* and draw *ab* equal to force *AB* and parallel to it to some suitable scale.
2. Similarly, draw *bc*, *cd* and *de* equal to and parallel to the forces *BC*, *CD* and *DE* respectively.
3. Now take suitable point *o* and join *oa*, *ob*, *oc* and *oe*.
4. Extend the lines of action of the forces *AB*, *BC*, *CD* and *DE*.
5. Select some suitable point p_1 on the line of action of force at *AB*. Through p_1 draw a line Lp_1 parallel to *ao* ; and then p_1p_2 parallel to *bo* meeting the line of action of the force at p_2.
6. Similarly, draw lines p_2p_3, p_3p_4 and p_4M parallel to *co*, *do* and *eo* respectively.
7. Extend the lines Lp_1 and Mp_4 to meet at *k*.
8. Through *k* draw a line parallel to *ea* which gives the inclination and position of the resultant force.
9. By measurement we find that magnitude of the resultant force *ae*

$$ae = 3760 \text{ N} \quad \textbf{Ans.}$$

Direction of the resultant force *ae*

$$= 62° \text{ with the horizontal} \quad \textbf{Ans.}$$

and distance between *P* and line of action of the resultant force *i.e.*

$$PK = 4.25 \text{ m} \quad \textbf{Ans.}$$

EXERCISE 2

1. Find the magnitude and direction of the resultant of the concurrent forces of 8 N, 12 N, 15 N and 20 N making angles of 30°, 70°, 120°.25 and 155° respectively with a fixed line. [**Ans.** 39.5 N ; 111.7°]
2. Find magnitude of the resultant force, if 30, 40, 50 and 60 N forces are acting along the lines joining the centre of a square to its vertices. [**Ans.** 28.3 N]
3. Four forces of 25 N, 20 N, 15 N and 10 N are acting simultaneously along straight lines *OA*, *OB*, *OC* and *OD* such that

$$\angle AOB = 45°;\ \angle BOC = 100° \text{ and } \angle COD = 125°.$$

Find graphically magnitude and direction of the resultant force. Also check the answer analytically. [**Ans.** 29.5 N ; 25.4° with *OA*]

QUESTIONS

1. Define the term 'force' and state clearly the effects of force.
2. What are the various characteristics of a force?
3. Distinguish clearly between resolution of forces and composition of forces.
4. What are the methods for finding out the resultant force for a given system of forces?
5. State and prove parallelogram law of forces.
6. State triangle law of forces and polygon law of forces.
7. Show that the algebraic sum of the resolved part of a number of forces in a given direction, is equal to the resolved part of their resultant in the same direction.
8. Explain clearly the procedure for finding out the resultant force analytically as well as graphically.

OBJECTIVE TYPE QUESTIONS

1. Which of the following statement is correct?
   (*a*) A force is an agent which produces or tends to produce motion.
   (*b*) A force is an agent which stops or tends to stop motion.
   (*c*) A force may balance a given number of forces acting on a body.
   (*d*) Both (*a*) and (*b*).
2. In order to determine the effects of a force acting on a body, we must know
   (*a*) Its magnitude and direction of the line along which it acts.
   (*b*) Its nature (whether push or pull).
   (*c*) Point through which it acts on the body.
   (*d*) All of the above.
3. If a number of forces are acting simultaneously on a particle, then the resultant of these forces will have the same effect as produced by the all the forces. This is known as
   (*a*) Principle of physical independence of forces.
   (*b*) Principle of transmissibility of forces.
   (*c*) Principle of resolution of forces.
   (*d*) None of the above.
4. The vector method, for the resultant force, is also called polygon law of forces
   (*a*) Correct (*b*) Incorrect
5. The resultant of two forces P and Q acting at an angle θ is equal to
   (*a*) $\sqrt{P^2 + Q^2 + 2PQ\sin\theta}$ (*b*) $\sqrt{P^2 + Q^2 + 2PQ\cos\theta}$
   (*c*) $\sqrt{P^2 + Q^2 - 2PQ\sin\theta}$ (*d*) $\sqrt{P^2 + Q^2 - 2PQ\cos\theta}$
6. If the resultant of two forces P and Q acting at an angle (α) with P, then
   (*a*) $\tan\alpha = \dfrac{P\sin\theta}{P + Q\cos\theta}$ (*b*) $\tan\alpha = \dfrac{P\cos\theta}{P + Q\cos\theta}$
   (*c*) $\tan\alpha = \dfrac{Q\sin\theta}{P + Q\cos\theta}$ (*d*) $\tan\alpha = \dfrac{Q\cos\theta}{P + Q\cos\theta}$

ANSWERS

1. (*d*) 2. (*d*) 3. (*a*) 4. (*b*) 5. (*b*) 6. (*c*)

CHAPTER

3 Moments

Contents

1. INTRODUCTION

In the previous chapter, we have been discussing the effects of forces, acting on a body, through their lines of action or at the point of their intersection. But in this chapter, we shall discuss the effects of these forces, at some other point, away from the point of intersection or their lines of action.

2. *MOMENT OF A FORCE

[GTU, WBUT]

It is the turning effect produced by a force, on the body, on which it acts. The **moment of a force** is equal to the product of the force and the perpendicular distance of the point, about which the moment is required and the line of action of the force.

Mathematically,

Moment, $M = P \times l$

where P = Force acting on the body, and

l = Perpendicular distance between the point, about which the moment is required and the line of action of the force.

3. GRAPHICAL REPRESENTATION OF MOMENT

Consider a force P represented, in magnitude and direction, by the line AB. Let O be a point, about which the moment of this force is required to be found out, as shown in Fig. 1. From O, draw OC perpendicular to AB. Join OA and OB.

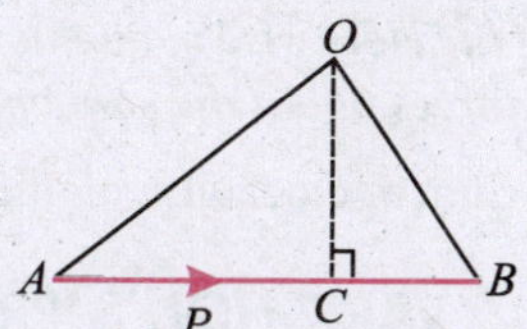

Fig. 1. Representation of moment

Now moment of the force P about O

$$= P \times OC = AB \times OC$$

But $AB \times OC$ is equal to twice the area of triangle ABO.

Thus the moment of a force, about any point, is equal to twice the area of the triangle, whose base is the line to some scale representing the force and whose vertex is the point about which the moment is taken.

4. UNITS OF MOMENT

Since the moment of a force is the product of force and distance, therefore the units of the moment will depend upon the units of force and distance. Thus, if the force is in Newton and the distance is in meters, then the units of moment will be Newton-meter (briefly written as N-m). Similarly, the units of moment may be kN-m (*i.e.* kN × m), N-mm (*i.e.* N × mm) etc.

5. TYPES OF MOMENTS

Broadly speaking, the moments are of the following two types:

1. Clockwise moments. 2. Anticlockwise moments.

6. CLOCKWISE MOMENT

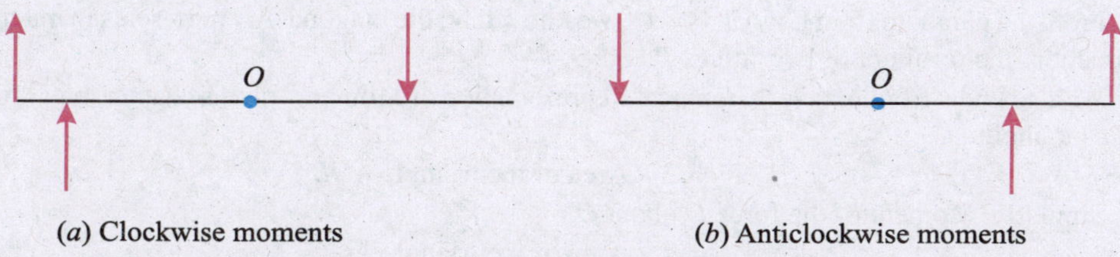

(*a*) Clockwise moments (*b*) Anticlockwise moments

Fig. 2.

It is the moment of a force, whose effect is to turn or rotate the body, about the point in the same direction in which hands of a clock move as shown in Fig. 2 (*a*).

7. ANTICLOCKWISE MOMENT

It is the moment of a force, whose effect is to turn or rotate the body, about the point in the opposite direction in which the hands of a clock move as shown in Fig. 2 (*b*).

Note. The general convention is to take clockwise moment as positive and anticlockwise moment as negative.

8. LAWS RELATED TO MOMENTS

In this section, we deal with the following two laws of moments which are important from the subject point of view :

1. Law of Moments. **2.** Varignon's principle of moments.

8.1 LAW OF MOMENTS

It states, ***"If a body is in rotational equilibrium under the action of a number of forces, the sum of the clockwise moments of the forces about any point is equal to the anti-clockwise moments of the forces about the same point."*** Mathematically,

Sum of clockwise moments = Sum of anti-clockwise moments

8.2. **VARIGNON'S PRINCIPLE OF MOMENTS

[RTU, RGPV, MU, WBUT, AUT]

It states, ***"If a number of coplanar forces are acting simultaneously on a particle, the algebraic sum of the moments of all the forces about any point is equal to the moment of their resultant force about the same point."***

Proof

Let us consider, for the sake of simplicity, two concurrent forces P and Q represented in magnitude and direction by AB and AC as shown in Fig. 3.

Let O be the point, about which the moments are taken. Through O, draw a line OD parallel to the direction of force P, to meet the line of action of the force Q at C. Now, with AB and AC as two adjacent sides, complete the parallelogram $ABDC$ as shown in Fig. 3. Join the diagonal AD of the parallelogram and OA and OB.

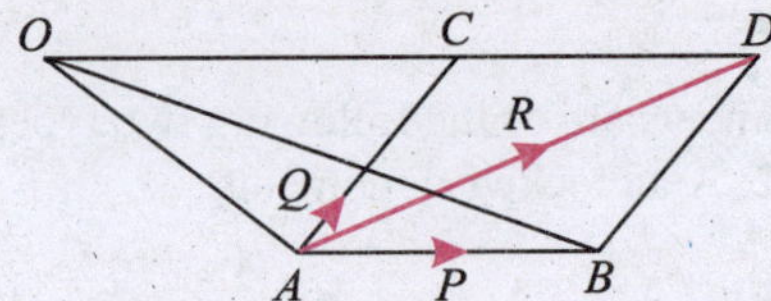

Fig. 3. Proof of Varignon's principle.

From the parallelogram Law of Forces, we know that, the diagonal AD represents in magnitude and direction, the resultant of two forces P and Q.

We also know from Article "Graphical Representation of Moment" of this chapter that, Moment of forces P about O,

$$= 2 \times \text{Area of the triangle } AOB \qquad \ldots(i)$$

Similarly, Moment of the force Q about O,

$$= 2 \times \text{Area of the triangle } AOC \qquad \ldots(ii)$$

And Moment of the resultant force R about O,

$$= 2 \times \text{Area of the triangle } AOD \qquad \ldots(iii)$$

But from the geometry of the figure, we find that,

$$\text{Area of } \Delta AOD = \text{Area of } \Delta AOC + \text{Area of } \Delta ACD$$

But area of ΔACD = Area of ΔADB = Area of ΔAOB

...($\because$ the two Δs AOB and ADB are on the same base AB and between the same parallel lines)

$$\therefore \quad \text{Area of } \Delta AOD = \text{Area of } \Delta AOC + \text{Area of } \Delta AOB$$

Multiplying both sides by 2, we get

$$2 \times \text{ Area of } \Delta AOD = 2 \times \text{Area of } \Delta AOC + 2 \times \text{Area of } \Delta AOB$$

i.e., Moment of force R about O

$$= \text{Moment of force } P \text{ about } O + \text{Moment of force } Q \text{ about } O$$

Note. We have considered, for the sake of simplicity only, two forces P and Q. But this principle may be extended for any number of forces.

Examples based on Law of Moments

Example 1. *A force of 15 N is applied perpendicular to the edge of a door 0.8 m wide as shown in Fig. 4 (a). Find the moment of the force about the hinge.*

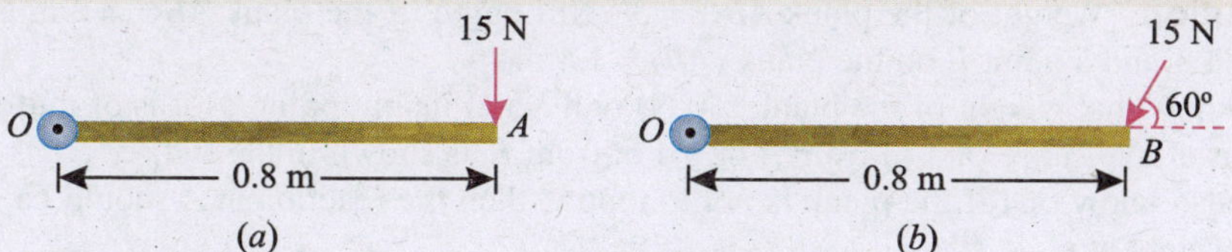

Fig. 4.

If this force is applied at an angle of 60° to the edge of the same door, as shown in Fig. 4 (b), find the moment of this force.

Solution. Given : Force applied (P) = 15 N and width of the door (l) = 0.8 m

Moment when the force acts perpendicular to the door

We know that, Moment of the force about the hinge,

$$= P \times l = 15 \times 0.8 = 12.0 \text{ N-m} \qquad \textbf{Ans.}$$

Moment when the force acts at an angle of 60° to the door

This part of the example may be solved either by finding out the perpendicular distance between the hinge and the line of action of the force as shown in Fig. 5 (*a*) or by finding out the vertical component of the force as shown in Fig. 5 (*b*).

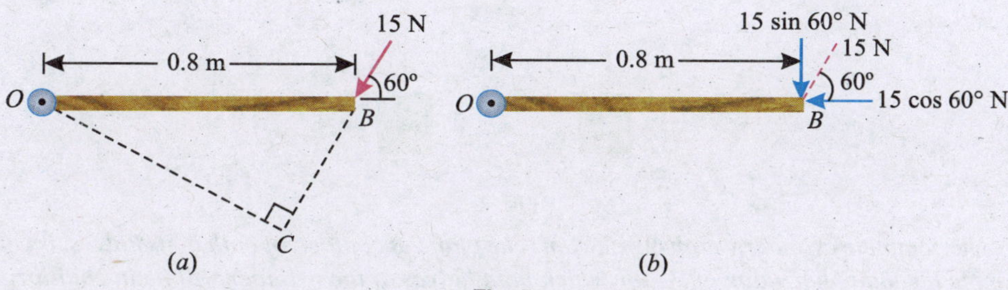

Fig. 5.

From the geometry of Fig. 5 (*a*), we find that the perpendicular distance between the line of action of the force and hinge,

$$OC = OB \sin 60° = 0.8 \times 0.866 = 0.693 \text{ m}$$

$\therefore$ Moment = $15 \times 0.693 = 10.4$ N-m **Ans.**

In the second case, we know that the vertical component of the force

$$= 15 \sin 60° = 15 \times 0.866 = 13.0 \text{ N}$$

$\therefore$ Moment = $13 \times 0.8 = 10.4$ N-m **Ans.**

Note. Since distance between the horizontal component of force (15 cos 60°) and the hinge is zero, therefore moment of horizontal component of the force about the hinge is also zero.

***Example 2.** *A uniform plank ABC of weight 30 N and 2 m long is supported at one end A and at a point B 1.4 m from A as shown in Fig. 6.*

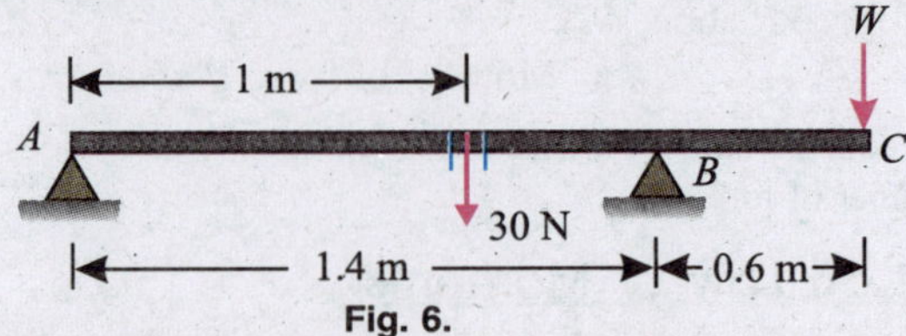

Fig. 6.

Find the maximum weight W, that can be placed at C, so that the plank does not topple.

Solution. Weight of the plank ABC = 30 N; Length of the plank ABC = 2 m and distance between end A and a point B on the plank (AB) = 1.4 m.

We know that weight of the plank (30 N) will act at its midpoint, as it is of uniform section. This point is at a distance of 1 m from A or 0.4 m from B as shown in the figure.

We also know that if the plank is not to topple, then the reaction at A should be zero for the maximum weight at C.

Taking moments about B, and equating the anti-clocwise moments with clockwise moments,

$$30 \times 0.4 = W \times 0.6$$

or

$$W = \frac{30 \times 0.4}{0.6} = 20 \text{ N} \quad \textbf{Ans.}$$

****Example 3.** *Two halves of a round homogeneous cylinder are held together by a thread wrapped round the cylinder with two weights each equal to P attached to its ends as shown in Fig. 7.*

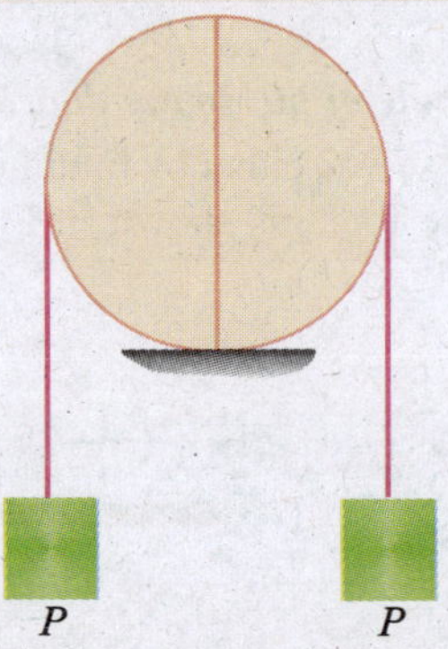

Fig. 7.

The complete cylinder weighs W newton. The plane of contact, of both of its halves, is vertical. Determine the minimum value of P, for which both halves of the cylinder will be in equilibrium on a horizontal plane. **[BPUT]**

Solution. The free body diagram of one of the halves of the cylinder with minimum value of P is shown in Fig. 8

We know that the centre of gravity of a semi-circle is at a distance of $4r/3\pi$ from its base measured along the radius.

Taking moments about A, and equating the anti-clockwise moments with clockwise moments,

$$(P \times r) + \left(\frac{W}{2} \times \frac{4r}{3\pi}\right) = P \times 2r$$

Fig. 8.

or $$P \times r = \frac{2W}{3\pi} \times r$$

or $$P = \frac{2W}{3\pi} \quad \textbf{Ans.}$$

***Example 4.** *A uniform wheel of 600 mm diameter, weighing 5 kN rests against a rigid rectangular block of 150 mm height as shown in Fig. 9.*

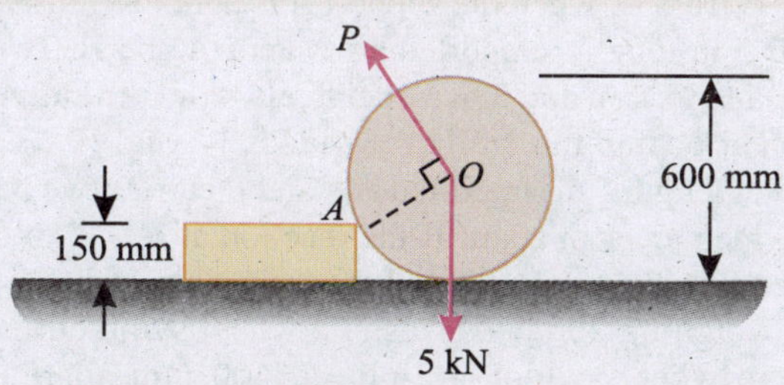

Fig. 9.

Find the least pull, through the centre of the wheel, required just to turn the wheel over the corner A of the block. Also find the reaction on the block. Take all the surfaces to be smooth.

[DTU, PU, RGPV, AUT, WBUT]

Solution. Given : Diameter of wheel = 600 mm; Weight of wheel = 5 kN and height of the block = 150 mm.

Least pull required just to turn the wheel over the corner

Let P = Least pull required just to turn the wheel in kN.

A little consideration will show that for the least pull, it must be applied normal to AO. The system of forces is shown in Fig. 10. From the geometry of the figure, we find that

$$\sin\theta = \frac{150}{300} = 0.5$$

or $$\theta = 30°$$

and $$AB = \sqrt{(300)^2 - (150)^2} = 260 \text{ mm}$$

P
O
300 mm
150 mm
A
θ
B
R
5 kN

Fig. 10.

Taking moments about A, and equating the anti-clockwise moments with clockwise moments,

$$P \times 300 = 5 \times 260$$

or $$P = \frac{1300}{300} = 4.33 \text{ kN} \quad \textbf{Ans.}$$

Reaction on the block

Let R = Reaction on the block in kN.

Resolving the forces horizontally and equating forces acting in left direction with forces acting in right direction,

$$R \cos 30° = P \sin 30°$$

or $$R = \frac{P \sin 30°}{\cos 30°} = \frac{4.33 \times 0.5}{0.866} = 2.5 \text{ kN} \quad \textbf{Ans.}$$

EXERCISE 1

1. A rod AB 2·5 m long is supported at A and B. The rod is carrying a point load of 5 kN at a distance of 1 m from A. What are the reactions at A and B ? **[Ans.** 2 kN ; 3 kN]
2. A beam AB of length 1 m is supported horizontally at A and B. A weight of 500 N is attached on the rod at C. If the support at A can not bear a pressure more than 300 N, find the distance of C from A, when the support A is about to fail. **[Ans.** 0.4 m]

*3. A beam AB 5 m long is supported at its ends A and B. Two point loads W_1 and W_2 are placed at C and D, 1 m and 3 m respectively from the end A. If the reaction at A is twice the reaction at B, find the ratio of the loads W_1 and W_2. **[Ans.** $W_1 : W_2 = 2 : 1$]

4. A beam AB of length 5 m supported at A and B carries two point loads W_1 and W_2 of 3 kN and 5 kN which are 1 m apart. If the reaction at B is 2 kN more than that at A, find the distance between the support A and the load 3 kN. **[Ans.** 2.5 m]
5. A tricycle weighing 200 N has a small wheel symmetrically placed 500 mm in front of two large wheels which are placed 400 mm apart. If centre of gravity of the cycle be at a horizontal distance of 150 mm from the rear wheels and that of the rider, whose weight is 150 N, be 100 mm from the rear wheels, find the thrust on the ground under the different wheels. **[Ans.** 90 N ; 130 N ; 130 N]

*6. Two identical prismatic bars PQ and RS each weighing 75 N are welded together to form a Tee and are suspended in a vertical plane as shown in Fig. 11.

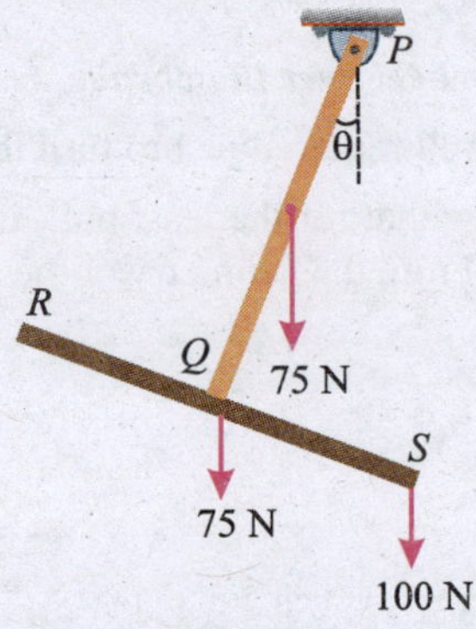

Fig. 11.

Calculate the value of θ, that the bar PQ will make with vertical when a load of 100 N is applied at S. **[PU, RGPV, BPUT]** **[Ans.** 13.25°]

9. APPLICATIONS OF MOMENTS

Though the moments have a number of applications, in the field of Engineering science, yet the following are important from the subject point of view :

1. Position of Resultant Force **2. Lever** **3. Steel Yard**

4. Lever Safety Valve **5. Balance** **6. Finding Support Reactions**

10. *POSITION OF THE RESULTANT FORCE BY MOMENTS

It is also known as analytical method for the resultant force. The position of a resultant force may be found out by moments as discussed below :

1. First of all, find out the magnitude and direction of the resultant force by the method of resolution as discussed earlier in chapter 'Composition and Resolution of Forces'.

An engineer designing a suspension bridge like one above, takes account of forces acting at points within the structure and the turning moment of forces.

2. Now equate the moment of the resultant force with the algebraic sum of moments of the given system of forces about any point. This may also be found out by equating the sum of clockwise moments and that of the anticlockwise moments about the point, through which the resultant force will pass.

Examples based on Varignon's Principle of Moments

Example 5. *Three forces of 2P, 3P and 4P act along the three sides of an equilateral triangle of side 100 mm taken in order. Find the magnitude and position of the resultant force.*

Solution. The system of given forces is shown in Fig. 12.

Magnitude of the resultant force

Resolve all the forces horizontally and find the algebraic sum of all the horizontal components,

$$\Sigma H = 2P + 3P \cos 120° + 4P \cos 240°$$
$$= 2P + 3P(-0.5) + 4P(-0.5)$$
$$= -1.5\,P \qquad \text{...}(i)$$

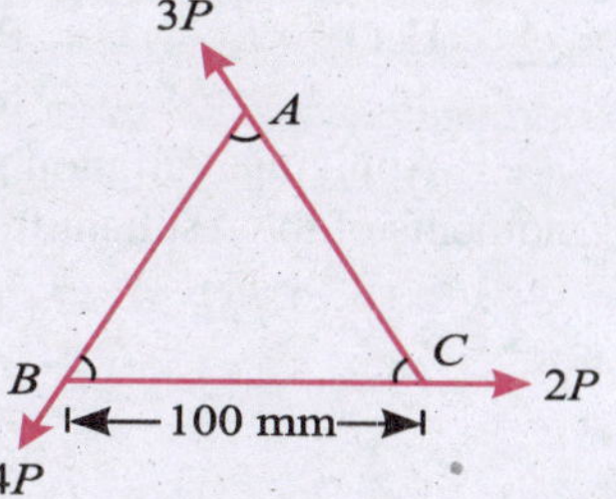

Fig. 12.

Resolve all the forces vertically and find the algebraic sum of all the vertical components,

$$\Sigma V = 3P \sin 60° - 4P \sin 60°$$
$$= (3P \times 0.866) - (4P \times 0.866)$$
$$= -0.866\,P \qquad \text{...}(ii)$$

We know that magnitude of the resultant force

$$R = \sqrt{(\Sigma H)^2 + (\Sigma V)^2} = \sqrt{(-1.5\,P)^2 + (-0.866\,P)^2} = 1.732\,P \quad \textbf{Ans.}$$

Position of the resultant force

Let x = Perpendicular distance between B and the line of action of the resultant force.

Applying Varignon's principle. Now taking moments about B and equating the moments of the resultant force with the moments of component forces,

$$1.732\,P \times x = 3P \times 100 \sin 60°$$
$$= 3P \times (100 \times 0.866) = 259.8\,P$$

or
$$x = \frac{259.8}{1.732} = 150 \text{ mm} \qquad \textbf{Ans.}$$

Note. The moment of the force $2P$ and $4P$ about the point B will be zero, as they pass through it.

Example 6. *Four forces equal to P, 2P, 3P and 4P are respectively acting along the four sides of square ABCD taken in order. Find the magnitude, direction and position of the resultant force.*

Solution. The system of given forces is shown in Fig. 13.

Magnitude of the resultant force

Resolve all the forces horizontally and find the algebraic sum of all the horizontal components,

$$\Sigma H = P - 3P = -2P \qquad ...(i)$$

Resolve all the forces vertically and find the algebraic sum of all the vertical components,

$$\Sigma V = 2P - 4P = -2P \qquad ...(ii)$$

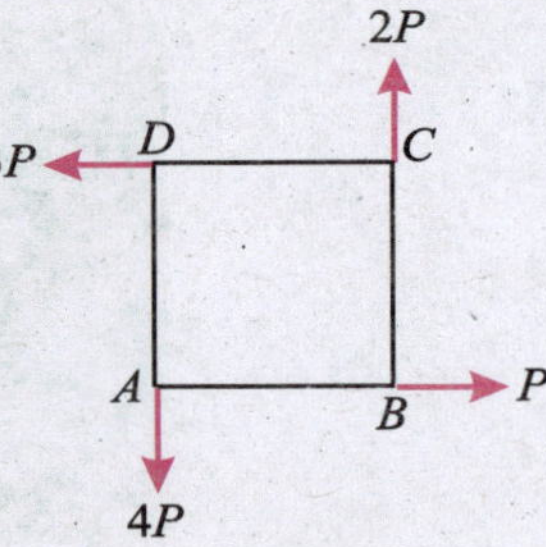

Fig. 13.

We know that magnitude of the resultant forces,

$$R = \sqrt{(\Sigma H)^2 + (\Sigma V)^2} = \sqrt{(-2P)^2 + (-2P)^2}$$

$$= 2\sqrt{2}P \quad \textbf{Ans.}$$

Direction of the resultant force

Let θ = Angle, which the resultant makes with the horizontal.

$$\therefore \qquad \tan\theta = \frac{\Sigma V}{\Sigma H} = \frac{-2P}{-2P} = 1 \quad \text{or} \quad \theta = 45^\circ$$

Since ΣH as well as ΣV are –ve, therefore resultant lies between 180°and 270°. Thus actual angle of the resultant force = 180° + 45° = 225° **Ans.**

Position of the resultant force

Let x = Perpendicular distance between A and the line of action of the resultant force.

Applying Varignon's principle. Now taking moments about A and equating equating the moments of the resultant force with the moments of component forces,

$$2\sqrt{2}P \times x = (2P \times a) + (3P \times a) = 5P \times a$$

or $$x = \frac{5a}{2\sqrt{2}} \quad \textbf{Ans.}$$

Note. The moment of the forces P and $4P$ about the point A will be zero, as they pass through it.

Example 7. *ABCD is a square, each side being 20 cm and E is the middle point of AB. Forces of 7, 8, 12, 5, 9 and 6* kN *act on the lines of directions AB, EC, BC, BD, CA and DE respectively. Find the magnitude, direction and position of the resultant force.*

Solution. The system of the given forces is shown in Fig. 14

Magnitude of resultant force

Let $\angle BEC = \alpha$

We know that,

$$\tan\alpha = \frac{20}{10} = 2$$

$$\sin\alpha = \frac{2}{\sqrt{5}} = 0.894$$

and $$\cos\alpha = \frac{1}{\sqrt{5}} = 0.447$$

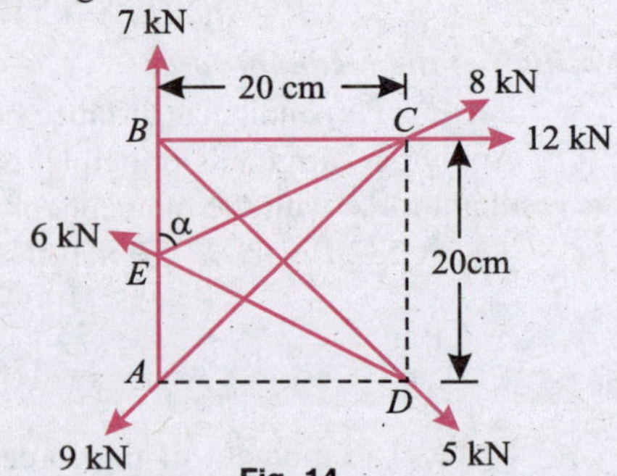

Fig. 14.

Resolve all the forces horizontally and find the algebraic sum of all the horizontal components,

$$\Sigma H = 8 \sin \alpha + 12 + 5 \sin 45° - 9 \sin 45° - 6 \sin \alpha$$
$$= (8 \times 0.894) + (12) + (5 \times 0.707) - (9 \times 0.707) - (6 \times 0.894) \text{ kN}$$
$$= 10.96 \text{ kN} \quad ...(i)$$

Resolve all the forces vertically and find the algebraic sum of all the vertical components,

$$\Sigma V = 7 + 8 \cos \alpha - 5 \cos 45° - 9 \cos 45° + 6 \cos \alpha$$
$$= 7 + (8 \times 0.447) - (5 \times 0.707) - (9 \times 0.707) + (6 \times 0.447) \text{ kN}$$
$$= 3.36 \text{ kN} \quad ...(ii)$$

We know that magnitude of the resultant force,

$$R = \sqrt{(\Sigma H)^2 + (\Sigma V)^2} = \sqrt{(10.96)^2 + (3.36)^2} = 11.46 \text{ kN} \quad \textbf{Ans.}$$

Direction of the resultant force

Let θ = Angle, which the resultant force makes with *BC* *i.e.*, with the horizontal.

$$\therefore \quad \tan \theta = \frac{\Sigma V}{\Sigma H} = \frac{3.36}{10.96} = 0.3066 \quad \text{or} \quad \theta = 17.05° \quad \textbf{Ans.}$$

Note : Since both the values of ΣH and ΣV are + ve, therefore resultant actual angle of the resultant force lies between 0° and 90°.

Position of the resultant force

Let x = Perpendicular distance between the point *E* and the line of action of the resultant force.

Applying Varignon's principle. Now taking moments about *E* and equating equating the moments of the resultant force with the moments of component forces,

... (The point *E* has been intentionally selected as the moments of the forces 7 kN, 8 kN and 6 kN about this point are zero, because these forces pass through *E*.)

$$11.46\, x = (7 \times 0) + (8 \times 0) + (12 \times 10) + (5 \times 0.707) + (9 \times 0.707) + (6 \times 0)$$
$$= 129.9$$

or
$$x = \frac{129.9}{11.46} = 11.33 \text{ cm} \quad \textbf{Ans.}$$

Example 8. *ABCD is a square. Forces of 10, 8 and 4 units act at A in the directions AD, AC and AB respectively. Using the analytical method, determine*

(*i*) *resultant force in magnitude and direction ;*

(*ii*) *magnitude and sense of two forces along the directions AJ and AH, where J and H are mid-points of CD and BC respectively, which together will balance the above resultant.*

Solution. The system of forces is shown in Fig. 15

Resultant force in magnitude and direction

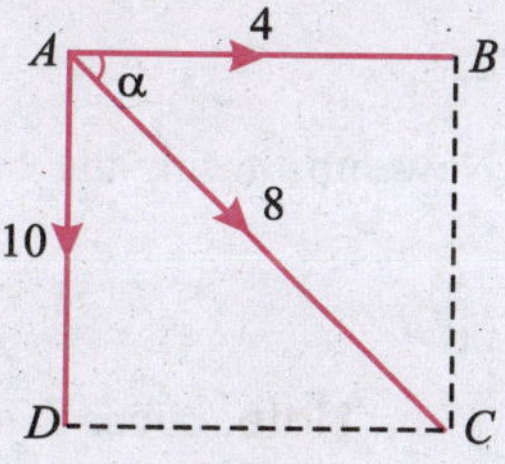

Fig. 15.

Resolve all the forces horizontally and find the algebraic sum of all the horizontal components,

$$\Sigma H = 4 + 8 \cos 45° = 4 + (8 \times 0.707) \text{ units}$$
$$= 9.656 \text{ units}$$

Resolve all the forces vertically and find the algebraic sum of all the vertical components,

$$\Sigma V = -10 + (-8 \cos 45°) = -10 - (8 \times 0.707)$$
$$= -15.656 \text{ units}$$

We know that magnitude of the resultant force,

$$R = \sqrt{(\Sigma H)^2 + (\Sigma V)^2} = \sqrt{(9.656)^2 + (-15.656)^2} = 18.39 \text{ units} \quad \textbf{Ans.}$$

Let α = Angle which the resultant force makes with the horizontal.

$$\therefore \quad \tan\alpha = \frac{\Sigma V}{\Sigma H} = \frac{-15.656}{9.656} = -1.6214 \quad \text{or} \quad \alpha = 58.3° \textbf{ Ans.}$$

Since ΣH is positive and ΣV is negative, therefore resultant lies between 270° and 360°. Thus actual angle of the resultant force = 360° – 58.3° = 301.7° **Ans.**

Magnitude and sense of two forces along AJ and JH

Let P_1 = Force in units along *AH*, and

P_2 = Force in units along *AJ*.

The system of forces along with the resultant (*R*) is shown in Fig. 16. From the geometry of the figure, we find that

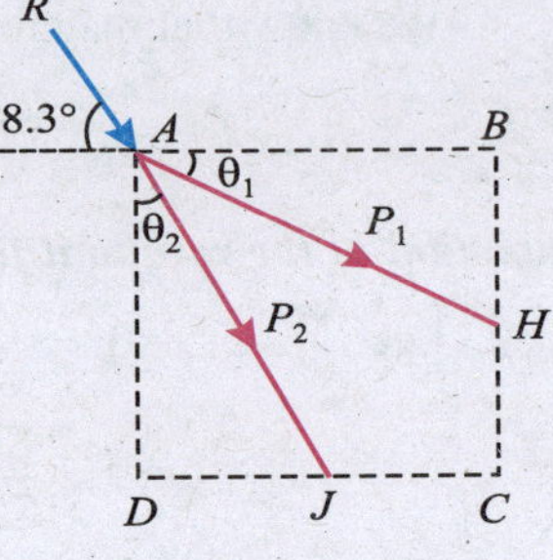

Fig. 16.

$$\tan\theta_1 = \tan\theta_2 = \frac{0.5}{1} = 0.5$$

or $$\theta_1 = \theta_2 = 26.6°$$

Resolving the forces P_1 and P_2 horizontally, and equating it with the horizontal component of the resultant force,

$$P_1 \cos\theta_1 + P_2 \sin\theta_2 = 9.656$$

or $$P_1 \cos 26.6° + P_2 \sin 26.6° = 9.656$$

or $$P_1 \times 0.8944 + P_2 \times 0.4472 = 9.656$$

or $$2P_1 + P_2 = 21.59 \quad ...(i)$$

Resolving the forces P_1 and P_2 vertically and equating them with the vertical component of the resultant force,

$$P_1 \sin\theta_1 + P_2 \cos\theta_2 = 15.656$$

or $$P_1 \sin 26.6° + P_2 \cos 26.6° = 15.656$$

or $$P_1 \times 0.4472 + P_2 \times 0.8944 = 15.656$$

or $$P_1 + 2P_2 = 35.01$$

or $$0.5\,P_1 + P_2 = 17.505 \quad ...(ii)$$

Subtracting equation (*ii*) from equation (*i*),

$$1.5\,P_1 = 4.085$$

or $$P_1 = \frac{4.085}{1.5} = 2.72 \text{ units} \quad \textbf{Ans.}$$

Now substituting the value of P_1 in equation (*i*),

$$(2 \times 2.72) + P_2 = 21.59$$

or $$P_2 = 21.59 - 5.44 = 16.15 \text{ units} \quad \textbf{Ans.}$$

Note. Since the two forces P_1 and P_2 together will balance the resultant force, therefore their directions will be in the opposite directions as assumed *i.e.* along *HA* and *JA*.

11. LEVERS

A lever is a rigid bar (straight, curved or bent) and is hinged at one point. It is free to rotate about the hinged end called fulcrum. The common examples of the use of lever are crow bar, pair of scissors, fire tongs, etc.

It may be noted that there is a point for effort (called effort arm) and another point for overcoming resistance or lifting load (called load arm).

12. TYPES OF LEVERS

Though the levers are of many types, yet the following are important from the subject point of view.

1. Simple levers. 2. Compound levers.

13. SIMPLE LEVERS

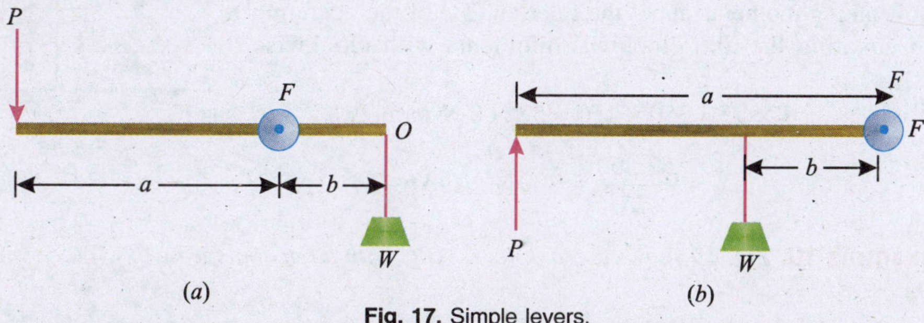

Fig. 17. Simple levers.

A lever, which consists of one bar having one fulcrum is known as simple lever as shown in Fig. 17 (*a*) and (*b*).

The simple levers are classified as

(a) ***Lever of first order.*** In this type of lever, the effort and load act on the opposite sides of the fulcrum.

(b) ***Lever of second order.*** In this type of lever, the effort and load act on the same side of the fulcrum. But effort acting at a greater distance than the load. Or in other words, load is acting between the fulcrum and the effort.

(c) ***Lever of third order.*** In this type of lever, the effort and load act on the same side of the fulcrum. But load acting at a greater distance than the effort. Or in other words, effort is acting between the fulcrum and the load.

Let us consider simple lever shown in Fig. 17 (*a*) and (*b*).

Where P = Effort applied

W = Weight lifted

a = Length between fulcrum and effort, and

b = Length between fulcrum and weight.

Now taking moments of the effort and load about the fulcrum (F) and equating the same,

$$P.a = W.b \qquad \text{or} \qquad \frac{W}{P} = \frac{a}{b}$$

The terms $\frac{W}{P}$ and $\frac{a}{b}$ are commonly known as mechanical advantage and leverage. These will be discussed in more details in the chapter 'Simple Lifiting Machines'.

A little consideration will show that in order to increase the mechanical advantage, either length of the effort arm a is to be increased or length of the load arm b is to be reduced.

Note. A simple lever may be straight, curved or even bent.

Example 9. *Find the tension required in the operating wire to raise the signal through the system of levers as shown in Fig. 18. All dimensions are in mm.*

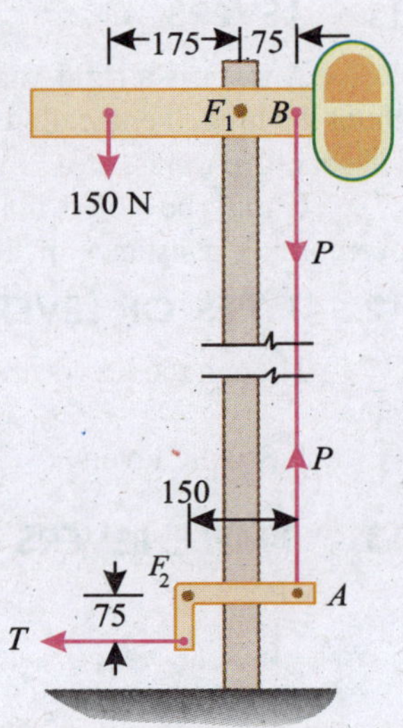

Fig. 18.

Solution. Given : Weight of signal arm (W) = 150 N

Let T = Tension required in operating wire, and

P = Tension in wire AB.

First of all, taking moment about the fulcrum (F_1) of the signal and equating the anti-clockwise moments with clockwise moments,

$$P \times 75 = 150 \times 175 = 26\ 250 \text{ N-mm}$$

or
$$P = \frac{26\ 250}{75} = 350 \text{ N}$$

Now taking moments about the fulcrum (F_2) of the operating wire and equating the anti-clockwise moments with clockwise moments,

$$T \times 75 = 350 \times 150 = 52\ 500 \text{ N-mm}$$

or
$$T = \frac{52\ 500}{75} = 700 \text{ N} \quad \textbf{Ans.}$$

Example 10. *Fig. 19 shows a crank-lever ABC with a tension spring (T). The lever weighs 0.2 N/mm.* **[PU]**

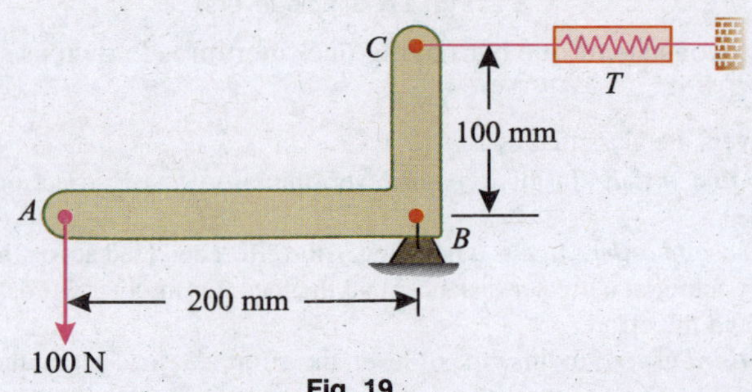

Fig. 19.

Determine the tension developed in the spring, when a load of 100 N is applied at A.

Solution. Given : Weight of lever = 0.2 N/ mm ; Force applied on the effort arm (P) = 100 N; Length of the effert arm (a) = 200 mm and length of the load arm (b) = 100 mm.

Let T = Tension developed in the spring.

We know that weight of the lever AB,

$$= 200 \times 0.2 = 40 \text{ N}$$

... (The weight of lever BC will have no moment about the hinge B. Therefore its weight has been ignored.)

and it is acting at the mid-point of AB *i.e.* 100 mm from A or B.

Taking moments about the hinge B and equating the clockwise moments with anti-clockwise moments,.

$$T \times 100 = (100 \times 200) + (40 \times 100) = 24\ 000 \text{ N-mm}$$

or
$$T = \frac{24\ 000}{100} = 24 \text{ N} \quad \textbf{Ans.}$$

Example 11. *The lever ABC of a component of a machine is hinged at B, and is subjected to a system of coplaner forces as shown in Fig. 20.*

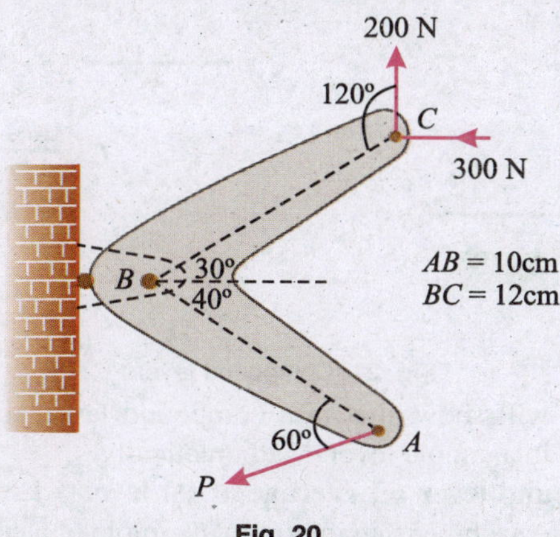

Fig. 20.

Neglecting friction, find the magnitude of the force (P) to keep the lever in equilibrium. Also determine the magnitude and direction of the reaction at B.

Solution. Given : Vertical force at $C = 200$ N and horizontal force at $C = 300$ N.

Magnitude of the force (P)

Taking moments about the hinge B and equating the clockwise moments with ant-clockwise moments,,

$$P \times 10 \sin 60° = (200 \times 12 \cos 30°) + (300 \times 12 \cos 60°)$$

or $$P \times 10 \times 0.866 = (200 \times 12 \times 0.866) + (300 \times 12 \times 0.5)$$

or $$8.66\, P = 2078 + 1800 = 3878$$

or $$P = \frac{3878}{8.66} = 447.8 \text{ N} \quad \textbf{Ans.}$$

Magnitude of the reaction at B

Resolve all the forces horizontally and find the algebraic sum of all the horizontal components,

$$\Sigma H = 300 + P \cos 20° = 300 + 447.8 \times 0.9397 = 720.8 \text{ N}$$

Resolve all the forces vertically and find the algebraic sum of all the vertical components,

$$\Sigma V = 200 - P \sin 20° = 200 - 447.8 \times 0.3420 = 46.85 \text{ N}$$

$\therefore$ Magnitude of the reaction at B,

$$R = \sqrt{(\Sigma H)^2 + (\Sigma V)^2} = \sqrt{(720.8)^2 + (46.85)^2} = 722.3 \text{ N} \quad \textbf{Ans.}$$

Direction of the reaction at B

Let θ = Angle, which the reaction at B makes with the horizontal.

$$\therefore \quad \tan\theta = \frac{\Sigma V}{\Sigma H} = \frac{46.85}{720.8} = 0.0650 \quad \text{or} \quad \theta = 3.7° \textbf{ Ans.}$$

Note. Since both the values of ΣH and ΣV are +ve. therefore resultant lies between 0° and 90°.

14. COMPOUND LEVERS

A lever, which consists of a number of simple levers is known as a compound lever, as shown in Fig. 21 (*a*) and (*b*).

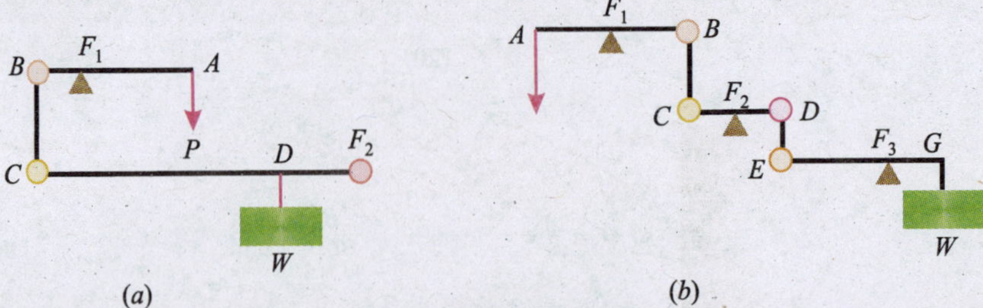

Fig. 21. Compound levers

A little consideration will show, that in a compound lever, the mechanical advantage (or leverage) is greater than that in a simple lever. Mathematically,

Leverage in a compound lever = Leverage of 1st lever × Leverage of 2nd lever × ...

The platform weighing machine is an important example of a compound lever. This machine is used for weighing heavy loads such as trucks, wagons alongwith their contents. On smaller scales, these machines are used in godowns and parcel offices of transport companies for weighing consignment goods.

Example 12. *A compound lever shown in Fig. 22 is required to lift a heavy load W*

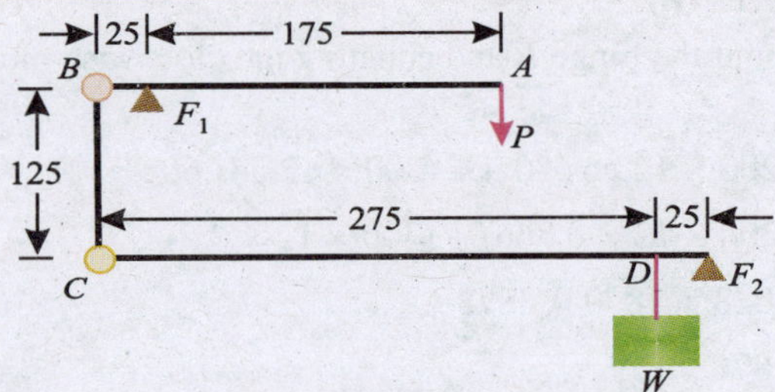

Fig. 22.

Find the value of W, if an effort (P) of 100 N is applied at A.

Solution. Given : Effort, $(P) = 100$ N

From the geometry of the lever, we find that,

$$\text{the Leverage of the upper lever } AB = \frac{AF_1}{BF_1} = \frac{175}{25} = 7 \qquad ...(i)$$

$$\text{and Leverage of the lower lever } CF_2 = \frac{CF_2}{DF_2} = \frac{275 + 25}{25} = 12 \qquad ...(ii)$$

$\therefore$ Total leverage of the compound lever $= 7 \times 12 = 84$ ...(*iii*)

We know that, the weight (W) can be found out from the relation for total leverage,

$$84 = \frac{W}{P} = \frac{W}{100}$$

or $W = 100 \times 84 = 8400 \text{ N} = 8.4 \text{ kN}$ **Ans.**

EXERCISE 2

1. An oil drum of 500 mm diameter and 1.5 long is to be rolled across a footstep of 100 mm high. Find the minimum push required at the top of the drum. Take density of the oil as 1 kg/litre. Neglect weight of the drum. **[Ans.** 1444.5 N**]**

2. Three forces equal to 3P, 5P and 7P act simultaneously along the three sides AB, BC, and CA of an equilateral triangle ABC of side a. Find the magnitude, direction and position of the resultant once. **[Ans.** 3.46 P; 210° ; 1.25 a]
3. A square $ABCD$ of 60 mm side is subjected to force of 10 N, 20 N, 30 N and 40 N along the sides AB, AC, BD and AD respectively. Find magnitude. direction and position of the resultant force with respect to A. **[Ans.** 75.4 N; 87.7° ; 16.88 mm]
4. A compound lever shown in Fig. 23 is required to lift a load of 9 kN with an effort P.

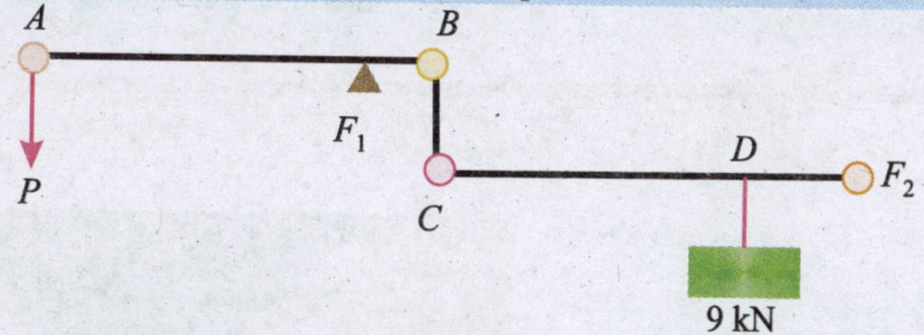

Fig. 23.

The dimensions are $AF_1 = 450$ mm ; $F_1B = 50$ mm ; $BC = 100$ mm ; $CD = 300$ mm ; $DF_2 = 75$ mm. Find the effort required to lift the load. **[Ans.** 200 N]

QUESTIONS

1. What is meant by moment of a force? How will you explain it mathematically?
2. How will you represent the moment of a force geometrically?
3. Explain clearly the difference between clockwise moments and anticlockwise moments.
4. State clearly the law of moments.
5. State the Varignon's principle of moments.
6. What is a lever? Distinguish clearly between a simple lever and a compound lever.
7. What do you understand by the term 'leverage'? Explain its usefulness.

OBJECTIVE TYPE QUESTIONS

1. The moment of a force about any point is geometrically equal to... area of the triangle whose base is the line representing the force and vertex is the point about which the moment is taken.
   (*a*) Half (*b*) Same (*c*) Twice (*d*) None of these
2. In a clockwise moment, we actually use wall clock in order to know the time for which the moment is applied.
   (*a*) Right (*b*) Wrong
3. If a number of coplaner forces are acting simultaneously on a particle, the algebraic sum of the moments of all forces about any point is equal to the moment of their resultant force about the same point. This principle is known as
   (*a*) Principle of moments. (*b*) Principle of levers (*c*) None of them
4. In the study of levers, we ..... the principle of moments.
   (*a*) Use (*b*) Do not use
5. In a compound lever, the leverages of all the simple levers is
   (*a*) Added (*b*) Subtracted (*c*) Multiplied (*d*) Divided
6. The term 'leverage' and 'mechanical advantage' in a compound lever have got the same meaning.
   (*a*) Agree (*b*) Disagree

ANSWERS

1. (*a*) 2. (*b*) 3. (*a*) 4. (*a*) 5. (*c*) 6. (*a*)

CHAPTER

4 Parallel Forces and Couples

Contents

1. PARALLEL FORCES

[UPTU]

In the previous chapters, we have been studying forces acting at one point. But, sometimes, the given forces have their lines of action parallel to each other. A little consideration will show, that such forces do not meet at any point, though they do have some effect on the body on which they act. The forces, whose lines of action are parallel to each other, are known as **parallel forces**.

2. CLASSIFICATION OF PARALLEL FORCES

The parallel forces may be, broadly, classified into the following two categories, depending upon their directions :

1. **Like parallel forces.**
2. **Unlike parallel forces.**

3. LIKE PARALLEL FORCES

The forces, whose lines of action are parallel to each other and all of them act in the same direction as shown in Fig. 1 (*a*) are known as like parallel forces.

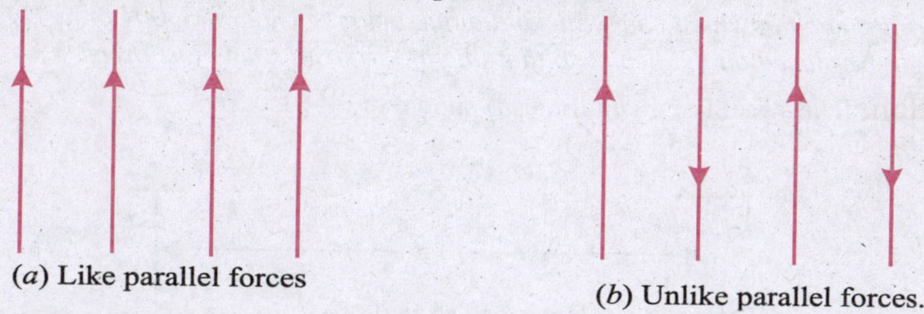

(*a*) Like parallel forces (*b*) Unlike parallel forces.

Fig. 1.

4. UNLIKE PARALLEL FORCES

The forces, whose lines of action are parallel to each other and all of them do not act in the same direction as shown in Fig. 1 (*b*) are known as unlike parallel forces.

5. METHODS TO FIND MAGNITUDE AND POSITION OF THE RESULTANT OF PARALLEL FORCES

The magnitude and position of the resultant force, of a given system of parallel forces (like or unlike) may be found out analytically or graphically. Here we shall discuss both the methods one by one.

6. ANALYTICAL METHOD TO FIND THE RESULTANT OF PARALLEL FORCES

In this method, the sum of clockwise moments is equated with the sum of anti-clockwise moments about a point.

Example 1. *Two like parallel forces of 50 N and 100 N act at the ends of a rod 360 mm long. Find the magnitude of the resultant force and the point where it acts.*

Solution. Given : The system of given forces is shown in Fig. 2

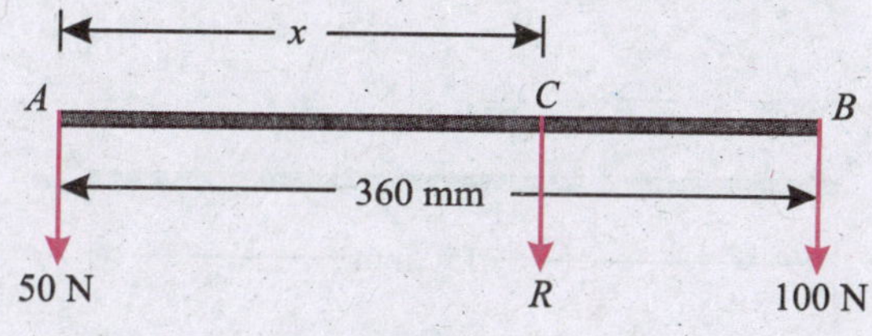

Fig. 2.

Magnitude of the resultant force

Since the given forces are like and parallel, therefore magnitude of the resultant force,

$$R = 50 + 100 = 150 \text{ N} \quad \text{Ans.}$$

Point where the resultant force acts

Let x = Distance between the line of action of the resultant force (R) and A (*i.e.* AC) in mm.

Now taking clockwise and anti-clockwise moments of the forces about C and equating the same,

$$50 \times x = 100\,(360 - x) = 36\,000 - 100\,x$$

or

$$150\,x = 36\,000$$

or $$x = \frac{36\,000}{150} = 240 \text{ mm} \quad \textbf{Ans.}$$

Example 2. *A beam 3 m long weighing 400 N is suspended in a horizontal position by two vertical strings, each of which can withstand a maximum tension of 350 N only. How far a body of 200 N weight be placed on the beam, so that one of the strings may just break ?*

Solution. The system of given forces is shown in Fig. 3.

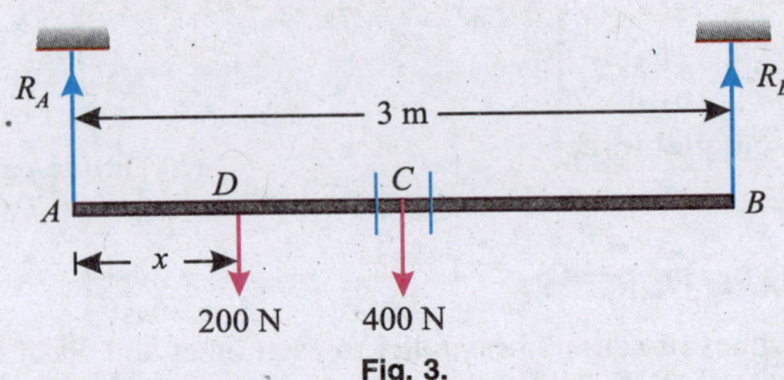

Fig. 3.

Let x = Distance between the body of weight 200 N and support A.

We know that one of the string (say A) will just break, when the tension will be 350 N. (*i.e.*, R_A = 350 N). Now taking clockwise and anti-clockwise moments about B and equating the same,

$$350 \times 3 = 200\,(3 - x) + 400 \times 1.5$$

or $$1\,050 = 600 - 200\,x + 600 = 1200 - 200\,x$$

or $$200\,x = 1\,200 - 1\,050 = 150$$

or $$x = \frac{150}{200} = 0.75 \text{ m} \quad \textbf{Ans.}$$

Example 3. *Two unlike parallel forces of magnitude 400 N and 100 N are acting in such a way that their lines of action are 150 mm apart. Determine the magnitude of the resultant force and the point at which it acts.*

Solution. Given : The system of given force is shown in Fig. 4

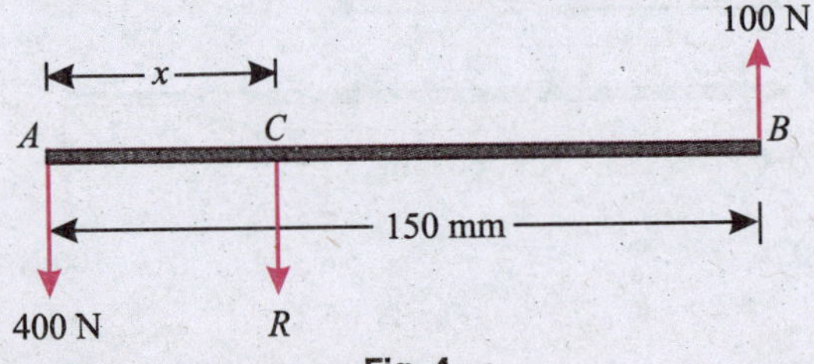

Fig. 4.

Magnitude of the resultant force

Since the given forces are unlike and parallel, therefore magnitude of the resultant force,

$$R = 400 - 100 = 300 \text{ N} \quad \textbf{Ans.}$$

Point where the resultant force acts

Let x = Distance between the lines of action of the resultant force and A in mm.

Now taking clockwise and anti-clockwise moments about A and equating the same,

$$300 \times x = 100 \times 150$$
$$= 15\,000$$

or $$x = \frac{15\ 000}{300} = 50 \text{ mm} \quad \textbf{Ans.}$$

Example 4. *A uniform beam AB of weight 100 N and 6 m long had two bodies of weights 60 N and 80 N suspended from its two ends as shown in Fig. 5.*

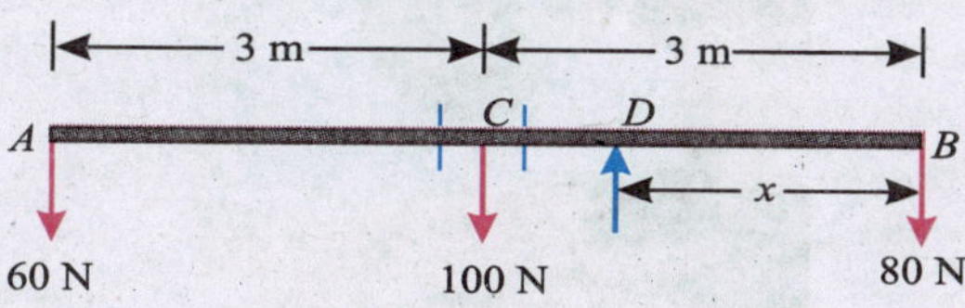

Fig. 5.

Find analytically at what point the beam should be supported, so that it may rest horizontally.

Solution. Given : Weight of rod AB = 100 N ; Length of rod AB = 6 mm and weight of the bodies supported at A and B = 60 N and 80 N.

Let x = Distance between B and the point where the beam should be supported.

We know that for the beam to rest horizontally, the moments of the weights should be equal. Now taking moments of the weights about D and equating the anti-clockwise moments with clock-wise moments,,

$$80x = 60\,(6 - x) + 100\,(3 - x)$$
$$= 360 - 60x + 300 - 100x = 660 - 160x$$
$$240x = 660$$

or $$x = \frac{660}{240} = 2.75 \text{ m} \quad \textbf{Ans.}$$

EXERCISE 1

1. Two like parallel forces of 10 N and 30 N act at the ends of a rod 200 mm long. Find magnitude of the resultant force and the point where it acts. [**Ans.** 40 N ; 150 mm]
2. Find the magnitude of two like parallel forces acting at a distance of 240 mm, whose resultant is 200 N and its line of action is at a distance of 60 mm from one of the forces. [**Ans.** 50 N ; 150 N]

   Hint.
   $$P + Q = 200$$
   $$Q \times 240 = 200 \times 60 = 12\ 000$$
   $$\therefore \quad Q = 50 \text{ N and } P = 200 - 50 = 150 \text{ N}$$
3. Two unlike parallel forces are acting at a distance of 450 mm from each other. The forces are equivalent to a single force of 90 N, which acts at a distance of 200 mm from the greater of the two forces. Find the magnitude of the forces. [**Ans.** 40 N ; 130 N]

7. COUPLE

A pair of two equal and unlike parallel forces (*i.e.* forces equal in magnitude, with lines of action parallel to each other and acting in opposite directions) is known as a couple.

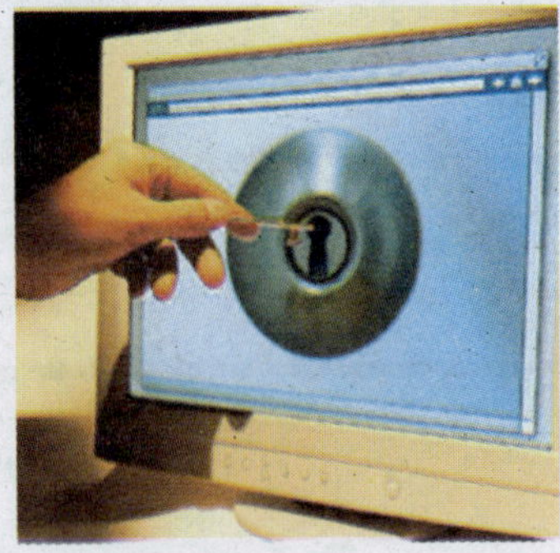

A couple is a pair of forces applied to the key of a lock.

As a matter of fact, a couple is unable to produce any translatory motion (*i.e.*, motion in a straight line). But it produces a motion of rotation in the body, on which it acts. The simplest example of a couple is the forces applied to the key of a lock, while locking or unlocking it.

8. ARM OF A COUPLE

The perpendicular distance (a), between the lines of action of the two equal and opposite parallel forces, is known *as arm of the couple* as shown in Fig. 6.

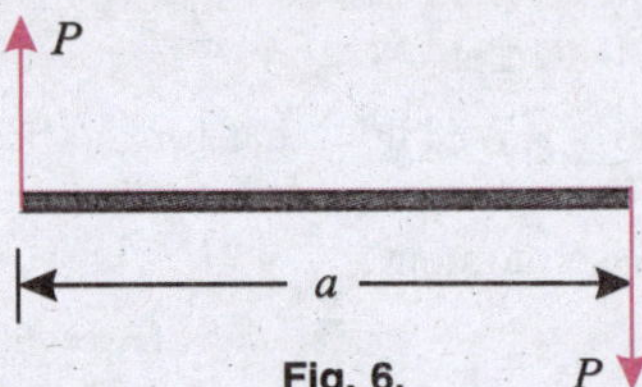

Fig. 6.

9. MOMENT OF A COUPLE

[GTU]

The moment of a couple is the product of the force (*i.e.*, one of the forces of the two equal and opposite parallel forces) and the arm of the couple. Mathematically:

Moment of a couple $= P \times a$

where P = Magnitude of the force, and

a = Arm of the couple.

10. CLASSIFICATION OF COUPLES

The couples may be, broadly, classified into the following two categories, depending upon their direction, in which the couple tends to rotate the body, on which it acts :

1. Clockwise couple, and 2. Anti-clockwise couple.

11. CLOCKWISE COUPLE

A couple, whose tendency is to rotate the body, on which it acts, in a clockwise direction, is known as a clockwise couple as shown in Fig. 7 (*a*). Such a couple is also called positive couple.

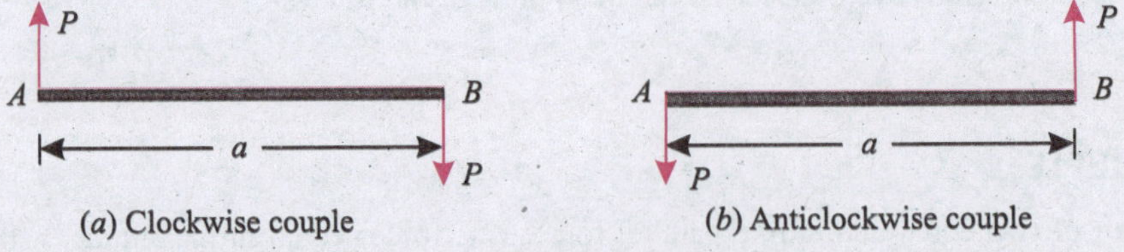

(*a*) Clockwise couple (*b*) Anticlockwise couple

Fig. 7.

12. ANTI-CLOCKWISE COUPLE

A couple, whose tendency is to rotate the body, on which it acts, in an anti-clockwise direction, is known as an anti-clockwise couple as shown in Fig. 7 (*b*). Such a couple is also called a negative couple.

13. CHARACTERISTICS OF A COUPLE

A couple (whether clockwise or anti-clockwise) has the following characteristics :

1. The algebraic sum of the forces, constituting the couple, is zero.
2. The algebraic sum of the moments of the forces, constituting the couple, about any point is the same, and equal to the moment of the couple itself.
3. A couple cannot be balanced by a single force. But it can be balanced only by a couple of opposite sense.
4. Any no. of coplaner couples can be reduced (*i.e.* added or subtracted) to a resultant single couple, whose magnitude will be equal to the algebraic sum of the moments of all the component couples.

Example 5. *A square ABCD has forces acting along its sides as shown in Fig. 8. Find the values of P and Q, if the system reduces to a couple. Also find magnitude of the couple, if the side of the square is 1 m.*

Solution. Given : Length of square = 1 m

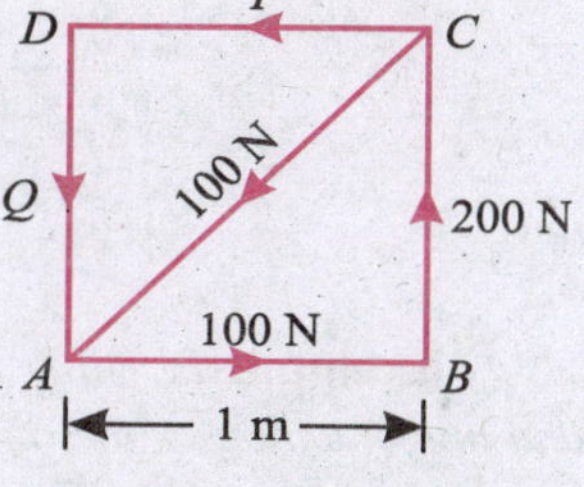

Fig. 8.

Values of P and Q

We know that if the system reduces to a couple, the resultant force in horizontal and vertical directions must be zero.

To find the force (P), Resolving the forces horizontally and equating their algebraic sum to zero,

$$100 - 100 \cos 45° - P = 0$$

or
$$P = 100 - 100 \cos 45° \text{ N}$$
$$= 100 - (100 \times 0.707) = 29.3 \text{ N} \quad \textbf{Ans.}$$

To find the force (Q), Resolving the forces vertically and equating their algebraic sum to zero,

$$200 - 100 \sin 45° - Q = 0$$

or
$$Q = 200 - (100 \times 0.707) = 129.3 \text{ N} \quad \textbf{Ans.}$$

Magnitude of the couple

We know that moment of the couple is equal to the algebraic sum of the moments about any point. Therefore moment of the couple (taking moments about A)

$$= (-200 \times 1) + (-P \times 1)$$
$$= -200 - (29.3 \times 1) \text{ N-m} = -229.3 \text{ N-m} \quad \textbf{Ans.}$$

Since the value of moment is negative, therefore the couple is anti-clockwise.

Example 6. *ABCD is a rectangle, such that AB = CD = a and BC = DA = b. A pair of forces equal to P act along AD and CB and another pair of forces equal to Q act along AB and CD respectively. Prove that the perpendicular distance between the resultants of P and Q at A and that of P and Q at C*

$$= \frac{(P \times a) - (Q \times b)}{\sqrt{(P^2 + Q^2)}}$$

Solution. Given : The system of forces is shown in Fig. 9.

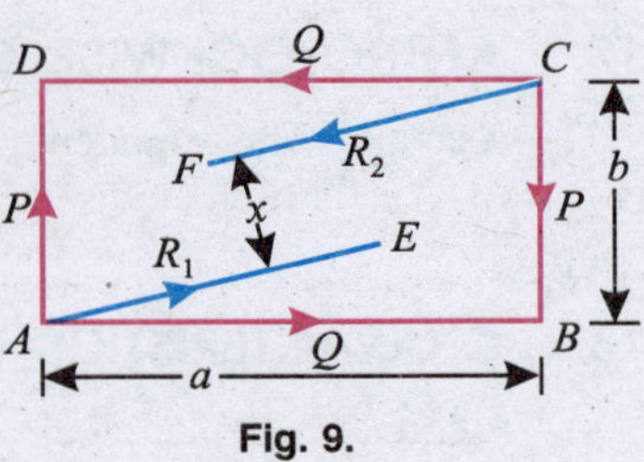

Fig. 9.

Let x = Perpendicular distance between the two resultants.

We know that the resultant of the forces P and Q at A,

$$R_1 = \sqrt{P^2 + Q^2} \quad ...(i)$$

and resultant of the forces P and Q at C,

$$R_2 = \sqrt{P^2 + Q^2} \quad ...(ii)$$

$\therefore$ Resultant $R = R_1 = R_2$ ...[from equations (*i*) and (*ii*)]

We know that moment of the force (P) about A,

$$M_1 = P \times a \quad ...(+ \text{ Due to clockwise})$$

and moment of the force (Q) about A,

$$M_2 = -Q \times b \quad ...(- \text{ Due to anti-clockwise})$$

$\therefore$ Net moment of the two couples

$$= (P \times a) - (Q \times b) \quad ...(iii)$$

and moment of the couple formed by the resultants

$$= R \times x = \sqrt{P^2 + Q^2} \times x \quad ...(iv)$$

Equating the moments (*iii*) and (*iv*),

$$\sqrt{P^2 + Q^2} \times x = (P \times a) - (Q \times b)$$

or

$$x = \frac{(P \times a) - (Q \times b)}{\sqrt{P^2 + Q^2}} \quad \textbf{Ans.}$$

Example 7. *Three forces, acting on a rigid body, are represented in magnitude, direction and line of action by the three sides of a triangle taken in order as shown in Fig.10.*

Fig. 10.

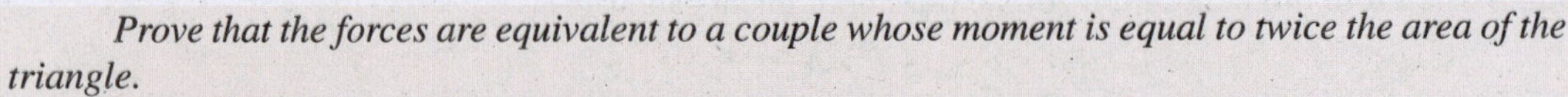

Prove that the forces are equivalent to a couple whose moment is equal to twice the area of the triangle.

Solution. The system of forces on the triangle ABC is shown in Fig. 11. Now complete the figure as discussed below :

1. Through A draw a line EF parallel to BC.
2. Extend CA to D, such that AD is equal to Q (*i.e.* CA).
3. Now apply two equal and opposite forces (P) at A represented by AE and AF.
4. Complete the parallelogram $ABED$ with adjacent sides AB and AD.

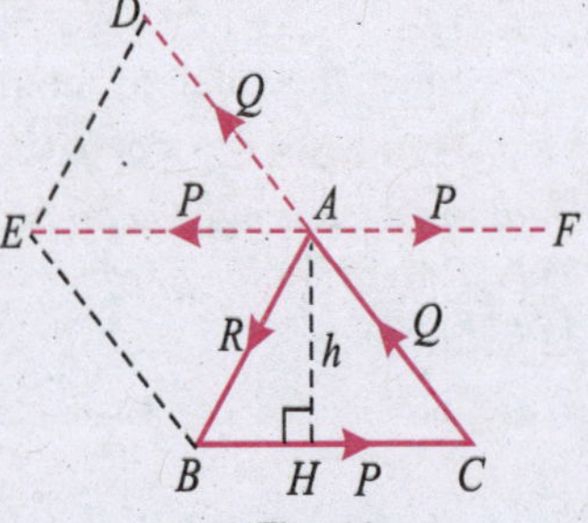

Fig. 11.

We know that the diagonal AE represents in magnitude and direction of the resultant of the two forces R and Q.

Thus the force AF (equal to P) will set the forces Q and R in equilibrium. Thus we are left with only two forces BC (equal to P) and AE (equal to P) which will constitute a couple. Now from A, draw AH perpendicular to BC. Let this perpendicular be equal to h.

We know that moment of the couple,

$$M = P \times a = P \times h \quad ...(i)$$

and area of triangle

$$= \frac{1}{2} \times \text{Base} \times \text{Height} = \frac{1}{2} \times P \times h \quad ...(ii)$$

From equations (*i*) and (*ii*), we find that moment of the couple

= Twice the area of triangle. **Ans.**

Example 8. *A machine component of length 2.5 metres and height 1 metre is carried upstairs by two men, who hold it by the front and back edges of its lower face.*

If the machine component is inclined at 30° to the horizontal and weighs 100 N, find how much of the weight each man supports ?

Solution. Given : Length of machine component = 2.5 m; Height of the component = 1 m ; Inclination = 30° and weight of component = 100 N

Let P = Weight supported by the man at A.

Q = Weight supported by the man at B.

C = Point where the vertical line through the centre of gravity cuts the lower face.

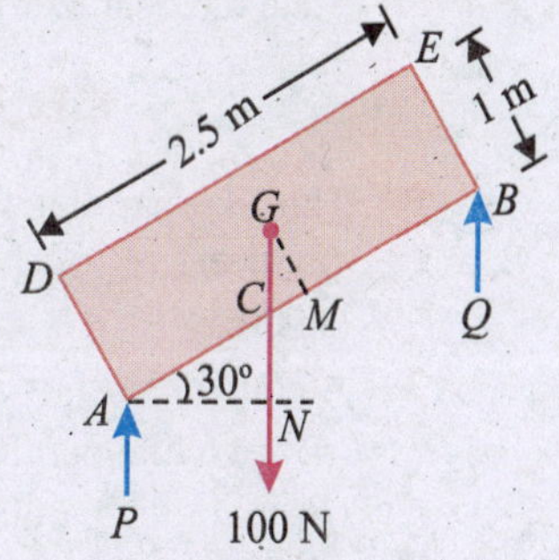

Fig. 12.

Now join G (*i.e.*, centre of gravity) with M (*i.e.*, mid-point of AB) as shown in Fig. 12.

From the geometry of the figure, we find that

$$GM = \frac{1}{2} = 0.5 \text{ m}$$

and

$$AM = \frac{2.5}{2} = 1.25 \text{ m}$$

$$\therefore \quad AC = AM - CM = 1.25 - GM \tan 30° \quad ...\left(\because \frac{CM}{GM} = \tan CGM = \tan 30°\right)$$

$$= 1.25 - (0.5 \times 0.577) = 0.96 \text{ m}$$

and

$$CB = AB - AC = 2.5 - 0.96 = 1.54 \text{ m}$$

We know that, the system is in equilibrium, therefore equations of equilibrium must be satisfied. Taking moments about point C and equting clockwise moment with anti-clockwise moment.

$$P \times CA = Q \times CB$$

or

$$P \times 0.96 = (100 - P)\, 1.54 = 154 - 1.54\, P \quad ...(\because P + Q = 100)$$

or

$$2.5\, P = 154$$

or

$$P = 61.6 \text{ N} \quad \textbf{Ans.}$$

and

$$Q = 100 - 61.6 = 38.4 \text{ N} \quad \textbf{Ans.}$$

EXERCISE 2

1. *ABCD* is rectangle, in which *AB* = *CD* = 100 mm and *BC* = *DA* = 80 mm. Forces of 100 N each act along *AB* and *CD* and forces of 50 N each at along *BC* and *DA*. Find the resultant moment of the two couples. **[Ans.** – 13 000 N-mm]
2. A square *ABCD* has sides equal to 200 mm. Forces of 150 N each act along *AB* and *CD* and 250 N act along *CB* and *AD*. Find the moment of the couple, which will keep the system in equilibrium. **[Ans.** – 20 000 N-mm]

QUESTIONS

1. What do you understand by the term 'parallel forces' ? Discuss their classifications.
2. Distinguish clearly between like forces and unlike forces.
3. What is a couple ? What is the arm of a couple and its moment ?
4. Discuss the classification of couples and explain clearly the difference between a positive couple and a negative couple.
5. State the characteristics of a couple.

OBJECTIVE TYPE QUESTIONS

1. The like parallel forces are those parallel forces, which are liked by the scientist and engineers.

   (*a*) Yes (*b*) No
2. A couple consists of

   (*a*) two like parallel forces of same magnitude.

   (*b*) two like parallel forces of different magnitudes.

   (*c*) two unlike parallel forces of same magnitude.

   (*d*) two unlike parallel forces of different magnitudes.
3. If the arm of a couple is doubled, its moment will

   (*a*) be halved (*b*) remain the same (*c*) be doubled
4. A couple can be balanced by a force equal to its magnitude.

   (*a*) Agree (*b*) Disagree
5. One of the characteristics of a couple is that it can cause a body to move in the direction of the greater force.

   (*a*) True (*b*) False
6. In a couple, the lines of action of the forces are

   (*a*) parallel (*b*) inclined (*c*) none of the two

ANSWERS

1. (*b*) 2. (*c*) 3. (*c*) 4. (*b*) 5. (*b*) 6. (*a*)

CHAPTER

5

Equilibrium of Forces

Contents

1. EQUILIBRIUM FORCES

[RTU, GTU]

In this chapter we shall study the different bodies and structures in the state of rest and under the condition of equilibrium of forces acting on it.

If the resultant of a number of forces, acting on a Stationary Rigid Body is Zero, the **body will be in equilibrium**. Such a set of forces whose resultant is zero are known as **equilibrium forces**.

1.1. RESULTANT AND EQUILIBRANT

When a body is acted upon by a number of forces, then the single force which will produce the same effect as poduced by all the given forces is known as **resultant force**.

A little consideration will show, that if we apply a force equal, opposite and collinear with the resultant force acting on the body, the body will come in equilibrium. The force, which brings the body in equilibrium, when applied to the body acted upon by set of forces is known as an **equilibrant**.

As a matter of fact, the equilibrant is equal to the resultant force in magnitude, but opposite in direction.

1.2. FREE BODY DIAGRAM

[DTU, UPTU, JNTU, AUT, IGNOU]

A structure may consist of more than one elements, portions and supports. Each portion can be isolated from the rest of the structure by incorporating the net effect of the remaining structure through a set of forces.

A force analysis of a structure is made in a simplified way by considering the equilibrium of a portion of the structure. To do this the portion is drawn seperately showing applied forces, self weight, and reactions at the pont of contact with other bodies of the structure.

The diagram of the isolated portion of the structure, along with the net effects of the different types forces on that portion, is known a **free-body diagram** of the portion.

We have to consider External and Internal Forces while drawing free body diagrams.

Free-body diagrams are useful in solving the forces and deformation of the system.

We shall be evaluating in detail the free body diagrams of the following bodies at respective places where ever they are discussed in different articles of this book.

1. A sphere or cylinder resting on frictionless surface.
2. A sphere attached to a string and resting against a vertical wall.
3. A sphere resting on V shaped groove.
4. A rod resting on V shaped groove.
5. A beam resting on supports.
6. A body hanging from spring.
7. A rod subjected to axial forces (tension or compression).

Example : Consider a sphere as shown in Fig. 1 (a). The sphere is supported by a string AB and is resting against a smooth vertical wall at C and also resting against a smooth horizontal floor at D.

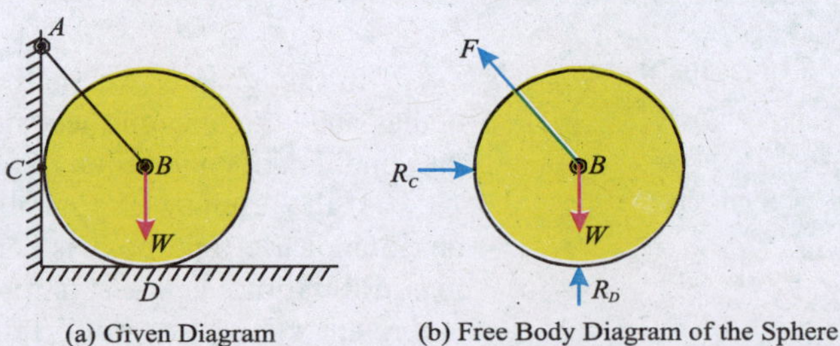

(a) Given Diagram (b) Free Body Diagram of the Sphere

Fig. 1.

Free-body diagram of the Sphere

Let us isolate the sphere completely from its supports and forces. Replace the vertical support with reaction R_C, horizontal support with reaction R_C and string AB with force F. Now the sphere is in equilibrium under action of following forces, which must pass through the centre of the sphere as shown in Fig. 1 (b):

(*i*) Reaction R_C at point C, normal to wall AC.

(ii) Force F in the direction of string.

(iii) Weight W of the sphere.

(iv) Reaction R_D at point D, normal to horizontal surface.

The Fig. 1 (b) illustrates the complete free-body diagram of the sphere.

2. *PRINCIPLES OF EQUILIBRIUM

[BPUT, AM-IETE]

Though there are many principles of equilibrium, yet the following three are important from the subject point of view :

1. *Two force principle.* As per this principle, if a body in equilibrium is acted upon by two forces, then they must be equal, opposite and collinear.
2. *Three force principle.* As per this principle, if a body in equilibrium is acted upon by three forces, then the resultant of any two forces must be equal, opposite and collinear with the third force.
3. *Four force principle.* As per this principle, if a body in equilibrium is acted upon by four forces, then the resultant of any two forces must be equal, opposite and collinear with the resultant of the other two forces.

2.1 CONDITIONS & EQUATIONS OF STATIC EQUILIBRIUM FOR COPLANAR CONCURRENT FORCES

If a number of **coplanar concurrent forces** act on a body, the body is said to be at rest or in equilibrium, if the resultant of all the forces acting on the body is zero. Or in other words, the horizontal component of all the forces (ΣH) and vertical component of all the forces ($\sum V$) must be zero. Mathematically, **conditions for equilibrium for coplanar concurrent forces** are:

$$\Sigma H = 0 \qquad \text{and} \qquad \Sigma V = 0 \qquad \ldots(i)$$

Thus, if a body is in equilibrium under the action of a number of coplanar concurrent forces, the equation (*i*) must be satisfied.

3. METHODS FOR THE EQUILIBRIUM OF COPLANAR FORCES

Though there are many methods of studying the equilibrium of forces, yet the following are important from the subject point of view :

1. Analytical method. **2.** Graphical method.

4. *ANALYTICAL METHOD FOR THE EQUILIBRIUM OF COPLANAR FORCES

The equilibrium of coplanar forces may be studied, analytically, by Lami's theorem as discussed below :

5. *LAMI'S THEOREM

[MU, JNTU, AUT, WBUT]

It states, "*If three coplanar forces acting at a point be in equilibrium, then each force is proportional to the sine of the angle between the other two.*" Mathematically,

$$\frac{P}{\sin\alpha} = \frac{Q}{\sin\beta} = \frac{R}{\sin\gamma}$$

P Q γ β α R

Fig. 2. Lami's theorem

where, P, Q, and R are three forces and α, β, γ are the angles as shown in Fig. 2.

Proof

Consider three coplanar forces P, Q, and R acting at a point O. Let the opposite angles to three forces be α , β and γ as shown in Fig. 3.

Now let us complete the parallelogram $OACB$ with OA and OB as adjacent sides as shown in the figure. We know that the resultant of two forces P and Q will be given by the diagonal OC both in magnitude and direction of the parallelogram $OACB$.

Since these forces are in equilibrium, therefore the resultant of the forces P and Q must be in line with OD and equal to R, but in opposite direction.

Fig. 3. Proof of Lami's theorem

From the geometry of the figure, we find

$$BC = P \quad \text{and} \quad AC = Q$$

$$\therefore \quad \angle AOC = (180° - \beta)$$

and $$\angle ACO = \angle BOC = (180° - \alpha)$$

$$\therefore \quad \angle CAO = 180° - (\angle AOC + \angle ACO)$$

$$= 180° - [(180° - \beta) + (180° - \alpha)]$$

$$= 180° - 180° + \beta - 180° + \alpha$$

$$= \alpha + \beta - 180°$$

But $$\alpha + \beta + \gamma = 360°$$

Subtracting 180° from both sides of the above equation,

$$(\alpha + \beta - 180°) + \gamma = 360° - 180° = 180°$$

or $$\angle CAO = 180° - \gamma$$

We know that in triangle AOC,

$$\frac{OA}{\sin\angle ACO} = \frac{AC}{\sin\angle AOC} = \frac{OC}{\sin\angle CAO}$$

or $$\frac{OA}{\sin(180° - \alpha)} = \frac{AC}{\sin(180° - \beta)} = \frac{OC}{\sin(180° - \gamma)}$$

or $$\frac{P}{\sin\alpha} = \frac{Q}{\sin\beta} = \frac{R}{\sin\gamma} \qquad \ldots[\because \sin(180° - \theta) = \sin\theta]$$

***Example 1.** *An electric light fixture weighting 15 N hangs from a point C, by two strings AC and BC. The string AC is inclined at 60° to the horizontal and BC at 45° to the horizontal as shown in Fig. 4*

Using Lami's theorem, or otherwise, determine the forces in the strings AC and BC. **[IGNOU]**

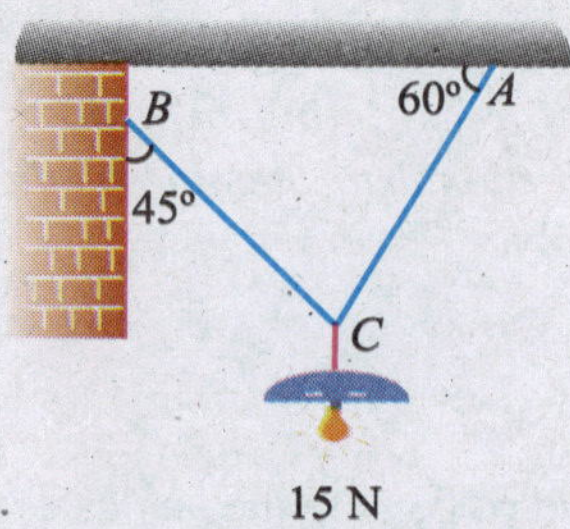

Fig. 4.

Solution. Given : Weight at C = 15 N

Let T_{AC} = Force in the string AC, and

T_{BC} = Force in the string BC.

The system of forces is shown in Fig. 5. From the geometry of the figure, we find that angle between T_{AC} and 15 N is 150° and angle between T_{BC} and 15 N is 135°.

Fig. 5.

$\therefore \quad \angle ACB = 180° - (45° + 60°) = 75°$

Applying Lami's equation at C,

$$\frac{15}{\sin 75°} = \frac{T_{AC}}{\sin 135°} = \frac{T_{BC}}{\sin 150°}$$

or

$$\frac{15}{\sin 75°} = \frac{T_{AC}}{\sin 45°} = \frac{T_{BC}}{\sin 30°} \qquad ...[\because \sin(180° - \theta) = \sin\theta]$$

Now

$$T_{AC} = \frac{15 \sin 45°}{\sin 75°} = \frac{15 \times 0.707}{0.9659} = 10.98 \text{ N} \quad \textbf{Ans.}$$

and

$$T_{BC} = \frac{15 \sin 30°}{\sin 75°} = \frac{15 \times 0.5}{0.9659} = 7.76 \text{ N} \quad \textbf{Ans.}$$

***Example 2.** *A string ABCD, attached to fixed points A and D has two equal weights of 1000 N attached to it at B and C. The weights rest with the portions AB and CD inclined at angles as shown in Fig. 6.*

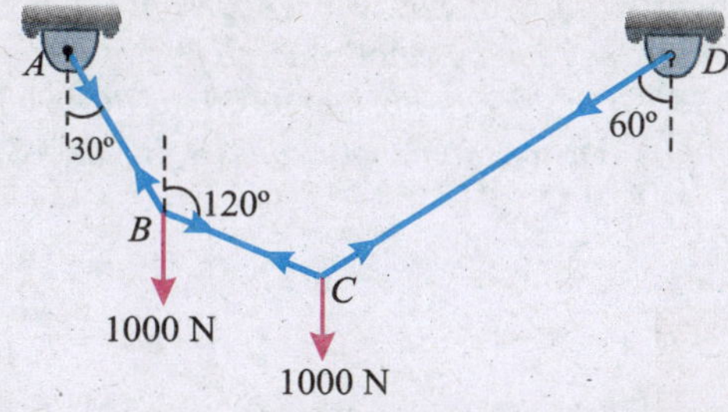

Fig. 6.

Find the tensions in the portions AB, BC and CD of the string, if the inclination of the portion BC with the vertical is 120°. **[WBUT]**

Solution. Given : Load at B = Load at C = 1000 N

For the sake of convenience, let us split up the string *ABCD* into two parts. The system of forces at joints *B* and *C* is shown in Fig. 7 (*a*) and (*b*).

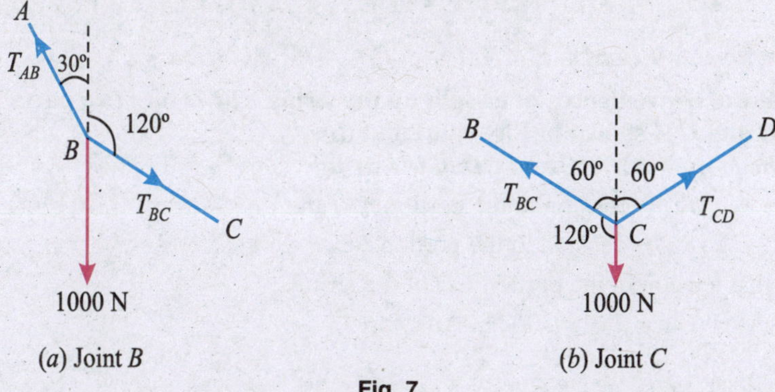

Fig. 7.

Let T_{AB} = Tension in the portion AB of the string,

T_{BC} = Tension in the portion BC of the string, and

T_{CD} = Tension in the portion CD of the string.

Applying Lami's equation at joint B,

$$\frac{T_{AB}}{\sin 60°} = \frac{T_{BC}}{\sin 150°} = \frac{1000}{\sin 150°}$$

or
$$\frac{T_{AB}}{\sin 60°} = \frac{T_{BC}}{\sin 30°} = \frac{1000}{\sin 30°} \quad \text{...}[\because \sin(180° - \theta) = \sin\theta]$$

Now
$$T_{AB} = \frac{1000 \sin 60°}{\sin 30°} = \frac{1000 \times 0.866}{0.5} = 1732 \text{ N} \quad \textbf{Ans.}$$

and
$$T_{BC} = \frac{1000 \sin 30°}{\sin 30°} = 1000 \text{ N} \quad \textbf{Ans.}$$

Again applying Lami's equation at joint C,

$$\frac{T_{BC}}{\sin 120°} = \frac{T_{CD}}{\sin 120°} = \frac{1000}{\sin 120°}$$

Now
$$T_{CD} = \frac{1000 \sin 120°}{\sin 120°} = 1000 \text{ N} \quad \textbf{Ans.}$$

****Example 3.** *A light string ABCDE whose extremity A is fixed, has weights W_1 and W_2 attached to it at B and C. It passes round a small smooth peg at D carrying a weight of 300 N at the free end E as shown in Fig. 8. If in the equilibrium position, BC is horizontal and AB and CD make 150° and 120° with BC, find (i) Tensions in the portion AB, BC and CD of the string and (ii) Magnitudes of W_1 and W_2.* **[RGPV, AUT]**

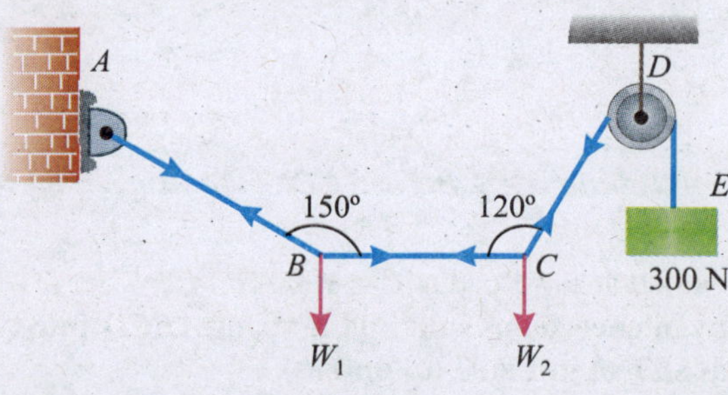

Fig. 8.

Solution. Given : Weight at E = 300 N

For the sake of convenience, let us split up the string *ABCD* into two parts. The system of forces at joints *B* and *C* is shown in Fig. 9. (*a*) and (*b*).

(*i*) *Tensions is the portion AB, BC and CD of the string*

Let T_{AB} = Tension in the portion AB, and

T_{BC} = Tension in the portion BC,

We know that tension in the portion CD of the string.

$$T_{CD} = T_{DE} = 300 \text{ N} \quad \textbf{Ans.}$$

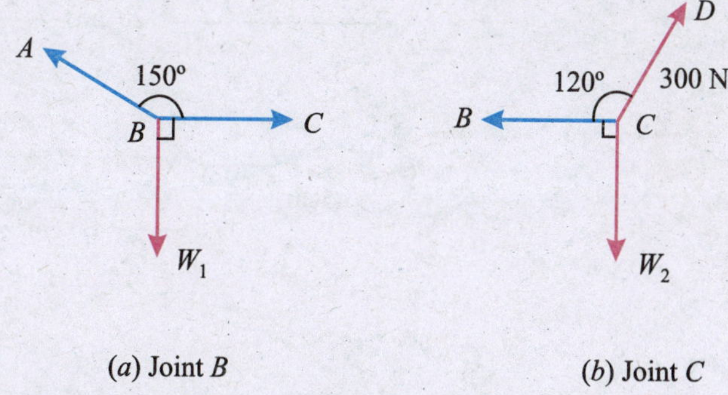

(a) Joint B (b) Joint C

Fig. 9.

Applying Lami's equation at C,

$$\frac{T_{BC}}{\sin 150^\circ} = \frac{W_2}{\sin 120^\circ} = \frac{300}{\sin 90^\circ}$$

or $$\frac{T_{BC}}{\sin 30^\circ} = \frac{W_2}{\sin 60^\circ} = \frac{300}{1} \qquad ...[\because \sin (180^\circ - \theta) = \sin \theta]$$

Now $$T_{BC} = 300 \sin 30^\circ = 300 \times 0.5 = 150 \text{ N}$$ **Ans.**

and $$W_2 = 300 \sin 60^\circ = 300 \times 0.866 = 259.8 \text{ N}$$

Again applying Lami's equation at B,

$$\frac{T_{AB}}{\sin 90^\circ} = \frac{W_1}{\sin 150^\circ} = \frac{T_{BC}}{\sin 120^\circ}$$

or $$\frac{T_{AB}}{1} = \frac{W_1}{\sin 30^\circ} = \frac{150}{\sin 60^\circ} \qquad ...[\because \sin (180^\circ - \theta) = \sin \theta]$$

Now $$T_{AB} = \frac{150}{\sin 60^\circ} = \frac{150}{0.866} = 173.2 \text{ N}$$ **Ans.**

and $$W_1 = \frac{150 \sin 30^\circ}{\sin 60^\circ} = \frac{150 \times 0.5}{0.866} = 86.6 \text{ N}$$

(ii) Magnitudes of W_1 and W_2

From the above calculations, we find that the magnitudes of W_1 and W_2 are 86.6 N and 259.8 N respectively. **Ans.**

EXERCISE 1

1. Two men carry a weight of 2 kN by means of two ropes fixed to the weight. One rope is inclined at 45° and the other at 30° with their vertices. Find the tension in each rope.
[**Ans.** 1.04 kN ; 1.46 kN]

2. Three forces acting on a particle are in equilibrium. The angles between the first and second is 90° and that between the second and third is 120°. Find the ratio of the forces.
[**Ans.** 1.73 : 1 : 2]

**3. A smooth sphere of weight W is supported by a string fastened to a point A on the smooth vertical wall, the other end is in contact with point B on the wall as shown in Fig. 10.

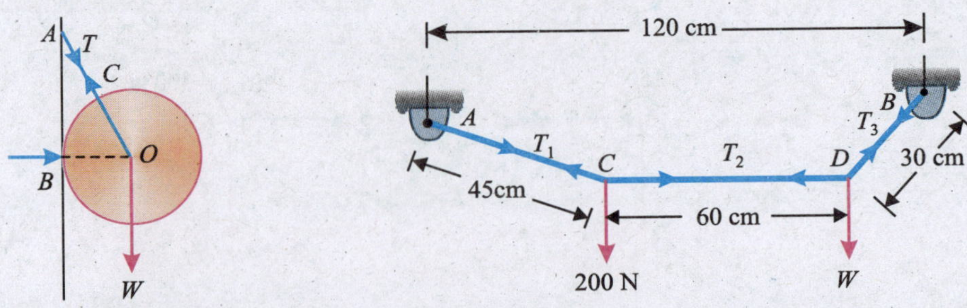

Fig. 10.   Fig. 11.

If length of the string AC is equal to radius of the sphere, find tension (T) in the string and reaction of the wall. **[Ans.** 1.155 W ; 0.577 W]

Hint. Since $AO = 2\ OB$, therefore $\angle AOB = 60°$ **[WBUT, BPUT, AMIE]**

4. A rope is connected between two points A and B 120 cm apart at the same level. A load of 200 N is suspended from a point C on the rope 45 cm from A as shown in Fig. 11. Find the load, that should be suspended from the rope D 30 cm from B, which will keep the rope CD horizontal. **[Ans.** 400 N]

5. A uniform sphere of weight W rests between a smooth vertical plane and a smooth plane inclined at an angle θ with the vertical plane. Find the reaction at the contact surfaces. **[Ans.** $W \cot \theta$; $W \operatorname{cosec} \theta$]

Example 4. *Two equal heavy spheres of 50 mm radius are in equilibrium within a smooth cup of 150 mm radius. Show that the reaction between the cup of one sphere is double than that between the two spheres.*

Solution. Given : Radius of spheres = 50 mm and radius of the cup = 150 mm.

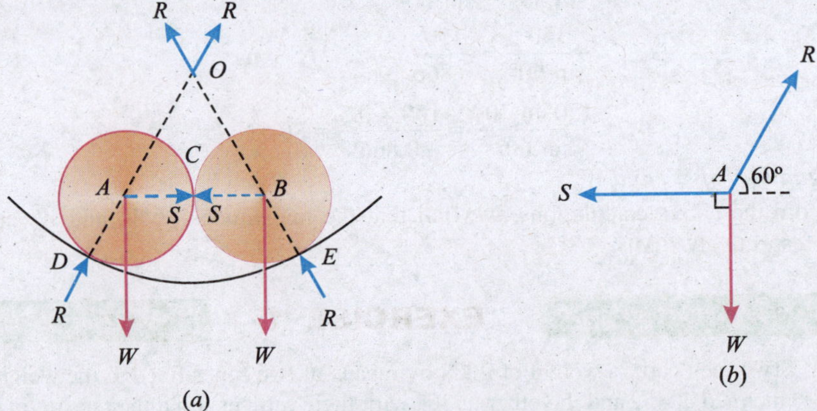

Fig. 12.

The two spheres with centres A and B, lying in equilibrium, in the cup with O as centre are shown in Fig. 12 (*a*). Let the two spheres touch each other at C and touch the cup at D and E respectively.

Let R = Reactions between the spheres and cup, and

S = Reaction between the two spheres at C.

From the geometry of the figure, we find that OD = 150 mm and AD = 50 mm. Therefore OA = 100 mm. Similarly OB = 100 mm. We also find that AB = 100 mm. Therefore OAB is an equilateral triangle. The system of forces at A is shown in Fig. 12 (*b*).

Applying Lami's equation at A,

$$\frac{R}{\sin 90°} = \frac{W}{\sin 120°} = \frac{S}{\sin 150°}$$

or
$$\frac{R}{1} = \frac{W}{\sin 60°} = \frac{S}{\sin 30°}$$

Now
$$R = \frac{S}{\sin 30°} = \frac{S}{0.5} = 2S$$

Hence the reaction between the cup and the sphere is double than that between the two spheres. **Ans.**

***Example 5.** *A smooth circular cylinder of radius 1.5 meter is lying in a triangular groove, one side of which makes 15° angle and the other 40° angle with the horizontal. Find the reactions at the surfaces of contact, if there is no friction and the cylinder weights 100 N.* **[MU, WBUT, AMIE]**

Solution. Given : Weight of cylinder = 100 N

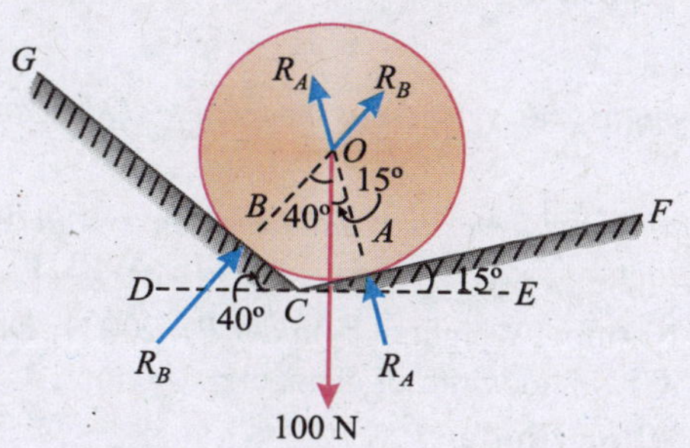

(*a*) Free Body Diagram of Cylinder

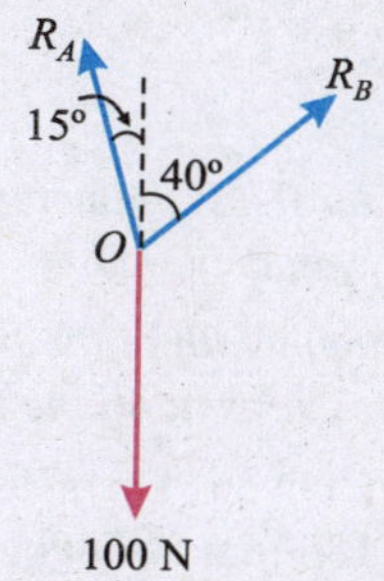

(*b*) Force Diagram of Cylinder

Fig. 13.

Let R_A = Reaction at A, and

R_B = Reaction at B.

First of all , consider the equlibrium of the cylinder.

Let us draw the free body diagram for the cylinder.

The cylinder is in equilibrium under the action of the following three forces which must pass through the centre of of cylinder as shown in Fig. 13 (*a*).

1. Weight of the cylinder (100 N) acting downwards.
2. Reaction (R_A) of the cylinder at the left side.
3. Reaction (R_B) of the cylinder at the right side.

Moreover, as there is no friction, the reactions R_A and R_B must be normal to the surfaces as shown in Fig. 13 (*a*). The system of forces is shown in Fig. 13 (*b*).

Applying Lami's equation, at O,

$$\frac{R_A}{\sin(180° - 40°)} = \frac{R_B}{\sin(180° - 15°)} = \frac{100}{\sin(15° + 40°)}$$

or
$$\frac{R_A}{\sin 40°} = \frac{R_B}{\sin 15°} = \frac{100}{\sin 55°}$$

Now $$R_A = \frac{100 \times \sin 40°}{\sin 55°} = \frac{100 \times 0.6428}{0.8192} = 78.5 \text{ N}$$ **Ans.**

and $$R_B = \frac{100 \times \sin 15°}{\sin 55°} = \frac{100 \times 0.2588}{0.8192} = 31.6 \text{ N}$$ **Ans.**

***Example 6.** *Two cylinders P and Q rest in a channel as shown in Fig. 14.*

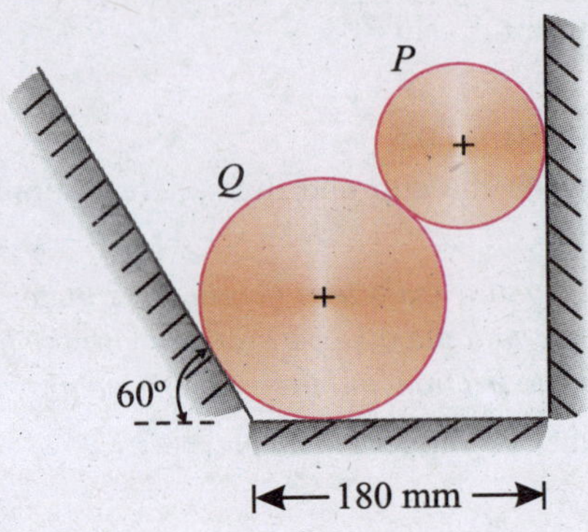

Fig. 14.

The cylinder P has diameter of 100 mm and weighs 200 N, whereas the cylinder Q has diameter of 180 mm and weighs 500 N.

If the bottom width of the box is 180 mm, with one side vertical and the other inclined at 60°, determine the pressures at all the four points of contact. **[MU, WBUT, JNTU]**

Solution. Given : Diameter of cylinder P = 100 mm ; Weight of cylinder P = 200 N ; Diameter of cylinder Q = 180 mm ; Weight of cylinder Q = 500 N and width of channel = 180 mm.

First of all, consider the equilibrium of the cylinder *P*.

Let us draw the free body diagram for the cylinder *P*.

The cylinder P is in equilibrium under the action of the following three forces which must pass through A *i.e.*, the centre of the cylinder P as shown in Fig. 15 (*a*).

1. Weight of the cylinder (200 N) acting downwards.
2. Reaction (R_1) of the cylinder P at the vertical side.
3. Reaction (R_2) of the cylinder P at the point of contact with the cylinder Q.

From the geometry of the figure, we find that

$$ED = \text{Radius of cylinder } P = \frac{100}{2} = 50 \text{ mm}$$

Similarly, $$BF = \text{Radius of cylinder } Q = \frac{180}{2} = 90 \text{ mm}$$

and $$\angle BCF = 60°$$

$\therefore$ $$CF = BF \cot 60° = 90 \times 0.577 = 52 \text{ mm}$$

$\therefore$ $$FE = BG = 180 - (52 + 50) = 78 \text{ mm}$$

and $$AB = 50 + 90 = 140 \text{ mm}$$

$\therefore$ $$\cos \angle ABG = \frac{BG}{AB} = \frac{78}{140} = 0.5571$$

or $$\angle ABG = 56.1°$$

The system of forces at A is shown in Fig. 15 (b).

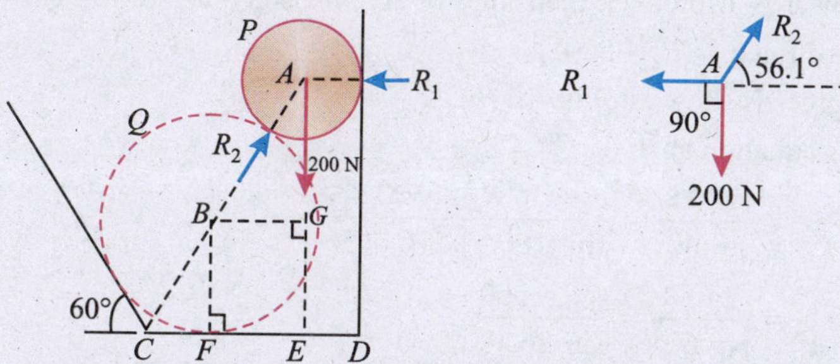

(a) Free body diagram of cylinder P (b) Force diagram of cylinder P

Fig. 15. For cylinder P

Applying Lami's equation at A,

$$\frac{R_1}{\sin(90° + 56.1°)} = \frac{R_2}{\sin 90°} = \frac{200}{\sin(180° - 56.1°)}$$

or
$$\frac{R_1}{\cos 56.1°} = \frac{R_2}{1} = \frac{200}{\sin 56.1°}$$

Now
$$R_1 = \frac{200 \cos 56.1°}{\sin 56.1°} = \frac{200 \times 0.5571}{0.830} = 134.2 \text{ N} \quad \textbf{Ans.}$$

and
$$R_2 = \frac{200}{\sin 56.1°} = \frac{200}{0.8300} = 240.8 \text{ N} \quad \textbf{Ans.}$$

Now consider the equilibrium of the cylinder Q.

Let us draw the free body diagram for the cylinder Q.

The cylinder Q is in equilibrium under the action of the following four forces, which must pass through the centre of the cylinder as shown in Fig. 16 (a).

1. Weight of the cylinder Q (500 N) acting downwards.
2. Reaction R_2 equal to 240.8 N of the cylinder P on cylinder Q.
3. Reaction R_3 of the cylinder Q on the inclined surface.
4. Reaction R_4 of the cylinder Q on the base of the channel.

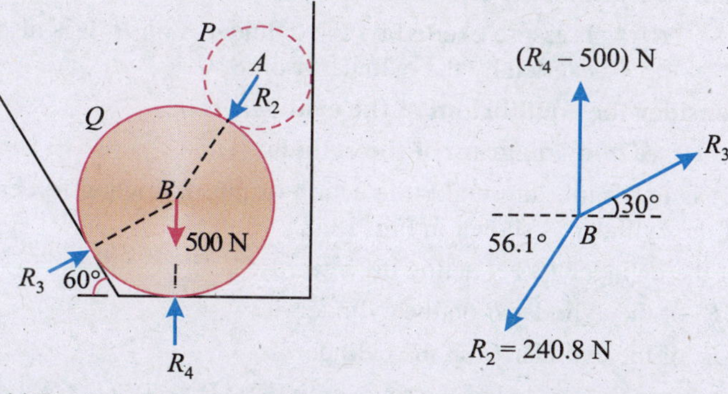

(a) Free body diagram of cylinder Q (b) Force diagram of cylinder Q

Fig. 16. For cylinder Q

A little consideration will show, that the weight of the cylinder Q is acting downwards and the reaction R_4 is acting upwards. Moreover, their lines of action also coincide with each other.

$\therefore$ Net downward force = $(R_4 - 500)$ N

The system of forces is shown in Fig. 16 (*b*).

Applying Lami's equation at B,

$$\frac{R_3}{\sin(90° + 56.1°)} = \frac{240.8}{\sin 60°} = \frac{R_4 - 500}{\sin(180° + 30° - 56.1°)}$$

or
$$\frac{R_3}{\cos 56.1°} = \frac{240.8}{\sin 60°} = \frac{R_4 - 500}{\sin 26.1°}$$

Now
$$R_3 = \frac{240.8 \times \cos 56.1°}{\sin 60°} = \frac{240.8 \times 0.5577}{0.866} = 155 \text{ N} \quad \textbf{Ans.}$$

and
$$R_4 - 500 = \frac{240.8 \times \sin 26.1°}{\sin 60°} = \frac{240.8 \times 399}{0.866} = 122.3 \text{ N}$$

$\therefore$
$$R_4 = 122.3 + 500 = 622.3 \text{ N} \quad \textbf{Ans.}$$

****Example 7.** *Three cylinders weighting 100 N each and of 80 mm diameter are placed in a channel of 180 mm width as shown in Fig. 17.*

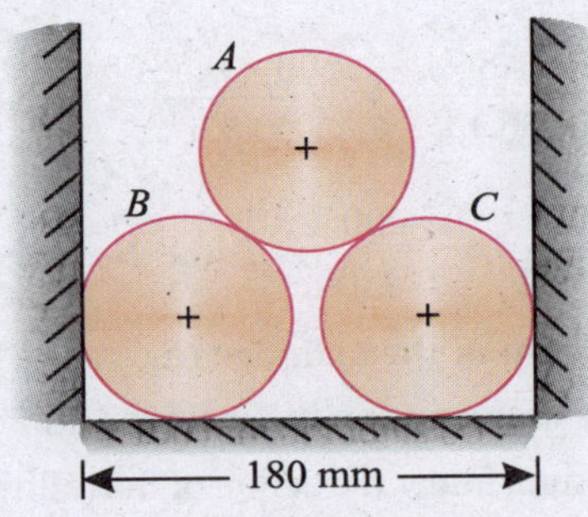

Fig. 17.

Determine the pressure exerted by (i) the cylinder A on B at the point of contact (ii) the cylinder B on the base and (iii) the cylinder B on the wall. **[AM-IETE]**

Solution. Given : Weight of each cylinder = 100 N ; Dia. of each cylinder = 80 mm and width of channel = 180 mm

(i) Pressure exerted by the cylinder A on the cylinder B

Let R_1 = Pressure exerted by the cylinder A on B. It is also equal to pressure exerted by the cylinder A on B.

First of all, consider the equilibrium of the cylinder *A*.

Let us draw the free body diagram of the cylinder *A*.

The cylinder A is in equilibrium under the action of the following forces, which must pass through the centre of the cylinder as shown in Fig. 18 (*a*).

1. Weight of the cylinder 100 N acting downwards.
2. Reaction R_1 of the cylinder B on the cylinder A.
3. Reaction R_2 of the cylinder C on the cylinder A.

Now join the centres O, P and Q of the three cylinders. Bisect PQ at S and join OS as shown in Fig. 18 (*b*).

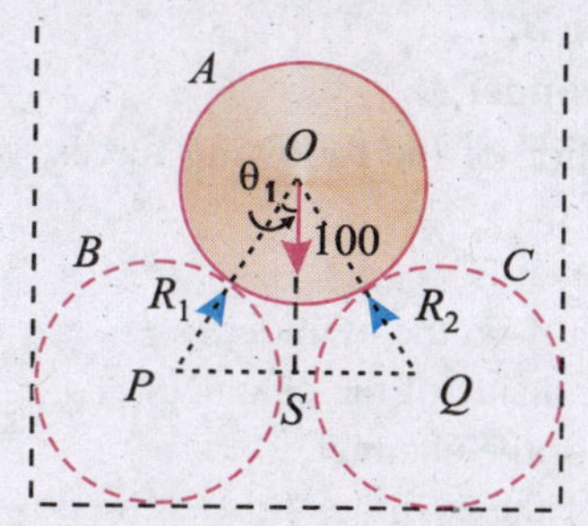

(*a*) Free body diagram of cylinder *A*

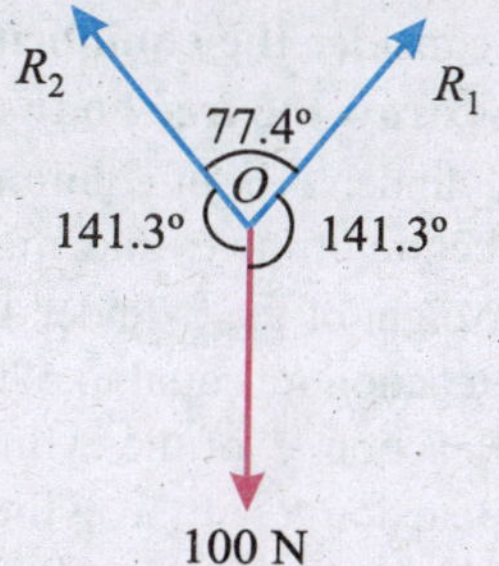

(*b*) Force diagram of cylinder *A*

Fig. 18. For cylinder *A*

From the geometry of the triangle *OPS*, we find that

$$OP = 40 + 40 = 80 \text{ mm}$$

and $$PS = 90 - 40 = 50 \text{ mm}$$

$$\therefore \quad \sin \angle POS = \frac{PS}{OP} = \frac{50}{80} = 0.625$$

or $$\angle POS = 38.7°$$

Since the triangle *OSQ* is similar to the triangle *OPS*, therefore $\angle SOQ$ is also equal to 38.7°. Thus the angle between R_1 and R_2 is 2 × 38.7° = 77.4°.

And angle between R_1 and *OS* (also between R_2 and *OS*) = 180° – 38.7° = 141.3°

The system of forces at *O* is shown in Fig. 17 (*b*). Applying Lami's equation at *O*,

$$\frac{R_1}{\sin 141.3°} = \frac{R_2}{\sin 141.3°} = \frac{100}{\sin 77.4°}$$

or $$\frac{R_1}{\sin 38.7°} = \frac{R_2}{\sin 38.7°} = \frac{100}{\sin 77.4°} \qquad ...[\because \sin (180° - \theta) = \sin \theta]$$

Now $$R_1 = \frac{100 \times \sin 38.7°}{\sin 77.4°} = \frac{100 \times 0.6252}{0.9759} = 64.0 \text{ N} \quad \textbf{Ans.}$$

Similarly $$R_2 = R_1 = 64.0 \text{ N} \quad \textbf{Ans.}$$

(ii) Pressure exerted by the cylinder B on the base

Let R_3 = Pressure exerted by the cylinder *B* on the wall, and

R_4 = Pressure exerted by the cylinder *B* on the base.

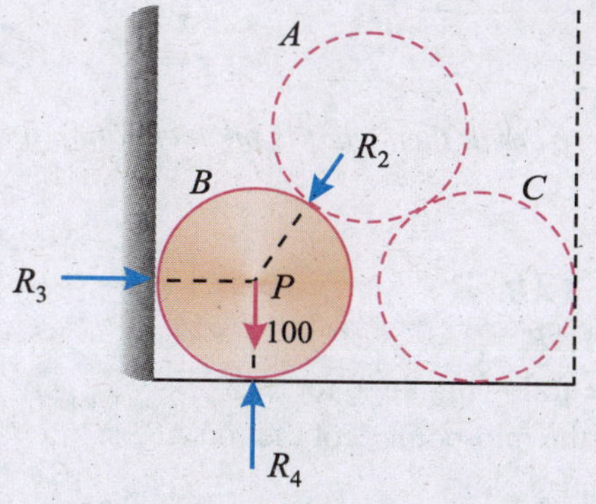

(*a*) Free body diagram of cylinder *B*

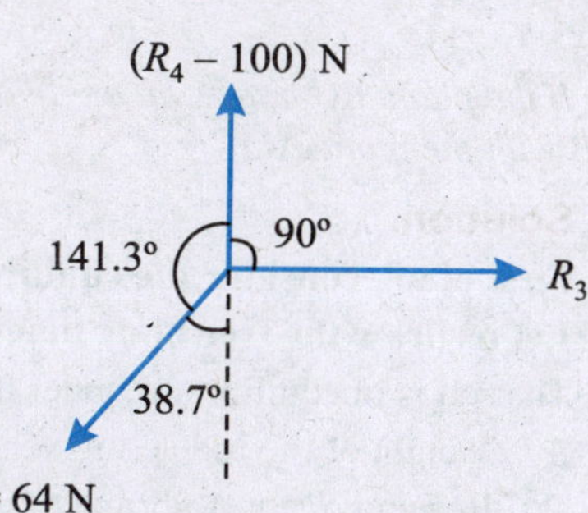

(*b*) Force diagram of cylinder *B*

Fig. 19. For cylinder *B*

Now consider the equilibrium of the cylinder *B*.

Let us draw the free body diagram of the cylinder *B*.

The cylinder *B* is in equilibrium under the action of the following forces, which must pass through the centre of the cylinder as shown in Fig. 19 (*a*).

1. Weight of the cylinder 100 N acting downwards.
2. Reaction R_2 equal to 64.0 N of the cylinder *A* on the cylinder *B*.
3. Reaction R_3 of the cylinder *B* on the vertical side of the channel.
4. Reaction R_4 of the cylinder *B* on the base of the channel.

A little consideration will show that weight of the cylinder *B* is acting downwards and the reaction R_4 is acting upwards. Moreover, their lines of action also coincide with each other. Therefore net downward force will be equal to $(R_4 - 100)$ N.

The system of forces is shown in Fig. 19 (*b*). Applying Lami's equation at *P*,

$$\frac{64}{\sin 90°} = \frac{R_3}{\sin (180° - 38.7°)} = \frac{(R_4 - 100)}{\sin (90° + 38.7°)}$$

or
$$\frac{64}{1} = \frac{R_3}{\sin 38.7°} = \frac{R_4 - 100}{\cos 38.7°}$$

Now
$$R_4 - 100 = 64 \cos 38.7° = 64 \times 0.7804 = 50 \text{ N}$$

or
$$R_4 = 50 + 100 = 150 \text{ N} \quad \textbf{Ans.}$$

(*iii*) *Pressure exerted by the cylinder B on the wall*

From the above Lami's equation, we also find that

$$R_3 = 64 \sin 38.7° = 64 \times 0.6252 = 40 \text{ N} \quad \textbf{Ans.}$$

Note. Since the cylinders *B* and *C* are symmetrically placed, therefore pressures exerted by the cylinder *C* on the wall as well as channel will be the same as those exerted by the cylinder *B*.

Example 8. *A uniform rod AB remains in equilibrium position resting on a smooth inclined planes AC and BC, which are at an angle of 90° as shown in figure given below :*

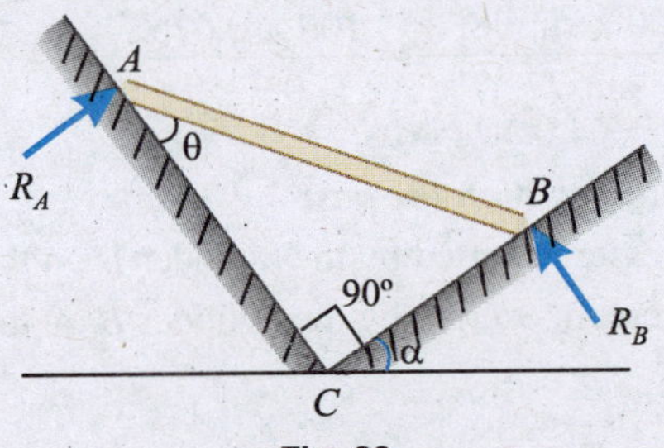

Fig. 20.

If the plane BC makes an angle of α with the horizontal, then what is the inclination θ of the rod AB with the plane AC.

Solution.

First of all, consider the equilibrium of the rod *AB*.

Let us draw the free body diagram of the rod *AB*.

The rod is in equilibrium under the action of the following three forces,

1. Weight of the rod acting vertically through the mid-point *G* of the rod *AB*.
2. Reaction R_A at *A* normal to the plane *AC*, and
3. Reaction R_B at *B* normal to the plane *BC*.

Let these three forces meet at point *D* as shown in Fig. 21

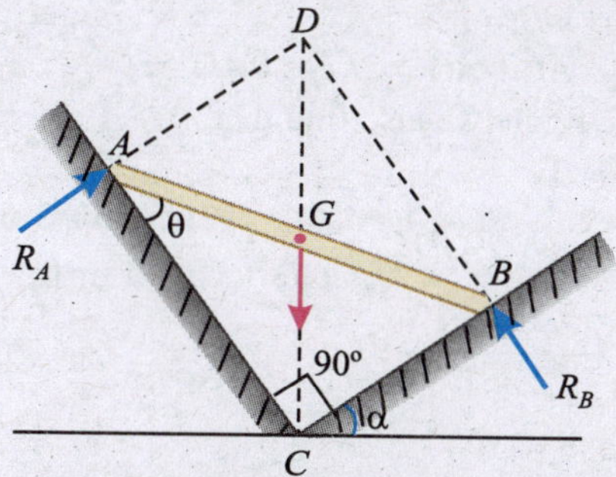

Fig. 21. Free body diagram of rod *AB*

Since *AD* is perpendicular to *AC* and *BD* is perpendicular to *BC*, therefore *AD* is parallel to *BC* and *BD* is parallel to *AC*.

and $\angle ADB = 90°$

The figure *ADBC* is a rectangle whose diagonal *DGC* is vertical

$$GA = GC$$

$$\angle GAC = \angle GCA$$

$\therefore$ $\theta = \alpha$ **Ans.**

Example 9. *A uniform rod AB of length 3r remains in equilibrium on a hemispherical bowl of radius r as shown in Fig. 22.*

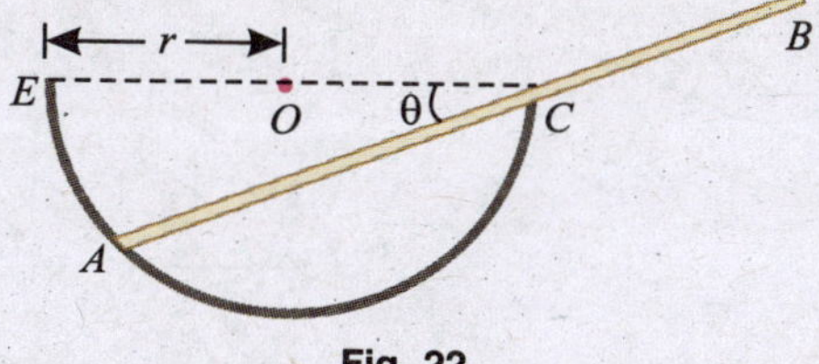

Fig. 22.

Ignoring friction find the inclination of the rod (θ) *with the horizontal.*

Solution. Given : Length of the rod $AB = 3r$ and radius of hemispherical ball $= r$

First of all, consider the equilibrium of the rod *AB*.

Let us draw the free body diagram of the rod *AB*.

The rod is in equilibrium under the action of the following three forces as shown in Fig. 23.

1. Weight of the rod (*W*) acting vertically downwards through the mid-point *G* of the rod *AB*
2. Reaction at *A* acting in the direction *AO*
3. Reaction at *C* acting at the right angle to *AB*

From the geometry of the figure we know that

$AD = 2r$

$AC = AD \cos\theta = 2r \cos\theta$

$CD = AD \sin\theta = 2r \sin\theta$

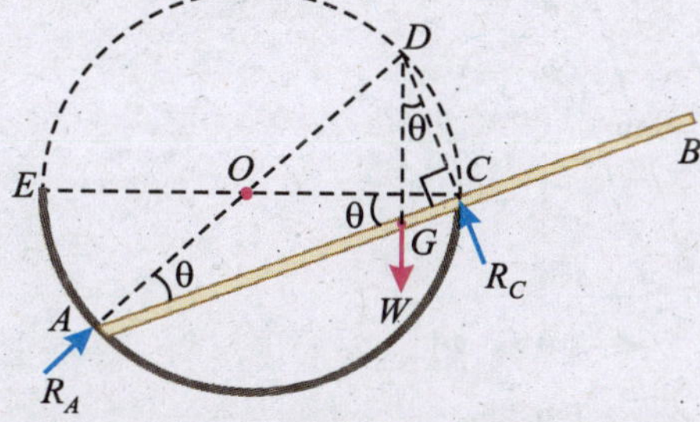

Fig. 23. Free body diagram of rod *AB*

$$AG = GB = 1.5r$$
$$GC = AC - AG = 2r\cos\theta - 1.5r$$

From the geometry of the figure we also find that

$$\angle GDC = \theta$$

$$\therefore \quad \tan\theta = \tan\angle GDC = \frac{GC}{CD} = \frac{2r\cos\theta - 1.5r}{2r\sin\theta} = \frac{r(2\cos\theta - 1.5)}{2r\sin\theta}$$

or $$\frac{\sin\theta}{\cos\theta} = \frac{2\cos\theta - 1.5}{2\sin\theta}$$

or $$2\sin^2\theta = 2\cos^2\theta - 1.5\cos\theta$$

or $$2(\cos^2\theta - \sin^2\theta) = 1.5\cos\theta$$

or $$2(2\cos^2\theta - 1) = 1.5\cos\theta$$

or $$4\cos^2\theta - 1.5\cos\theta - 2 = 0$$

Solving it as a quadratic equation,

$$\cos\theta = \frac{1.5 + \sqrt{2.25 + 32}}{8} = 0.9 \quad \text{or} \quad \theta = 25.8° \quad \textbf{Ans.}$$

Example 10. *Fig. 24 shows a shear leg crane lifting a load of 250 kN.*

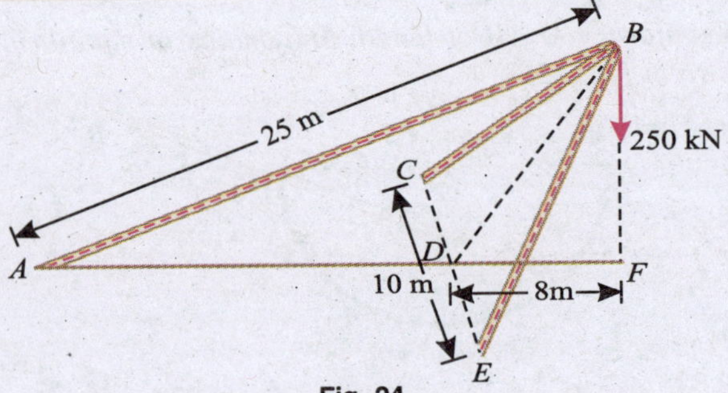

Fig. 24.

The legs BC and BE are 20 m long and 10 m apart at the base. The back stay AB is 25 m long. If all the members are pin-jointed at A, C and E, at the same level, find the forces in all the three members of the crane.

Solution. Given : Weight at $B = 250$ kN

Let P = Force in each members BC and BE, and

T = Force in the member AB.

From the geometry of the figure, we find that the points $ABDF$ lie in one vertical plane, in which $\angle AFB$ is a right angle. Moreover, the points $BCDE$ also lie in one plane, in which $\angle BDC$ and $\angle BDE$ are also right angles and D is in the mid point of C and E.

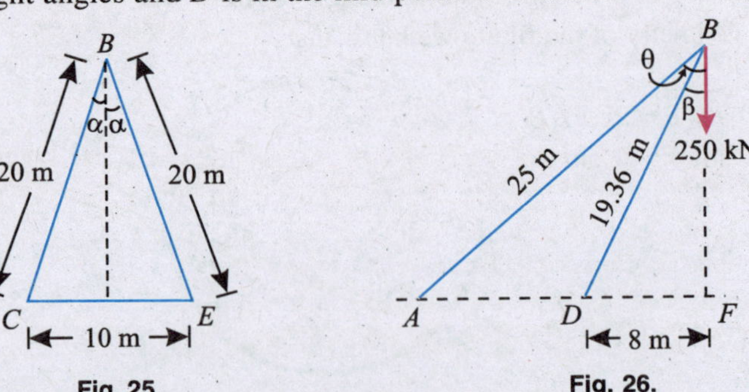

Fig. 25. **Fig. 26.**

First of all, draw the isosceles triangle *BCE* with *BC* and *BE* each equal to 20 m and *CE* equal to 10 m with *D* as mid point of *C* and *E* as shown in Fig. 25.

Now in triangle *BCD*, we find that

$$\sin \alpha = \frac{5}{20} = 0.25 \quad \text{or} \quad \alpha = 14.5°$$

and
$$BD = \sqrt{(20)^2 - (5)^2} = 19.36 \text{ m}$$

Now draw the triangle *ABF* with *DF* equal to 8 m, *AFB* equal to 90°, *DB* equal to 19.36 m and *AB* equal to 25 m as shown in Fig. 26.

From the geometry of the triangle *BDF*, we find that

$$\sin \beta = \frac{DF}{BD} = \frac{8}{19.36} = 0.4132 \quad \text{or} \quad \beta = 24.4°$$

and
$$BF = \sqrt{(19.36)^2 - (8)^2} = 17.63 \text{ m}$$

From the geometry of the triangle *ABF*, we also find that

$$\cos \angle ABF = \frac{BF}{AB} = \frac{17.63}{25} = 0.7052 \quad \text{or} \quad \angle ABF = 45.1°$$

$$\therefore \quad \theta = 45.1° - 24.4° = 20.7°$$

We know that resultant of the forces in members *BC* and *BE* (acting along *BD*)

$$R = 2P \cos \alpha = 2P \cos 14.5°$$
$$= 2P \times 0.9680 = 1.936\, P$$

The system of forces acting at *B* is shown in Fig 27.

Applying Lami's equation at *B*,

$$\frac{T}{\sin(180° - 24.4°)} = \frac{1.936\,P}{\sin 45.1°} = \frac{250}{\sin(180° - 20.7°)}$$

or
$$\frac{T}{\sin 24.4°} = \frac{1.936\,P}{\sin 45.1°} = \frac{250}{\sin 20.7°}$$

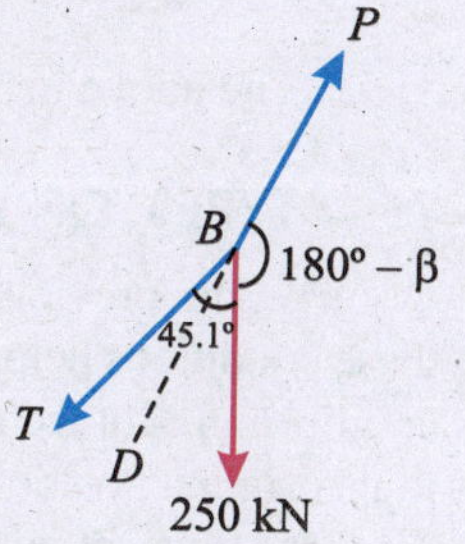

Fig. 27.

Now
$$T = \frac{250 \times \sin 24.4°}{\sin 20.7°} = \frac{250 \times 0.4131}{0.3543} = 291.5 \text{ kN} \quad \textbf{Ans.}$$

and
$$P = \frac{250 \times \sin 45.1°}{1.936 \times \sin 20.7°} = \frac{250 \times 0.7090}{1.936 \times 0.3543} = 258.4 \text{ kN} \quad \textbf{Ans.}$$

6. **CONDITIONS & EQUATIONS OF STATIC EQUILIBRIUM FOR COPLANAR NON-CONCURRENT FORCES

[RTU, AM-IETE]

Consider a body acted upon by a number of coplaner non-concurrent forces. A little consideration will show, that as a result of these forces, the body may have any one of the following states:

1. The body may move in any one direction.
2. The body may rotate about itself without moving.
3. The body may move in any one direction and at the same time it may also rotate about itself.
4. The body may be completely at rest.

Now we shall study the above mentioned four states one by one.

1. If the body moves in any direction, it means that there is a resultant force acting on it. A little consideration will show, that if the body is to be at rest or in equilibrium, the resultant

force causing movement must be zero. Or in other words, the horizontal component of all the forces (ΣH) and vertical component of all the forces (ΣV) must be zero. Mathematically,

$$\Sigma H = 0 \qquad \text{and} \qquad \Sigma V = 0$$

2. If the body rotates about itself, without moving, it means that there is a single resultant couple acting on it with no resultant force. A little consideration will show, that if the body is to be at rest or in equilibrium, the moment of the couple causing rotation must be zero. Or in other words, the resultant moment of all the forces (ΣM) must be zero. Mathematically,

$$\Sigma M = 0$$

3. If the body moves in any direction and at the same time it rotates about itself, if means that there is a resultant force and also a resultant couple acting on it. A little consideration will show, that if the body is to be at rest or in equilibrium, the resultant force causing movements and the reusltant moment of the couple causing rotation must be zero. Or in other words, horizontal component of all the forces (ΣH), vertical component of all the forces (ΣV) and resultant moment of all the forces (Σ M) must be zero. Mathematically,

$$\Sigma H = 0, \quad \Sigma V = 0 \qquad \text{and} \qquad \Sigma M = 0$$

4. If the body is completely at rest, it necessarily means that there is neither a resultant force nor a couple acting on it. A little consideration will show, that in this case the following conditions are already satisfied :

$$\Sigma H = 0, \quad \Sigma V = 0 \qquad \text{and} \qquad \Sigma M = 0$$

The above mentioned three equations are known as the conditions of equilibrium.

7. TYPES OF EQUILIBRIUM

In the previous article, we have discussed the conditions of equilibrium. As a matter of fact, these conditions help us in finding out the reactions or forces at a particular point, when the body is in equilibrium. But from practical point of view, a body is said to be in equilibrium when it comes back to its original position, after it is slightly displaced from its position of rest. In general, following are the three types of equilibrium :

1. Stable equilibrium

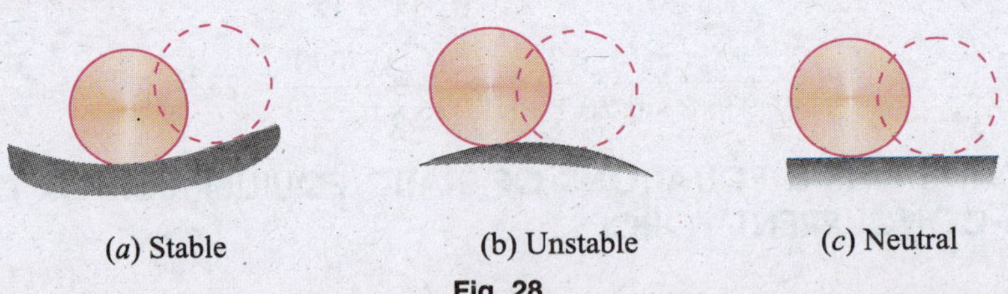

(*a*) Stable (b) Unstable (*c*) Neutral

Fig. 28.

A body is said to be in stable equilibrium, if it returns back to its original position, after it is slightly displaced from its position of rest. This happens when some additional force sets up due to displacement and brings the body back to its original position. A smooth cylinder, lying in a curved surface, is in stable equilibrium. If we slightly displace the cylinder from its position of rest (as shown by dotted lines), it will tend to return back to its original position in order to bring its weight normal to horizontal axis as shown in Fig. 28 (*a*).

2. Unstable equilibrium

A body is said to be in an unstable equilibrium, if it does not return back to its original position, and heels farther away, after slightly displaced from its position of rest. This happens when the addtional force moves the body away from its position of rest. This happens when the additional force moves the body away from its position of rest. A smooth cylinder lying on a convex surface is in unstable equilibrium. If we slightly displace the cylinder from its position of rest (as shown by dotted lines) the body will tend to move away from its original position as shown in Fig. 28 (*b*).

3. *Neutral equilibrium*

A body is said to be in a neutral equilibrium, if it occupies a new position (and remains at rest in this position) after slightly displaced from its position of rest. This happens when no additional force sets up due to the displacement. A smooth cylinder lying on a horizontal plane is in neutral equilibrium as shown in Fig. 28 (*c*).

Example 11. *A revolving crane is supported by a point at C and rollers at A and B. The crane carries a load P applied at D in addition to its own weight W at E as shown in Fig. 29.*

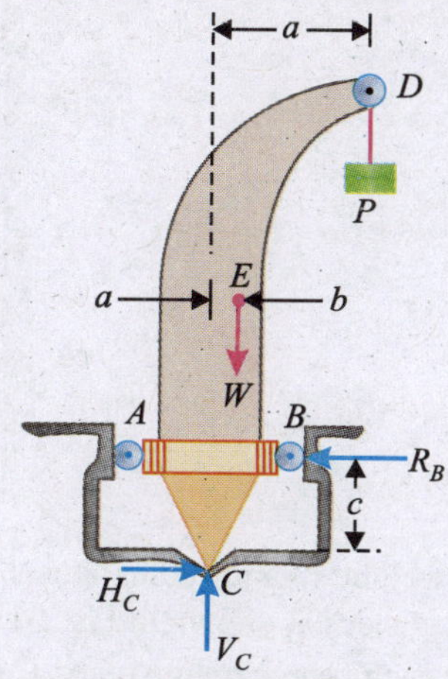

Fig. 29.

Determine the reactions R_B and R_C at the points B and C, if P = 4 kN, W = 2 kN, a = 3·0 m, b = 0·9 m and c = 1·8 m. Neglect friction.

Solution. Given : Load at *D* (*P*) = 4 kN ; Load at *E* (*W*) = 2 kN; *a* = 3·0 m ; *b* = 0·9 m and *c* = 1.8 m

Now let us use the conditions of equilibrium one by one.

$$H_C = R_B \quad \text{...(For } \Sigma H = 0)$$

and

$$V_C = P + W = 4 + 2 = 6 \text{ kN} \quad \text{...(For } \Sigma V = 0)$$

Now

$$R_B \times C = Pa + Wb \quad \text{...(For } \Sigma M = 0 \text{ about } C)$$

or

$$R_B \times 1.8 = (4 \times 3.0) + (2 \times 0.9) = 13.8$$

or

$$R_B = \frac{13.8}{1.8} = 7.67 \text{ kN} \quad \textbf{Ans.}$$

and

$$H_C = R_B = 7.67 \text{ kN} \quad \textbf{Ans.}$$

Now the reaction at *C*,

$$R_C = \sqrt{H_C^2 + V_C^2} = \sqrt{7.67^2 + 6^2} = 9.74 \text{ kN} \quad \textbf{Ans.}$$

EXERCISE 2

1. A spherical ball of weight 50 N is suspended vertically by a string 500 mm long. Find the magnitude and direction of the least force, which can hold the ball 100 mm above the lowest point. Also find tension in the string at that point. **[Ans.** 30 N ; 40 N]

   Hint. The force will be least, when it is applied at an angle of 90° with the string.

*2. A jib crane shown in Fig. 30 is required to lift a load of 5 kN. Find, graphically, the forces in the jib and tie. Also check the answer analytically.

[Ans. 13.7 kN (tension) ; 9.7 kN (compression)]

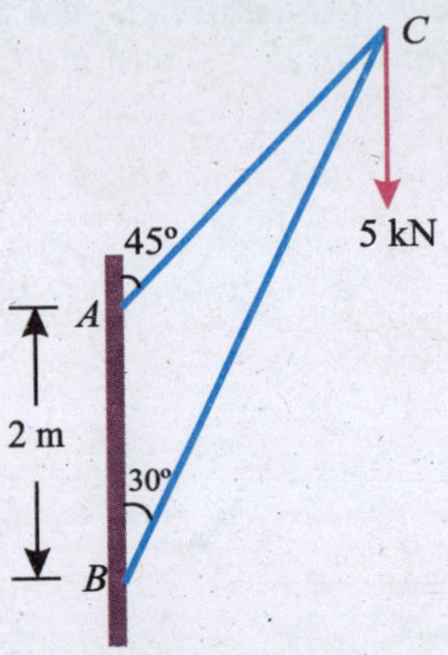

Fig. 30.

Fig. 31.

*3. Two smooth spheres of weight W and radius r each are in equilibrium in a horizontal channel of A and B vertical sides as shown in Fig. 31.

Find the force exerted by each sphere on the other. Calculate these values, if $r = 250$ mm, $b = 900$ mm and $W = 100$ N.**[AMIE, IGNOU]** **[Ans.** 133.3 N ; 166.7 N ; 133.3 N ; 200 N]

*4. A ladle is lifted by means of three chains each 2 m in length. The upper ends of the chains are attached to a ring, while the lower ends are attached to three hooks, fixed to the ladle forming an equilateral triangle of 1.2 metre side as shown in Fig. 32.

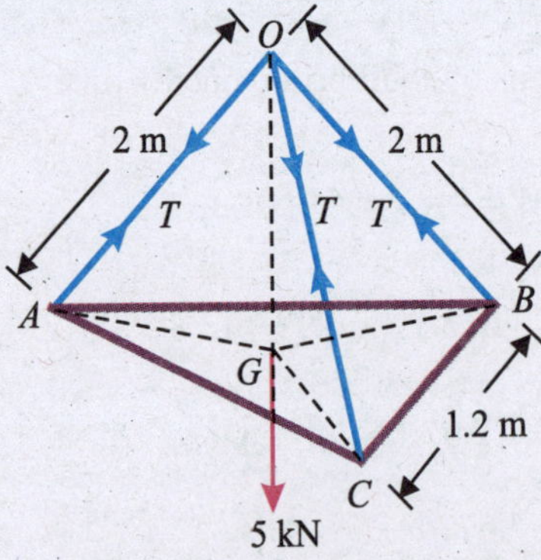

Fig. 32.

If weight of the ladle and its contents is 5 kN, find tension in each rope. **[Ans.** 1.78 kN]

Hint. First of all locate the point G of the equilateral triangle ABC. Now consider the right angled triangle AOG and resolve the forces vertically. **[PU]**

*5. A spherical ball, of weight W, rests in a triangular groove whose sides are inclined at angles α and β to the horizontal. Find the reactions at the surfaces of contact.

If a similar ball is now placed, so as to rest above the first ball, one side of which is inclined at angle α, find the reaction of the lower ball, on the surface inclined at an angle β.

$$\left[\textbf{Ans. } \frac{W \sin \alpha}{\sin(\alpha+\beta)}; \frac{W \sin \beta}{\sin(\alpha+\beta)}; \frac{2W \sin \alpha}{\sin(\alpha+\beta)}\right]$$

8. EQUIVALENT FORCE SYSTEM

For a given system of coplanar forces, the combination of resultant force passing through a point and a moment about a given point is called *equivalent system*. The force is the resultant of all forces acting on the body. And the moment is the sum of all the moments about that point. Hence equivalent system consists of the following.

(*i*) A resultant force R passing through a point and

(*ii*) A single moment M about a given point,

where R = the resultant of all forces acting on the body

M = sum of moments, of all the forces, about a given point

QUESTIONS

1. Enunciate any two principles of equilibrium.
2. State and prove Lami's Theorem.
3. Show that if three coplaner forces, acting at a point be in equilibrium, then, each force is proportional to the sine of the angle between the other two.
4. What are different methods of studying the equilibrium of coplaner forces ? Describe any one of them.
5. How would you find out the equilibrium of non-coplaner forces ?
6. Explain the conditions of equilibrium.
7. Discuss the various types of equilibrium.

OBJECTIVE TYPE QUESTIONS

1. According to Lami's Theorem, the three forces

   (*a*) Must be equal.

   (*b*) Must be at 120° to each other.

   (*c*) Must be both of above.

   (*d*) May not be any of the two.

2. The Lami's Theorem is applicable only for

   (*a*) Coplaner forces (*b*) Concurrent forces

   (*c*) Coplaner and concurrent forces (*d*) Any type of forces

3. If a body is in equilibrium. We may conclude that

   (*a*) No force is acting on the body

   (*b*) The resultant of all the forces acting on it is zero.

   (*c*) The moments of the forces about any point is zero.

   (*d*) Both (*b*) and (*c*)

4. If the sum of all the forces acting on a body is zero, then the body may be in equilibrium provided the forces are
   (*a*) Concurrent (*b*) Parallel
   (*c*) Like parallel (*d*) Unlike parallel
5. A body is said to be in equilibrium, if it has no linear motion.
   (*a*) True (*b*) False
6. Lami's Theorem can not be applied in case of concurrent forces
   (*a*) Agree (*b*) Disagree

ANSWERS

1. (*d*) 2. (*a*) 3. (*d*) 4. (*a*) 5. (*b*) 6. (*b*)

CHAPTER

Centre of Gravity

Contents

Note :

Important Results related to this chapter are given at the end of this book.

See Appendix - Properties of Plane Areas

1. CENTRE OF GRAVITY (C.G.)

[JNTU]

It has been established, since long, that every particle of a body is attracted by the earth towards its centre. The force of attraction, which is proportional to the mass of the particle, acts vertically downwards and is known as weight of the body. As the distance between the different particles of a body and the centre of the earth is taken as same, therefore these forces may be taken to act along parallel lines.

A point may be found out in a body, through which the resultant of all such parallel forces act.

This point, through which the whole weight of the body acts is known as **centre of gravity** (briefly written as **C.G.**). It may be noted that every body has one and only one centre of gravity. The centre of gravity depends upon the shape of the body.

2. CENTROID

[JNTU], [RGPV]

The plane figures (like triangle, quadrilateral, circle etc.) have only areas, but no mass. The **centre of area of plane figures** is known as **centroid**. The method of finding out the centroid of a figure is the same as that of finding out the centre of gravity of a body. In many books, the authors also write centre of gravity for centroid and vice versa.

3. METHODS FOR FINDING CENTRE OF GRAVITY

The centre of gravity (or centroid) may be found out by any one of the following two methods:

1. By geometrical considerations
2. By moments
3. By graphical method

As a matter of fact, the graphical method is a tedious and cumbersome method for finding out the centre of gravity of simple figures. That is why, it has academic value only. But in this book, we shall discuss the procedure for finding out the centre of gravity of simple figures by geometrical considerations and by moments one by ones.

4. CENTRE OF GRAVITY BY GEOMETRICAL CONSIDERATIONS

The centre of gravity of simple figures may be found out from the geometry of the figure as given below.

Plane Area Figures

1. The centre of gravity of a **rectangle (or a parallelogram)** is at the point, where its diagonals meet each other. It is also a middle point of the length as well as the breadth of the rectangle as shown in Fig. 1.

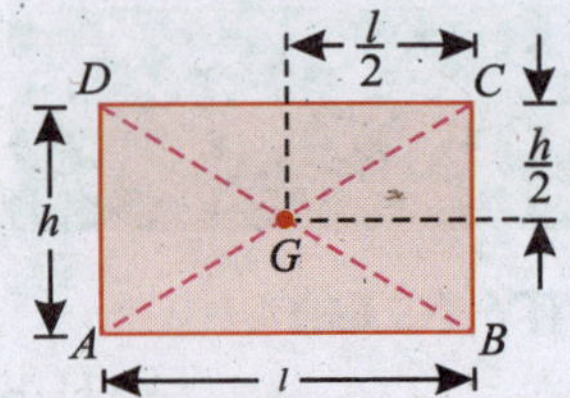

Fig. 1. Rectangle

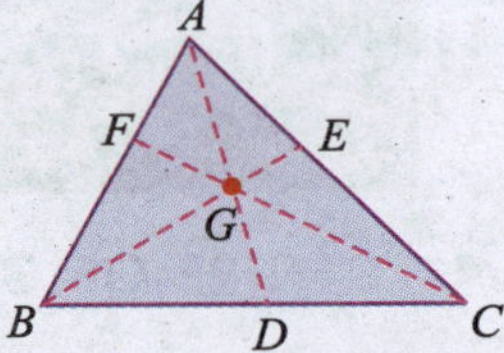

Fig. 2. Triangle

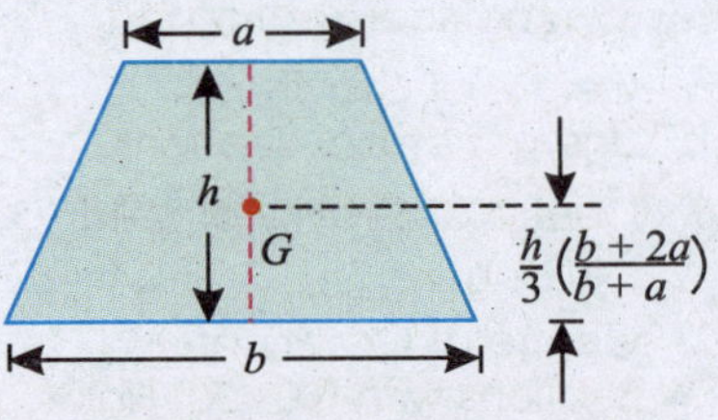

Fig. 3. Trapezium

2. The centre of gravity of a **triangle** is at the point, where the three medians (a median is a line connecting the vertex and middle point of the opposite side) of the triangle meet as shown in Fig. 2.
3. The centre of gravity of a **trapezium** with parallel sides a and b is at a distance of $\frac{h}{3} \times \left(\frac{b+2a}{b+a}\right)$ measured form the side b as shown in Fig. 3.
4. The centre of gravity of a **semicircle** is at a distance of $\frac{4r}{3\pi}$ from its base measured along the vertical radius as shown in Fig. 4 (a).

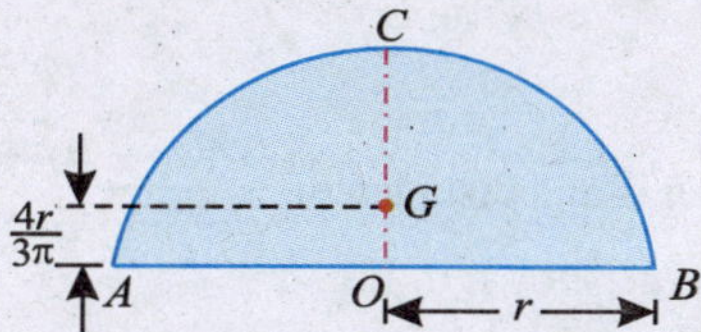

Fig. 4 (a). Semicircle

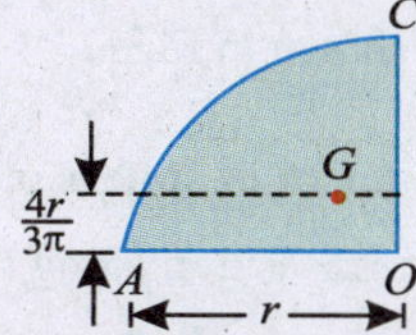

Fig. 4 (b). Quarter-circle

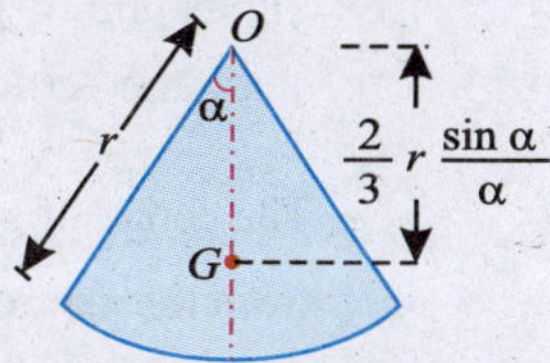

Fig. 4 (c). Circular sector

4. The centre of gravity of a **quarter-circle** is at a distance of $\frac{4r}{3\pi}$ from its base, measured along the vertical radius, as shown in Fig. 4 (b).
5. The centre of gravity of a **circular sector** making semi-vertical angle α is at a distance of $\frac{2r}{3}\frac{\sin\alpha}{\alpha}$ from the centre of the sector, measured along the central axis, as shown in Fig. 4 (c).

Solid Figures

6. The centre of gravity of a **cube** is at a distance of $\frac{l}{2}$ from every face (where l is the length of each side).
7. The centre of gravity of a **cylinder** is at a distance of $\frac{h}{2}$ from from its base (where h is the hieght of cylinder), as shown in Fig. 5.

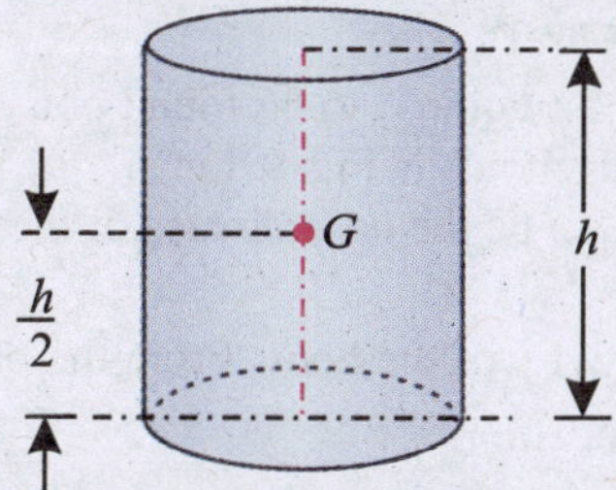

Fig. 5. Cylinder

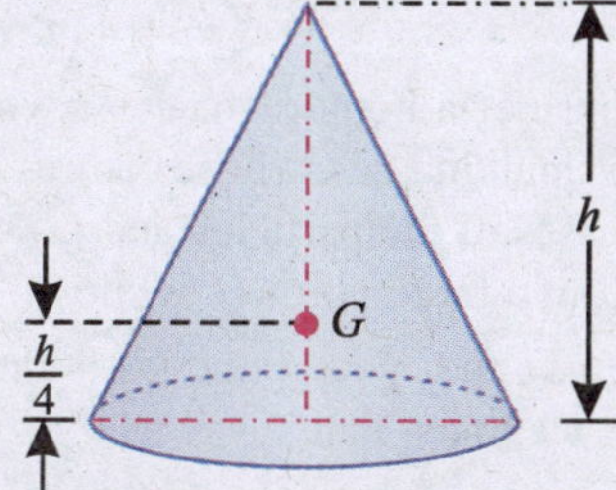

Fig. 6. Right circular solid cone

8. The centre of gravity of **right circular cone** is at a distance of $\frac{h}{4}$ from its base, measured along the vertical axis (where h is the hieght of cylinder), as shown in Fig. 6.
9. The centre of gravity of a **sphere** is at a distance of $\frac{d}{2}$ from every point (where d is the diameter of the sphere).

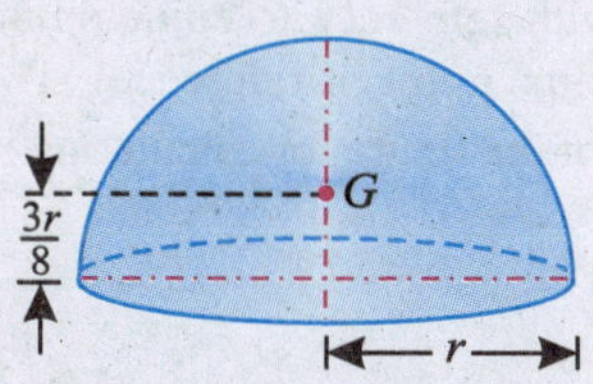

Fig. 7. Hemisphere

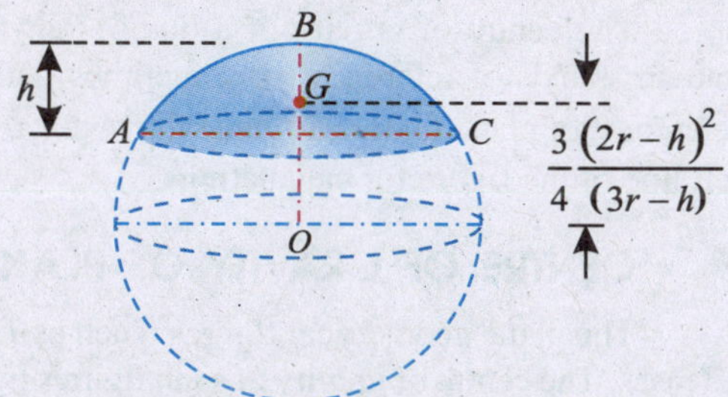

Fig.8. Segment of a sphere

10. The centre of gravity of a **hemisphere** is at a distance of $\frac{3r}{8}$ from its base, measured along the vertical radius, as shown in Fig. 7.

11. The centre of gravity of a **segment of sphere** of a height h is at a distance of $\frac{3}{4}\frac{(2r-h)^2}{(3r-h)}$ from the centre of the sphere, measured along the height, as shown in Fig. 8.

5. CENTRE OF GRAVITY BY MOMENTS

The centre of gravity of a body may also be found out by moments as discussed below:

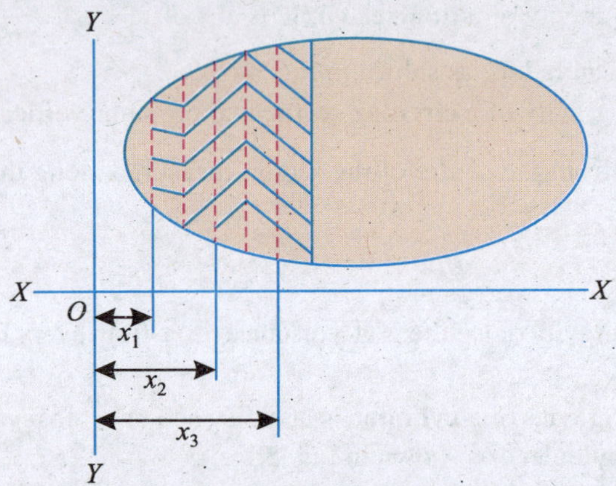

Fig. 9. Centre of gravity by moments

Consider a body of mass M whose centre of gravity is required to be found out. Divide the body into small masses, whose centres of gravity are known as shown in Fig. 9. Let m_1, m_2, m_3....; etc. be the masses of the particles and (x_1, y_1), (x_2, y_2), (x_3, y_3), ... be the co-ordinates of the centres of gravity from a fixed point O as shown in Fig. 9.

Let $\bar{x}$ and $\bar{y}$ be the co-ordinates of the centre of gravity of the body. From the principle of moments, we know that

$$M\bar{x} = m_1 x_1 + m_2 x_2 + m_3 x_3 \; .....$$

or
$$\bar{x} = \frac{\Sigma mx}{M}$$

Similarly
$$\bar{y} = \frac{\Sigma my}{M}$$

where
$$M = m_1 + m_2 + m_3 + \; .....$$

6. AXIS OF REFERENCE

The centre of gravity of a body is always calculated with reference to some assumed axis known as axis of reference (or sometimes with reference to some point of reference). The axis of reference, of plane figures, is generally taken as the lowest line of the figure for calculating $\bar{y}$ and the left line of the figure for calculating $\bar{x}$.

7. CENTRE OF GRAVITY OF PLANE FIGURES

The plane geometrical figures (such as T-section, I-section, L-section etc.) have only areas but no mass. The centre of gravity of such figures is found out in the same way as that of solid bodies. The centre of area of such figures is known as centroid, and coincides with the centre of gravity of the figure. It is a common practice to use centre of gravity for centroid and vice versa.

Let $\bar{x}$ and $\bar{y}$ be the co-ordinates of the centre of gravity with respect to some axis of reference, then

$$\bar{x} = \frac{a_1 x_1 + a_2 x_2 + a_3 x_3 + ...}{a_1 + a_2 + a_3}$$

and
$$\bar{y} = \frac{a_1 y_1 + a_2 y_2 + a_3 y_3 + ...}{a_1 + a_2 + a_3 + ...}$$

where a_1, a_2, a_3........ etc., are the areas into which the whole figure is divided x_1, x_2, x_3 ..... etc., are the respective co-ordinates of the areas a_1, a_2, a_3....... on *X-X* axis with respect to same axis of reference.

y_1, y_2, y_3....... etc., are the respective co-ordinates of the areas a_1, a_2, a_3....... on *Y-Y* axis with respect to same axis of the reference.

Note. While using the above formula, x_1, x_2, x_3 ..... or y_1, y_2, y_3 or $\bar{x}$ and $\bar{y}$ must be measured from the same axis of reference (or point of reference) and on the same side of it. However, if the figure is on both sides of the axis of reference, then the distances in one direction are taken as positive and those in the opposite directions must be taken as negative.

8. CENTRE OF GRAVITY OF SYMMETRICAL SECTIONS

Sometimes, the given section, whose centre of gravity is required to be found out, is symmetrical about *X-X* axis or *Y-Y* axis. In such cases, the procedure for calculating the centre of gravity of the body is very much simplified; as we have only to calculate either $\bar{x}$ or $\bar{y}$. This is due to the reason that the centre of gravity of the body will lie on the axis of symmetry.

Example 1. *Find the centre of gravity of a 100 mm × 150 mm × 30 mm T-section.*

Solution. As the section is symmetrical about *Y-Y* axis, bisecting the web, therefore its centre of gravity will lie on this axis. Split up the section into two rectangles *ABCH* and *DEFG* as shown in Fig 10.

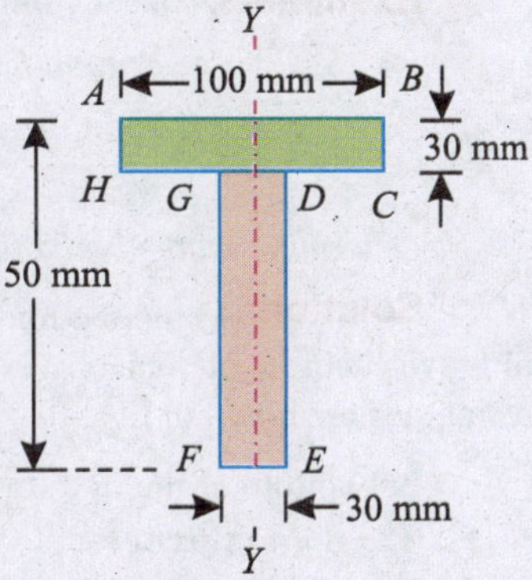

Fig. 10.

Let bottom of the web *FE* be the axis of reference.

(*i*) ***Rectangle ABCH***

Area, $a_1 = 100 \times 30 = 3000 \text{ mm}^2$

and $y_1 = \left(150 - \frac{30}{2}\right) = 135 \text{ mm}$

(*ii*) ***Rectangle DEFG***

Area, $a_2 = 120 \times 30 = 3600 \text{ mm}^2$

and $y_2 = \frac{120}{2} = 60 \text{ mm}$

We know that distance between centre of gravity of the section and bottom of the flange *FE*,

$$\bar{y} = \frac{a_1 y_1 + a_2 y_2}{a_1 + a_2} = \frac{(3000 \times 135) + (3600 \times 60)}{3000 + 3600} \text{ mm}$$

$= 94.1$ mm **Ans.**

Example 2. *Find the centre of gravity of a channel section 100 mm × 50 mm × 15 mm.*

Solution. As the section is symmetrical about *X-X* axis, therefore its centre of gravity will lie on this axis. Now split up the whole section into three rectangles *ABFJ*, *EGKJ* and *CDHK* as shown in Fig. 11.

Let the face *AC* be the axis of reference.

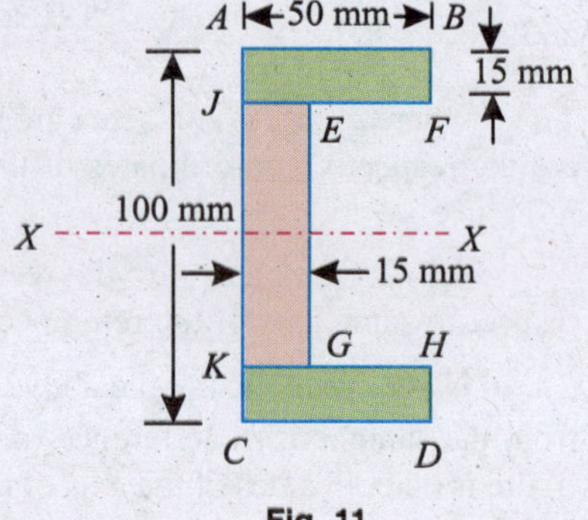

Fig. 11.

(i) Rectangle ABFJ

Area, $a_1 = 50 \times 15 = 750 \text{ mm}^2$

and $x_1 = \dfrac{50}{2} = 25 \text{ mm}$

(ii) Rectangle EGKJ

Area $a_2 = (100 - 30) \times 15 = 1050 \text{ mm}^2$

and $x_2 = \dfrac{15}{2} = 7.5 \text{ mm}$

(iii) Rectangle CDHK

Area, $a_3 = 50 \times 15 = 750 \text{ mm}^2$

and $x_3 = \dfrac{50}{2} = 25 \text{ mm}$

We know that distance between the centre of gravity of the section and left face of the section *AC*,

$$\bar{x} = \frac{a_1 x_1 + a_2 x_2 + a_3 x_3}{a_1 + a_2 + a_3}$$

$$= \frac{(750 \times 25) + (1050 \times 7.5) + (750 \times 25)}{750 + 1050 + 750} = 17.8 \text{ mm} \quad \textbf{Ans.}$$

Example 3. *An I-section has the following dimensions in mm units :*

Bottom flange = *300 × 100*

Top flange = *150 × 50*

Web = *300 × 50*

Determine mathematically the position of centre of gravity of the section. **[AMIE]**

Solution. As the section is symmetrical about *Y-Y* axis, bisecting the web, therefore its centre of gravity will lie on this axis. Now split up the section into three rectangles as shown in Fig. 12.

Let bottom of the bottom flange be the axis of reference.

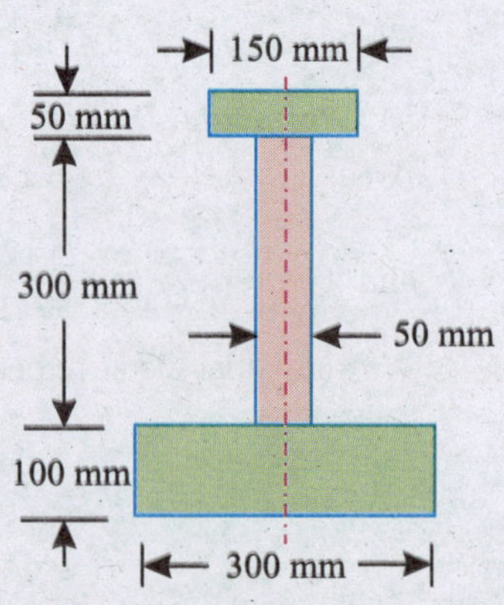

Fig. 12.

(i) Bottom flange

Area, $a_1 = 300 \times 100 = 30\,000 \text{ mm}^2$

and $y_1 = \dfrac{100}{2} = 50 \text{ mm}$

(ii) Web

Area, $a_2 = 300 \times 50 = 15\,000 \text{ mm}^2$

and $y_2 = 100 + \dfrac{300}{2} = 250 \text{ mm}$

(iii) Top flange

Area, $a_3 = 150 \times 50 = 7500 \text{ mm}^2$

and $y_3 = 100 + 300 + \dfrac{50}{2} = 425 \text{ mm}$

We know that distance between centre of gravity of the section and bottom of the flange,

$$\bar{y} = \frac{a_1 y_1 + a_2 y_2 + a_3 y_3}{a_1 + a_2 + a_3}$$

$$= \frac{(30\,000 \times 50) + (15\,000 \times 250) + (7500 \times 425)}{30\,000 + 15\,000 + 7500} = 160.7 \text{ mm} \qquad \textbf{Ans.}$$

9. CENTRE OF GRAVITY OF UNSYMMETRICAL SECTIONS

Sometimes, the given section, whose centre of gravity is required to be found out, is not symmetrical either about *X-X* axis or *Y-Y* axis. In such cases, we have to find out both the values of $\bar{x}$ and $\bar{y}$

Example 4. *Find the centroid of an unequal angle section 100 mm × 80 mm × 20 mm.*

Solution. As the section is not symmetrical about any axis, therefore we have to find out the values of $\bar{x}$ and $\bar{y}$ for the angle section. Split up the section into two rectangles as shown in Fig. 13.

Let left face of the vertical section and bottom face of the horizontal section be axes of reference.

(i) Rectangle 1

Area, $a_1 = 100 \times 20 = 2000 \text{ mm}^2$

$x_1 = \dfrac{20}{2} = 10 \text{ mm}$

and $y_1 = \dfrac{100}{2} = 50 \text{ mm}$

(ii) Rectangle 2

Area, $a_2 = (80 - 20) \times 20 = 1200 \text{ mm}^2$

$x_2 = 20 + \dfrac{60}{2} = 50 \text{ mm}$.

Fig. 13.

and $y_2 = \dfrac{20}{2} = 10 \text{ mm}$

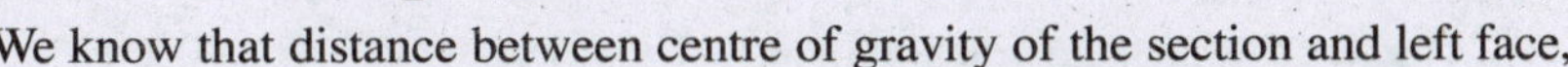

We know that distance between centre of gravity of the section and left face,

$$\bar{x} = \frac{a_1 x_1 + a_2 x_2}{a_1 + a_2} = \frac{(2000 \times 10) + (1200 \times 50)}{2000 + 1200} = 25 \text{ mm} \qquad \textbf{Ans.}$$

Similarly, distance between centre of gravity of the section and bottom face,

$$\bar{y} = \frac{a_1 y_1 + a_2 y_2}{a_1 + a_2} = \frac{(2000 \times 50) + (1200 \times 10)}{2000 + 1200} = 35 \text{ mm} \qquad \textbf{Ans.}$$

Example 5. *A uniform lamina shown in Fig. 14 consists of a rectangle, a circle and a triangle.*

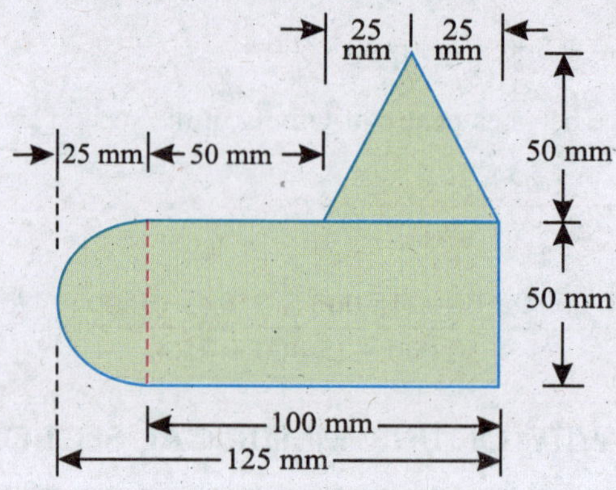

Fig. 14.

Determine the centre of gravity of the lamina. All dimensions are in mm.

Solution. As the section is not symmetrical about any axis, therefore we have to find out the values of both $\bar{x}$ and $\bar{y}$ for the lamina.

Let left edge of circular portion and bottom face rectangular portion be the axes of reference.

(i) Rectangular portion

Area, $a_1 = 100 \times 50 = 5000 \text{ mm}^2$

$$x_1 = 25 + \frac{100}{2} = 75 \text{ mm}$$

and $y_1 = \frac{50}{2} = 25 \text{ mm}$

(ii) Semicircular portion

Area, $a_2 = \frac{\pi}{2} \times r^2 = \frac{\pi}{2}(25)^2 = 982 \text{ mm}^2$

$$x_2 = 25 - \frac{4r}{3\pi} = 25 - \frac{4 \times 25}{3\pi} = 14.4 \text{ mm}$$

and $y_2 = \frac{50}{2} = 25 \text{ mm}$

(iii) Triangular portion

Area, $a_3 = \frac{50 \times 50}{2} = 1250 \text{ mm}^2$

$$x_3 = 25 + 50 + 25 = 100 \text{ mm}$$

and $y_3 = 50 + \frac{50}{3} = 66.7 \text{ mm}$

We know that distance between centre of gravity of the section and left edge of the circular portion,

$$\bar{x} = \frac{a_1 x_1 + a_2 x_2 + a_3 x_3}{a_1 + a_2 + a_3} = \frac{(5000 \times 75) + (982 \times 14.4) + (1250 \times 100)}{5000 + 982 + 1250}$$

$= 71.1 \text{ mm}$ **Ans.**

Similarly, distance between centre of gravity of the section and bottom face of the rectangular portion,

$$\bar{y} = \frac{a_1 y_1 + a_2 y_2 + a_3 y_3}{a_1 + a_2 + a_3} = \frac{(5000 \times 25) + (982 \times 25) + (1250 \times 66.7)}{5000 + 982 + 1250} \text{ mm}$$

$= 32.2$ mm **Ans.**

Example 6. *A plane lamina of 220 mm radius is shown in figure given below.*

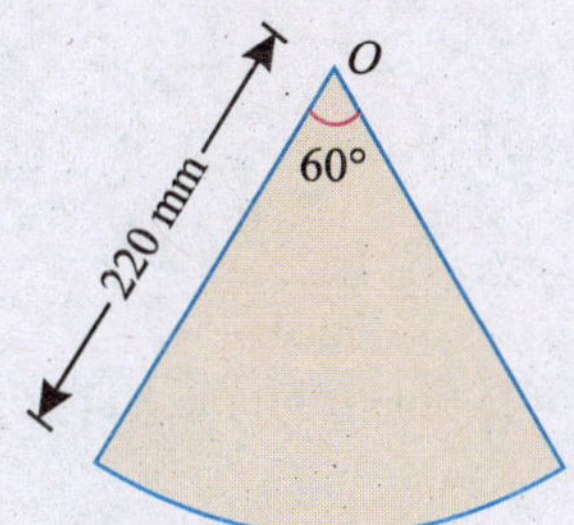

Fig. 15.

Find the centre of gravity of lamina from the point O.

Solution. As the lamina is symmetrical about *y-y* axis, bisecting the lamina, therefore its centre of gravity lies on this axis. Let *O* be the reference point. From the geometry of the lamina. We find that semi-vertical angle of the lamina,

$$\alpha = 30° = \frac{\pi}{6} \text{ rad}$$

We know that distance between the reference point *O* and centre of gravity of the lamina,

$$\bar{y} = \frac{2r}{3} \cdot \frac{\sin\alpha}{\alpha} = \frac{2 \times 220}{3} \times \frac{\sin 30°}{\left(\frac{\pi}{6}\right)} = \frac{440}{3} \times \frac{0.5}{\left(\frac{\pi}{6}\right)} = 140 \text{ mm} \quad \textbf{Ans.}$$

EXERCISE 1

1. Find the centre of gravity of a *T*-section with flange 150 mm × 10 mm and web also 150 mm × 10 mm. [**Ans.** 115 mm for bottom of the web]
2. Find the centre of gravity of an inverted *T*-section with flange 60 mm × 10 mm and web 50 mm × 10 mm [**Ans.** 18.6 mm from bottom of the flange]
3. A channel section 300 mm × 10 mm is 20 mm thick. Find the centre of gravity of the section from the back of the web. [**Ans.** 27.4 mm]
4. Find the centre of gravity of an *T*-section with top flange 100 mm × 20 mm, web 200 mm × 30 mm and bottom flange 300 mm × 40 mm. [**Ans.** 79 mm from bottom of lower flange]
5. Find the position of the centre of gravity of an unequal angle section 10 cm × 16 cm × 2cm. [**Ans.** 5.67 cm and 2.67 cm]

6. A figure consists of a rectangle having one of its sides twice the other, with an equilateral triangle described on the larger side. Show that centre of gravity of the section lies on the line joining the rectangle and triangle.
7. A plane lamina of radius 100 mm as shown in fig 16 given below:

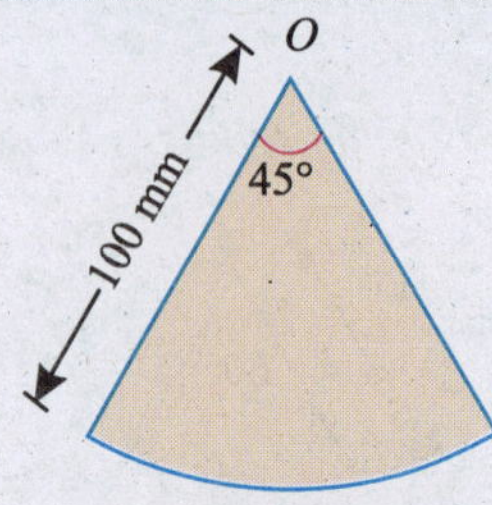

Fig. 16.

Find the centre of gravity of lamina from the point O.

[**Ans.** 65 mm]

10. CENTRE OF GRAVITY OF SOLID BODIES

The centre of gravity of solid bodies (such as hemispheres, cylinders, right circular solid cones etc.) is found out in the same way as that of plane figures. The only difference, between the plane figures and solid bodies, is that in the case of solid bodies, we calculate volumes instead of areas. The volumes of few solid bodies are given below :

1. Volume of cylinder $= \pi \times r^2 \times h$
2. Volume of hemisphere $= \dfrac{2\pi}{3} \times r^3$
3. Volume of right circular solid cone $= \dfrac{\pi}{3} \times r^2 \times h$

where r = Radius of the body, and
h = Height of the body.

Note. Sometimes the densities of the two solids are different. In such a case, we calculate the weights instead of volumes and the centre of gravity of the body is found out as usual.

Example 7. *A solid body formed by joining the base of a right circular cone of height H to the equal base of a right circular cylinder of height h. Calculate the distance of the centre of mass of the solid from its plane face, when H = 120 mm and h = 30 mm.*

Solution. As the body is symmetrical about the vertical axis, therefore its centre of gravity will lie on this axis as shown in Fig. 17. Let r be the radius of the cylinder base in cm. Now let base of the cylinder be the axis of reference.

(i) Cylinder

Volume, $v_1 = \pi \times r^2 \times 30 = 30\,\pi\, r^2 \text{ mm}^3$

and $y_1 = \dfrac{30}{2} = 15 \text{ mm}$

(ii) Right circular cone

Volume, $v_2 = \dfrac{\pi}{3} \times r^2 \times h = \dfrac{\pi}{3} \times r^2 \times 120 \text{ mm}^3$

$= 40\,\pi r^2 \text{ mm}^3$

and $y_2 = 30 + \dfrac{120}{4} = 60 \text{ mm}$

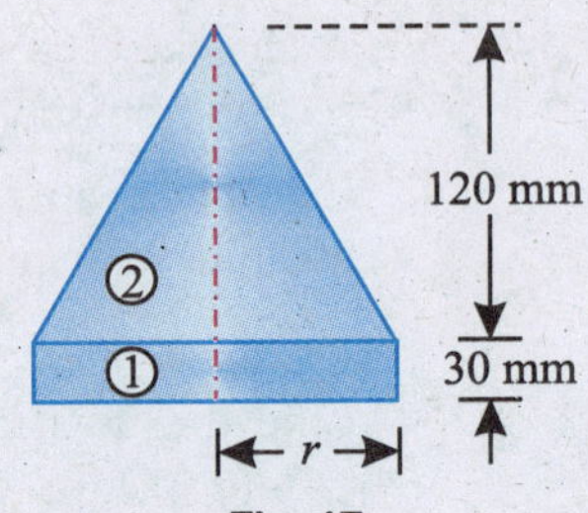

Fig. 17.

We know that distance between centre of gravity of the section and base of the cylinder,

$$\bar{y} = \frac{v_1 y_1 + v_2 y_2}{v_1 + v_2} = \frac{(30\pi r^2 \times 15) + (40\pi r^2 \times 60)}{30\pi r^2 + 40\pi r^2} = \frac{2850}{70} \text{ mm}$$

$$= 40.7 \text{ mm} \quad \textbf{Ans.}$$

Example 8. *A body consists of a right circular solid cone of height 40 mm and radius 30 mm placed on a solid hemisphere of radius 30 mm of the same material. Find the position of centre of gravity of the body.*

Solution. As the body is symmetrical about *Y-Y* axis, therefore its centre of gravity will lie on this axis as shown in Fig. 18. Let bottom of the hemisphere (*D*) be the point of reference.

(i) Hemisphere

Volume, $$v_1 = \frac{2\pi}{3} \times r^3 = \frac{2\pi}{3}(30)^3 \text{ mm}^3$$

$$= 18\,000\,\pi \text{ mm}^3$$

and $$y_1 = \frac{5r}{8} = \frac{5 \times 30}{8} = 18.75 \text{ mm}$$

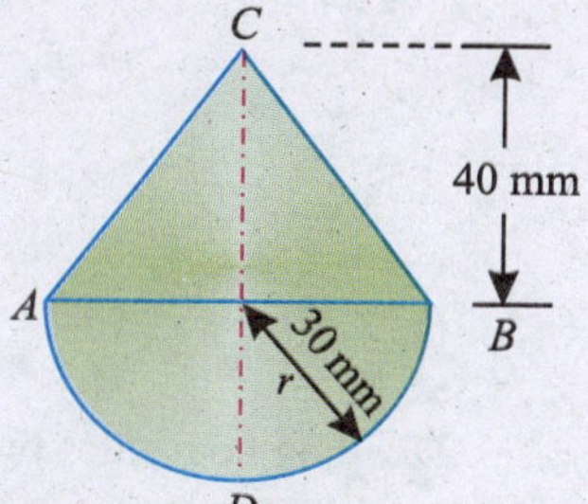

Fig. 18.

(ii) Right circular cone

Volume, $$v_2 = \frac{\pi}{3} \times r^2 \times h = \frac{\pi}{3} \times (30)^2 \times 40 \text{ mm}^3$$

$$= 12\,000\,\pi \text{ mm}^3$$

and $$y_2 = 30 + \frac{40}{4} = 40 \text{ mm}$$

We know that distance between centre of gravity of the body and bottom of hemisphere *D*,

$$\bar{y} = \frac{v_1 y_1 + v_2 y_2}{v_1 + v_2} = \frac{(18\,000\pi \times 18.75) + (12\,000\pi \times 40)}{18\,000\pi + 12\,000\pi} \text{ mm}$$

$$= 27.3 \text{ mm} \quad \textbf{Ans.}$$

Example 9. *A body consisting of a cone and hemisphere of radius r fixed on the same base rests on a table, the hemisphere being in contact with the table. Find the greatest height of the cone, so that the combined body may stand upright.*

Solution. As the body is symmetrical about *Y-Y* axis, therefore its centre of gravity will lie on this axis as shown in Fig. 19. Now consider two parts of the body *viz.*, hemisphere and cone. Let bottom of the hemisphere (*D*) be the axis of reference.

(i) Hemisphere

Volume, $$v_1 = \frac{2\pi}{3} \times r^3$$

and $$y_1 = \frac{5r}{8}$$

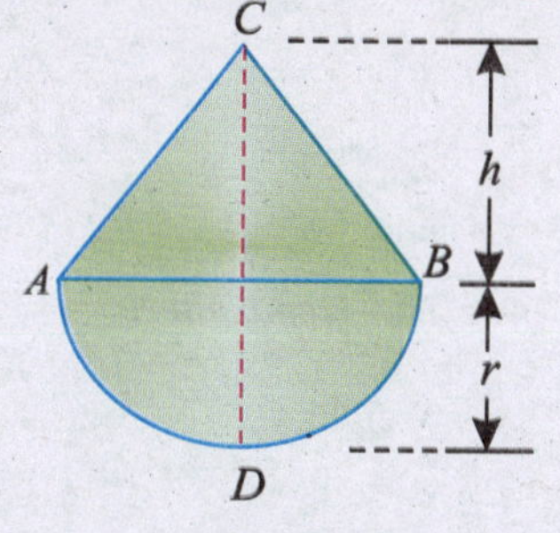

Fig. 19.

(ii) Cone

Volume, $$v_2 = \frac{\pi}{3} \times r^2 \times h$$

and $$y_2 = r + \frac{h}{4}$$

We know that distance between centre of gravity of the body and bottom of hemisphere D,

$$\bar{y} = \frac{v_1 y_1 + v_2 y_2}{v_1 + v_2} = \frac{\left(\frac{2\pi}{3} \times r^3 \times \frac{5r}{8}\right) + \left(\frac{\pi}{3} \times r^2 \times h\right)\left(r + \frac{h}{4}\right)}{\left(\frac{2\pi}{3} \times r^3\right) + \left(\frac{\pi}{3} \times r^2 \times h\right)}$$

Now for stable equilibrium, we know that the centre of gravity of the body should preferably be below the common face AB or maximum may coincide with it. Therefore substituting $\bar{y}$ equal to r in the above equation,

$$r = \frac{\left(\frac{2\pi}{3} \times r^3 \times \frac{5r}{8}\right) + \left(\frac{\pi}{3} \times r^2 \times h\right)\left(r + \frac{h}{4}\right)}{\left(\frac{2\pi}{3} \times r^3\right) + \left(\frac{\pi}{3} \times r^2 \times h\right)}$$

or
$$\left(\frac{2\pi}{3} \times r^4\right) + \left(\frac{\pi}{3} \times r^3 h\right) = \left(\frac{5\pi}{12} \times r^4\right) + \left(\frac{\pi}{3} \times r^3 \times h\right) + \left(\frac{\pi}{12} \times r^2 \times h^2\right)$$

Dividing both sides by πr^2,

$$\frac{2r^2}{3} + \frac{rh}{3} = \frac{5r^2}{12} + \frac{rh}{3} + \frac{h^2}{12} \quad \text{or} \quad \frac{3r^2}{12} = \frac{h^2}{12}$$

$$3r^2 = h^2 \quad \text{or} \quad h = 1.732\ r \qquad \textbf{Ans.}$$

Example 10. *A right circular cylinder of 12 cm diameter is joined with a hemisphere of the same diameter face to face. Find the greatest height of the cylinder, so that centre of gravity of the composite section coincides with the plane of joining the two sections. The density of the material of hemisphere is twice that the material of cylinder.*

Solution. As the body is symmetrical about the vertical axis, therefore its centre of gravity will lie on this axis. Now let the vertical axis cut the plane joining the two sections at O as shown in Fig. 20. Therefore centre of gravity of the section is at a distance of 60 mm from P *i.e.*, bottom of the hemisphere.

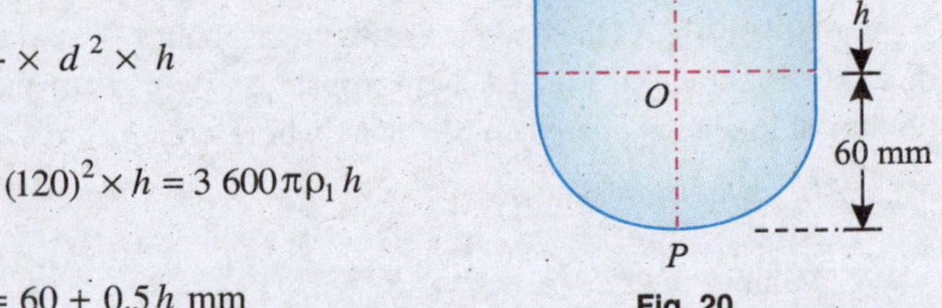

Fig. 20.

Let h = Height of the cylinder in mm.

(i) Right circular cylinder

Weight, $$w_1 = \rho_1 \times \frac{\pi}{4} \times d^2 \times h$$

$$= \rho_1 \times \frac{\pi}{4} \times (120)^2 \times h = 3\,600\pi\rho_1 h$$

and $$y_1 = 60 + \frac{h}{2} = 60 + 0.5h \text{ mm}$$

(ii) Hemisphere

Weight, $$w_2 = \rho_2 \times \frac{2\pi}{3} \times r^3 = 2\rho_1 \times \frac{2\pi}{3} \times (60)^3 \quad ...(\because \rho_2 = 2\rho_1)$$

$$= 288\,000\,\pi\,\rho_1$$

and $$y_2 = \frac{5r}{8} = \frac{5 \times 60}{8} = \frac{300}{8} = 37.5 \text{ mm}$$

We know that distance between centre of gravity of the combined body from $P(\bar{y})$,

$$60 = \frac{w_1 y_1 + w_2 y_2}{w_1 + w_2} = \frac{3\,600\pi\rho_1 h(60 + 0.5h) + (288\,000\pi\rho_1 \times 37.5)}{3\,600\pi\rho_1 h + 288\,000\pi\rho_1}$$

$$= \frac{216\,000h + 1800\,h^2 + 10\,800\,000}{3\,600h + 288\,000}$$

or $\quad 216\,000\,h + 17\,280\,000 = 216\,000\,h + 1\,800\,h^2 + 10\,800\,000$

or $\quad 1\,800\,h^2 = 17\,280\,000 - 10\,800\,000 = 6\,480\,000$

or $\quad h = \sqrt{\dfrac{6\,480\,000}{1\,800}} = \sqrt{3\,600} = 60 \text{ mm}$ **Ans.**

Example 11. *Find the centre of gravity of a segment of height 30 mm of a sphere of radius 60 mm.*

Solution. Let O be the centre of the given sphere and ABC is the segment of this sphere as shown in Fig. 21

As the section is symmetrical about X-X axis, therefore its centre of gravity lies on this axis.

Let O be the reference point.

We know that centre of gravity of the segment of sphere

$$\bar{x} = \frac{3(2r - h)^2}{4(3r - h)} = \frac{3(2 \times 60 - 30)^2}{4(3 \times 60 - 30)}$$

$$= \frac{3 \times (90)^2}{4 \times 150} = 40.5 \text{ mm.} \quad \textbf{Ans.}$$

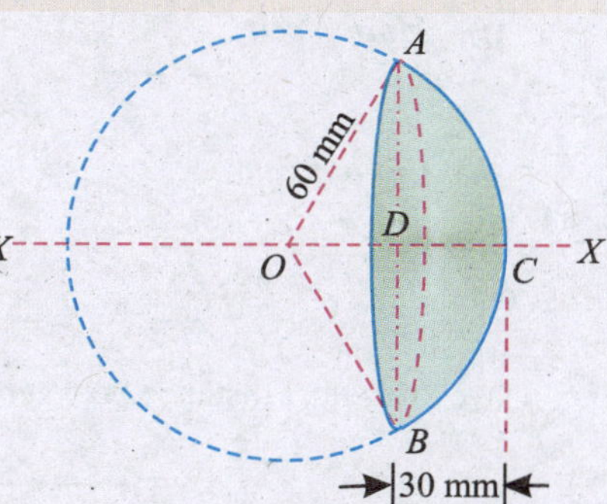

Fig. 21.

EXERCISE 2

1. A hemisphere of 60 mm diameter is placed on the top of the cylinder having 60 mm diameter. Find the common centre of gravity of the body from the base of cylinder, if its height is 100 mm. **[Ans.** 60.2 mm]

2. A solid consists of a cylinder and a hemisphere of equal radius fixed base to base. Find the ratio of the radius to the height of the cylinder, so that the solid has its centre of gravity at the common face. **[Ans.** $\sqrt{2} : 1$]

   Hint. For stable equilibrium, the centre of the body should be below the common face or maximum lie on it. So take the centre of gravity of the body at a distance (a) from the bottom of the hemisphere.

3. A body consisting of a cone and hemisphere of radius (r) on the same base rests on a table, the hemisphere being in contact with the table. Find the greatest height of the cone, so that the combined solid may be in stable equilibrium. **[Ans.** 1.732 r]

4. Find the centre of gravity of a segment of height 77 mm of a sphere of radius 150 mm. **[Ans.** 100 mm]

11. CENTRE OF GRAVITY OF SECTIONS WITH CUT OUT HOLES

The centre of gravity of such a section is found out by considering the main section, first as a complete one, and then deducting the area of the cut out hole *i.e.*, by taking the area of the cut out hole as negative. Now substituting a_2 (*i.e.*, the area of the cut out hole) as negative, in the general equation for the centre of gravity, we get

$$\bar{x} = \frac{a_1 x_1 - a_2 x_2}{a_1 - a_2} \quad \text{and} \quad \bar{y} = \frac{a_1 y_1 - a_2 y_2}{a_1 - a_2}$$

Note. In case of circle the section will be symmeterical along the line joining the centres of the bigger and the cut out circle.

Example 12. *A square hole is punched out of circular lamina, the digonal of the square being the radius of the circle as shown in Fig.22. Find the centre of gravity of the remainder, if r is the radius of the circle.* **[AMIE]**

Solution. As the section is symmetrical about *X-X* axis, therefore its centre of gravity will lie on this axis. Let *A* be the point of reference.

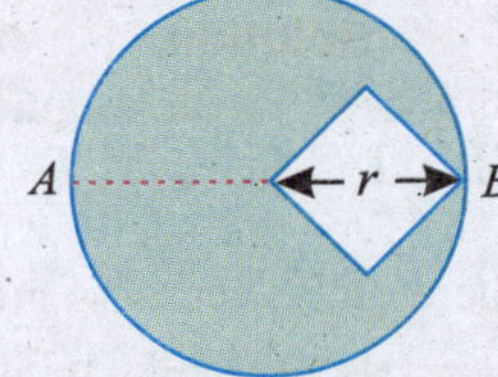

Fig. 22.

(i) Main circle

Area, $a_1 = \pi r^2$

and $x_1 = r$

(ii) Cut out square

Area, $a_2 = \dfrac{r \times r}{2} = 0.5\, r^2$

and $x_2 = r + \dfrac{r}{2} = 1.5\, r$

We know that distance between centre of gravity of the section and *A*,

$$\bar{x} = \frac{a_1 x_1 - a_2 x_2}{a_1 - a_2} = \frac{(\pi r^2 \times r) - (0.5 r^2 \times 1.5 r)}{\pi r^2 - 0.5 r^2}$$

$$= \frac{r^3(\pi - 0.75)}{r^2(\pi - 0.5)} = \frac{r(\pi - 0.75)}{\pi - 0.5} = 0.905 r \quad \textbf{Ans.}$$

Example 13. *A semicircle of 90 mm radius is cut out from a trapezium as shown in Fig 23*

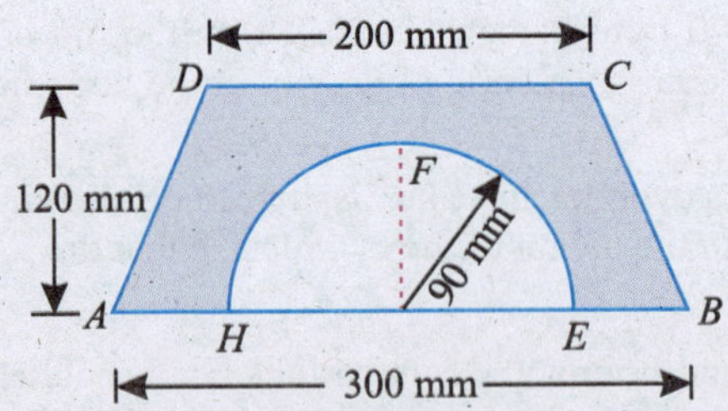

Fig. 23.

Find the position of the centre of gravity of the figure. **[MU, JNTU, PU]**

Solution. As the section is symmetrical about *Y-Y* axis, therefore its centre of gravity will lie on this axis. Now consider two portions of the figure *viz.*, trapezium *ABCD* and semicircle *EFH*.

Let base of the trapezium *AB* be the axis of reference.

(i) Trapezium ABCD

Area, $a_1 = 120 \times \dfrac{200 + 300}{2} = 30\,000 \text{ mm}^2$

and $$y_1 = \frac{h}{3} \times \left(\frac{a + 2b}{a + b}\right) = \frac{120}{3} \times \left(\frac{300 + (2 \times 200)}{300 + 200}\right) = 56 \text{ mm}$$

(ii) Semicircle

Area, $$a_2 = \frac{1}{2} \times \pi r^2 = \frac{1}{2} \times \pi \times (90)^2 = 4050\pi \text{ mm}^2$$

and $$y_2 = \frac{4r}{3\pi} = \frac{4 \times 90}{3\pi} = \frac{120}{\pi} \text{ mm}$$

We know that distance between centre of gravity of the section and *AB*,

$$\bar{y} = \frac{a_1 y_1 - a_2 y_2}{a_1 - a_2} = \frac{(30\,000 \times 56) - \left(4050\pi \times \frac{120}{\pi}\right)}{30\,000 - 4050\pi} \text{ mm}$$

$= 69.1$ mm **Ans.**

Example 14. *A semicircular area is removed from a trapezium as shown in Fig.24 (dimensions in mm)*

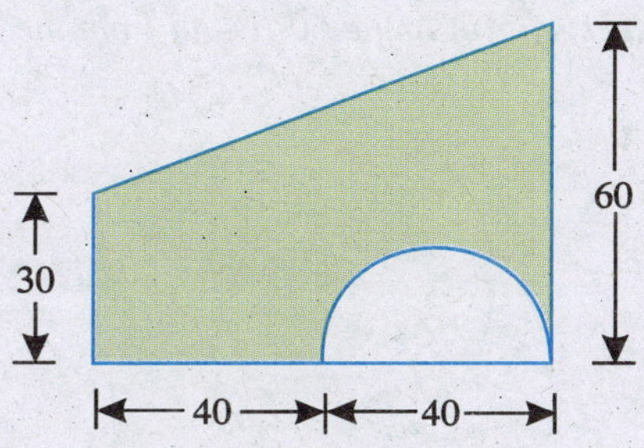

Fig. 24.

Determine the centroid of the remaining area (shown hatched).

[MU, PU, DTU, WBUT, IGNOU]

Solution. As the section in not symmetrical about any axis, therefore we have to find out the values of $\bar{x}$ and $\bar{y}$ for the area. Split up the area into three parts as shown in Fig. 25. Let left face and base of the trapezium be the axes of reference.

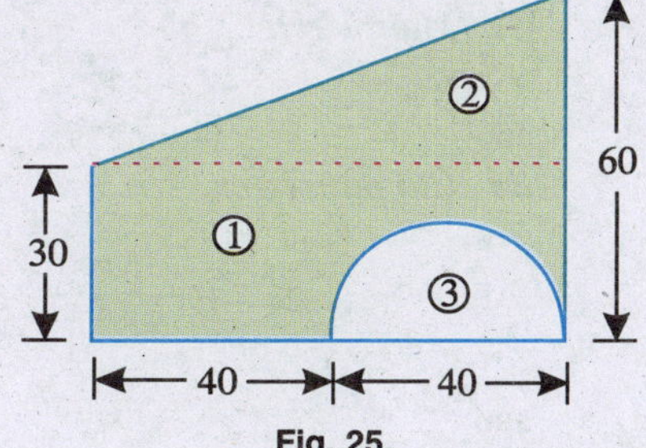

Fig. 25.

(i) Rectangle

Area, $$a_1 = 80 \times 30 = 2400 \text{ mm}^2$$

$$x_1 = \frac{80}{2} = 40 \text{ mm}$$

and $$y_1 = \frac{30}{2} = 15 \text{ mm}$$

(ii) Triangle

Area, $$a_2 = \frac{80 \times 30}{2} = 1200 \text{ mm}^2$$

$$x_2 = \frac{80 \times 2}{3} = 53.3 \text{ mm}$$

and $$y_2 = 30 + \frac{30}{3} = 40 \text{ mm}$$

(iii) Semicircle

Area, $a_3 = \frac{\pi}{2} \times r^2 = \frac{\pi}{2}(20)^2 = 628.3 \text{ mm}^2$

$x_3 = 40 + \frac{40}{2} = 60 \text{ mm}$

and $y_3 = \frac{4r}{3\pi} = \frac{4 \times 20}{3\pi} = 8.5 \text{ mm}$

We know that distance between centre of gravity of the area and left face of trapezium,

$$\bar{x} = \frac{a_1 x_1 + a_2 x_2 - a_3 x_3}{a_1 + a_2 - a_3} = \frac{(2400 \times 40) + (1200 \times 53.3) - (628.3 \times 60)}{2400 + 1200 - 628.3}$$

$= 41.1$ mm **Ans.**

Similarly, distance between centre of gravity of the area and base of the trapezium,

$$\bar{y} = \frac{a_1 y_1 + a_2 y_2 - a_3 y_3}{a_1 + a_2 - a_3} = \frac{(2400 \times 15) + (1200 \times 40) - (628.3 \times 8.5)}{2400 + 1200 - 628.3}$$

$= 26.5$ mm **Ans.**

Example 15. *A circular sector of angle 60° is cut from the circle of radius r as shown in Fig. 26 :*

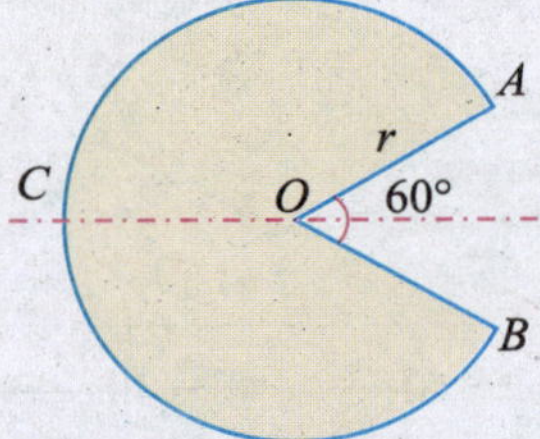

Fig. 26.

Determine the centre of gravity of the remainder.

Solution: As the section is symmetrical about *X-X* axis, therefore its centre of gravity will lie on this axis.

Let *C* be the reference point.

(i) *Main circle*

Area, $a_1 = \pi r^2$

and $x_1 = r$

(ii) *Cut out sector*

Area, $a_2 = \frac{\pi r^2 \theta}{360°} = \frac{\pi r^2 \times 60°}{360°} = \frac{\pi r^2}{6}$

and $x_2 = r + \frac{2r}{\pi}$

We know that distance between the centre of gravity of the section and *C*

$$\bar{x} = \frac{a_1 x_1 - a_2 x_2}{a_1 - a_2} = \frac{(\pi r^2 \times r) - \left[\frac{\pi r^2}{6} \times \left(r + \frac{2r}{\pi}\right)\right]}{\pi r^2 - \frac{\pi r^2}{6}}$$

$$= \frac{\pi r^2 \left[r - \frac{1}{6}\left(r + \frac{2r}{\pi} \right) \right]}{\pi r^2 \left(1 - \frac{1}{6} \right)} = \frac{r - \left[\frac{1}{6} \times \left(r + \frac{2r}{\pi} \right) \right]}{1 - \frac{1}{6}}$$

$$= \frac{6}{5}\left[r - \left(\frac{r}{6} + \frac{2r}{6\pi} \right) \right] = \frac{6}{5}\left[r - \frac{r}{6} - \frac{r}{3\pi} \right]$$

$$= \frac{6}{5}\left(\frac{5}{6}r - \frac{r}{3\pi} \right) = r - \frac{2r}{5\pi} \quad \textbf{Ans.}$$

Example 16. *A solid consists of a right circular cylinder and a hemisphere with a cone cut out from the cylinder as shown in Fig. 27.*

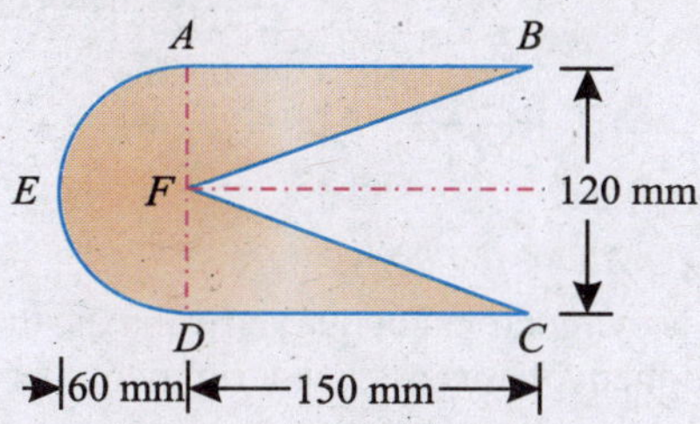

Fig. 27

Find the centre of gravity of the body.

Solution. As the solid is symmetrical about horizontal axis, therefore its centre of gravity lie on this axis.

Let the left edge of the hemispherical portion (E) be the axis of reference.

(i) Hemisphere ADE

Volume, $\quad v_1 = \frac{2\pi}{3} \times r^3 = \frac{2\pi}{3} \times (60)^3 = 144\,000\ \pi\ \text{mm}^3$

and $\quad x_1 = \frac{5r}{8} = \frac{5 \times 60}{8} = 37.5\ \text{mm}$

(ii) Right circular cylinder ABCD

Volume, $\quad v_2 = \pi \times r^2 \times h = \pi \times (60)^2 \times 150 = 540\,000\ \pi\ \text{mm}^3$

and $\quad x_2 = 60 + \frac{150}{2} = 135\,\text{mm}$

(iii) Cone BCF

Volume, $\quad v_3 = \frac{\pi}{3} \times r^2 \times h = \frac{\pi}{3} \times (60)^2 \times 150 = 180\,000\ \pi\ \text{mm}^3$

and $\quad x_3 = 60 + 150 \times \frac{3}{4} = 172.5\ \text{mm}$

We know that distance between centre of gravity of the solid and left edge of the hemisphere (E),

$$\bar{x} = \frac{v_1 x_1 + v_2 x_2 - v_3 x_3}{v_1 + v_2 - v_3}$$

$$= \frac{(144\ 000\ \pi \times 37.5) + (540\ 000\ \pi \times 135) - (180\ 000\ \pi \times 172.5)}{144\ 000\ \pi + 540\ 000\ \pi - 180\ 000\ \pi}$$

$= 93.75$ mm **Ans.**

Example 17. *A frustum of a solid right circular cone has an axial hole of 50 cm diameter as shown in Fig. 28.*

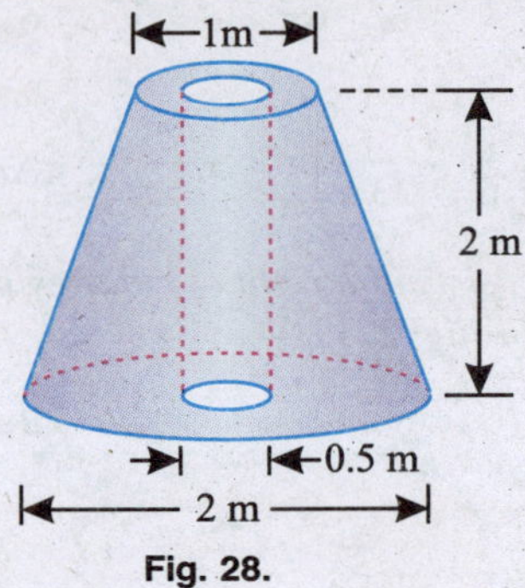

Fig. 28.

Determine the centre of gravity of the body.

Solution. As the body is symmetrical about vertical axis, therefore its centre of geravity lie on this axis. For the sake of simplicity, let us assume a right circular cone *OCD*, from which a right circulr cone *OAB* is cut off as shown in Fig. 29.

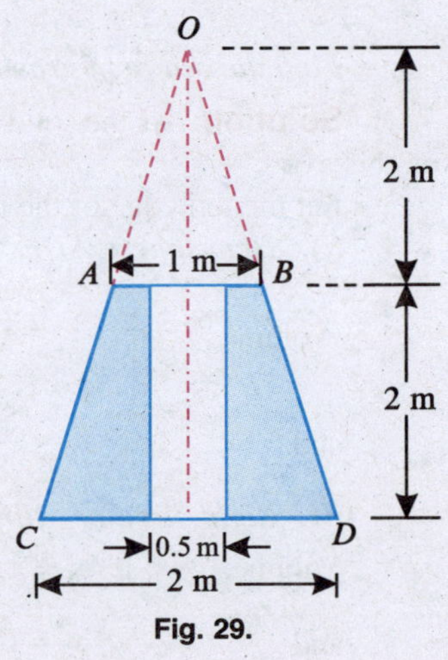

Fig. 29.

Let base of cone *CD* be the axis of reference.

(*i*) *Right circular cone OCD*

Volume, $v_1 = \frac{\pi}{3} \times R^2 \times H$

$= \frac{\pi}{3} \times (1)^2 \times 4 = \frac{4\pi}{3}$ m^3

and $y_1 = \frac{4}{4} = 1$m

(*ii*) *Right circular cone OAB*

Volume, $v_2 = \frac{\pi}{3} \times r^2 \times h$

$= \frac{\pi}{3} \times (0.5)^2 \times 2 = \frac{\pi}{6}$ m^3

and $y_2 = 2 + \frac{2}{4} = \frac{5}{2}$m

(*iii*) *Circular hole*

Volume, $v_3 = \frac{\pi}{4} \times d^2 \times h = \frac{\pi}{4} \times (0.5)^2 \times 2 = \frac{\pi}{8}$ m^3

and $y_2 = \frac{2}{2} = 1$ m

We know that distance between centre of gravity of the body and the base of the cone,

$$\bar{y} = \frac{v_1\, y_1 - v_2\, y_2 - v_3\, y_3}{v_1 - v_2 - v_3}$$

$$= \frac{\left(\frac{4\pi}{3} \times 1\right) - \left(\frac{\pi}{6} \times \frac{5}{2}\right) - \left(\frac{\pi}{8} \times 1\right)}{\frac{4\pi}{3} - \frac{\pi}{6} - \frac{\pi}{8}} = \frac{\frac{4}{3} - \frac{5}{12} - \frac{1}{8}}{\frac{4}{3} - \frac{1}{6} - \frac{1}{8}} = \frac{19}{25} = 0.76 \text{ m} \quad \textbf{Ans.}$$

Example 18. *A solid hemisphere of 20 mm radius supports a solid cone of the same base and 60 mm height as shown in Fig. 30. Locate the centre of gravity of the composite section.*

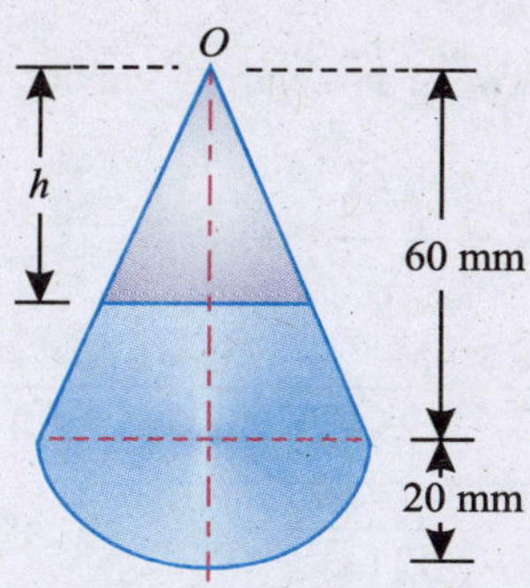

Fig. 30.

If the upper portion of the cone is removed by a certain section, the centre of gravity lowers down by 5 mm. Find the depth (h) of the section plane below the apex.

Solution. As the body is symmetrical about *Y-Y* axis, therefore its centre of gravity will lie on this axis.

Let apex of the cone (*O*) be the axis of reference.

Centre of gravity of the composite section

(i) Right circular cone

Volume, $v_1 = \frac{\pi}{3} \times r^2 \times h = \frac{\pi}{3} \times (20)^2\, 60 = 25\ 133 \text{ mm}^3$

and $y_1 = 60 \times \frac{3}{4} = 45 \text{ mm}$

(ii) Hemisphere

Volume, $v_2 = \frac{2\pi}{3} \times r^2 = \frac{2\pi}{3} \times (20)^3 = 16\ 755 \text{ mm}^3$

and $y_2 = 60 + \frac{3 \times 20}{8} = 67.5 \text{ mm}$

We know that distance between centre of gravity of the body and apex of the cone,

$$\bar{y} = \frac{v_1 y_1 + v_2 y_2}{v_1 + v_2} = \frac{(25\ 133 \times 45) + (16\ 755 \times 67.5)}{25\ 133 + 16\ 755} \text{ mm}$$

$$= \frac{2\ 261\ 950}{41\ 888} = 54 \text{ mm} \quad \textbf{Ans.}$$

Depth (h) of the section plane below the apex

We know that the radius of the cut out cone,

$$r = \frac{h}{3} \qquad \ldots \left(\because \frac{r}{20} = \frac{h}{60}\right)$$

∴ Volume of the cut out cone,

$$v_3 = \frac{\pi}{3} \times r^2 \times h = \frac{\pi}{3}\left(\frac{h}{3}\right)^2 \times h = 0.1164\ h^2\ \text{mm}^3$$

and distance between centre of gravity of the cut out cone and its apex,

$$y_3 = \frac{3h}{4} = 0.75\ h$$

We also know that distance between the centre of gravity of the body and apex of the cone (*i.e.* 54 + 5 = 59 mm),

$$\bar{y} = \frac{v_1 y_1 + v_2 y_2 - v_3 y_3}{v_1 + v_2 - v_2}$$

$$\therefore \qquad 59 = \frac{(25\,133 \times 45) + (16\,755 \times 67.5) - 0.1164\ h^3 \times 0.75\ h}{25\,133 + 16\,755 - 0.1164\ h^3}$$

$$= \frac{2\,261\,950 - 0.0873 h^4}{41\,888 - 0.1164\ h^3}$$

$$2\,471\,400 - 6.868\ h^3 = 2\,261\,950 - 0.0873\ h^4$$

$$0.0873\ h^4 - 6.868\ h^3 = -\ 209\,450$$

Dividing both sides by 0.0873,

$$h^4 - 78.67\ h^3 = -2\,399\,200 \qquad \text{...}(i)$$

We shall solve this equation by trial and error. First of all, let us substitute $h = 10$ mm in the left hand side of equation (*i*). We find

$$(10)^4 - 78.67\ (10)^3 = -\ 68\,670$$

We find that answer obtained does not tally with the value of right hand side of equation (*i*), and is much less than that. Now let us substitute $h = 20$ mm in the left hand side of equation (*i*),

$$(20)^4 - 78.67\ (20)^3 = -\ 469\,360$$

We again find that the answer obtained does not tally with the right hand side of equation (*i*), But it is closer to the value of right hand side than the first case (*i.e.* when we substituted $h = 10$ mm.) Or in other words, the value obtained is still less than the right hand side of equation (*i*). But the difference has reduced. Now let us substitute $h = 30$ mm in the left hand side of equation (*i*).

$$(30)^4 - 78.67\ (30)^3 = 1\,314\,100$$

We again find the answer obtained does not tally with the right hand side of equation (*i*), But it is more close to the right hand side than the previous case *i.e.* when we substituted $h = 20$ mm. Now let us substitute $h = 40$ mm in the left hand side of the equation (*i*).

$$(40)^4 - 78.67\ (40)^3 = 2474900$$

Now we find that the answer obtained does not tally with the right hand side of equation (*i*). But its value is more than the right hand side of equation (*i*), In the previous cases, the value of the answer obtained was less. Thus we find that the value of (h) is less than 40 mm.

A little consideration will show, that as the value of the answer is slightly more than the right hand side of equation (*i*). (as compared to the previous answers), the value of (h) is slightly less than 40 mm. Now let us substitude $h = 39$ mm in the left hand side of the equation (*i*).

$$(39)^4 - 78.67\ (39)^3 = -\ 2\,153\,200$$

Now we find that the answer obtained is less than the right hand side of equation (*i*). Thus the value of (h) is more than 39 mm. Or in other words it is within 39 and 40 mm. This is due to the reason that when we substitude $h = 39$ mm, the answer is less and when we substitute $h = 40$ mm, answer is more than the right hand side of equation (*i*), Now let us substitute $h = 39.5$ mm in the left hand side of the equation (*i*).

$$(39.5)^4 - 78.6\dot{7}\ (39.5)^3 = -2\ 414\ 000$$

Now we find that the answer obtained is more than the right hand side of equation (*i*). Thus the value of (h) is less than 39.5 mm. Now let us substitute the $h = 39.4$ mm in the left hand side of equation, (*i*).

$$(39.4)^4 - 78.67\ (39.4)^3 = -2\ 401\ 900$$

We find that is answer is very close to the right hand side of the equation and there is no need of further calculations. Thus the value of $h = 39.4$ mm **Ans.**

EXERCISE 3

*1. A circular hole of 50 mm diameter is cut out from a circular disc of 100 mm diameter as shown in Fig. 31. Find the centre of gravity of the section from *A*. [**Ans.** 41.7 mm] **[IGNOU]**

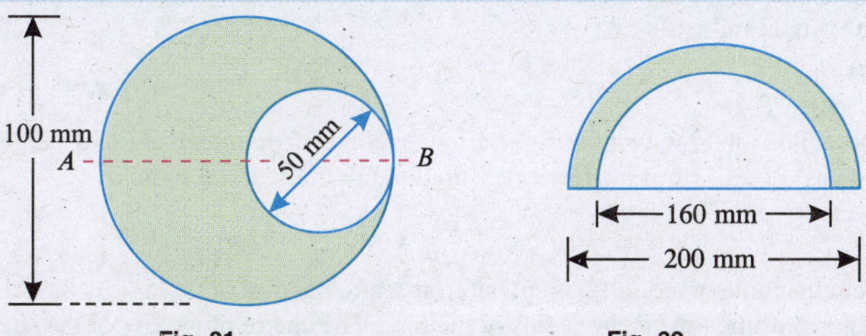

Fig. 31. Fig. 32.

*2. Find the centre of gravity of a semicircular section having outer and inner diameters of 200 mm and 160 mm respectively as shown in Fig. 32. [**Ans.** 57.5 mm from the base]

3. A circular sector of angle 45° is cut from the circle of radius 220 mm Determine the centre of gravity of the remainder from the centre of the sector. [**Ans.** 200 mm]

*4. A hemisphere of radius 80 mm is cut out from a right circular cylinder of diameter 80 mm and height 160 mm as shown in Fig. 33. Find the centre of gravity of the body from the base *AB*. [**Ans.** 77.2 mm]

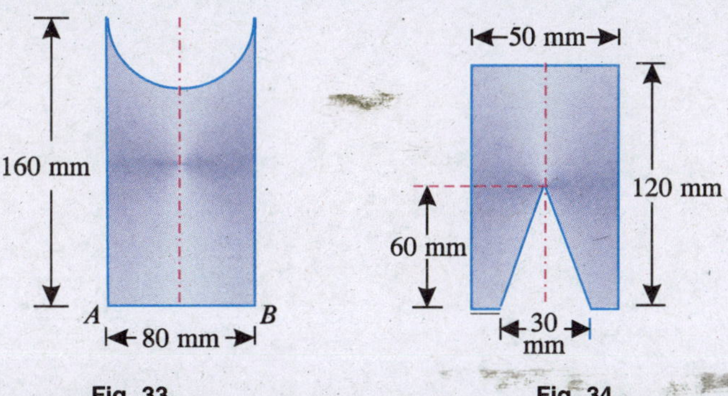

Fig. 33. Fig. 34.

*5. A right circular cone of 30 mm diameter and 60 mm height is cut from a cylinder of 50 mm diameter at 120 mm height as shown in Fig. 34. Find the position of the centre of gravity of the body from its base. [**Ans.** 60.7 mm]

QUESTIONS

1. Define the terms 'centre of gravity'.
2. Distinguish between centre of gravity and centroid.
3. How many centres of gravity a body has?
4. Describe the various methods of finding out the centre of gravity of a body.
5. How would you find out the centre of gravity of a section, with a cut out hole?

OBJECTIVE TYPE QUESTIONS

1. The centre of gravity of an equilateral triangle with each side (a) is ...... from any of the three sides.

   (*a*) $\frac{a\sqrt{3}}{2}$ (*b*) $\frac{a\sqrt{2}}{3}$ (*c*) $\frac{a}{2\sqrt{3}}$ (*d*) $\frac{a}{3\sqrt{2}}$

2. The centre of gravity of hemisphere lies at a distance of ......form its base measured along the vertical radius.

   (*a*) $\frac{3r}{8}$ (*b*) $\frac{3}{8r}$ (*c*) $\frac{8r}{3}$ (*d*) $\frac{8}{3r}$

3. The centre of gravity of a right circular cone of diameter (d) and height (h) lies at a distance of ...... from the base measured along the vertical radius.

   (*a*) $\frac{h}{2}$ (*b*) $\frac{h}{3}$ (*c*) $\frac{h}{4}$ (*d*) $\frac{h}{6}$

4. A circular hole of radius (r) is cut out from a circular disc of radius ($2r$) in such a way that the diagonal of the hole is the radius of the disc. The centre of gravity of the section lies at

   (*a*) Centre of a disc (*b*) Centre of the hole
   (*c*) Somewhere in the disc (*d*) Somewhere in the hole

ANSWERS

1. (*c*) 2. (*a*) 3. (*c*) 4. (*c*)

CHAPTER

7 Moment of Inertia

Contents

Note :

Important Results related to this chapter are given at the end of this book.

See Appendix - Properties of Plane Areas

1. MOMENT OF INERTIA (M.I.)

[PTU, RGPV, AUT]

The **moment of inertia** of an object is defined by the distribution of mass around an axis.

It depends not only on the total mass of the object, but also on the square of the perpendicular distance from the axis to each element of mass. This means the moment of inertia increases rapidly as masses are distributed more distant from the axis.

We know that the moment of a force (P) about a point, is the product of the force and perpendicular distance (x) between the point and the line of action of the force (*i.e.* $P.x$). This moment is also called **first moment of force**. If this moment is again multiplied by the perpendicular distance (x) between the point and the line of action of the force *i.e.* $P.x\ (x) = Px^2$, then this quantity is called **moment of the moment**

of a force or **second moment of force** or **moment of inertia** (briefly written as **M.I.**).

Sometimes, instead of force, area or mass of a figure or body is taken into consideration. Then the second moment is known as second moment of area or second moment of mass. But all such second moments are broadly termed as moment of inertia. In this chapter, we shall discuss the moment of inertia of plane areas only.

2. MOMENT OF INERTIA OF A PLANE AREA

Consider a plane area, whose moment of inertia is required to be found out. Split up the whole area into a number of small elements.

Let $a_1, a_2, a_3, ... =$ Areas of small elements, and

$r_1, r_2, r_3, ... =$ Corresponding distances of the elements from the line about which the moment of inertia is required to be found out.

Now the moment of inertia of the area,

$$I = a_1 r_1^2 + a_2 r_2^2 + a_3 r_3^2 + ...$$
$$= \Sigma a r^2$$

3. UNITS OF MOMENT OF INERTIA

As a matter of fact the units of moment of inertia of a plane area depend upon the units of the area and the length. *e.g.*,

1. If area is in m^2 and the length is also in m, the moment of inertia is expressed in m^4.
2. If area in mm^2 and the length is also in mm, then moment of inertia is expressed in mm^4.

4. METHODS FOR FINDING MOMENT OF INERTIA

The moment of inertia of a plane area (or a body) may be found out by any one of the following two methods :

1. By Routh's rule **2. By Integration.**

Note : The Routh's Rule is used for finding the moment of inertia of a plane area or a body of uniform thickness.

5. MOMENT OF INERTIA BY ROUTH'S RULE

The Routh's Rule states, if a body is symmetrical about three mutually perpendicular axes (*i.e.*, *X-X* axis, *Y-Y* axis and *Z-Z* axis), then the moment of inertia, about any one axis passing through its centre of gravity is given by:

$$I = \frac{A \text{ (or } M) \times S}{3}$$ ... (For a Square or Rectangular Lamina)

$$I = \frac{A \text{ (or } M) \times S}{4}$$ ... (For a Circular or Elliptical Lamina)

$$I = \frac{A \text{ (or } M) \times S}{5}$$ ... (For a Spherical Body)

where $A =$ Area of the plane area

$M =$ Mass of the body, and

$S =$ Sum of the squares of the two semi-axis, other than the axis, about which the moment of inertia is required to be found out.

Note : This method has only academic importance and is rarely used in the field of science and technology these days. The reason for the same is that it is equally convenient to use the method of integration for the moment of inertia of a body.

6. *MOMENT OF INERTIA BY INTEGRATION

The moment of inertia of an area may also be found out by the method of integration as discussed below:

Consider a plane figure, whose moment of inertia is required to be found out about X-X axis and Y-Y axis as shown in Fig 1. Let us divide the whole area into a no. of strips. Consider one of these strips.

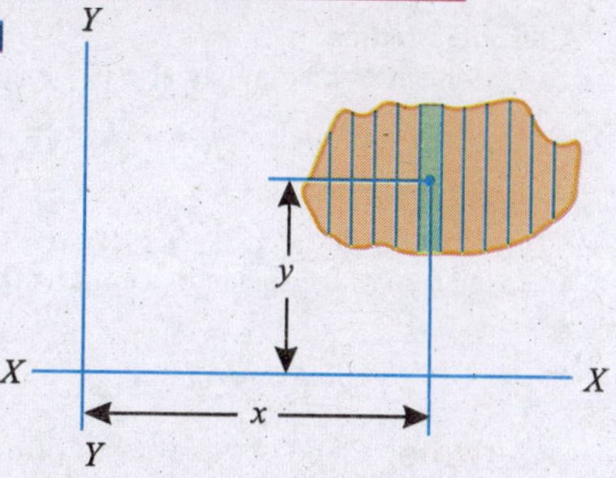

Fig. 1. Moment of inertia by integration.

Let dA = Area of the strip

x = Distance of the centre of gravity of the strip on X-X axis and

y = Distance of the centre of gravity of the strip on Y-Y axis.

We know that the moment of inertia of the strip about Y-Y axis

$$= dA \cdot x^2$$

Now the moment of inertia of the whole area about Y-Y and X-X axis respectively may be found out by integrating above equation. *i.e.*,

$$I_{YY} = \Sigma\, dA \cdot x^2$$

Similarly $$I_{XX} = \Sigma\, dA \cdot y^2$$

In the following pages, we shall discuss the applications of this method for finding out the moment of inertia of various cross-sections.

7. *MOMENT OF INERTIA OF A RECTANGULAR SECTION

[GJU, AMIE]

Consider a rectangular section $ABCD$ as shown in Fig. 2 whose moment of inertia is required to be found out.

Let b = Width of the section and

d = Depth of the section.

Now consider a strip PQ of thickness dy parallel to X-X axis and at a distance y from it as shown in the figure

∴ Area of the strip

$$= b \cdot dy$$

We know that moment of inertia of the strip about X-X axis,

$$= \text{Area} \times y^2 = (b.\ dy)\ y^2 = b \cdot y^2 .\ dy$$

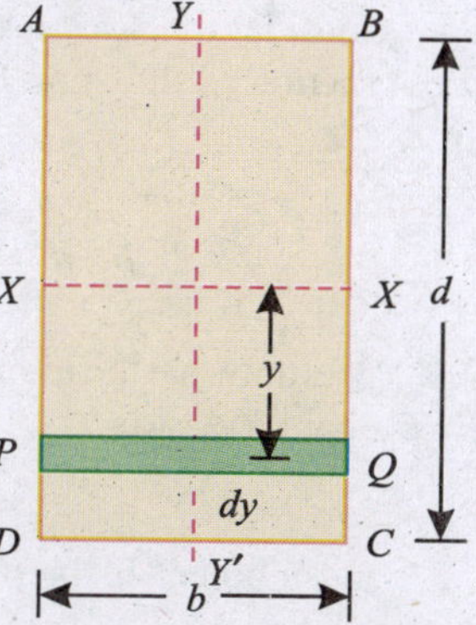

Fig. 2. Rectangular section.

Now *moment of inertia of the whole section may be found out by integrating the above equation for the whole length of the lamina *i.e.* from $-\dfrac{d}{2}$ to $+\dfrac{d}{2}$,

$$I_{XX} = \int_{-\frac{d}{2}}^{+\frac{d}{2}} b.y^2.dy = b\int_{-\frac{d}{2}}^{+\frac{d}{2}} y^2 \cdot dy = b\left[\frac{y^3}{3}\right]_{-\frac{d}{2}}^{+\frac{d}{2}} = b\left[\frac{(d/2)^3}{3} - \frac{(-d/2)^3}{3}\right]$$

or $$I_{XX} = \frac{bd^3}{12}$$

Similarly, $$I_{YY} = \frac{db^3}{12}$$

* **Alternate Method:**

Moment of Inertia may also be obtained by Routh's rule as discussed below:

$$I_{XX} = \frac{AS}{3} \qquad \text{...(for rectangular section)}$$

where area, $A = b \times d$,

and sum of the square of semi axes Y-Y and Z-Z, $S = \left(\frac{d}{2}\right)^2 + 0 = \frac{d^2}{4}$

$$\therefore \qquad I_{XX} = \frac{AS}{3} = \frac{(b \times d) \times \frac{d^2}{4}}{3} = \frac{bd^3}{12}$$

***Note.** Cube is to be taken of the side, which is at right angles to the line of reference.

Example 1. *Find the moment of inertia of a rectangular section 30 mm wide and 40 mm deep about X-X axis and Y-Y axis.*

Solution. Given: Width of the section (b) = 30 mm and depth of the section (d) = 40 mm.

We know that, moment of inertia of the section about an axis passing through its centre of gravity and parallel to X-X axis,

$$I_{XX} = \frac{bd^3}{12} = \frac{30 \times (40)^3}{12} = 160 \times 10^3 \text{ mm}^4 \qquad \textbf{Ans.}$$

Similarly, moment of inertia of the section about an axis passing through its centre of gravity and parallel to Y-Y axis,

$$I_{YY} = \frac{db^3}{12} = \frac{40 \times (30)^3}{12} = 90 \times 10^3 \text{ mm}^4 \qquad \textbf{Ans.}$$

8. MOMENT OF INERTIA OF A HOLLOW RECTANGULAR SECTION

Consider a hollow rectangular section, in which *ABCD* is the main section and *EFGH* is the cut out section as shown in Fig 3

Let b = Breadth of the outer rectangle,

d = Depth of the outer rectangle and

b_1, d_1 = Corresponding values for the cut out rectangle.

We know that the moment of inertia, of the outer rectangle *ABCD* about X-X axis

$$= \frac{bd^3}{12} \qquad \text{...(i)}$$

and moment of inertia of the cut out rectangle *EFGH* about X-X axis

$$= \frac{b_1 d_1^3}{12} \qquad \text{...(ii)}$$

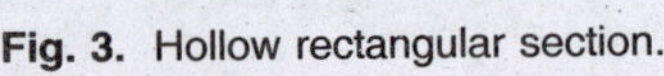

Fig. 3. Hollow rectangular section.

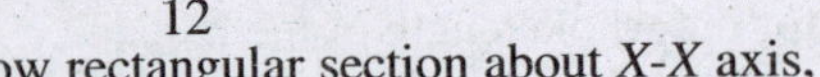

$\therefore$ M.I. of the hollow rectangular section about X-X axis,

I_{XX} = M.I. of rectangle *ABCD* – M.I. of rectangle *EFGH*

$$I_{XX} = \frac{bd^3}{12} - \frac{b_1 d_1^3}{12}$$

Similarly,

$$I_{YY} = \frac{db^3}{12} - \frac{d_1 b_1^3}{12}$$

Note : This relation holds good only if the centre of gravity of the main section as well as that of the cut out section coincide with each other.

Example 2. *Find the moment of inertia of a hollow rectangular section about its centre of gravity if the external dimensions are breadth 60 mm, depth 80 mm and internal dimensions are breadth 30 mm and depth 40 mm respectively.*

Solution. Given: External breadth (b) = 60 mm; External depth (d) = 80 mm ; Internal breadth (b_1) = 30 mm and internal depth (d_1) = 40 mm.

We know that moment of inertia of hollow rectangular section about an axis passing through its centre of gravity and parallel to X-X axis,

$$I_{XX} = \frac{bd^3}{12} - \frac{b_1 d_1^3}{12} = \frac{60\ (80)^3}{12} - \frac{30\ (40)^3}{12} = 2400 \times 10^3\ \text{mm}^4 \qquad \textbf{Ans.}$$

Similarly,
$$I_{YY} = \frac{db^3}{12} - \frac{d_1 b_1^3}{12} = \frac{80\ (60)^3}{12} - \frac{40\ (30)^3}{12} = 1350 \times 10^3\ \text{mm}^4 \qquad \textbf{Ans.}$$

9. *THEOREM OF PERPENDICULAR AXIS

[WBUT, IGNOU]

It states that, ***If I_{XX} and I_{YY} be the moments of inertia of a plane section about two perpendicular axis meeting at O, the moment of inertia I_{ZZ} about the axis Z-Z, perpendicular to the plane and passing through the intersection of X-X and Y-Y is given by***

$$I_{ZZ} = I_{XX} + I_{YY}$$

Proof :

Consider a small lamina (P) of area da having co-ordinates as x and y along OX and OY two mutually perpendicular axes on a plane section as shown in Fig. 4.

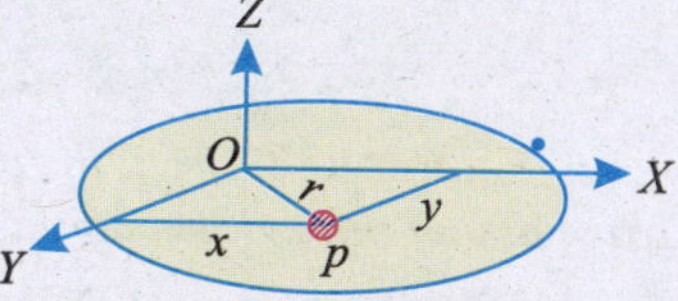

Fig. 4. Theorem of perpendicular axis.

Now consider a plane OZ perpendicular to OX and OY. Let (r) be the distance of the lamina (P) from Z-Z axis such that $OP = r$.

From the geometry of the figure, we find that

$$r^2 = x^2 + y^2$$

We know that the moment of inertia of the lamina P about X-X axis,

$$I_{XX} = da.\ y^2 \qquad \ldots[\because I = \text{Area} \times (\text{Distance})^2]$$

Similarly,
$$I_{YY} = da.\ x^2$$

and
$$I_{ZZ} = da.\ r^2 = da\ (x^2 + y^2) \qquad \ldots(\because r^2 = x^2 + y^2)$$
$$= da.\ x^2 + da.\ y^2 = I_{YY} + I_{XX}$$

10. MOMENT OF INERTIA OF A CIRCULAR SECTION

[PTU, AUT]

Consider a circle $ABCD$ of radius (r) with centre O and X-X′ and Y-Y′ be two axes of reference through O as shown in Fig. 5.

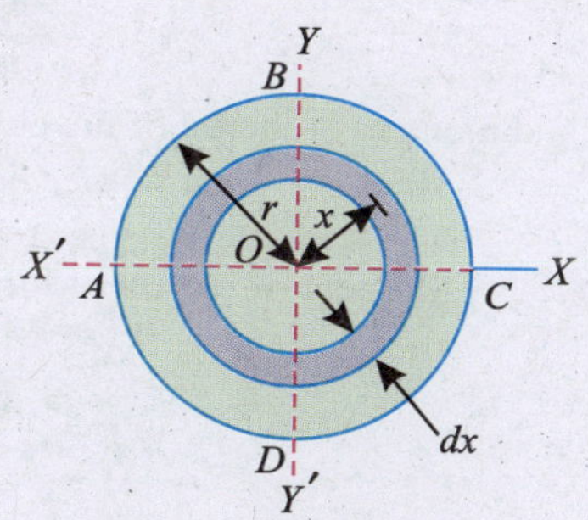

Fig. 5. Circular section.

Now consider an elementary ring of radius x and thickness dx. Therefore area of the ring,

$$da = 2\ \pi\ x.\ dx$$

and moment of inertia of ring, about X-X axis or Y-Y axis

$$= \text{Area} \times (\text{Distance})^2$$
$$= 2\ \pi\ x.\ dx \times x^2$$
$$= 2\ \pi\ x^3.\ dx$$

Now moment of inertia of the whole section, about the central axis, can be found out by integrating the above equation for the whole radius of the circle *i.e.*, from 0 to *r*.

$$\therefore \qquad I_{ZZ} = \int_0^r 2\pi x^3 \,.\, dx = 2\pi \int_0^r x^3 \,.\, dx = 2\pi \left[\frac{x^4}{4}\right]_0^r = \frac{\pi}{2}(r)^4$$

$$I_{ZZ} = \frac{\pi}{32}(d)^4 \qquad \ldots\left(\text{substituting } r = \frac{d}{2}\right)$$

We know from the Theorem of Perpendicular Axis that

$$I_{XX} + I_{YY} = I_{ZZ}$$

$$\therefore \qquad I_{XX} = I_{YY} = \frac{I_{ZZ}}{2} = \frac{1}{2} \times \frac{\pi}{32}(d)^4 = \frac{\pi}{64}(d)^4$$

* **Alternate Method:**

Moment of Inertia may also be obtained by Routh's rule as discussed below:

$$I_{XX} = \frac{AS}{4} \qquad \text{(for circular section)}$$

where area, $A = \frac{\pi}{4} \times d^2$

and sum of the square of semi axis *Y-Y* and *Z-Z*, $S = \left(\frac{d}{2}\right)^2 + 0 = \frac{d^2}{4}$

$$\therefore \qquad I_{XX} = \frac{AS}{4} = \frac{\left[\frac{\pi}{4} \times d^2\right] \times \frac{d^2}{4}}{4} = \frac{\pi}{64}(d)^4$$

Example 3. *Find the moment of inertia of a circular section of 50 mm diameter about an axis passing through its centre.*

Solution. Given: Diameter (*d*) = 50 mm

Moment of inertia of the circular section about an axis passing through its centre,

$$I_{XX} = \frac{\pi}{64}(d)^4 = \frac{\pi}{64} \times (50)^4 = 307 \times 10^3 \text{ mm}^4 \qquad \textbf{Ans.}$$

11. *MOMENT OF INERTIA OF A HOLLOW CIRCULAR SECTION

Consider a hollow circular section as shown in Fig.6, whose moment of inertia is required to be found out.

Let D = Diameter of the main circle, and

d = Diameter of the cut out circle.

We know that the moment of inertia of the main circle about *X-X* axis

$$= \frac{\pi}{64}(D)^4$$

and moment of inertia of the cut-out circle about *X-X* axis

$$= \frac{\pi}{64}(d)^4$$

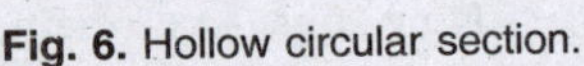

Fig. 6. Hollow circular section.

∴ Moment of inertia of the hollow circular section about *X-X* axis,

I_{XX} = Moment of inertia of main circle – Moment of inertia of cut out circle,

$$\text{or} \qquad I_{XX} = \frac{\pi}{64}(D)^4 - \frac{\pi}{64}(d)^4 = \frac{\pi}{64}(D^4 - d^4)$$

$$\text{Similarly,} \qquad I_{YY} = \frac{\pi}{64}(D^4 - d^4)$$

Note : This relation holds good only if the centre of the main circular section as well as that of the cut out circular section coincide with each other.

Example 4. *A hollow circular section has an external diameter of 80 mm and internal diameter of 60 mm. Find its moment of inertia about the horizontal axis passing through its centre.*

Solution. Given : External diameter (D) = 80 mm and internal diameter (d) = 60 mm.

We know that moment of inertia of the hollow circular section about the horizontal axis passing through its centre,

$$I_{XX} = \frac{\pi}{64}(D^4 - d^4) = \frac{\pi}{64}[(80)^4 - (60)^4] = 1374 \times 10^3 \text{ mm}^4 \qquad \textbf{Ans.}$$

12. *THEOREM OF PARALLEL AXIS

[UPTU, PUT, AUT, WBUT]

It states, ***If the moment of inertia of a plane area about an axis through its centre of gravity is denoted by I_G, then moment of inertia of the area about any other axis AB, parallel to the first, and at a distance h from the centre of gravity is given by:***

$$I_{AB} = I_G + ah^2$$

where I_{AB} = Moment of inertia of the area about an axis AB,

I_G = Moment of Inertia of the area about its centre of gravity

a = Area of the section, and

h = Distance between centre of gravity of the section and axis AB.

Proof

Consider a strip of a circle, whose moment of inertia is required to be found out about a line AB as shown in Fig. 7.

Let δa = Area of the strip

y = Distance of the strip from the centre of gravity the section and

h = Distance between centre of gravity of the section and the axis AB.

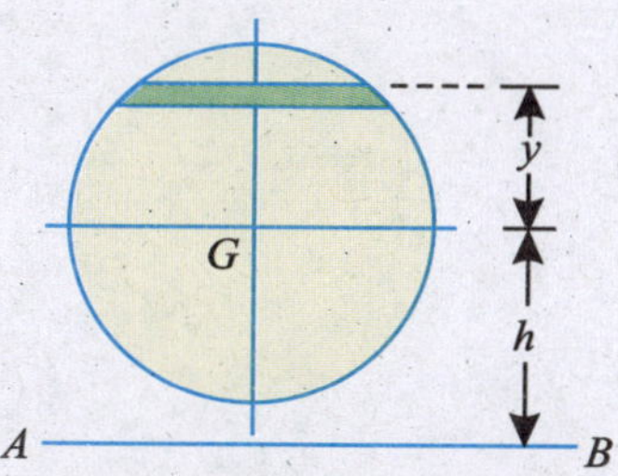

Fig. 7. Theorem of parallel axis.

We know that moment of inertia of the whole section about an axis passing through the centre of gravity of the section

$$= \delta a.\ y^2$$

and moment of inertia of the whole section about an axis passing through its centre of gravity,

$$I_G = \Sigma\ \delta a.\ y^2$$

∴ Moment of inertia of the section about the axis AB,

$$I_{AB} = \Sigma\ \delta a\ (h + y)^2 = \Sigma\ \delta a\ (h^2 + y^2 + 2\ h\ y)$$

$$= (\Sigma\ h^2.\ \delta a) + (\Sigma\ y^2.\ \delta a) + (\Sigma\ 2\ h\ y\ .\ \delta a)$$

$$I_{AB} = a\ h^2 + I_G + 0$$

It may be noted that $\Sigma\ h^2 .\ \delta a = a\ h^2$ and $\Sigma\ y^2 .\ \delta a = I_G$ [as per equation (*i*) above] and $\Sigma\ \delta a.y$ is the algebraic sum of moments of all the areas, about an axis through centre of gravity of the section and is equal to $a.\bar{y}$, where $\bar{y}$ is the distance between the section and the axis passing through the centre of gravity, which obviously is zero.

13. MOMENT OF INERTIA OF A TRIANGULAR SECTION

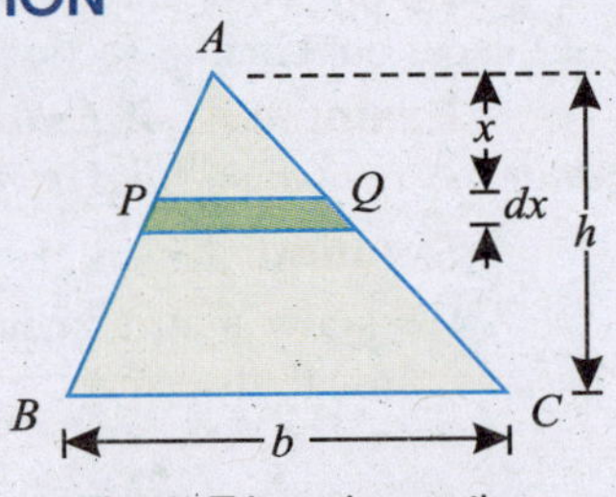

Fig. 8. Triangular section.

Consider a triangular section *ABC* whose moment of inertia is required to be found out.

Let b = Base of the triangular section and

h = Height of the triangular section.

Now consider a small strip *PQ* of thickness *dx* at a distance of *x* from the vertex *A* as shown in Fig. 8. From the geometry of the figure, we find that the two triangles *APQ* and *ABC* are similar. Therefore

$$\frac{PQ}{BC} = \frac{x}{h} \quad \text{or} \quad PQ = \frac{BC.x}{h} = \frac{bx}{h} \qquad (\because BC = \text{base} = b)$$

We know that area of the strip *PQ*

$$= \frac{bx}{h}.dx$$

and moment of inertia of the strip about the base *BC*

$$= \text{Area} \times (\text{Distance})^2 = \frac{bx}{h} dx\,(h-x)^2 = \frac{bx}{h}(h-x)^2\,dx$$

Now moment of inertia of the whole triangular section may be found out by integrating the above equation for the whole height of the triangle *i.e.*, from 0 to *h*.

$$I_{BC} = \int_0^h \frac{bx}{h}(h-x)^2\,dx$$

$$= \frac{b}{h}\int_0^h x\,(h^2 + x^2 - 2hx)\,dx$$

$$= \frac{b}{h}\int_0^h (xh^2 + x^3 - 2hx^2)\,dx$$

$$= \frac{b}{h}\left[\frac{x^2h^2}{2} + \frac{x^4}{4} - \frac{2hx^3}{3}\right]_0^h = \frac{bh^3}{12}$$

We know that distance between centre of gravity of the triangular section and base *BC*,

$$d = \frac{h}{3}$$

∴ Moment of inertia of the triangular section about an axis through its centre of gravity and parallel to *X-X* axis,

$$I_G = I_{BC} - ad^2 \qquad ...(\because I_{XX} = I_G + a\,h^2)$$

or
$$I_G = \frac{bh^3}{12} - \left(\frac{bh}{2}\right)\left(\frac{h}{3}\right)^2 = \frac{bh^3}{36}$$

Notes : 1. The moment of inertia of section about an axis through its vertex and parallel to the base

$$= I_G + ad^2 = \frac{bh^3}{36} + \left(\frac{bh}{2}\right)\left(\frac{2h}{3}\right)^2 = \frac{9bh^3}{36} = \frac{bh^3}{4}$$

2. This relation holds good for any type of triangle.

Example. 5. *An isosceles triangular section ABC has base width 80 mm and height 60 mm. Determine the moment of inertia of the section about the centre of gravity of the section and the base BC.*

Solution. Given : Base width (b) = 80 mm and height (h) = 60 mm.

Moment of inertia about the centre of gravity of the section

We know that moment of inertia of triangular section about its centre of gravity,

$$I_G = \frac{b h^3}{36} = \frac{80 \times (60)^3}{36} = 480 \times 10^3 \text{ mm}^4$$

Moment of inertia about the base BC

We also know that moment of inertia of triangular section about the base BC,

$$I_{BC} = \frac{b h^3}{12} = \frac{80 \times (60)^3}{12} = 1440 \times 10^3 \text{ mm}^4$$

***Example 6.** *A hollow triangular section shown in Fig. 9 is symmetrical about its vertical axis.*

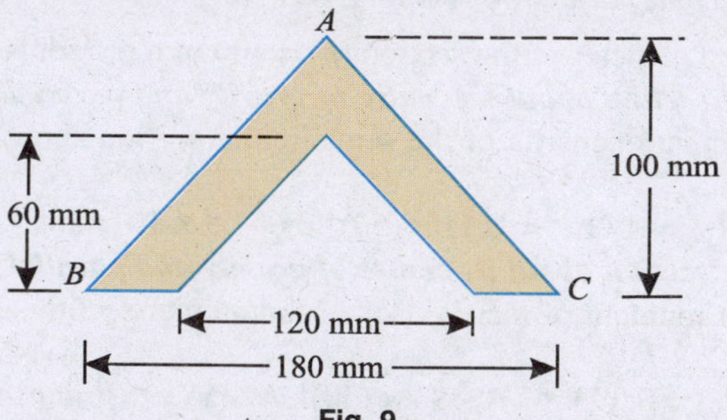

Fig. 9.

Find the moment of inertia of the section about the base BC.

Solution. Given : Base width of main triangle (B) = 180 mm; Base width of cut out triangle (b) = 120 mm; Height of main triangle (H) = 100 mm and height of cut out triangle (h) = 60 mm.

We know that moment of inertia of the triangular, section about the base BC,

$$I_{BC} = \frac{BH^3}{12} - \frac{bh^3}{12} = \frac{180 \times (100)^3}{12} - \frac{120 \times (60)^3}{12} \text{ mm}^4$$

$$= (15 \times 10^6) - (2.16 \times 10^6) = 12.84 \times 10^6 \text{ mm}^4 \quad \textbf{Ans.}$$

14. *MOMENT OF INERTIA OF A SEMICIRCULAR SECTION

[PTU]

Consider a semicircular section ABC whose moment of inertia is required to be found out as shown in Fig. 10.

Let r = Radius of the semicircle.

We know that moment of inertia of the semicircular section about the base AC is equal to half the moment of inertia of the circular section about AC. Therefore moment of inertia of the semicircular section ABC about the base AC,

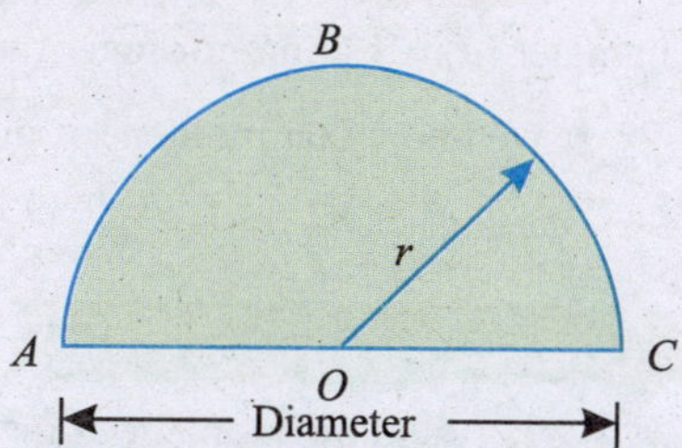

Fig. 10. Semicircular section *ABC*.

$$I_{AC} = \frac{1}{2} \times \frac{\pi}{64} \times (d)^4 = 0.393\, r^4$$

We also know that area of semicircular section,

$$a = \frac{1}{2} \times \pi r^2 \frac{\pi r^2}{2}$$

and distance between centre of gravity of the section and the base AC,

$$h = \frac{4r}{3\pi}$$

∴ Moment of inertia of the section through its centre of gravity and parallel to x-x axis,

$$I_G = I_{AC} - ah^2 = \left[\frac{\pi}{8} \times (r)^4\right] - \left[\frac{\pi r^2}{2}\left(\frac{4r}{3\pi}\right)^2\right] = \left[\frac{\pi}{8} \times (r)^4\right] - \left[\frac{8}{9\pi} \times (r)^4\right]$$

or $$I_G = 0.11\, r^4$$

Note. The moment of inertia about y-y axis will be the same as that about the base AC *i.e.*, $0.393\, r^4$.

Example 7. *Determine the moment of inertia of a semicircular section of 100 mm diameter about its centre of gravity and parallel to X-X and Y-Y axes.*

Solution. Given: Diameter of the section (d) = 100 mm or radius (r) = 50 mm

Moment of inertia of the section about its centre of gravity and parallel to X-X axis

We know that moment of inertia of the semicircular section about its centre of gravity and parallel to X-X axis,

$$I_{XX} = 0.11\, r^4 = 0.11 \times (50)^4 = 687.5 \times 10^3 \text{ mm}^4 \qquad \textbf{Ans.}$$

Moment of inertia of the section about its centre of gravity and parallel to Y-Y axis

We also know that moment of inertia of the semicircular section about its centre of gravity and parallel to Y-Y axis.

$$I_{YY} = 0.393\, r^4 = 0.393 \times (50)^4 = 2456 \times 10^3 \text{ mm}^4 \qquad \textbf{Ans.}$$

Example 8. *A hollow semicircular section has its outer and inner diameter of 200 mm and 120 mm respectively as shown in Fig. 11.*

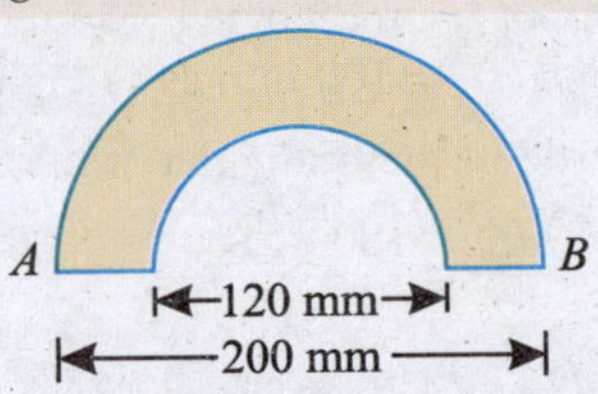

Fig. 11.

What is its moment of inertia about the base AB ?

Solution. Given: Outer diameter (D) = 200 mm or Outer Radius (R) = 100 mm and inner diameter (d) = 120 mm or inner radius (r) = 60 mm.

We know that moment of inertia of the hollow semicircular section about the base AB,

$$I_{AB} = 0.393\,(R^4 - r^4) = 0.393\,[(100)^4 - (60)^4] = 34.21 \times 10^6 \text{ mm}^4 \qquad \textbf{Ans.}$$

EXERCISE 1

1. Find the moment of inertia of a rectangular section 60 mm wide and 40 mm deep about its centre of gravity. **[Ans.** $I_{XX} = 320 \times 10^3 \text{ mm}^4$; $I_{YY} = 720 \times 10^3 \text{ mm}^4$]
2. Find the moment of inertia of a hollow rectangular section about its centre of gravity, if the external dimensions are 40 mm deep and 30 mm wide and internal dimensions are 25 mm deep and 15 mm wide. **[Ans.** $I_{XX} = 140\,470 \text{ mm}^4$: $I_{YY} = 82\,970 \text{ mm}^4$]

3. Find the moment of inertia of a circular section of 20 mm diameter through its centre of gravity. [**Ans.** 7854 mm^4]
4. Calculate the moment of inertia of a hollow circular section of external and internal diameters 100 mm and 80 mm respectively about an axis passing through its centroid. [**Ans.** 2.898×10^6 mm^4]
5. Find the moment of inertia of a triangular section having 50 mm base and 60 mm height about an axis through its centre of gravity and base. [**Ans.** 300×10^3 mm^4: 900×10^3 mm^4]
6. Find the moment of inertia of a semicircular section of 30 mm radius about its centre of gravity and parallel to X-X and Y-Y axes. [**Ans.** 89 100 mm^4 : 381 330 mm^4]

15. MOMENT OF INERTIA OF A COMPOSITE SECTION

The moment of inertia of a composite section may be found out by the following steps :

1. First of all, split up the given section into plane areas (*i.e.*, rectangular, triangular, circular etc., and find the centre of gravity of the section).
2. Find the moments of inertia of these areas about their respective centres of gravity.
3. Now transfer these moment of inertia about the required axis (AB) by the Theorem of Parallel Axis, *i.e.*,

$$I_{AB} = I_G + ah^2$$

where I_G = Moment of inertia of a section about its centre of gravity and parallel to the axis.

a = Area of the section,

h = Distance between the required axis and centre of gravity of the section.

4. The moments of inertia of the given section may now be obtained by the algebraic sum of the moment of inertia about the required axis.

Example 9. *Figure 12 shows an area ABCDEF.*

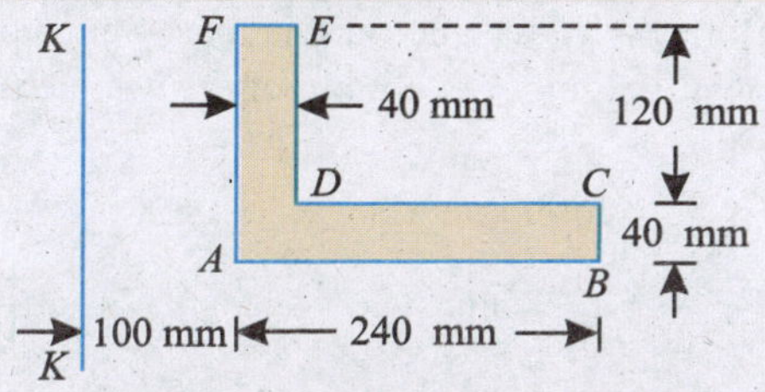

Fig. 12.

Compute the moment of inertia of the above area about axis K-K.

Solution. As the moment of inertia is required to be found out about the axis K-K, therefore there is no need of finding out the centre of gravity of the area.

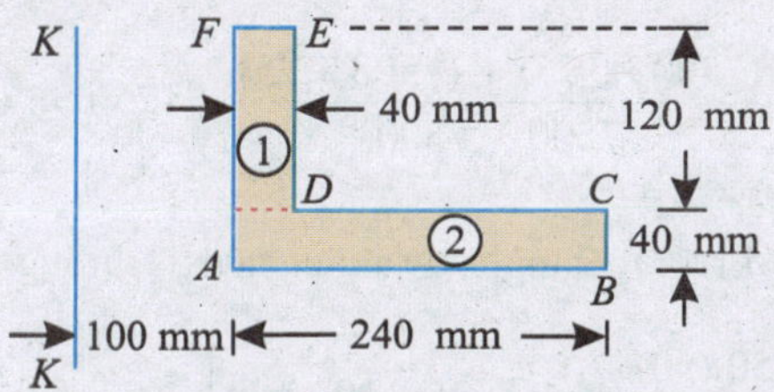

Fig. 13.

Let us split up the area into two rectangles 1 and 2 as shown in Fig. 13.

We know that moment of inertia of section (1) about its centre of gravity and parallel to axis *K-K*,

$$I_{G1} = \frac{120 \times (40)^3}{12} = 640 \times 10^3 \text{ mm}^4$$

and distance between centre of gravity of section (1) and axis *K-K*,

$$h_1 = 100 + \frac{40}{2} = 120 \text{ mm}$$

∴ Moment of inertia of section (1) about axis *K-K*

$$= I_{G1} + a_1 h_1^2 = (640 \times 10^3) + [(120 \times 40) \times (120)^2] = 69.76 \times 10^6 \text{ mm}^4$$

Similarly, moment of inertia of section (2) about its centre of gravity and parallel to axis *K-K*,

$$I_{G2} = \frac{40 \times (240)^3}{12} = 46.08 \times 10^6 \text{ mm}^4$$

and distance between centre of gravity of section (2) and axis *K-K*,

$$h_2 = 100 + \frac{240}{2} = 220 \text{ mm}$$

∴ Moment of inertia of section (2) about the axis *K-K*,

$$= I_{G2} + a_2 h_2^2 = (46.08 \times 10^6) + [(240 \times 40) \times (220)^2] = 510.72 \times 10^6 \text{ mm}^4$$

Now moment of inertia of the whole area about axis *K-K*,

$$I_{KK} = (69.76 \times 10^6) + (510.72 \times 10^6) = 580.48 \times 10^6 \text{ mm}^4$$ **Ans.**

***Example 10.** *Find the moment of inertia of a T-section with flange as 150 mm × 50 mm and web as 150 mm × 50 mm about X-X and Y-Y axes through the centre of gravity of the section.* **[WBUT, IGNOU]**

Solution. The given *T*-section is shown in Fig. 14.

First of all, let us find out centre of gravity of the section. As the section is symmetrical about *Y-Y* axis, therefore its centre of gravity will lie on this axis. Split up the whole section into two rectangles *viz.*, 1 and 2 as shown in figure. Let bottom of the web be the axis of reference.

***(i) Rectangle* (1)**

Area, $a_1 = 150 \times 50 = 7500 \text{ mm}^2$

and $y_1 = 150 + \frac{50}{2} = 175 \text{ mm}$

***(ii) Rectangle* (2)**

Area, $a_2 = 150 \times 50 = 7500 \text{ mm}^2$

and $y_2 = \frac{150}{2} = 75 \text{ mm}$

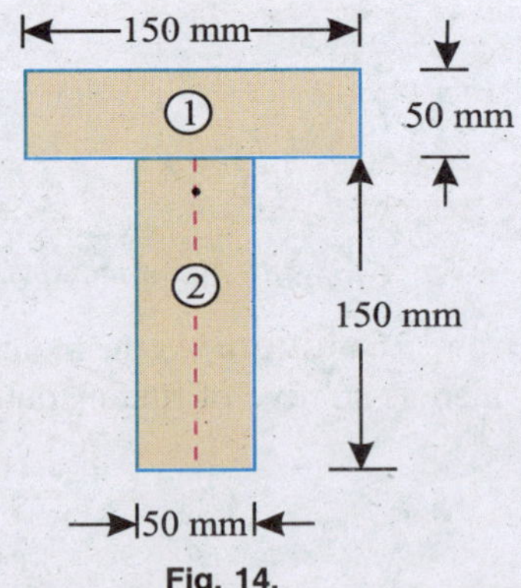

Fig. 14.

We know that distance between centre of gravity of the section and bottom of the web,

$$\bar{y} = \frac{a_1 y_1 + a_2 y_2}{a_1 + a_2} = \frac{(7500 \times 175) + (7500 \times 75)}{7500 + 7500} = 125 \text{ mm}$$

Moment of inertia about X-X axis

We also know that M.I. of rectangle (1) about an axis through its centre of gravity and parallel to *X-X* axis.

$$I_{G1} = \frac{150\,(50)^3}{12} = 1.5625 \times 10^6 \text{ mm}^4$$

and distance between centre of gravity of rectangle (1) and *X-X* axis,

$$h_1 = 175 - 125 = 50 \text{ mm}$$

∴ Moment of inertia of rectangle (1) about X-X axis

$$I_{G1} + a_1 h_1^2 = (1.5625 \times 10^6) + [7500 \times (50)^2] = 20.3125 \times 10^6 \text{ mm}^4$$

Similarly, moment of inertia of rectangle (2) about an axis through its centre of gravity and parallel to X-X axis,

$$I_{G2} = \frac{50\,(150)^3}{12} = 14.0625 \times 10^6 \text{ mm}^4$$

and distance between centre of gravity of rectangle (2) and X-X axis,

$$h_2 = 125 - 75 = 50 \text{ mm}$$

∴ Moment of inertia of rectangle (2) about X-X axis

$$= I_{G2} + a_2 h_2^2 = (14.0625 \times 10^6) + [7500 \times (50)^2] = 32.8125 \times 10^6 \text{ mm}^4$$

Now moment of inertia of the whole section about X-X axis,

$$I_{XX} = (20.3125 \times 10^6) + (32.8125 \times 10^6) = 53.125 \times 10^6 \text{ mm}^4$$ **Ans.**

Moment of inertia about Y-Y axis

We know that M.I. of rectangle (1) about Y-Y axis

$$= \frac{50\,(150)^3}{12} = 14.0625 \times 10^6 \text{ mm}^4$$

and moment of inertia of rectangle (2) about Y-Y axis,

$$= \frac{150\,(50)^3}{12} = 1.5625 \times 10^6 \text{ mm}^4$$

Now moment of inertia of the whole section about Y-Y axis,

$$I_{YY} = (14.0625 \times 10^6) + (1.5625 \times 10^6) = 15.625 \times 10^6 \text{ mm}^4$$ **Ans.**

***Example 11.** *An I-section is made up of three rectangles as shown in Fig. 15. Find the moment of inertia of the section about the horizontal axis passing through the centre of gravity of the section.* **[RGPV, BPUT, AMIE]**

Solution. First of all, let us find out centre of gravity of the section. As the section is symmetrical about Y-Y axis, therefore its centre of gravity will lie on this axis. Split up the whole section into three rectangles 1, 2 and 3 as shown in Fig. 15, Let bottom face of the bottom flange be the axis of reference.

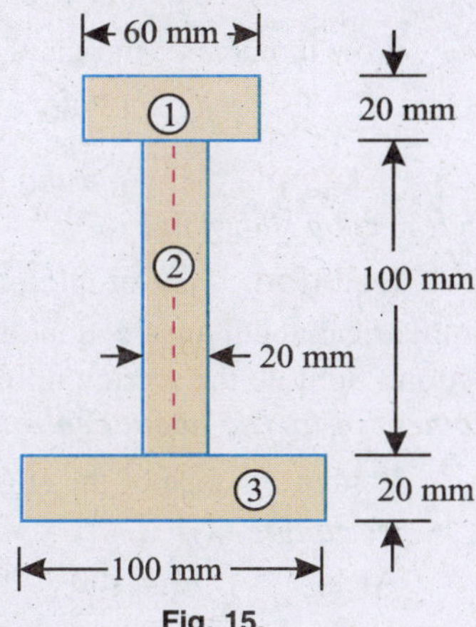

Fig. 15.

***(i) Rectangle* 1**

Area, $a_1 = 60 \times 20 = 1200$ mm

and $y_1 = 20 + 100 + \frac{20}{2} = 130$ mm

***(ii) Rectangle* 2**

Area, $a_2 = 100 \times 20 = 2000 \text{ mm}^2$

and $y_2 = 20 + \frac{100}{2} = 70$ mm

***(iii) Rectangle* 3**

Area, $a_3 = 100 \times 20 = 2000 \text{ mm}^2$

and $y_3 = \frac{20}{2} = 10$ mm

We know that the distance between centre of gravity of the section and bottom face,

$$\bar{y} = \frac{a_1 y_1 + a_2 y_2 + a_3 y_3}{a_1 + a_2 + a_3} = \frac{(1200 \times 130) + (2000 \times 70) + (2000 \times 10)}{1200 + 2000 + 2000} \text{ mm}$$

$$= 60.8 \text{ mm}$$

We know that moment of inertia of rectangle (1) about an axis through its centre of gravity and parallel to X-X axis,

$$I_{G1} = \frac{60 \times (20)^3}{12} = 40 \times 10^3 \text{ mm}^4$$

and distance between centre of gravity of rectangle (1) and X-X axis,

$$h_1 = 130 - 60.8 = 69.2 \text{ mm}$$

∴ Moment of inertia of rectangle (1) about X-X axis,

$$= I_{G1} + a_1 h_1^2 = (40 \times 10^3) + [1200 \times (69.2)^2] = 5786 \times 10^3 \text{ mm}^4$$

Similarly, moment of inertia of rectangle (2) about an axis through its centre of gravity and parallel to X-X axis,

$$I_{G2} = \frac{20 \times (100)^3}{12} = 1666.7 \times 10^3 \text{ mm}^4$$

and distance between centre of gravity of rectangle (2) and X-X axis,

$$h_2 = 70 - 60.8 = 9.2 \text{ mm}$$

∴ Moment of inertia of rectangle (2) about X-X axis,

$$= I_{G2} + a_2 h_2^2 = (1666.7 \times 10^3) + [2000 \times (9.2)^2] = 1836 \times 10^3 \text{ mm}^4$$

Now moment of inertia of rectangle (3) about an axis through its centre of gravity and parallel to X-X axis,

$$I_{G3} = \frac{100 \times (20)^3}{12} = 66.7 \times 10^3 \text{ mm}^4$$

and distance between centre of gravity of rectangle (3) and X-X axis,

$$h_3 = 60.8 - 10 = 50.8 \text{ mm}$$

∴ Moment of inertia of rectangle (3) about X-X axis,

$$= I_{G3} + a_3 h_3^2 = (66.7 \times 10^3) + [2000 \times (50.8)^2] = 5228 \times 10^3 \text{ mm}^4$$

Now moment of inertia of the whole section about X-X axis,

$$I_{XX} = (5786 \times 10^3) + (1836 \times 10^3) + (5228 \times 10^3) = 12\,850 \times 10^3 \text{ mm}^4$$ **Ans.**

***Example 12.** *Find the moment of inertia about the centroidal X-X and Y-Y axes of the angle section shown in Fig. 16.* **[UPTU, GTU]**

Solution. First of all, let us find the centre of gravity of the section. As the section is not symmetrical about any section, therefore we have to find out the values of $\bar{x}$ and $\bar{y}$ for the angle section. Split up the section into two rectangles (1) and (2) as shown in Fig. 16.

Moment of inertia about centroidal X-X axis

Let bottom face of the angle section be the axis of reference.

***Rectangle* (1)**

Area, $a_1 = 100 \times 20 = 2000 \text{ mm}^2$

and $y_1 = \frac{100}{2} = 50 \text{ mm}$

***Rectangle* (2)**

Area, $a_2 = (80 - 20) \times 20 = 1200 \text{ mm}^2$

and $y_2 = \frac{20}{2} = 10 \text{ mm}$

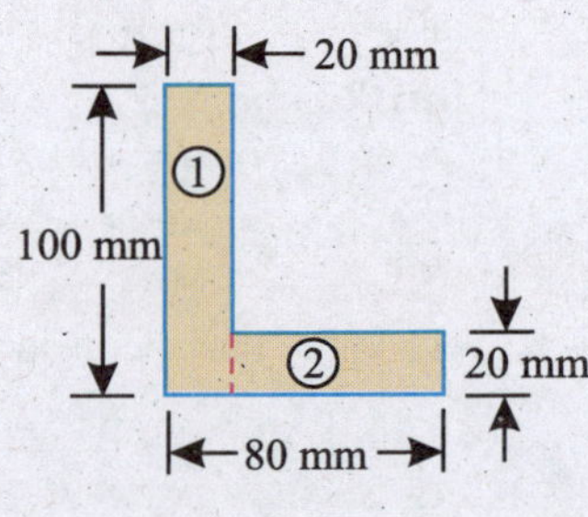

Fig. 16.

We know that distance between the centre of gravity of the section and bottom face,

$$\bar{y} = \frac{a_1 y_1 + a_2 y_2}{a_1 + a_2} = \frac{(2000 \times 50) + (1200 \times 10)}{2000 + 1200} = 35 \text{ mm}$$

We know that moment of inertia of rectangle (1) about an axis through its centre of gravity and parallel to *X-X* axis,

$$I_{G1} = \frac{20 \times (100)^3}{12} = 1.667 \times 10^6 \text{ mm}^4$$

and distance of centre of gravity of rectangle (1) from *X-X* axis,

$$h_1 = 50 - 35 = 15 \text{ mm}$$

∴ Moment of inertia of rectangle (1) about *X-X* axis

$$= I_{G1} + a h_1^2 = (1.667 \times 10^6) + [2000 \times (15)^2] = 2.117 \times 10^6 \text{ mm}^4$$

Similarly, moment of inertia of rectangle (2) about an axis through its centre of gravity and parallel to *X-X* axis,

$$I_{G2} = \frac{60 \times (20)^3}{12} \; 0.04 \times 10^6 \text{ mm}^4$$

and distance of centre of gravity of rectangle (2) from *X-X* axis,

$$h_2 = 35 - 10 = 25 \text{ mm}$$

∴ Moment of inertia of rectangle (2) about *X-X* axis

$$= I_{G2} + a h_2^2 = (0.04 \times 10^6) + [1200 \times (25)^2] = 0.79 \times 10^6 \text{ mm}^4$$

Now moment of inertia of the whole section about *X-X* axis,

$$I_{XX} = (2.117 \times 10^6) + (0.79 \times 10^6) = 2.907 \times 10^6 \text{ mm}^4$$ **Ans.**

Moment of inertia about centroidal Y-Y axis

Let left face of the angle section be the axis of reference.

***Rectangle* (1)**

Area, $a_1 = 2000 \text{ mm}^2$ ...(As before)

and $x_1 = \frac{20}{2} = 10 \text{ mm}$

***Rectangle* (2)**

Area, $a_2 = 1200 \text{ mm}^2$ ...(As before)

and $x_2 = 20 + \frac{60}{2} = 50 \text{ mm}$

We know that distance between the centre of gravity of the section and left face,

$$\bar{x} = \frac{a_1 x_1 + a_2 x_2}{a_1 + a_2} = \frac{(2000 \times 10) + (1200 \times 50)}{2000 + 1200} = 25 \text{ mm}$$

We know that moment of inertia of rectangle (1) about an axis through its centre of gravity and parallel to *Y-Y* axis,

$$I_{G1} = \frac{100 \times (20)^3}{12} = 0.067 \times 10^6 \text{ mm}^4$$

and distance of centre of gravity of rectangle (1) from *Y-Y* axis,

$$h_1 = 25 - 10 = 15 \text{ mm}$$

∴ Moment of inertia of rectangle (1) about *Y-Y* axis

$$= I_{G1} + a_1 h_1^2 = (0.067 \times 10^6) + [2000 \times (15)^2] = 0.517 \times 10^6 \text{ mm}^4$$

Similarly, moment of inertia of rectangle (2) about an axis through its centre of gravity and parallel to *Y-Y* axis,

$$I_{G2} = \frac{20 \times (60)^3}{12} = 0.36 \times 10^6 \text{ mm}^4$$

and distance of centre of gravity of rectangle (2) from *Y-Y* axis,

$$h_2 = 50 - 25 = 25 \text{ mm},$$

∴ Moment of inertia of rectangle (2) about *Y-Y* axis

$$= I_{G2} + a_2 h_2^2 = 0.36 \times 10^6 + [1200 \times (25)^2] = 1.11 \times 10^6 \text{ mm}^4$$

Now moment of inertia of the whole section about *Y-Y* axis,

$$I_{YY} = (0.517 \times 10^6) + (1.11 \times 10^6) = 1.627 \times 10^6 \text{ mm}^4$$ **Ans.**

Example 13. *Figure 17 shows the cross-section of a cast iron beam.*

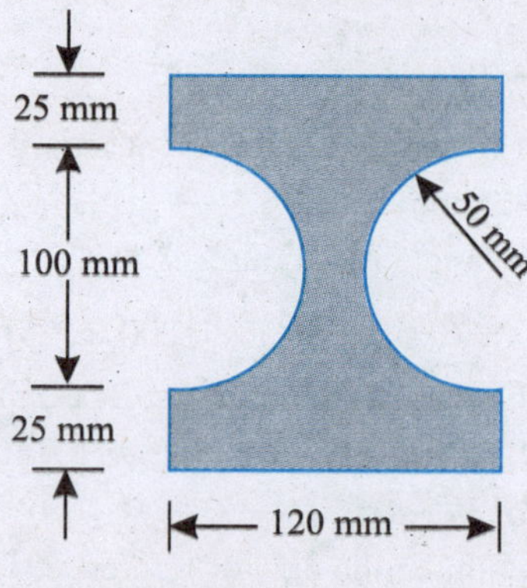

Fig. 17.

Determine the moments of inertia of the section about horizontal and vertical axes passing through the centroid of the section.

Solution. As the section is symmetrical about its horizontal and vertical axes, therefore centre of gravity of the section will lie at the centre of the rectangle. A little consideration will show that when the two semicircles are placed together, it will form a circular hole with 50 mm radius or 100 mm diameter.

Moment of inertia of the section about horizontal axis passing through the centroid of the section

We know that moment of inertia of the rectangular section about its horizontal axis passing through its centre of gravity,

$$= \frac{bd^3}{12} = \frac{120 \times (150)^3}{12} = 33.75 \times 10^6 \text{ mm}^4$$

and moment of inertia of the circular section about a horizontal axis passing through its centre of gravity,

$$= \frac{\pi}{4}(r)^4 = \frac{\pi}{4}(50)^4 = 4.91 \times 10^6 \text{ mm}^4$$

∴ Moment of inertia of the whole section about horizontal axis passing through the centroid of the section,

$$I_{XX} = (33.75 \times 10^6) - (4.91 \times 10^6) = 28.84 \times 10^6 \text{ mm}^4$$ **Ans.**

Moment of inertia of the section about vertical axis passing through the centroid of the section

We know that moment of inertia of the rectangular section about the vertical axis passing through its centre of gravity,

$$I_{G1} = \frac{db^3}{12} = \frac{150 \times (120)^3}{12} = 21.6 \times 10^6 \text{ mm}^4 \qquad ...(i)$$

and area of one semicircular section with 50 mm radius,

$$a = \frac{\pi r^2}{2} = \frac{\pi (50)^2}{2} = 3927 \text{ mm}^2$$

We also know that moment of inertia of a semicircular section about a vertical axis passing through its centre of gravity,

$$I_{G2} = 0.11\, r^4 = 0.11 \times (50)^4 = 687.5 \times 10^3 \text{ mm}^4$$

and distance between centre of gravity of the semicircular section and its base

$$= \frac{4r}{3\pi} = \frac{4 \times 50}{3\pi} = 21.2 \text{ mm}$$

∴ Distance between centre of gravity of the semicircular section and centre of gravity of the whole section,

$$h_2 = 60 - 21.2 = 38.8 \text{ mm}$$

and moment of inertia of one semicircular section about centre of gravity of the whole section,

$$= I_{G2} + a_2 h_2^2 = (687.5 \times 10^3) + [3927 \times (38.8)^2] = 6.6 \times 10^6 \text{ mm}^4$$

∴ Moment of inertia of both the semicircular sections about centre of gravity of the whole section,

$$= 2 \times (6.6 \times 10^6) = 13.2 \times 10^6 \text{ mm}^4 \qquad ...(ii)$$

and moment of inertia of the whole section about a vertical axis passing through the centroid of the section,

$$= (21.6 \times 10^6) - (13.2 \times 10^6) = 8.4 \times 10^6 \text{ mm}^4$$ **Ans.**

***Example 14.** *Find the moment of inertia of a hollow section shown in Fig. 18. about an axis passing through its centre of gravity or parallel X-X axis.* **[RTU, IGNOU]**

Solution. As the section is symmentrical about *Y-Y* axis, therefore centre of a gravity of the section will lie on this axis. Let $\bar{y}$ be the distance between centre of gravity of the section from the bottom face.

200 mm
100 mm
150 mm
300 mm

Fig. 18.

(i) Rectangle

Area, $a_1 = 300 \times 200 = 60\ 000 \text{ mm}^2$

and $y_1 = \frac{300}{2} = 150 \text{ mm}$

(ii) Circular hole

Area, $a_2 = \frac{\pi}{4} \times (150)^2 = 17\ 670 \text{ mm}^2$

and $y_2 = 300 - 100 = 200 \text{ mm}$

We know that distance between the centre of gravity of the section and its bottom face,

$$\bar{y} = \frac{a_1 y_1 - a_2 y_2}{a_1 - a_2} = \frac{(60000 \times 150) - (17670 \times 200)}{60000 - 17670} = 129.1 \text{ mm}$$

∴ Moment of inertia of rectangular section about an axis through its centre of gravity and parallel to *X-X* axis,

$$I_{G1} = \frac{200 \times (300)^3}{12} = 450 \times 10^6 \text{ mm}^4$$

and distance of centre of gravity of rectangular section and X-X axis,

$$h_1 = 150 - 129.1 = 20.9 \text{ mm}$$

∴ Moment of inertia of rectangle about X-X axis

$$= I_{G1} + ah^2 = (450 \times 10^6) + [(300 \times 200) \times (20.9)]^2 = 476.21 \times 10^6 \text{ mm}^4$$

Similarly, moment of inertia of circular section about an axis through its centre of gravity and parallel to X-X axis,

$$I_{G2} = \frac{\pi}{64} \times (150)^4 = 24.85 \times 10^6 \text{ mm}^4$$

and distance between centre of gravity of the circular section and X-X axis,

$$h_2 = 200 - 129.1 = 70.9 \text{ mm}$$

∴ Moment of inertia of the circular section about X-X axis,

$$= I_{G2} + ah^2 = (24.85 \times 10^6) + [(17\,670) \times (70.9)^2] = 113.67 \times 10^6 \text{ mm}^4$$

Now moment of inertia of the whole section about X-X axis

$$= (476.21 \times 10^6) - (113.67 \times 10^6) = 362.54 \times 10^6 \text{ mm}^4$$ **Ans.**

***Example 15.** *A rectangular hole is made in a triangular section as shown in Fig. 19.*

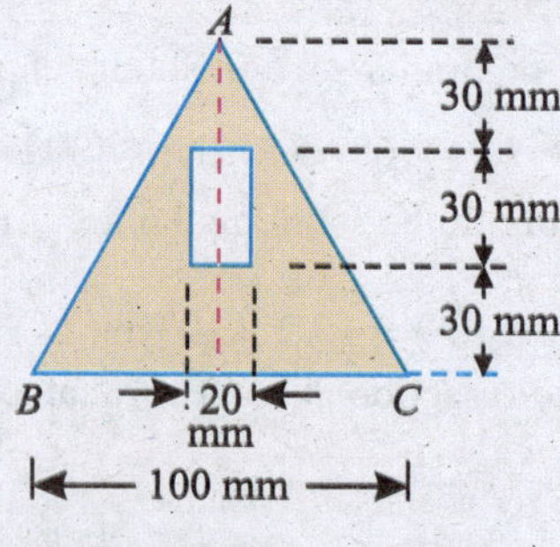

Fig. 19.

Determine the moment of inertia of the section about X-X axis passing through its centre of gravity and the base BC. **[AUT]**

Solution. As the section is symmetrical about Y-Y axis, therefore centre of gravity of the section will lie on this axis. Let $\bar{y}$ be the distance between the centre of gravity of the section and the base BC.

(i) Triangular section

Area, $$a_1 = \frac{100 \times 90}{2} = 4500 \text{ mm}^2$$

and $$y_1 = \frac{90}{3} = 30 \text{ mm}$$

(ii) Rectangular hole

Area, $$a_2 = 30 \times 20 = 600 \text{ mm}^2$$

and $$y_2 = 30 + \frac{30}{2} = 45 \text{ mm}$$

We know that distance between the centre of gravity of the section and base BC of the triangle,

$$\bar{y} = \frac{a_1 y_1 - a_2 y_2}{a_1 - a_2} = \frac{(4500 \times 30) - (600 \times 45)}{4500 - 600} = 27.7 \text{ mm}$$

Moment of inertia of the section about X-X axis

We also know that moment of inertia of the triangular section through its centre of gravity and parallel to *X-X* axis,

$$I_{G1} = \frac{bd^3}{36} = \frac{100 \times (90)^3}{36} = 2025 \times 10^3 \text{ mm}^4$$

and distance between the centre of gravity of the section and *X-X* axis,

$$h_1 = 30 - 27.7 = 2.3 \text{ mm}$$

∴ Moment of inertia of the triangular section about *X-X* axis

$$= I_{G1} + a_2 h_1^2 = 2025 \times 10^3 + [4500 \times (2.3)^2] = 2048.8 \times 10^3 \text{ mm}^4$$

Similarly moment of inertia of the rectangular hole through its centre of gravity and parallel to the *X-X* axis

$$I_{G2} = \frac{bd^3}{12} = \frac{20 \times (30)^3}{12} = 45 \times 10^3 \text{ mm}^4$$

and distance between the centre of gravity of the section and *X-X* axis

$$h_2 = 45 - 27.7 = 17.3 \text{ mm}$$

∴ Moment of inertia of rectangular section about *X-X* axis

$$= I_{G2} + a_2 h_2^2 = (45 \times 10^3) + [600 \times (17.3)^2] = 224.6 \times 10^3 \text{ mm}^4$$

Now moment of inertia of the whole section about *X-X* axis.

$$I_{xx} = (2048.8 \times 10^3) - (224.6 \times 10^3) = 1824.2 \times 10^3 \text{ mm}^4$$ **Ans.**

Moment of inertia of the section about the base BC

We know that moment of inertia of the triangular section about the base *BC*

$$I_{G1} = \frac{bd^3}{12} = \frac{100 \times (90)^3}{12} = 6075 \times 10^3 \text{ mm}^4$$

Similarly moment of inertia of the rectangular hole through its centre of gravity and parallel to *X-X* axis,

$$I_{G2} = \frac{bd^3}{12} = \frac{20 \times (30)^3}{12} = 45 \times 10^3 \text{ mm}^4$$

and distance between the centre of gravity of the section about the base *BC*,

$$h_2 = 30 + \frac{30}{2} = 45 \text{ mm}$$

∴ Moment of inertia of rectangular section about the base *BC*,

$$= I_{G2} + a_2 h_2^2 = (45 \times 10^3) + [600 \times (45)^2] = 1260 \times 10^3 \text{ mm}^4$$

Now moment of inertia of the whole section about the base *BC*,

$$I_{BC} = (6075 \times 10^3) - (1260 \times 10^3) = 4815 \times 10^3 \text{ mm}^4$$ **Ans.**

16. MOMENT OF INERTIA OF A BUILT-UP SECTION

A built-up section consists of a number of sections such as rectangular sections, channel sections, I-sections etc., A built-up section is generally made by symmetrically placing and then fixing these section by welding or riveting. It will be interesting to know that a built-up section behaves as one unit. The moment of inertia of such a section is found out by the following steps.

1. Find out the moment of inertia of the various sections about their respective centres of gravity as usual.
2. Now transfer these moments of inertia about the required axis (say *X-X* axis or *Y-Y* axis) by the Theorem of Parallel Axis.

Note. In most of the standard sections, their moments of inertia of about their respective centres of gravity is generally given. However, if it is not given then we have to calculate it before transferring it to the required axis.

Example 16. *A compound beam is made by welding two steel plates 160 mm × 12 mm one on each flange of an ISLB 300 section as shown in Fig 20.*

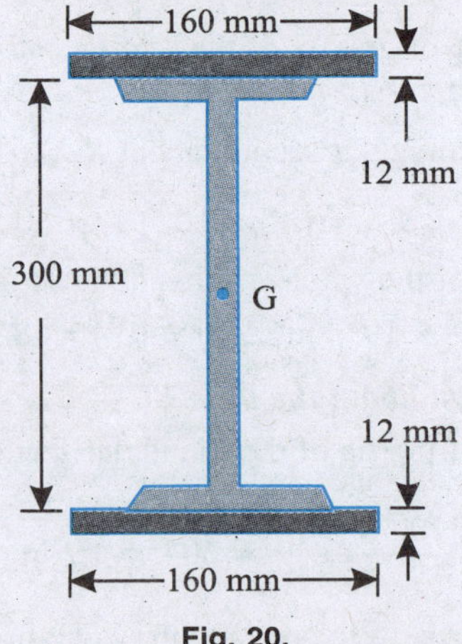

Fig. 20.

Find the moment of inertia the beam section about an axis passing through its centre of gravity and parallel to X-X axis. Take moment of inertia of the ISLB 300 section about X-X axis as 73.329×10^6 *mm*^{4}.

Solution. Given: Size of two steel plates = 160 mm × 12 mm and moment of inertia of ISLB 300 section about *X-X* axis = 73.329

From the geometry of the compound section, we find that it is symmetrical about both the *X-X* and *Y-Y* axes. Therefore centre of gravity of the section will lie at *G i.e.* centre of gravity of the beam section.

We know that moment of inertia of one steel plate section about an axis passing through its centre of gravity and parallel to *X-X* axis.

$$I_G = \frac{160 \times (12)^3}{12} = 0.023 \times 10^6 \text{ mm}^4$$

and distance between the centre of gravity of the plate section and *X-X* axis,

$$h = 150 + \frac{12}{2} = 156 \text{ mm}$$

∴ Moment of inertia of one plate section about X-X axis,

$$= I_G + a\,h^2 = (0.023 \times 10^6) + [(160 \times 12) \times (156)^2] = 46.748 \times 10^6 \text{ mm}^4$$

and moment of inertia of the compound beam section about X-X axis,

I_{XX} = Moment of inertia of ISLB section

+ Moment of inertia of two plate sections.

$$= (73.329 \times 10^6) + 2\,(46.748 \times 10^6) = 166.825 \times 10^6 \text{ mm}^4$$ **Ans.**

Example 17. *A compound section is built-up by welding two plates 200 mm × 15 mm on two steel beams ISJB 200 placed symmetrically side by side as shown in Fig. 21.*

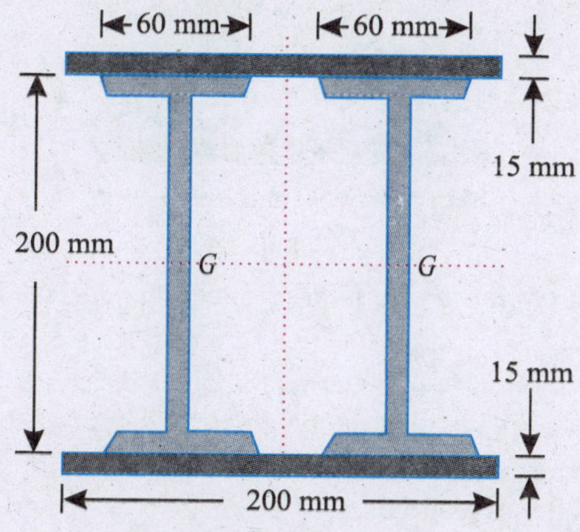

Fig. 21.

What is the moment of inertia of the compound section about an axis passing through its centre of gravity and parallel to X-X axis ? Take I_{XX} for the ISJB section as 7.807×10^6 mm⁴.

Solution. Given: Size of two plates = 200 mm × 15 mm and moment of inertia of ISJB 200 section about X-X axis $= 7.807 \times 10^6 \text{ mm}^4$.

From the geometry of the compound section, we find that it is symmetrical about both the X-X and Y-Y axis. Therefore centre of gravity of the section will lie at G *i.e.*, centre of gravity of the beam sections.

We know that moment of inertia of one plate section about an axis passing through its centre of gravity and parallel to X-X axis,

$$I_G = \frac{200 \times (15)^3}{12} = 0.056 \times 10^6 \text{ mm}^4$$

and distance between the centre of gravity of the plate section and X-X axis,

$$h = 100 + \frac{15}{2} = 107.5 \text{ mm}$$

∴ Moment of inertia of the plate section about x-x axis

$$= I_G + a\,h^2 = (0.056 \times 10^6) + (200 \times 15) \times (107.5)^2 = 34.725 \times 10^6 \text{ mm}^4$$

and moment of inertia of the compound section about x-x axis,

I_{XX} = Moment of inertia of two ISJB sections

+ Moment of inertia of two plate sections

$$= [2 \times (7.807 \times 10^6) + 2 \times (34.725 \times 10^6)] = 85.064 \times 10^6 \text{ mm}^4$$ **Ans.**

Example 18. *A built up section is made by needing too stable and two channel sections as shown in Fig. 22.*

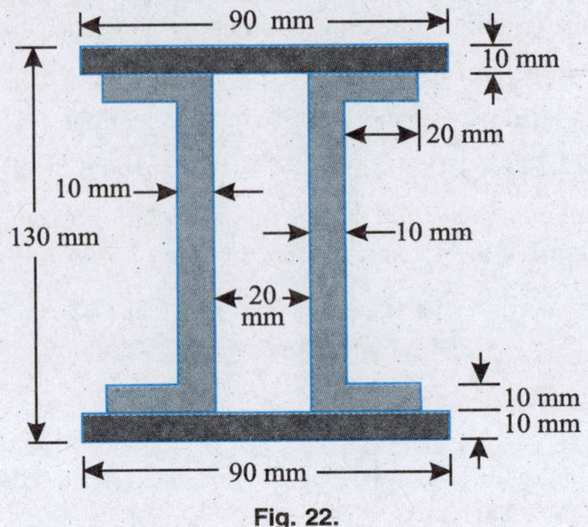

Fig. 22.

Determine moment of inertia of a built up section about X-X axis passing through centre of gravity of the section.

Solution. As the section is symmetrical about X-X axis and Y-Y axis therefore centre of gravity of the section will coincide with the geometrical centre of section.

We know that the moment of inertia of one top or bottom plate about an axis through its centre os gravity and parallel to X-X axis,

$$I_{G1} = \frac{90 \times (10)^3}{12} = 7500 \text{ mm}^4$$

and distance between centre of gravity of the plates from X-X axis,

$$h_1 = 65 - 5 = 60 \text{ mm}$$

∴ Moment of inertia of top and bottom plates about X-X axis,

$$= I_{G1} + a\,h^2 = 2\,[7500 + (90 \times 10) \times (60)^2] \text{ mm}^4$$

(because of two plates)

$$= 6.5 \times 10^6 \text{ mm}^4$$

Now moment of inertia of part (1) of one channel section about an axis through its centre of gravity and parallel to X-X axis,

Fig. 23.

$$I_{G2} = \frac{30 \times (10)^3}{12} = 2500 \text{ mm}^4$$

and distance of centre of gravity of this part from X-X axis,

$$h_2 = 55 - 5 = 50 \text{ mm}$$

∴ Moment of inertia of part (1) about X-X axis,

$$= I_{G2} + a\,h^2 = 4\,[2500 + (30 \times 10) \times (50)^2 \text{ mm}^4 \quad \text{...(because of four plates)}$$

$$= 3.0 \times 10^6 \text{ mm}^4$$

Similarly moment of inertia of part (2) of the channel about an axis through its centre of gravity and parallel to X-X axis,

$$I_{G3} = 2\left[\frac{10 \times (90)^3}{12}\right] = 0.6 \times 10^6 \text{ mm}^4 \quad \text{...(because of two plates)}$$

Now moment of inertia of the whole built-up section about an axis through its centre of gravity and parallel to X-X axis,

$$I_{XX} = (6.5 \times 10^6) + (3.0 \times 10^6) + (0.6 \times 10^6) = 10.1 \times 10^6 \text{ mm}^4 \qquad \textbf{Ans.}$$

17. *PRODUCT OF INERTIA, PRINCIPAL MOMENT OF INERTIA AND PRINCIPAL AXES

We shall study this topic in detail in the chapter of Product of Inertia.

18. *MASS MOMENT OF INERTIA AND RADIUS OF GYRATION

We shall study this topic in detail in the chapter of Kinetics of Motion of Rotation and Mass Moment of Inertia.

EXERCISE 2

1. Find the moment of inertia of a *T*-section having flange and web both 120 mm × 30 mm about *X-X* axis passing through the centre of gravity of the section. **[Ans.** $14\,715 \times 10^3$ mm^4]

2. Calculate the moment of inertia of an I-section having equal flanges 30 mm × 10 mm and web also 30 mm × 10 mm about an axis passing through its centre of gravity and parallel to *X-X* and *Y-Y* axes. **[Ans.** 267.5×10^3 mm^4; 47×10^3 mm^4]

*3. Find the moment of inertia of the lamina with a circular hole of 30 mm diameter about the axis *AB* as shown in Fig. 24. **[AM-IETE, BPUT]** **[Ans.** 638.3×10^3 mm^4]

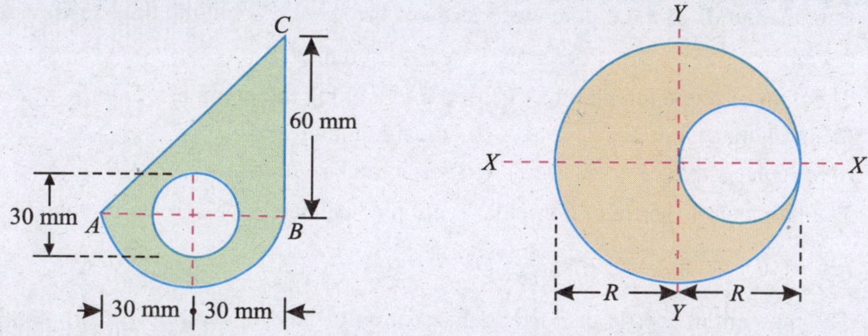

Fig. 24. Fig. 25.

*4. A circular hole of diameter *R* is punched out from a circular plate of radius *R* shown in Fig. 25. Find the moment of inertia about both the centroidal axes.

$$\left[\textbf{Ans. } I_{XX} = \frac{15\pi R^4}{64};\ I_{YY} = \frac{29\pi R^4}{192}\right]$$

*5. The cross-section of a beam is shown in Fig. 26. Find the moment of inertia of the section about the horizontal centroidal axis. **[RGPV, JNTU]** **[Ans.** 1.354×10^6 mm^4]

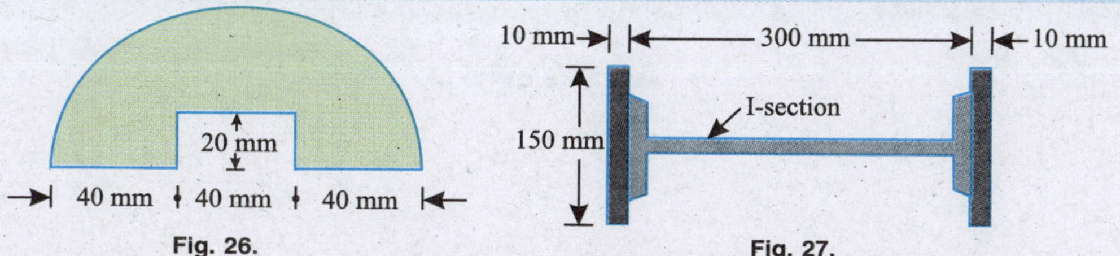

Fig. 26. Fig. 27.

6. A built-up section consists of an I-section and two plates as shown in Fig 27. Find values of I_{XX} and I_{YY} of the section. Take values of I_{XX} as 3.762×10^6 mm^4 and I_{YY} as 73.329×10^6 mm^6 respectively for the I-section.

[**Ans.** $I_{XX} = 17.095 \times 10^6$ mm^4 ; $I_{YY} = 169.46 \times 10^6$ mm^4]

QUESTIONS

1. How would you find out the moment of inertia of a plane area ?
2. What is Routh's rule for finding out the moment of inertia of an area ? Explain where it is used and why ?
3. Derive an equation for moment of inertia of the following sections about centroidal axis:
   (*a*) a rectangular section, (*b*) a hollow rectangular section,
   (*c*) a circular section, and (*d*) a hollow circular section.
4. State and prove the theorem of perpendicular axis applied to moment of inertia.
5. Prove the parallel axis theorem in the determination of moment of inertia of areas with the help of a neat sketch.
6. Describe the method of finding out the moment of inertia of a composite section.

OBJECTIVE TYPE QUESTIONS

1. If the area of a section is in mm^2 and the distance of the centre of area from a lines is in mm, then units of the moment of inertia of the section about the line is expressed in
   (*a*) mm^2 (*b*) mm^3 (*c*) mm^4 (*d*) mm^5
2. Theorem of perpendicular axis is used in obtaining the moment of inertia of a
   (*a*) triangular lamina (*b*) square lamina
   (*c*) circular lamina (*d*) semicircular lamina
3. The moment of inertia of a circular section of diameter (*d*) is given by the relation
   (*a*) $\frac{\pi}{16}(d)^4$ (*b*) $\frac{\pi}{32}(d)^4$ (*c*) $\frac{\pi}{64}(d)^4$ (*d*) $\frac{\pi}{96}(d)^4$
4. The moment of inertia of a triangular section of base (*b*) and height (*h*) about an axis through its *c.g.* and parallel to the base is given by the relation.
   (*a*) $\frac{bh^3}{12}$ (*b*) $\frac{bh^3}{24}$ (*c*) $\frac{bh^3}{36}$ (*d*) $\frac{bh^3}{48}$
5. The moment of inertia of a triangular section of base (*b*) and height (*h*) about an axis passing through its vertex and parallel to the base is ... as that passing through its C.G. and parallel to the base.
   (*a*) twelve times (*b*) nine times
   (*c*) six times (*d*) four times

ANSWERS

1. (*c*) 2. (*b*) 3. (*c*) 4. (*c*) 5. (*b*)

CHAPTER

8 Principles of Friction

Contents

1. FRICTION OR FORCE OF FRICTION

[GTU, AM-IETE]

It has been established since long that all surfaces of the bodies are never perfectly smooth. It has been observed that whenever, even a very smooth surface is viewed under a microscope, it is found to have some roughness and irregularities, which may not be detected by an ordinary touch.

It will be interesting to know that if a block of one substance is placed over the level surface of the same or different material, a certain degree of interlocking of the minutely projecting particles takes place. This does not involve any force, so long as the block does not move or tends to move. But whenever one of the blocks moves or tends to move tangentially with respect to the surface, on which it rests, the interlocking property of the projecting particles opposes the motion. This opposing force, which acts in the opposite direction of the movement of the block, is called **force of friction** or simply **friction**. It is of the following two types:

1. **Static friction.**
2. **Dynamic friction.**

2. *STATIC FRICTION

It is the friction experienced by a body when it is at rest. Or in other words, it is the friction when the body tends to move.

3. *DYNAMIC FRICTION (OR KINETIC FRICTION)

It is the friction experienced by a body when it is in motion. It is also called kinetic friction. The dynamic friction is of the following two types :

1. **Sliding friction.** It is the friction, experienced by a body when it slides over another body.
2. **Rolling friction.** It is the friction, experienced by a body when it rolls over another body.

4. *LIMITING FRICTION

[JNTU]

It has been observed that when a body, lying over another body, is gently pushed, it does not move because of the frictional force, which prevents the motion. It shows that the force of the hand is being exactly balanced by the force of friction, acting in the opposite direction. If we again push the body, a little harder, it is still found to be in equilibrium. It shows that the force of friction has increased itself so as to become equal and opposite to the applied force. Thus the force of friction has a remarkable property of adjusting its magnitude, so as to become exactly equal and opposite to the applied force, which tends to produce motion.

There is, however, a limit beyond which the force of friction cannot increase. If the applied force exceeds this limit, the force of friction cannot balance it and the body begins to move, in the direction of the applied force. This maximum value of frictional force, which comes into play, when a body just begins to slide over the surface of the other body, is known as limiting friction. It may be noted that when the applied force is less than the limiting friction, the body remains at rest, and the friction is called static friction, which may have any value between zero and limiting friction.

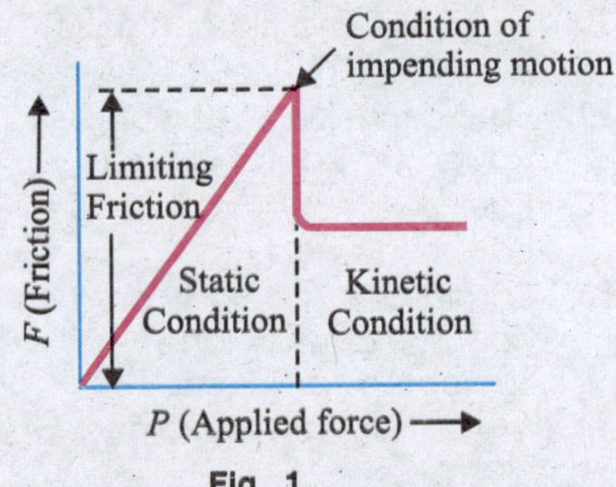

Fig. 1.

5. *NORMAL REACTION

It has been experienced that whenever a body, lying on a horizontal or an inclined surface, is in equilibrium, its weight acts vertically downwards through its centre of gravity. The surface, in turn, exerts an upward reaction on the body. This reaction, which is taken to act perpendicular to the plane, is called normal reaction and is, generally, denoted by R. It will be interesting to know that the term 'normal reaction' is very important in the field of friction, as force of friction is directly proportional to it.

6. *ANGLE OF FRICTION

[RTU, JNTU, AM-IETE]

Consider a body of weight W resting on an inclined plane as shown in Fig. 2. We know that the body is in equilibrium under the action of the following forces :

Fig. 2. Angle of friction.

1. Weight (W) of the body, acting vertically downwards,
2. Friction force (F) acting upwards along the plane, and
3. Normal reaction (R) acting at right angles to the plane.

Let the angle of inclination (α) be gradually increased, till the body just starts sliding down the plane. This angle of inclined plane, at which a body just begins to slide down the plane, is called the **angle of friction**. This is also equal to the angle, which the normal reaction makes with the vertical.

7. *COEFFICIENT OF FRICTION

[GTU]

It is the ratio of limiting friction to the normal reaction, between the two bodies, and is generally denoted by μ.

Mathematically, coefficient of friction,

$$\mu = \frac{F}{R} = \tan \phi \qquad \text{or} \qquad F = \mu R$$

where ϕ = Angle of friction,
F = Limiting friction, and
R = Normal reaction between the two bodies.

7.1. *CONE OF FRICTION

[PU, JNTU, BPUT]

The Fig. 3 shows a body of weight W acted upon by a force P. If the line OA making the maximum angle of friction ϕ with the normal, is revolved about OB as an axis, the cone generated is called the **cone of friction**.

If the resultant R of the normal reaction N and the force of friction F, falls within the cone of friction, the forces acting on the body are not great enough to cause the motion.

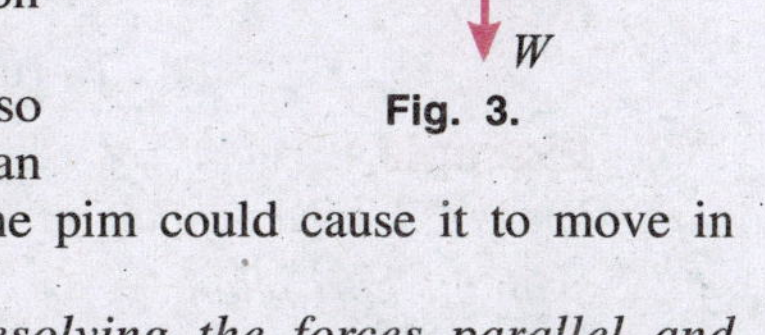

Fig. 3.

This principle is used in **self locking mechanisms** and also **in taperpins**. For example, if the angle of taper pin is less than the angle of friction, no force at right angle to the axis of the pim could cause it to move in the direction of its axis.

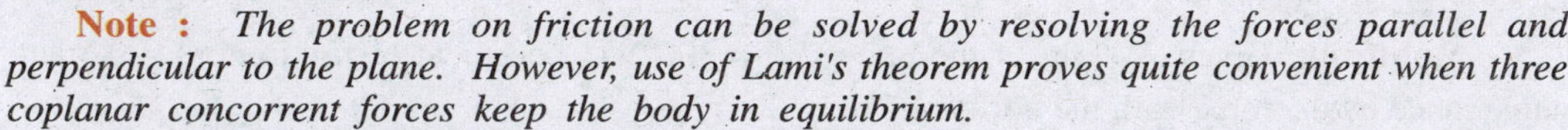
Note : *The problem on friction can be solved by resolving the forces parallel and perpendicular to the plane. However, use of Lami's theorem proves quite convenient when three coplanar concorrent forces keep the body in equilibrium.*

8. *LAWS OF FRICTION

Prof. Coulomb, after extensive experiments, gave some laws of friction, which may be grouped under the following heads :

1. Laws of static friction, and
2. Laws of dynamic or kinetic friction.

9. *LAWS OF STATIC FRICTION

[WBUT, BPUT]

Following are the laws of static friction :

1. The force of friction always acts in a direction, opposite to that in which the body tends to move, if the force of friction would have been absent.
2. The magnitude of the force of friction is exactly equal to the force, which tends to move the body.
3. The magnitude of the limiting friction bears a constant ratio to the normal reaction between the two surfaces. Mathematically :

This rock climber uses the static frictional force between his hands and feet and the vertical rock walls.

$$\frac{F}{R} = \text{Constant}$$

where F = Limiting friction, and

R = Normal reaction.

4. The force of friction is independent of the area of contact between the two surfaces.
5. The force of friction depends upon the roughness of the surfaces.

10. LAWS OF DYNAMIC OR KINETIC FRICTION

Following are the laws of dynamic or kinetic friction :

1. The force of friction always acts in a direction, opposite to that in which the body is moving.
2. The magnitude of kinetic friction bears a constant ratio to the normal reaction between the two surfaces. But this ratio is slightly less than that in case of limiting friction.
3. For moderate speeds, the force of friction remains constant. But it decreases slightly with the increase of speed.

11. EQUILIBRIUM OF A BODY ON A ROUGH HORIZONTAL PLANE

We know that a body, lying on a rough horizontal plane will remain in equilibrium. But whenever a force is applied on it, the body will tend to move in the direction of the force. In such cases, equilibrium of the body is studied first by resolving the forces horizontally and then vertically.

Now the value of the force of friction is obtained from the relation :

$$F = \mu R$$

where μ = Coefficient of friction, and

R = Normal reaction.

Example 1. *A body of weight 300 N is lying on a rough horizontal plane having a coefficient of friction as 0.3. Find the magnitude of the force, which can move the body, while acting at an angle of 25° with the horizontal.*

Solution. Given: Weight of the body (W) = 300 N; Coefficient of friction (μ) = 0.3 and angle made by the force with the horizontal (α) = 25°

Let P = Magnitude of the force, which can move the body,

F = Force of friction.

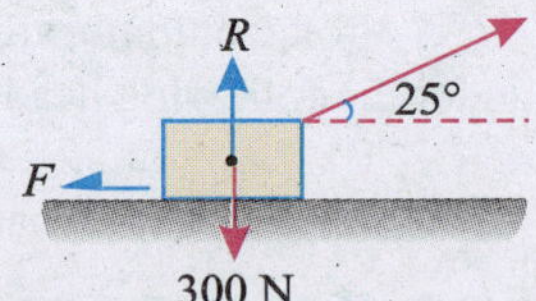

Fig. 4.

Resolving the forces horizontally and equating forces acting in left direction with forces acting in right direction,

$$F = P \cos \alpha = P \cos 25° = P \times 0.9063$$

and now resolving the forces vertically and equating forces acting in upward direction with forces acting in downward direction,

$$R = W - P \sin \alpha = 300 - P \sin 25° = 300 - P \times 0.4226$$

We know that the force of friction (F),

$$0.9063\,P = \mu R = 0.3 \times (300 - 0.4226\,P) = 90 - 0.1268\,P$$

or $$90 = 0.9063\,P + 0.1268\,P = 1.0331\,P$$

or $$P = \frac{90}{1.0331} = 87.1 \text{ N} \qquad \textbf{Ans.}$$

Example 2. *A body, resting on a rough horizontal plane, required a pull of 180 N inclined at 30° to the plane just to move it. It was found that a push of 220 N inclined at 30° to the plane just moved the body. Determine the weight of the body and the coefficient of friction.* **[RGPV]**

Solution. Given: Pull = 180 N; Push = 220 N and angle at which force is inclined with horizontal plane (α) = 30°

Let W = Weight of the body

R = Normal reaction, and

μ = Coefficient of friction.

First of all, consider a pull of 180 N acting on the body. We know that in this case, the force of friction (F_1) will act towards left as shown in Fig. 5. (*a*).

Resolving the forces horizontally and equating forces acting in left direction with forces acting in right direction,

$$F_1 = 180 \cos 30° = 180 \times 0.866 = 155.9 \text{ N}$$

and now resolving the forces vertically and equating forces acting in upward direction with forces acting in downward direction,

$$R_1 = W - 180 \sin 30° = W - 180 \times 0.5 = W - 90 \text{ N}$$

We know that the force of friction (F_1),

$$155.9 = \mu R_1 = \mu (W - 90) \qquad ...(i)$$

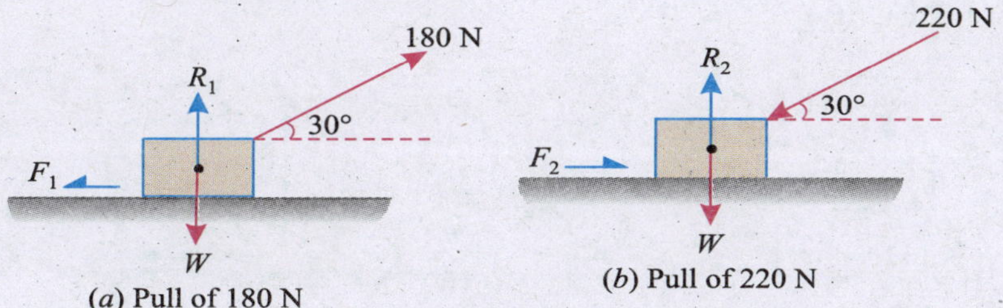

(*a*) Pull of 180 N

(*b*) Pull of 220 N

Fig. 5.

Now consider a push of 220 N acting on the body. We know that in this case, the force of friction (F_2) will act towards right as shown in Fig. 5 (*b*).

Resolving the forces horizontally and equating forces acting in right direction with forces acting in left direction,

$$F_2 = 220 \cos 30° = 220 \times 0.866 = 190.5 \text{ N}$$

and now resolving the forces vertically and equating forces acting in upward direction with forces acting in downward direction,

$$R_2 = W + 220 \sin 30° = W + 220 \times 0.5 = W + 110 \text{ N}$$

We know that the force of friction (F_2),

$$190.5 = \mu.R_2 = \mu (W + 110) \qquad ...(ii)$$

Dividing equation (*i*) by (*ii*)

$$\frac{155.9}{190.5} = \frac{\mu (W - 90)}{\mu (W + 110)} = \frac{W - 90}{W + 110}$$

or $\quad 155.9\, W + 17\,149 = 190.5\, W - 17\,145$

or $\quad 34.6\, W = 34\,294$

or $\quad W = \dfrac{34\,294}{34.6} = 991.2 \text{ N}$ **Ans.**

Now substituting the value of W in equation (i),

$$155.9 = \mu\,(991.2 - 90) = 901.2\,\mu$$

or $$\mu = \frac{155.9}{901.2} = 0.173 \quad \textbf{Ans.}$$

***Example 3.** *Two blocks A and B of weights 1 kN and 2 kN respectively are in equilibrium position as shown in Fig. 6.*

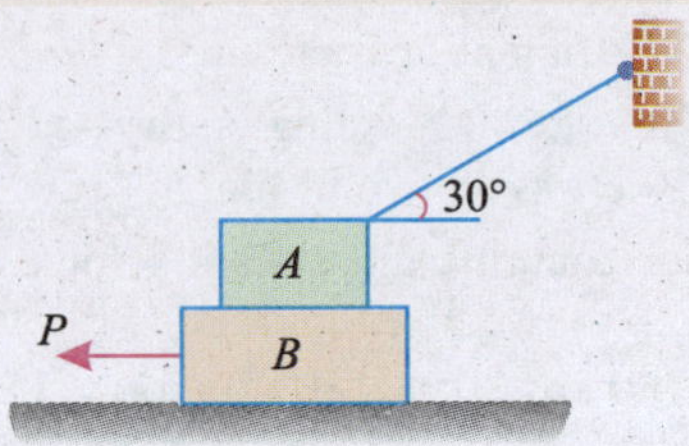

Fig. 6.

If the coefficient of friction between the two blocks as well as the block B and the floor is 0.3, find the force (P) required to move the block B. **[AUT, WBUT]**

Solution. Given: Weight of block A (W_A) = 1 kN; Weight of block B (W_B) = 2 kN and coefficient of friction (μ) = 0.3.

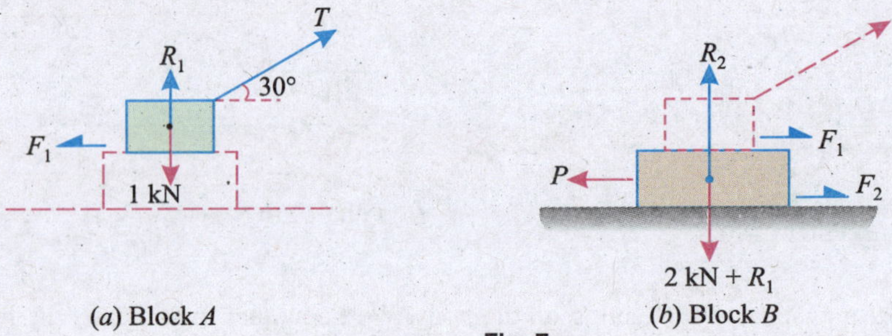

(a) Block A (b) Block B

Fig. 7.

Let T = Tension in the rope

The forces acting on the two blocks A and B are shown in Fig. 7 (a) and (b) respectively. First of all, consider the forces acting in the block A.

Resolving the forces vertically and equating forces acting in upward direction with forces acting in downward direction,

$$R_1 + T\sin 30° = 1 \text{ kN}$$

or $$T\sin 30° = 1 - R_1 \qquad ...(i)$$

and now resolving the forces horizontally and equating forces acting in left direction with forces acting in right direction,

$$T\cos 30° = F_1 = \mu R_1 = 0.3\,R_1 \qquad ...(ii)$$

Dividing equation (i) by (ii)

$$\frac{T\sin 30°}{T\cos 30°} = \frac{1 - R_1}{0.3\,R_1} \quad \text{or} \quad \tan 30° = \frac{1 - R_1}{0.3\,R_1}$$

or $$0.5774 = \frac{1 - R_1}{0.3\,R_1} \quad \text{or} \quad 0.5774 \times 0.3\,R_1 = 1 - R_1$$

or $$0.173\,R_1 = 1 - R_1 \quad \text{or} \quad 1.173\,R_1 = 1$$

or $$R_1 = \frac{1}{1.173} = 0.85 \text{ kN}$$

and $$F_1 = \mu.R_1 = 0.3 \times 0.85 = 0.255 \text{ kN} \quad ...(iii)$$

Now consider the block *B*. A little consideration will show that the downward force of the block *A* (equal to R_1) will also act alongwith the weight of the block *B*.

Resolving the forces vertically and equating forces acting in upward direction with forces acting in downward direction,

$$R_2 = 2 + R_1 = 2 + 0.85 = 2.85 \text{ kN}$$

and $$F_2 = \mu R_2 = 0.3 \times 2.85 = 0.855 \text{ kN} \quad ...(iv)$$

and now resolving the forces horizontally and equating forces acting in left direction with forces acting in right direction,

$$P = F_1 + F_2 = 0.255 + 0.855 = 1.11 \text{ kN} \quad \textbf{Ans.}$$

****Example 4.** *What is the maximum load (W) which a force P equal to 5 kN will hold up, if the coefficient of friction at C is 0.2 in the arrangement shown in Fig. 8. Neglect other friction and weight of the member.*

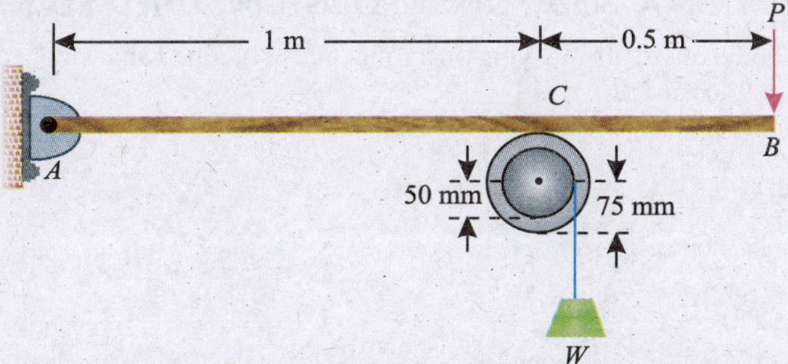

Fig. 8.

If W = 3 kN and P = 4.5 kN, what are the normal and tangential forces transmitted at C ?

[GTU, WBUT, AMIE]

Solution. Given: Force $(P) = 5$ kN and coefficient of friction at C $(\mu) = 0.2$

Maximum load W

Let R = Normal reaction of the pulley on the beam at C.

First of all, consider the equilibrium of the beam *AB*. Taking moments about the hinge *A* and equating the anti-clockwise moments with clockwise moments,

$$R \times 1 = 5 \times 1.5 = 7.5 \quad \text{or} \quad R = 7.5 \text{ kN}$$

Now consider the equilibrium of the pulley. It is subjected to a normal reaction of 7.5 kN (as calculated above). The load (*W*) tends to rotate it. A little consideration will show that the rotation of the pulley is prevented by the frictional force between the pulley and beam at *C*. We know that maximum force of friction at *C*

$$= \mu \,.\, R = 0.2 \times 7.5 = 1.5 \text{ kN}$$

Now taking moments about the centre of the pulley and equating the equating the clockwise moments with anti-clockwise moments,,

$$W \times 50 = 1.5 \times 75 = 112.5$$

or $$W = \frac{112.5}{50} = 2.25 \text{ kN} \quad \textbf{Ans.}$$

Normal and tangential forces transmitted at C

Now consider a weight W equal to 3 kN suspended from the pulley and a force P equal to 4.5 kN applied at B.

Let R_1 = Normal force or normal reaction at C, and

F_1 = Tangential force at C.

Again consider equilibrium of the beam. Taking moments about the hinge A and equating the anti-clockwise moments with clockwise moments,,

$$R_1 \times 1 = 4.5 \times 1.5 = 6.75$$

or $R_1 = 6.75$ kN **Ans.**

We know that the tangential force at C will be the frictional force between the pulley and beam. Again taking moments about the centre of the pulley and equating the anti-clockwise moments with clockwise moments,,

$$F_1 \times 75 = W \times 50 = 3 \times 50 = 150$$

or $F_1 = \frac{150}{75} = 2$ kN **Ans.**

12. EQUILIBRIUM OF A BODY ON A ROUGH INCLINED PLANE

Consider a body, of weight W, lying on a rough plane inclined at an angle α with the horizontal as shown in Fig. 9 (*a*) and (*b*).

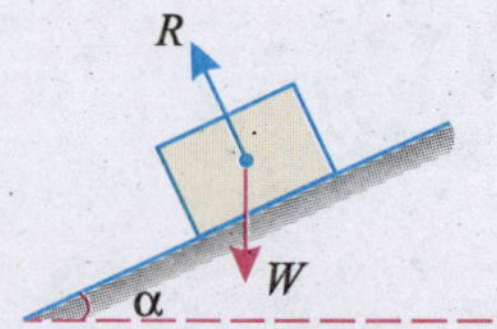

(*a*) Angle of inclination less than the angle of friction

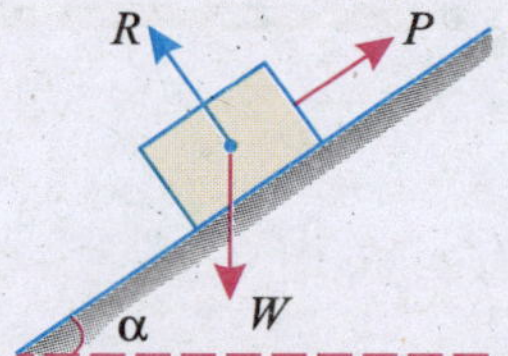

(*b*) Angle of inclination more than the angle of friction

Fig. 9.

A little consideration will show, that if the inclination of the plane, with the horizontal, is less the angle of friction, the body will be automatically in equilibrium as shown in Fig. 9 (*a*). If in this condition, the body is required to be moved upwards or downwards, a corresponding force is required, for the same. But, if the inclination of the plane is more than the angle of friction, the body will move down. And an upward force (P) will be required to resist the body from moving down the plane as shown in Fig. 9 (*b*).

Though there are many types of forces, for the movement of the body, yet the following are important from the subject point of view :

1. Force acting along the inclined plane.
2. Force acting horizontally.
3. Force acting at some angle with the inclined plane.

Note. In all the above mentioned three types of forces, we shall discuss the magnitude of force, which will keep the body in equilibrium, when it is at the point of sliding downwards or upwards.

13. EQUILIBRIUM OF A BODY ON A ROUGH INCLINED PLANE SUBJECTED TO A FORCE ACTING ALONG THE INCLINED PLANE

Consider a body lying on a rough inclined plane subjected force acting along the inclined plane, which keeps it in equilibrium as shown in Fig. 10. (*a*) and (*b*).

Let W = Weight of the body,

α = Angle, which the inclined plane makes with the horizontal,

R = Normal reaction,

μ = Coefficient of friction between the body and the inclined plane, and

ϕ = Angle of friction, such that $\mu = \tan \phi$.

A little consideration will show that if the force is not there, the body will slide down the plane. Now we shall discuss the following two cases :

1. *Minimum force (P_1) which will keep the body in equilibrium, when it is at the point of sliding downwards.*

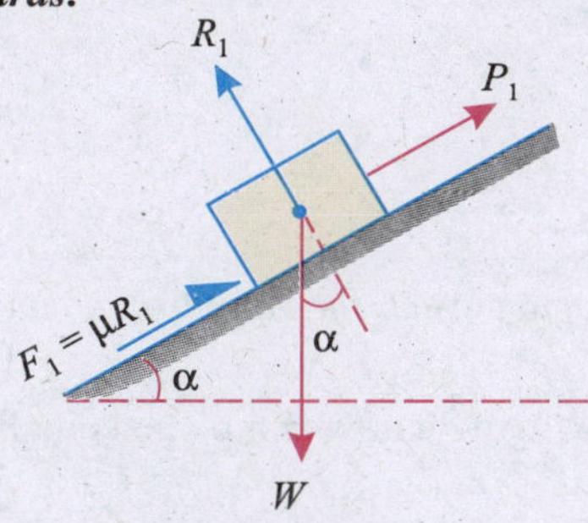

(*a*) Body at the point of sliding downwards

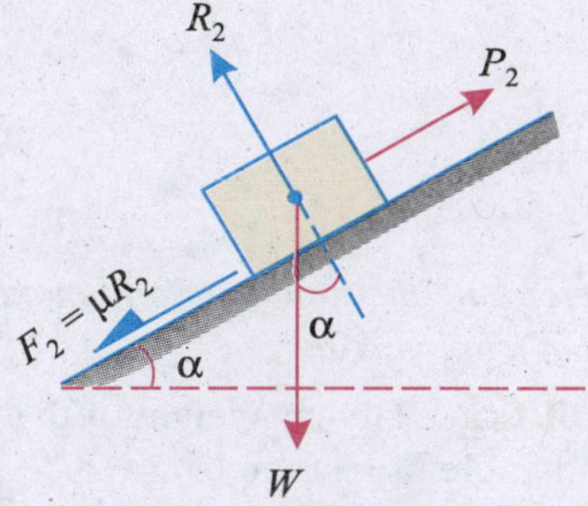

(*b*) Body at the point of sliding upwards

Fig. 10.

In this case, the force of friction ($F_1 = \mu.R_1$) will act upwards, as the body is at the point of sliding downwards as shown in Fig. 10 (*a*). Now resolving the forces along the plane,

$$P_1 = W \sin \alpha - \mu.R_1 \qquad ...(i)$$

and now resolving the forces perpendicular to the plane,

$$R_1 = W \cos \alpha \qquad ...(ii)$$

Substituting the value of R_1 in equation (*i*),

$$P_1 = W \sin \alpha - \mu\, W \cos \alpha = W (\sin \alpha - \mu \cos \alpha)$$

and now substituting the value of $\mu = \tan \phi$ in the above equation,

$$P_1 = W (\sin \alpha - \tan \phi \cos \alpha)$$

Multiplying both sides of this equation by $\cos \phi$,

$$P_1 \cos \phi = W (\sin \alpha \cos \phi - \sin \phi \cos \alpha) = W \sin (\alpha - \phi)$$

or
$$P_1 = W \times \frac{\sin (\alpha - \phi)}{\cos \phi}$$

2. *Maximum force (P_2) which will keep the body in equilibrium, when it is at the point of sliding upwards.*

In this case, the force of friction ($F_2 = \mu.R_2$) will act downwards as the body is at the point of sliding upwards as shown in Fig. 10 (*b*). Now resolving the forces along the plane,

$$P_2 = W \sin \alpha + \mu.R_2 \qquad ...(i)$$

and now resolving the forces perpendicular to the plane,

$$R_2 = W \cos \alpha \qquad ...(ii)$$

Substituting the value of R_2 in equation (*i*),

$$P_2 = W \sin \alpha + \mu\, W \cos \alpha = W (\sin \alpha + \mu \cos \alpha)$$

and now substituting the value of $\mu = \tan \phi$ in the above equation,

$$P_2 = W (\sin \alpha + \tan \phi \cos \alpha)$$

Multiplying both sides of this equation by cos ϕ,

$$P_2 \cos \phi = W (\sin \alpha \cos \phi + \sin \phi \cos \alpha) = W \sin (\alpha + \phi)$$

or
$$P_2 = W \times \frac{\sin (\alpha + \phi)}{\cos \phi}$$

***Example 5.** *A body of weight 500 N is lying on a rough plane inclined at an angle of 25° with the horizontal. It is supported by an effort (P) parallel to the plane as shown in Fig. 11.*

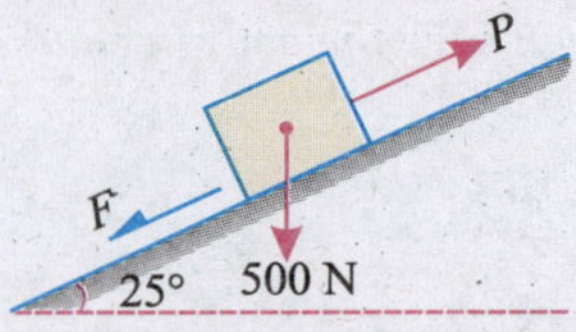

Fig. 11.

Determine the minimum and maximum values of P, for which the equilibrium can exist, if the angle of friction is 20°. **[RGPV]**

Solution. Given: Weight of the body (W) = 500 N ; Angle at which plane is inclined (α) = 25° and angle of friction (ϕ) = 20°.

Minimum value of P

We know that for the minimum value of P, the body is at the point of sliding downwards. We also know that when the body is at the point of sliding downwards, then the force

$$P_1 = W \times \frac{\sin (\alpha - \phi)}{\cos \phi} = 500 \times \frac{\sin (25° - 20°)}{\cos 20°} \text{ N}$$

$$= 500 \times \frac{\sin 5°}{\cos 20°} = 500 \times \frac{0.0872}{0.9397} = 46.4 \text{ N} \qquad \textbf{Ans.}$$

Maximum value of P

We know that for the maximum value of P, the body is at the point of sliding upwards. We also know that when the body is at the point of sliding upwards, then the force

$$P_2 = W \times \frac{\sin (\alpha + \phi)}{\cos \phi} = 500 \times \frac{\sin (25° + 20°)}{\cos 20°} \text{ N}$$

$$= 500 \times \frac{\sin 45°}{\cos 20°} = 500 \times \frac{0.7071}{0.9397} = 376.2 \text{ N} \qquad \textbf{Ans.}$$

Example 6. *An inclined plane as shown in Fig. 12. is used to unload slowly a body weighing 400 N from a truck 1.2 m high into the ground.*

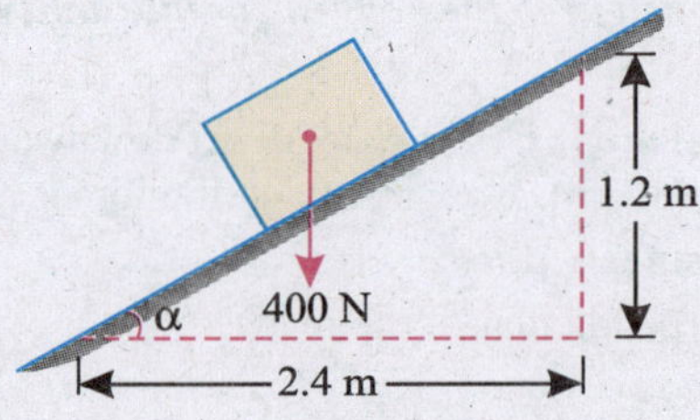

Fig. 12.

The coefficient of friction between the underside of the body and the plank is 0.3. State whether it is necessary to push the body down the plane or hold it back from sliding down. What minimum force is required parallel to the plane for this purpose ?

Solution. Given: Weight of the body (W) = 400 N and coefficient of friction (μ) = 0.3.

Whether it is necessary to push the body down the plane or hold it back from sliding down.

We know that

$$\tan \alpha = \frac{1.2}{2.4} = 0.5 \quad \text{or} \quad \alpha = 26.5°$$

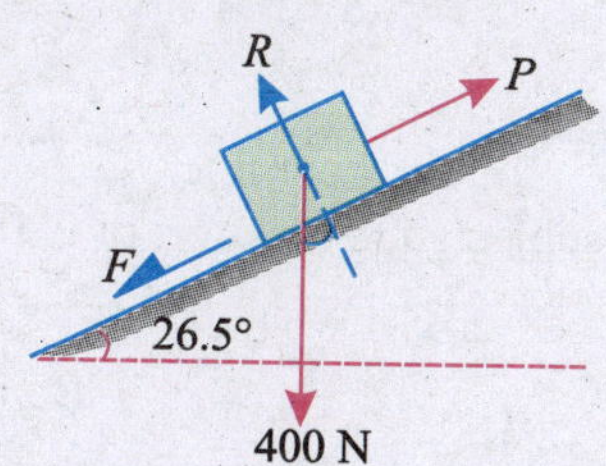

Fig. 13.

and normal reaction, $R = W \cos \alpha = 400 \cos 26.5°$ N

$= 400 \times 0.8949 = 357.9$ N

$\therefore$ Force of friction,

$$F = \mu R = 0.3 \times 357.9 = 107.3 \text{ N} \quad ...(i)$$

Now resolving the 400 N force along the plane

$= 400 \sin \alpha = 400 \times \sin 26.5°$ N

$= 400 \times 0.4462 = 178.5$ N ...(ii)

We know that as the force along the plane (which is responsible for sliding the body) is more than the force of friction, therefore the body will slide down. Or in other words, it is not necessary to push the body down the plane, rather it is necessary to hold it back from sliding down. **Ans.**

Minimum force required parallel to the plane

We know that the minimum force required parallel to the plane to hold the body back,

$$P = 178.5 - 107.3 = 71.2 \text{ N} \quad \textbf{Ans.}$$

Example 7. *An effort of 200 N is required just to move a certain body up an inclined plane of angle 15° the force acting parallel to the plane. If the angle of inclination of the plane is made 20° the effort required, again applied parallel to the plane, is found to be 230 N. Find the weight of the body and the coefficient of friction.*

Solution. Given: First case : When effort (P_1) = 200 N, then angle of inclination (α_1) = 15° and second case : When effort (P_2) = 230 N, then angle of inclination (α_2) = 20°.

Let μ = Coefficient of friction,

W = Weight of the body,

R = Normal reaction, and

F = Force of friction.

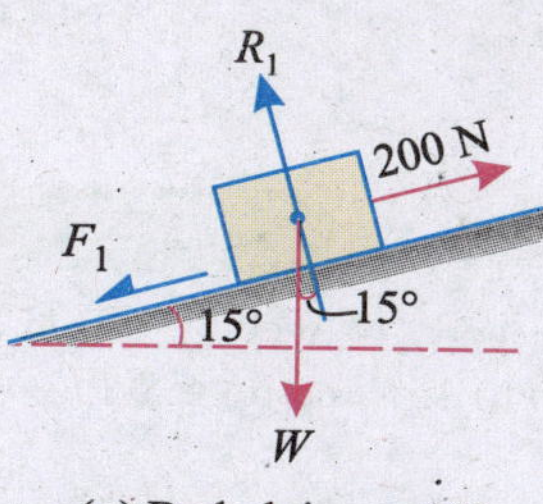

(*a*) Body lying at 15°

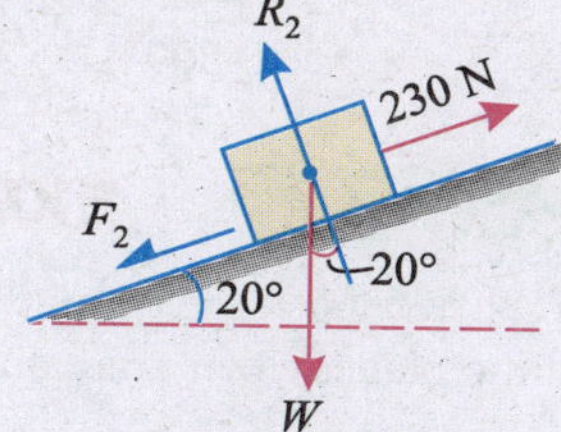

(*b*) Body lying at 20°

Fig. 14.

First of all, consider the body lying on a plane inclined at an angle of 15° with the horizontal and subjected to an effort of 200 N as shown in Fig. 14 (*a*).

Resolving the forces at right angles to the plane,

$$R_1 = W \cos 15° \quad ...(i)$$

and now resolving the forces along the plane,

$$200 = F_1 + W \sin 15° = \mu.R_1 + W \sin 15° \qquad ...(\because F = \mu.R)$$

$$= \mu\, W \cos 15° + W \sin 15° \qquad ...(\because R_1 = W \cos 15°)$$

$$= W (\mu \cos 15° + \sin 15°) \qquad ...(ii)$$

Now consider the body lying on a plane inclined at an angle of 20° with the horizontal and subjected to an effort of 230 N shown in Fig. 14 (*b*).

Resolving the forces at right angles to the plane,

$$R_2 = W \cos 20° \qquad ...(iii)$$

and now resolving the forces along the plane,

$$230 = F_2 + W \sin 20° = \mu\, R_2 + W \sin 20° \qquad ...(\because F = \mu.R)$$

$$= \mu\, W \cos 20° + W \sin 20° \qquad ...(\because R_2 = W \cos 20°)$$

$$= W (\mu \cos 20° + \sin 20°) \qquad ...(iv)$$

Coefficient of friction

Dividing equation (*iv*) by (*ii*),

$$\frac{230}{200} = \frac{W (\mu \cos 20° + \sin 20°)}{W (\mu \cos 15° + \sin 20°)}$$

or $$230\, \mu \cos 15° + 230 \sin 15° = 200\, \mu \cos 20° + 200 \sin 20°$$

or $$230\, \mu \cos 15° - 200\, \mu \cos 20° = 200 \sin 20° - 230 \sin 15°$$

or $$\mu\, (230 \cos 15° - 200 \cos 20°) = 200 \sin 20° - 230 \sin 15°$$

or $$\mu = \frac{200 \sin 20° - 230 \sin 15°}{230 \cos 15° - 200 \cos 20°} = \frac{(200 \times 0.3420) - (230 \times 0.2588)}{(230 \times 0.9659) - (200 \times 0.9397)} = 0.259$$ **Ans.**

Weight of the body

Substituting the value of μ in equation (*ii*),

$$200 = W (0.259 \cos 15° + \sin 15°)$$

$$= W (0.259 \times 0.9659 + 0.2588) = 0.509\, W$$

or $$W = \frac{200}{0.509} = 392.9 \text{ N}$$ **Ans.**

EXERCISE 1

1. Find the horizontal force required to drag a body of weight 100 N along a horizontal plane. If the plane, when gradually raised up to 15°, the body will begin to slide. [**Ans.** 26.79 N]

   Hint. $\phi = 15°$ or $\mu = \tan \phi = \tan 15° = 0.2679$

*2. A body of weight 50 N is hauled along a rough horizontal plane by a pull of 18 N acting at an angle of 14° with the horizontal. Find the coefficient of friction. [**Ans.** 0.383]

3. A man is walking over a dome of 10 m radius. How far he can descend from the top of the dome without slipping ? Take coefficient of friction between the surface of the dome and shoes of the man as 0.6. [**Ans.** 1.413 m]

Hint. Let the man start from the top of the dome (A) and reach a point (B), beyond which he can not walk. Now let (W) be the weight of the man and (θ) and angle subtended by the arc AB at the centre of the dome.

$\therefore$ Normal reaction at $B = W\cos\theta$

and force of friction, $F = W\sin\theta$

$\therefore$ $W\sin\theta = \mu.W\cos\theta$

$\tan\theta = \mu = 0.6$ or $\theta = 31°$

*4. A force of 250 N pulls a body of weight 500 N up an inclined plane, the force being applied parallel to the plane. If the inclination of the plane to the horizontal is 15°, find the coefficient of friction. **[IGNOU]** [**Ans.** 0.25]

14. EQUILIBRIUM OF A BODY ON A ROUGH INCLINED PLANE SUBJECTED TO A FORCE ACTING HORIZONTALLY

[UPTU, GTU]

Consider a body lying on a rough inclined plane subjected to a force acting horizontally, which keeps it in equilibrium as shown in Fig. 15. (*a*) and (*b*).

W = Weight of the body,

α = Angle, which the inclined plane makes with the horizontal,

R = Normal reaction,

μ = Coefficient of friction between the body and the inclined plane, and

ϕ = Angle of friction, such that $\mu = \tan\phi$.

A little consideration will show that if the force is not there, the body will slide down on the plane. Now we shall discuss the following two cases :

1. ***Minimum force (P_1) which will keep the body in equilibrium, when it is at the point of sliding downwards.***

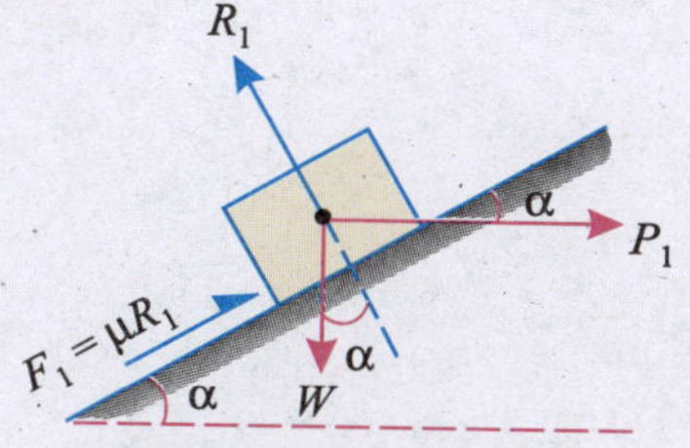

(*a*) Body at the point of sliding downwards

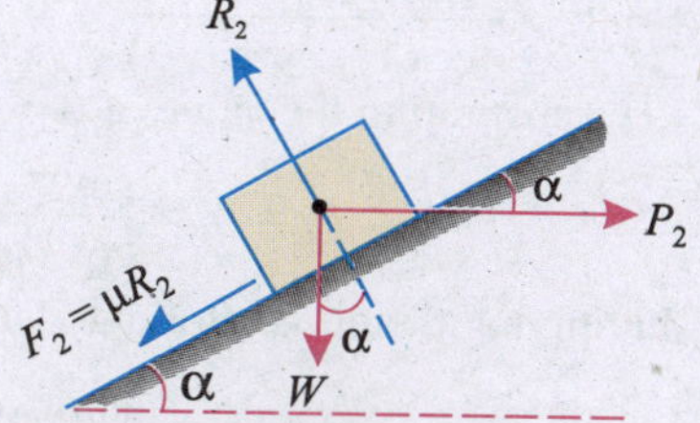

(*b*) Body at the point of sliding upwards

Fig. 15.

In this case, the force of friction ($F_1 = \mu.R_1$) will act upwards, as the body is at the point of sliding downwards as shown in Fig. 15. (*a*). Now resolving the forces along the plane,

$$P_1\cos\alpha = W\sin\alpha - \mu R_1 \quad ...(i)$$

and now resolving the forces perpendicular to the plane,

$$R_1 = W\cos\alpha + P_1\sin\alpha \quad ...(ii)$$

Substituting this value of R_1 in equation (i),

$$P_1\cos\alpha = W\sin\alpha - \mu(W\cos\alpha + P_1\sin\alpha)$$

$$= W\sin\alpha - \mu W\cos\alpha - \mu P_1\sin\alpha$$

or $P_1 \cos\alpha + \mu P_1 \sin\alpha = W\sin\alpha - \mu W\cos\alpha$

or $P_1(\cos\alpha + \mu\sin\alpha) = W(\sin\alpha - \mu\cos\alpha)$

or $$P_1 = W \times \frac{(\sin\alpha - \mu\cos\alpha)}{(\cos\alpha + \mu\sin\alpha)}$$

Now substituting the value of $\mu = \tan\phi$ in the above equation,

$$P_1 = W \times \frac{(\sin\alpha - \tan\phi\cos\alpha)}{(\cos\alpha + \tan\phi\sin\alpha)}$$

Multiplying the numerator and denominator by $\cos\phi$,

$$P_1 = W \times \frac{\sin\alpha\cos\phi - \sin\phi\cos\alpha}{\cos\alpha\cos\phi + \sin\alpha\sin\phi} = W \times \frac{\sin(\alpha - \phi)}{\cos(\alpha - \phi)}$$

$$= W\tan(\alpha - \phi) \qquad \text{...(when } \alpha > \phi)$$

$$= W\tan(\phi - \alpha) \qquad \text{...(when } \phi > \alpha)$$

2. *Maximum force (P_2) which will keep the body in equilibrium, when it is at the point of sliding upwards*

In this case, the force of friction ($F_2 = \mu R_2$) will act downwards, as the body is at the point of sliding upwards as shown in Fig.15. (*b*). Now resolving the forces along the plane,

$$P_2\cos\alpha = W\sin\alpha + \mu R_2 \qquad \text{...}(iii)$$

and now resolving the forces perpendicular to the plane,

$$R_2 = W\cos\alpha + P_2\sin\alpha \qquad \text{...}(iv)$$

Substituting this value of R_2 in the equation (*iii*),

$$P_2\cos\alpha = W\sin\alpha + \mu(W\cos\alpha + P_2\sin\alpha)$$

$$= W\sin\alpha + \mu W\cos\alpha + \mu P_2\sin\alpha$$

or $P_2\cos\alpha - \mu P_2\sin\alpha = W\sin\alpha + \mu W\cos\alpha$

or $P_2(\cos\alpha - \mu\sin\alpha) = W(\sin\alpha + \mu\cos\alpha)$

or $$P_2 = W \times \frac{(\sin\alpha + \mu\cos\alpha)}{(\cos\alpha - \mu\sin\alpha)}$$

Now substituting the value of $\mu = \tan\phi$ in the above equation,

$$P_2 = W \times \frac{\sin\alpha + \tan\phi\cos\alpha}{\cos\alpha - \tan\phi\sin\alpha}$$

Multiplying the numerator and denominator by $\cos\phi$,

$$P_2 = W \times \frac{\sin\alpha\cos\phi + \sin\phi\cos\alpha}{\cos\alpha\cos\phi - \sin\phi\sin\alpha} = W \times \frac{\sin(\alpha + \phi)}{\cos(\alpha + \phi)}$$

$$= W\tan(\alpha + \phi)$$

Example 8. *An object of weight 100 N is kept in position on a plane inclined 30° to the horizontal by a horizontally applied force (F). If the coefficient of friction of the surface of the inclined plane is 0.25, determine the minimum magnitude of the force (F).* **[WBUT]**

Solution. Given: Weight of the object (W) = 100 N; Angle at which plane is inclined (α) = 30° and coefficient of friction (μ) = 0.25 = $\tan\phi$ or $\phi = 14°$.

We know that the minimum magnitude of the force to kept the object in position (when it is at the point of sliding downwards),

$$F = W\tan(\alpha - \phi) = 100\tan(30° - 14°) = 100\tan 16°$$

$$= 100 \times 0.2867 = 28.67 \text{ N}$$ **Ans.**

Example 9. *A load of 1.5 kN, resting on an inclined rough plane, can be moved up the plane by a force of 2 kN applied horizontally or by a force 1.25 kN applied parallel to the plane. Find the inclination of the plane and the coefficient of friction.*

Solution. Given: Load (W) = 1.5 kN; Horizontal effort (P_1) = 2 kN and effort parallel to the inclined plane (P_2) = 1.25 kN.

Inclination of the plane

Let α = Inclination of the plane, and

ϕ = Angle of friction.

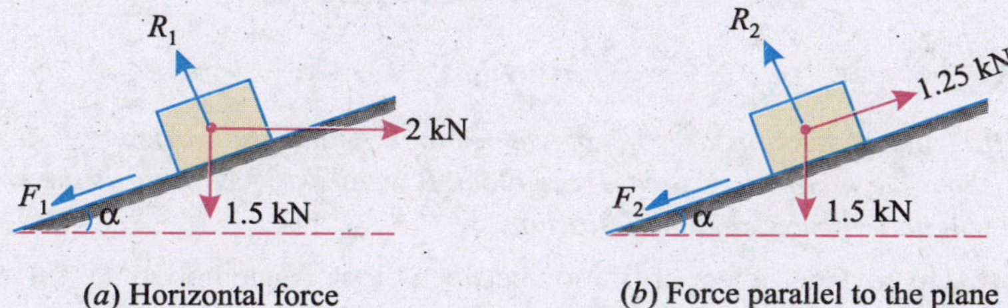

(*a*) Horizontal force (*b*) Force parallel to the plane

Fig. 16.

First of all, consider the load of 1.5 kN subjected to a horizontal force of 2 kN as shown in Fig. 16 (*a*). We know that when the force is applied horizontally, then the magnitude of the force, which can move the load up the plane.

$$P = W \tan(\alpha + \phi)$$

or $$2 = 1.5 \tan(\alpha + \phi)$$

or $$\tan(\alpha + \phi) = \frac{2}{1.5} = 1.333 \quad \text{or} \quad (\alpha + \phi) = 53.1°$$

Now consider the load of 1.5 kN subjected to a force of 1.25 kN along the plane as shown in Fig. 16 (*b*). We Know that when the force is applied parallel to the plane, then the magnitude of the force, which can move the load up the plane,

$$P = W \times \frac{\sin(\alpha + \phi)}{\cos \phi}$$

or $$1.25 = 1.5 \times \frac{\sin 53.1°}{\cos \phi} = 1.5 \times \frac{0.8}{\cos \phi} = \frac{1.2}{\cos \phi}$$

or $$\cos \phi = \frac{1.2}{1.25} = 0.96 \quad \text{or} \quad \phi = 16.3°$$

and $$\alpha = 53.1° - 16.3° = 36.8° \quad \textbf{Ans.}$$

Coefficient of friction

We know that the coefficient of friction,

$$\mu = \tan \phi = \tan 16.3° = 0.292 \quad \textbf{Ans.}$$

Example 10. *Two blocks A and B, connected by a horizontal rod and frictionless hinges are supported on two rough planes as shown in Fig. 17.*

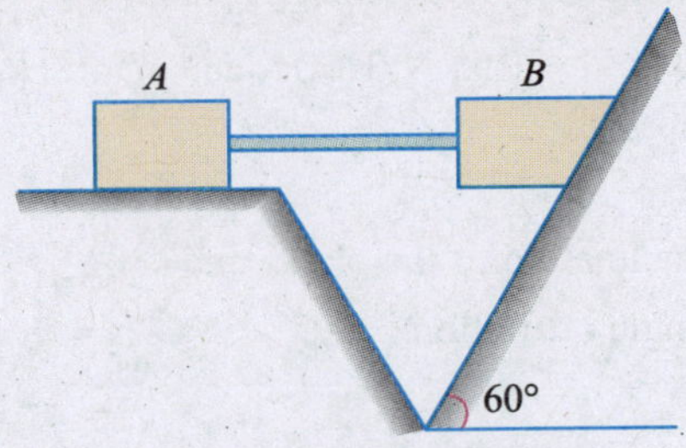

Fig. 17.

The coefficients of friction are 0.3 between block A and the horizontal surface, and 0.4 between block B and the inclined surface. If the block B weighs 100 N, what is the smallest weight of block A, that will hold the system in equilibrium? **[IGNOU, RGPV]**

Solution. Given: Coefficient of friction between block A and horizontal surface (μ_A) = 0.3; Coefficient of friction between block B and inclined surface (μ_B) = 0.4 and weight of block B (W_B) = 100 N.

Let W_A = Smallest weight of block A.

We know that force of friction of block A, which is acting horizontally on the block B,

$$P = \mu_A W_A = 0.3 \times W_A = 0.3\, W_A$$

and angle of friction of block B

$$\tan \phi = \mu_B = 0.4 \quad \text{or} \quad \phi = 21.8°$$

We also know that the smallest force, which will hold the system in equilibrium (or will prevent the block B from sliding downwards),

$$P = W_B \tan (\alpha - \phi) = 100 \tan (60° - 21.8°)$$

or $$0.3\, W_A = 100 \tan 38.2° = 100 \times 0.7869 = 78.69$$

or $$W_A = \frac{78.69}{0.3} = 262.3 \text{ N} \qquad \textbf{Ans.}$$

Alternative method

Consider the equilibrium of block B. We know that it is in equilibrium under the action of the following four forces as shown in Fig. 18.

1. Its own weight 100 N
2. Normal reaction R,
3. Force of friction of block A (acting horizontally on B),
$$F_A = \mu_A \times W_A = 0.3 \times W_A = 0.3\, W_A$$
4. Force of friction between the block B and inclined surface,
$$F = \mu_B \times R = 0.4\, R$$

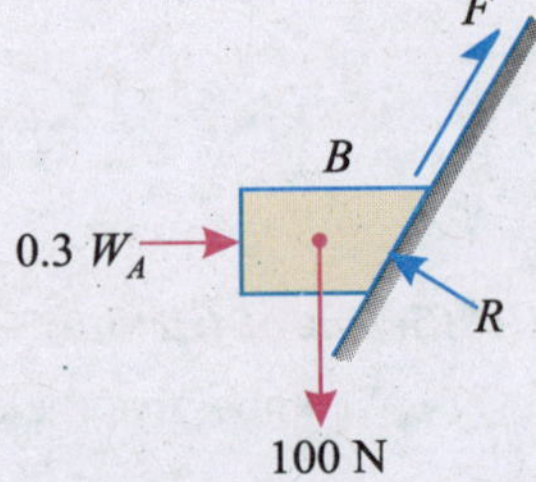

Fig. 18.

Resolving the forces along the plane,

$$F = 100 \cos 30° - 0.3\, W_A \cos 60°$$
$$= 100 \times 0.866 - 0.3\, W_A \times 0.5$$

or $$0.4\, R = 86.6 - 0.15\, W_A \qquad ...(i)$$

and now resolving the forces at right angles to the plane,

$$R = 0.3\ W_A \cos 30° + 100 \sin 30°$$
$$= 0.3\ W_A \times 0.866 + 100 \times 0.5$$
$$= 0.26\ W_A + 50 \quad \text{...}(ii)$$

Substituting the value of R in equation (*i*)

$$0.4\ (0.26\ W_A + 50) = 86.6 - 0.15\ W_A$$

or $$0.104\ W_A + 20 = 86.6 - 0.15\ W_A$$

or $$0.254\ W_A = 86.6 - 20 = 66.6$$

or $$W_A = \frac{66.6}{0.254} = 262.2 \text{ N} \quad \textbf{Ans.}$$

15. EQUILIBRIUM OF A BODY ON A ROUGH INCLINED PLANE SUBJECTED TO A FORCE ACTING AT SOME ANGLE WITH THE INCLINED PLANE

Consider a body lying on a rough inclined plane subjected to a force acting at some angle with the inclined plane, which keeps it in equilibrium as shown in Fig. 19 (*a*) and (*b*).

Let W = Weight of the body,

α = Angle which the inclined plane makes with the horizontal,

θ = Angle which the force makes with the inclined surface,

R = Normal reaction,

μ = Coefficient of friction between the body and the inclined plane, and

ϕ = Angle of friction, such that $\mu = \tan \phi$.

A little consideration will show that if the force is not there, the body will slide down the plane. Now we shall discuss the following two cases :

1. *Minimum force (P_1) which will keep the body in equilibrium when it is at the point of sliding downwards.*

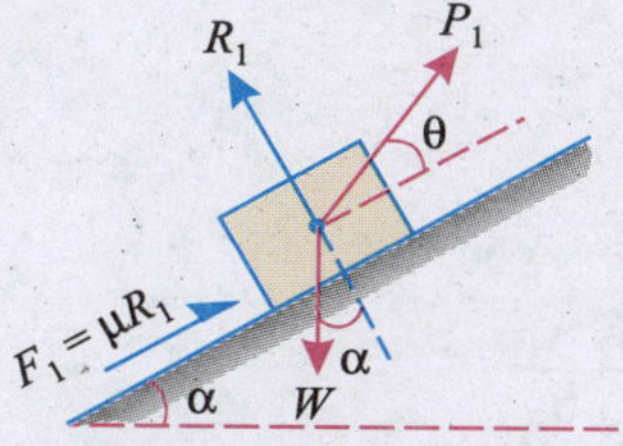

(*a*) Body at the point of sliding downwards

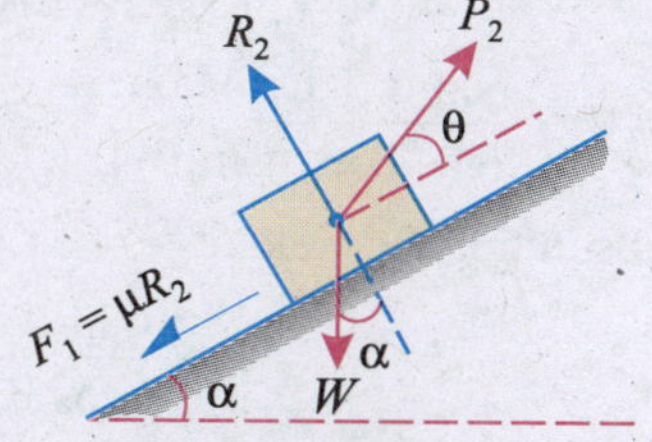

(*b*) Body at the point of sliding upwards

Fig. 19.

In this case, the force of friction ($F_1 = \mu R_1$) will act upwards, as the body is at the point of sliding downwards as shown in Fig. 19 (*a*). Now resolving the forces along the plane,

$$P_1 \cos \theta = W \sin \alpha - \mu R_1 \quad \text{...}(i)$$

and now resolving the forces perpendicular to the plane,

$$R_1 = W \cos \alpha - P_1 \sin \theta \quad \text{...}(ii)$$

Substituting the value of R_1 in equation (*i*),

$$P_1 \cos\theta = W\sin\alpha - \mu\,(W\cos\alpha - P_1\sin\theta)$$
$$= W\sin\alpha - \mu\,W\cos\alpha + \mu\,P_1\sin\theta$$

or $\quad P_1\cos\theta - \mu\,P_1\sin\theta = W\sin\alpha - \mu\,W\cos\alpha$

or $\quad P_1(\cos\theta - \mu\sin\theta) = W(\sin\alpha - \mu\cos\alpha)$

or $$P_1 = W \times \frac{(\sin\alpha - \mu\cos\alpha)}{(\cos\theta - \mu\sin\theta)}$$

and now substituting the value of $\mu = \tan\phi$ in the above equation,

$$P_1 = W \times \frac{(\sin\alpha - \tan\phi\cos\alpha)}{(\cos\theta - \tan\phi\sin\theta)}$$

Multiplying the numerator and denominator by $\cos\phi$,

$$P_1 = W \times \frac{(\sin\alpha\cos\phi - \sin\phi\cos\alpha)}{(\cos\theta\cos\phi - \sin\phi\sin\theta)} = W \times \frac{\sin(\alpha - \phi)}{\cos(\theta + \phi)}$$

2. *Maximum force (P_2) which will keep the body in equilibrium, when it is at the point of sliding upwards.*

In this case, the force of friction ($F_2 = \mu R_2$) will act downwards as the body is at the point of sliding upwards as shown in Fig. 19 (*b*). Now resolving the forces along the plane.

$$P_2\cos\theta = W\sin\alpha + \mu\,R_2 \qquad \ldots(iii)$$

and now resolving the forces perpendicular to the plane,

$$R_2 = W\cos\alpha - P_2\sin\theta \qquad \ldots(iv)$$

Substituting the value of R_2 in equation (*iii*),

$$P_2\cos\theta = W\sin\alpha + \mu\,(W\cos\alpha - P_2\sin\theta)$$
$$= W\sin\alpha + \mu\,W\cos\alpha - \mu\,P_2\sin\theta$$

or $\quad P_2\cos\theta + \mu\,P_2\sin\theta = W\sin\alpha + \mu\,W\cos\alpha$

or $\quad P_2(\cos\theta + \mu\sin\theta) = W(\sin\alpha + \mu\cos\alpha)$

or $$P_2 = W \times \frac{(\sin\alpha + \mu\cos\alpha)}{(\cos\theta + \mu\sin\theta)}$$

and now substituting the vaue of $\mu = \tan\phi$ in the above equation,

$$P_2 = W \times \frac{(\sin\alpha + \tan\phi\cos\alpha)}{(\cos\theta + \tan\phi\sin\theta)}$$

Multiplying the numerator and denominator by $\cos\phi$,

$$P_2 = W \times \frac{(\sin\alpha\cos\phi + \sin\phi\cos\alpha)}{(\cos\theta\cos\phi + \sin\phi\sin\theta)} = W \times \frac{\sin(\alpha + \phi)}{\cos(\theta - \phi)}$$

Example 11. *Find the force required to move a load of 300 N up a rough plane, the force being applied parallel to the plane. The inclination of the plane is such that when the same load is kept on a perfectly smooth plane inclined at the same angle, a force of 60 N applied at an inclination of 30° to the plane, keeps the same load in equilibrium.*

Assume coefficient of friction between the rough plane and the load to be equal to 0.3.

[IGNOU, BPUT, JNTU, AUT]

Solution. Given: Load (W) = 300 N; Force (P_1) = 60 N and angle at which force is inclined (θ) = 30°,

Let α = Angle of inclination of the plane.

First of all, consider the load lying on a smooth plane inclined at an angle (α) with the horizontal and subjected to a force of 60 N acting at an angle 30° with the plane as shown in Fig. 20 (*a*).

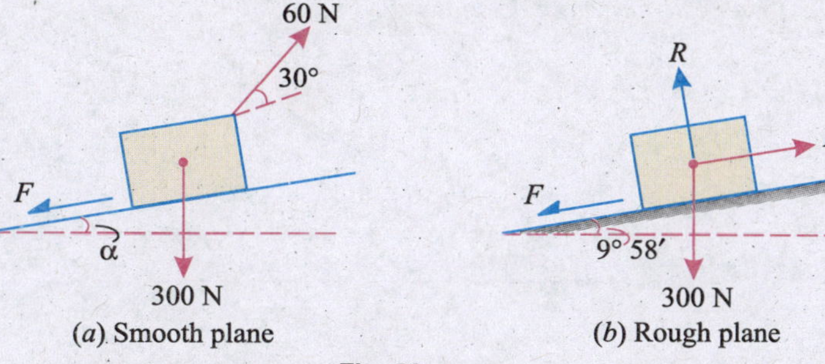

(*a*) Smooth plane (*b*) Rough plane

Fig. 20.

We know that in this case, because of the smooth plane $\mu = 0$ or $\phi = 0$. We also know that the force required, when the load is at the point of sliding upwards (P),

$$60 = W \times \frac{\sin(\alpha + \phi)}{\cos(\theta - \phi)} = 300 \times \frac{\sin \alpha}{\cos 30°} = 300 \times \frac{\sin \alpha}{0.866} = 346.4 \sin \alpha \quad ...(\because \phi = 0)$$

or $$\sin \alpha = \frac{60}{346.4} = 0.1732 \quad \text{or} \quad \alpha = 10°$$

Now consider the load lying on the rough plane inclined at an angle of 10° with the horizontal as shown in Fig. 20. (*b*). We know that in this case, $\mu = 0.3 = \tan \phi$ or $\phi = 16.7°$.

We also know that force required to move the load up the plane,

$$P = W \times \frac{\sin(\alpha + \phi)}{\cos \phi} = 300 \times \frac{\sin(10° + 16.7°)}{\cos 16.7°} \text{ N}$$

$$= 300 \times \frac{\sin 26.7°}{\cos 16.7°} = 300 \times \frac{0.4493}{0.9578} = 140.7 \text{ N} \quad \textbf{Ans.}$$

Alternative method

Ist case

Given: In this case load (P) = 60 N; Angle (θ) = 30° and force of friction $F = 0$ (because of smooth plane).

Resolving the forces along the inclined plane,

$$60 \cos 30° = 300 \sin \alpha$$

$$\therefore \quad \sin \alpha = \frac{60 \cos 30°}{300} = \frac{60 \times 0.866}{300} = 0.1732 \quad \text{or} \quad \alpha = 10°$$

2nd case

Given: In this case, coefficient of friction (μ) = 0.3 = $\tan \phi$ or $\phi = 16.7°$

Let P = Force required to move the load up the plane,

R = Normal reaction, and

F = Force of friction equal to 0.3 R.

Resolving the forces along the plane,

$$P = F + 300 \sin 10° = 0.3\,R + (300 \times 0.1732) = 0.3\,R + 51.96 \text{ N} \quad ...(i)$$

and now resolving the forces at right angles to the plane,

$$R = 300 \cos 10° = 300 \times 0.9849 = 295.5 \text{ N} \qquad ...(ii)$$

Substituting the value of R in equation (i),

$$P = (0.3 \times 295.5) + 51.96 = 140.7 \text{ N} \qquad \textbf{Ans.}$$

Example 12. *The upper half of an inclined having inclination θ with the horizontal is smooth, while the lower half in rough as shown in Fig 21.*

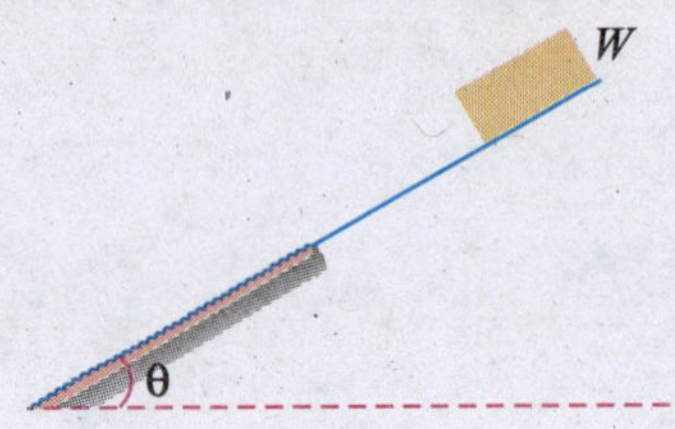

Fig. 21.

If a body of weight W slides down from rest at the top, again comes to rest at the bottom of the plane, then determine the value of coefficient of friction for the lower half of the plane.

Solution. Given: Angle of inclination = θ and weight of the body = W.

Let μ = Coefficient of friction for the lower half of the inclined alone.

We know that acceleration on the smooth surface of the plane

$$= g \sin \theta \qquad ...(i)$$

and retardation on the rough surface of the plane

$$= -g(\sin \theta - \mu \cos \theta) \qquad \text{...(Minus sign due to retardation)}$$

$$= g(\mu \cos \theta - \sin \theta) \qquad ...(ii)$$

Since body starts from rest at the top comes to rest at the bottom of therefore acceleration on the smooth surface is equal to retardation on the rough surface.

$\therefore \quad g \sin \theta = g(\mu \cos \theta - \sin \theta) \quad$ or $\quad \sin \theta = \mu \cos \theta - \sin \theta$

or $\quad \mu \cos \theta = 2 \sin \theta \quad$ or $\quad \mu = 2 \tan \theta \quad$ **Ans.**

Example 13. *Two loads, W_1 (equal to 1 kN) and W_2 resting on two inclined rough planes OA and OB are connected by a horizontal link PQ as shown in Fig. 22.*

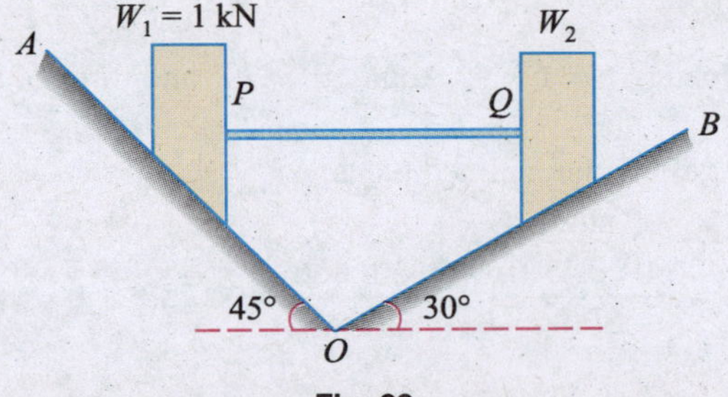

Fig. 22.

Find the maximum and minimum values of W_2 for which the equilibrium can exist. Take angle of friction for both the planes as 20°.

Solution. Given: First load (W_1) = 1 kN ; Angle made by inclined plane OA with the horizontal (α_1) 45°; Angle made by inclined plane OB with the horizontal (α_2) = 30° and Angle of friction for both the planes (ϕ) = 20°.

Maximum value of W_2

We know that for maximum value of W_2, the load W_2 will be at the point of sliding downwards whereas the load W_1 will be at the point of sliding upwards. We also know that when the load W_1 is at the point of sliding upwards on the plane *OA*, the horizontal thrust in the link *PQ*,

$$P = W_1 \tan(\alpha_1 + \phi) = 1 \times \tan(45° + 20°) \text{ kN}$$
$$= 1 \tan 65° = 1 \times 2.1445 = 2.1445 \text{ kN} \quad ...(i)$$

and when the load W_2 is at the point of sliding downwards on the plane *OB*, the horizontal thrust in the link *PQ*

$$P = W_2 \tan(\alpha_2 - \phi) = W_2 \tan(30° - 20°) \text{ kN}$$
$$= W_2 \tan 10° = W_2 \times 0.1763 \text{ kN} \quad ...(ii)$$

Since the values of the horizontal thrusts in the link *PQ*, obtained in both the above equations is the same, therefore equating equations (*i*) and (*ii*),

$$2.1445 = W_2 \times 0.1763$$

or $$W_2 = \frac{2.1445}{0.1763} = 12.16 \text{ kN} \quad \textbf{Ans.}$$

Minimum value of W_2

We know that for maximum value of W_2, the load W_2 will be at the point of sliding upwards whereas the load W_1 will be at the point of sliding downwards. We also know that when the load W_1 is at the point of sliding downwards on the plane *OA*, the horizontal thrust in the link *PQ*,

$$P = W_1 \tan(\alpha_1 - \phi) = 1 \times \tan(45° - 20°) \text{ kN}$$
$$= 1 \times \tan 25° = 1 \times 0.4663 = 0.4663 \text{ kN} \quad ...(iii)$$

and when the load W_2 is at the point of sliding upwards on the plane *OB*, the horizontal thrust in the link *PQ*,

$$P = W_2 \tan(\alpha_2 + \phi) = W_2 (30° + 20°) \text{ kN}$$
$$= W_2 \tan 50° = W_2 \times 1.1918 \text{ kN} \quad ...(iv)$$

Since the values of the horizontal thrust in the link *PQ*, obtained in the above equations is the same, therefore, equating (*iii*) and (*iv*),

$$0.4663 = W_2 \times 1.1918$$

or $$W_2 = \frac{0.4663}{1.1918} = 0.391 \text{ kN} = 391 \text{ N} \quad \textbf{Ans.}$$

Example 14. *A block (A) weighing 1 kN rests on a rough inclined plane whose inclination to the horizontal is 45°. This block is connected to another block (B) weighing 3 kN rests on a rough horizontal plane by a weightless rigid bar inclined at an angle of 30° to the horizontal as shown in Fig. 23.*

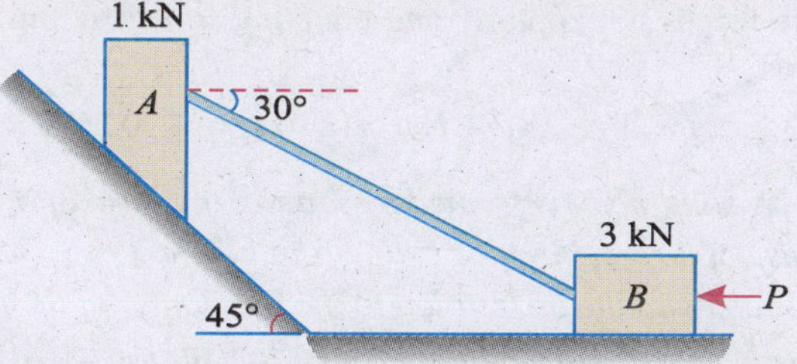

Fig. 23.

Find horizontal force (P) required to be applied to the block (B) just to move the block (A) in upward direction. Assume angle of limiting friction as 15° at all surface where there is sliding.

[RGPV, JNTU]

Solution. Given: Weight of block A (W_A) = 1 kN; Weight of block B (W_B) = 3 kN; Angle of inclination of plane with horizontal (α) = 45° or coefficient of friction (μ) = tan ϕ = tan 15° = 0.2679 or angle between rod and inclined plane (θ) = 45° – 30° = 15° and angle of limiting friction (ϕ) = 15°.

First of all, consider the equilibrium of the block (A), which is subjected to the forces as shown in Fig. 24 (a).

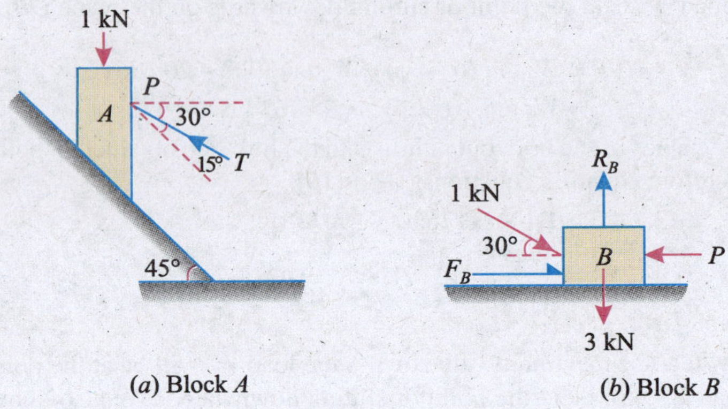

(a) Block A (b) Block B

Fig. 24.

We know that the thrust or force in the rigid bar just to move the block (A) in the upward direction,

$$T = W_A \times \frac{\sin(\alpha + \phi)}{\cos(\theta - \phi)} = 1 \times \frac{\sin(45° + 15°)}{\cos(-15° - 15°)} \text{ kN}$$

...(The value of θ is taken as negative)

$$= 1 \times \frac{\sin 60°}{\cos(-30°)} = 1 \times \frac{\sin 60°}{\cos 30°} = 1 \times \frac{0.866}{0.866} = 1 \text{ kN} \quad ...[\because \cos(-\theta) = \cos\theta]$$

Now consider the equilibrium of the block (B), which is subjected to the forces as shown in Fig. 24 (b). We know that as the block is at the point of sliding towards left, therefore, the force of friction ($F_B = \mu R_B$) will act towards right as shown in the figure.

Now resolving the forces vertically and equating forces acting in upward direction with forces acting in downward direction,

$$R_B = 3 + 1 \sin 30° = 3 + (1 \times 0.5) = 3.5 \text{ kN}$$

and now resolving the forces horizontally and equating forces acting in left direction with forces acting in right direction,

$$P = 1 \cos 30° + F_B = (1 \times 0.866) + (0.2679 \times 3.5) = 1.8 \text{ kN} \quad \textbf{Ans.}$$

Example 15. *A solid body is formed by joining the base of a right circular cone of height 12 cm to the equal base of a right circular cylinder of height 3 cm. The solid is placed with its face on a rough inclined plane, and the inclination to the horizontal of the plane is gradually increased. Show that if radius r of the base is 2 cm, and the coefficient of friction $\mu = 0.5$, the body will topple over before it begins to slide.*

If the heights are so chosen that the centre of mass of the solid is at the centre of the common base, show that if $\mu H < r\sqrt{6}$, the solid will slide before it topples.

Solution. Given: Height of right circular cone (H) = 12 cm; Height of right circular cylinder (h) = 3 cm; Radius of the base (r) = 2 cm and coefficient of friction (μ) = 0.5.

We have already found out in example 6.6 that the centre of gravity of the body is at a height of 4.07 cm from the base. A little consideration will show, that when the body is at the point of toppling, the weight of the body will pass through the extreme point B as shown in Fig. 25. Thus in the limiting case, the angle (θ_1) at which the body is inclined with the horizontal is given by:

$$\tan \theta_1 = \frac{BD}{DG} = \frac{2}{4.07} = 0.49$$

Thus we see that the angle (θ_1) is less than the angle (θ). It is so, as the value of $\tan \theta_1$ (equal to 0.49) is less than $\tan \theta$ (equal to 0.5). Or in other words, the angle of inclination is less than the angle of friction. Thus the body will topple over before it begins to slide. **Ans.**

Now in the second case, let us first find out the relation between h and H from the given condition that the centre of the body is at the centre of the common base (G) as shown in Fig. 26.

(i) Cylinder

$$v_1 = \pi r^2 h$$

and $$y_1 = \frac{h}{2}$$

(ii) Cone

$$v_2 = \frac{1}{3} \pi r^2 H$$

and $$y_2 = \frac{H}{4}$$

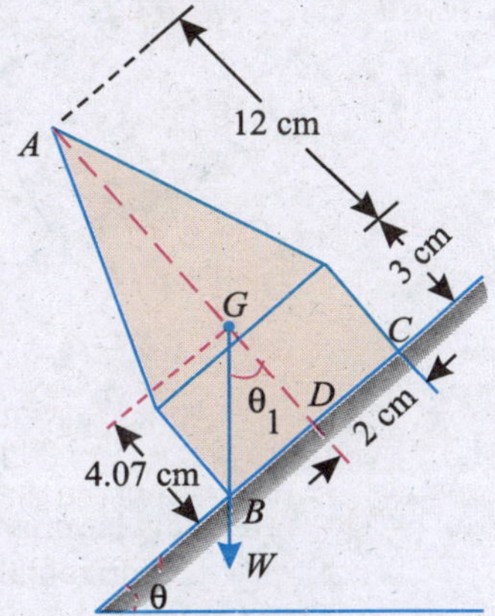

Fig. 25.

We know that the distance between centre of gravity of the body from the base of the cylinder $(\bar{y})$

$$h = \frac{v_1 y_1 + v_2 y_2}{v_1 + v_2} = \frac{\left[\pi r^2 h \times \frac{h}{2}\right] + \left[\frac{1}{3} \pi r^2 H \left(h + \frac{H}{4}\right)\right]}{\pi r^2 h + \frac{1}{3} \pi r^2 H}$$

$$\pi r^2 h^2 + \frac{\pi r^2 H h}{3} = \frac{\pi r^2 h^2}{2} + \frac{\pi r^2 Hh}{3} + \frac{\pi r^2 H^2}{12}$$

$$\frac{\pi r^2 h^2}{2} = \frac{\pi r^2 H^2}{12}$$

or $$h^2 = \frac{H^2}{6} \quad \text{or} \quad h = \frac{H}{\sqrt{6}}$$

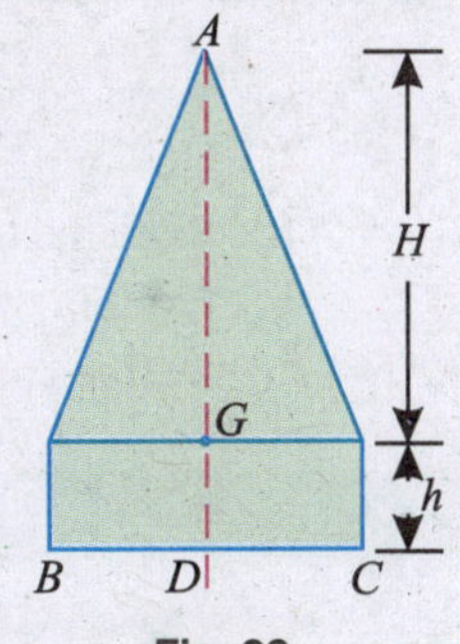

Fig. 26.

In this case, when the body is at the point of toppling, the angle (θ_2) at which it is inclined with the horizontal is given by :

$$\tan \theta_2 = \frac{r}{h} = \frac{r}{H/\sqrt{6}} = \frac{r\sqrt{6}}{H}$$

Now if the body is to slide before toppling, the angle of friction should be less than the angle of friction. Or in other words, $\tan \theta$ should be less than $\tan \theta_2$, Mathematically,

$$\mu < \frac{r\sqrt{6}}{H} \quad \text{or} \quad \mu H < r\sqrt{6} \quad \textbf{Ans.}$$

EXERCISE 2

1. A load of 500 N is lying on an inclined plane, whose inclination with the horizontal is 30°. If the coefficient of friction between the load and the plane is 0.4, find the minimum and maximum horizontal force, which will keep the load in equilibrium.

[**Ans.** 72.05 N, 635.4 N]

*2. Two blocks *A* and *B* of weight 100 N and 300 N respectively are resting on a rough inclined plane as shown in Fig. 27.

Find the value of the angle (θ) when the block *B* is about to slide. Take coefficient of friction between the two blocks as well as block *B* and the inclined plane as 0.25.

[IGNOU] [**Ans.** 22.6°]

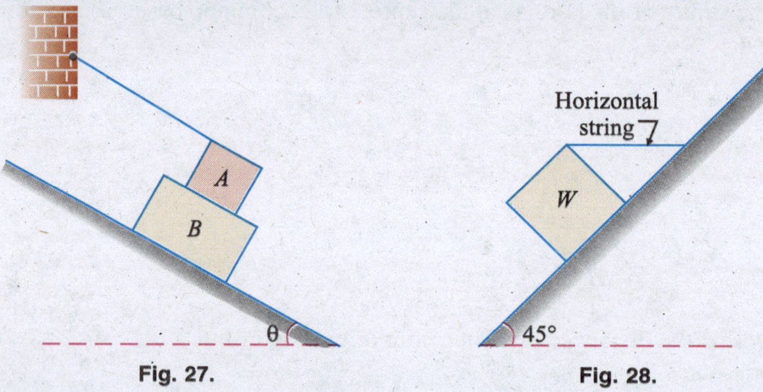

Fig. 27. **Fig. 28.**

*3. A rectangular prism (*W*) weighing 150 N, is lying on an inclined plane whose inclination with the horizontal is shown in Fig. 28.
The block is tied up by a horizontal string, which has a tension of 50 N. From fundamentals find (*i*) the frictional force on the block (*ii*) the normal reaction of the inclined plane, (*iii*) the coefficient of friction between the surface of contact.

[**Ans.** 70.7 N ; 141.4 N ; 0.5 ;]

QUESTIONS

1. What do you understand by the term friction ? Explain clearly why it comes into play ?
2. How will you distinguish between static friction and dynamic friction ?
3. State the laws of friction.
4. Explain the term angle of friction.
5. Define coefficient of friction and limiting friction.

OBJECTIVE TYPE QUESTIONS

1. The force of friction between two bodies in contact

   (*a*) Depends upon the area of their contact

   (*b*) Depends upon the relative velocity between them

   (*c*) Is always normal to the surface of their contact

   (*d*) All of the above

2. The magnitude of the force of friction between two bodies, one lying above the other, depends upon the roughness of the
   (*a*) Upper body (*b*) Lower body
   (*c*) Both the bodies (*d*) The body having more roughness
3. The force of friction always acts in a direction opposite to that
   (*a*) In which the body tends to move
   (*b*) In which the body is moving
   (*c*) Both (*a*) and (*b*)
   (*d*) None of the two
4. Which of the following statement is correct ?
   (*a*) The force of friction does not depend upon the area of contact
   (*b*) The magnitude of limiting friction bears a constant ratio to the normal reaction between the two surfaces
   (*c*) The static friction is slightly less than the limiting friction.
   (*d*) All (*a*), (b) and (*c*)

ANSWERS

1. (*c*)	2. (*c*)	3. (*c*)	4. (*d*)

CHAPTER

Applications of Friction

Contents

1. APPLICATIONS OF FRICTION

In the last chapter, we have discussed the principles of friction of various types. Though these principles have a number of applications in Engineering-science, yet the following are important from the subject point of view:

1. Ladder friction
2. Wedge friction
3. Screw friction.

2. LADDER FRICTION

The ladder is a device for climbing or scaling on the roofs or walls. It consists of two long uprights of wood, iron or rope connected by a number of cross pieces called rungs. These runing serve as steps.

Consider a ladder AB resting on the rough ground and leaning against a wall, as shown in Fig. 1.

As the upper end of the ladder tends to slip downwards, therefore the direction of the force of friction between the ladder and the wall (F_w) will be upwards as shown in the figure. Similarly, as the lower end of the ladder tends to slip away from the wall, therefore the direction of the force of friction between the ladder and the floor (F_f) will be towards the wall as shown in the figure.

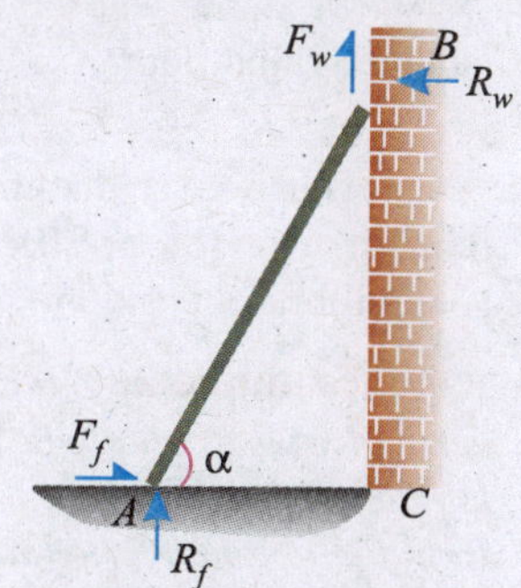

Fig. 1. Ladder friction

Since the system is in equilibrium, therefore the algebraic sum of the horizontal and vertical components of the forces must also be equal to zero.

Note: The normal reaction at the floor (R_f) will act perpendicular of the floor. Similarly, normal reaction of the wall (R_w) will also act perpendicular to the wall.

Example 1. *A uniform ladder of length 3.25 m and weighing 250 N is placed against a smooth vertical wall with its lower end 1.25 m from the wall. The coefficient of friction between the ladder and floor is 0.3.*

What is the frictional force acting on the ladder at the point of contact between the ladder and the floor? Show that the ladder will remain in equilibrium in this position. **[BPUT]**

Solution. Given: Length of the ladder (l) = 3.25 m; Weight of the ladder (w) = 250 N; Distance between the lower end of ladder and wall = 1.25 m and coefficient of friction between the ladder and floor (μ_f) = 0.3.

Frictional force acting on the ladder.

The forces acting on the ladder are shown in Fig. 2.

let F_f = Frictional force acting on the ladder at the Point of contact between the ladder and floor, and

R_f = Normal reaction at the floor.

Fig. 2.

Since the ladder is placed against a smooth vertical wall, therefore there will be no friction at the point of contact between the ladder and wall. Resolving the forces vertically and equating forces acting in upward direction with forces acting in downward direction,

$$R_f = 250 \text{ N}$$

From the geometry of the figure, we find that

$$BC = \sqrt{(3.25)^2 - (1.25)^2} = 3.0 \text{ m}$$

Taking moments about B and equating the anti-clockwise moments with clockwise moments,

$$(F_f \times 3) + (250 \times 0.625) = (R_f \times 1.25)$$

or
$$F_f \times 3 = (250 \times 1.25) - (250 \times 0.625) = 156.2 \text{ N}$$

or
$$F_f = \frac{156.2}{3} = 52.1 \text{N} \quad \textbf{Ans.}$$

Equilibrium of the ladder

We know that the maximum force of friction available at the point of contact between the ladder and the floor

$$= \mu R_f = 0.3 \times 250 = 75 \text{ N}$$

Thus we see that the amount of the force of friction available at the point of contact (75 N) is more than the force of friction required for equilibrium (52.1 N). Therefore the ladder will remain in an equilibrium position. **Ans.**

Example 2. *A ladder 5 meters long rests on a horizontal ground and leans against a smooth vertical wall at an angle 70° with the horizontal. The weight of the ladder is 900 N and acts at its middle. The ladder is at the point of sliding, when a man weighing 750N stands on a rung 1.5 metre from the bottom of the ladder.*

Calculate the coefficient of friction between the ladder and the floor. **[AUT, GTU]**

Solution. Given: Length of the ladder (l) = 5 m; Angle which the ladder makes with the horizontal (α) = 70°; Weight of the ladder (w_1) = 900 N; Weight of man (w_2) = 750 N and distance between the man and bottom of ladder = 1.5 m.

Forces acting on the ladder are shown in Fig. 3.

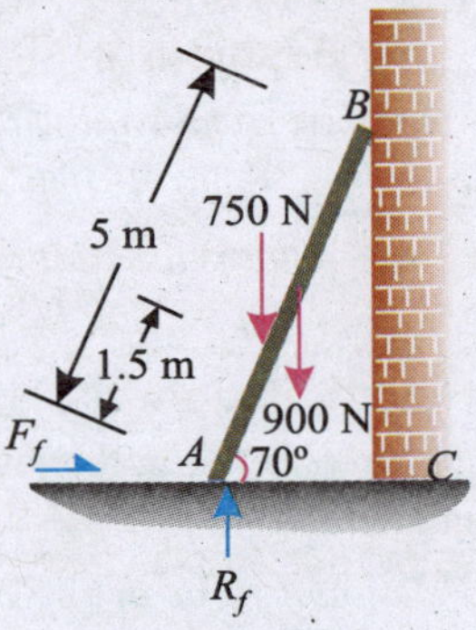

Fig. 3.

Let μ_f = Coefficient of friction between ladder and floor and

R_f = Normal reaction at the floor.

Resolving the forces vertically and equating forces acting in upward direction with forces acting in downward direction,

$$R_f = 900 + 750 = 1650 \text{ N} \quad ...(i)$$

∴ Force of friction at A

$$F_f = \mu_f \times R_f = \mu_f \times 1650 \quad ...(ii)$$

Now taking moments about B and equating the clockwise moments with anti-clockwise moments,

$$R_f \times 5 \sin 20° = (F_f \times 5 \cos 20°) + (900 \times 2.5 \sin 20°) + (750 \times 3.5 \sin 20°)$$

$$= (F_f \times 5 \cos 20°) + (4875 \sin 20°)$$

$$= (\mu_f \times 1650 \times 5 \cos 20°) + 4875 \sin 20°$$

and now substituting the values of R_f and F_f from equations (i) and (ii)

$$1650 \times 5 \sin 20° = (\mu_f \times 1650 \times 5 \cos 20°) + (4875 \sin 20°)$$

Dividing both sides by 5 sin 20°,

$$1650 = (\mu_f \times 1650 \cot 20°) + 975$$

$$= (\mu_f \times 1650 \times 2.7475) + 975 = 4533\,\mu_f + 975$$

or $$\mu_f = \frac{1650 - 975}{4533} = 0.15$$ **Ans.**

Example 3. *A uniform ladder of 4 m length rests against a vertical wall with which it makes an angle of 45°. The coefficient of friction between the ladder and the wall is 0.4 and that between ladder and the floor is 0.5. If a man, whose weight is one-half of that of the ladder ascends it, how high will it be when the ladder slips?* **[JNTU, GTU, IGNOU, AMIE]**

Solution. Given: Length of the ladder (l) = 4 μ; Angle which the ladder makes with the horizontal (α) = 45°; Coefficient of friction between the ladder and the wall (μ_w) = 0.4 and coefficient of friction between the ladder and the floor (μ_f) = 0.5.

The forces acting on the ladder are shown in Fig. 4.

Let x = Distance between A and the man, when the ladder is at the point of slipping.

W = Weight of the ladder, and

R_f = Normal reaction at floor.

$\therefore$ Weight of the man $= \dfrac{W}{2} = 0.5\ W$

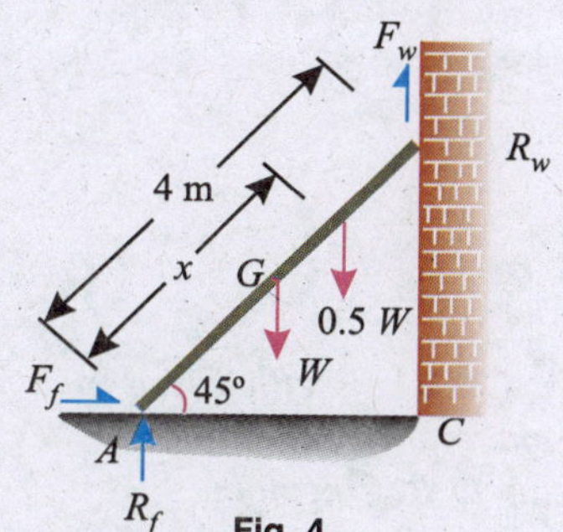

Fig. 4.

We know that frictional force at the floor,

$$F_f = \mu_f\, R_f = 0.5\ R_f \qquad ...(i)$$

and frictional force at the wall,

$$F_w = \mu_w\ R_w = 0.4\ R_w \qquad ...(ii)$$

Resolving the forces vertically and equating forces acting in upward direction with forces acting in downward direction,

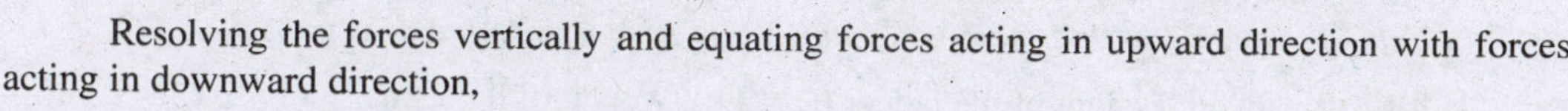

$$R_f + F_w = W + 0.5W = 1.5\ W \qquad ...(iii)$$

and now resolving the forces horizontally and equating forces acting in left direction with forces acting in right direction,

$$R_w = F_f = 0.5\ R_f \quad \text{or} \quad R_f = 2R_w$$

Now substituting the values of R_f and F_w in equation (*iii*),

$$2R_w + 0.4\ R_w = 1.5\ W$$

or $$R_w = \frac{1.5\ W}{2.4} = 0.625\ W$$

and $$F_w = 0.4\ R_w = 0.4 \times 0.625\ W = 0.25\ W \qquad ...(iv)$$

Taking moments about A and equating the clockwise moments with anti-clockwise moments,

$$(W \times 2 \cos 45°) + (0.5\ W \times x \cos 45°)$$
$$= (R_w \times 4 \sin 45°) + (F_w \times 4 \cos 45°)$$

Substituting values of R_w *and* F_w from equations (*iii*) and (*iv*) in the above equation,

$$(W \times 2 \cos 45°) + (0.5\ W \times x \cos 45°)$$
$$= (0.625\ W \times 4 \sin 45°) + (0.25\ W \times 4 \cos 45°)$$

Dividing both sides by ($W \sin 45°$),

$$2 + 0.5\,x = 2.5 + 1 = 3.5 \qquad ...(\because \sin 45° = \cos 45° = 0.707)$$

or $$x = \frac{3.5 - 2}{0.5} = 3.0\,\text{m} \qquad \textbf{Ans.}$$

Example 4. *A uniform ladder 3 m long weighs 200 N. It is placed against a wall making an angle of 60° with the floor as shown in Fig. 5.*

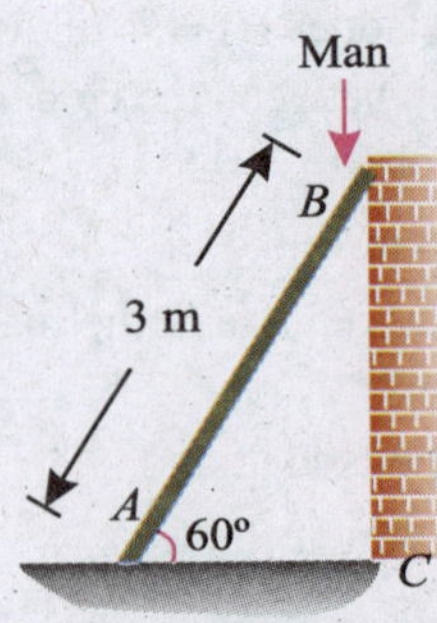

Fig. 5.

The coefficient of friction between the wall and the ladder is 0.25 and that between the floor and ladder is 0.35. The ladder, in addition to its own weight, has to support a man of 1000 N at its top at B. Calculate:

(*i*) *The horizontal force P to be applied to ladder at the floor level to prevent slipping .*

(*ii*) *If the force P is not applied, what should be the minimum inclination of the ladder with the horizontal, so that there is no slipping of it with the man at its top.*

Solution. Given: Length of the ladder (l) = 3 μ ; Weight of the ladder (W) = 200 N; Coefficient of friction between the wall and the ladder (μ_w) = 0.25 and coefficient of friction between the floor and ladder (μ_f) = 0.35.

The forces acting in both the cases are shown in Fig. 6 (*a*) and (*b*).

First of all, cosider the ladder inclined at an angle of 60° and subjected to a horizontal force (P) at the floor as shown in Fig. 6 (*a*).

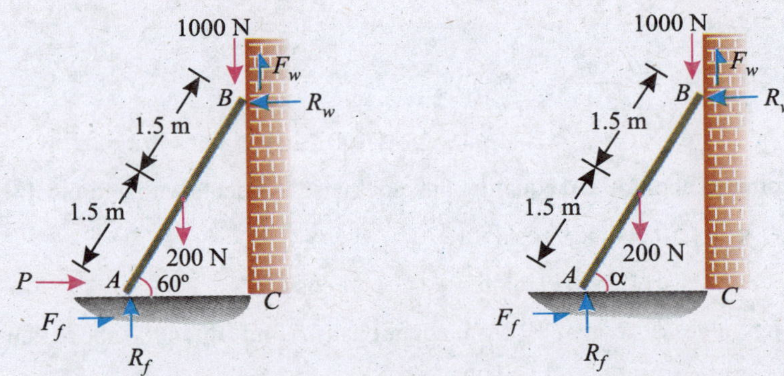

Fig. 6.

(*i*) *Horizontal force (P) applied to the ladder at floor level to prevent slipping*

Resolving the forces horizontally and equating forces acting in left direction with forces acting in right direction,

$$P + F_f = R_w \qquad \text{... (i)}$$

and now resolving the forces vertically and equating forces acting in upward direction with forces acting in downward direction,

$$R_f + F_w = 1000 + 200 = 1200 \text{ N} \qquad \text{...(ii)}$$

Taking moments about A and equating the clockwise moments with anti-clockwise moments,

$$(200 \times 1.5 \cos 60°) + 1000 \times 3 \cos 60°)$$
$$= (F_w \times 3 \cos 60°) + (R_w \times 3 \sin 60°)$$

Dividing both sides by the cos 60°,

$$300 + 3000 = (3 \times F_w) + (3 \times R_w \tan 60°)$$

or $$1100 = F_w + R_w \tan 60° \quad ...(iii)$$

We know that $F_w = \mu_w \times R_w = 0.25\ R_w$ ...($\because \mu_w = 0.25$)

Substituting this value of F_w in equation (*iii*),

$$1100 = (0.25\ R_w) + (R_w \tan 60°) = R_w\,(0.25 + 1.732) = R_w \times 1.982$$

or $$R_w = \frac{1100}{1.982} = 555 \text{ N}$$

and $$F_w = 0.25\ R_w = 0.25 \times 555 = 138.7 \text{ N}$$

Now substituting the value of F_w in equation (*ii*),

$$R_f + 138.7 = 1200$$

or $$R_f = 1200 - 138.7 = 1061.3 \text{ N}$$

and $$F_f = \mu_f R_f = 0.35 \times 1061.3 = 371.5 \text{ N}$$

Now substituting the value of F_f in equation (*i*),

$$P + 371.5 = 555$$

or $$P = 555 - 371.5 = 183.5 \text{ N} \quad \textbf{Ans.}$$

(*ii*) *Inclination of the ladder with the horizontal for no slipping*

Now consider the ladder inclined at angle (α) and without any horizontal force acting at the floor as shown in Fig. 6 (*b*).

Resolving the forces horizontally and equating forces acting in left direction with forces acting in right direction,

$$R_w = F_f = \mu_f \times R_f = 0.35 \times R_f \quad ...(iv)$$

and now resolving the forces vertically and equating forces acting in upward direction with forces acting in downward direction,

$$R_f + F_w = 1000 + 200 = 1200 \text{ N}$$

We know that, $F_w = \mu_w \times R_w = 0.25(0.35\ R_f) = 0.09\ R_f$ ...($\because R_w = 0.35\ R_f$)

or $$R_f + 0.09\ R_f = 1200$$

or $$R_f = \frac{1200}{1.09} = 1101 \text{ N}$$

and $$R_w = 0.35 R_f = 0.35 \times 1101 = 385.4 \text{ N}$$

Similarly $$F_w = 0.09\ R_f = 0.09 \times 1101 = 99.1 \text{ N}$$

Taking moments about A and equating the clockwise moments with anti-clockwise moments,

$$(1000 \times 3 \cos \alpha) + (200 \times 1.5 \cos \alpha)$$
$$= (F_w \times 3 \cos \alpha) + (R_w \times 3 \sin \alpha)$$

Dividing both sides by 3 cos α,

$$1000 + 100 = F_w + R_w \tan\alpha$$

or $$1100 = 99.1 + 385.4 \tan\alpha$$

or $$385.4 \tan\alpha = 1100 - 99.1 = 1000.9$$

or $$\tan\alpha = \frac{1000.9}{385.4} = 2.5970 \quad \text{or} \quad \alpha = 68.9° \quad \textbf{Ans.}$$

***Example 5.** *Two identical blocks of weight W are supported by a rod inclined at 45° with the horizontal as shown in Fig. 7.*

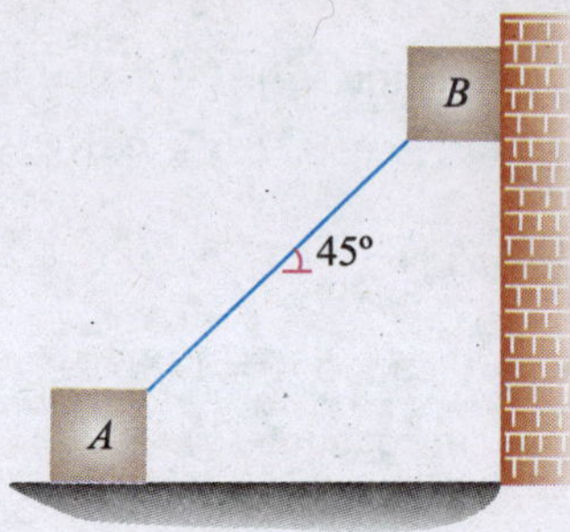

Fig. 7.

If both the blocks are in limiting equilibrium, find the coefficient of friction (μ), assuming it to be the same at floor as well as at wall. **[WBUT]**

Solution. Given: Weight of blocks A and $B = W$ and inclination of rod with the horizontal $(\alpha) = 45°$.

Let μ = Coefficient of friction, and

l = Length of the rod.

The forces acting on both the blocks are shown in Fig. 8.

Resolving the forces vertically and equating forces acting in upward direction with forces acting in downward direction,

$$F_w + R_f = 2W$$

or $$\mu R_w + R_f = 2W \qquad ...(\because F_w = \mu R_w) \; ...(i)$$

and now resolving the forces horizontally and equating forces acting in left direction with forces acting in right direction,

$$R_w = F_f = \mu R_f \qquad ...(ii)$$

Now substituting this value of R_w in equation (*i*),

$$\mu(\mu R_f) + R_f = 2W$$

or $$\mu^2 R_f + R_f = 2W$$

or $$R_f(\mu^2 + 1) = 2W$$

or $$R_f = \frac{2W}{\mu^2 + 1} \qquad ...(iii)$$

Fig. 8.

and now substituting this value of R_f in equation (*ii*),

$$R_w = \mu \times \frac{2W}{\mu^2 + 1} \qquad ...(iv)$$

Taking moments of the forces about the block A and equating the anti-clockwise moments with clockwise moments,

$$(R_w \times l \cos 45°) + (F_w \times l \cos 45°) = W \times l \cos 45°$$

$$R_w + F_w = W$$

or $$R_w + \mu R_w = W$$

or $$R_w (1 + \mu) = W$$

Substituting the value of R_w from equation (*iv*),

$$\frac{\mu \times 2W}{\mu^2 + 1}(1 + \mu) = W$$

or $$2\mu\,(1 + \mu) = \mu^2 + 1$$

or $$2\mu + 2\mu^2 = \mu^2 + 1$$

or $$\mu^2 + 2\mu - 1 = 0$$

Solving it as quadratic equation for μ.

$$\mu = \frac{-2 \pm \sqrt{(2)^2 + 4}}{2} = 0.414 \qquad \textbf{Ans.}$$

EXERCISE 1

1. A 4 m ladder weighing 250 N is placed against a smooth vertical wall with its lower end 1.5 m away from the wall. If the coefficient of friction between the ladder and the floor is 0.3, show that the ladder will remain in equilibrium in this position.
2. A ladder shown in Fig. 9 is 4 m long and is supported by a horizontal floor and vertical wall. The coefficient of friction at the wall is 0.25 and that at the floor is 0.5. The weight of the ladder is 30 N and is considered to be concentrated at *G*. The ladder also supports a vertical load of 150 at *C*.

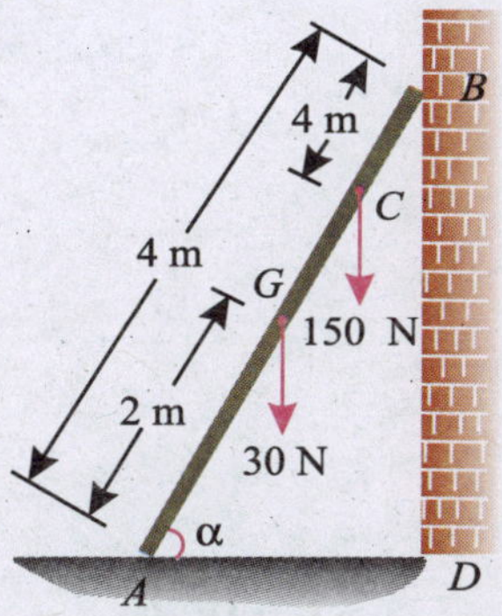

Fig. 9.

Determine the reactions at *A* and *B* and compute the least value of (α) at which the ladder may be placed without slipping to the left.

[Ans. $R_A = 178.9$ N; $R_B = 82.46$ N; $\alpha = 53.3°$]

3. WEDGE FRICTION

[BPUT]

A wedge is, usually, of a triangular or trapezoidal in cross-section. It is, generally, used for slight adjustements in the position of a body *i.e.* for tightening fits or keys for shafts. Sometimes, a wedge is also used for lifting heavy weights as shown in Fig. 10.

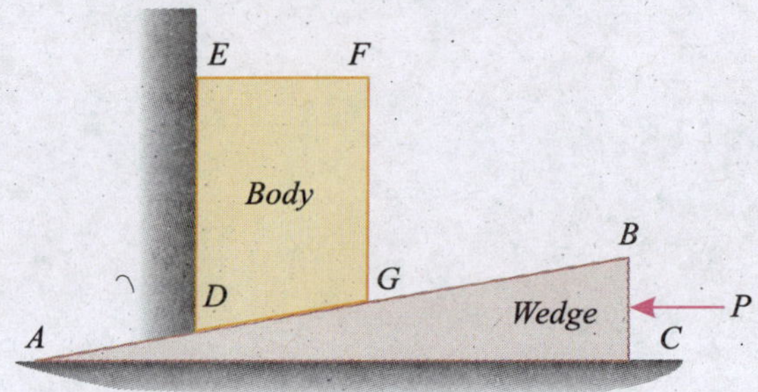

Fig. 10.

It will be interesting to know that the problems on wedges are basically the problems of equilibrium on inclined planes. Thus these problems may be solved either by the equilibrium method or by applying Lami's theorem. Now consider a wedge *ABC*, which is used to lift the body *DEFG*.

Let W = Weight fo the body *DEFG*,

P = Force required to lift the body, and

μ = Coefficient of friction on the planes *AB*, *AC* and *DE* such that $\tan \phi = \mu$.

A little consideration will show that when the force is sufficient to lift the body, the sliding will take place along three planes *AB*, *AC* and *DE* will also occur as shown in Fig. 11 (*a*) and (*b*).

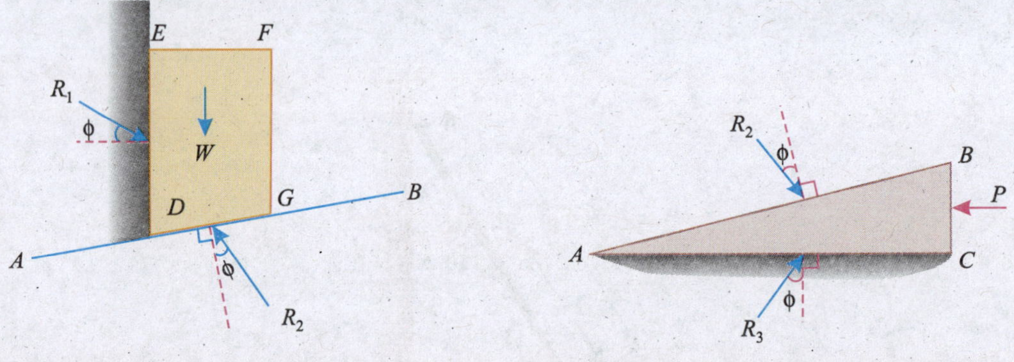

(*a*) Forces on the body *DEFG* (*a*) Forces on the wedge *ABC*

Fig. 11.

The three reactions and the horizontal force (*P*) may now be found out either by graphical method or analytical method as discussed below:

Graphical method

1. First of all, draw the space diagram for the body *DEFG* and the wedge *ABC* as shown in Fig. 12 (*a*). Now draw the reactions R_1, R_2 and R_3 at angle ϕ with normal to the faces *DE*, *AB* and *AC* respectively (such that $\tan \phi = \mu$).
2. Now consider the equilibrium of the body *DEFG*. We know that the body is in equilibrium under the action of

(a) Its own weight (W) acting downwards

(b) Reaction R_1 on the face DE, and

(c) Reaction R_2 on the face AB.

Now, in order to draw the vector diagram for the above mentioned three forces, take some suitable point l and draw a vertical line lm parallel to the line of action of the weight (W) and cut off lm equal to the weight of the body to some suitable scale. Through l draw a line parallel to the reaction R_1. Similarly, through m draw a line parallel to the reaction R_2, meeting the first line at n as shown in Fig. 12 (b).

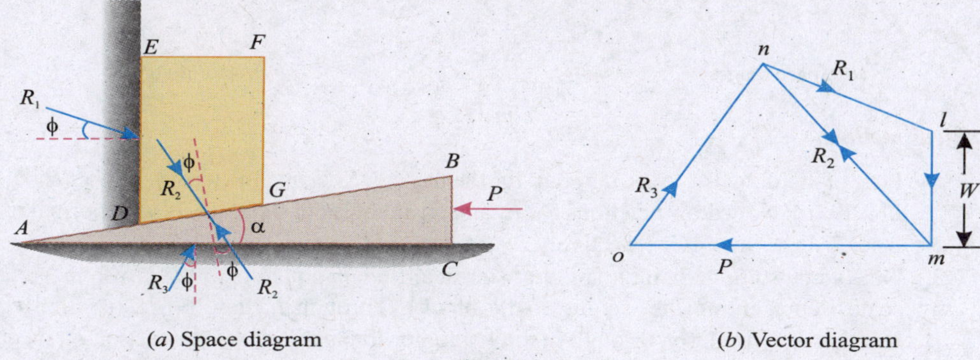

(a) Space diagram (b) Vector diagram

Fig. 12.

3. Now consider the equilibrium of the wedge ABC. We know that it is equilibrium under the action of

   (a) Force acting on the wedge (P),

   (b) Reaction R_2 on the face AB, and

   (c) Reaction R_3 on the face AC.

   Now, in order to draw the vector diagram for the above mentioned three forces, through m draw a horizontal line parallel to the force (P) acting on the wedge. Similarly, through n draw a line parallel to the reaction R_3 meeting the first line at O as shown in Fig. 12 (b).

4. Now the force (P) required on the wedge to raise the load will be given by mo to the scale.

Analytical method

1. First of all, consider the equilibrium of the body $DEFG$. And resolve the forces W, R_1 and R_2 horizontally as well as vertically.

2. Now consider the equilibrium of the wedge ABC. And resolve the forces P, R_2 and R_3 horizontally as well as vertically.

***Example 6.** *A block weighing 1500 N, overlying a 10° wedge on a horizontal floor and leaning against a vertical wall, is to be raised by applying a horizontal force to the wedge.*

Assuming the coefficient of friction between all the surface in contact to be 0.3, determine the minimum horizontal force required to raise the block. **[WBUT, JNTU, MU]**

Solution. Given: Weight of the block (W) = 1500 N; Angle of the wedge (α) = 10° and coefficient of friction between all the four surfaces of contact (μ) = 0.3 = tan ϕ or ϕ = 16.7°.

Let P = Minimum horizontal force required to raise the block.

The example may be solved graphically or analytically. But we shall solve it by both the methods.

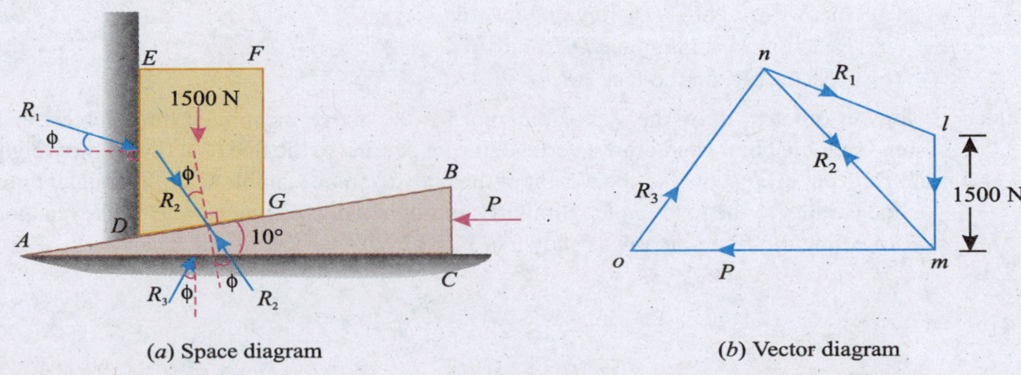

(a) Space diagram (b) Vector diagram

Fig. 13.

Graphical method

1. First of all, draw the space diagram for the block *DEFG* and the wedge *ABC* as shown in Fig. 13 (*a*). Now draw reactions R_1, R_2 and R_3 at angles of ϕ (*i.e.* 16.7° with normal to the faces *DE*, *AB* and *AC* respectively.
2. Take some suitable point *l*, and draw vertical line *lm* equal to 1500 N to some suitable scale (representing the weight of the block). Through *l*, draw a line parallel to the reaction R_1. Similarly, through *m* draw another line parallel to the reaction R_2 meeting the first line at *n*.
3. Now through *m*, draw a horizontal line (representing the horizontal force *P*). Similarly,through *n* draw a line parallel to the reaction R_3 meeting the first line at *O* as shown in Fig. 13(*b*).
4. Now measuring *mo* to the scale, we find that the required horizontal force

$$P = 1420 \text{ N.} \quad \textbf{Ans.}$$

Analytical method

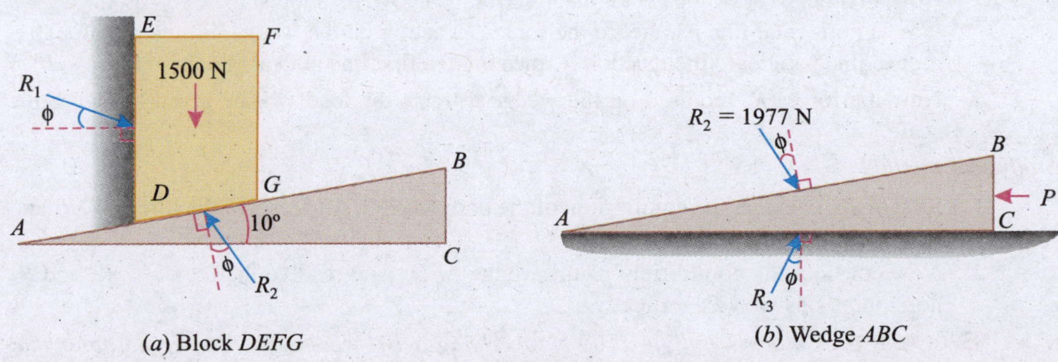

(a) Block *DEFG* (b) Wedge *ABC*

Fig. 14.

First of all, consider the equilibrium of the block. We know that it is in equilibrium under the action of the following forces as shown in Fig. 14 (*a*).

1. Its own weight 1500 N acting downwards.
2. Reaction R_1 on the face *DE*.
3. Reaction R_2 on the face *DG* of the block.

Resolving the forces horizontally and equating forces acting in left direction with forces acting in right direction,

$$R_1 \cos (16.7°) = R_2 \sin (10 + 16.7°) = R_2 \sin 26.7°$$

or $\quad R_1 \times 0.9578 = R_2 \times 0.4493$

or $\quad R_2 = 2.132\ R_1$

and now resolving the forces vertically and equating forces acting in downward direction with forces acting in upward direction,

$$(R_1 \times \sin(16.7°) + 1500 = R_2 \cos(10° + 16.7°) = R_2 \cos 26.7°$$

or $\quad (R_1 \times 0.2874) + 1500 = R_2 \times 0.8934$

$= (2.132\ R_1)\ 0.8934 = 1.905\ R_1$ $\quad$ ...(because $R_2 = 2.132\ R_1$)

or $\quad R_1(1.905 - 0.2874) = 1500$

or $\quad R_1 = \dfrac{1500}{1.6176} = 927.3$ N

and $R_2 = 2.132\ R_1 = 2.132 \times 927.3 = 1977$ N

Now consider the equilibrium of the wedge. We know that it is in equilibrium under the action of the following forces as shown in Fig. 14 (*b*).

1. Reaction R_2 of the block on the wedge.
2. Force (P) acting horizontally, and
3. Reaction R_3 on the face AC of the wedge.

Resolving the forces vertically and equating forces acting in upward direction with forces acting in downward direction,

$$R_3 \cos 16.7° = R_2 \cos(10° + 16.7°) = R_2 \cos 26.7°$$

or $\quad R_3 \times 0.9578 = R_2 \times 0.8934 = 1977 \times 0.8934 = 1766.2$

or $\quad R_3 = \dfrac{1766.2}{0.9578} = 1844$ N

and now resolving the forces horizontally and equating forces acting in left direction with forces acting in right direction,

$$P = R_2 \sin(10° + 16.7°) + R_3 \sin 16.7° = 1977 \sin 26.7° + 1844 \sin 16.7° \text{ N}$$

$$= (1977 \times 0.4493) + (1844 \times 0.2874) = 1418.3 \text{ N}$$ **Ans.**

***Example 7.** *A 15° wedge (A) has to be driven for tightening a body (B) loaded with 1000 N weight as shown in Fig. 15.*

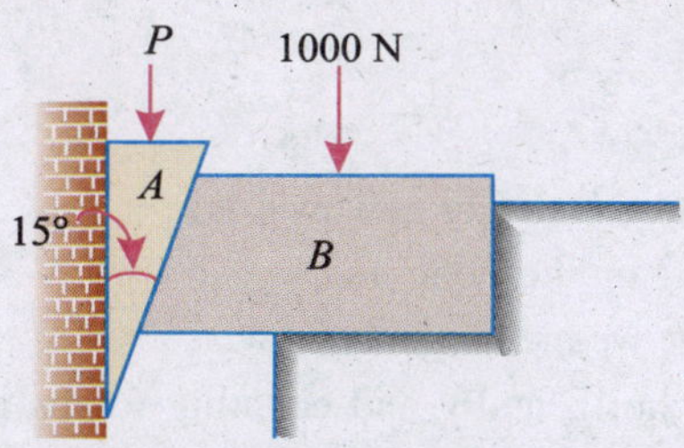

Fig. 15.

If the angle of friction for all the surfaces is 14°, find graphically the force (P), which should be applied to the wedge. Also check the answer analytically. **[GTU, MU]**

Solution. Given: Angle of the Wedge (α) = 15°; Weight acting on the body (W) = 1000 N and angle of friction for all the surfaces of contact (ϕ) = 14°.

Graphical solution

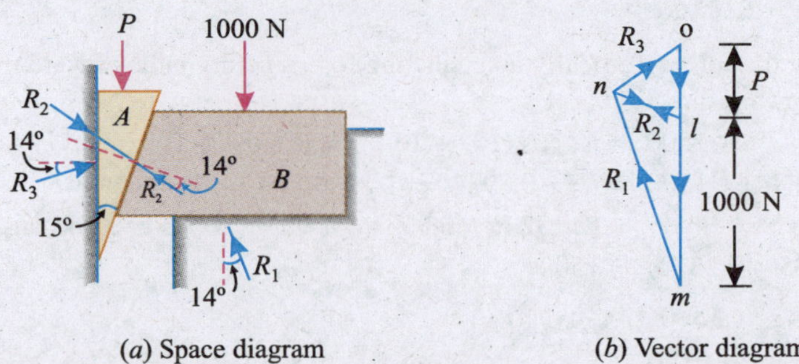

(*a*) Space diagram (*b*) Vector diagram

Fig. 16.

1. First of all, draw the space diagram for the body (*B*) and wedge (*A*) as shown in Fig. 16 (*a*). Now draw the reactions R_1, R_2 and R_3 at angles of 14° with normal to the faces.
2. Take some suitable point *l* and draw a vertical line *lm* equal to 1000 N to some suitable scale, representing the weight of the body. Through *l* draw a line parallel to the reaction R_2. Similarly, through *m* draw another line parallel to the reaction R_1 meeting first line at *n*.
3. Now through *l* draw a vertical line representing the vertical force (*P*). Similarly, through *n* draw a line parallel to the reaction R_3 meeting the first line at *O* as shown in Fig. 16 (*b*).
4. Now measuring *ol* to the scale, we find that the required vertical force, $P = 232$ N **Ans.**

Analytical check

First of all, consider equilibrium of the body. We know that it is in equilibrium under the action of the following forces as shown in Fig. 17 (*a*).

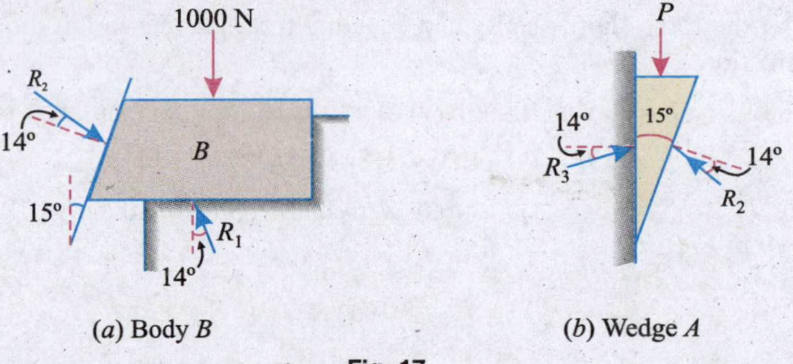

(*a*) Body *B* (*b*) Wedge *A*

Fig. 17.

1. Its own weight 1000 N acting downwards
2. Reaction R_1 acting on the floor, and
3. Reaction R_2 of the wedge on the body.

Resolving the forces horizontally and equating forces acting in left direction with forces acting in right direction,

$$R_1 \sin 14° = R_2 \cos (15° + 14°) = R_2 \cos 29°$$

or $$R_1 \times 0.2419 = R_2 \times 0.8746$$

$$\therefore \qquad R_1 = \frac{0.8746}{0.2419} R_2 = 3.616\ R_2$$

and now resolving the forces vertically and equating forces acting in downward direction with forces acting in upward direction,

$$R_2 \sin (15° + 14°) + 1000 = R_1 \cos 14°$$

or $$(R_2 \times 0.4848) + 1000 = R_1 \times 0.9703 = (3.616\ R_2)\ 0.9703 = 3.51\ R_2 \quad ...(\because R_1 = 3.616\ R_2)$$

or $$1000 = R_2\ (3.51 - 0.4848) = 3.0252\ R_2$$

or $$R_2 = \frac{1000}{3.0252} = 330.6 \text{ N}$$

Now consider equilibrium of the wedge. We know that it is in equilibrium under the action of the following forces as shown in Fig. 17. (*b*) :

1. Reaction R_2 of the body on the wedge,
2. Force (P) acting vertically downwards, and
3. Reaction R_3 on the vertical surface.

Resolving the forces horizontally and equating forces acting in left direction with forces acting in right direction,

$$R_3 \cos 14° = R_2 \cos (14° + 15°) = R_2 \cos 29°$$

or $$R_3 \times 0.9703 = R_2 \times 0.8746 = 330.6 \times 0.8746 = 289.1$$

or $$R_3 = \frac{289.1}{0.9703} = 297.9 \text{ N}$$

and now resolving the forces vertically and equating forces acting in downward direction with forces acting in upward direction,

$$P = R_3 \sin 14° + R_2 \sin (14° + 15°)$$

$$= (297.9 \times 0.2419) + (330.6 \times 0.4848) = 232.3 \text{ N} \qquad \textbf{Ans.}$$

EXERCISE 2

1. A block (A) of weight 5 kN is to be raised by means of a 20° wedge (B) by the application of a horizontal force (P) as shown in Fig. 18. The block A is constrained to move vertically by the application of a horizontal force (S). Find the magnitude of the forces F and S, when the coefficient of friction at the contact surfaces is 0.25.

[**Ans.** 4.62 kN; 3.77 kN]

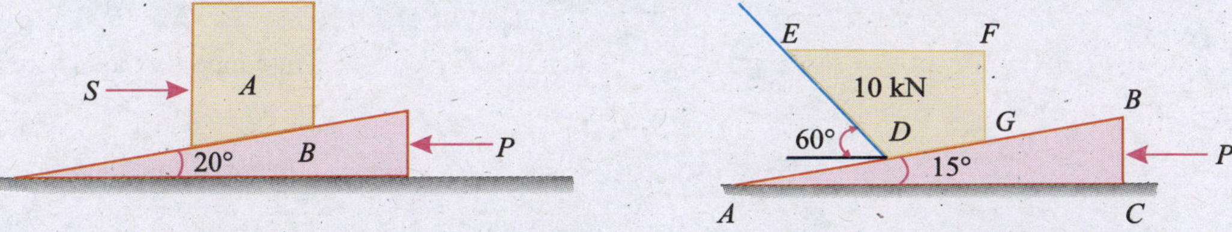

Fig. 18. **Fig. 19.**

*2. A block weighing 10 kN is to be raised against a surface, which is inclined at 60° with the horizontal by means of a 15° wedge as shown in Fig. 19.

Find graphically the horizontal force (P) which will just start the block to move, if the coefficient of friction between all the surfaces of contact be 0.2. Also check the answer analytically.

[MU] [**Ans.** 6 kN]

4. *SCREW FRICTION

The screws, bolts, studs, nuts etc. are widely used in various machines and structures for fastenings. These fastenings have screw threads, which are made by cutting a continuous helical groove on a cylindrical surface. If the threads are cut on the outer surface of a solid rod, these are known as external threads. But if the threads are cut on the internal surface of a hollow rod these are known as internal threads.

The screw threads are mainly of two types *viz.* *V*-threads and square threads. The *V*-threads are stronger and offer more frictional resistance to motion than square threads. Moreover, the *V*-threads have an advantage of preventing the nut from slackening. It will be interesting to know that the V-threads are used for the purpose of tightening pieces together (*e.g.* bolts and nuts etc.). Square threads are used in screw jacks, vice screws etc. which are used for lifting heavy loads. The following terms are important for the study of screws:

1. ***Helix.*** It is the curve traced by a particle, while describing a circular path at a uniform speed and advancing in the axial direction at a uniform rate. Or in other words, it is the curve traced by a particle while moving along a screw thread.
2. ***Pitch.*** It is the distance from one point of a thread to the corresponding point on the next thread. It is measured parallel to the axis of the screw.
3. ***Lead.*** It is the distance through which a screw thread advances axially in one turn.
4. ***Depth of thread.*** It is the distance between the top and bottom surfaces of a thread (also known as crest and root of thread).
5. ***Single-threaded screw.*** If the lead of a screw is equal to its pitch, it is known as single-threaded screw.
6. ***Multi-threaded screw.*** If more than one threads are cut in one lead distance of a screw, it is known as multi-threaded screw *e.g.* in a double-threaded screw, two threads are cut in one lead length. In such cases, all the threads run independently along the length of the rod. Mathematically,

$$\text{Lead} = \text{Pitch} \times \text{No. of threads.}$$

7. ***Slope of the thread.*** It is the inclination of the thread with horizontal. Mathematically,

$$\tan\alpha = \frac{\text{Lead of screw}}{\text{Circumference of screw}}$$

$$= \frac{p}{\pi d} \qquad \text{...(In single-threaded screw)}$$

$$= \frac{np}{\pi d} \qquad \text{...(In multi-threaded screw)}$$

where α = Angle of inclination of the thread,

p = Pitch of the screw,

d = Mean diameter of the screw, and

n = No. of threads in one lead.

5. RELATION BETWEEN EFFORT AND WEIGHT LIFTED BY A SCREW JACK

The screw jack is a device for lifting heavy loads, by applying a comparatively smaller effort at its handle. The principle, on which a screw jack works, is similar to that of an inclined plane.

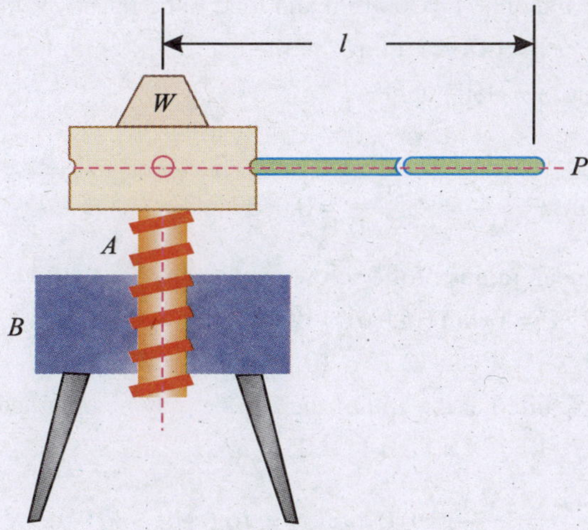

Fig. 20. Screw jack

Fig. 20 shows common form of a screw jack, which consists of a threaded rod A, called screw rod or simply screw. The screw has square threads, on its outer surface, which fit into the inner threads of the jack B. The load, to be raised or lowered, is placed on the head of the screw, which is rotated by the application of an effort at the end of the lever for lifting or lowering the load.

If one complete turn of a screw thread, be imagined to be unwound, from the body of the screw and developed, it will form an inclined plane as shown in Fig. 21

Let p = Pitch of the screw,

d = Mean diameter of the screw

r = Mean radius of the screw, and

α = Helix angle.

From the geometry of the figure, we find that

$$\tan \alpha = \frac{p}{\pi d} = \frac{p}{2\pi r} \qquad \text{...(where } d = 2r\text{)}$$

Fig. 21. Helix angle

Now let P = Effort applied at the mean radius of the screw jack to lift the load,

W = Weight of the body to be lifted,

and

μ = Coefficient of friction, between the screw and nut.

Let ϕ = Angle of friction, such that $\mu = \tan \phi$.

As a matter of fact, the principle, on which a screw jack works, is similar to that of an inclined plane. Thus the force applied on the lever of a screw jack is considered to be horizontal. We have already discussed (in Article 8.14 in chapter of Principles of Friction), that the horizontal force required to lift a load on an inclined rough plane

$$P = W \tan (\alpha + \phi)$$

***Example 8.** *A screw jack has mean diameter of 50 mm and pitch 10 mm. If the coefficient of friction between its screw and nut is 0.15, find the effort required at the end of 700 mm long handle to raise a load of 10 kN.*

Solution. Given: Mean diameter of screw jack (d) = 50 mm or radius (r) = 25 mm; Pitch of the screw (p) = 10 mm; Coefficient of friction between screw and nut (μ) = 0.15 = $\tan \phi$ or

$\phi = 8.5°$; Length of the handle (l) = 700 mm and load to be raised (W) = 10 kN.

Let P_1 = Effort required at the end of 700 mm long handle to raise the load,

and α = Helix angle

We know that,

$$\tan\alpha = \frac{p}{\pi d} = \frac{10}{\pi \times 50} = 0.0637 \quad \text{or} \quad \alpha = 3.6°$$

and effort required at mean radius of the screw to raise the load,

$$P = W\tan(\alpha + \phi) = W\tan(3.6° + 8.5°)$$
$$= W\tan 12.1° = 10 \times 0.2144 = 2.144 \text{ kN}$$

Now the effort required at the end of the handle may be obtained from the relation.

$$P_1 \times 700 = P \times r = 2.144 \times 25 = 53.6$$

or $$P_1 = \frac{53.6}{700} = 0.0766 \text{ kN} = 76.6 \text{ N}$$ **Ans.**

***Example 9.** *The mean radius of the screw of a square threaded screw jack is 25 mm. The pitch of thread is 7.5 mm. If the coefficient of friction is 0.12, what effort applied at the end of lever 60 cm length is needed to raise a weight of 2 kN.* **[AMIE]**

Solution. Given: Mean radius of the screw (r) = 25 mm; Pitch of the thread (p) = 7.5 mm; Coefficient of friction (μ) = 0.12 = $\tan\phi$; Length of the lever (l) = 60 cm and weight to be raised = 2 kN = 2000 N.

Let P_1 = Effort required at the end of the 60 cm long handle to raise the weight,

and α = Helix angle.

We know that

$$\tan\alpha = \frac{p}{2\pi r} = \frac{0.75}{2\pi \times 2.5} = 0.048$$

and effort required at mean radius of the screw to raise the weight,

$$P = W\tan(\alpha + \phi) = W \times \frac{\tan\alpha + \tan\phi}{1 - \tan\alpha . \tan\phi} \quad P = 2000 \times \frac{0.048 + 0.12}{1 - 0.048 \times 0.12}$$
$$= 2000 \times 0.169 = 338 \text{ N}$$

Now the effort applied at the end of the lever, may be found out from the relation,

or $$P_1 \times 60 = P \times 2.5 = 338 \times 2.5 = 845$$

or $$P_1 = \frac{845}{60} = 14.1 \text{ N}$$ **Ans.**

Example 10. *A screw press is used to compress books. The thread is a double thread (square head) with a pitch of 4 mm and a mean radius of 25 mm. The coefficient of the friction* (μ) *for the contact surface of the thread is 0.3. Find the torque for a pressure of 500 N.*

Solution. Given: No. of threads (n) = 2; Pitch (p) = 4 mm; Mean radius (r) = 25 mm ; Coefficient of friction (μ) = 0.3 = $\tan\phi$ or $\phi = 16.7°$ and pressure (W) = 500 N

Let α = Helix angle.

We know that, $$\tan\alpha = \frac{np}{2\pi r} = \frac{2 \times 4}{2\pi \times 25} = 0.0509 \quad \text{or} \quad \alpha = 2.9°$$

∴ Effort required at the mean radius of the screw to press the books

$$P = W \tan(\alpha + \phi) = 500 \tan(2.9° + 16.7°) \text{ N}$$
$$= 500 \tan 19.6° = 500 \times 0.356 = 178 \text{ N}$$

and torque required to press the books,

$$\text{Torque} = \text{Force} \times \text{Radius}$$

or $$T = P \times r = 178 \times 25 = 4450 \text{ N-mm} \quad \textbf{Ans.}$$

6. RELATION BETWEEN EFFORT AND WEIGHT LOWERED BY A SCREW JACK

We have already discussed in the last article that the principle, on which a screw works, is similar to that of an inclined plane. And force applied on the lever of a screw jack is considered to be horizontal. We have also discussed (in Article 8.14 in chapter of Principles of Friction) that the horizontal force required to lower a load on an inclined plane,

$$P = W \tan(\alpha - \phi) \quad \text{...(when } \alpha > \phi)$$
$$= W \tan(\phi - \alpha) \quad \text{...(when } \phi > \alpha)$$

Note. All the notations have the usual values as discussed in the last article.

Example 11. *A screw Jack has a square thread of 75 mm mean diameter and a pitch of 15 mm. Find the force, which is required at the end of 500 mm long lever to lower a load of 25 kN. Take coefficient of friction between the screw and thread as 0.05.*

Solution. Given: Mean diameter of thread (d) = 75 mm or radius (r) = 37.5 mm; Pitch of thread (p) = 15mm; Length of lever (l) = 500 mm; load to be lowered (W) = 25 kN and coefficient of friction between the screw and thread (μ) = 0.05 = $\tan \phi$ or $\phi = 2.9°$

Let P_1 = Effort required at end of 500 mm long handle to lower the load,

and α = Helix angle.

We know that, $$\tan \alpha = \frac{p}{\pi d} = \frac{15}{\pi \times 75} = 0.0637 \quad \text{or} \quad \alpha = 3.6°$$

and effort required at the mean radius of the screw to lower the load,

$$P = W \tan(\alpha - \phi) = W \tan(3.6° - 2.9°)$$
$$= W \tan 0.7° = 25 \times 0.0122 = 0.305 \text{ kN} = 305 \text{ N}$$

Now the effort required at the end of the handle may be found out from the relation,

or $$P_1 \times 500 = P \times r = 305 \times 37.5 = 11438$$

or $$P_1 = \frac{11438}{500} = 23 \text{ N} \quad \textbf{Ans.}$$

***Example 12.** *The screw of a jack is square threaded with two threads in a centimeter. The outer diameter of the screw is 5 cm. If the coefficient of friction is 0.1, calculate the force required to be applied at the end of the lever, which is 70 cm long (a) to lift a load of 4 kN, and (ii) to lower it.*

Solution. Given: Outer diameter of the screw (D) = 5 cm; Coefficient of friction (μ) = 0.1 = $\tan \phi$; Length of the lever (l) = 70 cm and load to be lifted (W) = 4 kN = 4000 N.

We know that as there are two threads in a cm, (*i.e.* $n = 2$) therefore pitch of the screw,

$$p = 1/2 = 0.5 \text{ cm}$$

and internal diameter of the screw,

$$= 5 - (2 \times 0.5) = 4 \text{ cm}$$

∴ Mean diameter of the screw,

$$d = \frac{5+4}{2} = 4.5 \text{ cm}$$

Let α = Helix angle.

We know that $\tan\alpha = \dfrac{p}{\pi d} = \dfrac{0.5}{\pi \times 4.5} = 0.0353$

(i) *Force required at the end of 70 cm long lever to lift the load*

Let P_1 = Force required at the end of the lever to lift the load.

We know that the force required to be applied at the mean radius to lift the load,

$$P = W \tan(\alpha + \phi) = W \times \frac{\tan\alpha + \tan\phi}{1 - \tan\alpha . \tan\phi}$$

$$= 4000 \times \frac{0.0353 + 0.1}{1 - 0.0353 \times 0.1} = 543.1 \text{ N}$$

Now the force required at the end of the lever may be found out from the relation,

$$P_1 \times 70 = P \times \frac{d}{2} = 543.1 \times \frac{4.5}{2} = 1222$$

or $$P_1 = \frac{1222}{7.0} = 17.5 \text{ N} \quad \textbf{Ans.}$$

(ii) *Force required at the end of 70 cm long lever to lower the load*

Let P_2 = Force required at the end of the lever to lower the load.

We know that the force required at the mean radius to lower the load,

$$P = W \tan(\phi - \alpha) = 4000 \times \frac{\tan\phi - \tan\alpha}{1 + \tan\phi \tan\alpha}$$

$$= 4000 \times \frac{0.1 - 0.0353}{1 + 0.1 \times 0.0353} = 257.9 \text{ N}$$

Now the force required at the end of the lever may be found out from the relation:

$$P_2 \times 70 = P \times \frac{d}{2} = 257.9 \times \frac{4.5}{2} = 580.3$$

$\therefore$ $$P_2 = \frac{580.3}{70} = 8.3 \text{ N} \quad \textbf{Ans.}$$

7. *EFFICIENCY OF A SCREW JACK

We have seen in Article 6 of this chapter that the effort (P) required at the mean radius of a jack to lift the load (W),

$$P = W \tan(\alpha + \phi) \qquad ...(i)$$

where α = Helix angle, and

μ = Coefficient of friction between the screw and the nut,

$= \tan\phi$ ...(where ϕ = Angle of friction)

If there would have been no friction between the screw and the nut, then ϕ will be zero. In such a case, the value of effort (P_0) necessary to raise the same load, will be given by the equation :

$$P_0 = W \tan\alpha \qquad \text{...[substituting } \phi = 0 \text{ in equation } (i)]$$

$$\therefore \quad \text{Efficiency } (\eta) = \frac{\text{Ideal effort}}{\text{Actual effort}} = \frac{P_0}{P} = \frac{W\tan\alpha}{W\tan(\alpha + \phi)} = \frac{\tan\alpha}{\tan(\alpha + \phi)}$$

It shows that the efficiency of a screw jack is independent of the weight lifted or effort applied.

The above equation for the efficiency of a screw jack may also be written as:

$$\eta = \frac{\dfrac{\sin\alpha}{\cos\alpha}}{\dfrac{\sin(\alpha+\phi)}{\cos(\alpha+\phi)}} = \frac{\sin\alpha\times\cos(\alpha+\phi)}{\cos\alpha\times\sin(\alpha+\phi)}$$

or

$$1-\eta = 1-\frac{\sin\alpha\times\cos(\alpha+\phi)}{\cos\alpha\times\sin(\alpha+\phi)} = \frac{\cos\alpha\,\sin(\alpha+\phi)-\sin\alpha\,\cos(\alpha+\phi)}{\cos\alpha\sin(\alpha+\phi)}$$

$$= \frac{\sin\phi}{\cos\alpha\,\sin(\alpha+\phi)} \quad [\because \sin(A-B) = \sin A\cos B - \cos A\sin B]$$

$$1-\eta = \frac{2\sin\phi}{2\cos\alpha\times\sin(\alpha+\phi)} \quad \text{[Multiplying and dividing by 2]}$$

$$= \frac{2\sin\phi}{\sin(2\alpha+\phi)+\sin\phi}$$

$$\ldots[\because\ 2\sin A\cos B = \sin(A+B)+\sin(A-B)]$$

Now for the efficiency to be maximum, the term $(1-\eta)$ should be the least. Or in other words, the value of $\sin(2\alpha+\phi)$ should be the greatest. This is only possible, when

$$2\alpha+\phi = 90° \quad \text{or} \quad 2\alpha = 90°-\phi$$

or

$$\alpha = 45° - \frac{\phi}{2}$$

It shows that the maximum efficiency of a screw jack is also independent of the weight lifted or effort applied.

***Example 13.** *A load of 2.5 kN is to be raised by a screw jack with mean diameter of 75 mm and pitch of 12 mm. Find the efficiency of the screw jack, if the coefficient of friction between the screw and nut is 0.075.* **[AMIE]**

Solution. Given: Load (W) = 2.5 kN ; Mean diameter of the screw (d) = 75 mm; Pitch of the screen (p) = 12 mm and coefficient of friction between the screw and nut $(\mu) = 0.075 = \tan\phi$.

We know that $\tan\alpha = \dfrac{p}{\pi d} = \dfrac{12}{\pi\times 75} = 0.051$

and efficiency of the screw jack,

$$\eta = \frac{\tan\alpha}{\tan(\alpha+\phi)} = \frac{\tan\alpha}{\dfrac{\tan\alpha+\tan\phi}{1-\tan\alpha\tan\phi}} = \frac{0.051}{\dfrac{0.051+0.075}{1-(0.051\times0.075)}} = \frac{0.051}{0.1265}$$

$$= 0.403 = 40.3\% \quad \textbf{Ans.}$$

Example 14. *A screw jack has a square thread of 75 mm mean diameter and 15 mm pitch. The load on the jack revolves with the screws. The coefficient of friction at the screw thread is 0.05. (i) Find the tangential force to be applied to the jack at 360 mm radius, so as to lift a load of 6 kN weight. (ii) State whether the jack is selflocking. If it is, find the torque necessary to lower the load. If not, find the torque which must be applied to keep the load from descending.*

Solution. Given: Mean diameter of square thread (d) = 75 mm or mean radius (r) = 37.5 mm; Pitch (p) = 15 mm; Coefficient of friction $(\mu) = 0.05 = \tan\phi$; Radius of effort arm = 360 mm and load lifted = 6 kN = 6000 N.

(i) ***Tangential force to be applied at the jack.***

Let P_1 = Tangential force to be applied at 36 cm radius to lift the load, and

α = Helix angle.

We know that, $$\tan\alpha = \frac{p}{\pi d} = \frac{15}{\pi \times 75} = 0.064$$

and tangential force required at the mean radius to lift the load,

$$P = W\tan(\alpha + \phi) = W \times \frac{\tan\alpha + \tan\phi}{1 - \tan\alpha.\tan\phi}$$

$$= 6000 \times \frac{0.064 + 0.05}{1 - 0.064 \times 0.05} = 686.2 \text{ N}$$

Now the effort applied at a radius of 36 cm may be found out from the relation

$$P_1 \times 360 = P \times r = 686.2 \times 37.5 = 25\,732$$

or $$P_1 = \frac{25\,732}{360} = 71.48 \text{ N} \quad \textbf{Ans.}$$

(ii) ***Self-locking of the screw jack***

We know that efficiency of the screw jack,

$$\eta = \frac{\tan\alpha}{\tan(\alpha + \phi)} = \frac{\tan\alpha}{\dfrac{\tan\alpha + \tan\phi}{1 - \tan\alpha.\tan\phi}} = \frac{0.064}{\dfrac{0.064 + 0.05}{1 - (0.064 \times 0.05)}} = \frac{0.064}{0.1144}$$

$$= 0.559 = 55.9\%$$

Since efficiency of the jack is more than 50%, therefore, it is not *self-locking. **Ans.**

(*Note: For details, please refer to article of Self Locking Machine (10.14) in chapter of Principles of Lifting Machines.)

Torque, which must be applied to keep the load from descending

We know that the force which must be applied at the mean radius to keep the load from descending (*i.e.* to prevent the load from descending).

$$P_2 = W\tan(\alpha - \phi) = W \times \frac{\tan\alpha - \tan\phi}{1 + \tan\alpha.\tan\phi}$$

$$= 6000 \times \frac{0.064 - 0.05}{1 + 0.064 \times 0.05} = 83.73$$

∴ Torque, which must be applied to keep the load from descending

$$= P_2 \times r = 83.73 \times 37.5 = 3140 \text{ N-mm} \quad \textbf{Ans.}$$

EXERCISE 3

1. A square threaded screw jack of mean diameter 25 mm and a pitch of 6 mm is used to lift the load of 1500 N. Find the force required at the mean circumference if the coefficient of friction between the screw and nut is 0.02. **[Ans.** 144.2 N]
2. A square threaded screw jack of mean diameter 50 mm has 3° angle of inclination of the thread and coefficient of friction 0.06. Find the effort required at the end of handle 450 mm long lever *(i)* to raise a load of 20 kN, and *(ii)* to lower the same load. **[Ans.** 126 N; 8.4 N]

QUESTIONS

1. What is a wedge? State its uses and the method of solving the problems on wedge friction.
2. What is a screw jack? Explain the principle, on which it works.
3. Establish a relation between the effort and load, when a square threaded screw is used for lifting purposes, taking friction into account.
4. Derive a relation for the efficiency of a screw jack, taking friction into account.
5. In a screw jack, the helix angle is α and the angle of friction is ϕ. Show that its efficiency is maximum, when $2\alpha = 90° - \phi$.

OBJECTIVE TYPE QUESTIONS

1. If a ladder is not in equilibrium against a smooth vertical wall, then it can be made in equilibrium by
   (*a*) increasing the angle of inclination.
   (*b*) decreasing the angle of inclination.
   (*c*) increasing the length of the ladder.
   (*d*) decreasing the length of the ladder.
2. The efficiency of a screw jack may be increased by
   (*a*) increasing its pitch.
   (*b*) decreasing its pitch.
   (*c*) increasing the load to be lifted.
   (*d*) decreasing the load to be lifted.
3. The efficiency of a screw jack is maximum when the helix angle is equal to

   (*a*) $45° + \frac{\phi}{2}$ (*b*) $45° - \frac{\phi}{2}$ (*c*) $\frac{\phi}{2} + 30°$ (*d*) $\frac{\phi}{2} - 30°$

ANSWERS

1. (*a*) 2. (*a*) 3. (*b*)

CHAPTER

10 Principles of Lifting Machines

Contents

1. INTRODUCTION

In olden times, thousands of slaves had to be arranged, whenever a heavy load had to be lifted or dragged. Even today, in the absence of a suitable device, many people have to be arranged to lift a motor car so that its tyres can be changed. In order to overcome such difficulties, a few simple machines were invented, which could save the man power *i.e.*, a single man can do the same work as many could do, though at a lesser speed.

Before entering into the details of simple machines and their working, following terms should be clearly understood at this stage :

2. SIMPLE MACHINE

In a broad sense, a simple machine may be defined as a device, which enables us to do some useful work at some point or to overcome some resistance, when an effort or force is applied to it, at some other convenient point.

3. COMPOUND MACHINE

A compound machine may be defined as a device, consisting of a number of simple machines, which enables us to do some useful work at a faster speed or with a much less effort as compared to a simple machine.

4. LIFTING MACHINE

It is a device, which enables us to lift a heavy load (W) by applying a comparatively smaller effort (P).

5. MECHANICAL ADVANTAGE

The mechanical advantage (briefly written as M.A.) is the ratio of weight lifted (W) to the effort applied (P) and is always expressed in pure number. Mathematically,

Mechanical Advantage,

$$\text{M.A.} = \frac{W}{P}$$

6. INPUT OF A MACHINE

The input of a machine is the work done on the machine. In a lifting machine, it is measured by the product of effort and the distance through which it has moved.

7. OUTPUT OF A MACHINE

The output of a machine is the actual work done by the machine. In a lifting machine, it is measured by the product of the weight lifted and the distance through which it has been lifted.

8. EFFICIENCY OF A MACHINE

It is the ratio of output to the input of a machine and is generally expressed as a percentage. Mathematically, Efficiency,

$$\eta = \frac{\text{Output}}{\text{Input}} \times 100$$

9. IDEAL MACHINE

If the efficiency of a machine is 100% *i.e.*, if the output is equal to the input, the machine is called as a perfect or an *ideal machine.*

10. VELOCITY RATIO

[RGPV]

The velocity ratio (briefly written as V.R.) is the ratio of distance moved by the effort (y) to the distance moved by the load (x) and is always expressed in pure number. Mathematically,

Velocity Ratio,

$$\text{V.R.} = \frac{y}{x}$$

11. RELATION BETWEEN EFFICIENCY, MECHANICAL ADVANTAGE AND VELOCITY RATIO OF A LIFTING MACHINE

It is an important relation of a lifting machine, which throws light on its mechanism. Now consider a lifting machine, whose efficiency is required to be found out.

Let W = Load lifted by the machine,

P = Effort required to lift the load,

Y = Distance moved by the effort, in lifting the load, and

x = Distance moved by the load.

We know that $\text{M.A.} = \dfrac{W}{P} = W/P$ and $\text{V.R.} = \dfrac{y}{x} = y/x$

We also know that input of a machine

= Effort applied × Distance through which the effort has moved

$= P \times y$ ...(*i*)

and output of a machine = Load lifted × Distance through which the load has been lifted

$= W \times x$ ...(*ii*)

$\therefore$ Efficiency, $$\eta = \frac{\text{Output}}{\text{Input}} = \frac{W \times x}{P \times y} = \frac{W/P}{y/x} = \frac{\text{M.A.}}{\text{V.R.}}$$

Note. It may be seen from the above relation that the values of M.A. and V.R. are equal only in case of a machine whose efficiency is 100%. But in actual practice, it is not possible.

Example 1. *In a certain weight lifting machine, a weight of 1 kN is lifted by an effort of 25 N. While the weight moves up by 100 mm, the point of application of effort moves by 8 m. Find mechanical advantage, velocity ratio and efficiency of the machine.* **[DTU]**

Solution. Given: Weight (W) = 1 kN = 1000 N ; Effort (P) = 25 N ; Distance through which the weight is moved (x) = 100 mm = 0.1 m and distance through which effort is moved (y) = 8 m.

Mechanical advantage of the machine

We know that mechanical advantage of the machine

$$\text{M.A.} = \frac{W}{P} = \frac{1000}{25} = 40 \quad \textbf{Ans.}$$

Velocity Ratio of the machine

We know that velocity ratio of the machine

$$\text{V.R.} = \frac{y}{x} = \frac{8}{0.1} = 80 \quad \textbf{Ans.}$$

Efficiency of the machine

We also know that efficiency of the machine,

$$\eta = \frac{\text{M.A.}}{\text{V.R.}} = \frac{40}{80} = 0.5 = 50\% \quad \textbf{Ans.}$$

12. REVERSIBILITY OF A MACHINE

Sometimes, a machine is also capable of doing some work in the reversed direction, after the effort is removed. Such a machine is called a reversible machine and its action is known as *reversibility of the machine.*

13. CONDITION FOR THE REVERSIBILITY OF A MACHINE

[GTU, DTU]

Consider a reversible machine, whose condition for the reversibility is required to be found out.

Let W = Load lifted by the machine,

P = Effort required to lift the load,

y = Distance moved by the effort, and

x = Distance moved by the load.

Input of the machine $= P \times y$ ...(*i*)

Output of the machine $= W \times x$ ...(*ii*)

Machine friction $= \text{Input} - \text{Output} = (P \times y) - (W \times x)$ ...(*iii*)

A little consideration will show that in a reversible machine, the *output of the machine should be more than the machine friction, when the effort (P) is zero. *i.e.*,

$$W \times x > P \times y - W \times x$$

or $$2\,W \times x > P \times y$$

or $$\frac{W \times x}{P \times y} > \frac{1}{2}$$

or $$\frac{\frac{W}{P}}{\frac{y}{x}} > \frac{1}{2}$$

or $$\frac{\text{M.A.}}{\text{V.R.}} > \frac{1}{2} \qquad \ldots\left(\because \frac{W}{P} = \text{M.A.} \quad \text{and} \quad \frac{y}{x} = \text{V.R.}\right)$$

$\therefore$ $$\eta > \frac{1}{2} = 0.5 = 50\%$$

Hence the condition for a machine, to be reversible, is that its efficiency should be *more than 50%.*

***Note:** As a matter of fact, a machine will work only when the input is more than the machine friction. When the machine has to work in the reverse order (as per condition of a reversible machine) its output (which will act as an input, $P = 0$) should be more than the machine friction.

14. SELF-LOCKING MACHINE

[RGPV, GTU]

Sometimes, a machine is not capable of doing any work in the reversed direction, after the effort is removed. Such a machine is called a non-reversible or *self-locking machine.* A little consideration will show, that the condition for a machine to be non-reversible or self-locking is that its efficiency should *not be more than* 50%.

Example 2. *A certain weight lifting machine of velocity ratio 30 can lift a load of 1500 N with the help of 125 N effort. Determine if the machine is reversible.*

Solution. Given: Velocity ratio (V.R.) = 30; Load (W) = 1500 N and effort (P) = 125 N.

Mechanical Advantage,

$$\text{M.A.} = \frac{W}{P} = \frac{1500}{125} = 12$$

Efficiency, $$\eta = \frac{\text{M.A.}}{\text{V.R.}} = \frac{12}{30} = 0.4 = 40\%$$

Since efficiency of the machine is less than 50%, therefore the machine is non-reversible. **Ans.**

Example 3. *In a lifting machine, whose velocity ratio is 50, an effort of 100 N is required to lift a load of 4 kN. Is the machine reversible ? If so, what effort should be applied, so that the machine is at the point of reversing ?*

Solution. Given: Velocity ratio (V.R.) = 50 ; Effort (P) = 100 N and load (W) = 4 kN = 4000 N.

Reversibility of the machine

Mechanical Advantage,

$$\text{M.A.} = \frac{W}{P} = \frac{4000}{100} = 40$$

and Efficiency,

$$\eta = \frac{\text{M.A.}}{\text{V.R.}} = \frac{40}{50} = 0.8 = 80\%$$

Since efficiency of the machine is more than 50%, therefore the machine is reversible. **Ans.**

Effort to be applied

A little consideration will show that the machine will be at the point of reversing, when its efficiency is 50% or 0.5.

Let P_1 = Effort required to lift a load of 4000 N when the machine is at the point of reversing.

Mechanical Advantage, $$\text{M.A.} = \frac{W}{P_1} = \frac{4000}{P_1} = 4000 / P_1$$

Efficiency, $$0.5 = \frac{\text{M.A.}}{\text{V.R.}} = \frac{4000/P_1}{50} = \frac{80}{P_1}$$

$\therefore$ $$P_1 = \frac{80}{0.5} = 160 \text{ N} \qquad \textbf{Ans.}$$

15. FRICTION IN A MACHINE

In previous articles, we have not taken the friction of the lifting machines into account. Or in other words, we have assumed every machine to be frictionless. But this is a rare phenomenon, as every machine cannot be frictionless. It has been observed that there is always some amount of friction present in every machine, which can be expressed on a graph of effort (P) and load or weight lifted (W). If we record the various values of efforts required to raise the corresponding loads or weights and plot a graph between effort and load, we shall obtain a straight line AB as shown in Fig. 1.

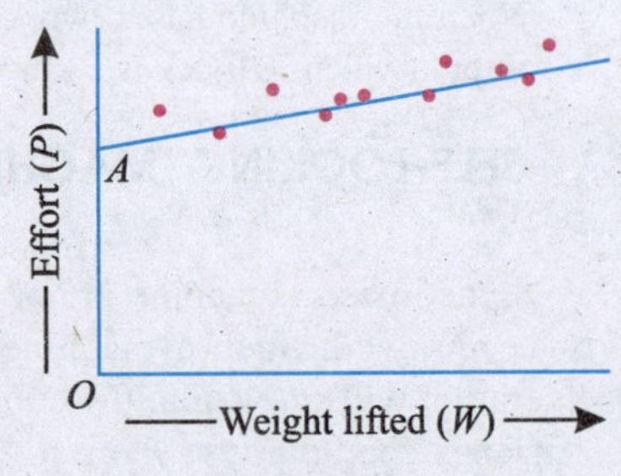

Fig. 1.

It may be noted from the graph that the intercept OA represents the amount of friction present in the machine. The machine friction may be expressed either on the effort side or on the load side. If expressed on the effort side, the friction may be defined as an additional effort required to overcome the frictional froce. But if expressed on the load side, the friction may be defined as the additional load that can be lifted or the additional resistance that can be overcome. Now consider a lifting machine having some amount of friction.

Let P = Actual effort (considering the machine friction) required to lift a weight,

P' = Ideal weight (neglecting the machine friction) required to lift the same weight,

W = Actual weight (considering machine friction) lifted by an effort and

W' = Ideal weight (neglecting machine friction) lifted by the same effort.

A little consideration will show, that in the above equations (P) is greater than (P'). Similarly, (W') is greater than (W). It is thus obvious, from the above equations, that ($P - P'$) is the amount of effort required to overcome the machine friction and ($W' - W$) is the load equivalent to the machine friction.

It may be noted, that in the case of an ideal machine, to lift a weight W, the effort required is P' only; whereas to lift a load W' effort required is P. But in the case of an actual machine to lift a weight W, effort required is P.

We know that, Efficiency of the machine,

$$\eta = \frac{\text{M.A.}}{\text{V.R.}} = \frac{W / P}{\text{V.R.}} \qquad ...(i)$$

In case of an ideal machine, the Efficiency is equal to 1. Substituting this value of Efficiency equal to 1 equation (i),

$$\frac{W}{P} = \text{V.R.} \qquad ...(ii)$$

As already discussed, to lift a load W, in the case of an ideal machine, the effort required is P'. Now substituting P' instead of P in equation (ii),

$$\frac{W}{P'} = \text{V.R.}$$

or

$$P' = \frac{W}{\text{V.R.}} \qquad ...(iii)$$

Now the effort (P) is required to lift the load (W), when the friction is considered and the effort (P') is required to lift the same load when the mchine is considered to be ideal *i.e.*, when the friction is neglected.

$$\therefore \quad P - P' = P - \frac{W}{\text{V.R.}}$$

We know that ($P - P'$) is the friction, if expressed in terms of effort.

$$\therefore \quad F_{(\text{effort})} = P - \frac{W}{\text{V.R.}} \qquad ...(iv)$$

Now the effort (P) will lift a load (W') if the mchine friction is neglected and the same effort will lift a load (W) if the machine friction is considered. Therefore substituting W' for W (considering the machine to be ideal, *i.e.*, neglecting the machine friction) in equation (ii),

$$\frac{W'}{P} = \text{V.R.} \qquad \text{or} \qquad W' = P \times \text{V.R.}$$

$$\therefore \quad W' - W = (P \times \text{V.R.}) - W$$

We know that ($W' - W$) is the friction, if expressed in terms of load.

$$\therefore \quad F_{(\text{load})} = (P \times \text{V.R.}) - W \qquad ...(v)$$

Example 4. *In a certain machine, an effort of 100 N is just able to lift a load of 840 N, Calculate efficiency and friction both on effort and load side, if the velocity ratio of the machine is 10.*

Solution. Given: Effort (P) = 100 N ; Load (W) = 840 N and velocity ratio (V.R.) = 10.

Efficiency of the machine

Mechanical Advantage, $\text{M.A.} = \frac{W}{P} = \frac{840}{100} = 8.4$

and Efficiency, $\eta = \frac{\text{M.A.}}{\text{V.R.}} = \frac{8.4}{10} = 0.84 = 84\%$ **Ans.**

Friction of the machine

Friction of the machine in terms of effort,

$$F_{(effort)} = P - \frac{W}{\text{V.R.}} = 100 - \frac{840}{10} = 16 \text{ N} \qquad ...(i)$$

Friction of the machine in terms of load,

$$F_{(load)} = (P \times \text{V.R.}) - W = (100 \times 10) - 840 = 160 \text{ N} \qquad ...(ii)$$

It may be noted from equations (*i*) and (*ii*) that an effort of 16 N is required to overcome the friction. Or in other words, this effort can lift an additional load of 160 N **Ans.**

EXERCISE 1

1. A load of 120 N is raised by means of a certain weight lifting machine through a distance of 200 mm. If the effort applied is 20 N and has moved through a distance of 1.5 m, find the efficiency of the machine. **[Ans.** 80%]
2. In a weight lifting machine, an effort of 50 N is required to lift a load (*W*). The distances moved by the load and effort are 20 mm and 500 mm respectively. Determine the magnitude of the load (*W*), if the efficiency of the machine is 80%. **[Ans.** 1 kN]
3. In a weight lifting machine, whose velocity ratio is 20, a weight of 1 kN can be raised by an effort of 80 N. If the effort is removed, show that the machine can work in the reverse direction.

   Hint. M.A. = W/P = 1000/80 = 12.5

   and η = M.A./V.R. = 12.5/20 = 0.625 = 62.5%.

   Since efficiency is more than 50%, therefore the machine can work in the reverse direction. **Ans.**
4. In a certain weight lifting machine, an effort of 25 N can lift a load of 315 N. If the velocity ratio of the machine is 14, find the effort lost in friction and the frictional load. **[Ans.** 2.5 N; 3.5 N]

16. LAW OF A MACHINE

[RGPV, GTU]

The term 'law of a machine' may be defined as relationship between the effort applied and the load lifted. Thus for any machine, if we record the various efforts required to raise the corresponding loads, and plot a graph between effort and load, we shall get a straight line *AB* as shown in Fig. 2.

We also know that the intercept *OA* represents the amount of friction offered by the machine. Or in other words, this is the effort required by the machine to overcome the friction, before it can lift any load.

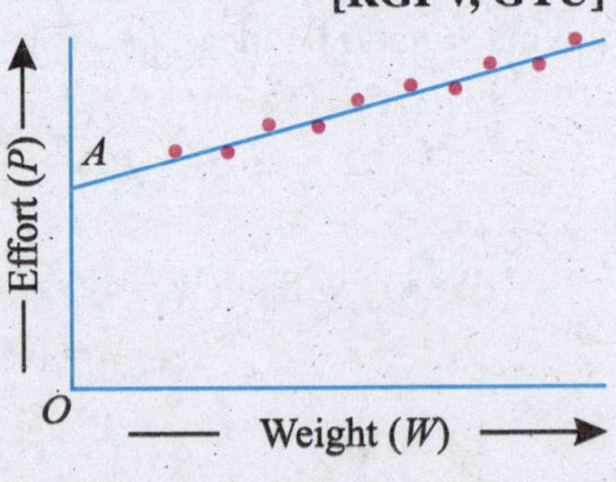

Fig. 2.

Mathematically, the law of a lifting machine is given by the relation :

$$P = mW + C$$

where P = Effort applied to lift the load,

m = A constant (called coefficient of friction) which is equal to the slope of the line *AB*,

W = Load lifted, and

C = Another constant, which represents the machine friction, (*i.e. OA*).

Example 5. *What load can be lifted by an effort of 120 N, if the velocity ratio is 18 and efficiency of the machine at this load is 60% ?*

Determine the law of the machine, if it is observed that an effort of 200 N is required to lift a load of 2600 N and find the effort required to run the machine at a load of 3.5 kN. **[UPTU]**

Solution. Given: Effort (P) = 120 N ; Velocity ratio (V.R.) = 18 and efficiency (η) = 60% = 0.6.

Load lifted by the machine

Let W = Load lifted by the machine.

$$\text{Mechanical Advantage, M.A.} = \frac{W}{P} = \frac{W}{120} = W/120$$

$$\text{and Efficiency, } 0.6 = \frac{\text{M.A.}}{\text{V.R.}} = \frac{W/120}{18} = \frac{W}{2160}$$

$$\text{or} \qquad W = 0.6 \times 2160 = 1296 \text{ N} \qquad \textbf{Ans.}$$

Law of the machine

In the second case, P = 200 N and W = 2600 N

Substituting the two values of P and W in the law of the machine, *i.e.*, $P = m\,W + C$,

$$120 = m \times 1296 + C \qquad ...(i)$$

$$\text{and} \qquad 200 = m \times 2600 + C \qquad ...(ii)$$

Subtracting equation (*i*) from (*ii*),

$$80 = 1304\, m \qquad \text{or} \qquad m = \frac{80}{1304} = 0.06$$

and now substituting the value of m in equation (*ii*)

$$200 = (0.06 \times 2600) + C = 156 + C$$

$$\text{or} \qquad C = 200 - 156 = 44$$

Now substituting the value of m = 0.06 and C = 44 in the law of the machine,

$$P = 0.06\, W + 44 \qquad \textbf{Ans.}$$

Effort required to run the machine at a load of 3.5 kN

Substituting the value of W = 3.5 kN or 3500 N in the law of machine,

$$P = (0.06 \times 3500) + 44 = 254 \text{ N} \qquad \textbf{Ans.}$$

Example 6. *In a lifting machine, an effort of 40 N raised a load of 1 kN. If efficiency of the machine is 0.5, what is its velocity ratio ? If on this machine, an effort of 74 N raised a load of 2 kN, what is now the efficiency ? What will be the effort required to raise a load of 5 kN ?*

Solution. Given: When Effort (P) = 40 N; Load (W) = 1 kN = 1000 N; Efficiency (η) = 0.5; When effort (P) = 74 N and load (W) = 2 kN = 2000 N.

Velocity Ratio when Efficiency is 0.5

$$\text{Mechanical Advantage, M.A.} = \frac{W}{P} = \frac{1000}{40} = 25$$

$$\text{and Efficiency, } 0.5 = \frac{\text{M.A.}}{\text{V.R.}} = \frac{25}{\text{V.R.}}$$

$$\text{and Velocity Ratio, V.R.} = \frac{25}{0.5} = 50 \qquad \textbf{Ans.}$$

Efficiency when P is 74 N and W is 2000 N

Mechanical Advantage. $\text{M.A.} = \frac{W}{P} = \frac{2000}{74} = 27$

and Efficiency, $\eta = \frac{\text{M.A.}}{\text{V.R.}} = \frac{27}{50} = 0.54 = 54\%$ **Ans.**

Effort required to raise a load of 5 kN or 5000 N

Substituting the two values of P and W in the law of the machine, *i.e.* $P = mW + C$

$$40 = m \times 1000 + C \quad \text{...(i)}$$

and

$$74 = m \times 2000 + C \quad \text{...(ii)}$$

Subtracting equation (*i*) from (*ii*),

$$34 = 1000\,m \qquad \text{or} \qquad m = \frac{34}{1000} = 0.034$$

and now substituting this value of m in equation (*i*),

$$40 = (0.034 \times 1000) + C = 34 + C$$

or

$$C = 40 - 34 = 6$$

Substituting these values of $m = 0.034$ and $C = 6$ in the law of machine,

$$P = 0.034\,W + 6 \quad \text{...(iii)}$$

∴ Effort required to raise a load of 5000 N,

$$P = (0.034 \times 5000) + 6 = 176 \text{ N}$$ **Ans.**

Example 7. *What load will be lifted by an effort of 12 N, if the velocity ratio is 18 and efficiency of the machine at this load is 60 % ?*

If the machine has a constant friction resistance, determine the law of the machine and find the effort required to run this machine at (i) *no load, and* (ii) *a load of 900 N.*

Solution. Given: Effort (P) = 12 N ; Velocity ratio (V.R.) = 18 and efficiency (η) = 60 % = 0.6.

Load lifted by the machine

Let W = Load lifted by the machine,

Mechanical Advantage, $\text{M.A.} = \frac{W}{P} = \frac{W}{12} = W/12$

and Efficiency, $0.6 = \frac{\text{M.A.}}{\text{V.R.}} = \frac{W/12}{18} = \frac{W}{216}$

or $W = 0.6 \times 216 = 129.6$ N **Ans.**

Law of the machine

We know that, Effort lost in friction,

$$F_{(\text{effort})} = P - \frac{W}{\text{V.R.}} = 12 - \frac{129.6}{18} = 4.8\text{N}$$

Since the frictional resistance is constant, therefore 4.8 N is the amount of friction offered by the machine. Now substituting the values of $P = 12$ and $C = 4.8$ in the law of the machine.

$$12 = m \times 129.6 + 4.8 \quad \text{...}(\because P = mW + C)$$

or

$$m = \frac{12 - 4.8}{129.6} = \frac{1}{18}$$

∴ Law of the machine will be given by the equation,

$$P = \frac{1}{18} W + 4.8$$ **Ans.**

Effort required to run the machine at no load

Substituting the value of $W = 0$ in the law of the machine (for no load condition),

$$P = 4.8 \text{ N} \quad \textbf{Ans.}$$

Effort required to run the machine at a load of 900 N

Substituting the value of $W = 900$ N in the law of machine,

$$P = \frac{1}{18} \times 900 + 4.8 = 54.8 \text{ N} \quad \textbf{Ans.}$$

17. MAXIMUM MECHANICAL ADVANTAGE OF A LIFTING MACHINE

We know that, Mechanical Advantage of a lifting machine,

$$\text{M.A.} = \frac{W}{P}$$

For Maximum Mechanical Advantage, substituting the value of $P = mW + C$ in the above equation,

$$\text{Max. M.A.} = \frac{W}{mW + C} = \frac{1}{m + \dfrac{C}{W}} = \frac{1}{m} \qquad \ldots\left(\text{Neglecting } \frac{C}{W}\right)$$

18. MAXIMUM EFFICIENCY OF A LIFTING MACHINE

We know that, Efficiency of a lifting machine,

$$\eta = \frac{\text{Mechanical advantage}}{\text{Velocity ratio}} = \frac{\dfrac{W}{P}}{\text{V.R.}} = \frac{W}{P \times \text{V.R.}}$$

For *Maximum Efficiency, substituting the value of $P = mW + C$ in the above equation,

$$\text{Max. } \eta = \frac{W}{(mW + C) \times \text{V.R}} = \frac{1}{\left(m + \dfrac{C}{W}\right) \times \text{V.R.}} = \frac{1}{m \times \text{V.R.}} \qquad \ldots\left(\text{Neglecting } \frac{C}{W}\right)$$

*Note: We know that, Efficiency of a lifting machine,

$$\eta = \frac{\text{M.A.}}{\text{V.R.}}$$

A little consideration will show that the efficiency will be maximum, when the mechanical advantage will be maximum.

or
$$\text{Max. } \eta = \frac{\text{Max. M.A.}}{\text{V.R.}}$$

$$= \frac{1}{m \times \text{V.R.}} \qquad \ldots\left(\because \text{ Max. M.A.} = \frac{1}{m}\right)$$

Example 8. *The law of a machine is given by the relation :*

$$P = 0.04\, W + 7.5$$

where (P) is the effort required to lift a load (W), both expressed in newtons. What is the mechanical advantage and efficiency of the machine, when the load is 2 kN and velocity ratio is 40 ? What is the maximum efficiency of the machine ?

If (F) is the effort lost in friction, find the relation between F and W. Also find the value of F, when W is 2 kN.

Solution. Given: Law of machine $P = 0.04\, W + 7.5$; Load $(W) = 2$ kN $= 2000$ N and velocity ratio (V.R.) = 40.

Mechanical advantage

Substituting the value of W in the law of the machine,

$$P = mW + C = 0.04 \times 2000 + 7.5 = 87.5 \text{ N}$$

$\therefore$ Mechanicsl Advantage, $\text{M.A.} = \dfrac{W}{P} = \dfrac{2000}{87.5} = 22.9$ **Ans.**

Efficiency of the machine

We know that, Effficiency,

$$\eta = \frac{\text{M.A.}}{\text{V.R.}} = \frac{22.9}{40} = 0.5725 = 57.25\% \quad \textbf{Ans.}$$

Maximum Efficiency of the machine

We know that, *Maximum Efficiency of the machine,

$$\text{Max.}\eta = \frac{1}{m \times \text{V.R.}} = \frac{1}{0.04 \times 40} = 0.625 = 62.5\% \quad \textbf{Ans.}$$

(*Note: In this example, the value of $m = 0.04$ and $C = 7.5$ as the law of the machine is given by the relation $P = mW + C$.)

Relation between F and W

We know that, Effort lost in friction,

$$F_{(\text{effort})} = P - \frac{W}{\text{V.R.}} = (0.04\,W + 7.5) - \frac{W}{\text{V.R.}}$$

$$= W\left(0.04 - \frac{1}{\text{V.R.}}\right) + 7.5 = W\left(0.04 - \frac{1}{40}\right) + 7.5$$

$$= W\,(0.04 - 0.025) + 7.5 = 0.015\,W + 7.5 \quad \textbf{Ans.}$$

Value of F when W is 2 kN

Substituting the value of W equal to 2 kN or 2000 N in the above equation,

$$F = (0.015 \times 2000) + 7.5 = 37.5 \text{ N} \quad \textbf{Ans.}$$

Example 9. *The law of a certain lifting machine is :*

$$P = \frac{W}{50} + 8$$

The velocity ratio of the machine is 100. Find the maximum possible mechanical advantage and the maximum possible efficiency of the machine. Determine the effort required to overcome the machine friction, while lifting a load of 600 N. Also calculate the efficiency of the machine at this load. **[AMIE]**

Solution. Given: Law of lifting machine $P = \dfrac{W}{50} + 8 = 0.02W + 8$; Velocity ratio (V.R.) = 100 and load (W) = 600 N.

Maximum Possible Mechanical Advantage

Comparing the given law of the machine with the standard relation for the law of the machine (*i.e.* $P = mW + C$) we find that in the given law of the machine, $m = 0.02$. We know that,

Maximum Possible Mechanical Advantage

$$\text{Max M.A.} = \frac{1}{m} = \frac{1}{0.02} = 50 \quad \textbf{Ans.}$$

Maximum Possible Efficiency

We know that, Maximum Possible Efficiency

$$\text{Max.}\eta = \frac{1}{m \times \text{V.R.}} = \frac{1}{0.02 \times 100} = \frac{1}{2} = 0.5 = 50\% \quad \textbf{Ans.}$$

Effort required to overcome the machine friction

Effort required to lift a load of 600 N,

$$P = mW + 8 = (0.02 \times 600) + 8 = 20 \text{ N}$$

Effort required to overcome the machine friction, while lifting a load of 600 N,

$$F_{(\text{effort})} = P - \frac{W}{\text{V.R.}} = 20 - \frac{600}{100} = 14 \text{ N} \quad \textbf{Ans.}$$

Efficiency of the machine

We know that, Mechanical advantage of the machine while lifting a load of 600 N.

$$\text{M.A.} = \frac{W}{P} = \frac{600}{20} = 30$$

and Efficiency, $\eta = \dfrac{\text{M.A.}}{\text{V.R.}} = \dfrac{30}{100} = 0.3 = 30\%$ **Ans.**

Example 10. *In an experiment of a weight lifting machine, with velocity ratio as 18, the values of effort required to lift various loads were as given in the table below :*

Load (W) in N	*250*	*500*	*750*	*1000*	*1500*	*2500*
Effort (P) in N	*42.5*	*62.5*	*82.5*	*105*	*142.5*	*220*

Plot a graph showing the relation between effort and load, and determine the law of the machine. Find the effort required and efficiency of the machine, when the load is 2 kN. Also find the maximum efficiency, this machine can attain.

Solution. Given: Velocity ratio (V. R.) = 18

Law of the machine

First of all, draw a suitable graph and plot the points 1, 2, 3, 4, 5, 6. From the geometry of the points, we find that points 1, 2, 3, and 5 lie on the same straight line, whereas the point 4 lies above the line and the point 6 below the line. Therefore let us ignore the points 4 and 6. Now draw a straight line *AB* passing through the points 1, 2, 3 and 5 as shown in Fig. 3. Now let us measure the intercept *OA* on *y-y* axis, which is equal to 22.5 N.

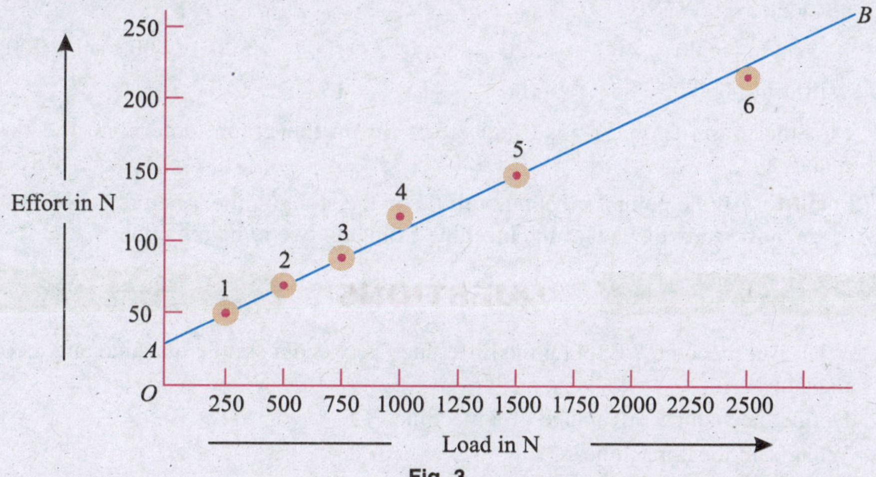

Fig. 3.

Now consider two points 1 and 5 (having maximum distance between them), which lie on the straight line AB. From the geometry of these two points, we find that the slope of the line AB,

$$m = \frac{142.5 - 42.5}{1500 - 250} = \frac{100}{1250} = 0.08$$

∴ Law of the machine is given by the relation,

$$P = 0.08\,W + 22.5 \qquad \textbf{Ans.}$$

Effort required when the load is 2 kN

Substituting the value of W equal to 2 kN or 2000 N in the above equation,

$$P = (0.08 \times 2000) + 22.5 = 182.5 \text{ N} \qquad \textbf{Ans.}$$

Efficiency of the machine when the load is 2 kN

Mechanical Advantage, $\text{M.A.} = \dfrac{W}{P} = \dfrac{2000}{182.5} = 10.96$

and Efficiency, $\eta = \dfrac{\text{M.A}}{\text{V.R.}} = \dfrac{10.96}{18} = 0.609 = 60.9\%$ **Ans.**

Maximum Efficiency the machine can attain

We also know that, Maximum Efficiency the machine can attain,

$$\text{Max. } \eta = \frac{1}{m \times \text{V.R.}} = \frac{1}{0.08 \times 18} = 0.694 = 69.4\% \qquad \textbf{Ans.}$$

EXERCISE 2

1. In a certain weight lifting machine, an effort of 15 N can lift a load of 300 N and an effort of 20 N can lift a load of 500 N. Find the law of the machine. Also find the effort required to lift a load of 880 N. **[Ans.** $P = 0.025\,W + 7.5$; 29.5 N]
2. In a weight lifting machine, an effort of 40 N can lift a load of 1000 N and an effort of 55 N can lift a load of 1500 N. Find the law of the machine. Also find maximum mechanical advantage and maximum efficiency of the machine. Take velocity ratio of the machine as 48. **[AMIE, AM-IETE]** [**Ans.** $P = 0.03\,W + 10$; 33.3 ; 69.4%]
3. The following results were obtained from a test on a certain weight lifting machine having a velocity ratio of 20 :

Load in N (W)	400	500	600	700	800	900	1000
Effort in N (P)	85	100	115	135	145	160	175

Plot the graph showing load and effort. From the graph, determine the law of the machine. **[Ans.** $P = 0.15\,W + 2.5$]

Hint. All the points, except 4, will lie on the straight line. Therefore ignore this point and now study the law of the machine from any two remaining points.

QUESTIONS

1. What is a machine ? Explain the difference between a simple machine and a compound machine.
2. Define mechanical advantage of a machine.
3. What is an ideal machine ?

4. Define velocity ratio of a machine.
5. Derive the relation between mechanical advantage, velocity ratio and efficiency of a machine.
6. Explain how the efficiency of a simple machine is determined ?
7. What do you understand by the term 'Reversibility' of a machine ? Explain the difference between a reversible machine and a self-locking machine.
8. What is law of a machine ? Derive an equation for the same.
9. Obtain an equation for the maximum mechanical advantage and maximum efficiency of a machine.
10. What is meant by 'friction in a machine'? In how many ways it can be expressed in terms of velocity ratio ?

OBJECTIVE TYPE QUESTIONS

1. The efficiency of a lifting machine is the ratio of
   (*a*) Its output to input
   (*b*) Work done by it to the work done on it
   (*c*) Its mechanical advantage to its velocity ratio
   (*d*) All of the above.
2. If efficiency of a lifting machine is kept constant, its velocity ratio is directly proportional to its
   (*a*) Mechanical advantage (*b*) Effort applied
   (*c*) Machine friction (*d*) All of the above
3. In an ideal machine, the mechanical advantage is .... velocity ratio
   (*a*) Equal to (*b*) Less than (*c*) Greater than
4. A lifting machine having an efficiency less than 50% is known as
   (*a*) Reversible machine (*b*) Non-reversible machine
   (*c*) Ideal machine (*d*) None of the above
5. A weight of 1000 N can be lifted by an effort of 80 N. If the velocity ratio of the machine is 20, then the machine is
   (*a*) Reversible (*b*) Non-reversible (*c*) Ideal
6. The maximum mechanical advantage of a lifting machine is
   (*a*) $1 + m$ (*b*) $1 - m$ (*c*) $1/m$ (*d*) m
7. The maximum efficiency of a lifting machine is
   (*a*) $1/m$ (*b*) V. R./m (*c*) m/V. R. (*d*) 1/(m × V. R.)

ANSWERS

1. (*d*) 2. (*a*) 3. (*a*) 4. (*b*) 5. (*a*) 6. (*c*) 7. (*d*)

CHAPTER

11 Simple Lifting Machines

Contents

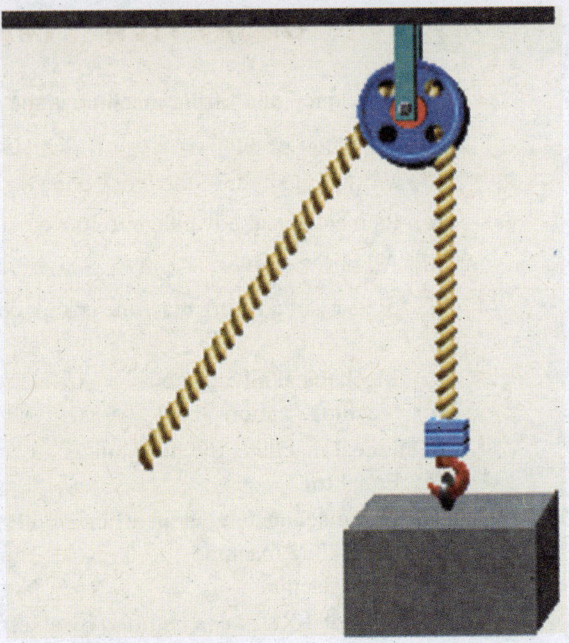

1. INTRODUCTION

In the last chapter, we have discussed the principles of lifting machines. Now in this chapter, we shall discuss the applications of these principles on a few lifting machines.

2. TYPES OF LIFTING MACHINES

These days, there are many types of lifting machines which are available in the market. But the basic principle, on which all these machines are based, is the same. It will be interesting to know that engineers who have designed (or invented) these machines have tried to increase the velocity ratio of their respective lifting machines. A little consideration will show, that if the efficiency of a machine remains almost the same, then increase in the velocity ratio must increase its mechanical advantage. The increased mechanical advantage, of a machine, means the application of a smaller force to lift the same load; or to lift a heavier load with the application of the same force.

The following lifting machines, which are important from the subject point of view, will be discussed in the following pages:

1. Simple wheel and axle.
2. Differential wheel and axle.
3. Weston's differential pulley block.
4. Geared pulley block.
5. Worm and worm wheel.
6. Worm geared pulley block.
7. Single purchase crab winch.
8. Double purchase crab winch.
9. Pulleys:
   (*a*) First system of pulleys.
   (*b*) Second system of pulleys.
   (*c*) Third system of pulleys.
10. Simple screw jack.
11. Differential screw jack.
12. Worm geared screw jack.

3. SIMPLE WHEEL AND AXLE

[WBUT]

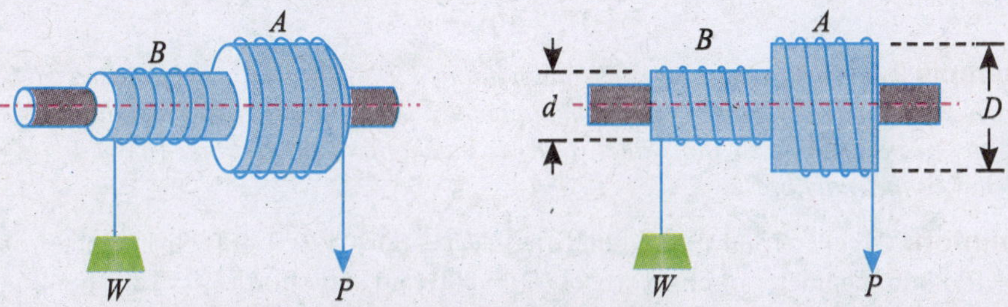

Fig. 1. Simple wheel and axle.

In Fig. 1 is shown a simple wheel and axle, in which the wheel *A* and axle *B* are keyed to the same shaft. The shaft is mounted on ball bearings, order to reduce the frictional resistance to a minimum. A string is wound round the axle *B*, which carries the load to be lifted. A second string is wound round the wheel *A* in the *opposite direction* to that of the string on *B*.

Let D = Diameter of effort wheel,

d = Diameter of the load axle,

W = Load lifted, and

P = Effort applied to lift the load.

One end of the string is fixed to the wheel, while the other is free and the effort is applied to this end. Since the two strings are wound in opposite directions, therefore a downward motion of the effort (P) will raise the load (W).

Since the wheel as well as the axle are keyed to the same shaft, therefore when the wheel rotates through one revolution, the axle will also rotate through one revolution.

We know that displacement of the effort in one revolution of effort wheel *A*,

$$= \pi D \quad ...(i)$$

and displacement of the load in one revolution

$$= \pi d \quad ...(ii)$$

Velocity Ratio, $\text{V.R.} = \dfrac{\text{Distance moved by the effort}}{\text{Distance moved by the load}} = \dfrac{\pi D}{\pi d} = \dfrac{D}{d}$

Mechanical advantage, $\text{M.A.} = \dfrac{\text{Load lifted}}{\text{Effort applied}} = \dfrac{W}{P}$ ..as usual

Efficiency, $\eta = \dfrac{\text{M.A}}{\text{V.R.}}$ ...as usual

Example 1. *A simple wheel and axle has wheel and axle of diameters of 300 mm and 30 mm respectively. What is the efficiency of the machine, if it can lift a load of 900 N by an effort of 100 N.*

Solution. Given: Diameter of wheel (D) = 300 mm; Diameter of axle (d) = 30 mm; Load lifted by the machine (W) = 900 N and effort applied to lift the load (P) = 100 N

Velocity ratio of the simple wheel and axle,

$$\text{V.R.} = \frac{D}{d} = \frac{300}{30} = 10$$

Mechanical advantage, $\text{M.A.} = \dfrac{W}{P} = \dfrac{900}{100} = 9$

$\therefore$ Efficiency, $\eta = \dfrac{\text{M.A.}}{\text{V.R.}} = \dfrac{9}{10} = 0.9 = 90\%$ **Ans.**

Example 2. *A drum weighing 60 N and holding 420N of water is to be raised from a well by means of wheel and axle. The axle is 100 mm diameter and the wheel is 500 mm diameter. If a force of 120 N has to be applied to the wheel, find (i) mechanical advantage, (ii) velocity ratio and (iii) efficiency of the machine.*

Solution. Given: Total load to be lifted (W) = 60 + 420 = 480 N; Diameter of the load axle (d) = 100 mm; Diameter of effort wheel (D) = 500 mm and effort (P) = 120 N.

Mechanical advantage,

$$\text{M.A.} = \frac{W}{P} = \frac{480}{120} = 4 \quad \textbf{Ans.}$$

Velocity ratio,

$$\text{V.R.} = \frac{D}{d} = \frac{500}{100} = 5 \quad \textbf{Ans.}$$

Efficiency of the machine,

$$\eta = \frac{\text{M.A.}}{\text{V.R.}} = \frac{4}{5} = 0.8 = 80\% \quad \textbf{Ans.}$$

Note : If we consider weight of the water only (*i.e.*, neglecting weight of the drum) then

Mechanical advantage, $\text{M.A.} = \dfrac{420}{120} = 3.5$ **Ans.**

Efficiency, $\eta = \dfrac{\text{M.A.}}{\text{V.R.}} = \dfrac{3.5}{5} = 0.7 = 70\%$ **Ans.**

4. DIFFERENTIAL WHEEL AND AXLE

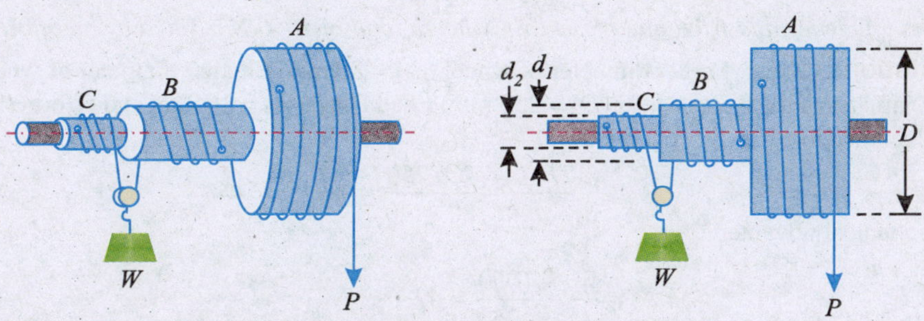

Fig. 2. Differential wheel and axle.

It is an improved form of simple wheel and axle, in which the velocity ratio is intensified with the help of load axle. In Fig. 2 is shown a differential wheel and axle. In this case, the load axle *BC* is made up of two parts of different diameters. Like simple wheel and axle, the wheel *A*, and the axles *B* and *C* are keyed to the same shaft, which is mounted on ball bearings in order to reduce the frictional resistance to a minimum.

The effort string is wound round the wheel *A*. Another string wound round the axle *B*, which after passing round the pulley (to which the weight *W* is attached) is wound round the axle *C* in opposite direction to that of the axle *B*; *care being taken to wind the string on the wheel A and axle C in the same direction.* As a result of this, when the string unwinds from the wheel *A*, the other string also unwinds from the axle *C*. But it winds on the axle *B* as shown in Fig. 2.

Let D = Diameter of the effort wheel *A*,

d_1 = Diameter of the axle *B*,

d_2 = Diameter fo the axle *C*,

W = Weight lifted by the machine, and

P = Effort applied to lift the weight.

We know that displacement of the effort in one revolution of effort wheel *A*

$$= \pi D \qquad ...(i)$$

∴ Length of string, which will wound on axle *B* in one revolution

$$= \pi d_1$$

and length of string, which will unwound from axle *C* in one revolution

$$= \pi d_2$$

∴ Length of string which will wound in one revolution

$$= \pi\, d_1 - \pi\, d_2 = \pi\, (d_1 - d_2)$$

and Displacement of weight $= \dfrac{1}{2} \times \pi(d_1 - d_2) = \dfrac{\pi}{2}(d_1 - d_2)$ ...(ii)

Velocity Ratio, $\text{V.R.} = \dfrac{\text{Distance moved by the effort}}{\text{Distance moved by the load}} = \dfrac{\pi D}{\dfrac{\pi}{2}(d_1 - d_2)} = \dfrac{2D}{d_1 - d_2}$

Mechanical Advantage, $\text{M.A.} = \dfrac{W}{P}$ ...as usual

Efficiency, $\eta = \dfrac{\text{M.A.}}{\text{V.R.}}$ ...as usual

Example 3. *The larger and smaller diameters of a differential wheel and axle are 80 mm and 70 mm respectively. The effort is applied to the wheel of diameter 250 mm. What is the velocity ratio?*

Find efficiency and frictional effort lost, when a load of 1050 N is lifted by an effort of 25 N.

Solution. Given: Larger diameter of wheel (d_1) = 80 mm; Smaller diameter of wheel (d_2) = 70 mm; Diameter of the effort wheel (D) = 250 mm; Load lifted (W) = 1050 N and effort (P) = 25 N.

Velocity ratio,

$$\text{V.R.} = \frac{2D}{d_1 - d_2} = \frac{2 \times 250}{80 - 70} = 48 \quad \textbf{Ans.} \quad ...(i)$$

Mechanical advantage

$$\text{M.A.} = \frac{W}{P} = \frac{1050}{25} = 42 \quad ...(ii)$$

Efficiency,

$$\eta = \frac{\text{M.A.}}{\text{V.R.}} = \frac{42}{50} = 0.84 = 84\% \quad \textbf{Ans.}$$

Frictional effort lost,

$$F_{(\text{effort})} = P - \frac{W}{\text{V.R.}} = 25 - \frac{1050}{50} = 4 \text{ N} \quad \textbf{Ans.}$$

Example 4. *With a differential wheel and axle, an effort of 6 N raised a load of 60 N. If the efficiency at this load is 80%, find the velocity ratio of the machine.*

If the diameter of the effort wheel is 300 mm, determine the difference betwen the diameters of the axles. If the sum of the diameters of the axles is 280 mm, determine the diameter of each axle.

[RTU]

Solution: Given: Effort (P) = 6 N; Load raised (W) = 60 N; Efficiency (η) = 80% = 0.8; Diameter of effort wheel (D) = 300 mm and sum of the diameters of axles ($d_1 + d_2$) = 280 mm.

Mechanical advantage,

$$\text{M.A.} = \frac{W}{P} = \frac{60}{6} = 10$$

Efficiency,

$$0.8 = \frac{\text{M.A.}}{\text{V.R.}} = \frac{10}{\text{V.R.}}$$

or Velocity Ratio,

$$\text{V.R.} = \frac{10}{0.8} = 12.5 \quad \textbf{Ans.}$$

Difference between the diameters of the axles

We know that velocity ratio of a differential wheel and axle,

$$12.5 = \frac{2D}{d_1 - d_2} = \frac{2 \times 300}{d_1 - d_2} = \frac{600}{d_1 - d_2}$$

or

$$(d_1 - d_2) = \frac{600}{12.5} = 48 \quad \textbf{Ans.}$$

Diameter of each axle

Solving $(d_1 - d_2) = 48$ and $(d_1 + d_2) = 280$ simultaneously, we get

$$d_1 = 164 \text{ mm and } d_2 = 116 \text{ mm} \quad \textbf{Ans.}$$

Example 5. *In a differential wheel and axle, the diameter of the effort wheel is 400 mm. The radii of the axles are 150 mm and 100 mm respectively. The diameter of the rope is 10 mm.*

Find the load which can be lifted by an effort of 25 N assuming the efficiency of the machine to be 84%.

Solution. Given: Diameter of effort wheel = 480 mm; Radii of axles = 150 mm and 100 mm or diameter of axles = 300 mm and 200 mm; Diameter of rope = 10 mm; Effort (P) = 25 N and efficiency (η) = 84% = 0.84.

Let W = Load that can be lifted by the machine.

We know that, effective diameter of the effort wheel,

$$D = 400 + 10 = 410 \text{ mm}$$

Effective diameters of axles,

$$d_1 = 300 + 10 = 310 \text{ mm and } d_2 = 200 + 10 = 210 \text{ mm}$$

Velocity ratio of a differential wheel and axle,

$$\text{V.R.} = \frac{2D}{d_1 - d_2} = \frac{2 \times 410}{310 - 210} = 8.2$$

Mechanical Advantage, $$\text{M.A.} = \frac{W}{P} = \frac{W}{25}$$

Efficiency,

$$84 = \frac{\text{M.A.}}{\text{V.R.}} = \frac{\frac{W}{25}}{8.2} = \frac{W}{205}$$

or $$W = 0.84 \times 205 = 172.2 \text{ N}$$ **Ans.**

5. WESTON'S DIFFERENTIAL PULLEY BLOCK

It consists of two blocks A and B. The upper block A has two pulley (P_1 and P_2) one having its diameter a little larger than that of the other. The Pulleys turn together as one pulley, *i.e.*, both of them behave as one pulley with two grooves. The lower block B also carries a pulley, to which the load W is attached.

An endless (*i.e.*, a continuous) chain passes round the pulleys then round the lower block pulley and finally round the pulley P_2 (*i.e.*, smaller of the upper pulleys). The remaining chain hangs slack and is joined to the first portion of the chain as shown in Fig. 3.

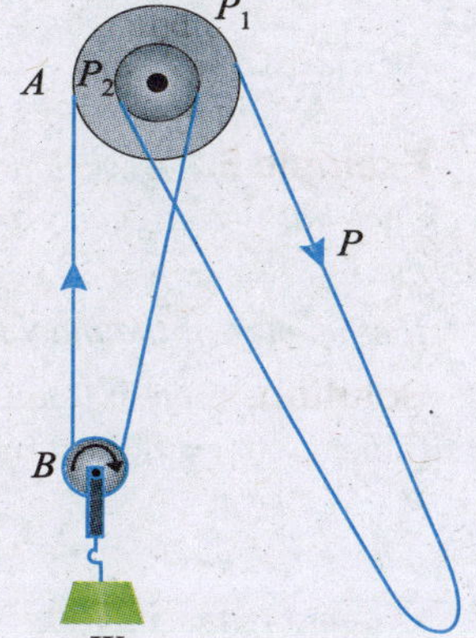

Fig. 3. Differential pulley block.

The effort P is applied to the chain passing over the pulley P_1 (*i.e.*, larger of upper pulleys) as shown in Fig. 3. To prevent the chain from slipping, projection are provided in the grooves of both the upper pulleys.

Let D = Diameter of the pulley P_1,
d = Diameter of the pulley P_2,
W = Weight lifted, and
P = Effort applied to lift the weight.

We know that displacement of the effort in one revolution of the upper pulley block,

$$= \pi D \qquad ...(i)$$

This is also equal to the length of the chain pulled over the larger pulley. Since the smaller pulley also turns with the larger one, therefore length of the chain released by the smaller pulley

$$= \pi d$$

$\therefore$ Net shortening of the chain $= \pi D - \pi d = \pi (D - d)$

This shortening of chain will be equally divided between the two portions of the chain, supporting the load. Therefore distance through which the load will move up

$$= \frac{1}{2} \times \pi(D - d) = \frac{\pi}{2}(D - d) \quad \text{...(ii)}$$

Velocity Ratio, $$\text{V.R.} = \frac{\text{Distance moved by the effort}}{\text{Distance moved by the load}}$$

$$= \frac{\pi D}{\frac{\pi}{2}(D - d)} = \frac{2D}{D - d} \quad \text{... (iii)}$$

Mechanical Advantage, $\text{M.A.} = \dfrac{W}{P}$ ... as usual

Efficiency, $\eta = \dfrac{\text{M.A.}}{\text{V.R.}}$ ...as usual

Notes: 1. Sometimes, the size of the upper pulleys are given in terms of radius (instead of diameters).

If R = Radius of the larger pulley, and

r = Radius of the smaller pulley.

Then $\text{V.R.} = \dfrac{2R}{R - r}$

2. Sometimes, the diameters of upper pulleys are not given and the relatives sizes are expressed in terms of number of teeth.

If T_1 = No. of teeth of the larger pulley, and

T_2 = No. of teeth of the smaller pulley

Then $\text{V.R.} = \dfrac{2T_1}{T_1 - T_2}$

Example 6. *A weston differential pulley consists of a lower block and an upper block. The upper block has two cogged grooves, one of which has a radius of 125 mm and the other has a radius of 115 mm.*

If efficiency of the machine is 80%, calculate the effort required to raise load of 1500 N.

Solution. Given: Radius of larger groove (R) = 125 mm; Radius of smaller groove (r) = 115 mm; Efficiency (η) = 80% = 0.8 and load to be raised = 1500 N.

Let P = Effort requried to raise the load.

Velocity ratio, $$\text{V.R.} = \frac{2R}{R - r} = \frac{2 \times 125}{125 - 115} = 25$$

Mechanical Advantage, $$\text{M.A.} = \frac{W}{P} = \frac{1500}{P}$$

Efficiency,

$$0.8 = \frac{\text{M.A.}}{\text{V.R.}} = \frac{\frac{1500}{P}}{25} = \frac{60}{P}$$

or $$P = \frac{60}{0.8} = 75 \text{ N} \quad \textbf{Ans.}$$

Example 7. *In a differential pulley block, a load of 1800 N is raised by an effort of 100 N. The number of teeth on the larger and smaller blocks are 12 and 11 respectively. Find the velocity ratio, mechanical advantage and efficiency of the machine.*

Solution. Given: Load (W) = 1800 N; Effort applied (P) = 180 N; Number of teeth on the larger block (T_1) = 12 and number of keeth on the smaller block (T_2) = 11.

Velocity ratio

Velocity ratio

$$\text{V.R.} = \frac{2T_1}{T_1 - T_2} = \frac{2 \times 12}{12 - 11} = 24 \quad \textbf{Ans.}$$

Mechanical advantage

$$\text{M.A.} = \frac{W}{P} = \frac{1800}{100} = 18 \quad \textbf{Ans.}$$

Efficiency of the machine,

$$\eta = \frac{\text{M.A.}}{\text{V.R.}} = \frac{18}{24} = 0.75 = 75\% \quad \textbf{Ans.}$$

EXERCISE 1

1. In a simple wheel and axle, radii of effort wheel and axle is 240 mm and 40 mm respectively. Find the efficiency of the machine, if a load of 600 N can be lifted by an effort of 120 N. **[Ans.** 83.3%]
2. In a simple wheel and axle, the diameter of the wheel is 150 mm and that of the axle is 30 mm. If efficiency of the machine is 60%, determine the effort required to lift a load of 500 N. **[Ans.** 166-7 N]
3. A differential wheel and axle, having velocity ratio as 12, is used to lift a load of 300 N. If the coefficient of friction for the pulley spindle is 0.2, find the effort required to lift the load. **[Ans.** 31.25 N]

   Hint : Efficiency = 1 – 0.2 = 0.8
4. In a differential wheel and axle, having a velocity ratio of 24, a load of 200 N can be lifted by and effort of 15N and a load of 250 N can be lifted by an effort of 18 N. Find (*a*) Effort required to lift a load 375 N; (*b*) Effort wasted in friction and (*c*) Maximum efficiency of the machine. **[Ans.** 25.5 N; 7.58 N; 69.4%]

6. GEARED PULLEY BLOCK

It is an improved form of a differential pulley block in which the velocity ratio is intensified with the help of gears. A geared pulley block consists of a cogwheel A, around which is passed an endless chain. A smaller gear wheel B, known as pinion, is keyed to the same shaft as that of A.

The wheel axle B is geared with another bigger wheel C, called the spur wheel. A cogwheel D is keyed to the same shaft as that of spur wheel C.

The load is attached to a chain, that passes over the cogwheel D. The effort is applied to the endless chain, which passes over the wheel A as shown in Fig. 4.

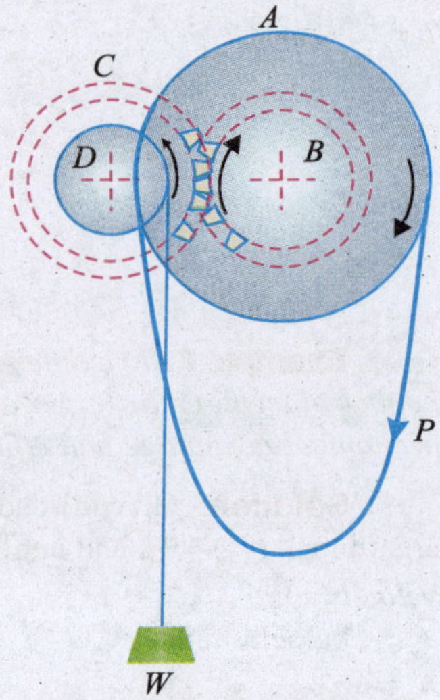

Let T_1 = No. of cogs on wheel A (known as effort wheel),

T_2 = No. of teeth on wheel B (known as pinion),

T_3 = No. of teeth on the wheel C. known as spur wheel, and

T_4 = No. of cogs on the wheel D known as load wheel.

Fig. 4. Geared pulley block.

We know that distance moved by the effort in one revolution of the cogwheel A,,

$$= T_1 \quad \text{...}(i)$$

No. of revolution made by the pinion B

$$= 1$$

∴ No. of revolutions made by the spur wheel C

$$= \frac{T_2}{T_3}$$

No. of revolutions made by the load wheel D

$$= \frac{T_2}{T_3}$$

∴ Distance moved by the load

$$= \frac{T_2}{T_3} \times T_4 \quad \text{...}(ii)$$

Velocity Ratio, $$\text{V.R.} = \frac{\text{Distance moved by the effort}}{\text{Distance moved by the load}} = \frac{T_1}{\frac{T_2}{T_3} \times T_4} = \frac{T_1}{T_2} \times \frac{T_3}{T_4} \quad \text{...}(iii)$$

Mechanical Advantage, $\text{M.A.} = \dfrac{W}{P}$ ...as usual

Efficiency, $\eta = \dfrac{\text{M.A}}{\text{V.R.}}$ ...as usual

Example 8. *A geared pulley block, used to lift a load, has the following dimensions:*

No. of cogs on the effort wheel = 90

No. of cogs on the load wheel = 8

No. of teeth on the pinion = 25

No. of teeth on the spur wheel = 40

Find the maximum load that can be lifted by an effort of 50 N on machine. Take efficiency of the pulley block as 75%.

Solution. Given: No. of cogs on the effort wheel (T_1) = 90; No. of cogs on the pinion (T_2) = 25; No. of teeth on the spur wheel (T_3) = 40; No. of teeth on the load wheel (T_4) = 8; Effort (P) = 50 N and efficiency (η) = 75% = 0.75.

Let W = Load that can be lifted by the machine.

Velocity ratio of a geared pulley block,

$$\text{V.R.} = \frac{T_1}{T_2} \times \frac{T_3}{T_4} = \frac{90}{25} \times \frac{40}{8} = 18$$

Mechanical Advantage, $\text{M.A.} = \frac{W}{P} = \frac{W}{50}$

Efficiency,

$$0.75 = \frac{\text{M.A.}}{\text{V.R.}} = \frac{\frac{W}{50}}{18} = \frac{W}{900}$$

or $W = 0.75 \times 900 = 675$ N **Ans.**

7. WORM AND WORM WHEEL

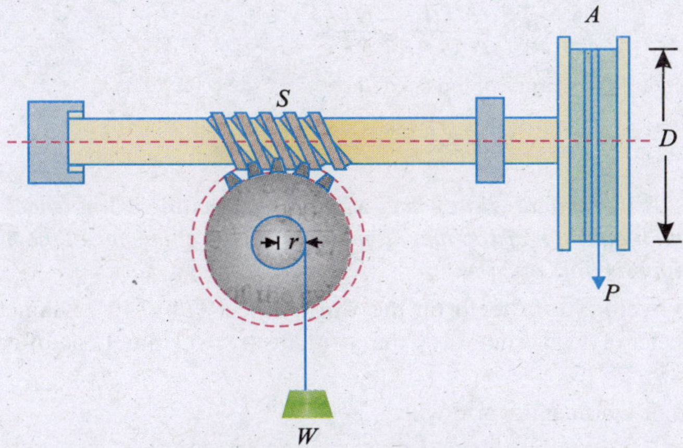

Fig. 5. Worm and worm wheel

It consists of a square threaded screw, S (known as worm) and a toothed wheel (known as worm wheel) geared with each other, as shown in Fig. 5. A wheel A is attached to the worm, over which passes a rope as shown in the figure. Sometimes, a handle is also fixed to the worm (instead of the wheel). A load drum is securely mounted on the worm wheel.

Let D = Diameter of the effort wheel,

r = Radius of the load drum

W = Load lifted,

P = Effort applied to lift the load, and

T = No. of teeth on the worm wheel.

We know that distance moved by the effort in one revolution of the wheel (or handle)

$$= \pi D \qquad ...(i)$$

If the worm is single-threaded (*i.e.*, for one revolution of the wheel A, the screw S pushes the worm wheel through one teeth), then the load drum will move through

$$= \frac{1}{T} \text{ revolution}$$

and distance, through which the load will move

$$= \frac{2\pi r}{T} \quad ...(ii)$$

Velocity Ratio,

Velocity Ratio, $\text{V.R.} = \dfrac{\text{Distance moved by the effort}}{\text{Distance moved by the load}}$

$$= \frac{\pi D}{\dfrac{2\pi r}{T}} = \frac{DT}{2r} \quad ...(iii)$$

Mechanical Advantage, $\text{M.A.} = \dfrac{W}{P}$ ...as usual

Efficiency, $\eta = \dfrac{\text{M.A.}}{\text{V.R.}}$ ...as usual

Notes : 1. If the worm is double-threaded *i.e.*, for one revolution of wheel *A*, the screw *S* pushes the worm wheel through two teeths, then

$$\text{V.R.} = \frac{DT}{2 \times 2r} = \frac{DT}{4r}$$

2. In general, if the worm is *n* threaded, then

$$\text{V.R.} = \frac{DT}{2nr}$$

Example 9. *A worm and worm wheel with 40 teeth on the worm wheel has effort wheel of 300 mm diameter and load drum of 100 mm diameter. Find the efficiency of the machine, if it can lift a load of 1800 N with an effort of 24 N.*

Solution. Given: No. of teeth on the worm wheel (T) = 40 ; Diameter of effort wheel (D) = 300 mm Diameter of load drum = 100 mm or radius (r) = 50 mm; Load lifted (W) = 1800 N and effort (P) = 24 N.

Velocity ratio of worm and worm wheel,

$$\text{V.R.} = \frac{DT}{2r} = \frac{300 \times 40}{2 \times 50} = 120$$

Mechanical Advantage, $\text{M.A.} = \dfrac{W}{P} = \dfrac{1800}{24} = 75$

Efficiency, $\eta = \dfrac{\text{M.A.}}{\text{V.R.}} = \dfrac{75}{120} = 0.625 = 62.5\%$ **Ans.**

Example 10. *In a double threaded worm and worm wheel, the number of teeth on the worm wheel is 60. The diameter of the effort wheel is 250 mm and that of the load drum is 100 mm. Calculate the velocity ratio. If the efficiency of the machine is 50%, determine the effort required to lift a load of 300 N.*

Solution. Given : No. of threads (n) = 2; No. of teeth on the worm wheel (T) = 60; Diameter of effort wheel = 250 mm; Diameter of load drum = 100 mm or radius (r) = 50 mm; Efficiency (η) = 50% = 0.5 and load to be lifted (W) = 300 N.

Velocity ratio of a worm and worm wheel,

$$\text{V.R.} = \frac{DT}{2nr} = \frac{250 \times 60}{2 \times 2 \times 50} = 75 \qquad \textbf{Ans.}$$

Effort required to lift the load

Let P = Effort required to lift the load.

Mechanical advantage,

$$\text{M.A.} = \frac{W}{P} = \frac{300}{P}$$

Efficiency,
$$0.5 = \frac{\text{M.A.}}{\text{V.R.}} = \frac{\frac{300}{P}}{75} = \frac{4}{P}$$

or
$$P = \frac{4}{0.5} = 8 \text{ N} \qquad \textbf{Ans.}$$

8. WORM GEARED PULLEY BLOCK

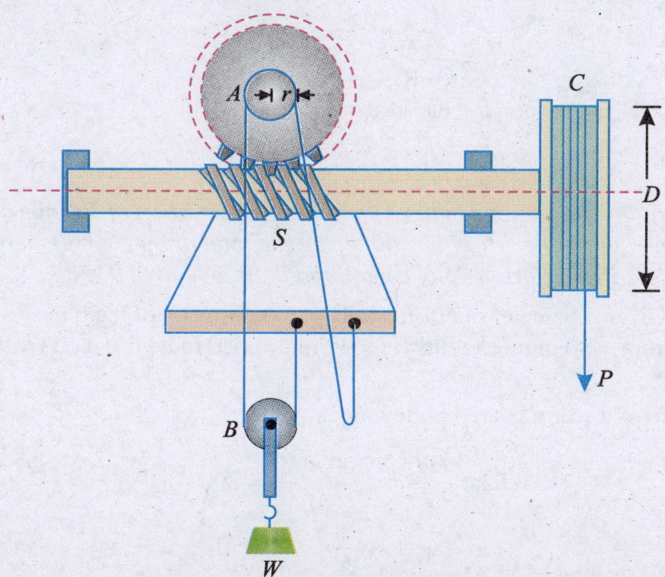

Fig. 6. Worm geared pulley block.

It is an improved form of worm and worm wheel, in which the velocity ratio is intensified with the help of a pulley block. It consists of a square threaded screw S and a toothed wheel. One end of the string is fastened to the frame. This string passes round the load drum B and then over the pulley A (which is keyed to the wheel). The other end of the string is also fixed to the frame. The effort is applied to the wheel C by a rope as shown in Fig. 6.

Let D = Diameter of effort wheel,

r = Radius of pulley A,

W = Load lifted,

P = Effort applied to lift the load, and

T = No. of teeth on the worm wheel.

We know that, Distance moved by effort in one revolution of wheel

$$= \pi D$$

If the worm is single threaded (*i.e.*, for one revolution of wheel C, the screw S pushes the worm wheel through one tooth) then the worm wheel and pulley A will move through.

$$= \frac{1}{T} \text{ revolution}$$

∴ Net shortening of the string

$$= \frac{2\pi r}{T} \qquad ...(i)$$

Since the load drum is supported by two strings, therefore the load will be lifted through a distance

$$= \frac{1}{2} \times \frac{2\pi r}{T} = \frac{\pi r}{T} \qquad ...(ii)$$

Velocity Ratio, $$\text{V.R.} = \frac{\text{Distance moved by the effort}}{\text{Distance moved by the load}} = \frac{\pi D}{\frac{\pi r}{T}} = \frac{DT}{r} \qquad ...(iii)$$

Mechanical Advantage, $$\text{M.A.} = \frac{W}{P} \qquad ...\text{as usual}$$

Efficiency, $$\eta = \frac{\text{M.A.}}{\text{V.R.}} \qquad ...\text{as usual}$$

Note: In general, if the worm is n threaded then

$$\text{V.R.} = \frac{DT}{nr}$$

Example 11. *A worm geared pulley block has its effort wheel of 200 mm diameter. The worm is single threaded and the worm wheel has 60 teeth. The load drum is of 80 mm diameter. Find the efficiency of the block, if an effort of 75 N is required to lift a load of 9* kN.

Solution. Given: Diameter of effort wheel = 200 mm; No. of teeth in worm wheel (T) = 60; Diameter of load drum = 80 mm or radius (r) = 40 mm; Load to be lifted (W) = 9 kN = 9000 N and effort (P) = 75 N.

Velocity ratio of a worm geared pulley block,

$$\text{V.R.} = \frac{DT}{r} = \frac{200 \times 60}{40} = 300$$

Mechanical Advantage, $$\text{M.A.} = \frac{W}{P} = \frac{9000}{75} = 120$$

Efficiency, $$\eta = \frac{\text{M.A.}}{\text{V.R.}} = \frac{120}{300} = 0.4 = 40\% \qquad \textbf{Ans.}$$

Example 12. *A double threaded worm geared pulley block has effort wheel of 400 mm diameter and load drum of 100 mm. If the machine is 35% efficient, find the load it can lift by an effort of 80 N. The worm wheel has 50 teeth.*

Solution. Given: No. of threads (n) = 2; Diameter of effort wheel = 400 mm; Diameter of load drum = 100 mm or radius (r) = 50 mm; Efficiency (η) = 35% = 0.35; Effort (P) = 80 N and no of teeth in worm wheel (T) = 50.

Let W = Load which can be lifted by the machine

Velocity ratio of a worm geared pulley block,

$$\text{V.R.} = \frac{DT}{nr} = \frac{400 \times 50}{2 \times 50} = 200$$

Mechanical Advantage, $\text{M.A.} = \frac{W}{P} = \frac{W}{80}$

Efficiency,

$$0.35 = \frac{\text{M.A.}}{\text{V.R.}} = \frac{\frac{W}{80}}{200} = \frac{W}{16000}$$

or $$W = 0.35 \times 16000 = 5600 \text{ N.} \quad \textbf{Ans.}$$

9. SINGLE PURCHASE CRAB WINCH

[RGPV]

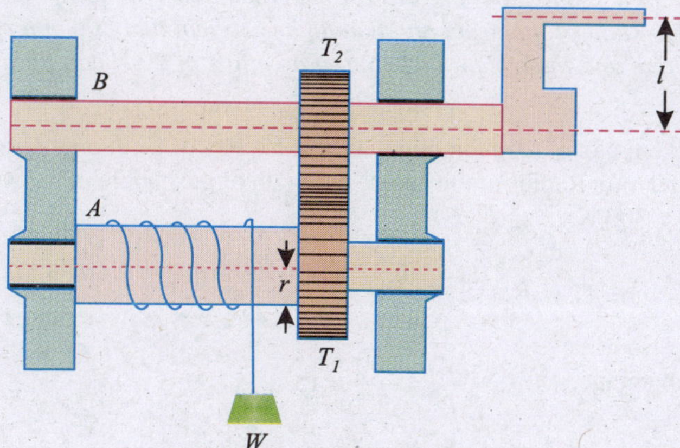

Fig. 7. Single purchase crab winch.

In single purchase crab winch, a rope is fixed to the drum and is wound a few turns round it. The free end of the rope carries the load W. A toothed wheel A is rigidly mounted on the load drum. Another toothed wheel B, called pinion, is geared with the toothed wheel A as shown in Fig. 7.

The effort is applied at the end of the handle to rotate it.

Let T_1 = No. of teeth on the main gear (or spur wheel) A,

T_2 = No. of teeth on the pinion B,

l = Length of the handle,

r = Radius of the load drum.

W = Load lifted, and

P = Effort applied to lift the load.

We know that distance moved by the effort in one revolution of the handle,

$$= 2\pi l \quad \text{...}(i)$$

No. of revolutions made by the pinion B

$$= 1$$

No. of revolutions made by the wheel A

$$= \frac{T_2}{T_1}$$

No. of revolutions made by the load drum

$$= \frac{T_2}{T_1}$$

Distance moved by the load $= 2\pi r \times \dfrac{T_2}{T_1}$ ...(*ii*)

Velocity Ratio, $\text{V.R.} = \dfrac{\text{Distance moved by the effort}}{\text{Distance moved by the load}} = \dfrac{2\pi l}{2\pi r \times \dfrac{T_2}{T_1}} = \dfrac{l}{r} \times \dfrac{T_1}{T_2}$ ...(*iii*)

Mechanical Advantage, $\text{M.A.} = \dfrac{W}{P}$ ...as usual

Efficiency, $\eta = \dfrac{\text{M.A.}}{\text{V.R.}}$ ...as usual

Example 13. *In a single purchase crab winch, the number of teeth on pinion is 25 and that on the spur wheel 100. Radii of the drum and handle are 50 mm and 300 mm respectively. Find the efficiency of the machine and the effect of friction, if an effort of 20 N can lift a load of 300 N.*

[GTU]

Solution. Given: No. of teeth on pinion (T_2) = 25; No. of teeth on the spur wheel (T_1) = 100; Radius of daum (r) = 50 mm; Radius of the handle or length of the handle (l) = 300 mm; Effort (P) = 20 N and load lifted (W) = 300 N.

Velocity ratio,

$$\text{V.R.} = \frac{1}{r} \times \frac{T_1}{T_2} = \frac{300}{50} \times \frac{100}{25} = 24$$

Mechanical Advantage, $\text{M.A.} = \dfrac{W}{P} = \dfrac{300}{20} = 15$

Efficiency, $\eta = \dfrac{\text{M.A.}}{\text{V.R.}} = \dfrac{15}{24} = 0.625 = 62.5\%$ **Ans.**

Effect of friction

Effect of friction in terms of load,

$$F_{(\text{load})} = (P \times \text{V.R.}) - W = (20 \times 24) - 300 = 180 \text{ N}$$

Effect of friction in terms of effort,

$$F_{(\text{effort})} = P - \frac{W}{\text{V.R.}} = 20 - \frac{300}{24} = 7.5 \text{ N}$$

It means that if the machine would have been ideal (*i.e.* without friction) then it could lift an extra load of 180 N with the same effort of 20 N. Or it could have required 7.5 N less force to lift the same load of 300 N. **Ans.**

Example 14. *A single purchase crab winch, has the following details:*

Length of lever = *700 mm*
Number of pinion teeth = *12*
Number of spur gear teeth = *96*
Diameter of load axle = *200 mm*

It is observed that an effort of 60 N can lift a load of 1800 N and an effort of 120 N can lift a load of 3960 N.

What is the law of the machine ? Also find efficiency of the machine in both the cases.

Solution. Given: Length of lever (l) = 700 *mm*; No. of pinion teeth (T_2) = 12; No. of spur geer teeth (T_1) = 96 and dia of load axle = 200 mm or radius (r) = 200/2 = 100 mm.

(i) Law of the machine

When P_1 = 60 N, W_1 = 1800 N and when P_2 = 120 N, W_2 = 3960 N.

Substituting the values of P and W in the law of the machine *i.e.*, $P = mW + C$

$$60 = (m \times 1800) + C \qquad ...(i)$$

and

$$120 = (m \times 3960) + C \qquad ...(ii)$$

Subtracting equation (*i*) from equation (*ii*)

$$60 = m \times 2160$$

or

$$m = \frac{60}{2160} = \frac{1}{36}$$

Now substituting this value of m in equation (*i*),

$$60 = \left(\frac{1}{36} \times 1800\right) + C = 50 + C$$

or

$$C = 60 - 50 = 10$$

and now substituting the value of m = 1/36 and C = 10 in the law of machine,

$$P = \frac{1}{36} W + 10 \qquad \textbf{Ans.}$$

(ii) Efficiencies of the machine in both the cases

Velocity ratio,

$$\text{V.R.} = \frac{l}{r} \times \frac{T_1}{T_2} = \frac{700}{100} \times \frac{96}{12} = 56$$

Mechanical Advantage in the first case,

$$\text{M.A.} = \frac{W_1}{P_1} = \frac{1800}{60} = 30$$

Efficiency,

$$\eta_1 = \frac{\text{M.A}}{\text{V.R.}} = \frac{30}{56} = 0.536 = 53.6\% \qquad \textbf{Ans.}$$

Similarly, Mechanical Advantage in the second case,

$$\text{M.A.} = \frac{W_2}{P_2} = \frac{3960}{120} = 33$$

Efficiency,

$$\eta_2 = \frac{\text{M.A.}}{\text{V.R.}} = \frac{33}{56} = 0.589 = 58.9\% \qquad \textbf{Ans.}$$

10. DOUBLE PURCHASE CRAB WINCH

A double purchase crab winch is an improved form of a single purchase crab winch, in which the velocity ratio is intensified with the help of one more spur wheel and a pinion. In a double purchase crab winch, there are two spur wheels of teeth T_1 and T_2 and T_3 as well as two pinions of teeth T_2 and T_4.

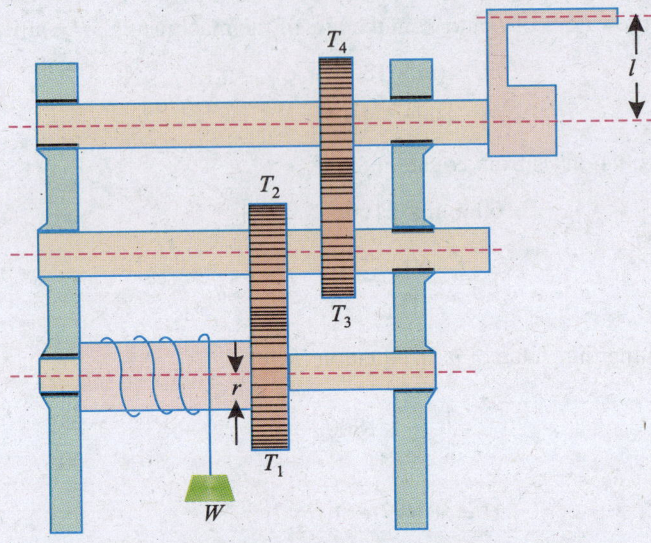

Fig. 8. Double purchase crab winch.

The arrangement of spur wheels and pinions are such that the spur wheel with T_1 gears with the pinion of teeth T_2. Similarly, the spur wheel with teeth T_3 gears with the pinion of the teeth T_4, The effort is applied to a handle as shown in Fig. 11.8.

Let T_1 and T_3 = No. of teeth of spur wheels,

T_2 and T_4 = No. of teeth of the pinions

l = Length of the handle,

r = Radius of the load drum,

W = Load lifted, and

P = Effort applied to lift the load, at the end of the handle.

We know that, Distance moved by the effort in one revolution of the handle,

$$= 2\pi l \qquad \text{...}(i)$$

No. of revolutions made by the pinion 4

$$= 1$$

No. of revolutions made by the wheel 3

$$= \frac{T_4}{T_3}$$

No. of revolutions made by the pinion 2

$$= \frac{T_4}{T_3}$$

No. of revolutions made by the wheel 1

$$= \frac{T_2}{T_1} \times \frac{T_4}{T_3}$$

∴ Distance moved by the load

$$= 2\pi r \times \frac{T_2}{T_1} \times \frac{T_4}{T_3} \qquad ...(ii)$$

Velocity Ratio, $$\text{V.R.} = \frac{\text{Distance moved by the effort}}{\text{Distance moved by the load}}$$

$$= \frac{2\pi l}{2\pi r \times \frac{T_2}{T_1} \times \frac{T_4}{T_3}} = \frac{l}{r}\left(\frac{T_1}{T_2} \times \frac{T_3}{T_4}\right)$$

Mechanical Advantage, $\text{M.A.} = \dfrac{W}{P}$ ...as usual

Efficiency, $\eta = \dfrac{\text{M.A.}}{\text{V.R.}}$ ...as usual

Example 15. *In a double purchase crab winch,* teeth of pinions are *20 and 25 and that of spur wheels are 50 and 60. Length of the handle is 0.5 metre and radius of the load drum is 0.25 metre. If efficiency of the machine is* 60%, *find the effort required to lift a load of 720 N.*

Solution. Given: No. of teeth of pinion (T_2) = 20 and (T_4) = 25; No. of teeth of spur wheel (T_1) = 501 and (T_3) = 60; Length of the handle (l) = 0.5 m; Radius of the load drum (r) = 0.25 m; Efficiency (η) = 60% = 0.6 and load to be lifted (W) = 720 N.

Let P = Effort required in newton to lift the load.

Velocity ratio, $$\text{V.R.} = \frac{1}{r}\left(\frac{T_1}{T_2} \times \frac{T_3}{T_4}\right) = \frac{0.5}{0.25}\left(\frac{50}{20} \times \frac{60}{25}\right) = 12$$

Mechnical Advantage, $\text{M.A.} = \dfrac{W}{P} = \dfrac{720}{P}$

Efficiency, $$0.6 = \frac{\text{M.A.}}{\text{V.R.}} = \frac{\frac{720}{P}}{12} = \frac{60}{P}$$

or $$P = \frac{60}{0.6} = 100 \text{ N} \qquad \textbf{Ans.}$$

Example 16. *A double purchase crab used in a laboratory has the following dimensions :*

Diameter of load drum = *160 mm*

Length of handle = *360 mm*

No. of teeth on pinions = *20* and *30*

No. of teeth on spur wheels = *75* and *90*

When tested, it was found that an effort of 90 N was required to lift a load of 1800 N and an effort of 135 N was required to lift a load of 3150 N. Determine :

(*a*) *Law of the machine,*

(*b*) *Probable effort to lift a load of 4500 N,*

(*c*) *Efficiency of the machine in the above case,*

(*d*) *Maximum efficiency of the machine.*

Solution. Given: Dia of load drum = 160 mm or radius (r) = 100/2 = 80 mm; Length of handle (l) = 360 mm; No. of teeth on pinions (T_2) = 20 and (T_4) = 30 and no. of teeth on spus wheels (T_1) = 75 and (T_3) = 90.

When P = 90 N, W = 1800 N when P = 135 N, W = 3150 N

(a) Law of the machine

Substituting the values of P and W in the law of the machine, *i.e.*, $P = m\,W + C$

$$90 = (m \times 1800) + C \qquad ...(i)$$

and

$$135 = (m \times 3150) + C \qquad ...(ii)$$

Subtracting equation (*i*) from equation (*ii*),

$$45 = m \times 1350$$

or

$$m = \frac{45}{1350} = \frac{1}{30}$$

Now substituting this value of m in equation (*i*),

$$90 = \frac{1}{30} \times 1800 + C = 60 + C$$

or

$$C = 90 - 60 = 30$$

Now substituting the value fo m and C in the law of the machine,

$$P = \frac{1}{30}W + 30 \qquad \textbf{Ans.}$$

(b) Effort to lift a load of 4500 N

Substituting the value of W equal to 4500 N in the law of the machine,

$$P = \left(\frac{1}{30} \times 4500\right) + 30 = 180 \text{ N} \qquad \textbf{Ans.}$$

(c) Efficiency of the machine in the above case

Velocity ratio,
$$\text{V.R.} = \frac{l}{r}\left(\frac{T_1 \times T_3}{T_2 \times T_4}\right) = \frac{360}{80}\left(\frac{75}{20} \times \frac{90}{30}\right) = 50.6$$

Mechnical advantage,
$$\text{M.A.} = \frac{W}{P} = \frac{4500}{180} = 25$$

Efficiency,
$$\eta = \frac{\text{M.A.}}{\text{V.R.}} = \frac{25}{50.6} = 0.494 = 49.4\% \qquad \textbf{Ans.}$$

(d) Maximum efficiency of the machine

We also know that maximum efficiency of the machine,

$$\eta_{max} = \frac{1}{m \times \text{V.R.}} = \frac{1}{\frac{1}{30} \times 50.6} = 0.593 = 59.3\% \qquad \textbf{Ans.}$$

EXERCISE 2

1. In a differential pulley block, diameters of the concentric pulleys are 300 mm and 250 mm respectively. It was found that an effort of 20 N just lifts a load of 200 N. Calculate (*i*) Efficiency of the machine; (*ii*) Effort lost in friction and (*iii*) Frictional load.

   [**Ans.** 83.3%; 3.33 N; 40 N]

2. In a Weston's differential pulley block, the difference in the number of teeth of the two pulleys is 3. If the efficiency of the machine is 60% and an effort of 100 N just lifts a load of 1 kN, find the number of teeth of the two pulleys. [**Ans.** 22 and 25]

   Hint :

   Given: $T_1 = T_2 + 3$

   Mechanical Advantage, $\text{M.A.} = \dfrac{W}{p} = \dfrac{1000}{100}$

   Velocity Ratio, $\text{V.R.} = \dfrac{2T_1}{T_1 - T_2} = \dfrac{2(T_2 + 3)}{(T_2 + 3) - T_2} = \dfrac{2T_2 + 6}{3}$

   Efficiency, $0.6 = \dfrac{\text{M.A.}}{\text{V.R.}} = \dfrac{10}{\dfrac{2T_2 + 6}{3}} = \dfrac{30}{2T_2 + 6}$

   or $2T_2 + 6 = \dfrac{30}{0.6}$

   or $T_2 = 22$ and $T_1 = 22 + 3 = 25$

3. A geared pulley block has the following dimensions :

   No. of cogs on the effort wheel = 100
   No. of cogs on the load wheel = 20
   No. of teeth on the pinion = 10
   No. of teeth on the spur wheel = 60

   If the machine has an efficiency of 60%, find the effort required to lift a load of 900 N.

   [**Ans.** 50 N]

4. In a worm and worm wheel, the number of teeth in the worm wheel is 25. The effort handle is 300 mm long and the load drum is 150 mm diameter. Find the efficiency of the machine, if an effort of 30 N can lift a load of 345 N and the worm is double threaded.

   [**Ans.** 23%]

5. A worm geared pulley block has its effort wheel of 250 mm diameter. The worm is double threaded and the worm wheel has 50 teeth. The load drum is of 100 mm diameter. Determine the effort required to lift a load of 5 kN, if efficiency of the machine is 40%.

   [**Ans.** 100 N]

6. A single purchase crab has 300 mm long handle, 120 mm diameter drum and diameter of the lifting rope is 10 mm. Number of teeth on the pinion are 25 and that on the wheel 130. Calculate the velocity ratio of the crab. If an effort of 20 N lifts a load of 300 N, what is the mechanical advantage and efficiency of the crab? [**Ans.** 24; 15; 62.5%]

   Hint : Effective diameter of the effort wheel

   $= 120 + 10 = 130$ mm

   $\therefore$ Effective radius, $r = 65$ mm

7. In a single purchase crab winch, length of the handle is 160 mm and the gear ratio is 5. Find the velocity ratio and efficiency of the machine, if a load of 1 kN is lifted by an effort of 50 N. Take diameter of the drum as 60 mm. **[Ans.** 26.67 ; 75%]

   Hint : $T_1/T_2 = 5$

8. In a double purchase crab winch the pinions have 15 and 20 teeth respectively, while the spur wheels have 45 and 40 teeth. The effort handle is 400 mm long and the effective diameter of the drum is 200 mm. If the efficiency of the winch is 50%, find the effort required to lift a load of 1500 N. **[Ans.** 125 N]

11. SIMPLE PULLEY

A simple pulley is a wheel of metal or wood, with a groove around its circumference, to receive rope or chain. The pulley rotates freely about its axle, which passes through its centre and is perpendicular to its surface plane. This axle is supported by a metal or a wooden frame, called block as show in Fig. 9. Following assumptions are made in the study of pulley system, which are quite reasonable from the practical point of view :

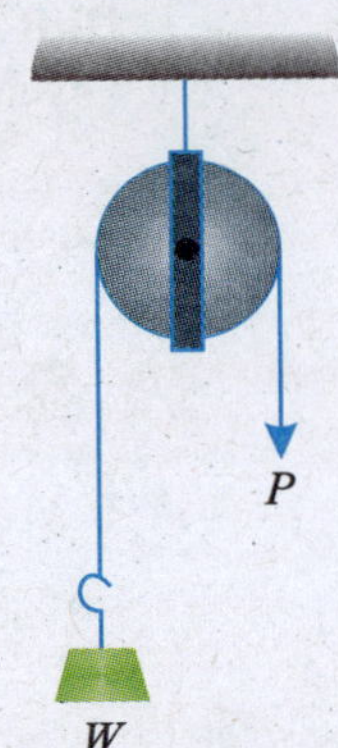

Fig. 9. Simple pulley.

1. The weight of the pulley block is small as compared to the weight to be lifted, and thus may be neglected in calculations.
2. The friction between the pulley surface and the string is negligible, and thus the tension in the two sides of the rope, passing round the pulley, may be taken to be equal.

A little consideration will show, that in a simple pulley, its mechanical advantage as well as velocity ratio is 1 under the assumed conditions mentioned above. The only advantage of a simple pulley is that the effort can be applied Fig. 10 (*a*), (*b*) and (*c*). Simple pulleys are generally used in certain mechanical advantage and efficiency.

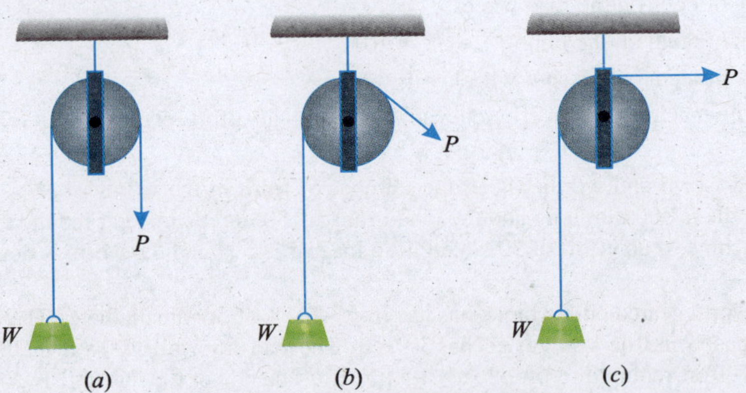

Fig. 10. Force applied in different directions.

Though there are many typed of pulleys used by engineers, yet the following system of pulleys are commonly used :

1. First system of pulleys.
2. Second system of pulleys.
3. Third system of pulleys.

12. FIRST SYSTEM OF PULLEYS

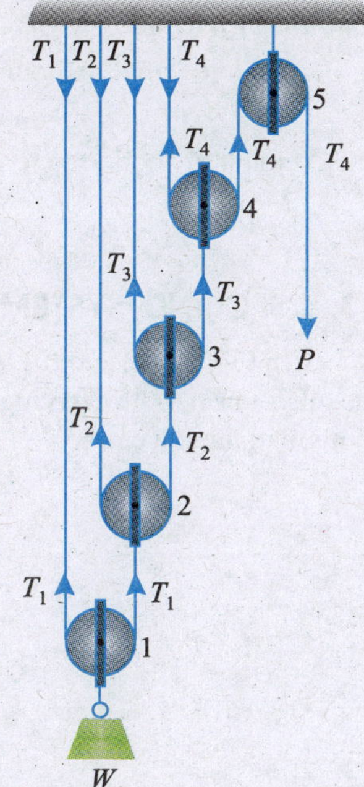

Fig. 11. First system of pulleys.

In Fig. 11 is shown the first system of pulleys. In this system, the pulleys are so arranged that there are as many strings as there are pulleys. The end of each string is fastened to a rigid ceiling; while the other end passing round the bottom periphery of the pulley, is fastened to the next higher pulley.

The load is attached to the bottom-most pulley ; whereas the effort is applied to the far end of the string passing round the last pulley. Another pulley (no. 5) is used just to change the direction of the effort.

The velocity ratio of the system may be obtained by considering a unit motion of the load. In this case, let the weight W be raised by x metres. Since the loads is supported on both sides of the string, thus this slackness of x metres will have to be taken up by the pulley 2. If the relative position of the pulley 2, with respect to the pulley 1, is to remain undisturbed, then the pulley 2 should move upwards through a distance of $2x$ metres.

Now this upward movement of pulley 2 through a distance of $2x$ metres will cause a total slackness of $2 \times (2x) = 2^2x$ metres in the string, which has to be taken up by the pulley 3. Thus the pulley 3 should move upwards through a distance of 2^2x metres, thus causing a slackness of $2 \times (2^2x) = 2^3x$ in the string passing round the pulley 4. Thus the pulley 4 should move upwards through a distance of 2^3x metres causing a slackness of $2 \times (2^3x) = 2^4x$ meters, which must be taken up by the free end of the string to which the effort is applied. Thus the effort must move through a distanece of 2^4x metres.

$$\therefore \qquad \text{V.R.} = \frac{\text{Distance moved by the effort}}{\text{Distance moved by the load}}$$

Thus in general, is there are n pulley in the system, then

Velocity Ratio, $\text{V.R.} = 2^n$

Mechanical Advantage, $\text{M.A.} = \dfrac{W}{P}$ ...as usual

Efficiency, $\eta = \dfrac{\text{M.A.}}{\text{V.R.}}$ ...as usual

Example. 17. *In a system of pulleys of the first type, there are three pulleys, and a weight of 320 N can be lifted by an effort of 50 N.*

Find the efficiency of the machine and the amount of friction.

Solution. Given : No. of pulleys (n) = 3 ; Weight lifted (W) = 320 N and effort (P) = 50 N.

Efficeincy of the machine

Velocity ratio of first system of pulleys.

Velocity Ratio, $\text{V.R.} = 2^n = 2^3 = 8$

Mechanical Advantage, $\text{M.A.} = \dfrac{W}{P} = \dfrac{320}{50} = 6.4$

Efficiency, $\eta = \dfrac{\text{M.A.}}{\text{V.R.}} = \dfrac{6.4}{8} = 0.80 = 80\%$ **Ans.**

Amount of friction

We know that, Amount of friction in terms of load,

$$F_{(load)} = (P \times \text{V.R.}) - W = (50 \times 8) - 320 = 80 \text{ N} \quad \textbf{Ans.}$$

and Amount of friction in terms of effort,

$$F_{(effort)} = P - \frac{W}{\text{V.R.}} = 50 - \frac{320}{8} = 10 \text{ N} \quad \textbf{Ans.}$$

13. SECOND SYSTEM OF PULLEYS

In Fig. 12 (*a*) and (*b*) is shown second system of pulleys containing two blocks, one upper and the other lower, both carrying either equal number of pulleys or the upper block may have one pulley more than the lower one.

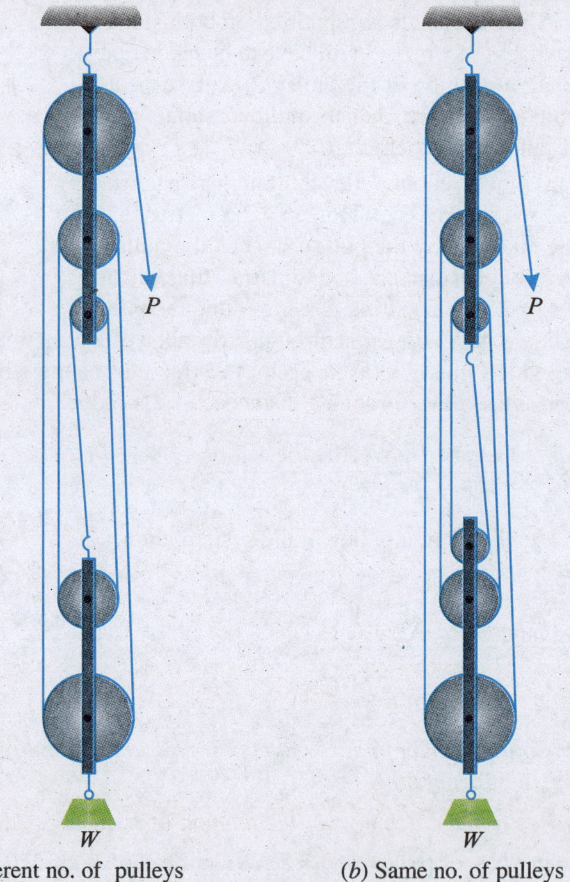

(*a*) Different no. of pulleys (*b*) Same no. of pulleys

Fig. 12. Second system of pulleys.

In both the cases, the upper block is fixed and the lower one is movable. There is obly one string, which passes round all the pulleys one end of which is fixed to the upper block (when both the blocks have the same no. of pulleys) or to the lower block (when the upper block has one pulley more than the lower one). The other end of the string is free and the effort is applied to this free end as shown in Fig. 12 (*a*) and (*b*). In both the cases the load is attached to lower block.

A little consideration will show, that for x displacement of the weight, the effort will move through a distance to nx, where n is the number of pulleys in both the blocks. Thus velocity ratio

Velocity Ratio, $\text{V.R.} = \dfrac{nx}{x} = n$

Mechanical Advantage, $\text{M.A.} = \dfrac{W}{P}$ ...as usual

Efficeincy, $\eta = \dfrac{\text{M.A.}}{\text{V.R.}}$ ...as usual

Example 18. *A weight of 1 kN is lifted by an effort of 125 N by second system of pulleys, having 5 pulleys in each block.*

Calculate the amount of effort wasted in friction and the frictional load.

Solution. Given: Weight lifted (W) = 1 kN = 1000 N ; Effort (P) = 125 N and no. of pulleys (n) = 2 × 5 = 10.

Amount of effort wasted in friction

Velocity ratio,

$$\text{V.R.} = n = 10$$

Amount of effort wasted in friction,

$$F_{(\text{effort})} = P - \frac{W}{\text{V.R.}} = 125 - \frac{1000}{10} = 25 \text{ N} \quad \textbf{Ans.}$$

Amount of frictional load

We also know that amount of frictional load,

$$F_{(\text{load})} = (P \times \text{V.R.}) - W = (125 \times 10) - 1000 = 240 \text{ N} \quad \textbf{Ans.}$$

14. THIRD SYSTEM OF PULLEYS

In Fig. 13, is shown a third system of pulleys. In this system, like the first system of pulleys, the pulleys are arranged in such a way that there are as many strings as there are pulleys. One end of each string is fixed to a block B –B, to which the load is attached. The other end of each string, passing round the upper periphery of the pulley, is fastened to the next lower pulley as shown in Fig. 13.

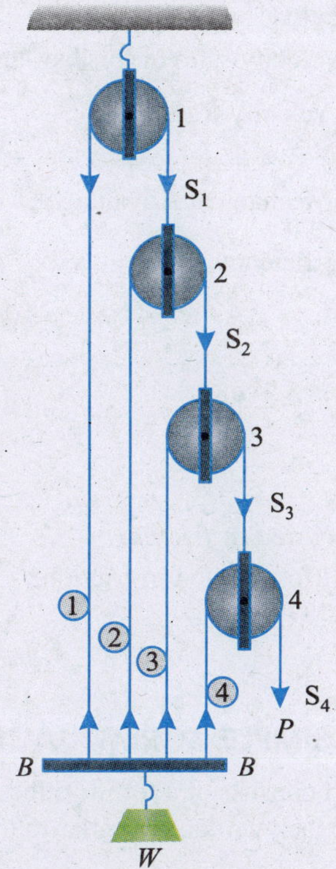

Fig. 13. Third system of pulleys.

The velocity ratio, of the system, may be obtained by considering a unit motion of the load. In this case, let the weight W be raised by x metres. Since the load is supported on all the strings, therefore all the strings will be slackened by x metres.

Now consider the pulley 1, which is fixed to the ceiling. The slackness of string s_1 equal to x metres will have to be taken up by the pulley 2, which should come down through a distance of $2x$ metres. But as the string s_2 also slacks by x metres, therefore the string s_1 will be pulled through a distance of $(2x - x) = x$ metre. Now consider the pulley 2. As the string s_1 has been pulled through a distance x metres, therefore the string s_2 will be pulled through a distance of $2x + x = 3x = (2^2 - 1)x$. Similarly, in order to keep the relative position of the pulley 3 undisturbed, the string s_3 will be pulled through a distance of $(2 \times 3x + x) = 7x = (2^3 - 1)\,x$ and the string s_4 *i.e.*, effort will be pulled through a distance of $(2 \times 7x + x) = 15x = (2^4 - 1)\,x$.

$$\therefore \qquad \text{V.R.} = \frac{\text{Distance moved by the effort}}{\text{Distance moved by load}}$$

$$= \frac{(2^4 - 1)\,x}{x} = (2^4 - 1)$$

Thus, in general, if there are n pulleys in this system, then

Velocity Ratio, $\quad \text{V.R.} = 2^n$

Mechanical advantage, $\quad \text{M.A.} = \dfrac{W}{P}$ ... as usual

Efficiency, $\quad \eta = \dfrac{\text{M.A.}}{\text{M.R.}}$ ... as usual

Example. 19. *In a third system of pulleys, there are 4 pulleys. Find the effort required to lift a load of 1800 N, if efficiency of the machine is 75%.*

Calculate the amount of effort wasted in friction.

Solution. Given: No. of pulleys $(n) = 4$; Load lifted $(W) = 1800$ N and efficiency $(\eta) = 75\% = 0.75$.

Effort required to lift the load

Let $\quad P$ = Effort required in newton to lift the load.

Velocity ratio of third system of pulleys.

Velocity Ratio, $\quad \text{V.R.} = 2^n - 1 = 2^4 - 1 = 15$

Mechanical Advantage, $\quad \text{M.A.} = \dfrac{W}{P} = \dfrac{1800}{P}$

Efficiency,

$$0.75 = \frac{\text{M.A.}}{\text{V.R.}} = \frac{\frac{1800}{P}}{15} = \frac{120}{P}$$

or

$$P = \frac{120}{0.75} = 160 \text{ N} \qquad \textbf{Ans.}$$

Effort wasted in friction

Effort wasted in friction,

$$F_{(\text{effort})} = P - \frac{W}{\text{V.R.}} = 160 - \frac{1800}{15} = 40 \text{ N} \qquad \textbf{Ans.}$$

15. SIMPLE SCREW JACK

It consists of a screw, fitted in a nut, which forms the body of the jack. The principle, on which a screw jack works, is similar to that of an inclined plane.

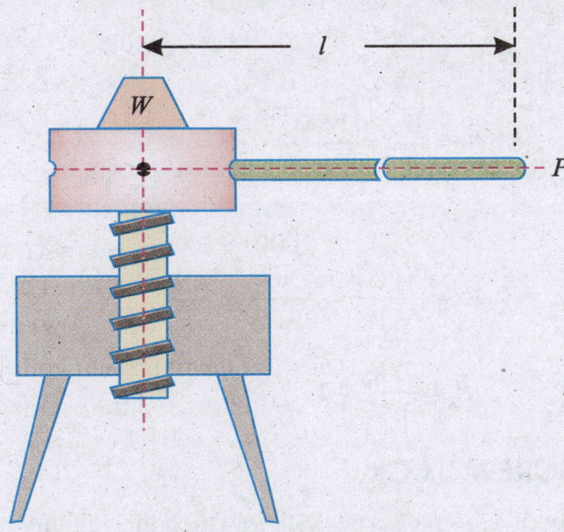

Fig. 14. Simple screw jack.

Fig. 14. shows a simple screw jack, which is rotated by the application of an effort at the end of the lever, for lifting the load. Now consider a single threaded simple screw jack.

Let l = Length of the effort arm,

p = Pitch of the screw,

W = Load lifted, and

P = Effort applied to lift the load at the end of the lever.

Distance moved by the effort in one revolution of screw,

$$= 2\pi l \quad ...(i)$$

Distance moved by the load $= p$ ...(*ii*)

$$\therefore \quad \text{Velocity ratio} = \frac{\text{Distance moved by the effort}}{\text{Distance moved by the load}} = \frac{2\pi l}{p} \quad ...(iii)$$

Mechanical Advantage, $\text{M.A.} = \dfrac{W}{P}$ ...as usual

Efficeincy, $\eta = \dfrac{\text{M.A.}}{\text{V.R.}}$ ...as usual

Note: The value of P *i.e.*, the effort applied may also found out by the relation :

$$^{*}P = W \tan(\alpha + \phi)$$

where W = Load lifted

$$\tan \alpha = \frac{p}{\pi d}$$

and $\tan \phi = \mu$ = Coefficient of friction.

Example 20. *A screw jack has a thread of 10 mm pitch. What effort applied at the end of a handle 400 mm long will be required to lift a load of 2 kN, if the efficiency at this load is 45%.*

[UPTU, IGNOU]

Solution. Given: Pitch of thread (p) = 10 mm; Length of the handle (l) = 400 mm; Load lifted (W) = 2 kN = 2000 N and efficiency (n) = 45% = 0.45.

Let P = Effort required to lift the load.

Velocity ratio,

$$\text{V.R.} = \frac{2\pi l}{p} = \frac{2\pi \times 400}{10} = 251.3$$

and

$$\text{M.A.} = \frac{W}{P} = \frac{2000}{P}$$

Efficiency,

$$0.45 = \frac{\text{M.A.}}{\text{V.R.}} = \frac{\frac{2000}{P}}{251.3} = \frac{7.96}{P}$$

or

$$P = \frac{7.96}{0.45} = 17.7 \text{ N} \qquad \textbf{Ans.}$$

16. DIFFERENTIAL SCREW JACK

It is an improved form of a simple screw Jack in which the velocity ratio is intensified with the help of a differential screw. In Fig. 15. is shown a jack, with a differential screw. The principle on which this machine works, is the same as that of any other differential machine, *i.e.*, action of one part of the machine is subtracted from the action of another part.

In this machine, the differential screw is in two parts, *A* and *B*. Part *A* is threaded both on inside and outside ; whereas the part *B* is threaded on the outside only. The external threads of *A* gear with the threads of the nut *C*, which forms the body of the differential screw jack. The internal threads of *A* gear with the external threads of the screw *B*. Thus the part *A* behaves as a screw for the nut *C* and as a nut for the screw *B*.

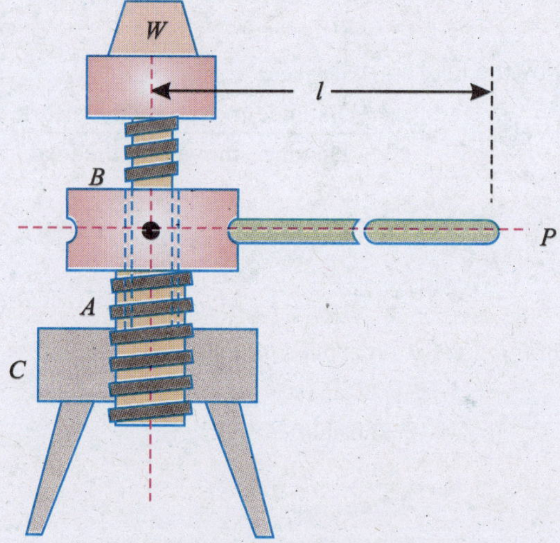

Fig. 15. Differential screw jack.

The screw *B* does not rotate, but moves in vertical direction only, and carries the load. When the effort is applied at the lever, the screw *A* rises up and simultaneously the screw *B* goes down. Thus the lift of the load is algebraic sum of the motions of the screw *A* and screw *B*.

Let

p_1 = Pitch of screw *A*,

p_2 = Pitch of the screw *B*,

l = Length of the lever arm,

W = Load lifted, and

P = Effort applied, at the end of the lever to lift the load.

Distance moved by the effort in one revoluton of the lever arm,

$$= 2\pi l \qquad ...(i)$$

∴ Upward distance moved by the screw A

$$= p_1$$

and Downward distance moved by the screw B

$$= p_2$$

∴ Distance through which the load is lifted

$$= p_1 - p_2 \qquad ...(ii)$$

∴ $$\text{Velocity ratio} = \frac{\text{Distance moved by the effort}}{\text{Distance moved by the load}} = \frac{2\pi l}{p_1 - p_2} \qquad ...(iii)$$

Mechanical Advantage, $\text{M.A.} = \dfrac{W}{P}$ ...as usual

Efficiency, $\eta = \dfrac{\text{M.A.}}{\text{V.R.}}$ ...as usual

Example 21. *A differential screw jack has pitch of 12 mm and 10 mm and 30 mm arm length. What will be the efficiency of the machine, if it can lift a load of 7.5 kN by an effort of 30 N.*

Solution. Given: Pitch of the screw (p_1) = 12 mm and (p_2) = 10 mm ; Arm length of screw jack (l) = 300 mm ; Load lifted (W) = 7.5 kN = 7500 N and effort (P) = 30 N.

Velocity ratio, $$\text{V.R.} = \frac{2\pi l}{p_1 - p_2} = \frac{2\pi \times 300}{12 - 10} = 942$$

Mechanical Advantage $$\text{M.A.} = \frac{W}{P} = \frac{7500}{30} = 250$$

Efficiency, $$\eta = \frac{\text{M.A.}}{\text{V.R.}} = \frac{250}{942} = 0.265 = 26.5\% \qquad \textbf{Ans.}$$

Example 22. *In a differential screw jack, the screw threads have pitch of 10 mm and 7 mm. If the efficiency of the machine is 28%, find the effort required at the end of an arm 360 mm long to lift a load of 5 kN.* **[IGNOU]**

Solution. Given: Pitch of the screw jack (p_1) = 10 mm and (p_2) = 7 mm ; Efficiency (η) = 28% = 0.28; Arm length of screw jack (l) = 360 mm and load lifted (W) = 5 kN = 5000 N.

Let P = Effort required to lift the load.

Velocity ratio of a Differential Screw Jack,

$$\text{V.R.} = \frac{2\pi l}{p_1 - p_2} = \frac{2\pi \times 360}{10 - 7} = 754$$

Mechanical Advantage, $$\text{M.A.} = \frac{W}{P} = \frac{5000}{P}$$

Efficiency,

$$0.28 = \frac{\text{M.A.}}{\text{V.R.}} = \frac{\frac{5000}{P}}{754} = \frac{6.63}{P}$$

or $$P = \frac{6.63}{0.28} = 23.7 \text{ N} \qquad \textbf{Ans.}$$

17. WORM GEARED SCREW JACK

It is a further improved form of differential screw jack, in which the velocity ratio is further intensified with the help of a geared screw jack.

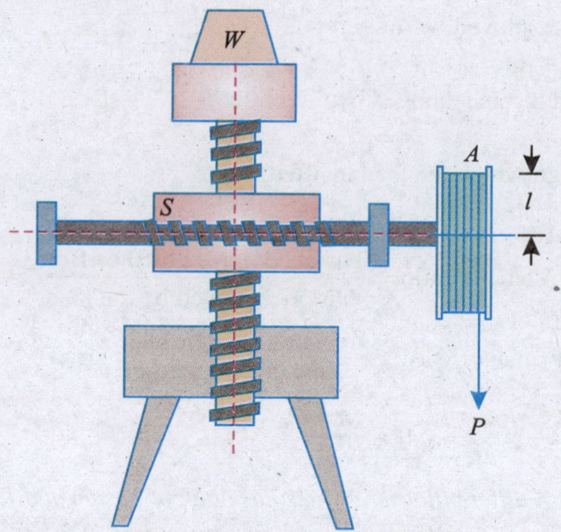

Fig. 16. Worm geared screw jack.

In Fig. 16. is shown a jack with worm geared screw. It is also an improved form of screw jack, in which the screw is lifted with the help of worm and worm wheel, instead of effort at the end of a lever. Now consider a worm geared screw jack.

Let l = Radius of the effort wheel (or length of the handle).

p = Pitch of the screw,

W = Load lifted,

P = Effort applied to lift the load, and

T = No. of teeth on the worm wheel.

Distance moved by effort in one revolution of wheel (or handle)

$$= 2\pi l \qquad \text{...}(i)$$

If the worm is single threaded (*i.e.* for one revolution of the wheel A, the screw S pushes the worm wheel through one teeth) then the worm wheel move through $1/T$ revolution. Therefore distance moved by the load

$$= \frac{p}{T} \qquad \text{...}(ii)$$

$$\therefore \quad \text{Velocity ratio} = \frac{\text{Distance moved by the effort}}{\text{Distance moved by the load}} = \frac{2\pi l}{\frac{p}{T}} = \frac{2\pi l T}{p}$$

Mechanical Advantage, $\text{M.A.} = \dfrac{W}{P}$ ...as usual

Efficiency, $\eta = \dfrac{\text{M.A.}}{\text{V.R.}}$ ...as usual

Notes : 1. If the worm wheel is double threaded *i.e.* for one revolution of wheel A, the screw S pushes the worm wheel through two teeths then

$$\text{V.R.} = \frac{2\pi l T}{2p}$$

2. In general, if the worm wheel is n threaded, then

$$\text{V.R.} = \frac{2\pi l T}{np}$$

Example 23. *A worm geared screw jack has the following particulars :*

Length of handle	*= 300 mm*
No. of teeth in the worm wheel	*= 50*
Pitch of screw	*= 10 mm*
Effort applied	*= 100 N*
Load lifted	*= 100 kN*

If the worm is double threaded, find the efficiency of the jack.

Solution. Given: Length of handle (l) = 300 mm ; No. of teeth in the worm wheel (T) = 50 ; Pitch of screw (p) = 10 mm ; Effort (P) = 100 N ; Load lifted (W) = 100 kN = 100 000 N and no. of threads (n) = 2.

Velocity ratio of a Worm Geared Screw Jack,

$$\text{V.R.} = \frac{2\pi l T}{n p} = \frac{2\pi \times 300 \times 50}{2 \times 10} = 4712$$

Mechanical Advantage, $\text{M.A.} = \frac{W}{P} = \frac{100\ 000}{100} = 1000$

Efficiency, $\eta = \frac{\text{M.A.}}{\text{V.R.}} = \frac{1000}{4712} = 0.212 = 21.2\%$ **Ans.**

EXERCISE 3

1. In a first system of pulleys, there are 5 movable pulleys. If the efficiency of the machine is 75%, what effort can lift a load of 4 kN ? [**Ans.** 166.7 N]
2. In a second system of pulleys, there are nine pulleys in the two blocks. Find the efficiency of the machine, if an effort of 50 N can lift a load of 300 N. [**Ans.** 66.7%]
3. There are three pulleys arranged in the third system of pulleys. Find the load that can be lifted by an effort of 50 N, if efficiency of the machine is 80%. [**Ans.** 280 N]
4. A simple screw jack has a thread of pitch 12 mm. Find the load that can be lifted by an effort of 20 N applied at the end of handle 500 mm long. Take efficiency of the machine as 50%. [**Ans.** 2.094 kN]
5. In a simple screw jack, the pitch of the screw is 10 mm and length of the handle is 450 mm. Find the velocity ratio. If an effort of 25 N applied at the end of the handle can lift a load of 3 kN, find the efficiency of the jack. Also calculate the amount of effort wasted in friction and the frictional load. [**Ans.** 282.7 ; 42.4% ; 14.4 N ; 4.068 kN]
6. A differential screw jack has threads of 10 mm and 6 mm. It was found that an effort of 14.7 N applied at the end of handle 400 mm long can lift a load of 2.5 kN. Find the efficiency of the screw jack. [**Ans.** 27.1%]
7. A worm geared screw jack has the following particulars :

Handle length	= 400 mm
No. of teeth in the worm wheel	= 45
Pitch of screw	= 12 mm

If the worm is single-threaded, find the load that can be lifted by an effort of 20 N. Take efficiency of the machine as 20%. [**Ans.** 37.7 kN]

QUESTIONS

1. Distinguish clearly between a 'simple wheel and axle' and a 'differential wheel and axle'.
2. Describe the working of a Weston's differential pulley block.
3. In a differential pulley lifting tackle, the upper block consists of two pulleys of diameter D and d rotating on a fixed axis, and a movable pulley below, to which is attached the load. The tackle is operated by an endless chain and the effort is applied to lengths coming off the larger pulley. Find the expression for the velocity ratio, if $D > d$.
4. State the difference between a Weston's differential pulley block and a geared pulley block.
5. Derive an equation for the velocity ratio of a worm and worm wheel when it is (*a*) single threaded and (*ii*) double threaded.
6. Distinguish clearly the difference between the working of a single purchase crab winch and a double purchase crab winch.
7. What is pulley ? State the working of first system, second system and third system of pulleys. Derive relations for their respective velocity ratios.
8. What is a screw jack ? On what principle does it work derive a relation for the velocity ratio of a simple screw jack and differential screw jack.

OBJECTIVE TYPE QUESTIONS

1. The velocity ratio of a simple wheel and axle with D and d as the diameters of effort wheel and load axle is :

   (*a*) $D + d$ (*b*) $D - d$ (*c*) $D \times d$ (*d*) D/d

2. The velocity ratio of a differential wheel and axle with D as the diameter effort wheel and d_1 and d_2 as the diameters of larger and smaller axles respectively is

   (*a*) $\frac{2D}{d_1 + d_2}$ (*b*) $\frac{2D}{d_1 - d_2}$ (*c*) $\frac{D}{d_1 + d_2}$ (*d*) $\frac{D}{d_1 - d_2}$

3. A differential pulley block has larger and smaller diameters of 100 mm and 80 mm respectively. Its velocity ratio is

   (*a*) 5 (*b*) 10 (*c*) 20 (*d*) 40

4. In a wormed geared pulley block, if we double the number of teeth on the worm wheel, its velocity ratio is

   (*a*) doubled (*b*) remains the same (*c*) halved

5. The velocity ratio of a single purchase crab winch can be increased by

   (*a*) increasing the length of the handle
   (*b*) increasing the radius of the load drum
   (*c*) increasing the number of teeth on the pinion
   (*d*) all of the above

6. The velocity ratio of a first system of pulleys with 4 pulleys is

(*a*) 4 (*b*) 8 (*c*) 16 (*d*) 15

7. If the number of pulleys in a system is equal to its velocity ratio, then it is a set of

(*a*) First system of pulleys (*b*) Second system of pulleys

(*c*) Third system of pulleys

8. In a simple screw jack, with (l) as the length of the effort wheel and (p) as pitch of the screw, its velocity ratio is

(*a*) $\frac{2\pi l}{p}$ (*b*) $\frac{\pi l}{2p}$ (*c*) $\frac{2\pi p}{l}$ (*d*) $\frac{\pi p}{2l}$

9. The velocity ratio of a double threaded worm geared screw jack is double than that of a single threaded screw jack provided other dimensions remain the same.

(*a*) True (*b*) False

ANSWERS

1. (*d*)	2. (*b*)	3. (*b*)	4. (*a*)
5. (*a*)	6. (*c*)	7. (*b*)	8. (*a*)
9. (*b*)			

CHAPTER

12 Support Reactions

Contents

1. SUPPORT REACTIONS

[UPTU]

In our day-to-day work, we see that whenever we apply a force on a body, it exerts a reaction, *e.g.*, when a ceiling fan is hung from a girder, it is subjected to the following two forces:

1. Weight of the fan, acting downwards, and
2. Reaction on the girder, acting upwards.

A little consideration will show, that as the fan is in equilibrium therefore, the above two forces must be equal and opposite. Similarly, if we consider the equilibrium of a girder supported on the walls, we see that the total weight of the fan and girder is acting through the supports of the girder on the walls. It is thus obvious, that walls must exert equal and upward reactions at the supports to maintain the equilibrium. The upward reactions, offered by the walls, are known as *support reactions.* As a mater of fact, the support reaction depends upon the type of loading and the support.

2. TYPES OF LOADING

Though there are many types of loading, yet the following are important from the subject point of view :

1. Concentrated or point load,
2. Uniformly distributed load,
3. Uniformly varying load.

3. CONCENTRATED OR POINT LOAD

A load, acting at a point on a beam is known as a *concentrated or a point load* as shown in Fig. 1.

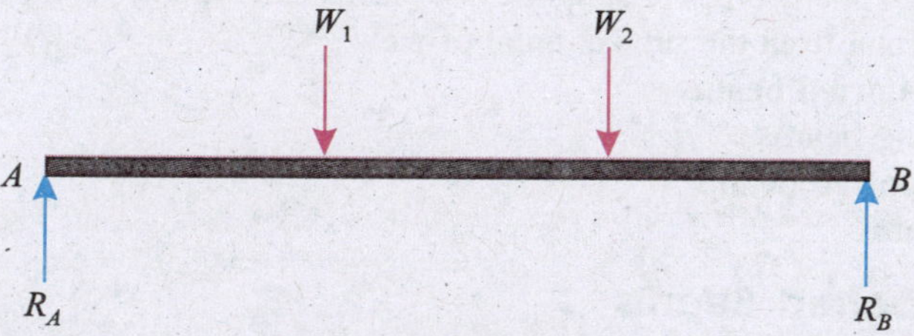

Fig. 1. Concentrated load.

In actual practice, it is not possible to apply a load at a point (*i.e.*, at a mathematical point), as it must have some contact area. But this area being so small, in comparison with the length of the beam, is negligible.

4. UNIFORMLY DISTRIBUTED LOAD

A load, which is spread over a beam, in such a manner that each unit length is loaded to the same extent, is known as *uniformly distributed load* (briefly written as U.D.L.) as shown in Fig. 2

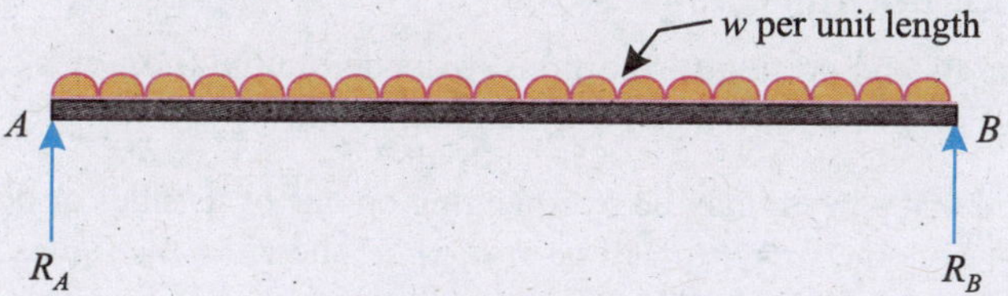

Fig. 2. Uniformly distributed load.

The total uniformly distributed load is assumed to act at the centre of gravity of the load for all sorts of calculations.

5. UNIFORMLY VARYING LOAD

A load, which is spread over a beam, in such a manner that its extent varies uniformly on each unit length (say from w_1 per unit length at one support to w_2 per unit length at the other support) is known as *uniformly varying load* as shown in Fig. 3.

Sometimes, the load varies from zero at one support to w at the other. Such a load is also called triangular load.

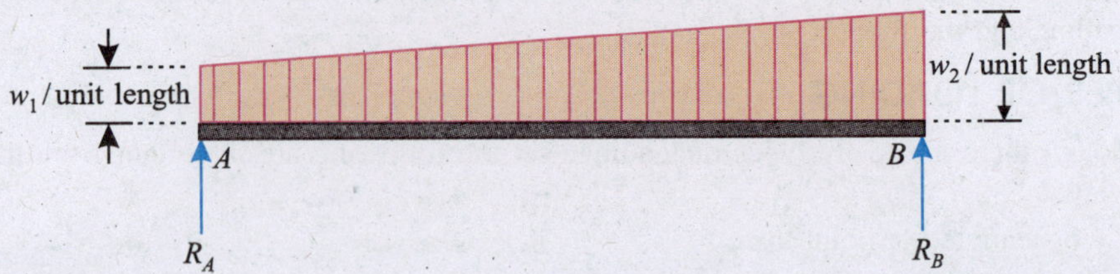

Fig. 3. Uniformly varying load.

Note : A beam may carry any one of the above-mentioned load system, or a combinations of the two or more.

5.1. TYPES OF END SUPPORTS OF BEAMS

Though there are many types of supports, for beams and frames, yet the following three types of supports are important from the subject point of view:

1. Simply supported beams
2. Overhanging beams
3. Roller supported beams
4. Hinged beams

Simply Supported Beams

It is a theoretical case, in which the end of a beam is simply supported over one of its support.

In such a case the reaction is always *vertical* as shown in Fig. 4.

Fig. 4. Simply supported beam

Overhanging Beams

A beam having its end portion (or portions) extended in the form of a cantilever, beyond its support, as shown in Fig. 5 is known as an overhanging beam.

It may be noted that a beam may be overhanging on one of its sides or both the sides. In such cases, the reactions at both the supports will be *vertical* as shown in the figure.

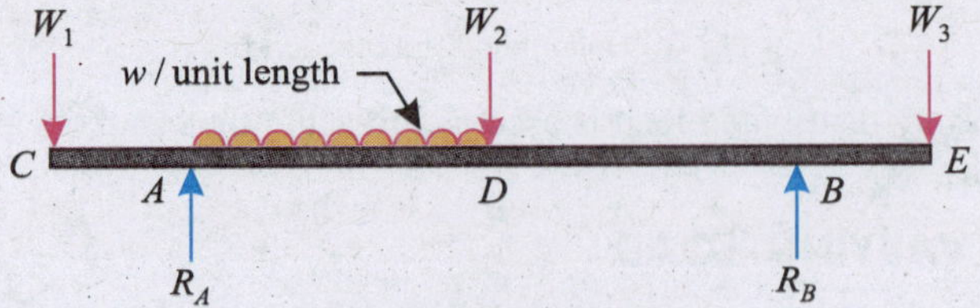

Fig. 5. Overhanging beam.

Roller Supported Beams

In such a case, the end of a beam is supported on rollers, and the reaction on such an end is always *normal to the support,* as shown in Fig. 6 (*a*) and (*b*). All the steel trusses, of the bridges, have one of their ends as supported on rollers.

The main advantage, of such a support, is that the beam can move easily towards left or right, on account of expansion or contraction due to change in temperature.

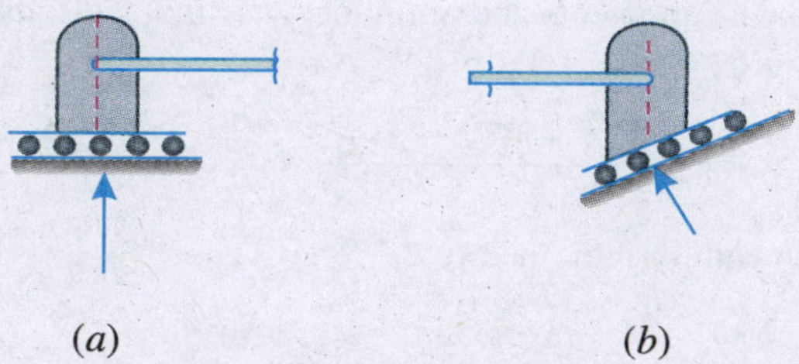

Fig. 6. Roller supported end

Hinged Beams

In such a case, the end of a beam is hinged to the support as shown in Fig. 7. The reaction on such an end may be *horizontal*, *vertical* or *inclined,* depending upon the type of loading. All the steel trusses of the bridges have one of their end roller supported, and the other hinged.

The main advantage of such a support is that the beam remains stable. A little consideration will show, that the beam cannot be stable, if both of its ends are supported on rollers. It is thus obvious, that one of the supports is made roller supported and the other hinged.

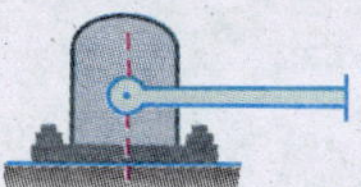

Fig. 7. Hinged end.

6. METHODS TO FIND THE REACTIONS OF A BEAM

The reactions at the two supports of a beam may be found out by any one of the following two methods:

1. Analytical method 2. Graphical method.

We shall discuss only analytical method in the following pages.

7. ANALYTICAL METHOD FOR THE REACTIONS OF A BEAM

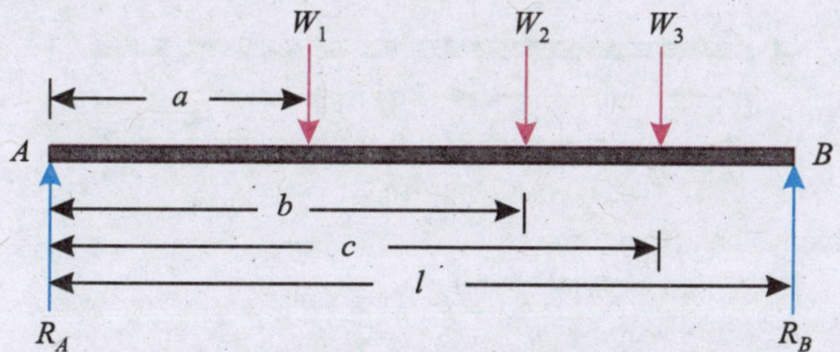

Fig. 8. Reactions of a beam.

Consider a simply supported beam *AB* of span *l*, subjected to point loads W_1, W_2 and W_3 at distances of *a*, *b* and *c*, respectively from the support *A*, as shown in Fig. 8

Let R_A = Reaction at *A*, and

R_B = Reaction at B.

We know that sum of the clockwise moments due to loads about A

$$= W_1 a + W_2 b + W_3 c \quad ...(i)$$

and anti-clockwise moment due to reaction R_B about A

$$= R_B\, l \quad ...(ii)$$

Now equating clockwise moments and anti-clockwise moments about A,

$$R_B\, l = W_1\, a + W_2\, b + W_3\, c \quad ...(\because \Sigma M = 0)$$

or

$$R_B = \frac{W_1 a + W_2 b + W_3 c}{l} \quad ..(iii)$$

Since the beam is in equilibrium, therefore

$$R_A + R_B = W_1 + W_2 + W_3 \quad ...(\because \Sigma V = 0)$$

and

$$R_A = (W_1 + W_2 + W_3) - R_B$$

8. NUMERICAL ON FINDING SUPPORT REACTIONS FOR DIFFERENT TYPES OF END SUPPORTS OF BEAMS

We shall solve the numerical based on following types of end suports:

1. Simply supported beams

2. Overhanging beams

3. Roller supported beams

4. Hinged beams

9. SIMPLY SUPPORTED BEAMS

It is a theoretical case, in which the end of a beam is simply supported over one of its support. In such a case the reaction is always *vertical* as shown in Fig. 9.

Fig. 9. Simply supported beam

Example 1. *A simply supported beam AB of span 5 m is loaded as shown in Fig. 10. Find the reactions at A and B.*

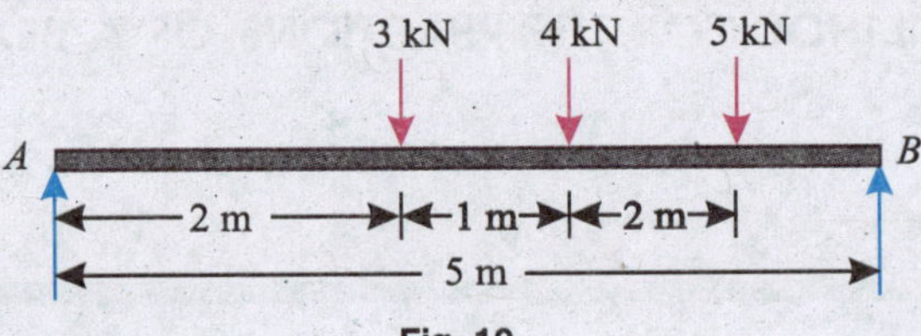

Fig. 10.

Solution. Given: Span (l) = 5 m

Let R_A = Reaction at A, and

R_B = Reaction at B.

The example may be solved either analytically or graphically. But we shall solve analytically only.

Anti-clockwise moment due to R_B about A,

$$= R_B \times l = R_B \times 5 = 5\, R_B \text{ kN-m} \quad ...(i)$$

Sum of the clockwise moments about A,

$$= (3 \times 2) + (4 \times 3) + (5 \times 4) = 38 \text{ kN-m} \quad ...(ii)$$

Equating anti-clockwise and clockwise moments about A given in (*i*) and (*ii*),

$$5\,R_B = 38$$

or $$R_B = \frac{38}{5} = 7.6 \text{ kN} \quad \textbf{Ans.}$$

and $$R_A = (3 + 4 + 5) - 7.6 = 4.4 \text{ kN} \quad \textbf{Ans.}$$

Example 2. *A simply supported beam, AB of span 6 m is loaded as shown in Fig. 11.*

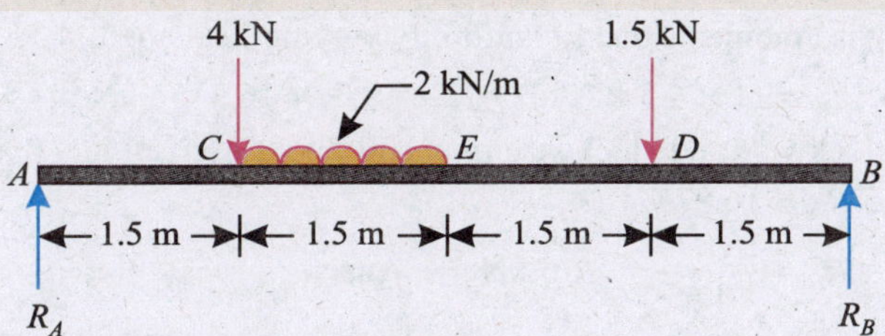

Fig. 11.

Determine the reactions R_A and R_B of the beam. **[PU]**

Solution. Given: Span (l) = 6m

Let R_A = Reaction at A, and

R_B = Reaction at B.

The example may be solved either analytically or graphically. But we shall solve it analytically.

Anti-clockwise moment due to the reaction R_B about A.

$$= R_B \times l = R_B \times 6 = 6\,R_B \text{ kN-m} \quad ...(i)$$

Sum* of the clockwise moments about A

$$= (4 \times 1.5) + (2 \times 1.5)\,2.25 + (1.5 \times 4.5) = 19.5 \text{ kN-m} \quad ...(ii)$$

Equating anti-clockwise and clockwise moments about A given in (*i*) and (*ii*),

$$6\,R_B = 19.5$$

or $$R_B = \frac{19.5}{6} = 3.25 \text{ k.N} \quad \textbf{Ans.}$$

and $$R_A = 4 + (2 \times 1.5) + 1.5 - 3.25 = 5.25 \text{ kN} \quad \textbf{Ans.}$$

***Note:** The uniformly distributed load of 2 kN/m for a length of 1.5 m (*i.e.*, between C and E) is assumed as an equivalent point load of 2 × 1.5 = 3 kN and acting at the centre of gravity of the load *i.e.*, at a distance of 1.5 + 0.75 = 2.25 m from A.

Example 3. *A simply supported beam AB of span 4.5 m is loaded as shown in Fig. 12.*

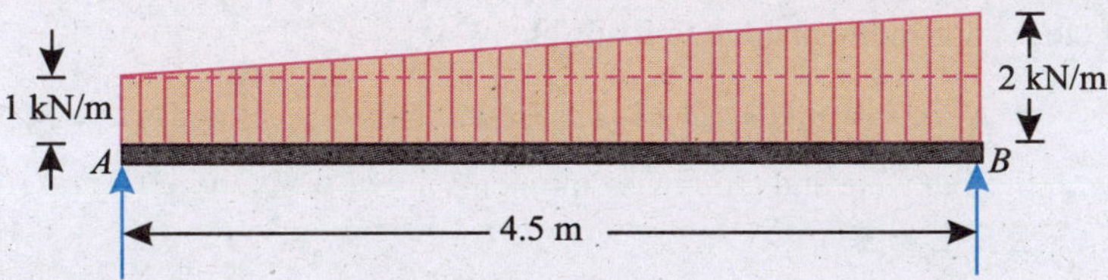

Fig. 12.

Find the support reactions at A and B.

Solution. Given: Span $(l) = 4.5$ m

Let R_A = Reaction at A, and

R_B = Reaction at B.

For the sake of simplicity, we shall assume the uniformly varying load to be split† up into (*a*) a uniformly distributed load of 1 kN/m over the entire span, and (*b*) triangular load of 0 at A to 1 kN/m at B.

Anti-clockwise moment due to R_B about A

$$= R_B \times l = R_B \times 4.5 = 4.5\, R_B \text{ kN-m} \quad ...(i)$$

Sum of clockwise moments due to uniformly varying load about A

$$= (1 \times 4.5 \times 2.25) + (2.25 \times 3) = 16.875 \text{ kN-m} \quad ...(ii)$$

Equating anti-clockwise and clockwise moments about A given in (*i*) and (*ii*),

$$4.5\, R_B = 16.875$$

or $$R_B = \frac{16.875}{4.5} = 3.75 \text{ kN} \quad \textbf{Ans.}$$

and $$R_A = [1 \times 4.5] + \left[4.5 \times \frac{0+1}{2}\right] - 3.75 = 3.0 \text{ kN} \quad \textbf{Ans.}$$

†**Note:** The uniformly distributed load of 1 kN/m over the entire span is assumed as an equivalent point load of $1 \times 4.5 = 4.5$ kN and acting at the centre of gravity of the load *i.e.* at a distance of 2.25 m from A. Similarly, the triangular load in assumed as an equivalent point load of $4.5 \times \frac{0+1}{2} = 2.25$ kN and acting at the centre of gravity of the load *i.e.*, distance of $4.5 \times \frac{2}{3} = 3$ m from A.

Example 4. *A simply supported beam AB of* 6 m *span is subjected to loading as shown in Fig. 13.*

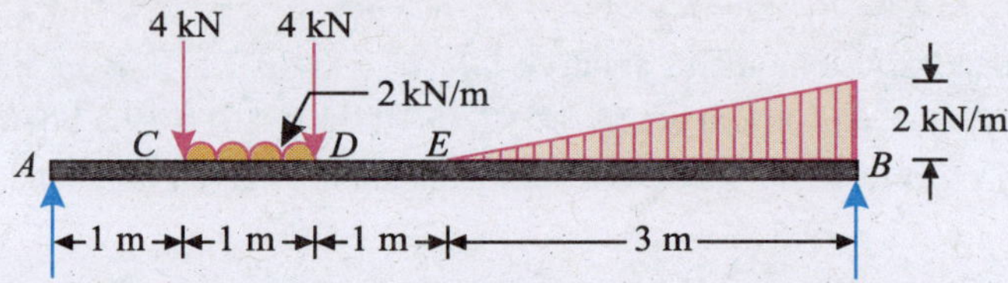

Fig. 13.

Find graphically or otherwise, the support reactions at A and B. **[MU]**

Solution. Given: Span $(l) = 6$ m

Let R_A = Reaction at A, and

R_B = Reaction at B.

Anti-clockwise moment due to R_B about A

$$= R_B \times l = R_B \times 6 = 6\, R_B \text{ kN-m} \quad ...(i)$$

*Sum of clockwise moments due to loads about A

$$= (4 \times 1) + (2 \times 1)\, 1.5 + (4 \times 2) + \frac{(0+2)}{2} \times 3 \times 5 = 30 \text{ kN-m} \quad ...(ii)$$

Equating anti-clockwise and clockwise moments about A given in (*i*) and (*ii*),

$$6\, R_B = 30$$

or $$R_B = \frac{30}{6} = 5 \text{ kN} \quad \textbf{Ans.}$$

and $$R_A = (4 + 2 + 4 + 3) - 5 = 8 \text{ kN} \quad \textbf{Ans.}$$

[*Note: It means converting the uniformly distributed load between C and D as well as triangular load between E and B into vertical loads as discussed below:

1. The uniformly distributed load is assumed as an equivalent point load of $2 \times 1 = 2$ kN acting at the centre of gravity of the load *i.e.*, at the mid point of C and D.
2. The triangular load is assumed as an equivalent point load of $\frac{0+2}{2} \times 3 = 3$ kN acting at the centre of gravity of the load *i.e.* at a distance of $\frac{2}{3} \times 3 = 2$ m from E or 5 m from A.]

10. OVERHANGING BEAMS

A beam having its end portion (or portions) extended in the form of a cantilever, beyond its support, as shown in Fig. 14 is known as an overhanging beam.

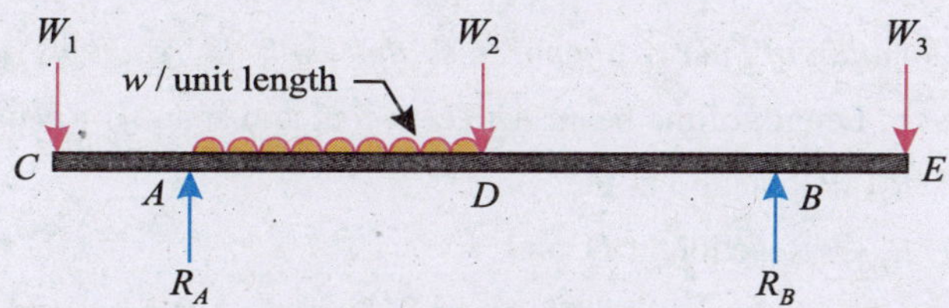

Fig. 14. Overhanging beam.

It may be noted that a beam may be overhanging on one of its sides or both the sides. In such cases, the reactions at both the supports will be vertical as shown in the figure.

Example 5. *A beam AB of span 3m, overhanging on both sides is loaded as shown in Fig. 15.*

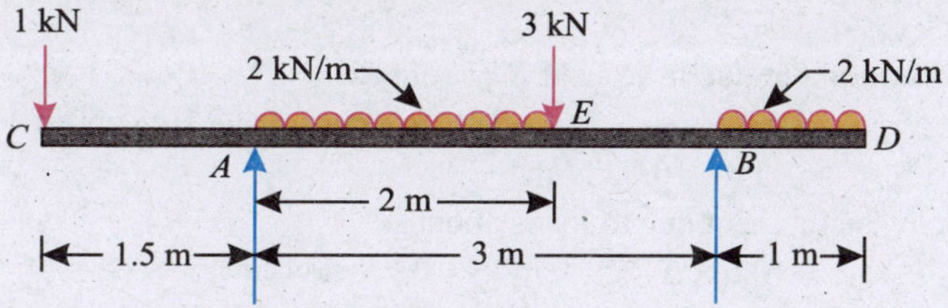

Fig. 15.

Determine the reactions at the supports A and B.

Solution. Given: Span (l) = 3 m

Let R_A = Reaction at A, and

R_B = Reaction at B.

Anti-clockwise moment due to R_B and load* at C about A,

...(*Note: The 1 kN load at C, will also cause an anti-clockwise moment about A.)

$$= R_B \times l + (1 \times 1.5) = R_B \times 3 + (1 \times 1.5) = 3R_B + 1.5 \text{ kN} \quad ...(i)$$

Sum of clockwise moments due to loads about A

$$= (2 \times 2)\, 1 + (3 \times 2) + (1 \times 1)\, 3.5 = 13.5 \text{ kN-m} \quad ...(ii)$$

Equating anti-clockwise and clockwise moments about A given in (*i*) and (*ii*),

$$3R_B + 1.5 = 13.5$$

or $$R_B = \frac{13.5 - 1.5}{3} = \frac{12}{3} = 4 \text{ kN} \quad \textbf{Ans.}$$

and $$R_A = 1 + (2 \times 2) + 3 + (1 \times 1) - 4 = 5 \text{ kN} \quad \textbf{Ans.}$$

Example 6. *A beam AB 5 m long, supported on two in termediate supports 3 m apart, carries a uniformly distributed load of 0.6 kN/m. The beam also carries two concentrated loads of 3 kN at left hand end A, and 5 kN at the right hand end B as shown in Fig. 16.*

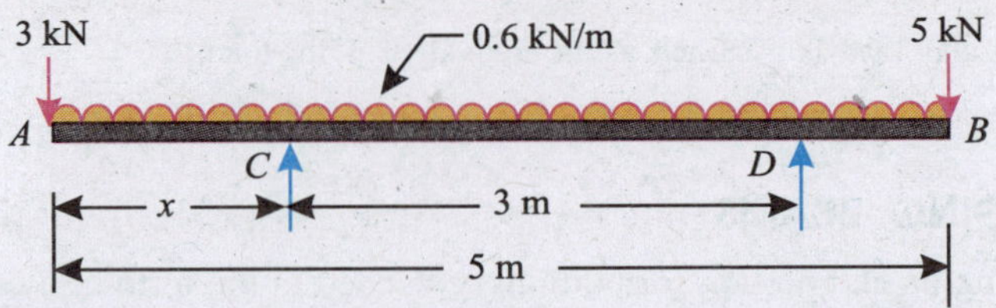

Fig. 16.

Determine the location of the two supports, so that both the reactions are equal.

Solution. Given: Length of the beam *AB* (*L*) = 5 m and span (*l*) = 3 m.

Let R_C = Reaction at C,

R_D = Reaction at *D*, and

x = Distance of the support *C* from the left hand end

Total load on the beam

$$= 3 + (0.6 \times 5) + 5 = 11 \text{ kN}$$

Since the reactions R_C and R_D are equal, therefore reaction at support

$$= \frac{11}{2} = 5.5 \text{ kN}$$

Anti-clockwise moment due to R_C and R_D about *A*

$$= 5.5 \times x + 5.5\,(x + 3) = 5.5\,x + 5.5\,x + 16.5 \text{ kN-m}$$

$$= 11x + 16.5 \text{ kN-m} \quad ...(i)$$

Sum of clockwise moment due to loads about *A*

$$= (0.6 \times 5)\,2.5 + 5 \times 5 = 32.5 \text{ kN-m} \quad ...(ii)$$

Distance of the support C from the left hand end (x) can be calculated by, Equating anti-clockwise and clockwise moments about *A* given in (*i*) and (*ii*)

$$11\,x + 16.5 = 32.5 \quad \text{or} \quad 11\,x = 16$$

or $$x = \frac{16}{11} = 1.45 \text{ m}$$

It is thus obvious that the first support will be located at distance of 1.45m from *A* and second support at a distance of 1.45 + 3 = 4.45 m from *A*. **Ans.**

Alternative Method

Sum of anti-clockwise moments due to R_D and point load at *A* about the support *C*

$$= (5.5 \times 3) + (3 \times x) = 16.5 + 3x$$

Sum of clockwise moments due to loads about *C*

$$= 5\,(5 - x) + 0.6 \times 5(2.5 - x) = 25 - 5x + 7.5 - 3x = 32.5 - 8x$$

Distance of the support C from the left hand end (x) can be calculated by, Equating anti-clockwise and clockwise moments about *A* given in (*iii*) and (*iv*)

$$16.5 + 3x = 32.5 - 8x \quad \text{or} \quad 11x = 16$$

or $$x = \frac{16}{11} = 1.45 \text{ m} \quad \textbf{Ans.}$$

Note: In the second method, the uniformly distributed load between A and C will cause anti-clockwise moment about C, while the load between C and B will cause clockwise moment. But for the sake of simplicity, we have taken the entire load from A to B (equal to 0.6×5) acting at its centre (*i.e.* 2.5 m from A or $2.5 - x$) from C.

EXERCISE 1

1. A simply supported beam AB of span 4 m is carrying a point loads of 5, 2 and 3 kN at 1, 2 and 3 m respectively from the support A. Calculate the reactions at the supports A and B.

   [Ans. 5.5 kN and 4.5 kN]

2. A simply supported beam of span 6 m is carrying a uniformly distributed load of 2 kN/m over a length of 3 m from the right end B. Calculate the support reactions.

   [Ans. R_A = 1.5 kN, R_B = 4.5 kN]

3. A simply supported beam AB of span 6 m is loaded as shown in Fig. 17.

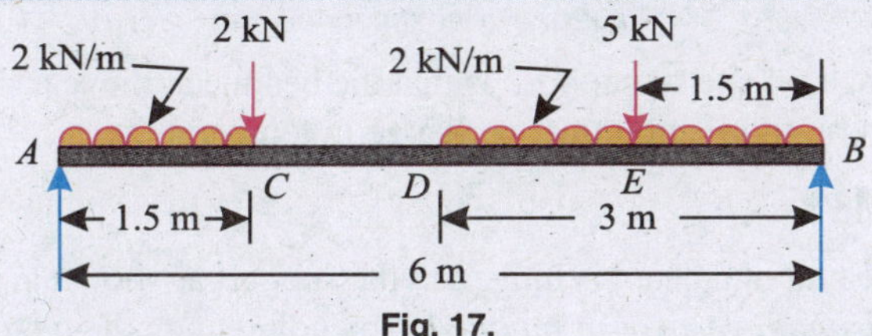

Fig. 17.

   Determine the reactions at A and B. **[PU] [Ans.** 6.875 kN, 9.125 kN]

4. A beam AB 6 m long rests on two supports 4 m apart, the right hand end is overhanging by 2 m. The beam carries a uniformly distributed load of 1 kN/m over the entire length of the beam.

   Determine the reactions at the two supports. **[Ans.** R_A = 1.5 kN, R_B = 4.5 kN]

5. A beam $ABCDEF$ of 7.5 m long and span 4.5 m is supported at B and E. The beam is loaded as shown in Fig. 18.

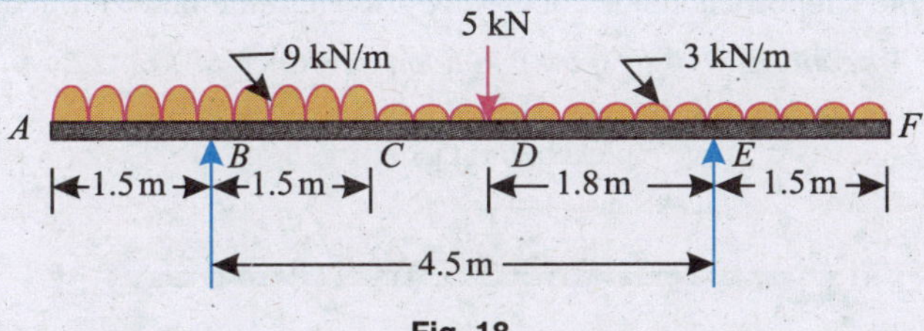

Fig. 18.

   Find graphically, or otherwise, the support reactions at the two supports.

   [Ans. R_B = 29.33 kN, R_E = 12.57 kN]

6. A beam $ABCDE$ hinged at A and supported on rollers at D, is loaded as shown in Fig. 19.

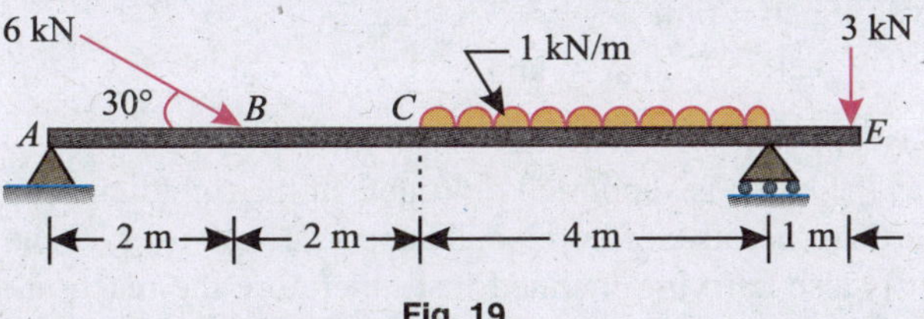

Fig. 19.

   Find the reactions at A and D. **[IGNOU] [Ans.** R_A = 5.94 kN, R_D = 7.125 kN, $\theta = 61°$]

11. ROLLER SUPPORTED BEAMS

In such a case, the end of a beam is supported on rollers, and the reaction on such an end is always *normal to the support,* as shown in Fig. 20 (*a*) and (*b*). All the steel trusses, of the bridges, have one of their ends as supported on rollers.

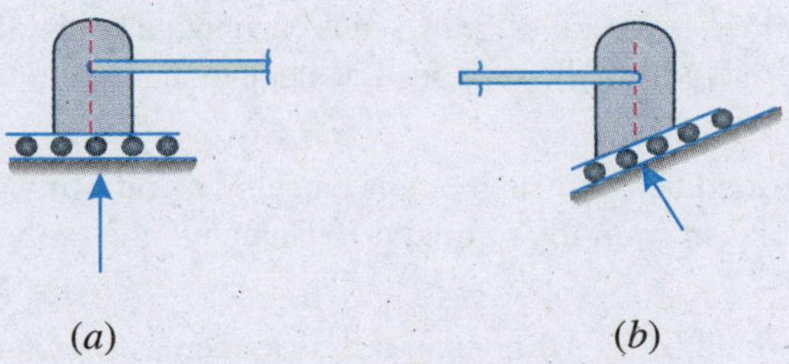

Fig. 20. Roller supported end

The main advantage, of such a support, is that the beam can move easily towards left or right, on account of expansion or contraction due to change in temperature.

12. HINGED BEAMS

In such a case, the end of a beam is hinged to the support as shown in Fig. 21. The reaction on such an end may be *horizontal*, *vertical* or *inclined,* depending upon the type of loading. All the steel trusses of the bridges have one of their end roller supported, and the other hinged.

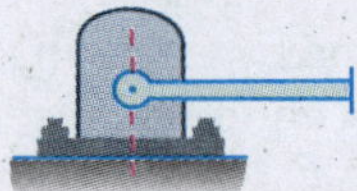

Fig. 21. Hinged end.

The main advantage of such a support is that the beam remains stable. A little consideration will show, that the beam cannot be stable, if both of its ends are supported on rollers. It is thus obvious, that one of the supports is made roller supported and the other hinged.

Example 7. *A beam AB of 6 m span is loaded as shown in Fig. 22.*

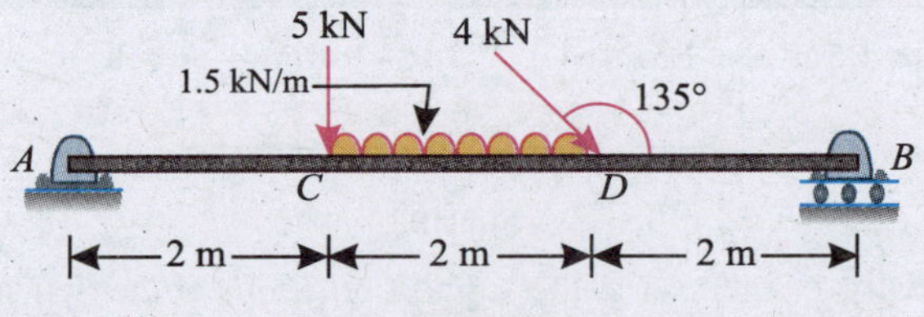

Fig. 22.

Determine the reactions at A and B. **[MU]**

Solution. Given: Span = 6 m

Let R_A = Reaction at *A*, and

R_B = Reaction at *B*.

We know that as the beam is supported on rollers at the right hand support (*B*), therefore the reaction R_B will be vertical (because of horizontal support). Moreover, as the beam is hinged at the left support (*A*) and it is also carrying inclined load, therefore the reaction at this end will be the resultant of horizontal and vertical forces, and thus will be inclined with the vertical.

The example may be solved either analytically or graphically, but we shall solve it by both the methods, one by one.

Analytical method

Resolving the 4 kN load at D vertically

$$= 4 \sin 45° = 4 \times 0.707 = 2.83 \text{ kN}$$

Resolving it horizontally

$$= 4 \cos 45° = 4 \times 0{\cdot}707 = 2{\cdot}83 \text{ kN}$$

Anti-clockwise moment due to R_B about A

$$= R_B \times 6 = 6\ R_B \text{ kN-m} \qquad ...(i)$$

*Sum of clockwise moments due to loads about A,

$$= (5 \times 2) + (1.5 \times 2)\ 3 + 2.83 \times 4 = 30.3 \text{ kN-m} \qquad ...(ii)$$

...(***Note:** Moment of horizontal component of 2.83 kN at D about A will be zero.)

Equating the anti-clockwise and clockwise moments about A given in (*i*) and (*ii*),

$$6\ R_B = 30.3$$

or $$R_B = \frac{30.3}{6} = 5.05 \text{ kN} \quad \textbf{Ans.}$$

Vertical component of the reaction R_A

$$= [5 + (1.5 \times 2) + 2.83] - 5.05 = 5.78 \text{ kN}$$

∴ Reaction at A,

$$R_A = \sqrt{(5.78)^2 + (2.83)^2} = 6.44 \text{ kN} \quad \textbf{Ans.}$$

Let θ = Angle, which the reaction at A makes with vertical.

∴ $$\tan\theta = \frac{2.83}{5.78} = 0.4896 \quad \text{or} \quad \theta = 26.1° \quad \textbf{Ans.}$$

Example 8. *A beam AB 8.5 m long is hinged at A and supported on rollers over a smooth surface inclined at 30° to the horizontal at B. The beam is loaded as shown in Fig. 23.*

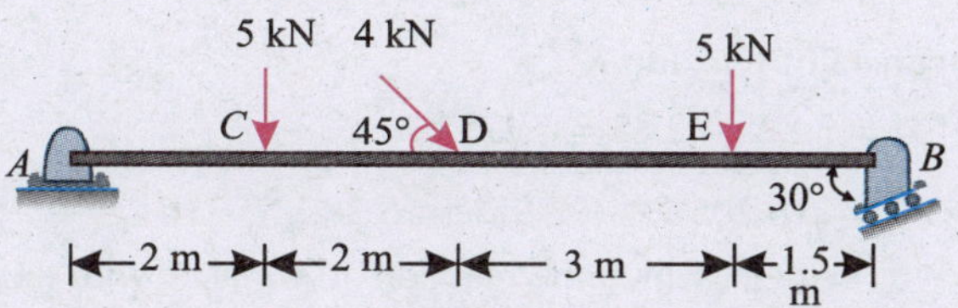

Fig. 23.

Determine graphically, or otherwise, the reactions at A and B. **[RGPV]**

Solution. Given: Span = 8.5 m

Let R_A = Reaction at A, and

R_B = Reactiion at B.

We know that, as the beam is supported on rollers at B, therefore the reaction at this end will be normal to the support *i.e.* inclined at an angle of 30° with the vertical (because the support is inclined at 30° with the horizontal) as shown in Fig. 24. Moreover, as the beam is hinged at A, therefore the reaction at this end will be the resultant of vertical and horizontal forces, and thus will be inclined with the vertical.

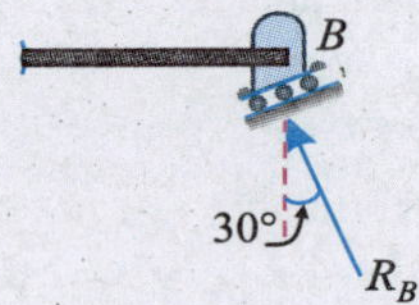

Fig. 24.

Resolving the 4 kN load at D vertically

$$= 4 \sin 45° = 4 \times 0.707 = 2.83 \text{ kN}$$

Resolving it horizontally

$$= 4 \cos 45° = 4 \times 0.707 = 2.83 \text{ kN}$$

Vertical component of reaction R_B

$$= R_B \cos 30° = R_B \times 0.866 = 0.866\, R_B$$

Anti-clockwise moment due to vertical component of reaction R_B about A

$$= 0.866\, R_B \times 8.5 = 7.361\, R_B \quad ...(i)$$

Sum of clockwise moments due to loads about A

$$= (5 \times 2) + (2.83 \times 4) + (5 \times 7) = 56.32 \text{ kN-m} \quad ...(ii)$$

*Equating anti-clockwise and clockwise moments about A given in (*i*) and (*ii*),

$$7.361\, R_B = 56.32$$

or $$R_B = \frac{56.32}{7.361} = 7.65 \text{ kN} \quad \textbf{Ans.}$$

...(***Note:** The moment of horizontal component of R_B and 2.83 kN at D about A will be zero.)

Vertical component of the reaction R_B

$$= 0.866\, R_B = 0.866 \times 7.65 = 6.625 \text{ kN}$$

Horizontal component of reaction R_B

$$= R_B \sin 30° = 7.65 \times 0.5 = 3.825 \text{ kN}$$

Vertical component of reaction R_A

$$= (5 + 2.83 + 5) - 6.625 = 6.205 \text{ kN}$$

Horizontal component of reaction R_A

$$= 3.825 - 2.83 = 0.995 \text{ kN}$$

$$\therefore \quad R_A = \sqrt{(6.205)^2 + (0.995)^2} = 6.28 \text{ kN}$$

Let θ = Angle, which the reaction at A makes with the vertical.

$$\therefore \quad \tan\theta = \frac{0.995}{6.205} = 0.1604 \quad \text{or} \quad \theta = 9.1°$$

Example 9. *A beam has hinged support at A and roller support at B as shown in Fig. 25.*

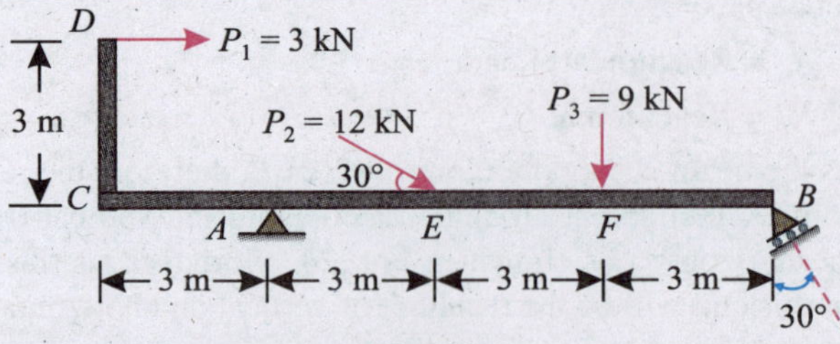

Fig. 25.

The beam is subjected to loads as shown. Determine analytically the reactions at A and B.

[JNTU]

Solution. Given: Span = 9 m

Let R_A = Reaction at A, and

R_B = Reaction at B.

The reaction at B, supported on rollers and inclined at an angle of 30° with the vertical is shown in Fig. 26. We know that as the beam is hinged at A, therefore the reaction at this end will be the resultant of vertical and horizontal forces, and thus will be inclined with the vertical.

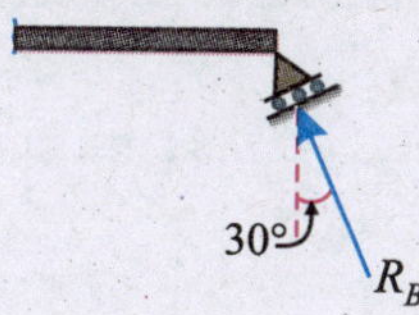

Fig. 26.

Resolving the 12 kN load at E vertically

$$= 12 \sin 30° = 12 \times 0.5 = 6 \text{ kN}$$

Resolving it horizontally

$$= 12 \cos 30° = 12 \times 0.866 = 10.4 \text{ kN}$$

Vertical component of reaction R_B.

$$= R_B \cos 30° = R_B \times 0.866 = 0.866\ R_B$$

Anti-clockwise moment due to vertical component of reaction R_B about A

$$= 0.866\ R_B \times 9 = 7.794\ R_B \qquad ...(i)$$

Sum of clockwise moments due to loads about A

$$= (6 \times 3) + (9 \times 6) + (3 \times 3) = 81 \text{ kN-m} \qquad ...(ii)$$

Equating the anticlockwise and clockwise moments about A given in (*i*) and (*ii*),

$$7.794\ R_B = 81 \quad \text{or} \quad R_B = \frac{81}{7.794} = 10.4 \text{ kN} \quad \textbf{Ans.}$$

Vertical component of the reaction R_B

$$= 0.866\ R_B = 0.866 \times 10.4 = 9.0 \text{ kN}$$

Horizontal component of reaction R_B

$$= R_B \sin 30° = 10.4 \times 0.5 = 5.2 \text{ kN}$$

Vertical component of reaction R_A

$$= (6 + 9) - 9 = 6 \text{ kN}$$

Horizontal component of reaction R_A

$$= (3 + 10.4) - 5.2 = 8.2 \text{ kN}$$

or $$R_B = \sqrt{(6)^2 + (8.2)^2} = 10.16 \text{ kN} \quad \textbf{Ans.}$$

Let θ = Angle, which the reaction at A makes with the vertical.

$$\therefore \quad \tan\theta = \frac{8.2}{6} = 1.3667 \quad \text{or} \quad \theta = 53.8° \quad \textbf{Ans.}$$

13. BEAMS SUBJECTED TO A MOMENT

Sometimes, a beam is subjected to a clockwise or anti-clockwise moment alongwith loads. In such a case, magnitude of the moment is taken into consideration while calculating the reactions. Since the moment does not involve any load, therefore it has no horizontal or vertical components.

Example 10. *Fig. 27 shows as beam ABCD simply supported on a hinged support at A and at D on a roller support inclined at 45° with the vertical.*

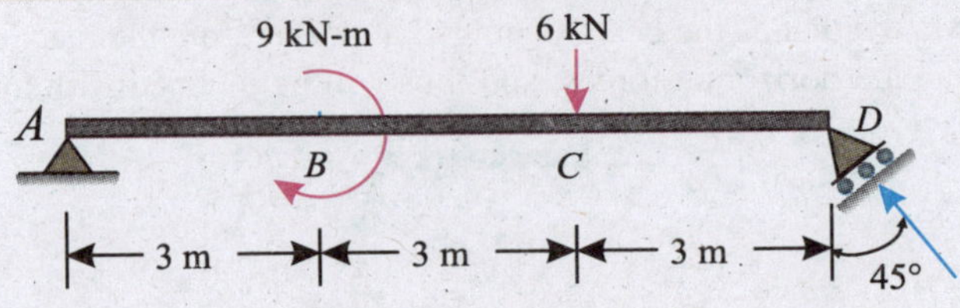

Fig. 27.

Determine the horizontal and vertical components of reaction at support A. Show clearly the direction as well as the magnitude of the resultant reaction at A.

Solution. Given: Span = 9 m

Let R_A = Reaction at A, and

R_D = Reaction at D.

The reaction R_D is inclined at 45° with the vertical as given in the example. We know that as the beam is hinged at A, therefore the reaction at this end will be the resultant of vertical and horizontal forces, and thus will be inclined with the vertical.

We know that, vertical component of reaction R_D

$$= R_D \cos 45° = R_D \times 0.707 = 0.707\ R_D$$

Anti-clockwise moment due to the vertical component of reaction R_D about A

$$= 0.707\ R_D \times 9 = 6.363\ R_D \qquad ...(i)$$

Sum of clockwise moments due to moment at B and Load at C about A.

$$= 9 + (6 \times 6) = 45 \text{ kN-m} \qquad ...(ii)$$

Equating the anti-clockwise and clockwise moments about A given in (*i*) and (*ii*),

$$6.363\ R_D = 45$$

or

$$R_D = \frac{45}{6.363} = 7.07 \text{ kN}$$

Vertical component of reaction R_D

$$= 7.07 \cos 45° = 7.07 \times 0.707 = 5 \text{ kN}$$

Horizontal component of R_D (this is also equal to horizontal component of reaction R_A as there is no inclined load on the beam)

$$= 7.07 \sin 45° = 7.07 \times 0.707 = 5 \text{ kN}$$

∴ Vertical component of reaction R_A

$$= 6 - 5 = 1 \text{ kN}$$

and

$$R_A = \sqrt{(5)^2 + (1)^2} = 5.1 \text{ kN} \qquad \textbf{Ans.}$$

Let θ = Angle, which the reaction at A makes with the vertical.

$$\therefore \quad \tan\theta = \frac{5}{1} = 5.0 \qquad \text{or} \qquad \theta = 78.7° \qquad \textbf{Ans.}$$

QUESTIONS

1. What are various type of loadings? Distinguish clearly between uniformly distributed load, uniformly varying load and triangular load.
2. Define the term 'support reaction'. Describe the analytical as well as graphical methods for finding out the support reactions of a beam carrying vertical loads only.
3. State the direction of a support reactions in the case of (*a*) simply supported end (*b*) hinged end and (*c*) roller supported end.
4. Write the assumptions, which are made, while finding out the reactions of a beam or a frame having both ends fixed.

OBJECTIVE TYPE QUESTIONS

1. In a simply supported beam carrying triangular load, the reactions can not be vertical
   (*a*) True (*b*) False
2. An overchanging beam with downward loads............have one of its reaction upward and the other downward.
   (*a*) can (*b*) can not
3. The reaction at the roller supported end of a beam is always
   (*a*) vertical (*b*) horizontal (*c*) none of the above
4. If the reaction of a beam, at one of its supports is the resultant of horizontal and vertical forces, then it is a
   (*a*) simply supported end (*b*) roller supported end (*c*) hinged end
5. A couple acting at the mid-point of a simply supported beam has some horizontal and vertical components.
   (*a*) Agree (*b*) Disagree.

ANSWERS

1. (*b*) **2.** (*a*) **3.** (*c*) **4.** (*c*) **5.** (*b*)

CHAPTER

13 Analysis of Perfect Frames (Analytical Method)

Contents

1. FRAME OR TRUSS

[AM-IETE, RGPV, BPUT]

A structure, made up of several bars, riveted or welded together is called a framed structure or **Frame** or **Truss**. The bars of frame are made up of angle irons or channel sections, and are called **members of the frame**.

Though these members are welded or riveted together, at their joints, yet for calculation purposes, the joints are assumed to be hinged or pin-jointed the determination of force in a frame is an important problem in engineering- science, which can be solved by the application of the principles of either statics or graphics. in this chapter, we shall be using the principles of statics for determining the forces in frames.

2. TYPES OF FRAMES

[RGPV, UPTU]

Though there are many types of frames, yet from the analysis point of view, the frames may be classified into the following two groups:

1. Perfect frame. **2. Imperfect frame.**

3. PERFECT FRAME

[AM-IETE, IGNOU, UPTU, RGPV, BPUT]

A perfect frame is that, which is made up of members just sufficient to keep it in equilibrium, when loaded, without any change in its shape.

The simplest perfect frame is a triangle, which contains three members and three joints as shown in Fig. 1. It will be intersting to know that if such a structure is loaded, its shape will not be distorted. Thus, for three jointed frame, there should be three members to prevent any distortion. It will be further noticed that if we want to increase a joint, to a triangular frame, we require two members as shown by dotted lines in Fig. 1. Thus we see that for every additional joint, to a triangular frame, two members are required.

The no. of members, in a perfect frame, may also be expressed by the relation :

$$n = (2j - 3)$$

n = No. of members, and

j = No. of joints.

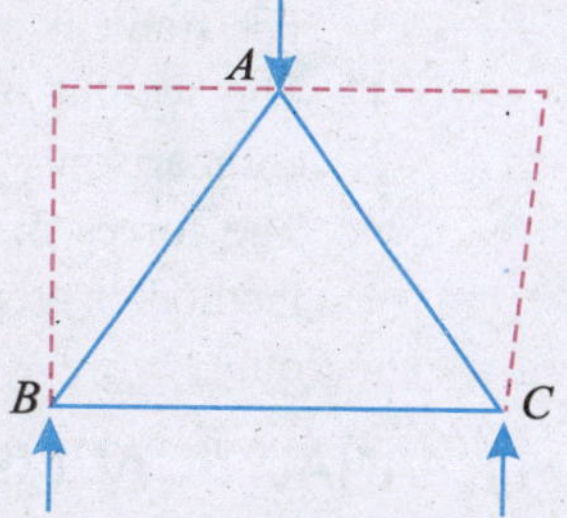

Fig. 1. Perfect Frame.

4. IMPERFECT FRAME

An imperfect frame is that which does not satisfy the equation :

$$n = (2j - 3)$$

Or in other words, it is a frame in which the no. of members are *more* or *less* than $(2j - 3)$. The imperfect frames may be further classified into the following two types :

1. Deficient frame. **2. Redundant frame.**

5. DEFICIENT FRAME

A deficient frame is an imperfect frame, in which the no. of members are less than $(2j - 3)$.

6. REDUNDANT FRAME

[AM-IETE, UPTU]

A redundant frame is an imperfect frame, in which the no. of members are more than $(2j - 3)$. In this chapter, we shall discuss only perfect frames.

7. STRESS

When a body is acted upon by a force, the internal force which is transmitted through the body is known as stress. Following two types of stress are important from the subject point of view :

1. Tensile stress. **2. Compressive stress.**

8. TENSILE STRESS

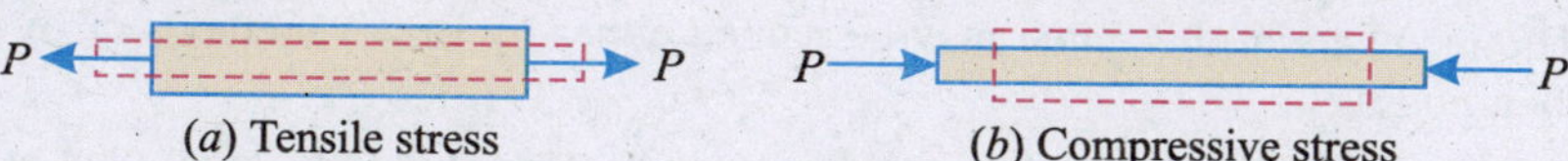

(*a*) Tensile stress (*b*) Compressive stress

Fig. 2.

Sometimes, a body is pulled outwards by two equal and opposite forces and the body tends to extend, as shown in Fig 2. (*a*). The stress induced is called tensile stress and corresponding force is called tensile force.

9. COMPRESSIVE STRESS

Sometimes, a body is pushed inwards by two equal and opposite forces and the body tends to shorten its length as shown in Fig. 2 (*b*). The stress induced is called compressive stress and the corresponding force is called compressive force.

10. ASSUMPTIONS FOR FORCES IN THE MEMBERS OF A PERFECT FRAME

[AM-IETE, JNTU, UPTU]

Following assumptions are made, while finding out the forces in the members of a perfect frame:

1. All the members are pin-jointed.
2. The frame is loaded only at the joints.
3. The frame is a perfect one.
4. The weight of the members, unless stated otherwise, is regarded as negligible in comparison with the other external forces or loads acting on the truss.

The forces in the members of a perfect frame may be found out either by analytical method or graphical method. But in this chapter, we shall discuss the analytical method only.

11. ANALYTICAL METHODS TO FIND FORCES IN MEMBERS OF PERFECT FRAME

The following two analytical methods for finding out the forces, in the members of a perfect frame, are important from the subject point of view :

1. Method of Joints. **2. Method of Sections.**

12. METHOD OF JOINTS

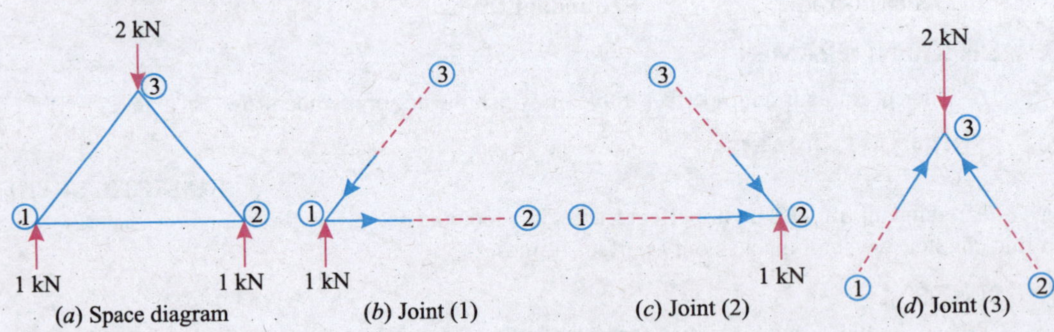

(*a*) Space diagram (*b*) Joint (1) (*c*) Joint (2) (*d*) Joint (3)

Fig. 3.

In this method, each and every joint is treated as a free body in equilibrium as shown in Fig. 3 (*a*), (*b*), (*c*) and (*d*). The unknown forces are then determined by equilibrium equations viz., $\Sigma V = 0$ and $\Sigma H = 0$. *i.e.,* Sum of all the vertical forces and horizontal forces is equated to zero.

Notes: 1. The members of the frame may be named either by Bow's methods or by the joints at their ends.

2. While selecting the joint, for calculation work, care should be taken that at any instant, the joint should not contain more than two members, in which the forces are unknown.

13. METHOD OF SECTIONS (OR METHOD OF MOMENTS)

This method is particularly convenient, when the forces in a few members of a frame are required to be found out. In this method, a section line is passed through the member or members, in which the forces are required to be found out as shown in Fig. 4 (*a*). A part of the structure, on any one side of the section line, is then treated as a free body in equilibrium under the action of external forces as shown in Fig. 4 (*b*) and (*c*).

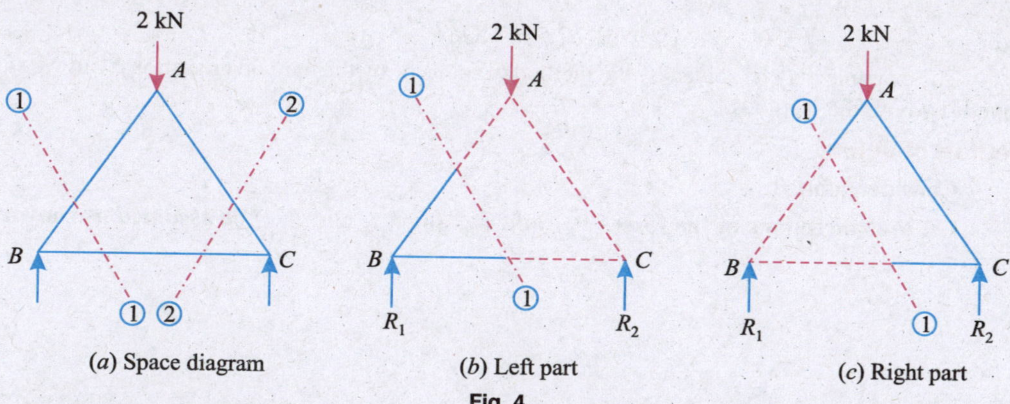

(*a*) Space diagram (*b*) Left part (*c*) Right part

Fig. 4.

The unknown forces are then found out by the application of equilibrium or the principles of statics *i.e.*, $\Sigma M = 0$.

Notes:1. To start with, we have shown section line 1-1 cutting the members *AB* and *BC*. Now in order to find out the forces in the member *AC*, section line 2-2 may be drawn.

2. While drawing a section line, care should always be taken not to cut more than three members, in which the forces are unknown.

14. FORCE TABLE

Finally, the results are tabulated showing the members, magnitudes of forces and their nature. Sometimes, tensile force is represented with a + ve sign and compressive force with a – ve sign.

Note: The force table is generally prepared, when force in all the members of a truss are required to be found out.

Example 1. *The truss ABC shown in Fig. 5 has a span of 5 metres. It is carrying a load of 10 kN at its apex.*

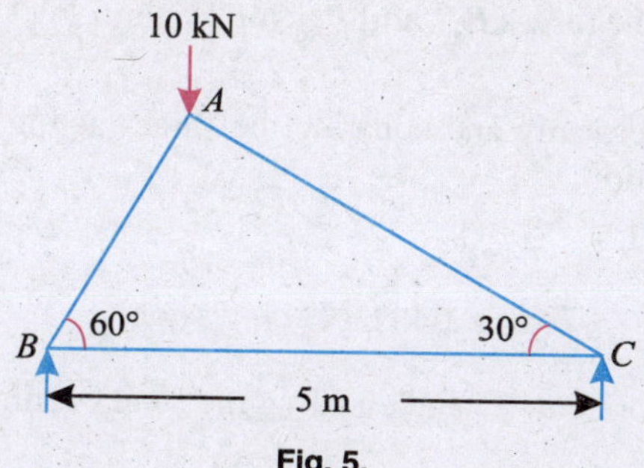

Fig. 5.

Find the forces in the members AB, AC and BC. **[UPTU]**

Solution. From the geometry of the truss, we find that the load of 10 kN is acting at a distance 1.25 m from the left hand support *i.e.*, *B* and 3.75 m from *C*.

Taking moments about *B* and equating the anti-clockwise moments with clockwise moments,

$$R_C \times 5 = 10 \times 1.25 = 12.5$$

or
$$R_C = \frac{12.5}{5} = 2.5 \text{ kN}$$

and
$$R_B = 10 - 2.5 = 7.5 \text{ kN}$$

The example may be solved by the method of joints or by the method of sections. But we shall solve it by both the methods.

Method of Joints

Consider joint *B*.

Let the *directions of the forces P_{AB} and P_{BC} (or P_{BA} and P_{CB}) be assumed as shown in Fig 6 (*a*).

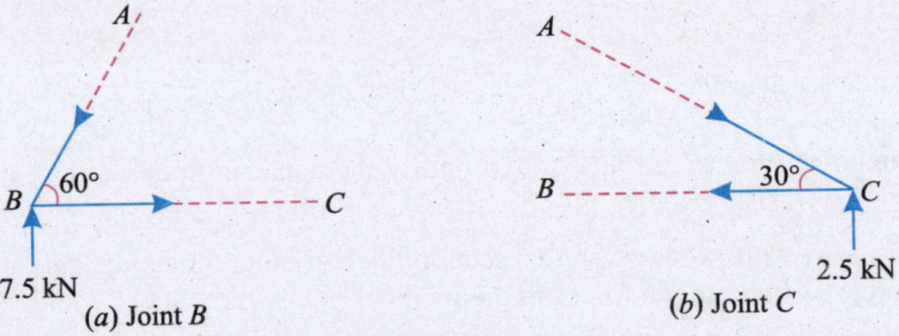

(*a*) Joint *B*

(*b*) Joint *C*

Fig. 6.

Resolving the forces vertically and equating the forces acting in downward direction with forces acting in upward direction,

$$P_{AB} \sin 60° = 7.5$$

or
$$P_{AB} = \frac{7.5}{\sin 60°} = \frac{7.5}{0.866} = 8.66 \text{ kN (Compression)}$$

Resolving the forces horizontally and equating the forces acting in left direction with forces acting in right direction,

$$P_{BC} = P_{AB} \cos 60° = 8.66 \times 0.5 = 4.33 \text{ kN (Tension)}$$

Consider the joint *C*.

Let the *directions of the forces P_{AC} and P_{BC} (or P_{CA} and P_{CB}) be assumed as shown in Fig. 6 (*b*).

Resolving the forces vertically and equating the forces acting in downward direction with forces acting in upward direction,

$$P_{AC} \sin 30° = 2.5$$

or
$$P_{AC} = \frac{2.5}{\sin 30°} = \frac{2.5}{0.5} = 5.0 \text{ kN (Compression)}$$

Resolving the forces horizontally and equating the forces acting in left direction with forces acting in right direction,

$$P_{BC} = P_{AC} \cos 30° = 5.0 \times 0.866 = 4.33 \text{ kN (Tension).}$$

...(As already obtained)

*__Note:__ The idea, of assuming the direction of the force P_{AB} to be downwards, is that the vertical component of the force P_{BC} is zero. Therefore in order to bring the joint B in equilibrium, the direction of the force P_{AB} must be downwards, or in other words, the direction of the force P_{AB} should be *opposite* to that of the reaction R_B. If, however the direction of the force P_{AB} is assumed to be upwards, then resolving the forces vertically and equating the forces acting in downward direction with forces acting in upward direction,

$$P_{AB} \sin 60° = -7.5 \text{ (Minus sign due to same direction of } R_B \text{ and } P_{AB}.)$$

or

$$P_{AB} = \frac{-7.5}{\sin 60°} = \frac{-7.5}{0.866} = -8.66 \text{ kN}$$

Minus sign means that the direction assumed is wrong. It should have been downwards instead of upwards. Similarly, the idea of assuming the direction of the force P_{BC} to be towards right is that the horizontal component of the reaction R_B is zero. Therefore in order to bring the joint B in equilibrium, the direction of the force P_{AB} must be towards right (because the direction of the horizontal component of the force P_{AB} is towards left).

Method of Sections

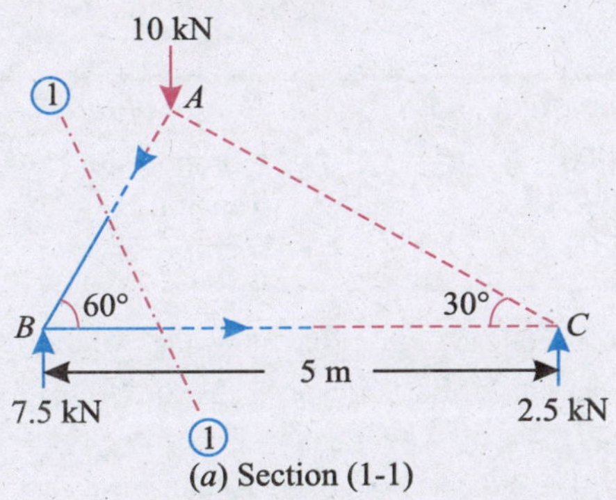

(*a*) Section (1-1)

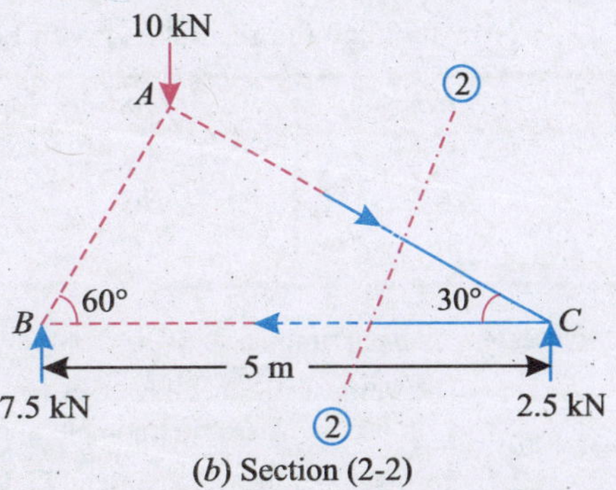

(*b*) Section (2-2)

Fig. 7.

First of all, pass section (1-1) cutting the truss into two parts (one part shown by firm lines and the other by dotted lines) through the members AB and BC of the truss as shown in Fig 7 (*a*).

Now consider equilibrium of the <u>Left Part</u> of the truss (because it is smaller than the right part).

Let the directions of the forces P_{AB} and P_{AC} be assumed as shown in Fig 7 (*a*).

Taking **moments of the forces acting in the left part of the truss only about the joint C and equating the anti-clockwise moments with clockwise moments,

$$P_{AB} \times 5 \sin 60° = 7.5 \times 5$$

or

$$P_{AB} = \frac{7.5 \times 5}{5 \sin 60°} = \frac{7.5}{0.866} = 8.66 \text{ kN (Compression)}$$

Taking moments of the forces acting in the left part of the truss only about the joint A and equating the anti-clockwise moments with clockwise moments,

$$P_{BC} \times 1.25 \tan 60° = 7.5 \times 1.25$$

or

$$P_{BC} = \frac{7.5 \times 1.25}{1.25 \tan 60°} = \frac{7.5}{1.732} = 4.33 \text{ kN (Tension)}$$

Now pass section (2-2) cutting the truss into two parts through the members AC and BC.

Now consider the equilibrium of the <u>Right Part</u> of the truss (because it is smaller than the left part).

Let the **direction of the forces P_{AC} and P_{BC} be assumed as shown in Fig 7 (*b*).

Taking moments of the force acting in the right part of the truss only about the joint *B* and equating the clockwise moments with anti-clockwise moments,

$$P_{AC} \times 5 \sin 30° = 2.5 \times 5$$

or $$P_{AC} = \frac{2.5}{\sin 30°} = \frac{2.5}{0.5} = 5 \text{ kN} \quad \text{(Compression)}$$

Taking moments of the forces acting in the right part of the truss only about the joint *A* and equating the clockwise moments with anti-clockwise moments,

$$P_{BC} \times 3.75 \tan 30° = 2.5 \times 3.75$$

or $$P_{BC} = \frac{2.5 \times 3.75}{3.75 \tan 30°} = \frac{2.5}{0.577} = 4.33 \text{ kN} \quad \text{(Tension)}$$

...(As already obtained)

Now tabulate the results as given below :

S.No.	*Member*	*Magnitude of force in kN*	*Nature of force*
1	*AB*	8.66	Compression
2	*BC*	4.33	Tension
3	*AC*	5.0	Compression

****Note:** The moment of the force P_{AB} about the joint *C* may be obtained in any one of the following two ways :

1. The vertical distance between the member *AB* and the joint *C* (*i.e.*, *AC* in this case) is equal to 5 sin 60° m. Therefore moment about *C* is equal to $P_{AB} \times 5 \sin 60°$ kN-m.
2. Resolve the force P_{AB} vertically and horizontally at *B*. The moment of horizontal component about *C* will be zero. The moment of vertical component (which is equal to $P_{AB} \times \sin 60°$) is equal to $P_{AB} \times \sin 60° \times 5 = P_{AB} \times 5 \sin 60°$ kN-m.

Example 2. *Fig 8 shows a Warren girder consisting of seven members each of 3 m length freely supported at its end points.*

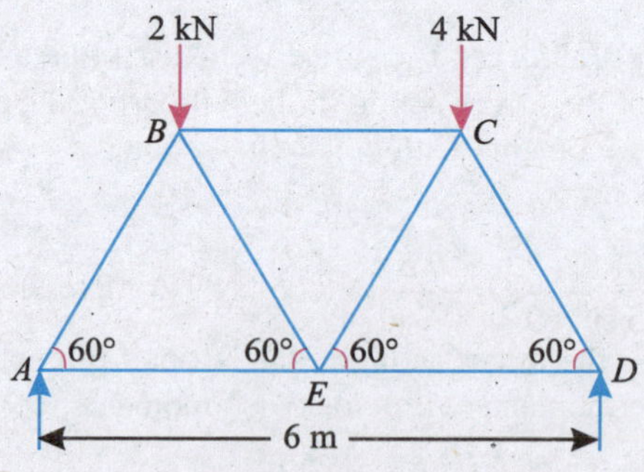

Fig. 8.

The girder is loaded at B and C as shown. Find the forces in all the members of the girder, indicating whether the force is compressive or tensile. **[AM-IETE, IGNOU]**

Solution. Taking moments about A and equating the anti-clockwise moments with clockwise moments,

$$R_D \times 6 = (2 \times 1.5) + (4 \times 4.5) = 21$$

or
$$R_D = \frac{21}{6} = 3.5 \text{ kN}$$

and
$$R_A = (2 + 4) - 3.5 = 2.5 \text{ kN}$$

The example may be solved by the method of joints or method of sections. But we shall solve it by both the methods.

Method of Joints

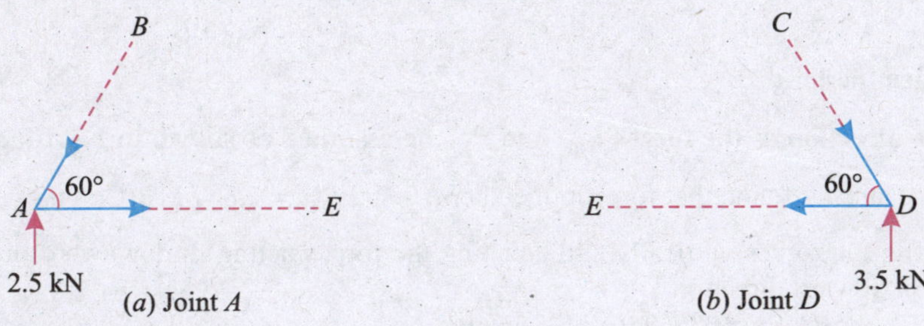

(a) Joint A (b) Joint D

Fig. 9.

Consider the joint A.

Let the directions of P_{AB} and P_{AE} be assumed as shown in Fig. 9 (a).

Resolving the forces vertically and equating the forces acting in downward direction with forces acting in upward direction,

$$P_{AB} \sin 60° = 2.5$$

or
$$P_{AB} = \frac{2.5}{\sin 60°} = \frac{2.5}{0.866} = 2.887 \text{ kN (Compression)}$$

Resolving the forces horizontally and equating the forces acting in right direction with forces acting in left direction,

$$P_{AE} = P_{AB} \cos 60° = 2.887 \times 0.5 = 1.444 \text{ kN (Tension)}$$

Consider the joint D.

Let the directions of the forces P_{CD} and P_{ED} be assumed as shown in Fig. 9 (b).

Resolving the forces vertically and equating the forces acting in downward direction with forces acting in upward direction,

$$P_{CD} \times \sin 60° = 3.5$$

or
$$P_{CD} = \frac{3.5}{\sin 60°} = \frac{3.5}{0.866} = 4.042 \text{ kN (Compression)}$$

Resolving the forces horizontally and equating the forces acting in left direction with forces acting in right direction,

$$P_{DE} = P_{CD} \cos 60° = 4.042 \times 0.5 = 2.021 \text{ kN (Tension)}$$

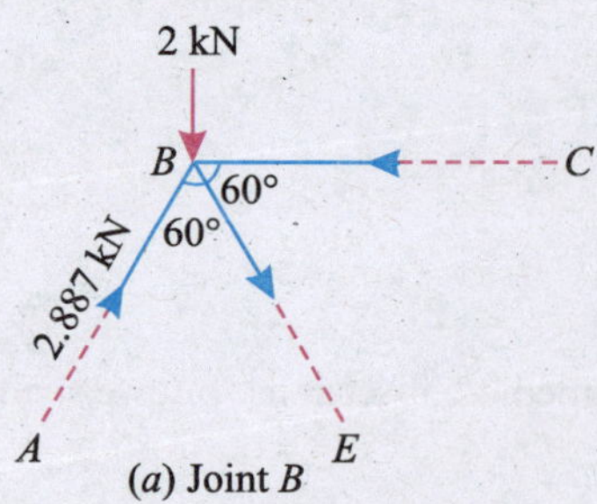

(a) Joint B

4 kN
1.732 kN
B
C
60°
4.042 kN
E
D

(b) Joint C

Fig. 10.

Consider the joint *B*.

Let the direction of the forces P_{BC} and P_{BE} be assumed as shown in Fig.10 (*a*).

We have already found that force in member *AB i.e.*, P_{AB} = 2.887 kN (Compression).

Resolving the forces vertically and equating the forces acting in downward direction with forces acting in upward direction,

$$P_{BE} \sin 60° + 2.0 = P_{AB} \sin 60°$$

or $$P_{BE} \sin 60° = P_{AB} \sin 60° - 2.0 = 2.887 \times 0.866 - 2.0 = 0.5 \text{ kN}$$

or $$P_{BE} = \frac{0.5}{\sin 60°} = \frac{0.5}{0.866} = 0.577 \text{ kN (Tension)}$$

Resolving the forces horizontally and equating the forces acting in left direction with forces acting in right direction,

$$P_{BC} = 2.887 \cos 60° + 0.577 \cos 60° \text{ kN}$$
$$= (2.887 \times 0.5) + (0.577 \times 0.5) \text{ kN} = 1.732 \text{ kN (Compression)}$$

Now consider joint *C*. We have already found out that the forces in the members *BC* and *CD* (*i.e.* P_{BC} = 1.732 kN (Compression) and P_{CD} = 4.042 kN (Compression).

Let the directions of P_{CE} be assumed as shown in Fig. 10 (*b*).

Resolving the forces vertically and equating the forces acting in upward direction with forces acting in downward direction,

$$P_{CE} \sin 60° + P_{CD} \sin 60° = 4$$

or $$P_{CE} \sin 60° = 4 - P_{CD} \sin 60° = 4 - (4.042 \times 0.866) = 0.5$$

or $$P_{CE} = \frac{0.5}{\sin 60°} = \frac{0.5}{0.866} = 0.577 \text{ kN (Compression)}$$

Method of Sections

First of all, pass section (1-1) cutting the truss through the members *AB* and *AE*.

Now consider equilibrium of the Left Part of the truss.

Let the directions of the forces P_{AB} and P_{AE} be assumed as shown in Fig. 11 (*a*).

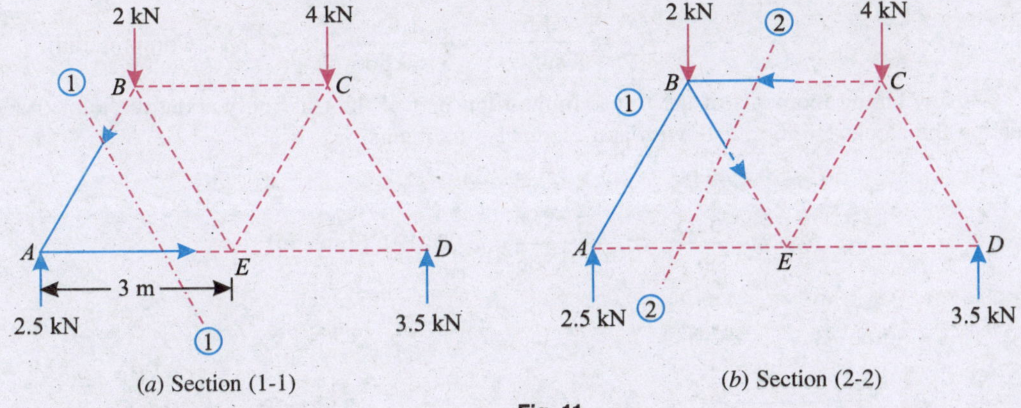

(*a*) Section (1-1)

(*b*) Section (2-2)

Fig. 11.

Taking moments of the forces acting in the left part of the truss only, about the joint *E* and equating the anti-clockwise moments with clockwise moments,

$$P_{AB} \times 3 \sin 60° = 2.5 \times 3$$

or
$$P_{AB} = \frac{2.5}{\sin 60°} = \frac{2.5}{0.866} = 2.887 \text{ kN (Compression)}$$

Now pass section (2-2) cutting the truss through the members *BC*, *BE* and *AE*.

Now consider equilibrium of the <u>Left Part</u> of the truss.

Let the directions of the forces P_{BC} and P_{BE} be assumed as shown in Fig. 11 (*b*).

Taking moments of the forces acting in left part of the truss only, about the joint *E* and equating the anti-clockwise moments with clockwise moments,

$$P_{BC} \times 3 \sin 60° + (2 \times 1.5) = (2.5 \times 3)$$

or
$$P_{BC} \times 3 \sin 60° = (2.5 \times 3) - (2 \times 1.5) = 4.5$$

or
$$P_{BC} = \frac{4.5}{3 \sin 60°} = \frac{4.5}{3 \times 0.866} = 1.732 \text{ kN (Compression)}$$

Now taking moments of the forces acting in the left part of the truss only about the joint *A* and equating the clockwise moments with anti-clockwise moments,

$$P_{BE} \times 3 \sin 60° + (2 \times 1.5) = (P_{BC} \times 3 \sin 60°)$$

or
$$P_{BE} \times 3 \sin 60° = (P_{BC} \times 3 \sin 60°) - (2 \times 1.5) = (1.732 \times 3 \times 0.866) - 3.0 = 1.5$$

or
$$P_{BE} = \frac{1.5}{3 \sin 60°} = \frac{1.5}{3 \times 0.866} = 0.577 \text{ kN (Tension)}$$

Now pass section (3-3) cutting the truss through the members *BC*, *CE* and *ED*.

Now consider the equilibrium of the <u>Right Part</u> of the truss.

Let the directions of the forces P_{CE} and P_{DE} be assumed as shown in Fig. 12 (*a*)

Taking moments of the forces in the right part of the truss only, about the joint *D* and equating the clockwise moments with anti-clockwise moments,

$$P_{CE} \times 3 \sin 60° + (P_{BC} \times 3 \sin 60°) = (4 \times 1.5)$$

or
$$P_{CE} \times 3 \sin 60° + (1.732 \times 3 \times 0.866) = 6.0$$

or
$$P_{CE} \times 3 \sin 60° = 6 - 4.5 = 1.5$$

or $$P_{CE} = \frac{1.5}{3 \sin 60°} = \frac{1.5}{3 \times 0.866} = 0.577 \text{ kN (Compression)}$$

Now taking moments of the forces in the right part of the truss only about the joint C and equating the clockwise moments with anti-clockwise moments,

$$P_{DE} \times 3 \sin 60° = 3.5 \times 1.5 = 5.25$$

or $$P_{DE} = \frac{5.25}{3 \sin 60°} = \frac{5.25}{3 \times 0.866} = 2.021 \text{ kN (Tension)}$$

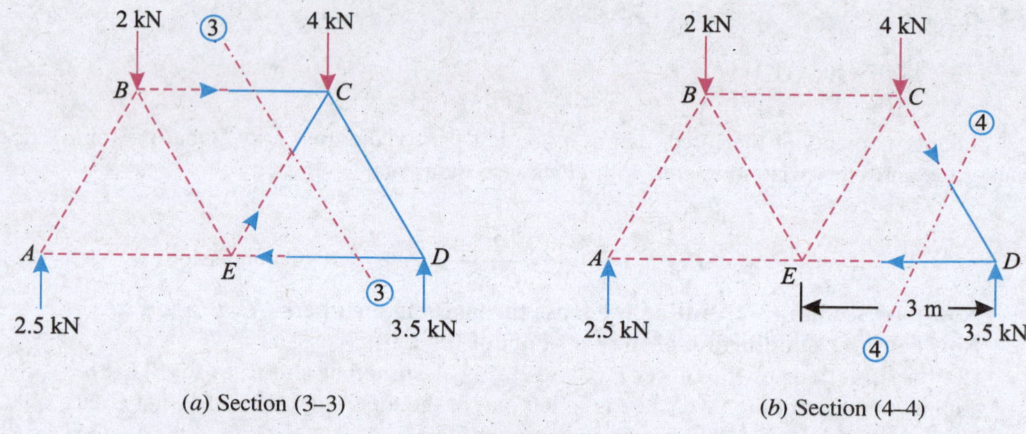

(*a*) Section (3–3)

(*b*) Section (4–4)

Fig. 12.

Now pass section (4-4) cutting the truss through the members *CD* and *DE*.

Now consider the equilibrium of the Right Part of the truss.

Let the directions of the forces P_{CD} be assumed as shown in Fig 12 (*b*).

Taking moments of the forces acting in the right part of the truss only about the joint E and equating the clockwise moments with anti-clockwise moments,

$$P_{CD} \times 3 \sin 60° = 3.5 \times 3$$

or $$P_{CD} = \frac{3.5}{\sin 60°} = \frac{3.5}{0.866} = 4.042 \text{ kN (Compression)}$$

Now tabulate the results as given below :

S.No.	*Member*	*Magnitude of force in kN*	*Nature of force*
1	*AB*	2.887	Compression
2	*AE*	1.444	Tension
3	*CD*	4.042	Compression
4	*DE*	2.021	Tension
5	*BE*	0.577	Tension
6	*BC*	1.732	Compression
7	*CE*	0.577	Compression

***Example 3.** *A plane is loaded and supported as shown in Fig 13.*

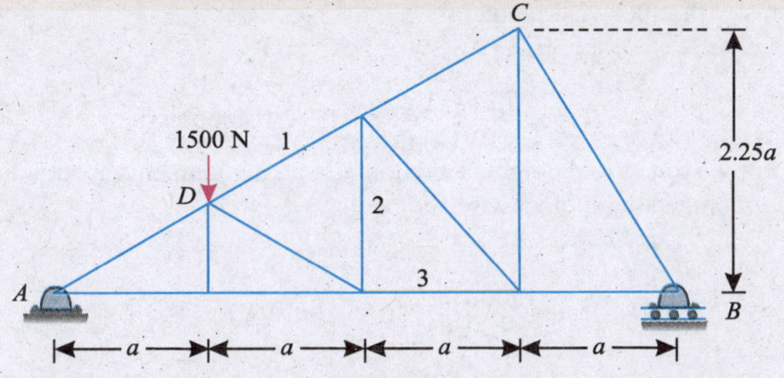

Fig. 13.

Determine the nature and magnitude of the forces in the members 1, 2 and 3. **[RGPV]**

Solution. Taking moments about A and equating the anti-clockwise moments with clock-wise moments,,

$$V_B \times 4\,a = 1500 \times a$$

or
$$V_B = \frac{1500}{4} = 375 \text{ N}$$

and
$$V_A = 1500 - 375 = 1125 \text{ N}$$

From the geometry of the figure, we find that

$$\tan\theta = \frac{2.25\,a}{3\,a} = 0.75$$

and
$$\sin\theta = \frac{3}{5} = 0.6 \text{ and } \cos\theta = \frac{4}{5} = 0.8$$

Method of Sections

The example may be solved by any method. But we shall solve it by the method of sections, as one section line can cut the members 1, 2 and 3 in which the forces are required to be found out.

Now let us pass section (1-1) cutting the truss into two parts as shown in Fig 14.

Now consider the equilibrium of the Right Part of the truss to the section (1-1).

Let the directions of P_1, P_2 and P_3 be assumed as shown in Fig. 14.

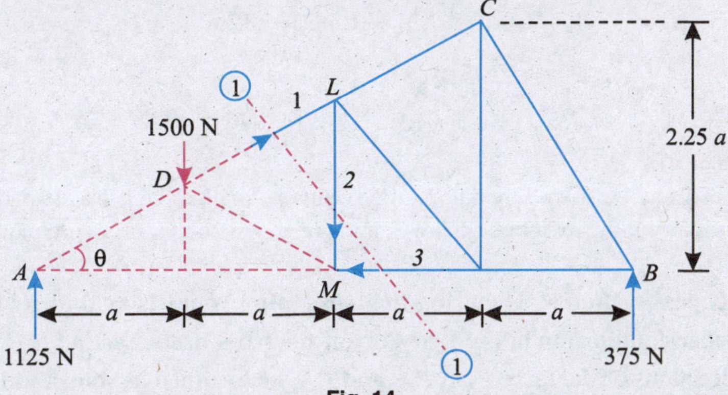

Fig. 14.

For the forces to the right of section 1-1, taking moments about joint M and equating the anti-clockwise moments with clockwise moments,

$$P_1 \times 2a \sin \theta = 375 \times 2a$$

or $$P_1 = \frac{375}{\sin \theta} = \frac{375}{0.6} = 625 \text{ N (Compression)}$$

Similarly, For the forces to the right of section 1-1, taking moments about joint A and equating the anti-clockwise moments with clockwise moments,

$$P_2 \times 2a = 375 \times 4a = 1500a$$

or $$P_2 = \frac{1500a}{2a} = 750 \text{ N (Tension)}$$

Similarly, For the forces to the right of section 1-1, taking moments about the joint L, and equating the anti-clockwise moments with clockwise moments,

$$P_3 \times \frac{3a}{2} = 375 \times 2a = 750a$$

or $$P_3 = \frac{750}{1.5} = 500 \text{ N (Tension)}$$

Example 4. *An inclined truss shown in Fig 15 is loaded as shown.*

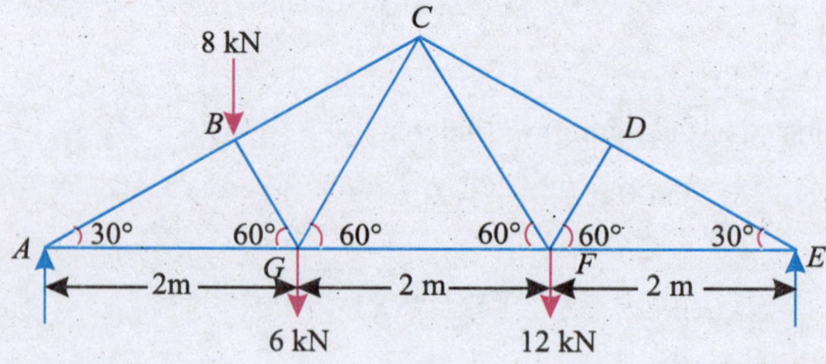

Fig. 15.

Determine the nature and magnitude of the forces in the members BC, GC and GF of the truss.

Solution. From the geometry of the figure, we find that the load 8 kN at B is acting at a distance of 1.5 m from the joint A.

Taking moments about A and equating the anti-clockwise moments with clockwise moments,

$$R_E \times 6 = (8 \times 1.5) + (6 \times 2) + (12 \times 4) = 72$$

or $$R_E = \frac{72}{6} = 12 \text{ kN}$$

and $$R_A = (8 + 6 + 12) - 12 = 14 \text{ kN}$$

Method of Sections

The example may be solved by any method. But we shall solve it by the method of sections, as one section line can cut the members BC, GC, and GF in which the forces are required to be found out.

Now let us pass section (1-1) cutting the truss into two parts as shown in Fig. 16.

Now consider equilibrium of the Left Part of the truss to the section (1-1).

Let the directions of the force P_{BC}, P_{GC} and P_{GF} be assumed as shown in Fig 16.

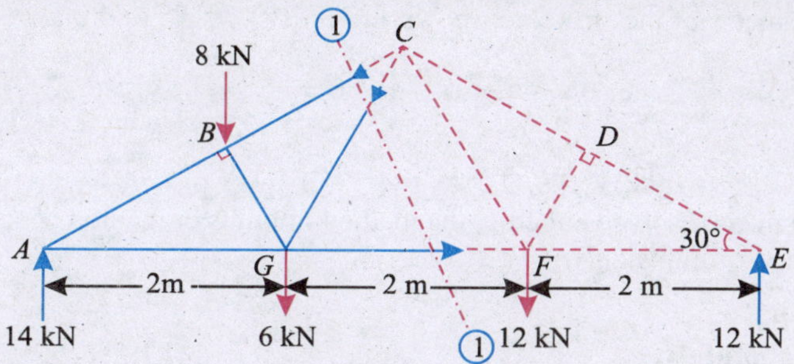

Fig. 16.

For the forces to the left of section 1-1, taking moments about the joint G and equating the anti-clockwise moments with clockwise moments,

$$P_{BC} \times 2 \sin 30° = (14 \times 2) - (8 \times 0.5) = 24$$

or
$$P_{BC} = \frac{24}{2 \sin 30°} = \frac{24}{2 \times 0.5} = 24 \text{ kN (Compression)}$$

Similarly, For the forces to the left of section 1-1, taking moments about the joint B and equating the anti-clockwise moments with clockwise moments,

$$P_{GC} \times 1 \cos 30° = (14 \times 1.5) + (6 \times 0.5) = 24 \text{ kN}$$

or
$$P_{GC} = \frac{24}{\cos 30°} = \frac{24}{0.866} = 27.7 \text{ kN (Compression)}$$

Similarly, For the forces to the left of section 1-1, taking moments about the joint C and equating the anti-clockwise moments with clockwise moments,

$$P_{GF} \times 3 \tan 30° = (14 \times 3) - (6 \times 1) = 36$$

or
$$P_{GF} = \frac{36}{3 \tan 30°} = \frac{12}{0.5774} = 20.8 \text{ kN (Tension)}$$

Example 5. *A framed of 6 m span is carrying a central load of 10 kN as shown in Fig. 17.*

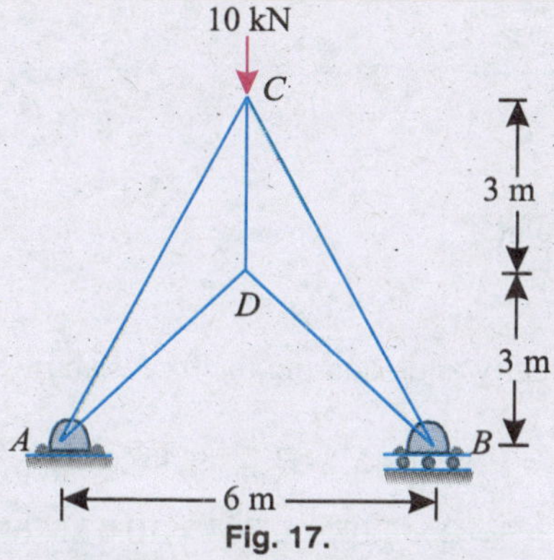

Fig. 17.

Find by any method, the magnitude and nature of forces in all members of the structure and tabulate the results.

Solution. Since the structure is symmetrical in geometry and loading, therefore reaction at A and B will be same and equal to half of the load,

$$R_A = R_B = 10/2 = 5 \text{ kN}$$

From the geometry of the structure, shown in Fig. 18 (*a*). we find that

$$\tan\theta = \frac{3}{3} = 1.0 \quad \text{or} \quad \theta = 45°$$

$$\tan\alpha = \frac{6}{3} = 2.0 \quad \text{or} \quad \alpha = 63.4°$$

The example may be solved either by the method of joints or method of sections. But we shall solve it by the method of joints only.

Method of Joints

Consider the joint *A*.

Let the directions of the forces P_{AC} and P_{AD} be assumed as shown in Fig 18 (*a*).

Resolving the forces horizontally and equating the forces acting in left direction with forces acting in right direction,

$$P_{AC}\cos 63.4° = P_{AD}\cos 45°$$

or
$$P_{AC} = \frac{P_{AD}\cos 45°}{\cos 63.4°} = \frac{P_{AD}\times 0.707}{0.4477} = 1.58\,P_{AD}$$

Resolving the forces vertically and equating the forces acting in downward direction with forces acting in upward direction,

$$P_{AC}\sin 63.4° = 5 + P_{AD}\sin 45°$$

or
$$1.58\,P_{AD}\times 0.8941 = 5 + P_{AD}\times 0.707 \qquad ...(\because P_{AC} = 1.58\,P_{AD})$$

or
$$0.7056\,P_{AD} = 5$$

or
$$P_{AD} = \frac{5}{0.7056} = 7.08 \text{ kN (Tension)}$$

Now
$$P_{AC} = 1.58\times P_{AD} = 1.58\times 7.08 = 11.19 \text{ kN (Compression)}$$

Consider the joint *D*.

Let the directions of the forces P_{CD} and P_{BD} be assumed as shown in Fig. 18 (*b*).

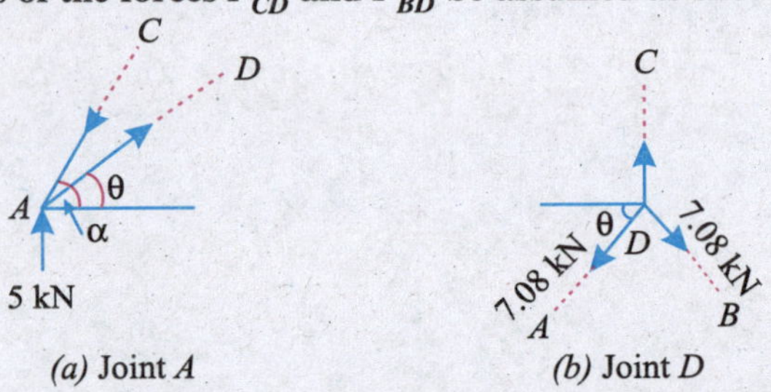

(*a*) Joint *A* (*b*) Joint *D*

Fig. 18.

Resolving the forces vertically and equating the forces acting in upward direction with forces acting in downward direction,

$$P_{CD} = P_{AD}\sin 45° + P_{BD}\sin 45° = 2\,P_{AD}\sin 45° \qquad ...(\because P_{BD} = P_{AD})$$

$$= 2\times 7.08\times 0.707 = 10.0 \text{ kN (Tension)}$$

Now tabulate these results as given below :

S.No.	*Member*	*Magnitude of force in kN*	*Nature of force*
1	*AD, DB*	7.08	Tension
2	*AC, CB*	11.19	Compression
3	*CD*	10.0	Tension

EXERCISE 1

1. A truss of span 10 meters is loaded as shown in Fig. 19. Find the forces in all the members of the truss. **[IGNOU]**

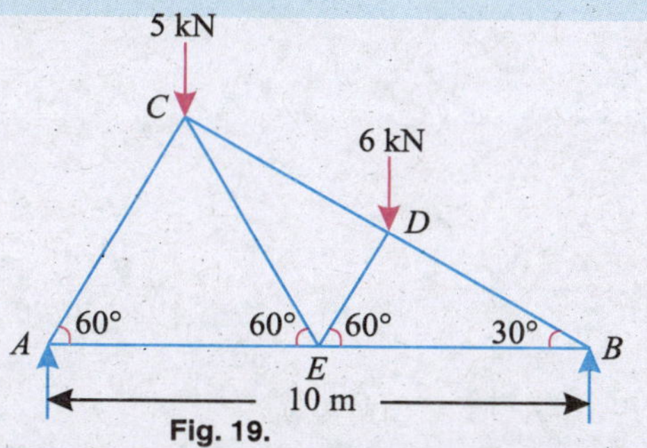

Fig. 19.

Ans. AC = 6.92 kN (Compression)
AE = 3.46 kN (Tension)
BD = 10.0 kN (Compression)
BE = 8.66 kN (Tension)
CD = 7.0 kN (Compression)
ED = 5.2 kN (Compression)
CE = 5.2 kN (Tension)

*2. A king post truss of 8 m span is loaded as shown in Fig 20. Find the forces in each member of the truss and tabulate the results. **[RGPV]**

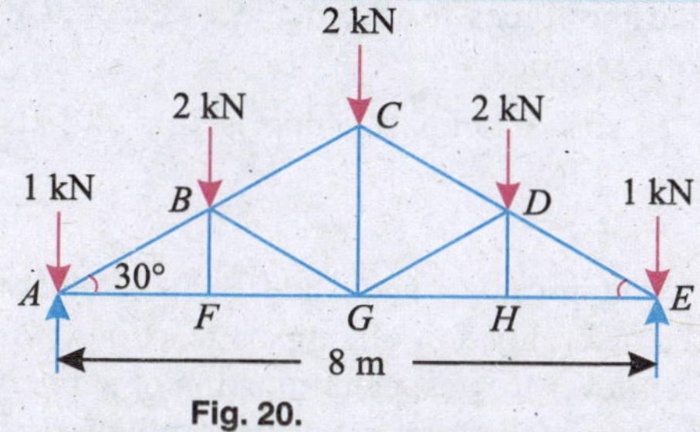

Fig. 20.

Ans. AB, DE = 6.0 kN (Compression)
AF, EH = 5.2 kN (Tension)
FG, GH = 5.2 kN (Tension)
BF, DH = 0
BG, DG = 2.0 kN (Compression)
BC, CD = 4.0 kN (Compression)
CG = 2.0 kN (Tension)

3. A plane truss of 6 m span is subjected to a point load of 30 kN as shown in the figure 21. Find graphically, or otherwise, the forces in all the members of the truss and tabulate the results.

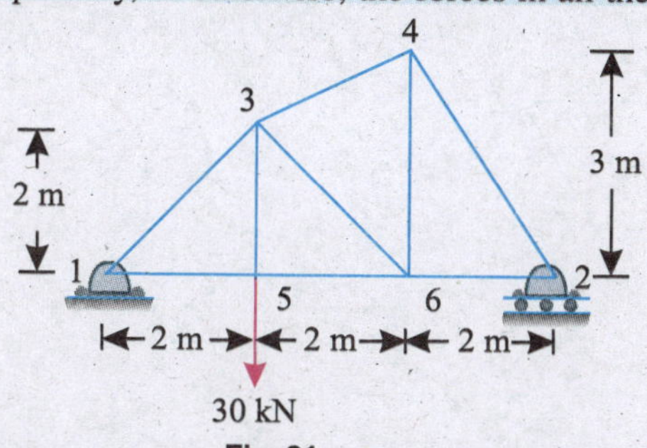

Fig. 21.

Ans. 1-3 = 28.3 kN (Compression)
1-5 = 20.0 kN (Tension)
2-4 = 12.0 kN (Compression)
2-6 = 6.7 kN (Tension)
1-5 = 20.0 kN (Tension)
3-5 = 30.0 kN (Tension)
3-6 = 18.8 kN (Compression)
4-6 = 13.3 kN (Tension)
3-4 = 7.5 kN (Compression)

4. A 9 m span truss is loaded as shown in Fig 22. Find the forces in the members BC, CH and HG of the truss.

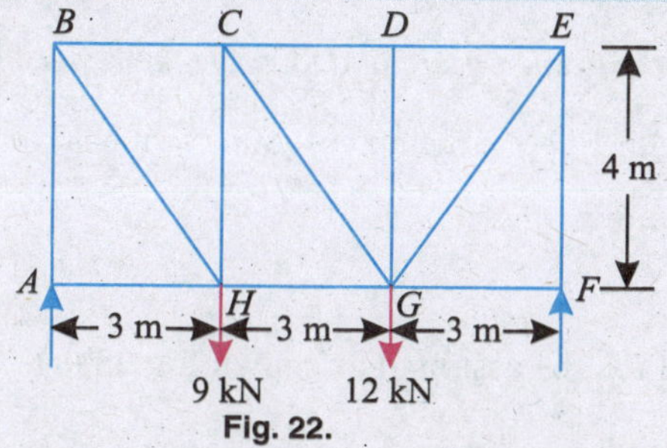

Fig. 22.

Ans. BC = 7.5 kN (Compression)
CH = 1.0 kN (Compression)
GH = 7.5 kN (Tension)

5. The roof truss shown in Fig. 23 is supported at A and B and carries vertical loads at each of the upper chord points.

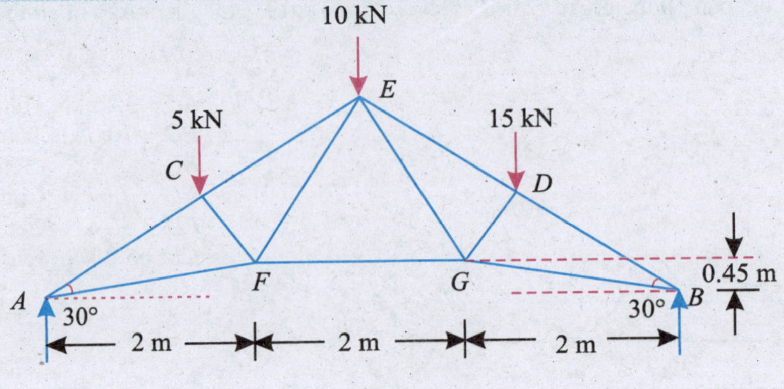

Fig. 23.

Using the method of sections, determine the forces in the members CE and FG of truss, stating whether they are in tension or compression.

[**Ans.** 38.5 kN (Compression); 24.2 kN (Tension)]

15. CANTILEVER TRUSSES

A truss, which is connected to a wall or a column at one end, and free at the other is known as a cantilever truss. In the previous examples, the determination of support reactions was absolutely essential to start the work. But in the case of cantilever trusses, determination of support reaction is not essential, as we can start the calculation work from the free end of the cantilever.

Example 6. *A cantilever truss of 3 m span is loaded as shown in Fig 24.*

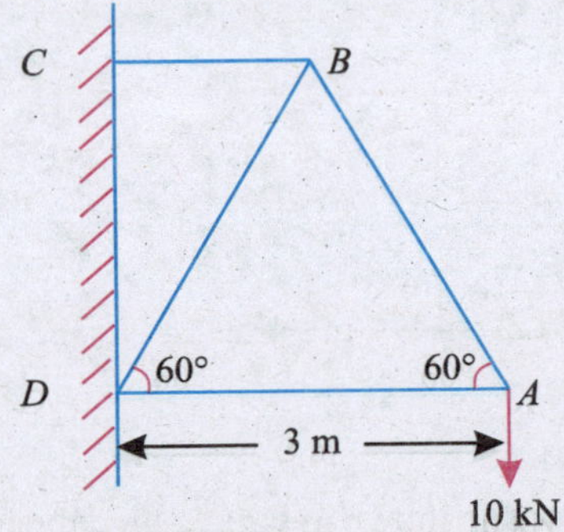

Fig. 24.

Find the forces in the various members of the framed truss, and tabulate the results. **[IGNOU]**

Solution. The example may be solved either by the method of joints or method of sections. But we shall solve it by both the methods one by one.

Method of Joints

Consider the joint A.

Let the directions of the forces P_{AB} and P_{AD} be assumed as shown Fig 25 (a).

Resolving the forces vertically and equating the forces acting in downward direction with forces acting in upward direction,

$$P_{AB} \sin 60° = 10$$

or $$P_{AB} = \frac{10}{\sin 60°} = \frac{10}{0.866} = 11.5 \text{ kN (Tension)}$$

Resolving the forces horizontally and equating the forces acting in left direction with forces acting in right direction,

$$P_{AD} = P_{AB} \cos 60° = 11.5 \times 0.5 = 5.75 \text{ kN (Compression)}$$

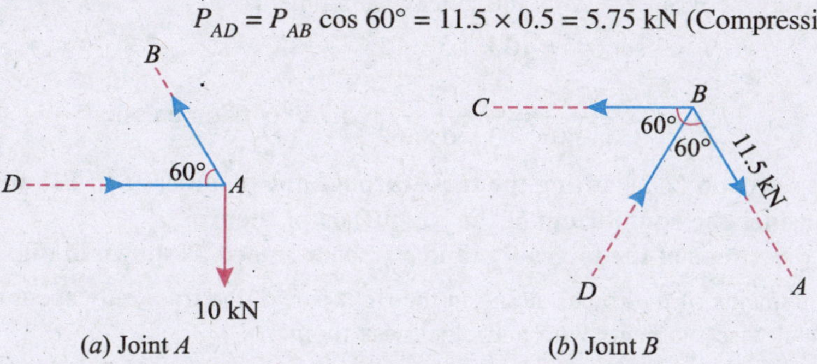

(*a*) Joint *A* (*b*) Joint *B*

Fig. 25.

Consider the joint *B*.

Let the directions of P_{BD} and P_{BC} be assumed as shown in Fig 25 (*b*).

We have already found out that the force in member *AB* is 11.5 kN (Tension) as shown in the figure 25 (*b*).

Resolving the forces vertically and equating the forces acting in downward direction with forces acting in upward direction,

$$P_{BD} \sin 60° = P_{AB} \sin 60° = 11.5 \sin 60°$$

$$\therefore \quad P_{BD} = P_{AB} = 11.5 \text{ kN (Compression)}$$

Resolving the forces horizontally and equating the forces acting in left direction with forces acting in right direction,

$$P_{BC} = P_{AB} \cos 60° + P_{BD} \cos 60°$$
$$= (11.5 \times 0.5) + (11.5 \times 0.5) = 11.5 \text{ kN (Tension)}$$

Method of Sections

First of all, pass section (1-1) cutting the truss through the members *AB* and *AD*.

Consider the equilibrium of the <u>Right Part</u> of the truss.

Let the directions of the forces P_{AB} and P_{AD} be assumed as shown in Fig 26 (*a*).

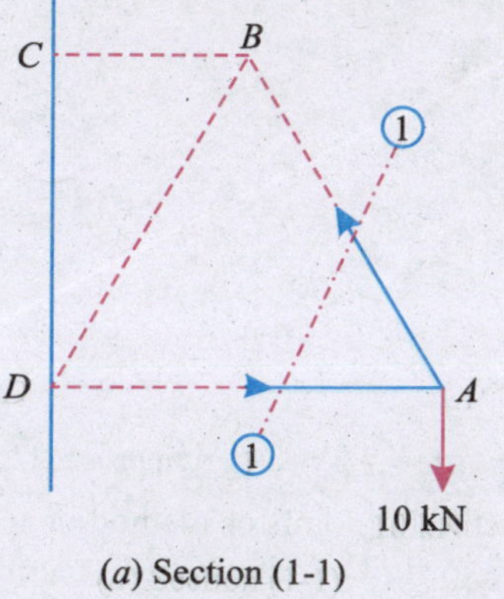

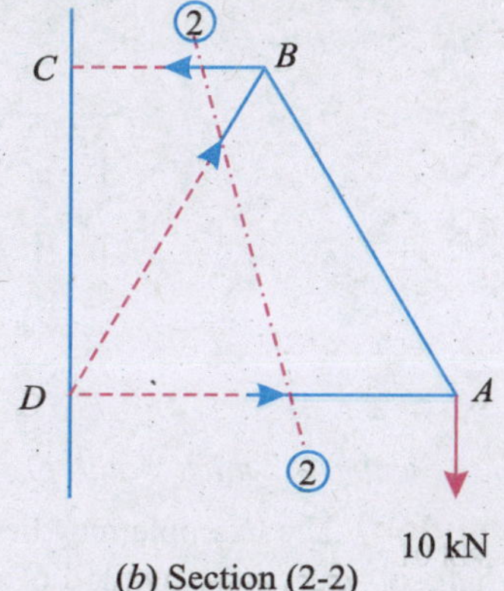

(*a*) Section (1-1) (*b*) Section (2-2)

Fig. 26.

Taking moments of the forces acting in the right part of the truss only about the joint D and equating the clockwise moments with anti-clockwise moments,

$$P_{AB} \times 3 \sin 60° = 10 \times 3$$

or $$P_{AB} = \frac{10}{\sin 60°} = \frac{10}{0.866} = 11.5 \text{ kN (Tension)}$$

Taking moments of the forces acting in the right part of the truss only about the joint B and equating the clockwise moments with anti-clockwise moments,

$$P_{AD} \times 3 \sin 60° = 10 \times 1.5 = 15$$

or $$P_{AD} = \frac{15}{3 \sin 60°} = \frac{15}{3 \times 0.866} = 5.75 \text{ kN (Compression)}$$

Now pass section (2-2) cutting the truss through the members *BC*, *BD* and *AD*.

Now consider the equilibrium of the Right Part of the truss.

Let the directions of the forces P_{BC} and P_{BD} be assumed as shown in Fig. 26 (*b*)

Taking moments of the forces acting in the right part of the truss only about the joint D and equating the clockwise moments with anti-clockwise moments,

$$P_{BC} \times 3 \sin 60° = 10 \times 3$$

or $$P_{BC} = \frac{10}{\sin 60°} = \frac{10}{0.866} = 11.5 \text{ kN (Tension)}$$

Taking moments of the forces acting in the right part of the truss only about the joint C and equating the clockwise moments with anti-clockwise moments,

$$P_{BD} \times 1.5 \sin 60° = (10 \times 3) - P_{AD} \times 3 \sin 60° = 30 - (5.75 \times 3 \times 0.866) = 15$$

or $$P_{BD} = \frac{15}{1.5 \sin 60°} = \frac{15}{1.5 \times 0.866} = 11.5 \text{ kN (Compression)}$$

Now tabulate the results as given below :

S.No.	*Members*	*Magnitude of force in kN*	*Nature of force*
1	*AB*	11.5	Tension
2	*AD*	5.75	Compression
3	*BD*	11.5	Compression
4	*BC*	11.5	Tension

Example 7. *A cantilever truss is loaded as shown in Fig 27.*

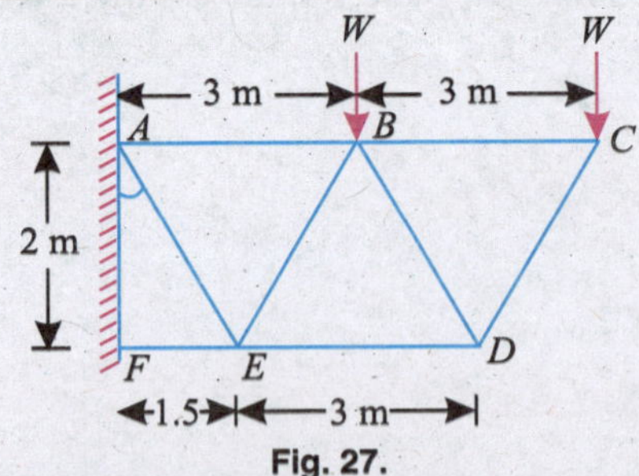

Fig. 27.

Find the value W, which would produce the force of magnitude 15 kN in the member AB.

Solution. The example may be solved either by the method of joints or method of sections. But we shall solve it by the method of section only as we have to find out the force in member *AB* only.

Method of Sections

First of all, let us find out the force in the member *AB* of the truss in terms of *W*.

Now pass section (1-1) cutting the truss through the members *AB, BE* and *ED* as shown in Fig. 28.

Consider the equilibrium of the <u>Right Part</u> of the truss.

Let the direction P_{AB} be assumed as shown in Fig 28.

Taking moments of the forces acting in the right part of the truss only about the joint *E* and equating the clockwise moments with anti-clockwise moments,

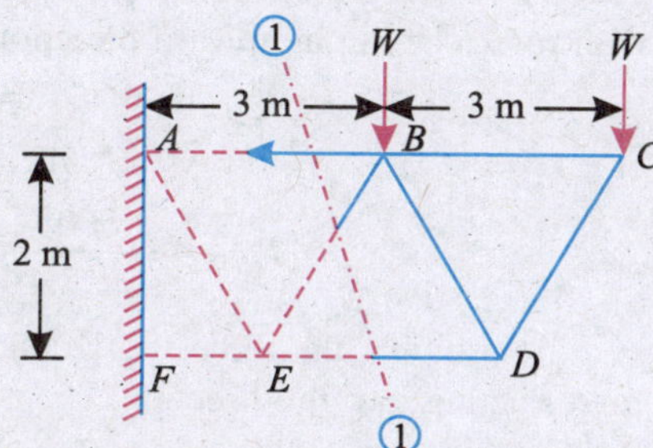

Fig. 28.

$$P_{AB} \times 2 = (W \times 1.5) + (W \times 4.5) = 6\,W$$

or $$P_{AB} = \frac{6W}{2} = 3W$$

Thus the value of *W*, which would produce the force of 15 kN in the member *AB*

$$= \frac{W}{3W} \times 15 = 5 \text{ kN} \qquad \textbf{Ans.}$$

Example 8. *Figure 29 shows a cantilever truss having a span of 4.5 meters. It is hinged at two joints to a wall and is loaded as shown.*

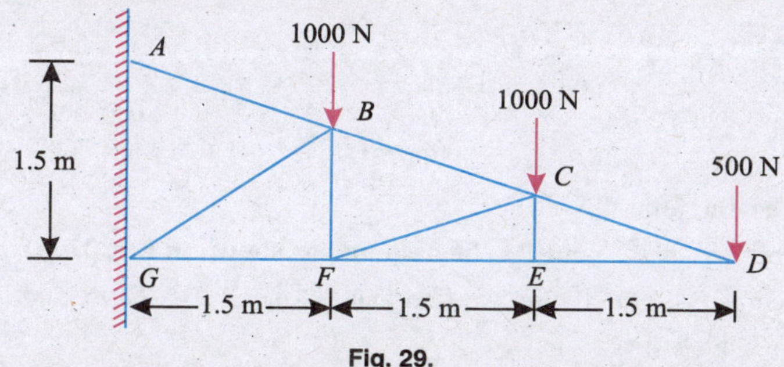

Fig. 29.

Find the forces in all the member of the truss. **[GTU]**

Solution. The example may be solved either by the method of joints or method of sections. But we shall solve it by the method of joints as we have to find out forces in all members of the truss.

Method of Joints

Force in all the members of the truss

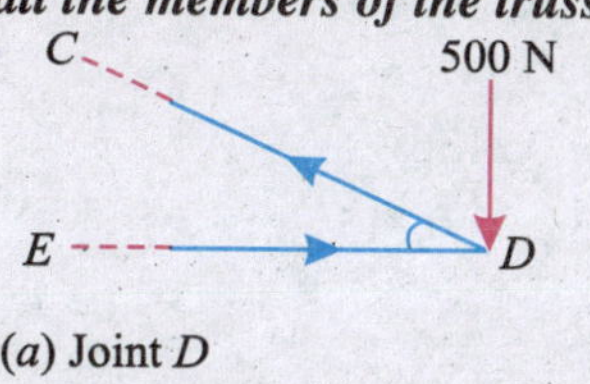

(*a*) Joint *D*

C *F* *E* *D*

(*b*) Joint *E*

Fig. 30.

Consider the joint *D*.

Let the directions of P_{CD} and P_{DE} be assumed as shown in Fig. 30 (*a*).

From the geometry of the figure, we find that

$$\tan \angle CDE = \frac{1.5}{4.5} = 0.3333 \qquad \text{or} \qquad \angle CDE = 18.4°$$

Resolving the forces vertically at D and equating the forces acting in downward direction with forces acting in upward direction,

$$P_{CD} \sin \angle CDE = 500$$

or $$P_{CD} \sin 18.4° = 500$$

or $$P_{CD} = \frac{500}{\sin 18.4°} = \frac{500}{0.3156} = 1584 \text{ N (Tension)}$$

Resolving the forces horizontally at D and equating the forces acting in left direction with forces acting in right direction,

$$P_{DE} = P_{CD} \cos \angle CDE = 1584 \cos 18.4°$$

$\therefore$ $$P_{DE} = 1584 \times 0.9488 = 1503 \text{ N (Compression)}$$

Now consider the joint E.

A little consideration will show that the value of the force P_{FE} will be equal to the force P_{ED} *i.e.*, 1503 N (Compression). Since the vertical components of the forces P_{FE} and P_{ED} are zero, therefore the value of the force P_{CE} will also be zero.

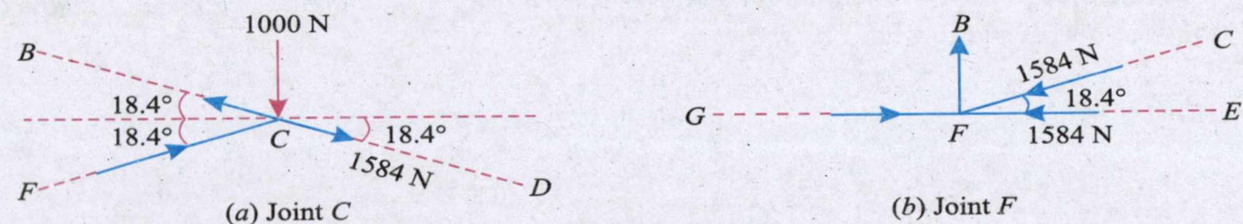

Fig. 31.

Now consider the joint C.

Let the directions of P_{BC} and P_{FC} be assumed as shown in Fig. 31 (*a*).

From the geometry of the figure, we find that the members CD, BC and FC make angle of 18.4° with the horizontal.

Resolving the forces horizontally and equating the forces acting in left direction with forces acting in right direction,

$$P_{BC} \cos 18.4° = 1584 \cos 18.4° + P_{FC} \cos 18.4°$$

or $$P_{BC} = 1584 + P_{FC} \qquad ...(i)$$

Resolving the forces vertically and equating the forces acting in downward direction with forces acting in upward direction,

$$1000 + 1584 \sin 18.4° = P_{FC} \sin 18.4° + P_{BC} \sin 18.4°$$

or $$1000 + (1584 \times 0.3156) = (P_{FC} \times 0.3156) + (P_{BC} \times 0.3156)$$

or $$1000 + (1581 \times 0.3156) = 0.3156\, P_{FC} + (1584 + P_{FC}) \times 0.3156$$

$$...(\because P_{BC} = 1584 + P_{FC})$$

or $$1000 + (1581 \times 0.3156) = 0.3156\, P_{FC} + (1584 \times 0.3156) + 0.3156\, P_{FC}$$

or $$P_{FC} = \frac{1000}{0.6312} = 1584 \text{ N (Compression)}$$

Substituting the value of P_{FC} in equation (*i*)

$$P_{BC} = 1584 + 1584 = 3168 \text{ N (Tension)}$$

Now consider the joint *F*.

Let the directions of the forces P_{GF} and P_{FB} be assumed as shown in Fig 31 (*b*).

Resolving the forces horizontally and equating the forces acting in left direction with forces acting in right direction,

$$P_{GF} = 1584 + 1584 \cos 18.4° = 1584 + (1584 \times 0.9488) \text{ N}$$
$$= 1584 + 1503 = 3087 \text{ N (Compression)}$$

Resolving the forces vertically and equating the forces acting in downward direction with forces acting in upward direction,

$$P_{BF} = 1584 \sin 18.4° = 1584 \times 0.3156 = 500 \text{ N (Tension)}$$

Now consider the joint *B*.

Let the direction of P_{BG} and P_{AB} be assumed as shown in Fig 32.

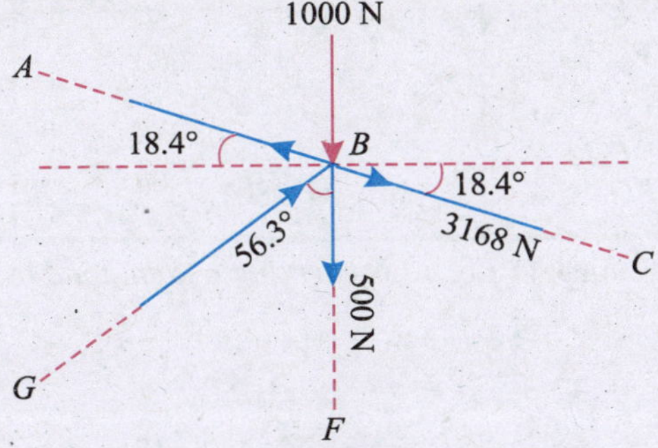

Fig. 32.

From the geometry of the figure, we find that

$$\tan \angle GBF = \frac{1.5}{1} = 1.5 \text{ or } \angle GBF = 56.3°$$

Resolving the forces horizontally at *B* and equating the forces acting in left direction with forces acting in right direction,

$$P_{AB} \cos 18.4° = P_{BG} \sin 56.3° + 3168 \cos 18.4°$$

or $$P_{AB} \times 0.9488 = P_{BG} \times 0.832 + 3168 \times 0.9488$$

or $$0.9488\, P_{AB} = 0.832\, P_{BG} + 3000 \quad \text{....}(ii)$$

Dividing the above equation by 3,

$$0.3156\, P_{AB} = 0.2773\, P_{BG} + 1000 \quad \text{....}(iii)$$

Resolving the forces vertically at *B* and equating the forces acting in downward direction with forces acting in upward direction,

$$P_{AB} \sin 18.4° + P_{BG} \cos 56.3° = 1000 + 500 + 3168 \sin 18.4°$$
$$= 1500 + (3168 \times 0.3156)$$

or $$P_{AB} \times 0.3156 + P_{BG} \times 0.5548 = 1500 + 1000$$

or $$0.3156\, P_{AB} + 0.5548\, P_{BG} = 2500 \quad \text{...}(iv)$$

Substracting equation (*iii*) from equation (*iv*),

$$0.8321\, P_{BG} = 1500$$

or $$P_{BG} = \frac{1500}{0.8321} = 1801 \text{ N (Compression)}$$

Substituting the value of P_{BG} in equation (*iii*)

$$0.3156\,P_{AB} = (0.2773 \times 1801) + 1000$$

$$0.3156\,P_{AB} = 500 + 1000 = 500$$

$$P_{AB} = \frac{1500}{0.3156} = 4753 \text{ N (Tension)}$$

Now tabulate the results as given below :

S.No.	*Member*	*Magnitude of force in kN*	*Nature of force*
1	*AB*	4753	Tension
2	*BC*	3168	Tension
3	*CD*	1584	Tension
4	*DE*	1503	Compression
5	*CE*	0	—
6	*FE*	1503	Compression
7	*FC*	1584	Compression
8	*BF*	500	Tension
9	*GF*	3087	Compression
10	*BG*	1801	Compression

Example 9. *A truss shown in Fig 33 is carrying a point load of 5 kN at E.*

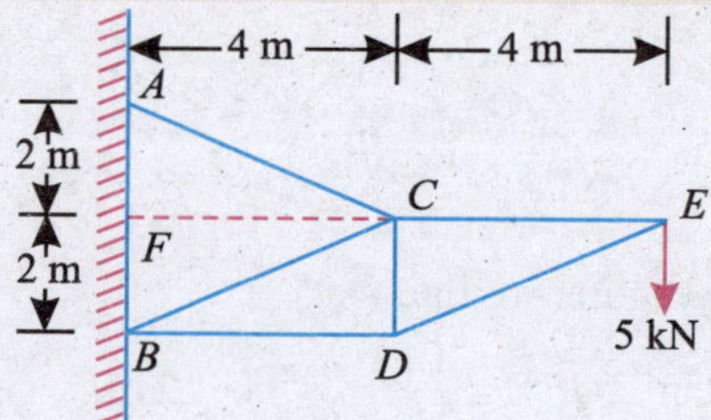

Fig. 33.

Find graphically, or otherwise, the force in the members CE, CD and BD of the truss.

[IGNOU, RGPV, BPUT]

Solution. The example may be solved either by the method of joints or method of sections. But we shall solve it by the method of sections, as one section line can cut the members *CE*, *CD* and *BD* in which the forces are required to be found out.

Method of Sections

Now let us pass section (1-1) cutting truss into two parts as shown in Fig. 34.

Now consider equilibrium of the <u>Right Part</u> of the truss.

Let the directions of the force P_{CE} P_{CD} and P_{BD} be assumed as shown in Fig. 34.

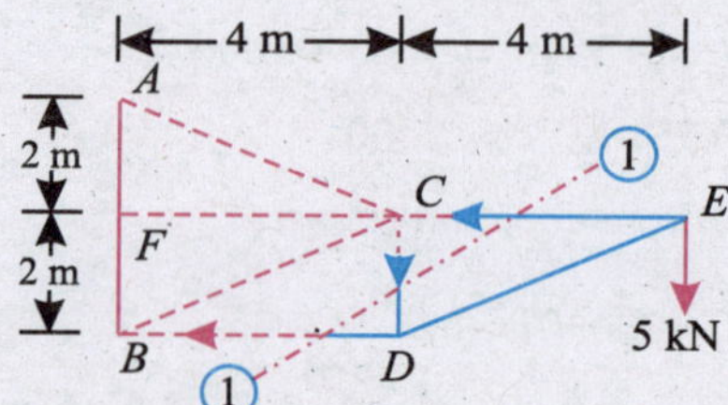

Fig. 34.

Taking moments about the joint D and equating the anti-clockwise moments with clockwise moments,

$$P_{CE} \times 2 = 5 \times 4 = 20$$

or $$P_{CE} = \frac{20}{2} = 10 \text{ kN (Tension)}$$

Taking moments about the joint B and equating the anti-clockwise moments with clockwise moments,

$$P_{CD} \times 4 = (5 \times 8) - (P_{CE} \times 2) = 40 - (10 \times 2) = 20$$

or $$P_{CD} = \frac{20}{4} = 5 \text{ kN (Compession)}$$

Taking moments about the joint C and equating the anti-clockwise moments with clockwise moments,

$$P_{BD} \times 2 = 5 \times 4 = 20$$

or $$P_{BD} = \frac{20}{2} = 10 \text{ kN (Tension)}$$

Example 10. *A pin-joined cantilever frame is hinged to a vertical wall at A and E and is loaded as shown in Fig 35.*

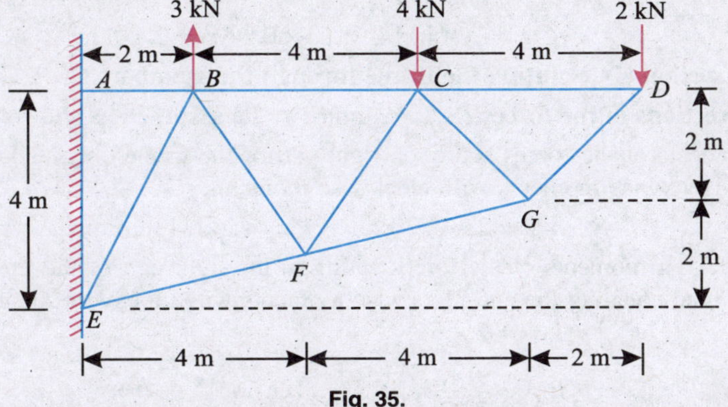

Fig. 35.

Determine the forces in the members CD, CG and FG.

Solution. First of all, extend the lines through the joints B, C and D as E, F and G meeting at O. Through G, draw GP perpendicular to CD. Similarly, through C, draw CQ perpendicular to FG.

Now extend the line of action of the member CG, and through O, draw a perpendicular to this line meeting at R as shown in Fig. 36.

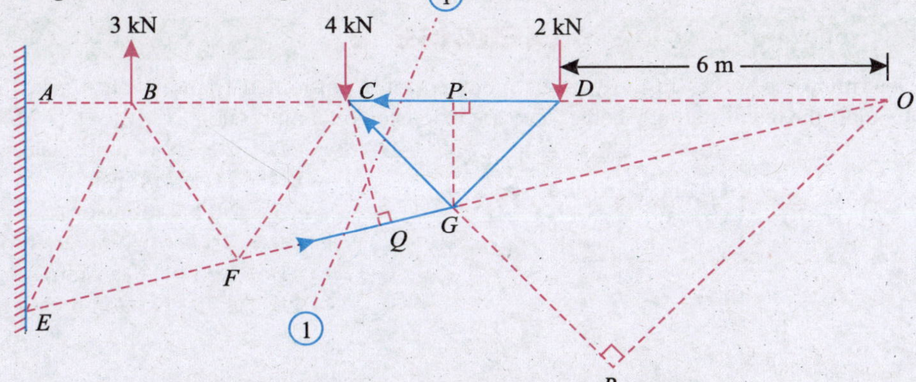

Fig. 36.

We know that in similar triangles OPG and OAE,

$$\frac{AO}{AE} = \frac{AP}{PG} \qquad \text{or} \qquad \frac{AO}{4} = \frac{8}{2} = 4$$

or $$AO = 4 \times 4 = 16 \text{ m}$$

and $$DO = 16 - 10 = 6 \text{ m}$$

Now in triangle CGP, we find that

$$\tan \angle GCP = \frac{2}{2} = 1 \qquad \text{or} \qquad \angle GCP = 45°$$

or $$\angle COR = 90° - 45° = 45°$$

and $$OR = OC \cos 45° = 10 \times 0.707 \text{ m} = 7.07 \text{ m}$$

From the geometry of the triangle OPG, we find that

$$\tan \angle GOP = \frac{2}{8} = 0.25 \qquad \text{or} \qquad \angle GOP = 14°$$

Similarly, in triangle OCQ, we find that

$$CQ = CO \sin 14° = 10 \times 0.2425 = 2.425 \text{ m}$$

Now pass section (1-1) cutting the frame through the members *CD*, *CG* and *FG*.

Let the directions of the forces P_{CD}, P_{CG} and P_{FG} be assumed as shown in Fig. 36.

Taking moments of the forces acting on right part of the frame only, about the joint G and equating the anti-clockwise moments with clockwise moments,

$$P_{CD} \times 2 = 2 \times 2 \qquad \text{or} \qquad P_{CD} = 2 \text{ kN (Tension)} \qquad \textbf{Ans.}$$

Similarly, taking moments of the forces acting in the right part of the truss only about the imaginary joint O and equating the anti-clockwise moments with clockwise moments,

$$P_{CG} \times 7.07 = 2 \times 6$$

or $$P_{CG} = \frac{12}{7.07} = 1.7 \text{ kN (Tension)} \qquad \textbf{Ans.}$$

Now taking moments of the forces acting in the right part of the truss only about the joint C and equating the anti-clockwise moments with clockwise moments,

$$P_{FG} \times 2.425 = 2 \times 4 = 8$$

or $$P_{FG} = \frac{8}{2.425} = 3.3 \text{kN (Compression)}$$

EXERCISE 2

1. Determine the forces in the various members of a pin-joined frame as shown in Fig. 37. Tabulate the result stating whether they are in tension or compression. **[AM-IETE]**

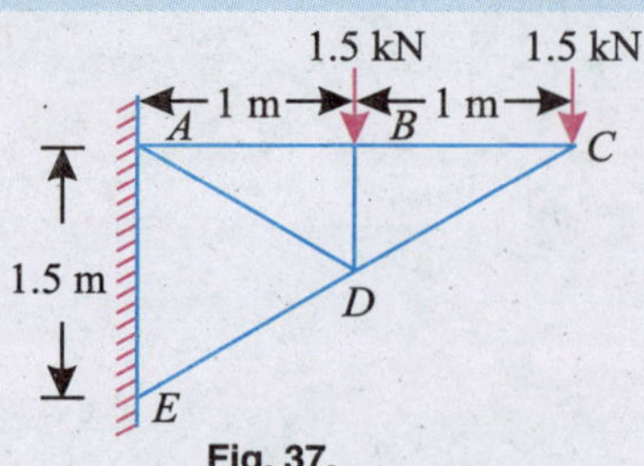

Fig. 37.

Ans. CD = 2.5 kN (Compression)
BC = 2.0 kN (Tension)
AB = 2.0 kN (Tension)
BD = 1.5 kN (Compression)
AD = 1.25 kN (Tension)
ED = 3.75 kN (Compression)

2. A cantilever truss of 4 m span is carrying two point loads of 1.5 kN each as shown in Fig. 38 Find the stresses in the members *BC* and *BD* of the truss.

Ans. 2.52 kN (Tension) ; Zero

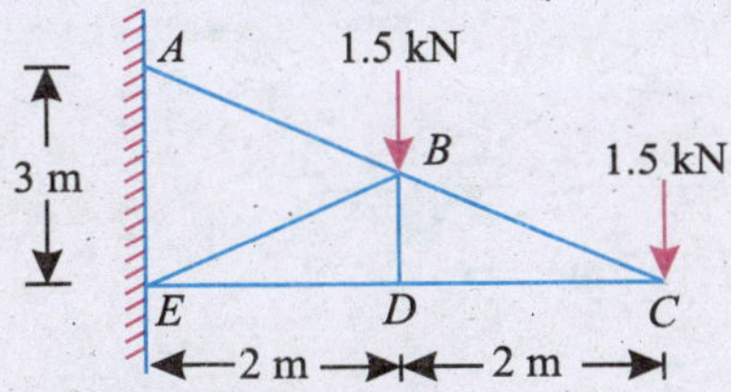

Fig. 38.

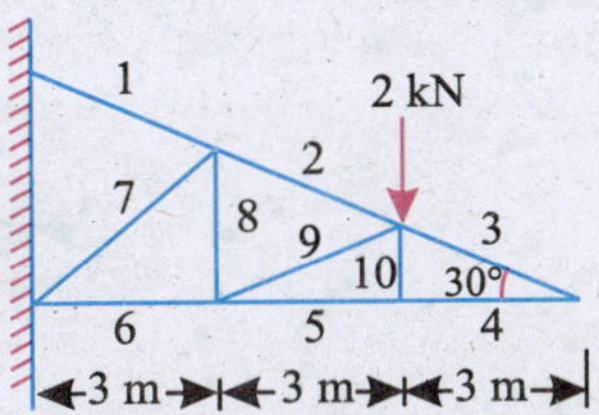

Fig. 39.

3. A cantilever truss carries two vertical load as shown in the Fig. 39. Find the magnitude and nature of stresses in the members 2, 9, 5 and 10 of the truss.

Ans. $P_2 = 6.0$ kN (Tension)
$P_9 = 2.9$ kN (Compression)
$P_5 = 3.46$ kN (Compression)
$P_{10} = 0$

4. A cantilever truss is subjected to two point loads of 3 kN each at *B* and *C* as shown in Fig 40. Find by any method the forces in the members *AB*. *BE* and *ED* of the truss.

Ans. $AB = 8.6$ kN (Tension)
$BE = 2.0$ kN (Tension)
$ED = 2.0$ kN (Compression)

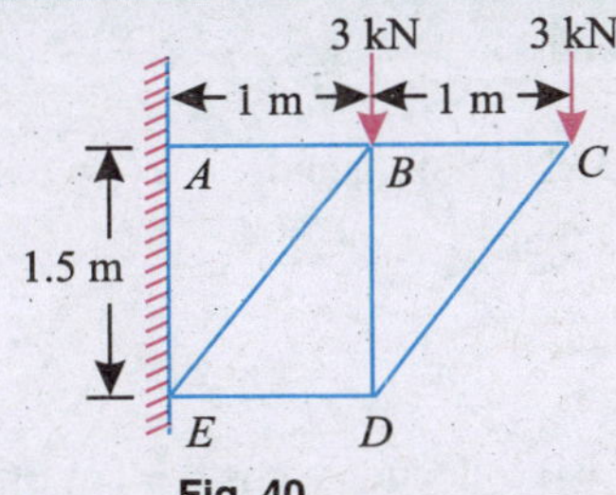

Fig. 40.

16. STRUCTURES WITH ONE END HINGED (OR PIN-JOINTED) AND THE OTHER FREELY SUPPORTED ON ROLLERS AND CARRYING HORIZONTAL LOADS

Sometimes, a structure is hinged or pin-jointed at one end, and freely supported on rollers at the other end. If such a truss carries vertical loads only, it does not present any special features. Such a structure may be solved just as a simply supported structure.

But, if such a structure carries horizontal loads (with or without vertical loads) the support reaction at the roller supported end will be normal to the support; where the support reaction at the hinged end will consist of :

1. Vertical reaction, which may be found out, by substracting the vertical support reaction at the roller supported end from the total vertical load.
2. Horizontal reaction, which may be found out, by algebraically adding all the horizontal loads.

After finding out the reactions, the forces in members of the frame may be found out as usual.

Example. 11. *Figure 41 shows a framed of 4 m span and 1.5 m height subjected to two point loads at B and D.*

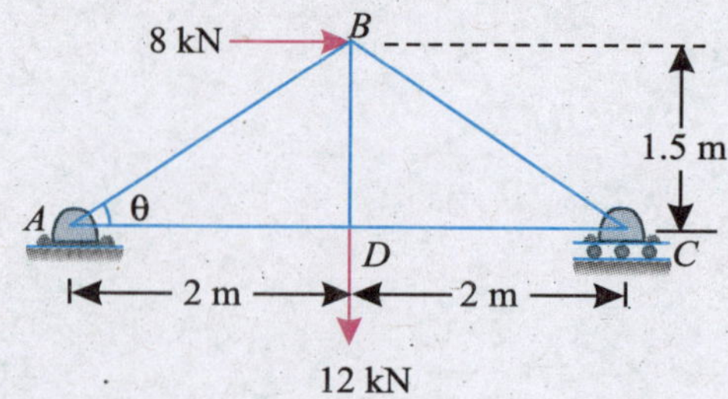

Fig. 41.

Find graphically, or otherwise, the forces in all the members of the structure.

[IGNOU, PU, DTU]

Solution. Since the structure is supported on rollers at the right hand support (*C*), therefore the reaction at this support will be vertical (because of horizontal support). The reaction at the left hand support (*A*) will be the resultant of vertical and horizontal forces and inclined with the vertical.

Taking moments about *A* and equating the anti-clockwise moments with clockwise moments,

$$V_C \times 4 = (8 \times 1.5) + (12 \times 2) = 36$$

or $$V_C = \frac{36}{4} = 9 \text{ kN} (\uparrow)$$

or $$V_A = 12 - 9 = 3 \text{ kN} (\uparrow) \quad \text{and} \quad H_A = 8 \text{ kN} (\leftarrow)$$

From the geometry of the figure, we find that

$$\tan \theta = \frac{1.5}{2} = 0.75 \quad \text{or} \quad \theta = 36.9°$$

Similarly, $$\sin \theta = \sin 36.9° = 0.6 \quad \text{and} \quad \cos \theta = \cos 36.9° = 0.8$$

The example may be solved either by the method of joints or method of sections. But we shall solve it by the method of joints as we have to find forces in all the members of the structure.

Method of Joints

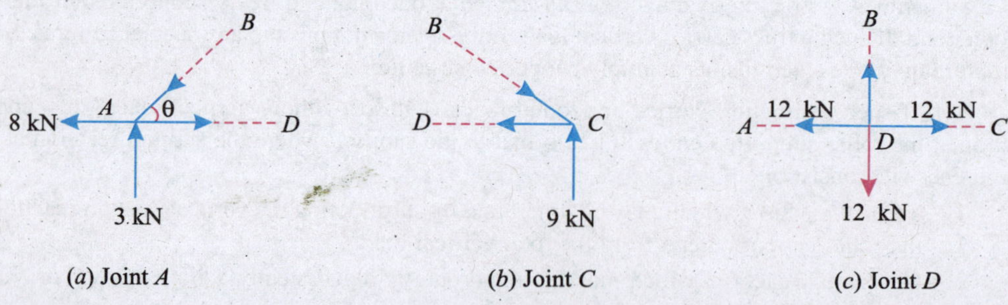

Fig. 42.

First of all, consider joint *A*.

Let directions of the forces P_{AB} and P_{AD} be assumed as shown in Fig. 42 (*a*).

We have already found that a horizontal force of 8 kN is acting at A as shown in Fig. 42 (*a*).

Resolving the forces vertically and equating the forces acting in downward direction with forces acting in upward direction,

$$P_{AB} \sin 36.9° = 3$$

or $$P_{AB} = \frac{3}{\sin 36.9°} = \frac{3}{0.6} = 5.0 \text{ kN (Compression)}$$

Resolving the forces horizontally and equating the forces acting in left direction with forces acting in right direction,

$$P_{AD} = 8 + P_{AB} \cos 36.9° = 8 + (5 \times 0.8) = 12.0 \text{ kN (Tension)}$$

Now consider the joint *C*.

Let the directions of the forces P_{BC} and P_{CD} be assumed as shown in Fig. 42 (*b*).

Resolving the forces vertically and equating the forces acting in downward direction with forces acting in upward direction,

$$P_{BC} \sin 36.9° = 9$$

or $$P_{BC} = \frac{9}{\sin 36.9°} = \frac{9}{0.6} = 15 \text{ kN (Compression)}$$

Resolving the forces horizontally and equating the forces acting in left direction with forces acting in right direction,

$$P_{CD} = P_{BC} \cos 36.9° = 15 \times 0.8 = 12.0 \text{ kN (Tension)}$$

Now consider the joint *D*.

A little consideration will show that the value of the force P_{BD} will be equal to the load 12 kN (Tension) as shown in Fig 42. (*c*). This will happen as the vertical components of the forces P_{AD} and P_{CD} will be zero.

Now tabulate the results as given below :

S.No.	*Member*	*Magnitude of force in kN*	*Nature of force*
1	*AB*	5.0	Compression
2	*AD*	12.0	Tension
3	*BC*	15.0	Compression
4	*CD*	12.0	Tension
5	*BD*	12.0	Tension

Example 12. 2 *A truss of 8 metres span, is loaded as shown in Fig. 43.*

Find the forces in the members CD, FD and FE of the truss. **[RGPV]**

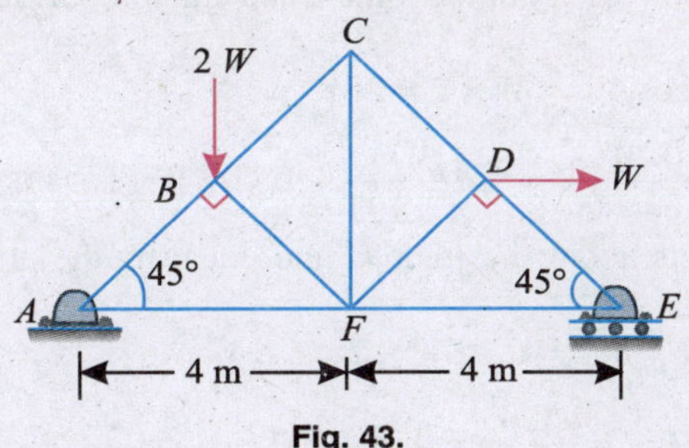

Fig. 43.

Solution. Since the truss is supported on rollers at the right hand support (E), therefore the reaction at this support will be vertical (because of horizontal support). The reaction at the left hand support (A) will be the resultant of vertical and horizontal forces and inclined with vertical.

Taking moments about A and equating same,

$$V_E \times 8 = (2\,W \times 2) + (W \times 2) = 6\,W$$

or
$$V_E = \frac{6W}{8} = 0.75W\ (\uparrow)$$

and
$$^*V_A = 2\,W - 0.75\,W = 1.25\,W\,(\uparrow) \quad \text{and} \quad H_A = W\,(\leftarrow)$$

*Note: There is no need of finding out the vertical and horizontal reaction at A, as we are not considering this part of the truss.

The example may be solved either by the method of joints or method of sections. But we shall solve it by the method of sections, as one section line can cut the members CD, FD and FE in which the forces are required to be found out.

Method of Sections

Now let us pass section (1-1) cutting the truss into two parts as shown in Fig. 44.

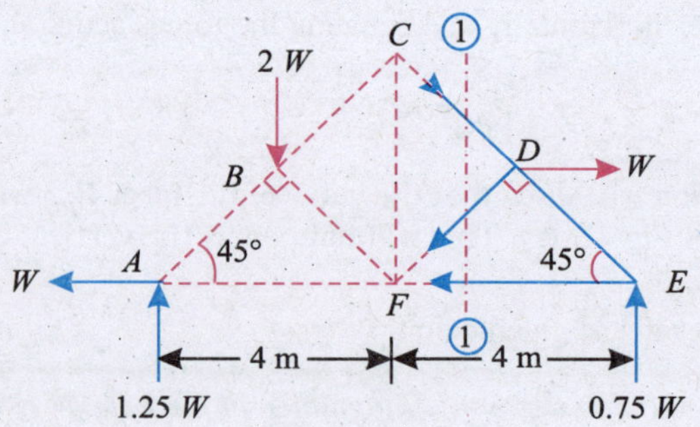

Fig. 44.

Now consider equilibrium of the Right Part of the truss.

Let the directions of the forces P_{CD}, P_{FD} and P_{FE} be assumed as shown in Fig. 44.

Taking moments about the joint F and equating the anti-clockwise moments with clockwise moments,

$$P_{CD} \times 4 \sin 45° = (0.75\,W \times 4) - (W \times 2) = W$$

or
$$P_{CD} = \frac{W}{4 \sin 45°} = \frac{W}{4 \times 0.707} = 0.354\,W \text{ (Compression)}$$

Similarly, taking moments about the joint E and equating the anti-clockwise moments with clockwise moments,

$$P_{FD} \times 4 \cos 45° = W \times 2 = 2\,W$$

or
$$P_{FD} = \frac{2W}{4 \cos 45°} = \frac{2W}{4 \times 0.707} = 0.707\,W \text{ (Tension)}$$

Now taking moments about the joint D and equating the anti-clockwise moments with clockwise moments,

$$P_{FE} \times 2 = 0.75\,W \times 2 = 1.5\,W$$

or
$$P_{FE} = \frac{1.5W}{2} = 0.75\,W \text{ (Tension)}$$

Example 13. *Figure 45 shows a pin-jointed frame carrying a vertical load at B and a horizontal load at D*

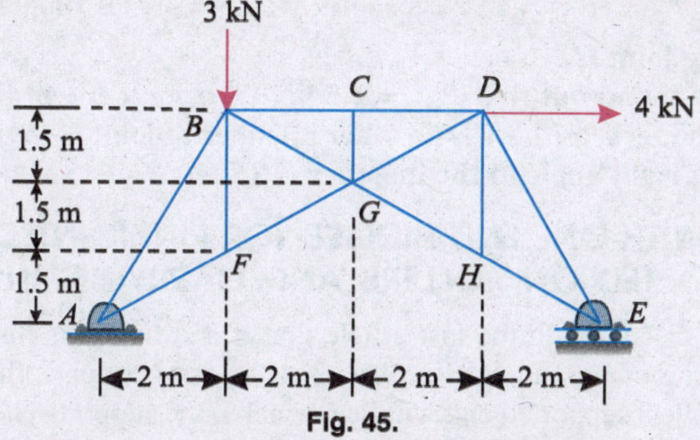

Fig. 45.

Find the forces in the members DF, HE and DH of the frame.

Solution. Since the frame is supported on rollers at the right hand support (*E*), therefore the reaction at this support will be vertical (because of horizontal support). The reaction at the left hand support (*A*) will be the resultant of vertical and horizontal forces and inclined with the vertical.

Taking moments about the joint* *A* and equating the anti-clockwise moments with clockwise moments,

$$R_E \times 8 = (3 \times 2) + (4 \times 4.5) = 24$$

or
$$R_E = \frac{24}{8} = 3 \text{ kN}$$

(*Note: There are no need of finding out the vertical and horizontal reaction at *A*, as we are not considering this part of the truss.)

From the geometry of the figure, we find that

$$\tan\theta = \frac{3}{4} = 0.75 \qquad \text{or} \qquad \theta = 36.9°$$

$$\tan\alpha = \frac{4.5}{2} = 2.25 \qquad \text{or} \qquad \alpha = 66°$$

The example may be solved either by the method of joints or method of sections. But we shall solve it by the method of joints, as we can resolve the forces in the members at joint *E* in which the force is required to be found out.

Method of Joints

Consider the joint *E*.

Let the directions of the forces P_{DE} and P_{HE} be assumed as shown in Fig. 46.

Resolving the forces horizontally and equating the forces acting in left direction with forces acting in right direction,

$$P_{DE} \cos 66° = P_{HE} \cos 36.9° = P_{HE} \times 0.8$$

or
$$P_{DE} = \frac{P_{HE} \times 0.8}{\cos 66°} = \frac{P_{HE} \times 0.8}{0.4062} = 1.97\ P_{HE}$$

Fig. 46.

Resolving the forces vertically and equating the forces acting in downward direction with forces acting in upward direction,

$$P_{DE} \sin 66° = P_{HE} \sin 36.9° + 3$$

or
$$1.97\ P_{HE} \times 0.9137 = (P_{HE} \times 0.6) + 3$$

$$1.2\ P_{HE} = 3$$

or $$P_{HE} = \frac{3}{1.2} = 2.5 \text{ kN (Tension)}$$

and $$P_{DE} = 1.97\, P_{HE} = 1.97 \times 2.5 = 4.93 \text{ (Compression)}$$

Now consider the joint *H*.

We have already found out that P_{HE} = 2.5 kN (Tension). It will be interesting to know that the force P_{DH} will be zero, as there is no other member at joint *H* to balance the component of this forces (if any) at right angle to the member *GHE*.

17. STRUCTURES WITH ONE END HINGED (OR PIN-JOINTED) AND THE OTHER FREELY SUPPORTED ON ROLLERS AND CARRYING INCLINED LOADS

We have already discussed in the last article that if a structure is hinged at one end, freely supported on rollers at the other, and carries horizontal loads (with or without vertical loads), the support reaction at the roller- supported end will be normal to the support. The same principle is used for structures carrying inclined loads also. In such a case, the support reaction at the hinged end will be the resultant of :

1. Vertical reaction, which may be found out by subtracting the vertical component of the support reaction at the roller supported end from the total vertical loads.
2. Horizontal reaction, which may be found out algebraically by adding all the horizontal loads.

Example 14. *Figure 47 represents a north-light roof truss with wind loads acting on it.*

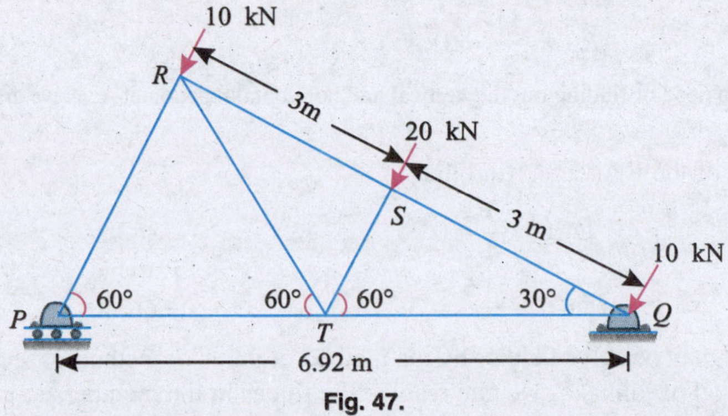

Fig. 47.

Find graphically, or otherwise, the forces in all the members of the truss Give your results in a tabulated form. **[MU]**

Solution. Since the truss is supported on rollers at *P*, therefore the reaction at this end will be vertical (because of horizontal support). Moreover, it is hinged at *Q*, therefore the reaction at this end will be the resultant of horizontal and vertical forces and inclined with the vertical.

Taking moments about *Q* and equating the anti-clockwise moments with clockwise moments,

$$V_P \times 6.92 = (20 \times 3) + (10 \times 6) = 120$$

or $$V_P = \frac{120}{6.92} = 17.3 \text{ kN}$$

Total wind loads on the truss = 10 + 20 + 10 = 40 kN

∴ Horizontal component of wind load,

$$H_Q = 40 \cos 60° = 40 \times 0.5 = 20 \text{ kN } (\rightarrow)$$

Vertical component of the wind load

$$= 40 \sin 60° = 40 \times 0.866 = 34.6 \text{ kN } (\downarrow)$$

∴ Vertical reaction at *Q*,

$$V_Q = 34.6 - 17.3 = 17.3 \text{ kN } (\uparrow)$$

The example may be solved either by the method of joints or method of sections. But we shall solve it by the method of joints, as we have to find out the forces in all the members of the truss.

Method of Joints

Consider the joint *P*.

Let the directions of the forces P_{PR} and P_{PT} be assumed as shown in Fig 48(*a*).

We know that a horizontal force of 20 kN is acting at *Q* as shown in Fig. 48 (*b*).

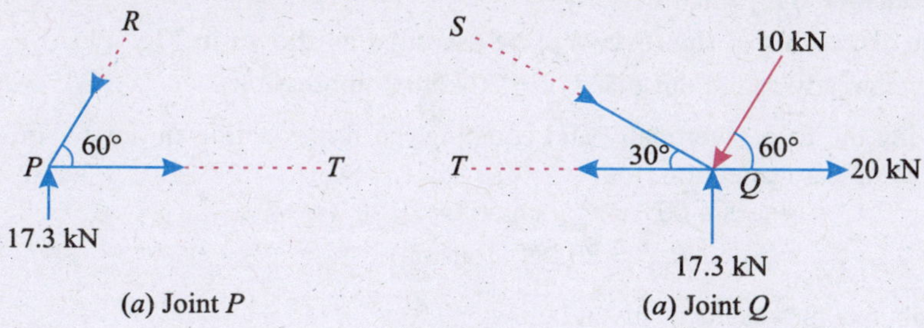

Fig. 48.

Resolving the forces vertically and equating the forces acting in downward direction with forces acting in upward direction,

$$P_{PR} \sin 60° = 17.3$$

or $$P_{PR} = \frac{17.3}{\sin 60°} = \frac{17.3}{0.866} = 20 \text{ kN (Compression)}$$

Resolving the forces horizontally and equating the forces acting in left direction with forces acting in right direction,

$$P_{PT} = P_{PR} \cos 60° = 20 \times 0.5 = 10 \text{ kN (Tension)}$$

Now consider the joint *Q*.

Let the directions of the forces P_{SQ} and P_{QT} be assumed as shown in Fig. 48 (*b*).

We know that a horizontal force of 20 kN is acting at *Q* as shown in Fig 48 (*b*).

Resolving the forces vertically and equating the forces acting in downward direction with forces acting in upward direction,

$$P_{SQ} \sin 30° = 17.3 - 10 \cos 30° = 17.3 - (10 \times 0.866) = 8.64$$

or $$P_{SQ} = \frac{8.64}{\sin 30°} = \frac{8.64}{0.5} = 17.3 \text{ kN (Compression)}$$

Resolving the forces horizontally and equating the forces acting in left direction with forces acting in right direction,

$$P_{QT} = P_{SQ} \cos 30° + 20 - 10 \sin 30°$$

$$= (17.3 \times 0.866) + 20 - (10 \times 0.5) = 30 \text{ kN (Tension)}$$

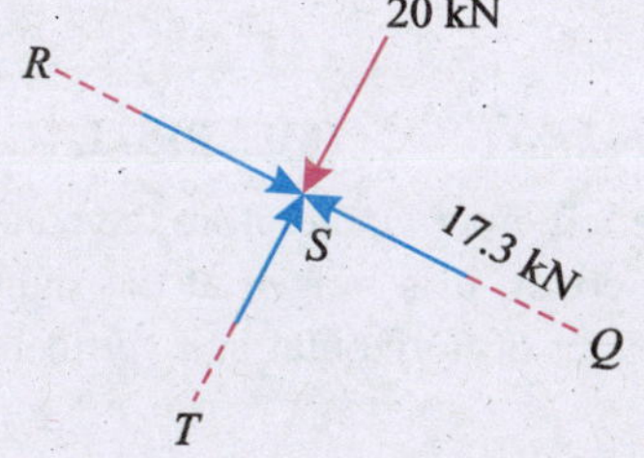

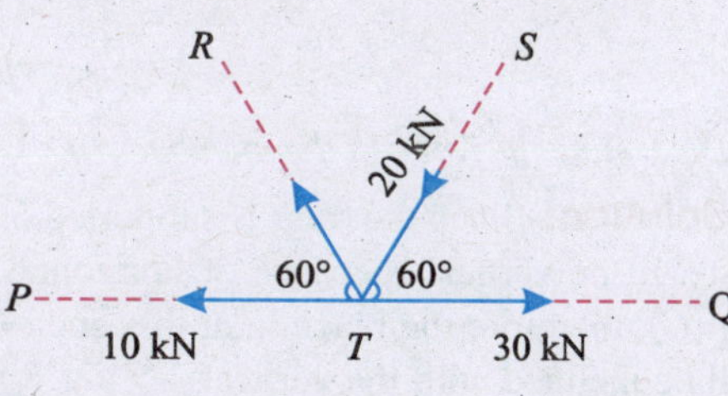

Fig. 49.

Now consider the joint *S*.

We have already found out that P_{SQ} = 17.3 kN (Compression). A little consideration will show that the value of the force P_{TS} will be equal to the force 20 kN (Compression). Similarly, the value of the force P_{RS} will be equal to P_{SQ} *i.e.*, 17.3 kN (Compression) as shown in Fig. 49 (*a*).

Now consider the joint *T*.

Let the directions of the force P_{RT} be assumed as shown in Fig. 49 (*b*).

We have already found out that P_{ST} = 20 kN (Compression).

Resolving the forces vertically and equating the forces acting in downward direction with forces acting in upward direction,

$$P_{RT} \sin 60° = P_{ST} \sin 60° = 20 \sin 60°$$

or
$$P_{RT} = 20 \text{ kN (Tension)}$$

Now tabulate the results as given below:

S.No.	*Member*	*Magnitude of force in kN*	*Nature of force*
1	*PR*	20.0	Compression
2	*PT*	10.0	Tension
3	*SQ*	17.3	Compression
4	*QT*	30.0	Tension
5	*ST*	20.0	Compression
6	*RS*	17.3	Compression
7	*RT*	20.0	Tension

Example 15. *A truss of 12 m span is loaded as shown in Fig 50.*

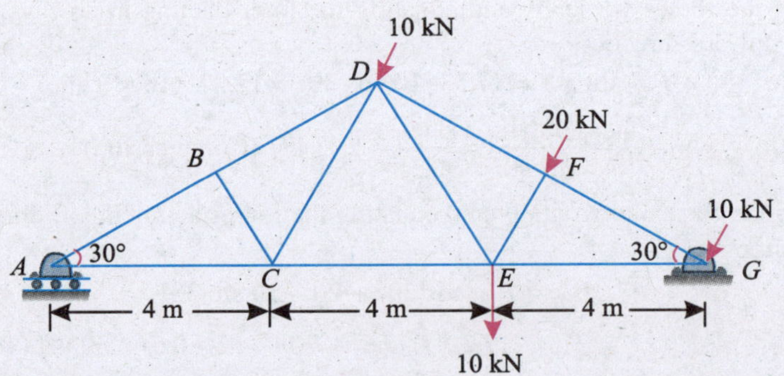

Fig. 50.

Determine the force in the members BD, CE and CD of the truss. **[MU, BPUT, IGNOU]**

Solution. Since the truss is supported on rollers on the left end (*A*), therefore the reaction at this end will be vertical (because of horizontal support). Moreover, it is hinged at the right hand support (*G*), therefore the reaction at this end will be the resultant of horizontal and vertical forces and will be inclined with the vertical.

Taking * moments about G and equating the anti-clockwise moments with clockwise moments,

$$V_A \times 12 = (10 \times 4) (20 \times 4 \cos 30°) + (10 \times 8 \cos 30°)$$

$$= 40 + (80 \times 0.866) + (80 \times 0.866) = 178.6$$

or $$V_A = \frac{178.6}{12} = 14.9 \text{ kN}$$

(*Note: There is no need of finding out the vertical and horizontal reaction at G, as we are not considering this part of the truss.)

The example may be solved either by the method of joints or method of sections. But we shall solve it by the method of sections, as one section line can cut the members BD, CE and CD in which forces are required to be found out.

Method of Sections

Now let us pass section (1-1) cutting the truss into two parts as shown in Fig 51.

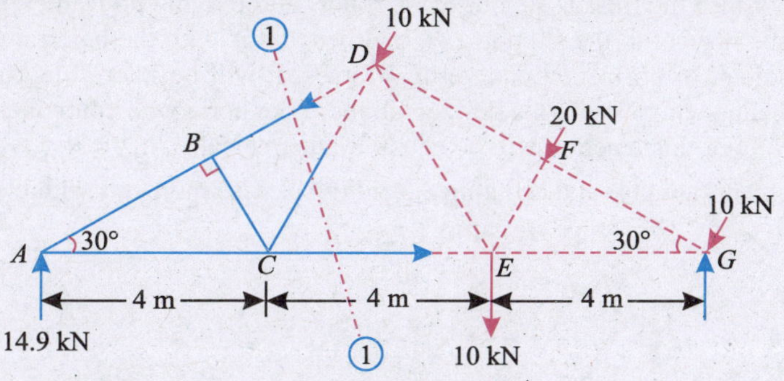

Fig. 51.

Now consider equilibrium of the <u>Left Part</u> of the truss.

Let the directions of the forces P_{BD}, P_{CE} and P_{CD} be assumed as shown in Fig 51.

Taking moments about the joint C and equating the anti-clockwise moments with clockwise moments,

$$P_{BD} \times 2 = 14.9 \times 4 = 59.6$$

or $$P_{BD} = \frac{59.6}{2} = 29.8 \text{ kN (Compression)}$$

Similarly taking moments about the joint D and equating the anti-clockwise moments with clockwise moments,

$$P_{CE} \times 6 \tan 30° = 14.9 \times 6 = 89.4$$

or $$P_{CE} = \frac{89.4}{6 \tan 30°} = \frac{89.4}{6 \times 0.5774} = 25.8 \text{ kN} \quad \text{(Tension)}$$

Now for finding out P_{CD}, we shall take moments about the A (where the other two members meet). Since there is no force in the lift of the truss (other than the reaction V_A, which will have zero moment about A), therefore the value of P_{CD} will be zero.

Note: The force P_{CD} may also be found out as discussed below :

At joint B, the force in member BC is zero, as there is no other member to balance the force (if any) in the member BC.

Now at joint C, since the force in member BC is zero, therefore the force in member CD is also equal to zero.

Example 16. *A truss hinged at A and supported on rollers at D, is loaded as shown in Fig. 52. Find the forces in the members BC, FC, FE of the truss.*

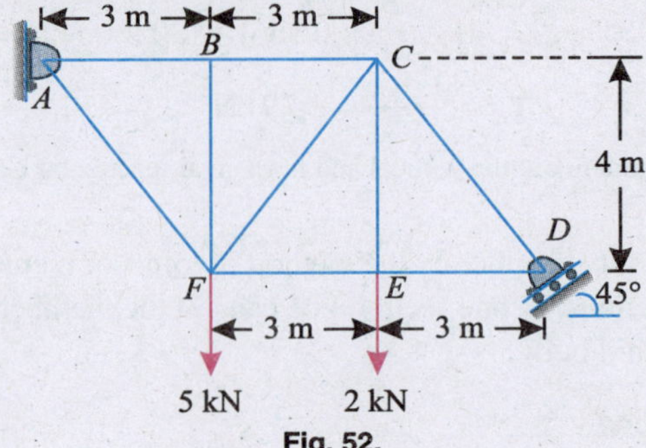

Fig. 52.

Solution. Since the truss is supported on rollers at the right end D, therefore the reaction at this support will be normal to the support *i.e.*, inclined at 45° with the horizontal. The reaction at A will be the resultant of horizontal and vertical forces. It will be interesting to know that as the reaction at D is inclined at 45° with the horizontal, therefore horizontal component (R_{DH}) and vertical component (R_{DV}) of this reaction will be equal. Mathematically $R_{DH} = R_{DV}$.

Taking moments about A and equating the anti-clockwise moments with clockwise moments,

$$(R_{DV} \times 9) - (R_{DH} \times 4) = (5 \times 3) + (2 \times 6)$$

or $$5\,R_{DH} = 27 \qquad [\because R_{DH} = R_{DV}]$$

or $$R_{DH} = \frac{27}{5} = 5.4 \text{ kN } (\leftarrow)$$

and $$R_{DV} = 5.4 \text{ kN } (\uparrow)$$

The example may be solved either by the method of joints or method of sections. But we shall solve it by the method of sections, as one section line can cut the members BC, FE and FC and in which forces are required to be found out.

Method of Sections

Now let us pass section (1-1) cutting the truss into two parts as shown in Fig. 53.

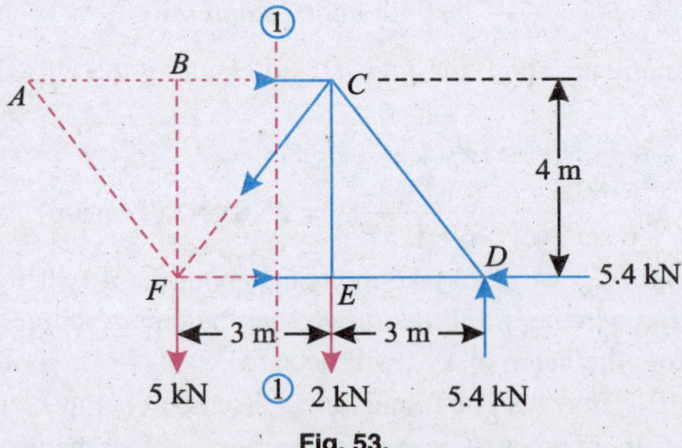

Fig. 53.

Now consider equilibrium of <u>Right Part</u> of the truss.

Let the directions of the forces P_{BC} and P_{FE} be assumed as shown in Fig 53.

Taking moments about the joint F and equating the anti-clockwise moments with clockwise moments,

$$P_{BC} \times 4 = (5.4 \times 6) - (2 \times 3) = 26.4$$

or $$P_{BC} = \frac{26.4}{4} = 6.6 \text{ kN (Compression)}$$

Similarly, taking moments about the joint C and equating the anti-clockwise moments with clockwise moments,

$$P_{FE} \times 4 = (5.4 \times 4) - (5.4 \times 3) = 5.4$$

$\therefore$ $$P_{FE} = \frac{5.4}{4} = 1.35 \text{ kN (Compression)}$$

Now taking moments about the joint B and equating the anti-clockwise moments with clockwise moments,

$$P_{FC} \times 2.4 = (P_{FE} \times 4) - (2 \times 3) + (5.4 \times 6) - (5.4 \times 4)$$
$$= (1.35 \times 4) - 6 + 32.4 - 21.6 = 10.2$$

or $$P_{FC} = \frac{10.2}{2.4} = 4.25 \text{ kN (Tension)}$$

18. MISCELLANEOUS STRUCTURES

In the previous articles we have been analysing the regular frames subjected to vertical, horizontal and inclined loads. We have been solving such examples by the methods of joints and sections. But sometimes we come across irregular structures.

Such structures may be analysed in the same way as that for regular structures. The casual look at such a structure, gives us a feeling that it is complicated problem. But a patient and thoughtful procedure helps us in solving such problems. The following examples will illustrate this point.

Example 17. *Figure 54 shows a bridge truss of 130 m span subjected to two points loads. Determine the forces in the members 1, 2 and 3 of the bridge truss by any suitable method.*

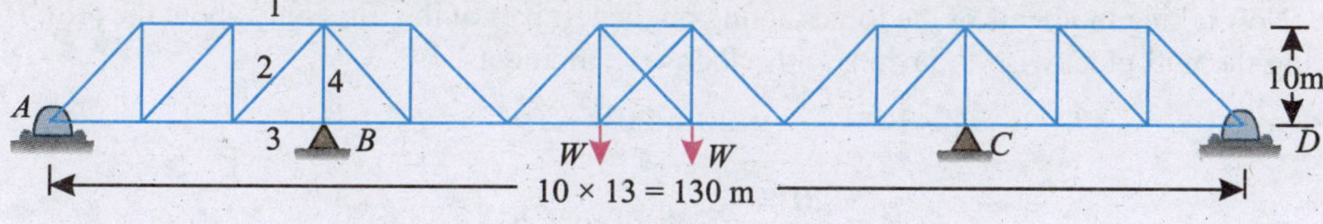

Fig. 54.

Solution. The whole structure may be considered to consist of two cantilever trusses supporting an intermediate truss. As a matter of fact, the two point loads acting at the intermediate truss are transferred to the ends of the cantilever trusses.

Since the two cantilever trusses are symmetrical and the point loads on the intermediate truss are also symmetrical, therefore each cantilever truss is subjected to a point load as shown in Fig. 55 (*a*).

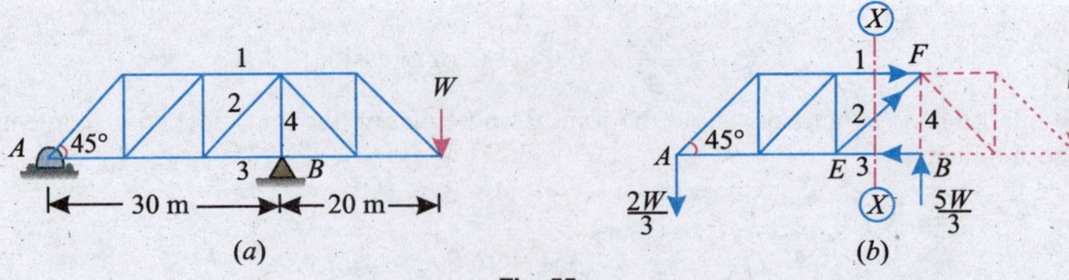

Fig. 55.

Let V_B = Vertical reaction at the support B.

Taking moments about the support A and equating the anti-clockwise moments with clock-wise moments,

$$V_B \times 30 = W \times 50 = 50\,W$$

or
$$V_B = \frac{50\,W}{30} = \frac{5\,W}{3}\ (\uparrow)$$

and
$$V_A = \frac{5\,W}{3} - W = \frac{2\,W}{3}\ (\downarrow)$$

First of all, pass section (X-X) cutting the truss into two parts and consider the equilibrium of the left part of the truss as shown in Fig. 55 (b). Now let the directions of the forces P_1, P_2, P_3 and P_4 be assumed as shown in Fig 55 (b).

First of all, let us consider the joint *B*.

A little consideration will show that the magnitude of the force P_4 will be equal and opposite to the reaction V_B *i.e.*, $5W/3$ (Compression). This will happen as the vertical components of the horizontal members at B will be zero.

Resolving the forces vertically and equating the forces acting in downward direction with forces acting in upward direction,

$$P_2 \times \cos 45° = \frac{2W}{3}$$

or
$$P_2 = \frac{2W}{3} \times \frac{1}{\cos 45°} = \frac{2W}{3 \times 0.707} = 0.943\ W\ \text{(Compression)}$$

Taking moments of the forces acting on the left part of the truss only about the joint E and equating the anti-clockwise moments with clockwise moments,

$$P_1 \times 10 = \frac{2W}{3} \times 20 = \frac{40W}{3}$$

or
$$P_1 = \frac{40W}{3} \times \frac{1}{10} = \frac{4W}{3}\ \text{(Compression)} \quad \textbf{Ans.}$$

Now taking moments of the forces acting on the left part of the truss only about the joint F and equating the anti-clockwise moments with clockwise moments,

$$P_3 \times 10 = \frac{2W}{3} \times 30 = 20W$$

or
$$P_3 = \frac{20W}{10} = 2W\ \text{(Tension)} \quad \textbf{Ans.}$$

Example 18. *A pin-jointed frame shown in Fig 56 is hinged at A and loaded at D. Horizontal chain is attached to C and pulled so that AD is horizontal. Determine the pull in the chain and also the force in each member. Tabulate the results.*

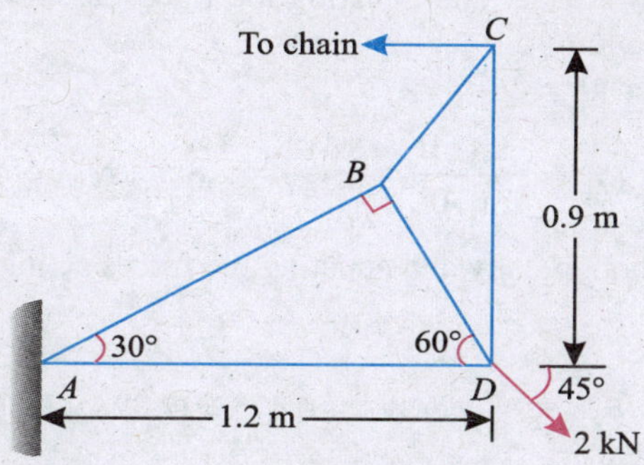

Fig. 56.

Solution. The example may be solved either by the method of joints or method of sections. But we shall solve it by the method of joints, as we have to find the force in each member.

Method of Joints

Pull in the chain

Let P = Pull in the chain.

Taking moments about the joint A and equating the anti-clockwise moments with clockwise moments,

$$P \times 0.9 = 2 \cos 45° \times 1.2 = 2 \times 0.707 \times 1.2 = 1.7$$

or $$P = \frac{1.7}{0.9} = 1.889 \text{ kN} \qquad \textbf{Ans.}$$

Force in each member

We know that, horizontal reaction at A,

$$H_A = 1.889 - (2 \cos 45°) = 1.889 - (2 \times 0.707) = 0.475 \text{ kN } (\rightarrow)$$

Vertical reaction at A,

$$V_A = 2 \sin 45° = 2 \times 0.707 = 1.414 \text{ kN } (\uparrow)$$

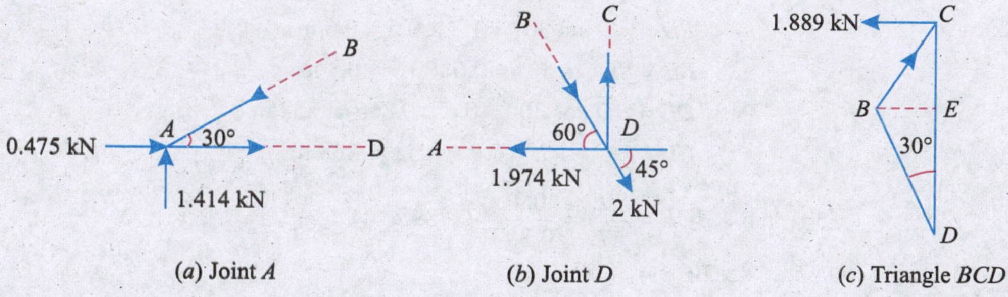

Fig. 57.

First of all, consider the joint A.

Let the directions of the forces P_{AB} and P_{AD} be assumed as shown in Fig 57 (*a*).

We have already found out that the horizontal and vertical reactions at A are 0.475 kN and 1.414 kN repectively as shown in the figure.

Resolving the forces vertically and equating the forces acting in downward direction with forces acting in upward direction,

$$P_{AB} \sin 30° = 1.414$$

or
$$P_{AB} = \frac{1.414}{\sin 30°} = \frac{1.414}{0.5} = 2.828 \text{ kN (Compression)}$$

Resolving the forces horizontally and equating the forces acting in left direction with forces acting in right direction,

$$P_{AD} = P_{AB} \cos 30° - 0.475 = (2.828 \times 0.866) - 0.475$$
$$= 1.974 \text{ kN (Tension)}$$

Now consider the joint D.

Let the directions of the forces P_{BD} and P_{CD} be assumed as shown in Fig 57 (*b*).

We have already found out that P_{AD} = 1.974 kN (Tension) as shown in the figure.

Resolving the forces horizontally and equating the forces acting in left direction with forces acting in right direction,

$$P_{BD} \cos 60° = 1.974 - 2 \cos 45° = 1.974 - (2 \times 0.707) = 0.56 \text{ kN}$$

or
$$P_{BD} = \frac{0.56}{\cos 60°} = \frac{0.56}{0.5} = 1.12 \text{ kN} \quad \text{(Compression)}$$

Resolving the forces vertically and equating the forces acting in downward direction with forces acting in upward direction,

$$P_{CD} = P_{BD} \sin 60° + 2 \sin 45°$$
$$= (1.12 \times 0.866) + (2 \times 0.707) = 2.384 \text{ kN (Tension)}$$

Now consider the triangle *BCD*.

From *B*, draw *BE* perpendicular to *CD*. Let the direction of P_{BC} be assumed as shown in Fig 57 (*c*).

From the geometry of this triangle, we find that

$$BD = AD \sin 30° = 1.2 \times 0.5 = 0.6 \text{ m}$$

and
$$BE = BD \sin 30° = 0.6 \times 0.5 = 0.3 \text{ m}$$

∴
$$DE = BD \cos 30° = 0.6 \times 0.866 = 0.52 \text{ m}$$

and
$$CE = DC - DE = 0.9 - 0.52 = 0.38 \text{ m}$$

∴
$$\tan \angle BCE = \frac{BE}{CE} = \frac{0.3}{0.38} = 0.7895$$

or
$$\angle BCE = 38.3°$$

Resolving the forces horizontally at *C* and equating the forces acting in left direction with forces acting in right direction,

$$P_{BC} \sin 38.3° = 1.889$$

or
$$P_{BC} = \frac{1.889}{\sin 38.3°} = \frac{1.889}{0.6196} = 3.049 \text{ kN (Compression)}$$

Now tabulate the results as given below :

S.No.	*Member*	*Magnitude of force in kN*	*Nature of force*
1	*AB*	2.828	Compression
2	*AD*	1.974	Tension
3	*BD*	1.12	Compression
4	*CD*	2.384	Tension
5	*BC*	3.049	Compression

Example 19. *The truss shown in the Fig. 58 is made up of three equilateral triangles loaded at each of the lower panel pains.*

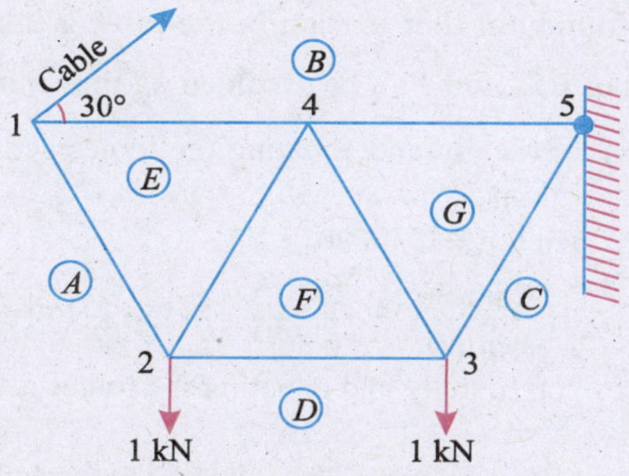

Fig. 58.

It is supported at the wall on the right hand side and by a cable on the left as shown. Determine (a) the tension in the cable (b) the reaction at the wall and (c) the nature and magnitude of the force in each bar.

Solution. The example may be solved either by the method of joints or method of sections. But we shall solve it by the method of joints, as we have to find out the forces in all the members of the truss.

(a) Tension in the cable

Let T = Tension in the cable and

a = Length of each side of the equilateral triangle.

Taking moments about the joint 5 and equating the anti-clockwise moments with clockwise moments,

$$(T \cos 60°) \times 2a = (1 \times 1.5\, a) + (1 \times 0.5\, a)$$

or $$(T \times 0.5)\, 2a = 2a$$

or $$T = 2 \text{ kN} \quad \textbf{Ans.}$$

(*b*) *Nature and magnitude of the force in each bar*

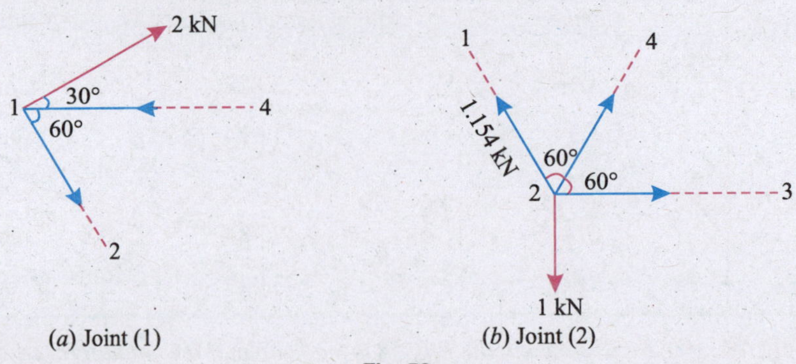

(*a*) Joint (1)

(*b*) Joint (2)

Fig. 59.

First of all consider the joint 1.

We have already found out that tension in the cable is 2 kN as shown in the figure.

Let the directions of P_{1-2} and P_{1-4} be assumed as shown in Fig. 59 (*a*).

Resolving the forces vertically and equating the forces acting in downward direction with forces acting in upward direction,

$$P_{1-2} \sin 60° = 2 \sin 30°$$

or $$P_{1-2} = \frac{2 \sin 30°}{\sin 60°} = \frac{2 \times 0.5}{0.866} = 1.154 \text{ kN (Tension)}$$

Resolving the forces horizontally and equating the forces acting in left direction with forces acting in right direction,

$$P_{1-4} = 2 \cos 30° + 1.154 \cos 60° \text{ kN}$$

$$= (2 \times 0.866) + (1.154 \times 0.5) = 2.309 \text{ kN (Compression)}$$

Now consider the joint 2.

We have already found out that the force in member 1-2 (*i.e.* P_{1-2}) is 1.54 kN (Tension).

Let the directions of the forces P_{2-4} and P_{2-3} be assumed as shown in Fig 59 (*b*).

Resolving the forces vertically and equating the forces acting in downward direction with forces acting in upward direction,

$$P_{2-4} \sin 60° = 1 - 1.154 \sin 60° = 1 - (1.154 \times 0.866) = 0$$

$$\therefore \quad P_{2-4} = 0$$

Resolving the forces horizontally and equating the forces acting in left direction with forces acting in right direction,

$$P_{2-3} = 1.154 \cos 60° = 1.154 \times 0.5 = 0.577 \text{ kN (Tension)}$$

Now consider the joint 4.

A little consideration will show that the force P_{3-4} will be zero.

This will happen as the force P_{2-4} is zero and the vertical components of forces P_{1-4} and P_{4-5} are also zero. Moreover, the force P_{4-5} will be equal to the force P_{1-4} *i.e.*, 2.309 kN (Compression). This will happen as the forces P_{2-4} and P_{2-5} (being zero) will have their vertical components as zero.

Now consider the joint 3.

Let the direction of the force P_{3-5} be assumed as shown in Fig. 60 (*b*).

We have already found out that the force P_{2-3} is 0.577 kN (Tension) and force P_{3-4} is zero.

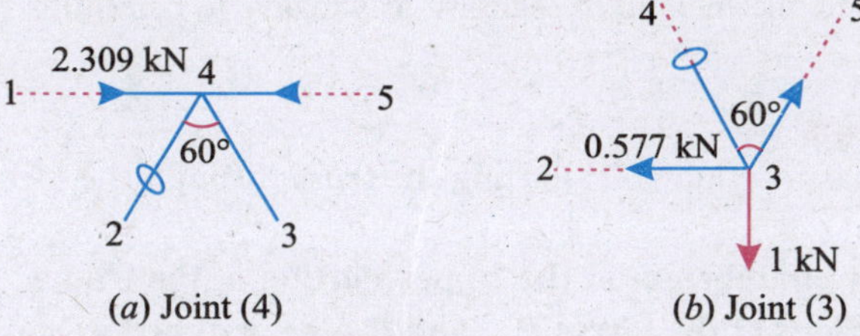

Fig. 60.

Resolving the forces vertically and equating the forces acting in downward direction with forces acting in upward direction,

$$P_{3-5} \cos 30° = 1$$

or
$$P_{3-5} = \frac{1}{\cos 30°} = \frac{1}{0.866} = 1.154 \text{ kN (Tension)}$$

Now tabulate the results as given below :

S.No.	*Member*	*Magnitude of force in kN*	*Nature of force*
1	1-2 (*AE*)	1.154	Tension
2	1-4 (*BE*)	2.309	Compression
3	2-4 (*EF*)	0	—
4	2-3 (*FD*)	0.577	Tension
5	3-4 (*FG*)	0	—
6	4-5 (*BG*)	2.309	Compression
7	3-5 (*GD*)	1.154	Tension

(*C*) *Reaction at the wall*

We know that the reaction at the wall will be the resultant of the forces P_{4-5} (*i.e.*, 2.309 kN Compression) and P_{3-5} (*i.e.*, 1.154 kN Tension).

This can be easily found out by the **parallelogram law of forces,**

$$R = \sqrt{(1.154)^2 + (2.309)^2 + 2 \times 1.154 \times 2.309 \cos 120°}$$

$$= \sqrt{1.332 + 5.331 + 5.329(-0.5)} = 2 \text{ kN} \quad \textbf{Ans.}$$

***Example 20.** *A frame ABCD is hinged at A and supported on rollers at D as shown in Fig. 61. Determine the forces in the member AB, CD and EF,.*

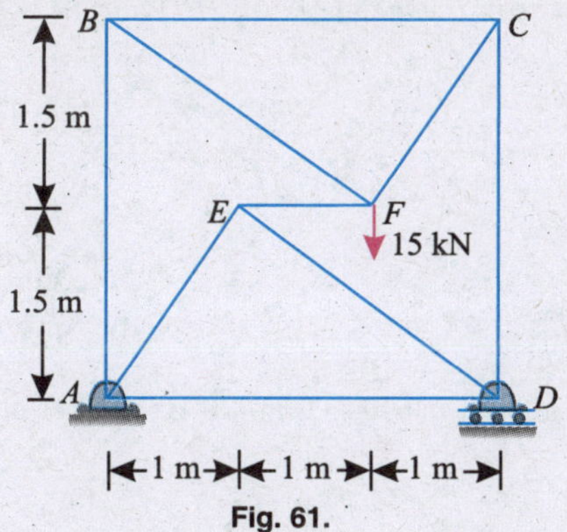

Fig. 61.

Solution. The example may be solved either by the method of joints or method of sections. But we shall solve it by the method of sections, as we have to determine forces in three members of the frame only.

Method of Sections

First of all pass section (1-1) cutting the truss through the members *AB*, *EF* and *CD* as shown in Fig 62.

Now consider equilibrium of the Upper Portion of the frame.

Let the directions of the forces P_{AB} and P_{CD} be assumed as shown in Fig 62.

Now consider the joint *F*.

We know that, Horizontal component of 15 kN load is zero. Therefore force in member *EF* is also zero. **Ans.**

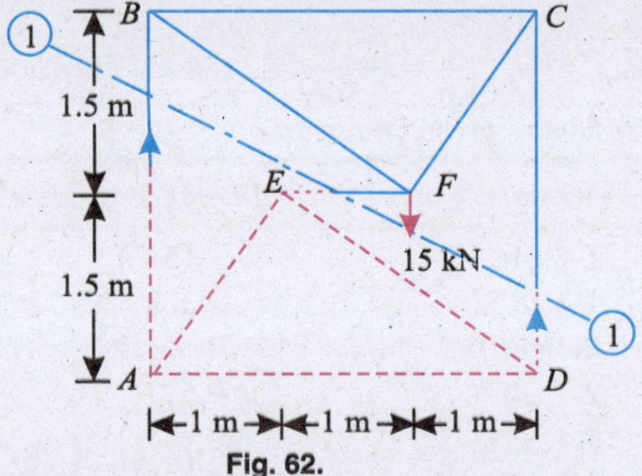

Fig. 62.

Taking moments of the forces acting on the upper portion of the frame about the joint *A* and equating the anti-clockwise moments with clockwise moments,

$$P_{CD} \times 3 = 15 \times 2 = 30$$

or
$$P_{CD} = \frac{30}{3} = 10\,\text{kN} \qquad \textbf{Ans.}$$

Now taking moments of the forces about the joint *D* and equating the anti-clockwise moments with clockwise moments,

$$P_{AB} \times 3 = 15 \times 1 = 15$$

or
$$P_{AB} = \frac{15}{3} = 5\,\text{kN} \qquad \textbf{Ans.}$$

Example 21. *A framed structure of 6 m span is carrying point loads as shown in Fig 63. Find by any method the forces in the members AC, BD and FG of the structure.*

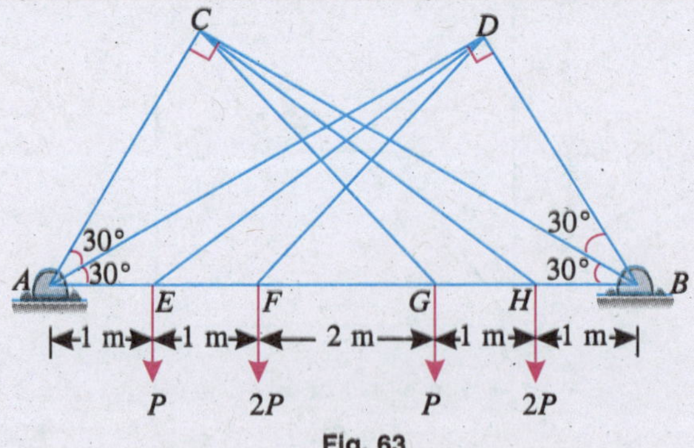

Fig. 63

Solution. First of all, from D draw DK perpendicular to AB as shown in Fig 63. From the geometry of the figure, we find that

$$AD = AB \cos 30° = 6 \times 0.866 = 5.196 \text{ m}$$

and $$DK = AD \sin 30° = 5.196 \times 0.5 = 2.598 \text{ m}$$

Similarly $$AK = AD \cos 30° = 5.196 \times 0.866 = 4.5 \text{ m}$$

$\therefore$ $$\tan \alpha = \frac{DK}{EK} = \frac{2.598}{3.5} = 0.7423 \quad \text{or} \quad \alpha = 36.6°$$

and $$\tan \beta = \frac{DK}{FK} = \frac{2.598}{2.5} = 1.0392 \quad \text{or} \quad \beta = 46.1°$$

Taking moments about B and equating the anti-clockwise moments with clockwise moments,

$$R_A \times 6 = (P \times 5) + (2\,P \times 4) + (P \times 2) + (2\,P \times 1) = 17\,P$$

or $$R_A = \frac{17\,P}{6} = 2.83\,P.$$

Let the directions of the various forces be assumed as shown in Fig 64.

Now resolving the forces vertically at E and equating the forces acting in downward direction with forces acting in upward direction,

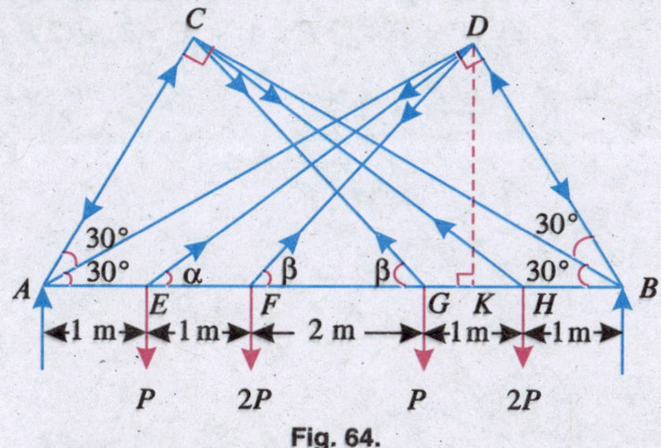

Fig. 64.

$$P_{ED} \sin 36.6° = P$$

or $$P_{ED} = \frac{P}{\sin 36.6°} = \frac{P}{0.5960} = 1.68\ P \text{ (Tension)}$$

Now resolving the forces vertically at F and equating the forces acting in downward direction with forces acting in upward direction,

$$P_{FD} \sin 46.1° = 2\,P$$

or $$P_{FD} = \frac{2\,P}{\sin 46.1°} = \frac{2\,P}{0.7206} = 2.78\ P \text{ (Tension)}$$

Similarly, resolving the forces vertically at G and equating the forces acting in downward direction with forces acting in upward direction,

$$P_{CG} \sin 46.1° = P$$

or $$P_{CG} = \frac{P}{\sin 46.1°} = \frac{P}{0.7206} = 1.39\ P \text{ (Tension)}$$

Now resolving the forces vertically at H and equating the forces acting in downward direction with forces acting in upward direction,

$$P_{CH} \sin 36.6° = 2P$$

or $$P_{CH} = \frac{2P}{\sin 36.6°} = \frac{2P}{0.5960} = 3.36\,P \text{ (Tension)}$$

From the geometry of the figure, we also find that

$$\angle EDB = \angle ACH = 180° - (36.6° + 60°) = 83.4°$$

and $$\angle FDB = \angle ACG = 180° - (46.1° + 60°) = 73.9°$$

Now at D, resolving the forces along BD and equating the same,

$$P_{BD} = P_{ED} \cos 83.4° + P_{FD} \cos 73.9°$$

....(The component of force P_{AD} about BD is zero)

$$= (1.68\,P \times 0.1146) + (2.78\,P \times 0.2773)$$

$$= 0.963\,P \text{ (Compression)}$$ **Ans.**

Now at C, resolving the forces along AC and equating the same,

$$P_{AC} = P_{CH} \cos 83.4° + P_{CG} \cos 73.9°$$

....(The component of force P_{BC} about AC is zero)

$$= (3.36\,P \times 0.1146) + (1.39\,P \times 0.2773)$$

$$= 0.772\,P \text{ (Compression)}$$ **Ans.**

Taking moments about B and equating the anti-clockwise moments with clockwise moments,

$$R_A \times 6 = (P \times 5) + (2P \times 4) + (P \times 2) + (2P \times 1) = 17P$$

or $$R_A = \frac{17P}{6} = 2.83\,P$$

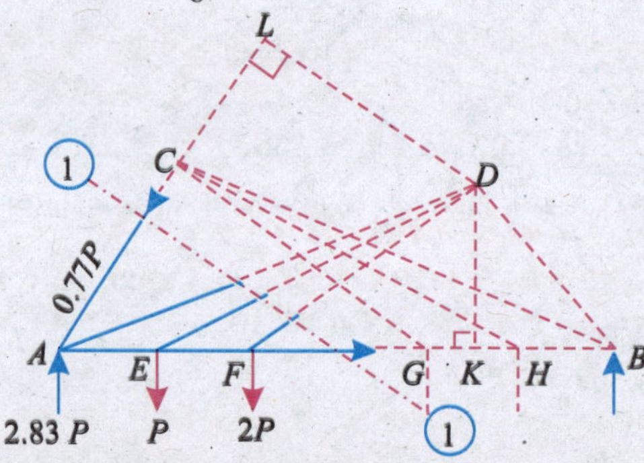

Fig. 65.

Now pass section (1-1) cutting the truss into two parts as shown in Fig 65.
Let us extend the line AC and through D draw DL perpendicular to AC.

From the geometry of the figure, we find that

$$DL = AD \sin 30° = 5.196 \times 0.5 = 2.598 \text{ m}$$

Taking moments of the forces in the left part of the truss about D and equating the anti-clockwise moments with clockwise moments,

$$2.83\,P \times 4.5 = (0.772\,P \times 2.598) + (P \times 3.5) + (2P \times 2.5) + (P_{FG} \times 2.598)$$

or $$12.74\,P = 10.5\,P + (P_{FG} \times 2.598)$$

or $$2.598\,P_{IG} = 12.74\,P - 10.5\,P = 2.24\,P$$

or $$P_{FG} = \frac{2.24P}{2.598} = 0.862\,P \text{ (Tension)}$$ **Ans.**

EXERCISE 3

1. A truss shown in Fig. 66 is subjected to two points loads at *B* and *F*. Find the forces in all the members of the truss and tabulate the results.

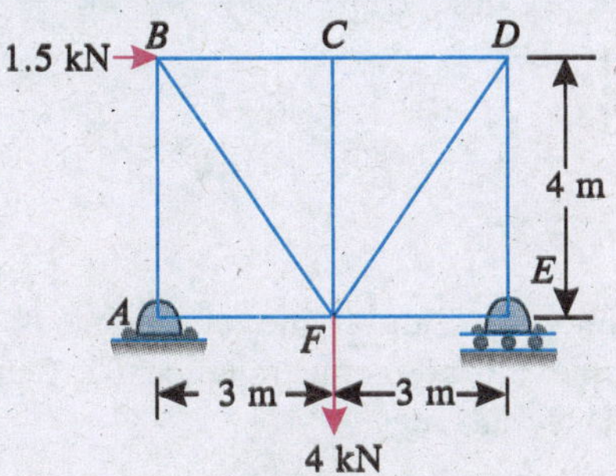

Fig. 66.

Ans. $AB = 1.0$ kN (Compression)
$AF = 1.5$ kN (Tension)
$AE = 3.0$ kN (Compression)
$EF \doteq 0$
$BF = 1.25$ kN (Tension)
$BC = 2.25$ kN (Compression)
$DF = 3.75$ kN (Tension)
$CD = 2.25$ kN (Compression)
$CF = 0$

*2. A cantilever braced truss supported on rollers at E and hinged at A is loaded as shown in Fig 67. Determine graphically or otherwise, the forces in the members of the truss, also determine the reactions at *A* and *E*.

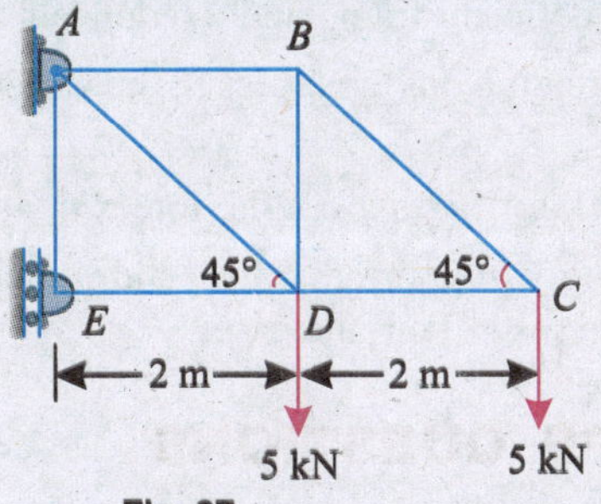

Fig. 67.

Ans. $BC = 7.1$ kN (Compression)
$CD = 5.0$ kN (Tension)
$AB = 5.0$ kN (Compression)
$BD = 5.0$ kN (Tension)
$AD = 14.1$ kN (Tension)
$ED = 15.0$ kN (Compression)
$R_E = 15$ kN
$R_E = 18$ kN

Note: Since the truss is freely supported on rollers at *E*, therefore the reaction at this support will be horizontal (because of vertical support).

3. A truss of 5 m span and 2.5 m height is subjected to wind load as shown in Fig. 68. Find by any method the magnitude of forces in all the members of the truss. Also state their nature.

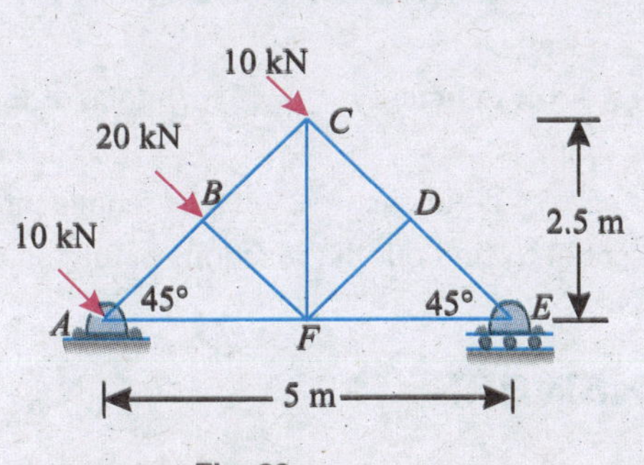

Fig. 68.

Ans. $AB = 10.0$ kN (Compression)
$AF = 28.28$ kN (Tension)
$DE = 20.0$ kN (Compression)
$EF = 14.14$ kN (Tension)
$BF = 20.0$ kN (Compression)
$BC = 10.0$ kN (Compression)
$CF = 14.11$ kN (Tension)
$CD = 20.0$ kN (Compression)
$DF = 0$

4. A truss 15 m long is subjected to a point load of 10 kN as shown in Fig. 69. Find the forces in the members 1, 2 and 3 of the truss.

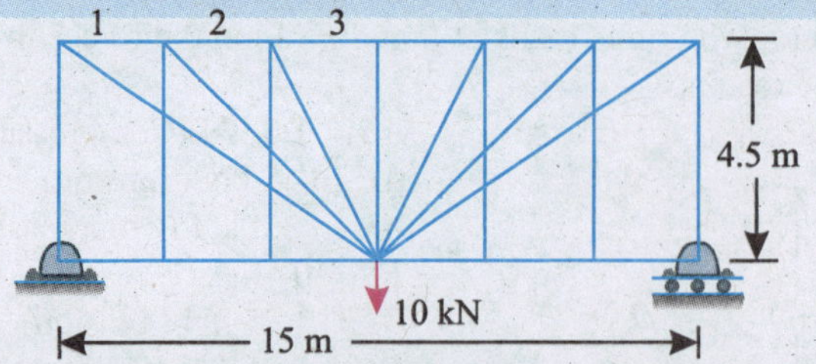

Fig. 69.

Ans. 1 = 40 kN (Compression)
2 = 10 kN (Compression)
3 = 10 kN (Compression)

Hint. Pass vertical sections cutting the members 1, 2 and 3 and take moments about the joint containing 100 kN load. Each time, all the members (except 1, 2 and 3) pass through the joint about which moments are taken.

QUESTIONS

1. What is a 'frame' ? Discuss its classification.
2. State clearly the difference between a perfect frame and an imperfect frame.
3. How would you distinguish between a deficient frame and a redundant frame ?
4. What are the assumptions made, while finding out the forces in the various members of a framed structure ?
5. Name the methods, which are employed, for finding out the forces in a frame.
6. What is the difference between a simply supported frame and a cantilever frame ? Discuss the method of finding out reactions in both the cases.

OBJECTIVE TYPE QUESTIONS

1. A framed structure is perfect, if the number of members are .....(2*j* – 3), where *j* is the number of joints.
   (*a*) less than (*b*) equal to (*c*) greater than (*d*) either (*a*) or (*c*)
2. A framed structure is imperfect, if the number of members are .....(2*j* – 3), where *j* is the number of joints.
   (*a*) less than (*b*) equal to (*c*) greater than (*d*) either (*a*) or (*c*)
3. A redundant frame is also called ......frame
   (*a*) perfect (*b*) imperfect (*c*) deficient (*d*) none of these
4. A framed structure of a triangular shape is
   (*a*) perfect (*b*) imperfect (*c*) deficient (*d*) redundant
5. In a cantilever truss, it is very essential to find out the reactions before analyzing it.
   (*a*) agree (*b*) disagree

ANSWERS

1. (*b*) 2. (*d*) 3. (*b*) 4. (*a*) 5. (*b*)

CHAPTER

14 Virtual Work

Contents

1. WORK

It has been observed that whenever a force acts on a body, and the body undergoes some displacement, some work is said to be done. Mathematically, if a force (P) acting on a body displaces it through a distance (s), then

$$\text{Workdone} = \text{Force} \times \text{Distance} = P \times s$$

But, sometimes, the body does not move in the direction of force (or in other words, the force does not act in the direction of motion of the body). In such a case,

$$\text{Workdone} = \text{Component of the force in the direction of motion} \times \text{Distance}$$

$$= P \cos \theta \times s$$

where θ is the inclination between the line of action of the force and the direction of the motion of the body.

A little consideration will show that

1. If the value of θ is between 0° and 90°, some work is done.
2. If the value of θ is 90°, then no work is done (because cos 90° = 0).
3. If the value of θ is between 90° and 180°, the body will move in the opposite direction and work is called as *negative.*

2. *VIRTUAL WORK

[AM-IETE, BPUT]

In the previous article, we have discussed that the work done by a force is equal to the force multiplied by the distance through which the body has moved in the direction of the force. But if the body is in equilibrium, under the action of a system of forces, the work done is zero. If we assume that the body, in equilibrium, undergoes an infinite small imaginary displacement (known as **virtual displacement**) some work will be *imagined* to be done. Such an imaginary work is called **virtual work**. This concept of **virtual work is used in finding out the unknown forces in structures**.

Note. The term 'virtual' is used to stress its purely hypothetical nature, as we do not actually displace the system. We only imagine, as to what would happen, if the system is displaced.

3. **PRINCIPLE OF VIRTUAL WORK

[AMIE, PTU, RTU, RGPV, WBUT]

It states, "*If a system of forces acting on a body or a system of bodies be in equilibrium, and the system be imagined to undergo a small displacement consistent with the geometrical conditions, then the algebraic sum of the virtual works done by all the forces of the system is zero.*"

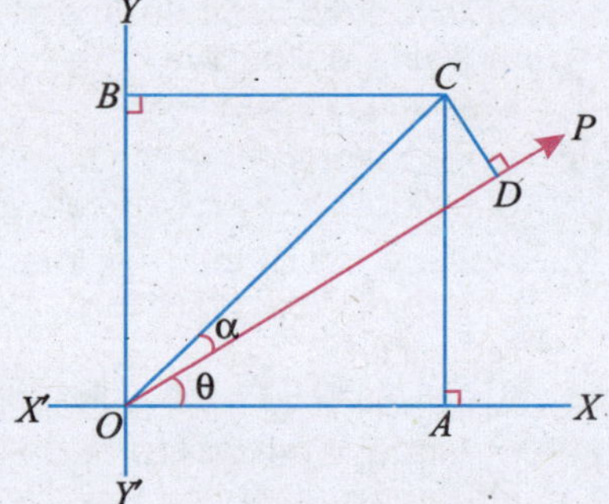

Fig. 1. Principle of virtual work.

Proof. Consider a body at O, subjected to a force P inclined at angle θ with X-X axis as shown in Fig. 1.

Let P_X = Component of the force along X-X axis, and

P_Y = Component of the force along Y-Y axis.

From the geometry of the figure, we find that

$$P_X = P \cos\theta \text{ and}$$
$$P_Y = P \sin\theta$$

Now consider the body to move from O to some other point C, under the action of the force P, such that the line OC makes an angle α with the direction of the force. Now draw CA and CB perpendiculars to OX and OY respectively as shown in Fig. 1.

From the geometry of the triangle OCA, we find that

$$\cos(\theta + \alpha) = \frac{OA}{OC}$$

$$\therefore \quad OA = OC \times \cos(\theta + \alpha)$$

Similarly, $OB = AC = OC \times \sin(\theta + \alpha)$

We know that the sum of the works done by the components P_X and P_Y of the force P

$$= P_X \times OA + P_Y \times OB$$
$$= [P \cos\theta \times OC \cos(\theta + \alpha)] + [P \sin\theta \times OC \sin(\theta + \alpha)]$$
$$= P \times OC\,[\cos\theta \times \cos(\theta + \alpha) + \sin\theta \times \sin(\theta + \alpha)]$$
$$= P \times OC \cos(\theta - \theta - \alpha) \quad \ldots (\because \cos A - B = \cos A \cos B + \sin A \sin B)$$

$= P \times OC \cos (-\alpha)$

$= P \times OC \cos \alpha$ ... ($\because \cos(-A) = \cos A$) ...(*i*)

We also know that the work done by the force P in moving the body from O to C.

$= P \times OC \cos \alpha$ ...(*ii*)

Since the equations (*i*) and (*ii*) are the same, therefore, work done by a force is equal to the sum of the works done by its resolved parts.

Note. For the sake of simplicity, we have considered only one force and its resolved parts. But it can be extended to any number of forces.

4. SIGN CONVENTIONS

Though there are different sign conventions for finding out the virtual works done in different books, yet we shall use the following sign conventions, which are internationally recognised.

1. Upward forces are considered as positive, whereas the downwards as negative.
2. Forces acting towards right are considered as positive, whereas those towards left as negative.
3. Forces acting in the clockwise direction are considered as positive, whereas the anticlockswise as negative.
4. Tensile forces are considered as positive whereas the compressive as negative.

5. APPLICATIONS OF THE PRINCIPLE OF VIRTUAL WORK

The principle of virtual work has very wide applications. But the following are important from the subject point of view;

1. Beams 2. Lifting machine. 3. Framed structures.

6. APPLICATION OF THE PRINCIPLE OF VIRTUAL WORK ON BEAMS CARRYING POINT LOAD

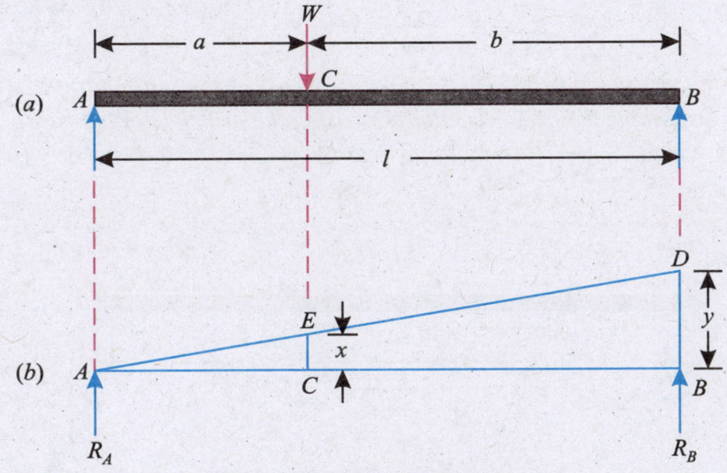

Fig. 2. Beam carrying point load.

Consider a beam AB, simply supported at its supports, and subjected to a point load W at C as shown in Fig. 2 (*a*)

Let R_A = Reaction at A, and

R_B = Reaction at B.

First of all, let us assume the beam to be hinged at *A*. Now consider an upward virtual displacement (y) of the beam at *B*. This is due to the reaction at *B* acting upwards as shown in Fig. 2 (*b*). Let x be the upward virtual displacement of the beam at *C* due to the point load.

Now in two similar triangles *ABD* and *ACE*,

$$\frac{x}{y} = \frac{a}{l} \qquad \text{or} \qquad x = \frac{a\,y}{l}$$

∴ Total virtual work done by the two reactions R_A and R_B

$$= + [(R_A \times 0) + (R_B \times y)] = + R_B \times y \qquad \text{...}(i)$$

... (Plus sign due to the reactions acting upwards)

and virtual work done by the point load*

$$= - W \times x \qquad \text{...}(ii)$$

... (Minus sign due to the load acting downwards)

(*Note: This may also be analysed by considering the downward vertical displacement of the beam at *C* (due to load *W*). In this case, the beam also undergoes a downward virtual displacement at *B*.)

We know, **from the principle of virtual work that, Algebraic sum of the virtual works done is zero**. Therefore

$$R_B \times y - W \times x = 0$$

or

$$R_B = \frac{W \times x}{y} = \frac{W}{y} \times \frac{a \times y}{l} = \frac{W \times a}{l}$$

Similarly, it can be proved that the vertical reaction at *A*,

$$R_A = \frac{W \times b}{l}$$

Notes : 1. For the sake of simplicity, we have taken only one point load *W* at *C*. But this principle may be extended for any number of loads.

2. The value of reaction at *A* (*i.e.*, R_A) may also be obtained by subtracting the value of R_B from the downward load *W*. Mathematically,

$$R_A = W - \frac{W a}{l} = W\left(1 - \frac{a}{l}\right) = W\left(\frac{l - a}{l}\right) = \frac{W b}{l}$$

Example 1. *A beam AB of span 5 metres is carrying a point load of 2 kN at a distance 2 metres from A. Determine the beam reactions, by using the principle of the virtual work.*

Solution. Given: Span (l) = 5 m; Point load (W) = 2 kN and distance between the point load and support *A* = 2 m.

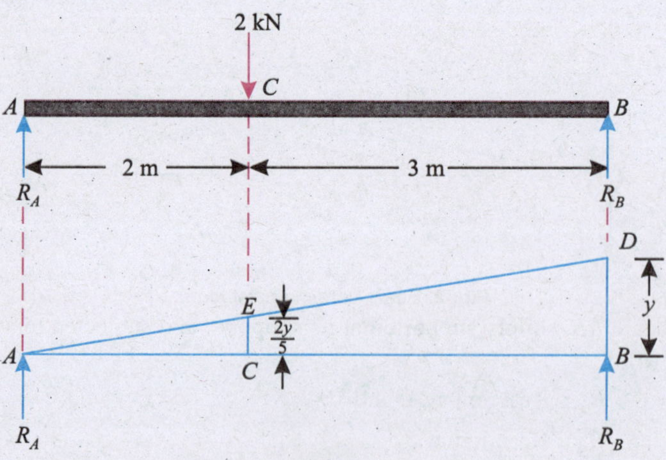

Fig. 3.

Let R_A = Reaction at A,

R_B = Reaction at B, and

y = Virtual upward displacement of the beam at B.

From the geometry of the figure, we find that when the virtual upward displacement of the beam at B is y, the virtual upward displacement of the beam at C is $\frac{2y}{5} = 0.4\ y$ as shown in Fig. 3.

∴ Total virtual work done by the two reactions R_A and R_B

$$= +[(R_A \times 0) + (R_B \times y)] = +R_B \times y \quad ...(i)$$

... (Plus sign due to the reactions acting upwards)

and virtual work done by the point load

$$= -Px = -2 \times 0.4 = -0.8\,y \quad ...(ii)$$

... (Minus sign due to the load acting downwards)

We know, **from the principle of virtual work that, Algebraic sum of the total virtual works done is zero**. Therefore

$$R_B \times y - 0.8\ y = 0$$

or $R_B = 0.8\ y/y = 0.8$ kN **Ans.**

and $R_A = 2 - 0.8 = 1.2$ kN **Ans.**

***Example 2.** *Two beams AC and CD of length 9 m and 10 m respectively are hinged at C. These are supported on rollers at the left and right ends (A and D). A hinged support is provided at B, 7m* from *A as shown in Fig. 4.*

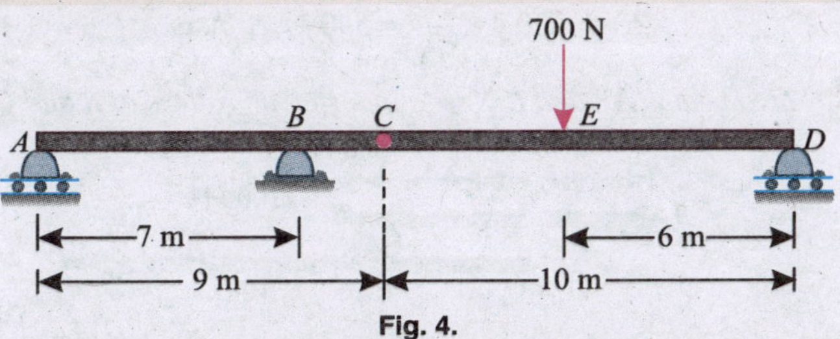

Fig. 4.

Using the principle of virtual work, determine the force transmitted by the hinge C and the reaction at the support B, when a load of 700 N acts at a point 6 m from D.

Solution. Given : Length of beam $AC = 9$ m; Length of beam $CD = 10$ m and load at $E = 700$ N.

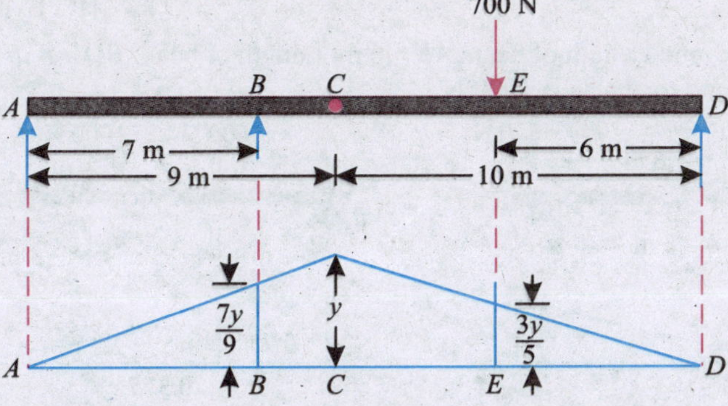

Fig. 5.

Let R_A = Reaction at A,

R_B = Reaction at B,

R_D = Reaction at D, and

y = Virtual upward displacement of the beam at the hinge (*i.e.* C).

From the geometry of the figure, we find that when the virtual upward displacement of the beam at the hinge (*i.e.* C) is y, then virtual displacement of B and E is $\frac{7y}{9}$ and $\frac{6y}{10} = \frac{3y}{5}$ respectively as shown in Fig. 5.

∴ Total work done by the three reactions (R_A, R_B and R_D)

$$= +\left[(R_A \times 0) + \left(R_B \times \frac{7y}{9}\right) + (R_D \times 0)\right]$$

$$= +R_B \times \frac{7y}{9} \qquad ...(i)$$

...(Plus sign due to reactions acting upwards)

and virtual work done by the load

$$= -700 \times \frac{3y}{5} = -420\,y \qquad ...(ii)$$

...(Minus sign due to load acting downwards)

We know, **from the principle of virtual work that, Algebraic sum of the total virtual works done is zero**. Therefore

$$R_B \times \frac{7y}{9} - 420\,y = 0$$

or $$R_B = 420\,y \times \frac{9}{7y} = 540 \text{ N} \quad \textbf{Ans.}$$

***Example 3.** *Two beams AE and BD are supported on rollers at B and C as shown in Fig. 6.*

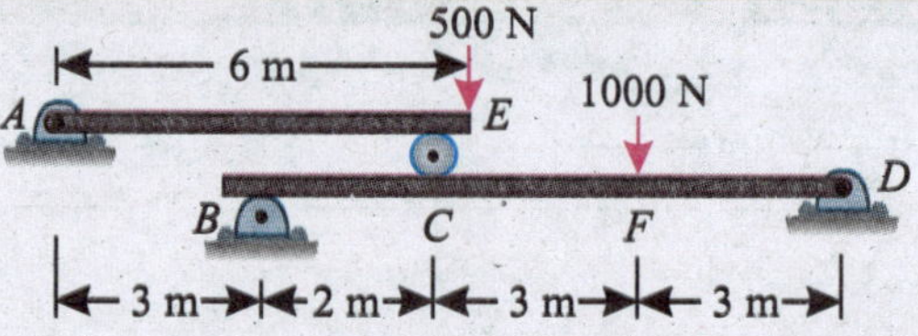

Fig. 6.

Determine the reactions at the rollers B and C, using the method of virtual work.

[RTU, MTU, AUT, AMIE]

Solution. Given : Length of beam AE = 6 m; Length of beam BD = 8 m; Distance AC = 5 m; Load at E = 500 N; and load at F = 1000 N.

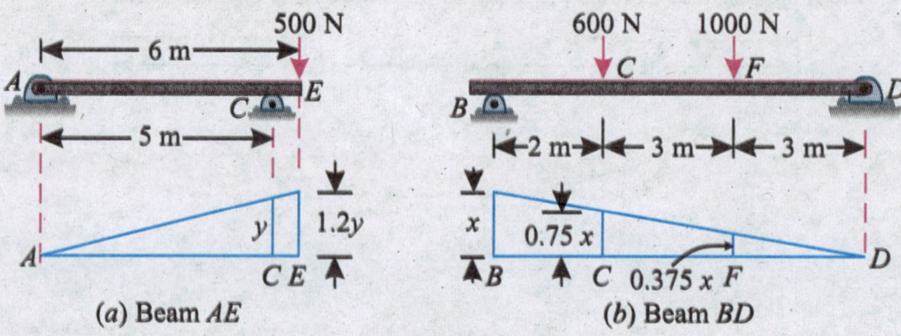

Fig. 7.

Let R_C = Reaction at the roller C, and

R_B = Reaction at the roller B.

First of all, let us consider the beam AE with roller support at C as shown in Fig. 7 (a).

Let y = Virtual upward displacement of the beam at C.

From the geometry of the figure, we find that when virtual upward displacement of the beam at C is y, then the virtual upward displacement of beam at E is $\frac{6y}{5} = 1.2\,y$ as shown in Fig. 7 (a)

Therefore total virtual work done by the two reactions R_A and R_C

$$= +\,[(R_A \times 0) + (R_C \times y)] = +\,R_C \times y \qquad ...(i)$$

...(Plus sign due to reaction at C acting upwards)

and virtual work done by the point load at E

$$= -\,(500 \times 1.2y) = 600y \qquad ...(ii)$$

...(Minus sign due to load acting downwards)

We know, **from principle of virtual work that, Algebraic sum of the total virtual work done is zero**. Therefore

$$\therefore \quad (R_C \times y) - 600\,y = 0$$

or $$R_C = 600 \text{ N} \quad \textbf{Ans.}$$

Now consider the beam BD with loads at C and F as shown in Fig. 7 (b)

Let x = Virtual upward displacement of the beam at B.

From the geometry of the figure, we find that when virtual upward displacement of the beam at B is x, then the virtual upward displacement of the beam at C and F is $\frac{6x}{8} = 0.75\,x$ and $\frac{3x}{8} = 0.375\,x$ respectively as shown in Fig. 7 (b).

Therefore total virtual work done by the two reactions R_B and R_D

$$= +\,(R_B \times x) + (R_D \times 0) = +\,R_B \times x \qquad ...(iii)$$

...(Plus sign due to reactions acting upwards)

and virtual work done by the point loads at C and F

$$= -\,[(600 \times 0.75\,x) + (1000 \times 0.375\,x)] = -\,825\,x$$

...(Minus sign due to loads acting downwards)

We know, **from principle of virtual work that, Algebraic sum of the total virtual works done is zero**. Therefore

$$R_B \times x - 825x = 0$$

or $$R_B = 825 \text{ N} \quad \textbf{Ans.}$$

Note. In this case, we have assumed the virtual upward displacement of the beam at B, because it is hinged at D. However, if we assume the virtual upward displacement at D, it is not wrong. In this case, we shall obtain the value of reaction at D.

7. APPLICATION OF THE PRINCIPLE OF VIRTUAL WORK FOR BEAMS CARRYING UNIFORMLY DISTRIBUTED LOAD

Consider a beam AB of length l simply supported at its both ends, and carrying a uniformly distributed load of w per unit length for the whole span from A to B as shown in Fig. 8 (a).

Let R_A = Reaction at A, and

R_B = Reaction at B.

First of all, let us assume the beam to be hinged at A. Now consider an upward virtual displacement (y) of the beam at B. This is due to the reaction at B acting upwards as shown in Fig. 8 (b)

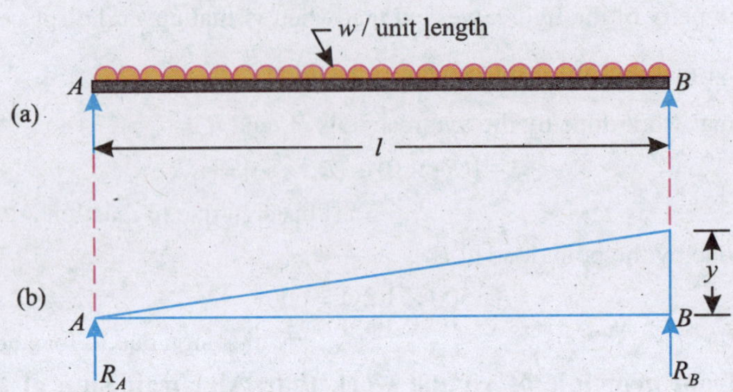

Fig. 8. Beam carrying uniformly distributed load.

∴ Total virtual work done by the two reactions R_A and R_B

$$= +[(R_A \times 0) + (R_B \times y)] = +R_B \times y \quad \text{...}(i)$$

...(Plus sign due to reaction acting upwards)

and virtual work done by the uniformly distributed load

$$= -w\left(\frac{0+y}{2} \times l\right) = -0.5\, wyl$$

...(Minus sign due to load acting downwards)

We know, **from principle of virtual work that, Algebraic sum of the total virtual work done is zero**. Therefore

$$(R_B \times y) - (0.5wl \times y) = 0$$

or $$R_B \times y = 0.5\, wl \times y$$

or $$R_B = 0.5\, wl$$

Note. For the sake of simplicity, we have taken the uniformly distributed load for the entire span from A to B. But this principle may be extended for any type of load on beam (*i.e.* simply supported or overhanging beam etc.)

***Example 4.** *A simply supported beam AB of span 5 m is loaded as shown in Fig. 9. Using the principle of virtual work, find the reactions at A and B.*

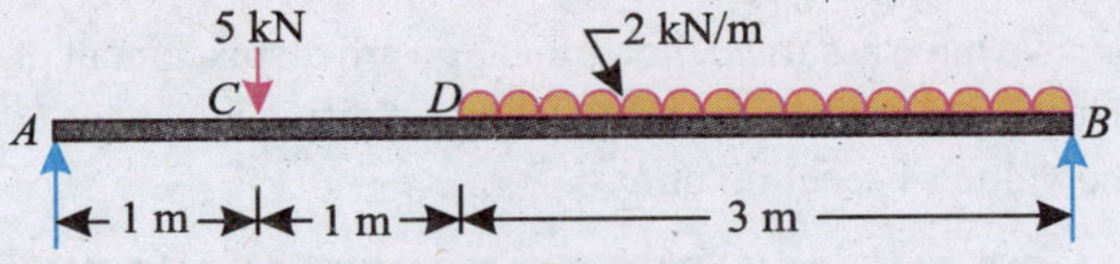

Fig. 9.

Solution. Given : Length of beam AB = 5 m; Point Load at C = 5 kN and uniformly distributed load between D and B = 2 kN/m

Let R_A = Reaction at A,

R_B = Reaction at B, and

y = Virtual upward displacement of the beam at B.

From the geometry of the figure, we find that when the virtual upward displacement of the beam at B is y, then the virtual upward displacement of the beam at C and D is $0.2\ y$ and $0.4\ y$ respectively as shown in Fig. 10.

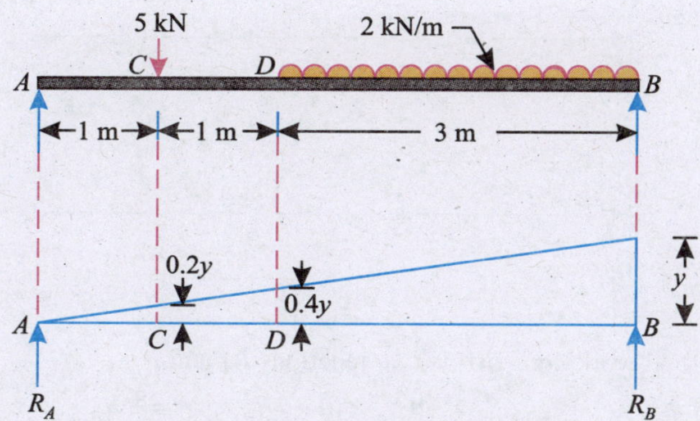

Fig. 10.

∴ Total virtual work done by the two reactions R_A and R_B

$$= + [(R_A \times 0) + (R_B \times y)] = + R_B \times y \qquad \text{...}(i)$$

...(Plus sign due to reactions acting upwards)

and total virtual work done by the point load at C and uniformly distributed load between D and B.

$$= -\left[(5 \times 0.2y) + 2\left(\frac{0.4y + y}{2}\right) \times 3\right] = -5.2\ y \qquad \text{...}(ii)$$

...(Minus sign due to loads acting downwards)

We know, **from principle of virtual work that, Algebraic sum of the total virtual work done is zero**. Therefore

$$R_B \times y - 5.2\ y = 0$$

or $\quad R_B = 5.2$ kN **Ans.**

and $\quad R_A = 5 + (2 \times 3) - 5.2 = 5.8$ kN **Ans.**

***Example 5.** *An overhanging beam ABC of span 3 m is loaded as shown in Fig. 11.*

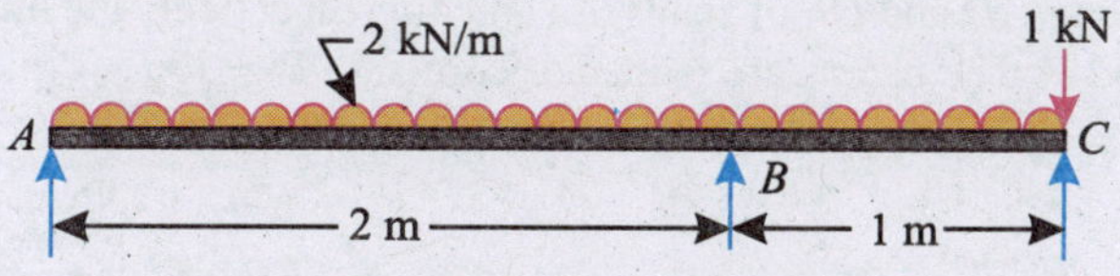

Fig. 11.

Using the principle of virtual work, find the reactions at A and B.

Solution. Given : Span AB = 2 m and span BC = 1 m

Let $\quad R_A$ = Reaction at A,

R_B = Reaction at B, and

y = Virtual upward displacement of beam at B.

From the geometry of the figure, we find that when the virtual upward displacement of the beam at B is y, then the virtual upward displacement of the beam at C is 1.5 y as shown in Fig. 12.

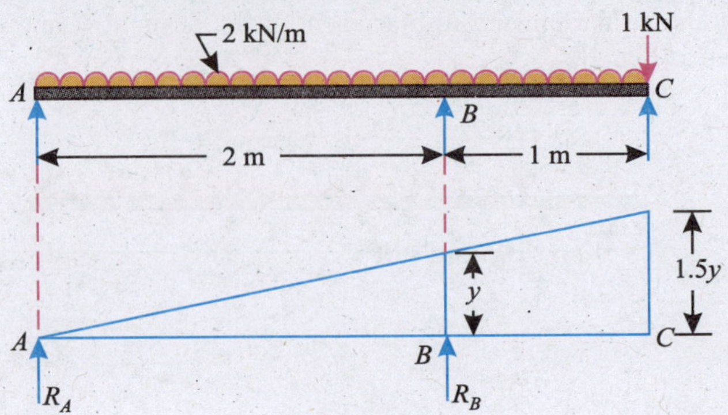

Fig. 12.

∴ Total virtual work done by the two reactions R_A and R_B

$$= +[(R_A \times 0) + (R_B \times y)] = + R_B \times y$$

...(Plus sign due to reactions acting upwards)

and total virtual work done by the point load at C and uniformly distributed load between A and C.

$$= -\left[(1 \times 1.5y) + 2\left(\frac{0 + 1.5y}{2} \times 3\right)\right]$$

$$= -(1.5y + 4.5\,y) = -6y$$

...(Minus sign due to loads acting downwards)

We know, **from principle of virtual work that, Algebraic sum of the total virtual work done is zero**. Therefore

$$(R_B \times y) - 6y = 0$$

or $\quad R_B = 6$ kN **Ans.**

and $\quad R_A = (2 \times 3) + 1 - 6 = 1$ kN **Ans.**

EXERCISE 1

1. A simply supported beam AB of span 4 m is subjected to a point load of 10 kN at a distance of 1.5 m from A. Using the principle of virtual work, determine the reactions at the two supports. (**Ans.** 3.75 kN ; 6.25 kN)
2. Two beams AD and DF of spans 6m and 4m respectively are hinged at C and supported at A, D and F. The beams are loaded as shown in Fig. 13.

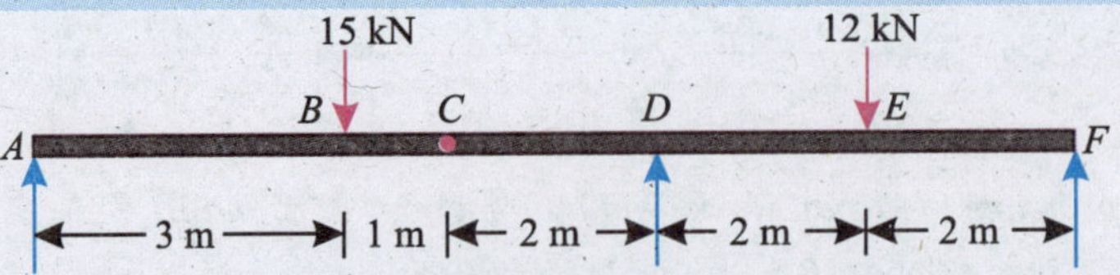

Fig. 13.

Using the principle of virtual work, find the reaction at D. (**Ans.** 22.9 kN)

3. A simply supported beam of span 4 m is carrying a uniformly distributed load of 5 kN/m as shown in Fig. 14.

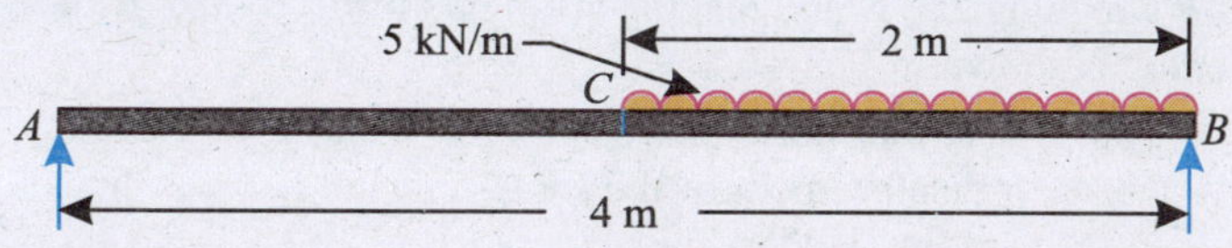

Fig. 14.

Using the principle of virtual work, find the reactions at *A* and *B*.

(**Ans.** 2.5 kN; 7.5 kN)

4. A beam of span 5 m is supported at *A* and *B*. It is subjected to a load system as shown in Fig. 15.

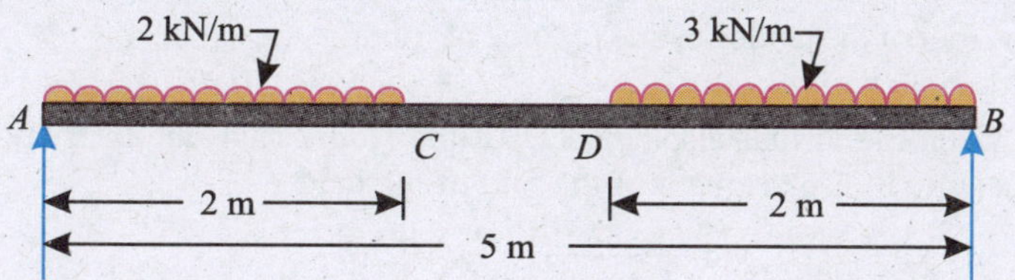

Fig. 15.

With the help of principle of virtual work, find the reactions at *A* and *B*.

(**Ans.** 4.4 kN; 5.6 kN)

8. APPLICATION OF THE PRINCIPLE OF VIRTUAL WORK ON LADDERS

We have already discussed (in article of Ladder Friction in chapter of Applications of Friction) that in case of a ladder, its foot moves on the floor towards or away from the wall. It is thus obvious, that no work is done by the normal reaction (R_f) at the foot of the ladder. However, some work is done by the frictional force (F_f) at the foot of the ladder. Similarly, top of the ladder moves up or down along the wall. Thus no work is done by the normal reaction (R_w) at the top of the ladder. However, some work is done by the frictional force (F_w) at the top of the ladder. This happens when the wall is not smooth, or in other words, the wall has some coefficient of friction.

Now the virtual works done by the frictional forces at the foot and top of the ladder are found out, and the principle of virtual work is applied as usual.

Note. If the vertical wall is smooth, then there is no frictional force at the top of the ladder. Thus no work is done at the top of the ladder.

****Example 6.** *A uniform ladder of weight 250 N rests against a smooth vertical wall and a rough horizontal floor making an angle of 45° with the horizontal. Find the force of friction at he floor using the method of virtual work.* **[MU, WBUT]**

Solution. Given : Weight of the ladder (W) = 250 N and inclination of the ladder with the horizontal (θ) = 45°

The ladder AB weighing 250 N and making an angle of 45° with the horizontal as shown in Fig. 16.

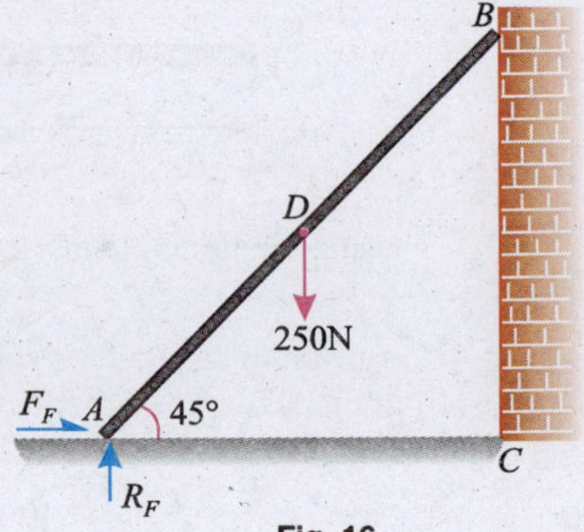

Fig. 16

Let x = Virtual horizontal displacement of the foot of the ladder, and

y = Virtual vertical displacement of the mid of the ladder at D.

From the geometry of the figure, we find that when mid point D of the ladder moves downwards (due to its weight) then bottom A of the ladder moves towards left, which is prevented by the force of friction. Or in other words, the virtual displacement of the foot of the ladder A, due to force of friction (F_f) will be towards right.

Moreover, when the virtual displacement of the ladder at A due to frictional force towards right is x. Then the virtual displacement of the mid of the ladder,

$$y = \frac{x}{2 \tan 45°} = \frac{x}{2} = 0.5\,x$$

∴ Virtual work done by the frictional force

$$= + F_f \times x = F_f x$$

...(Plus sign due to movement of force towards right)

and virtual work done by the 250 N weight of the ladder

$$= -(250 \times y) = -(250 \times 0.5\,x) = -125\,x$$

...(Minus sign due to downward movement of the weight)

We know, **from the principle of virtual work that, Algebraic sum of the total virtual works done is zero**. Therefore

$$F_f x - 125\,x = 0$$

or $$F_f = 125 \text{ N} \quad \textbf{Ans.}$$

***Example 7.** *A uniform ladder, 5 metres long and weighing 200 N, rests on a smooth floor at A and against a smooth wall at B as shown in Fig. 17. A horizontal rope PQ prevents the ladder from slipping. Using the method of virtual work, determine the tension in the rope.*

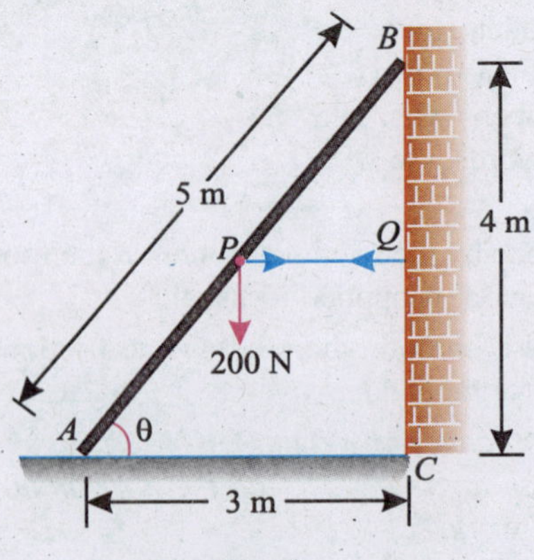

Fig. 17.

Solution. Given: Length of the ladder (l) = 5 m and weight of the ladder (W) = 200 N

Let T = Tension in the rope PQ,

θ = Angle, which the ladder makes with the horizontal.

x = Virtual vertical displacement of the mid of the ladder Q (or in other words, weight of the ladder), and

y = Virtual horizontal displacement of the rope PQ due to tension.

From the geometry of the figure, we find that

$$\tan\theta = \frac{4}{3}$$

We also find that when the mid point P of the ladder moves downwards (due to weight), it causes top of the ladder B to move downwards and bottom of the ladder A to move towards left. It causes tension (T) in the rope PQ. Moreover, when the virtual vertical displacement of the mid of the ladder P (or weight of the ladder) is x, then the virtual horizontal displacement of the ladder,

$$y = \frac{x}{\tan\theta} = \frac{x}{4/3} = \frac{3x}{4} = 0.75\,x$$

∴ Virtual work done by the tension in the rope

$$= +\,Tx \qquad \text{...}(i)$$

...(Plus sign due to tension)

and virtual work done by the 200 N weight of the ladder

$$= -\,200\,y \qquad \text{...}(ii)$$

...(Minus sign due to downward movement of the weight)

We know, **from the principle of virtual work that, Algebraic sum of the total virtual works done is zero.** Therefore

$$Tx - 200\,y = 0$$

or

$$T = \frac{200\,y}{x} = \frac{200 \times 0.75\,x}{x} = 150\text{ N} \quad \textbf{Ans.}$$

9. *APPLICATION OF THE PRINCIPLE OF VIRTUAL WORK ON LIFTING MACHINES

We know that in the case of lifting machines, the effort moves downwards, whereas the load moves upwards. In such cases, the virtual works done by the effort and that by the load are found out. Now apply the principle of virtual work as usual.

Example 8. *A weight (W) of 5 kN is raised by a system of pulleys as shown in Fig. 18*

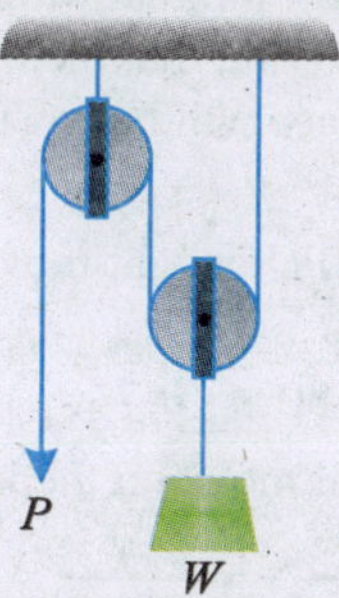

Fig. 18.

Using the method of virtual work, find the force P, which can hold the weight in equilibrium.

Solution. Given : Weight $(W) = 5$ kN

Let P = Force which can hold the weight in equilibrium, and

y = Virtual upward displacement of the weight.

From the geometry of the system of pulleys, we find that when the virtual upward displacement of the weight is y, the virtual downward displacement of the force is $2y$.

∴ Virtual work done by the load

$$= +Wy = 5y \qquad ...(i)$$

...(Plus sign due to upward movement of the load)

and virtual work done by the effort

$$= -P \times 2y = -2\,Py \qquad ...(ii)$$

... (Minus sign due to downward movement of the effort)

We know, **from the principle of virtual work that, Algebraic sum of the total virtual works done is zero**. Therefore

$$5y - 2Py = 0$$

or

$$P = \frac{5y}{2y} = 2.5 \text{ kN} \quad \textbf{Ans.}$$

Example 9. *A weight of 1000 N resting over a smooth surface inclined at 30° with the horizontal, is supported by an effort (P) resting on a smooth surface inclined at 45° with the horizontal as shown in Fig. 19.*

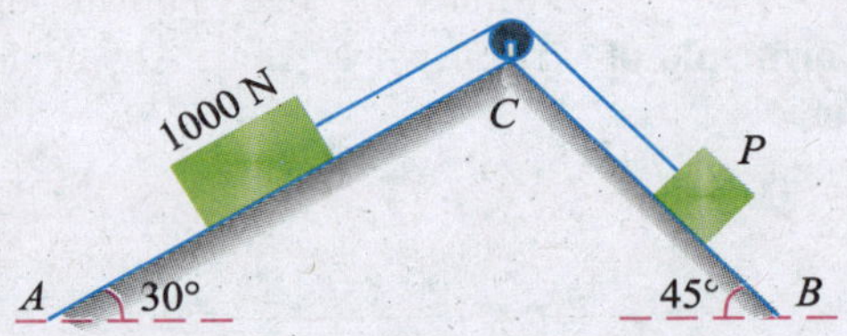

Fig. 19.

By using the principle of virtual work, calculate the value of effort (P). **[WBUT, BPUT]**

Solution. Given: Weight $(W) = 1000$ N; Inclination of weight surface $(\alpha_1) = 30°$ and inclination of effort surface $(\alpha_2) = 45°$

Let x = Virtual vertical displacement of 1000 N weight, and

y = Virtual vertical displacement of the effort (P).

From the geometry of the system, we find that when the weight (1000 N) moves downwards, the effort (P) will move upwards. We also find that the distance through which the 1000 N weight will move downwards on the inclined surface will be equal to the distance through which the load P will move upwards on the inclined surface.

∴ Distance through which the 1000 N weight moves on the inclined surface AC.

$$= \frac{x}{\sin 30°} = \frac{x}{0.5} = 2x \qquad ...(i)$$

and distance through which the load P will move on the inclined surface BC

$$= \frac{y}{\sin 45°} = \frac{y}{0.707} = 1.414\,y \qquad ...(ii)$$

Equating equations (*i*) and (*ii*),

$$2x = 1.414\,y \quad \text{or} \quad x = 0.707\,y$$

∴ Virtual work done by the effort (P)

$$= + P \times y \quad ...(i)$$

...(Plus sign due to upward movement of the effort)

and virtual work done by the 1000 N weight

$$= - 1000 \times x = - 1000 \times 0.707\, y$$

$$= - 707\, y \quad ...(\because x = 0.707\, y) \quad ...(ii)$$

...(Minus sign due to downward movement of the weight)

We know, **from the principle of virtual work that, Algebraic sum of the virtual works done is zero**. Therefore

$$Py - 707\, y = 0$$

or $$P = 707 \text{ N} \quad \textbf{Ans.}$$

Example 10. *A beam AB of 2 m length is held in equilibrium by the application of a force P as shown in Fig. 20.*

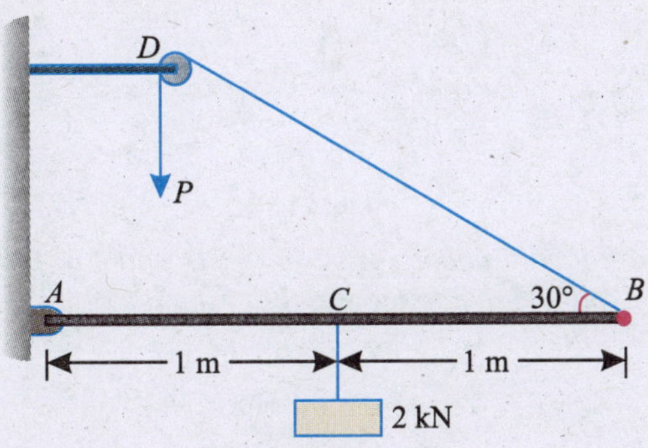

Fig. 20.

Using the principle of virtual work, find the magnitude of the force P when a weight of 2 kN is hung from the beam AB at its midpoint.

Solution. Given: Length of beam AB = 2 m; Span AC = 1m; Span CB = 1m and weight (W) = 2 kN

Let P = force required to keep the body in equilibrium, and

y = Virtual downward displacement of the effort.

From the geometry of the figure, we find that when the virtual downward displacement of the force (P) is y, then virtual upward displacement of the beam at B (or length of chord released)

$$= y$$

and virtual upward displacement of the load

$$= 0.5\, y$$

∴ Virtual work done by the load

$$= + (2 \times 0.5\, y) = + y \quad ...(i)$$

...(Plus sign due to upward movement of load)

and virtual work done by the effort

$$= - Py \quad ...(ii)$$

...(Minus sign due to downward movement of the effort)

We know, **from principle of virtual work that, Algebraic sum of the total virtual work done is zero**. Therefore

$$y - Py = 0$$

or $$P = y/y = 1 \text{ kN} \quad \textbf{Ans.}$$

10. APPLICATION OF THE PRINCIPLE OF VIRTUAL WORK ON FRAMED STRUCTURES

In a framed structure, first of all assume the member (in which force is required to be found out) to be removed. Now find out the virtual works done by all the remaining members of the frame and the force in the member, assumed to be removed. Now apply the principle of virtual work as usual.

Example 11. *A hexagonal frame is made up of six bars of equal length and cross-section as shown in Fig. 21. The bar ED is fixed in a horizontal position.*

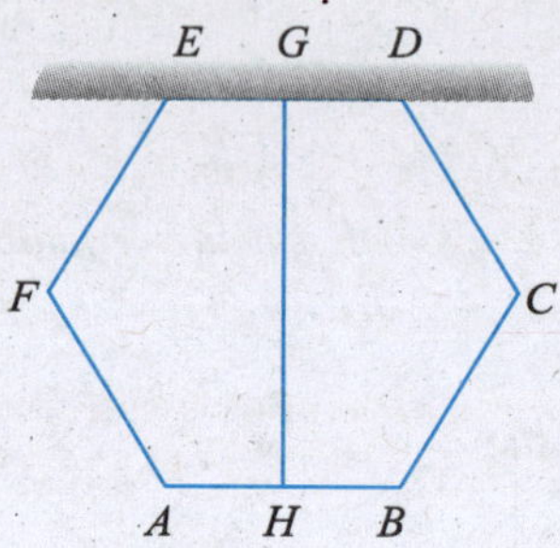

Fig. 21.

A rod GH is fixed at the mid-points of the bars ED and AB. Using the principle of virtual work, find the tension in the rod GH due to the weight of the bars.

Solution. Let W = Weight of each bar, and

T = Tension in the rod GH.

First of all, let us assume the rod GH to be removed.

Now let y = Virtual vertical upward displacement of the centre of gravity of the bars CD and EF.

From the geometry of the figure, we find that when the virtual vertical downward displacement of centre of gravity of the bars CD and EF is y (due to weight W), then the vertical downward virtual displacement of the bars BC and AF is $3y$; that of member AB is $4y$ and that of member ED is zero (because it is fixed in horizontal position). The vertical virtual displacement of member GH is $4y$.

$\therefore$ Virtual work done by the tension in rod GH

$$= + T \times 4y = + 4\,Ty \quad \text{...}(i)$$

...(Plus sign due to tension)

and virtual work done by the bars $= -\,[(2 \times W \times y) + (2 \times W \times 3y) + (W \times 4y)]$

$$= -\,12\,Wy \quad \text{...}(ii)$$

...(Minus sign due to downward movement of the bars)

We know, **from principle of virtual work that, Algebraic sum of the total virtual work done is zero**. Therefore

$$4\,Ty - 12Wy = 0$$

or

$$T = \frac{12Wy}{4\,y} = 3W \quad \textbf{Ans.}$$

***Example 12.** *A square pin-jointed frame ABCD with 500 mm side is subjected to a vertical force of 100 N at D and horizontal force of 50 N at C as shown in Fig. 22. All the pins are smooth and the bar weights are neglected.*

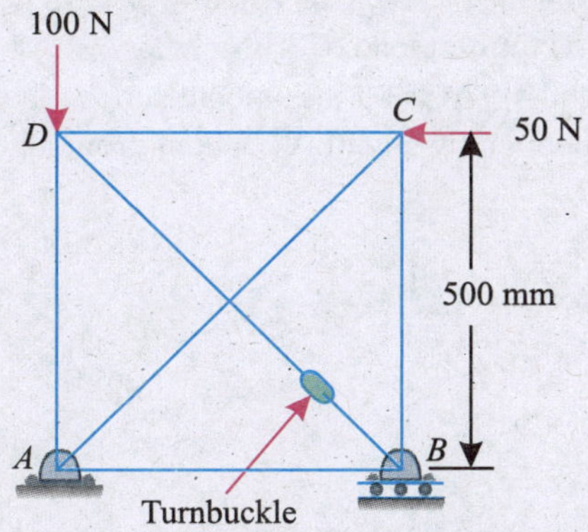

Fig. 22.

The force in the bar BD can be adjusted by means of a turnbuckle. Using the method of virtual work, determine the force in the turnbuckle, when the bar AC carries no load.

Solution. Given : Side of frame = 500 mm; force at D = 100 N and force at C = 50 N

Since the bar AC carries no load, therefore it may be assumed to be removed. Now let us assume the member BD (in which the force is required to be found out) to be removed as shown in Fig. 23.

Let T = Magnitude of force in the turnbuckle (*i.e.* in bar BD)

y = Virtual vertical displacement of the 100 N load, and

x = Virtual shortening of bar BD.

Fig. 23.

From the geometry of the figure, we find that

$$x = y\sqrt{2}$$

We know that virtual work done by the force in the turbuckle

$$= +T \times x = +T \times y\sqrt{2} \quad \text{...(i)}$$

...(Plus sign due to tension)

and virtual work done by the 100 N load at D

$$= -100 \times y = -100\, y \quad \text{...(ii)}$$

...(Minus sign due to downward movement of the load)

We know, **from principle of virtual work that, Algebraic sum of the total virtual work done is zero**. Therefore

$$Ty\sqrt{2} - 100\, y = 0$$

or

$$T = \frac{100\, y}{y\sqrt{2}} = 50\sqrt{2} = 70.7 \text{ N } \textbf{Ans.}$$

Note. During the displacement of the members, the forces at A and B do no work. Similarly, the horizontal force at C does not work in the vertical direction.

EXERCISE 2

1. The diameter of the pulleys in a differential pulley block are 300 mm and 250 mm respectively. Using the principle of virtual work and neglecting friction, find the value of the effort required to lift a load of 3 kN. **(Ans.** 250 N)

*2. A block of weight (W) rests on the smooth surface inclined at 20° with the horizontal. The block is supported by an effort (P) hung from a pulley as shown in Fig. 24.

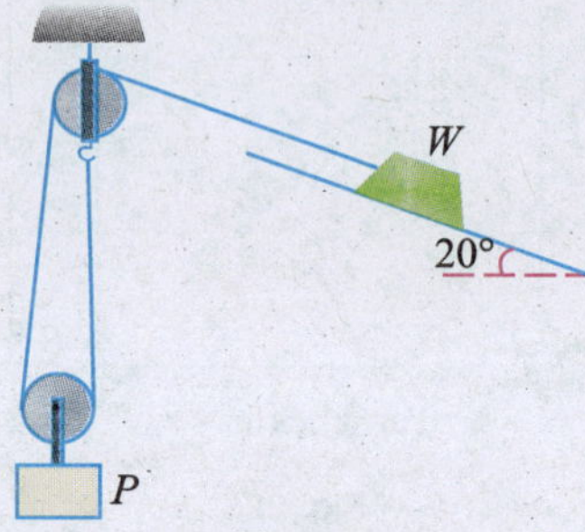

Fig. 24.

Using the principle of virtual work, obtain expression for (P) in terms of (W), when the system is at rest. **(Ans.** $P = 0.685\ W$)

[**Hint.**

Let x = Virtual vertical upward displacement of the load W, and

y = Virtual vertical downward displacement of the effort (P)

∴ Distance through which the load (W) moves along the inclined plane

$$= \frac{x}{\sin 20°} = \frac{x}{0.3420} = 2.92\,x \qquad ...(i)$$

and distance through which the effort (P) will move

$$= \frac{2.92x}{2} = 1.46x \qquad ..(ii)$$

∴ Virtual work done by the weight (W)

$$= +\,W\,x$$

and virtual work done by the effort (P)

$$= -\,P \times 1.46\,x = -\,1.46\,Px$$

$$\therefore\quad Wx - 1.46\,Px = 0 \quad \text{or} \quad P = \frac{W}{1.46} = 0.685\,W \quad \textbf{Ans.}$$

**3. Two weights W_1 and W_2 are resting on two smooth planes AB and BC at angles of θ_1 and θ_2 as shown in Fig. 25.

Fig. 25.

With the help of principle of virtual work, find the ratio of W_1 and W_2.

$$\left(\textbf{Ans.}\ \frac{W_1}{W_2} = \frac{\text{Sin}\,\theta_2}{\sin\theta_1}\right)$$

*4. A structure, with pin-jointed member is shown in Fig. 26.

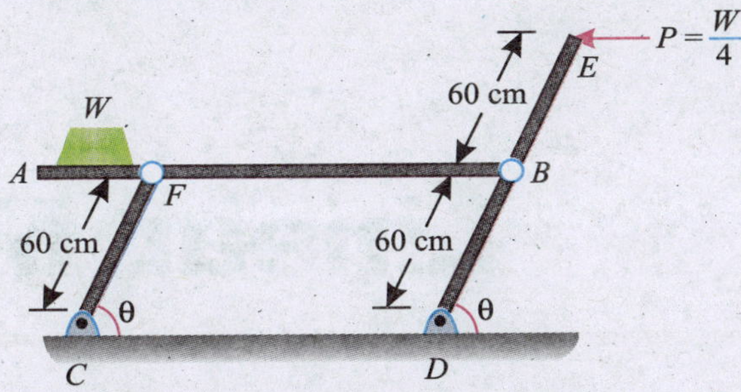

Fig. 26.

Determine the value of θ for equilibrium, using the principle of virtual work. Neglect friction and weight of the members. **(Ans.** 26.6°)

5. Five rods *AB*, *BC*, *CD*, *DA* and *DB* each of equal length and cross-section are pin-jointed together, so as to form a plane frame *ABCD*. The frame is suspended from the top most joint *A*.

A weight (*W*) is attached at the lower most joint *C*. Neglecting selfweight of the frame and using the method of virtual work, find magnitudes of the thrust in the member *BD*.

(Ans. 0.577 *W*)

QUESTIONS

1. Explain the principle of virtual work.
2. State the principle of virtual work, and explain how it can be used for solving problems in statics.
3. Explain the application of the principle of virtual work in case of lifting machines.
4. How will you apply the principle of virtual work in finding out the forces in a framed structure?

OBJECTIVE TYPE QUESTIONS

1. The term 'virtual work' refers to
   (*a*) actual work done by virtual forces
   (*b*) virtual work done by actual forces
   (*c*) virtual work done by virtual forces
2. The principle of virtual work can not be applied for finding out the reactions of a beam carrying simultaneously point loads and uniformly distributed loads.
   (*a*) Yes (*b*) No
3. The principle of virtual work is applicable for the bodies in equilibrium
   (*a*) Agree (*b*) Disagree
4. The principle of virtual work can be applied for all types of
   (*a*) possible displacements (*b*) impossible displacement (*c*) none of the two

ANSWERS

1. (*b*) **2.** (*b*) **3.** (*a*) **4.** (*a*).

CHAPTER

15 Linear Motion

Contents

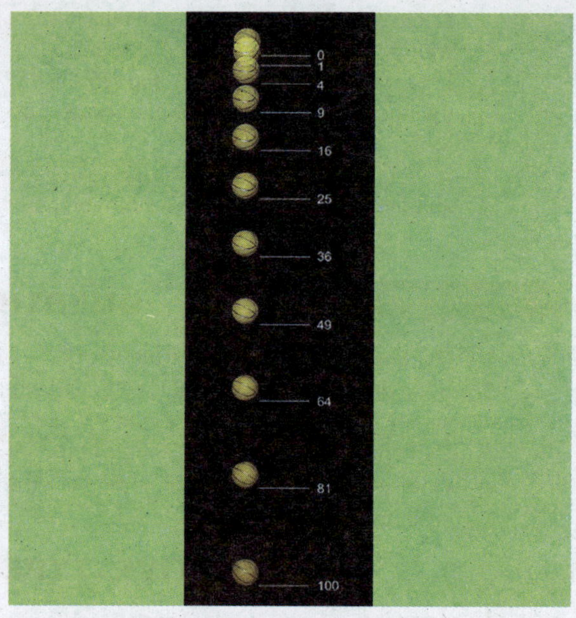

1. REST AND MOTION

A body is said to be at **rest**, if it occupies the same position with respect to its surroundings at all moments. But it is said to be in **motion**, if it changes its position, with respect to its surroundings.

2. IMPORTANT TERMS

The following terms, which will be frequently used in this chapter, should be clearly understood at this stage :

1. ***Speed.*** The speed of a body may be defined as its rate of change of displacement with respect to its surroundings. The speed of a body is irrespective of its direction and is, thus, a scalar quantity.

2. ***Velocity.*** The velocity of a body may be defined as its rate of change of displacement, with respect to its surroundings, in a particular direction. As the velocity is always expressed in particular direction, therefore it is a vector quantity.
3. ***Acceleration.*** The acceleration of a body may be defined as the rate of change of its velocity. It is said to be positive, when the velocity of a body increases with time, and negative when the velocity decreases with time. The negative acceleration is also called retardation. In general, the term acceleration is used to denote the rate at which the velocity is changing. It may be uniform or variable.
4. ***Uniform acceleration.*** If a body moves in such a way that its velocity changes in equal magnitudes in equal intervals of time, it is said to be moving with a uniform acceleration.
5. ***Variable acceleration.*** If a body moves in such a way, that its velocity changes in unequal magnitudes in equal intervals of time, it is said to be moving with a variable acceleration.
6. ***Distance traversed.*** It is the total distance moved by a body. Mathematically, if body is moving with a uniform velocity (v), then in (t) seconds, the distance traversed

$$s = vt$$

In this chapter, we shall discuss the motion under uniform acceleration only.

3. *MOTION UNDER UNIFORM ACCELERATION (EQUATIONS OF RECTILINEAR MOTION)

Fig. 1. Motion under uniform acceleration.

Consider *linear motion of a particle starting from O and moving along OX with a uniform acceleration as shown in Fig. 1. Let P be its position after t seconds.

Let u = Initial velocity,

v = Final velocity,

t = Time (in seconds) taken by the particle to change its velocity from u to v.

a = Uniform positive acceleration, and

s = Distance travelled in t seconds.

Since in t seconds, the velocity of the particle has increased steadily from (u) to (v) at the rate of a, therefore total increase in velocity

$$= at$$

∴ **Final velocity,**

$$v = u + at \qquad ...(i)$$

and average velocity

$$= \left(\frac{u+v}{2}\right)$$

We know that distance travelled by the particle,

$$s = \text{Average velocity} \times \text{Time}$$

$$= \left(\frac{u+v}{2}\right) \times t \qquad ...(ii)$$

Substituting the value of v from equation (*i*), we get

$$s = \left(\frac{u + u + at}{2}\right) \times t$$

∴ **Distance travelled in *t* seconds,**

$$s = ut + \frac{1}{2}at^2 \qquad ...(iii)$$

From equation (*i*), (*i.e.* $v = u + at$) we find that,

$$t = \frac{v - u}{a}$$

Now substituting this vlaue of t in equation (*ii*),

$$s = \left(\frac{u + v}{2}\right) \times \left(\frac{v - u}{a}\right) = \frac{v^2 - u^2}{2a}$$

or $$2as = v^2 - u^2$$

or $$v^2 = u^2 + 2as$$

Example 1. *A car starting from rest is accelerated at the rate of 0.4 m/s². Find the distance covered by the car in 20 seconds.*

Solution. Given : Initial velocity (u) = 0 (because, it starts from rest) ; Acceleration (a) = 0.4 m/s^2 and time taken (t) = 20 s

We know that, Distance covered by the car,

$$s = ut + \frac{1}{2}at^2 = (0 \times 20) + \frac{1}{2} \times 0.4 \times (20)^2 \text{ m} = 80 \text{ m} \quad \textbf{Ans.}$$

Example 2. *A train travelling at 27 km.p.h is accelerated at the rate of 0.5 m/s². What is the distance travelled by the train in 12 seconds ?*

Solution. Given : Initial velocity (u) = 27 km.p.h. = 7.5 m/s ; Accceleration (a) = 0.5 m/s^2 and time taken (t) = 12 s.

We know that, Distance travelled by the train,

$$s = ut + \frac{1}{2}at^2 = (7.5 \times 12) + \frac{1}{2} \times 0.5 \times (12)^2 \text{ m}$$

$$= 90 + 36 = 126 \text{ m} \quad \textbf{Ans.}$$

***Example 3.** *A scooter starts from rest and moves with a constant acceleration of 1.2 m/s². Determine its velocity, after it has travelled for 60 meters.*

Solution. Given : Initial velocity (u) = 0 (because it starts from rest) Acceleration (a) = 1.2 m/s^2 and distance travelled (s) = 60 m.

Let v = Final velocity of the scooter.

We know that, $v^2 = u^2 + 2as = (0)^2 + 2 \times 1.2 \times 60 = 144$

$$v = 12 \text{ m/s} = \frac{12 \times 3600}{1000} = 43.2 \text{ km.p.h.} \quad \textbf{Ans.}$$

***Example 4.** *On turning a corner, a motorist rushing at 20 m/s, finds a child on the road 50 m ahead. He instantly stops the engine and applies brakes, so as to stop the car within 10 m of the child. Calculate (i) retardation, and (ii) time required to stop the car.*

Solution. Given : Initial velocity (u) = 20 m/s ; Final velocity (v) = 0 (because the car is stopped) and distance travelled by the car (s) = 50 – 10 = 40 m

(i) Retardation

Let a = Acceleration of the motorist.

We know that, $v^2 = u^2 + 2as$

or $0 = (20)^2 + 2 \times a \times 40 = 400 + 80a$

or $80a = -400$

or $a = \dfrac{-400}{80} = -5 \text{ m/s}^2$ **Ans.**

...(Minus sign shows that the acceleration is negative *i.e.* retardation).

(ii) Time required to stop the car

Let t = Time required to stop the car in second.

We know that, Final velocity of the car (v),

$$0 = u + at = 20 - 5 \times t \qquad ...(\because a = -5 \text{ m/s}^2)$$

$\therefore$ $t = \dfrac{20}{5} = 4 \text{ s}$ **Ans.**

***Example 5.** *A motor car takes 10 seconds to cover 30 meters and12 seconds to cover 42 meters. Find the uniform acceleration of the car and its velocity at the end of 15 seconds.*

[JNTU, IGNOU]

Solution. Given : When t = 10 seconds, s = 30 m and when t = 12 seconds, s = 42 m.

Uniform acceleration

Let u = Initial velocity of the car, and

a = Uniform acceleration.

We know that, Distance travelled by the car in 10 seconds,

$$30 = ut + \frac{1}{2}at^2 = u \times 10 + \frac{1}{2} \times a(10)^2 = 10u + 50\,a$$

Multiplying the above equation by 6,

$$180 = 60u + 300a \qquad ...(i)$$

Similarly, Distance travelled by the car in 12 seconds,

$$42 = u \times 12 + \frac{1}{2} \times a(12)^2 = 12u + 72a$$

Mulitiplying the above equation by 5,

$$210 = 60u + 360a \qquad ...(ii)$$

Subtracting equation (*i*) from (*ii*),

$$30 = 60a \qquad \text{or} \qquad a = \frac{30}{60} = 0.5 \text{ m/s}^2 \text{ **Ans.**}$$

Velocity at the end of 15 seconds

Substituting the value of a in equation (*i*)

$$180 = 60u + (300 \times 0.5) = 60u + 150$$

or $u = \dfrac{(180 - 150)}{60} = 0.5 \text{ m/s}$

We know that, Velocity of the car after 15 seconds,

$$v = u + at = 0.5 + (0.5 \times 15) = 8 \text{ m/s}$$ **Ans.**

***Example 6.** *A burglar's car had a start with an acceleration of 2 m/s². A police vigilant party came after 5 seconds and continued to chase the burglar's car with a uniform velocity of 20 m/s. Find the time taken, in which the police van will overtake the burglar's car.*

Solution. Given : Acceleration of burglar's car = 2 m/s² and uniform velocity of the police party = 20 m/s.

Let t = Time taken by the police party, to overtake the burglar's car after reaching the spot.

First of all, consider the motion of the burglar's car. In this case, initial velocity $(u) = 0$ (because it starts from rest) ; acceleration = 2 m/s² and time taken by burglar's car to travel the distance of s m $(t_1) = (t + 5)$ second.

We know that, Distance travelled by the burglar's car,

$$s = ut + \frac{1}{2}at^2 = 0 + \frac{1}{2} \times 2(t+5)^2 = (t+5)^2 \quad ...(i)$$

Now consider the motion of the police party. In this case, uniform velocity, $v = 20$ m/s

∴ Distance travelled by the police party,

$$s = \text{Velocity} \times \text{Time} = 20 \times t \quad ...(ii)$$

For the police party to overtake the burglar's car, the two distances (i) and (ii) should be equal. Therefore equating equations (i) and (ii)

$$(t+5)^2 = 20 \times t$$

or $$t^2 + 25 + 10\,t = 20\,t$$

or $$t^2 + 25 - 10\,t = 0$$

or $$(t-5)^2 = 0$$

or $$(t-5) = 0 \quad \text{...(Taking square root)}$$

or $$t = 5 \text{ s}$$ **Ans.**

****Example 7.** *A train is uniformly accelerated and passes successive kilometre stones with velocities of 18 km.p.h. and 36 km.p.h. respectively. Calculate the velocity, when it passes the third kilometre stone. Also find the time taken for each of these two intervals of one kilometre.*

[JNTU]

Solution. First of all, consider the motion of the train between the first and second kilometre stones. In this case, distance (s) = 1 km = 1000 m ; initial velocity (u) = 18 km.p.h. = 5 m/s ; and final velocity (v) = 36 km.p.h. = 10 m/s

Velocity with which the train passes the third km stone

Let v = Velocity with which the train passes the third km, and

a = Uniform acceleration.

We know that, $$v^2 = u^2 + 2as$$

or $$(10)^2 = (5)^2 + (2a \times 1000) = 25 + 2000\,a$$

or $$a = \frac{100-25}{2000} = \frac{75}{2000} = 0.0375 \text{ m/s}^2$$

Now consider the motion of the train between the second and third kilometre stones. In this case, distance (s) = 1 km = 1000 m and initial velocity (u) = 36 km.p.h. = 10 m/s.

We know that, $$v^2 = u^2 + 2as = (10)^2 + (2 \times 0.0375 \times 1000) = 175$$

or $$v = 13.2 \text{ m/s} = 47.5 \text{ km.p.h.}$$ **Ans.**

Time taken for each of the two intervals of one kilometre

Let t_1 = Time taken by the train to travel the first one kilometre, and

t_2 = Time taken by the train to travel the second kilometre.

We know that, Velocity of the train after passing the first kilometre *i.e.*, in t_1 seconds (v_1),

$$10 = u + at_1 = 5 + 0.0375\ t_1$$

or $$t_1 = \frac{10 - 5}{0.0375} = 133.3 \text{ s} \quad \textbf{Ans.}$$

Similarly, Velocity of the train after passing the second kilometre *i.e.* in t_2 seconds,

$$13.2 = u + at_2 = 10 + 0.0375\ t_2$$

or $$t_2 = \frac{13.2 - 10}{0.0375} = 85.3 \text{ s} \quad \textbf{Ans.}$$

Example 8. *Two electric trains A and B leave the same station on parallel lines. The train A starts from rest with a uniform acceleration of 0.2 m/s² and attains a speed of 45 km.p.h., which is maintained constant afterwards. The train B leaves 1 minute after with a uniform acceleration of 0.4 m/s² to attain a maximum speed of 72 km.p.h., which is maintained constant afterwards. When will the train B overtake the train A ?* **[AUT]**

Solution. Given : Initial velocity of train A (u_A) = 0 (because it starts from rest) ; Uniform acceleration of train A (a_A) = 0.2 m/s² ; Final velocity of train A (v_A) = 45 km.p.h. = 12.5 m/s ; Initial velocity of train B (u_B) = 0 (because it also starts from rest) ; Uniform acceleration of train B (a_B) = 0.4 m/s² and final velocity of train B (v_B) = 72 km.p.h. = 20 m/s

Let t_A = Time taken by the train A to attain a speed of 12.5 m/s, and

T = Time in second when the train B will overtake the train A from its start.

Final velocity of the train A (v_A),

$$12.5 = u_A + a_A t_A = 0 + 0.2\ t_A$$

or $$t_A = \frac{12.5}{0.2} = 62.5 \text{ s}$$

Distance travelled by the train A to attain this speed,

$$s = u_A t_A + \frac{1}{2} a_A t_A^{\ 2} = 0 + \frac{1}{2} \times 0.2\ (62.5)^2 = 390.6 \text{ m}$$

Similarly, Final velocity of the train B (v_B),

$$20 = u_B + a_B t_B = 0 + 0.4\ t_B$$

or $$t_B = \frac{20}{0.4} = 50 \text{ s}$$

Distance travelled by the train B to attain this speed,

$$s = u_B t_B + \frac{1}{2} a_B t_B^{\ 2} = 0 + \frac{1}{2} \times 0.4 (50)^2 = 500 \text{ m}$$

Now we see that the train *A* has travelled for (*T* + 60) seconds. Therefore total distance travelled by the train *A* during this time,

$$s_A = 390.6 + 12.5\ [(T + 60) - 62.5\ ] \text{ m} \qquad ...(i)$$

and Total distance travelled by the Train B,

$$s_B = 500 + 20\,(T - 50)\text{ m} \quad \ldots(ii)$$

For the train B to overtake the train A, the two distances (i) and (ii) should be equal. Threfore equating equations (i) and (ii),

$$390.6 + 12.5\,[(T + 60) - 62.5] = 500 + 20\,(T - 50)$$

or $$12.5\,T - 31.3 = 109.4 + 20\,T - 1000$$

or $$7.5\,T = 1000 - 109.4 - 31.3 = 859.3$$

or $$T = \frac{859.3}{7.5} = 114.6\text{ s} \quad \textbf{Ans.}$$

EXERCISE 1

1. A body starts with a velocity of 3 m/s and moves in a straight line with a constant acceleration. If its velocity at the end of 5 seconds is 5.5 m/s, find (i) the uniform acceleration, and (ii) distance travelled in 10 seconds. [**Ans.** 0.5 m/s^2 ; 55 m]
2. Two cars start off to a race with velocities (u), and (v), and travel in a straight line with a uniform accelerations of (α) and (β). If the race ends in a dead beat, prove that the length of the race is:
$$\frac{2(u - v)\,(u\beta - v\alpha)}{(\beta - \alpha)^2}$$
Hint. Dead beat means that the distance travelled by both the cars is the same.
3. A car starts from rest and accelerates uniformly to a speed of 72 km.p.h. over a distance of 500 m. Find acceleration of the car and time taken to attain this speed.

   If a further acceleration rises the speed to 90 km.p.h. in 10 seconds, find the new acceleration and the further distance moved.

   [**Ans.** 0.4 m/s^2 ; 50 sec ; 0.5 m/s^2 ; 225 m ; 62.5 m]
4. A bullet moving at the rate of 300 m/s is fired into a thick target and penetrates up to 500 mm. If it is fired into a 250 mm thick target, find the velocity of emergence. Take the resistance to be uniform in both the cases. [**Ans.** 212.1 m/s]

4. *MOTION UNDER FORCE OF GRAVITY

It is a particular case of motion, under a constant acceleration of (g) where its value is taken as 9.8 m/s^2. **If there is a free fall under gravity, the expressions for velocity and distance travelled in terms of initial velocity, time and gravity acceleration will be :**

1. $v = u + gt$
2. $s = ut + \frac{1}{2}gt^2$
3. $v^2 = u^2 + 2\,gs$

But, **if the motion takes place against the force of gravity, *i.e.*, the particle is projected upwards, the corresponding equations will be :**

1. $v = -\,u + gt$
2. $s = -\,ut + \frac{1}{2}gt^2$
3. $v^2 = -\,u^2 + 2\,gs$

Notes : 1. In this case, the value of u is taken as negative due to upward motion.

2. In this case, the distances in upward direction are taken as negative, while those in the downward direction are taken as positive.

Example 9. *A stone is thrown upwards with a velocity of 4.9 m/s from a bridge. If it falls down in water after 2 s, then find the height of the bridge.*

Solution. Given : Initial velocity $(u) = -4.9$ m/s (*Minus sign due to upwards*) and time taken $(t) = 2$ s.

We know that, Height of the bridge,

$$h = ut + \frac{1}{2}gt^2 = (-4{\cdot}9 \times 2) + \frac{1}{2} \times 9{\cdot}8 \times (2)^2 \text{ m}$$

$$= -9.8 + 19.6 = 9.8 \text{ m} \quad \textbf{Ans.}$$

Example 10. *A packet is dropped from a balloon which is going upwards with a velocity 12 m/s. Calculate the velocity of the packet after 2 seconds.*

Solution. Given : Velocity of balloon when the packet is dropped $(u) = -12$ m/s (Minus sign due to upward motion) and time $(t) = 2$ s.

We know that, When the packet is dropped its initial velocity $(u) = -12$ m/s.

∴ Velocity of packet after 2 second,

$$v = u + gt = -12 + (9{\cdot}8 \times 2) = -12 + 19{\cdot}6 = 7{\cdot}6 \text{ m/s} \quad \textbf{Ans.}$$

Example 11. *A body is dropped from the top of a tall building. If it takes 2.8 seconds in falling on the ground, find the height of the building.*

Solution. Given : Initial velocity $(u) = 0$ (because it is dropped) and time taken $(t) = 2.8$ s.

We know that, Height of the building

$$s = ut + \frac{1}{2}gt^2 = (0 \times 2.8) + \frac{1}{2} \times 9.8 \times (2.8)^2 = 38.4 \text{ m} \quad \textbf{Ans.}$$

Example 12. *A stone is dropped from the top of a building, which is 65 m high. With what velocity will it hit the ground ?*

Solution. Given : Initial velocity $(u) = 0$ (because it is dropped) and height of the building $(s) = 65$ m

Let $\quad v$ = Final velocity of the stone with which it will hit the ground.

We know that, $\quad v^2 = u^2 + 2gs = (0)^2 + 2 \times 9.8 \times 65 = 1274$

or $\quad v = 35.7$ m/s **Ans.**

***Example 13.** *A body is thrown vertically upwards with a velocity of 28 m/s. Find the distance it will cover in 2 seconds.*

Solution. Given : Initial velocity $(u) = -28$ m/s (Minus sign due to upward motion) and time $(t) = 2$ s.

We know that, Distance covered by the body,

$$s = ut + \frac{1}{2}gt^2 = (-28 \times 2) + \frac{1}{2} \times 9.8 \times (2)^2 \text{ m}$$

$$= -56 + 19.6 = -36.4 \text{ m} \quad \textbf{Ans.}$$

....(Minus sign indicates that the body will cover the distance in upward direction)

Example 14. *A bullet is fired vertically upwards with a velocity of 80 m/s. To what height will the bullet rise above the point of projection ?*

Solution. Given : Initial velocity (u) = – 80 m/s (Minus sign due to upward motion) and final velocity (v) = 0 (because the bullet is at maximum rise)

Let s = Height to which the bullet will rise above the point of projection.

We know that, $v^2 = u^2 + 2\,gs$

or $(0)^2 = (-80)^2 + 2 \times 9.8 \times s = 6400 + 19.6\text{ s}$

or $s = \dfrac{6400}{-19.6} = -326.5\text{ m}$ **Ans.**

...(Minus sign indicates that the body will cover the distance in upward direction)

***Example 15.** *A body is released from a great height falls freely towards earth. Another body is released from the same height exactly one second later. Find the separation between both the bodies, after two seconds of the release of second body.*

Solution. Given : Initial velocity of both the bodies (u) = 0 (because they are released); Time taken by the first body (t_1) = 3 s and time taken by the second body (t_2) = 3 – 1 = 2 s (because the second body is released after 1 s of the release of first body).

Distance covered by the first body in 3 seconds

$$h_1 = ut_1 + \frac{1}{2}\,gt_1^2 = (0 \times 3) + \frac{1}{2} \times 9{\cdot}8 \times (3)^2 = 44.1\text{ m} \qquad ...(i)$$

Distance covered by the second body in 2 seconds

$$h_2 = ut_2 + \frac{1}{2}\,gt_2^2 = (0 \times 2) + \frac{1}{2} \times 9{\cdot}8 \times (2)^2 = 19.6\text{ m} \qquad ...(ii)$$

∴ Separation between the bodies

$= h_2 - h_1 = 44.1 - 19.6 = 24.5\text{ m}$ **Ans.**

***Example 16.** *A stone is thrown vertically upwards with a velocity of 29.4 m/s from the top of a tower 34.3 m high. Find the total time taken by the stone to reach the foot of the tower.*

Solution. Given : Initial velocity (u) = – 29.4 m/s (Minus sign due to upward motion) and height of tower (h) = 34.3 m

Let t = Time taken by the stone to reach the foot of the tower

Height of the tower (h),

$$34.3 = ut + \frac{1}{2}\,gt^2 = (-29.4 \times t) + \frac{1}{2} \times 9.8t^2$$

or $4.9\,t^2 - 29.4\,t - 34.3 = 0$

or $t^2 - 6t - 7 = 0$

This is a quadratic equation in (t).

$$\therefore \qquad t = \frac{+6 \pm \sqrt{(6)^2 + (4 \times 7)}}{2} = \frac{6 \pm 8}{2} = 7\text{ s} \quad \textbf{Ans.}$$

***Example 17.** *A stone is thrown vertically upwards, from the ground, with velocity 49 m/s. After 2 seconds, another stone is thrown vertically upwards from the same place. If both the stone strike the ground at the same time, find the velcocity, with which the second stone was thrown upwards.*

Solution. First of all, consider the upwards motion of the first stone. In this case, initial velocity $(u) = -49$ m/s (Minus sign due to upward motion) and final velocity $(v) = 0$ (because stone is at maximum height)

Let t = Time taken by the stone to reach maximum height.

We know that, Final velocity of the stone (v),

$$0 = u + gt = -49 + 9.8\,t \quad \text{...(Minus sign due to upwards motion)}$$

or $$t = \frac{49}{9.8} = 5 \text{ s}$$

It means that the stone will take 5 s to reach the maximum height and another 5 s to come back to the ground.

$\therefore$ Total time of flight $= 5 + 5 = 10$ s

Now consider the motion of second stone.

We know that, Time taken by the second stone for going upwards and coming back to the earth

$$= 10 - 2 = 8 \text{ s}$$

and Time taken by the second stone to reach maximum height

$$= \frac{8}{2} = 4 \text{ s}$$

Now consider the upward motion of the second stone.

Final velocity of the stone (v),

$$0 = u + gt = -u + 9.8 \times 4 = -u + 39.2$$

or $$u = 39.2 \text{ m/s} \quad \textbf{Ans.}$$

***Example 18.** *A stone is dropped from the top of a tower 50 m high. At the same time, another stone is thrown upwards from the foot of the tower with a velocity of 25 m/s. When and where the two stones cross each other ?* **[IGNOU]**

Solution. Given : Height of the tower = 50 m

Time taken by the stone to cross each other

First of all, consider downward motion of the first stone. In this case, initial velocity $(u) = 0$ (beacuse it is dropped)

Let t = Time taken for the stones to cross each other.

We know that, Distance traversed by the stone,

$$s = ut + \frac{1}{2}gt^2 = 0 + \frac{1}{2}gt^2 = 0.5\,gt^2 \quad \text{...(i)}$$

Now consider upward motion of the second stone. In this case, initial velocity = – 25 m/s (Minus sign due to upward) and distance traversed = $50 - s$,

We know that, Distance traversed,

$$50 - s = ut + \frac{1}{2}gt^2 = -25t + 0.5\,gt^2 \quad \text{...(ii)}$$

Adding equations (*i*) and (*ii*),

$$50 = 25\,t \quad \text{or} \quad t = \frac{50}{25} = 2 \text{ s} \quad \textbf{Ans.}$$

Point where the stones crosss each other

Substituting the value of $t = 2$ in equation (*i*),

$$s = 0.5 \times 9.8\,(2)^2 = 19.6 \text{ m} \quad \textbf{Ans.}$$

***Example 19.** *A stone is thrown vertically up from the top of a tower with a certain initial velocity. It reaches ground in 5.64 seconds. A second stone, thrown down from the same tower with the same initial velocity reaches ground in 3.6 seconds. Determine (i) the height of the tower, and (ii) the initial velocity of the stones.*

Solution. Given : Total time taken by first stone = 5.64 s ; Time taken by second stone = 3.6 s.

Initial velocity of the stones

Let u = Initial velocity of the stones.

First of all, consider upward motion of the first stone from the top of the tower. We know that it will first move upwards in the sky. And when its velocity becomes zero, it will start coming down. Its velocity at the top of the tower, while coming down, will be the same as that with which it was thrown upwards.

Therefore, Time taken by the first stone to reach maximum height and then to reach the top of the tower, from where it was thrown

$$= 5.64 - 3.6 = 2.04 \text{ s}$$

Thus, Time taken by the stone to reach maximum height (or in other words when its final velocity, $v = 0$)

$$= \frac{2.04}{2} = 1.02 \text{ s}$$

We know that, Final velocity of the stone (v),

$$0 = -u + gt = -u + (9.8 \times 1.02) = -u + 10$$

or $u = 10 \text{ m/s}$ **Ans.**

Height of the tower

Now consider downward motion of the second stone. In this case, initial velocity (u) = 10 m/s and time (t) = 3.6 s. We know that, Height of the tower (or distance travelled by the stone),

$$s = ut + \frac{1}{2} gt^2 = (10 \times 3.6) + \frac{1}{2} \times 9.8 (3.6)^2 = 99.5 \text{ m} \quad \textbf{Ans.}$$

***Example 20.** *A stone is thrown up with a velocity of 20 m/s. While coming down, it strikes a glass pan, held at half the height through which it has risen and loses half of its velocity in breaking the glass. Find the velocity with which it will strike the ground.*

Solution. First of all, consider upward motion of the stone. In this case, initial velocity (u_1) = – 20 m/s (Minus sign due to upward motion) and final velocity (v_1) = 0 (because it reaches maximum height).

Let s_1 = Maximum height reached by the stone.

We know that, $v_1^2 = u_1^2 + 2 g s_1$

or $0 = (-20)^2 + 2 \times 9.8 \times s_1 = 400 + 19.6\, s_1$

or $$s_1 = \frac{400}{-19.6} = -20.4 \text{ m} \quad ...(i)$$

....(Minus sign indicates that height reached by the stone is in upward direction)

Now consider downward motion of the stone up to the glass pan. In this case, initial velocity (u_2) = 0 (because it starts coming down after rising to the maximum height and distance covered by the stone (s_2) $= \dfrac{20.4}{2} = 10.2$ m

Let v_2 = Final velocity of the stone, with which it strikes the pan.

We know that, $v_2^2 = u_2^2 + 2\,gs_2 = 0 + (2 \times 9.8 \times 10.2) = 199.9$

or $v_2 = \sqrt{199.9} = 14.14$ m/s

Now consider motion of the stone after breaking the glass pan. In this case, initial velocity $(u_3) = \dfrac{14.14}{2} = 7.07$ m/s (because it looses half its velocity after striking the pan) and distance travelled by the stone $(s_3) = 20.4 - 10.2 = 10.2$ m.

Let v_3 = Final velocity of the stone, with which it strikes the ground.

We know that, $v_3^2 = u_3^2 + 2\,gs_3 = (7.07)^2 + (2 \times 9.8 \times 10.2) = 249.9$

or $v_3 = \sqrt{249.9} = 15.8$ m/s **Ans.**

****Example 21.** *A body falling freely, under the action of gravity passes two points 10 metres apart vertically in 0.2 second. From what height, above the higher point, did it start to fall ?*

[UPTU, BPUT, AMIE, IGNOU]

Solution. Let the body start from O and pass two points A and B 10 metres apart in 0.2 s as shown in Fig. 2.

First of all, consider the motion from O to A. In this case, initial velocity $(u) = 0$ (because it is falling freely) and distance travelled by body $(s) = OA = x$ m

Let t = Time taken by the body to travel from O to A.

We know that, $x = ut + \dfrac{1}{2}gt^2 = 0 + \dfrac{1}{2} \times 9.8t^2 = 4.9\,t^2$ ...(*i*)

Now consider motion of stone from O to B. In this case, Initial velocity $(u) = 0$ (because it is falling freely) ; Distance travelled $(s_1) = OB = (x + 10)$ m and time taken $(t_1) = (t + 0.2)$ s

We know that, $(x + 10) = ut_1 + \dfrac{1}{2}gt_1^2 = 0 + \dfrac{1}{2} \times 9.8(t + 0.2)^2$

$= 0 + 4.9\,(t^2 + 0.04 + 0.4\,t)$

$= 4.9\,t^2 + 0.196 + 1.96\,t$ ...(*ii*)

Fig. 2.

Subtracting equation (*i*) from (*ii*),

$10 = 0.196 + 1.96\,t$

or $t = \dfrac{10 - 0.196}{1.96} = 5$ s

Substituting the value of t in equation (*i*),

$x = 4.9\,t^2 = 4.9 \times (5)^2 = 122.5$ m **Ans.**

Alternative Method

First of all, consider the motion between A and B. In this case, distance travelled $(s) = 10$ m and time $(t) = 0.2$ s

Let u = Initial velocity of the body at A.

We know that, Distance travelled (s),

$$10 = ut + \frac{1}{2}gt^2 = u \times 0.2 + \frac{1}{2} \times 9.8(0.2)^2 = 0.2\,u + 0.196$$

or $u = \dfrac{10 - 0.196}{0.2} = 49$ m/s

Now consider the motion from O to A. In this case, initial velocity $(u) = 0$ (because it is falling freely) and final velocity $(v) = 49$ m/s

We know that, $$v^2 = u^2 + 2\,gs$$

or $$(49)^2 = 0 + 2 \times 9.8 \times x = 1.96\,x$$

or $$x = \frac{(49)^2}{19.6} = 122.5 \text{ m} \quad \textbf{Ans.}$$

****Example 22.** *A stone, dropped into a well, is heard to strike the water after 4 seconds. Find the depth of well, if velocity of the sound is 350 m/s.* **[JNTU, BPUT]**

Solution. Given: Time = 4 s and Velocity of sound = 300 m/s.

First of all, consider the downward motion of the stone. In this case, initial velocity $(u) = 0$ (because it is dropped)

Let t = Time taken by the stone to reach the bottom of the well.

We know that, Depth of the well,

$$s = ut + \frac{1}{2}\,gt^2 = 0 + \frac{1}{2} \times 9.8 \times t^2 = 4.9\,t^2 \qquad ...(i)$$

and Time taken by the sound to reach the top,

$$= \frac{\text{Depth of the well}}{\text{Velocity of sound}} = \frac{s}{350} = \frac{4.9t^2}{350} \qquad ...(ii)$$

Since the total time taken (*i.e.* stone to reach the bottom of the well and sound to reach the top of the well) is 4 seconds, therefore

$$t + \frac{4.9t^2}{350} = 4$$

or $$4.9\,t^2 + 350\,t = 1400$$

or $$4.9\,t^2 + 350\,t - 1400 = 0$$

This is a quadratic equation in t,

$$\therefore \quad t = \frac{-350 \pm \sqrt{(350)^2 + 4 \times 4.9 \times 1400}}{2 \times 4.9} = 3.8 \text{ s}$$

Now substituting the value of t in equation (*i*),

$$s = 4.9\,t^2 = 4.9\,(3.8)^2 = 70.8 \text{ m} \qquad \textbf{Ans.}$$

Example 23. *A particle, falling under gravity, falls 20 metres in a certain second. Find the time required to cover next 20 metres.*

Solution. Given : Distance travelled by particle $(s) = 20$ m and time $(t) = 1$ s

Let u = Initial velocity of particle at the time of starting.

We know that, Distance covered by the particle in one second (s),

$$20 = ut + \frac{1}{2}\,gt^2 = (u \times 1) + \frac{1}{2} \times 9.8(1)^2 = u + 4.9$$

or $$u = 20 - 4.9 = 15.1 \text{ m/s}$$

and Velocity of the particle after covering 20 metres (or after 1 second)

$$= u + g\,t = 15.1 + (9.8 \times 1) = 24.9 \text{ m/s}$$

Now let (t) be the time required to cover a distance of 20 metres when the particle has initial velocity of 24.9 m/s.

We know that, Distance covered by the particle in this time (s),

$$20 = ut + \frac{1}{2}\,gt^2 = 24.9t + \frac{1}{2} \times 9.8t^2 = 24.9t + 4.9t^2$$

or $$4.9\,t^2 + 24.9\,t - 20 = 0$$

This is a quadratic equation in t,

$$\therefore \quad t = \frac{-24.9 \pm \sqrt{(24.9)^2 - 4 \times 4.9 \times (-20)}}{2 \times 4.9} = 0.71 \text{ s} \qquad \textbf{Ans.}$$

Example 24. *Two particles A and B are dropped simultaneously from rest two points both 100 m above the ground. Particle A falls on the ground, while the particle B in its mid path, hits a fixed plane inclined to the horizontal as shown in the Fig. 3.*

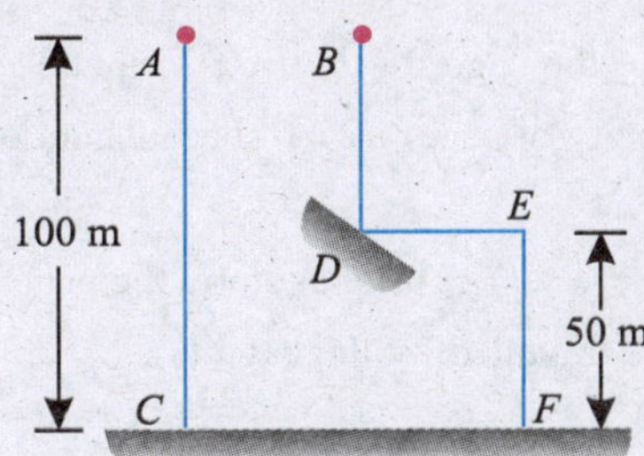

Fig. 3.

As a result of this impact, the direction of its velocity becomes horizontal. Compare the times of fall of the particles A and B to reach the ground.

Solution. First of all, consider motion of the stone A (*i.e.* from A to C). In this case, initial velocity $(u_1) = 0$ (because it is dropped) and distance $(s_1) = 100$ m

Let $\quad t_1$ = Time taken by the particle A to reach C.

We know that, Distance travelled by the stone A (s_1)

$$100 = ut + \frac{1}{2} gt^2 = 0 + \frac{1}{2} \times 9.8\, t_1^2 = 4.9\, t_1^2$$

or
$$t_1 = \sqrt{\frac{100}{4.9}} = 4.52 \text{ s}$$

Now consider motion of the stone B (first from B to D).

In this case, initial velocity $(u_2) = 0$ (because it is dropped) and distance $(s_2) = 50$ m

Let $\quad t_2$ = Time taken by the particle B to reach D.

We know that, Distance travelled by the stone B (s_2)

$$50 = ut + \frac{1}{2} gt^2 = 0 + \frac{1}{2} \times 9.8\, t_2^2 = 4.9\, t_2^2$$

or
$$t_2 = \sqrt{\frac{50}{4.9}} = 3.19 \text{ s} \qquad \text{...}(ii)$$

Now consider motion of the stone B (from D or E to F).

Since the direction of the particle B, after impact at D, becomes horizontal, therefore its velocity in the vertical direction becomes zero. A little consideration will show that the particle B will take another 3.19 sec to reach from D to F. Therefore total time taken by the particle B to reach F will be $T = 3.19 + 3.19 = 6.38$ sec. Therefore ratio of the two times,

$$\frac{T}{t_1} = \frac{6.38}{4.52} = 1.41 \qquad \textbf{Ans.}$$

***Example 25.** *A cage descends in a mine shaft with an acceleration of 0.5 m/s². After the cage has travelled 25 m, a stone is dropped from the top of the shaft. Determine the (a) time taken by the stone to hit the cage, and (b) distance travelled by the cage before impact.* **[BPUT]**

Solution. First of all, consider motion of the stone. In this case, initial velocity of stone $(u) = 0$ (because it is dropped).

(a) Time taken by the stone to hit the cage

Let t = Time taken by the stone to hit the cage

We know that, Distance travelled by the stone before the impact,

$$s = ut + \frac{1}{2}gt^2 = 0 + \frac{1}{2} \times 9.8t^2 = 4.9t^2 \quad ...(i)$$

Now consider motion of the cage for the first 25 metres.

In this case, initial velocity of cage $(u) = 0$ (because it descends); Accelaration $(a) = 0.5 \text{ m/s}^2$ and distance $(s) = 25$ m.

Let t = Time taken by the cage to travel 25 m.

We know that, Distance travelled by the cage (s),

$$25 = ut + \frac{1}{2} a.t^2 = 0 + \frac{1}{2} \times 0.5t^2 = 0.25t^2$$

or
$$t = \sqrt{\frac{25}{0.25}} = \sqrt{100} = 10 \text{ s}$$

It means that the cage has travelled for 10 sec, before the stone was dropped. Therefore total time taken by the cage before impact = $(10 + t)$ s.

We know that, Distance travelled by the cage in $(10 + t)$ s,

$$s = ut + \frac{1}{2} at^2 = 0 + \frac{1}{2} \times 0.5\,(10 + t)^2 = 0{\cdot}25\,(10 + t)^2 \quad ...(ii)$$

In order that the stone may hit the cage, the two distances must be equal. Therefore equating eqution (i) and (ii),

$$4{\cdot}9\,t^2 = 0.25\,(10 + t^2) = 0.25\,(100 + t^2 + 20\,t) = 25 + 0{\cdot}25\,t^2 + 5\,t$$

or
$$4{\cdot}65\,t^2 - 5\,t - 25 = 0$$

This is a quadratic equation in t,

$$\therefore \quad t = \frac{5 \pm \sqrt{(5)^2 + (4 \times 4{\cdot}65 \times 25)}}{2 \times 4{\cdot}65} = 2{\cdot}92 \text{ s} \quad \textbf{Ans.}$$

(b) Distance travelled by the cage before impact

Substituting the value of t in equation (ii),

$$s = 0{\cdot}25\,(10 + 2{\cdot}92)^2 = 41{\cdot}7 \text{ m} \quad \textbf{Ans.}$$

EXERCISE 2

1. A stone is thrown vertically upwards with a velocity of 40 m/s. Find its position after 5 seconds. **[Ans.** 77·5 m]
2. An elevator cage is going up with a velocity of 6 m/s. When the cage was 36 m above the bottom of the shaft, a bolt gets detached from the bottom of the cage floor. Find the velocity with which the bolt strikes the bottom of the shaft and the time that elapses. **[Ans.** 27·24 m/s ; 3·39 s]
3. A stone is dropped from the top of a cliff 120 metres high. After one second, another stone is thrown down and strikes the first stone when it has just reached the foot of the cliff. Find the velocity with which the second stone was thrown. **[Ans.** 11·0 m/s]

4. A body is projected upwards with a velocity of 30 m/s. Find (*a*) the time when its velocity will be 5 m/s ; (*b*) the time when it will be 20 metres above the point of projection and (*c*) the time when it will return to the point of projection.

[**Ans.** 2·55 s : 0·76 sec or 5·36 s ; 6·12 s]

5. A stone is dropped from the top of a tower 60 m high. Another stone is projected upwards at the same time from the foot of the tower, and meets the first stone at a height of 18 m. Find the velocity with which the second stone is projected upwards. [**Ans.** 20·48 m/s]

*6. A body, falling under the force of gravity from the top of a tower, covers 5/9 height of the tower in the last second of its motion. Find the height of the tower. [**Ans.** 44·1 m]

[RGPV]

7. A particle, starting from rest, falls 70 metres in the last second of its motion. Determine the total time taken by particle to fall, and the height from which it fell.

[**Ans.** 7·64 s ; 2·36 m]

5. *DISTANCE TRAVELLED IN THE *n*th SECOND

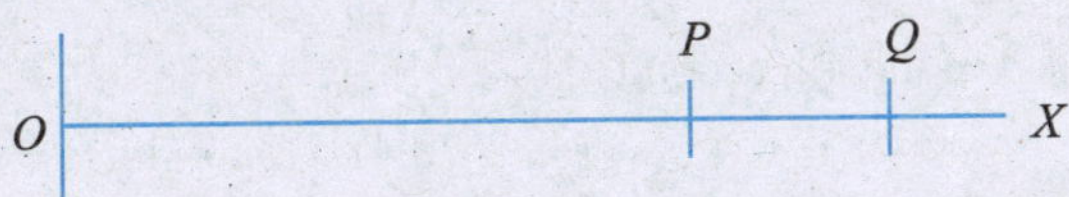

Fig. 4. Distance travelled in *n*th second.

Consider the motion of a particle, starting from O and moving along OX as shown in Fig. 4.

Let

u = Initial velocity of the particle,

v = Final velocity of the particle

a = Constant positive acceleration,

s_n = Distance (OQ) travelled in n sec,

s_{n-1} = Distance (OP) travelled in $(n-1)$ sec,

$s = (s_n - s_{n-1})$ = Distane (PQ) travelled in nth sec,

n = No. of second.

Substituting the values of $t = n$ and $t = (n-1)$ in the general equation of motion,

$$s_n = un + \frac{1}{2}a(n)^2 \qquad ...(i)$$

and

$$s_{n-1} = u(n-1) + \frac{1}{2}a(n-1)^2 \qquad ...(ii)$$

∴ Distance travelled in the nth sec,

$$s = s_n - s_{n-1}$$

$$= \left[un + \frac{1}{2}a(n)^2\right] - \left[u(n-1) + \frac{1}{2}a(n-1)^2\right]$$

$$= un + \frac{1}{2}an^2 - un + u - \frac{1}{2}a(n^2 + 1 - 2n)$$

$$= \frac{1}{2}an^2 + u - \frac{1}{2}an^2 - \frac{1}{2}a + an$$

$$= u - \frac{1}{2}a + an = u + a\left(n - \frac{1}{2}\right)$$

or

$$s = u + \frac{a}{2}(2n - 1)$$

Example 26. *A body was thrown vertically downwards from the top of a tower and traverses a distance of 40 metres during its 4th second of its fall. Find the initial velocity of the body.*

Solution. Given : Distance traversed (s) = 40 m ; No of second (n) = 4 and acceleration (a) = g = 9.8 m/s^2

Let u = Initial velocity of the body.

We know that, Distance traversed by the body in the 4th second (s_4),

$$40 = u + \frac{a}{2}(2n - 1) = u + \frac{9.8}{2}(2 \times 4 - 1) = u + 34.3$$

or $u = 40 - 34.3 = 5.7$ m/s **Ans.**

Alternative Method

Distance travelled in 3 seconds

$$s_3 = ut + \frac{1}{2}gt^2 = u \times 3 + \frac{1}{2} \times 9.8(3)^2 = 3u + 44.1 \text{ m}$$

Distance travelled in 4 seconds,

$$s_4 = ut + \frac{1}{2}gt^2 = u \times 4 + \frac{1}{2} \times 9.8(4)^2 = 4u + 78.4 \text{ m}$$

∴ Distance traversed in the 4th second

$$40 = s_4 - s_3 = (4u + 78.4) - (3u + 44.1) = u + 34.3$$

or $u = 40 - 34.3 = 5.7$ m/s **Ans.**

Example 27. *A particle starts from rest. Find the ratio of distances covered by it in the 3rd and 5th seconds of its motion.*

Solution. Given : Initial velocity (u) = 0 (because it starts from rest) ; Initial no. of second (n_1) = 3 and final no. of second (n_2) = 5.

Distance covered by the particle in the 3rd second after it starts,

$$s_3 = u + \frac{a}{2}(2n_1 - 1) = 0 + \frac{a}{2}[(2 \times 3) - 1] = \frac{5a}{2} \quad ...(i)$$

Distance covered by the particle in the 5th second after it starts,

$$s_5 = u + \frac{a}{2}(2n_2 - 1) = 0 + \frac{a}{2}[(2 \times 5) - 1] = \frac{9a}{2} \quad ...(ii)$$

∴ Ratio of distances $= s_3 : s_5 = \frac{5a}{2} : \frac{9a}{2} = 5 : 9.$ **Ans.**

***Example 28.** *A body, falling freely from rest travels in the first three seconds, a distance equal to the distance travelled by it in a certain second. Find the time of its travel for the body.*

Solution. Given : Initial velocity (u) = 0 (because it falls freely) and distance travelled in first three seconds = Distance travelled in nth second.

Let n = Time of travel for the body.

Distance travelled in the first 3 s,

$$s = ut + \frac{1}{2}gt^2 = (0 \times 3) + \frac{1}{2} \times g \times (3)^2 = 4{\cdot}5\ g \quad ...(i)$$

Distance travelled in the nth second after it starts,

$$s_n = u + \frac{g}{2}(2n - 1) = 0 + \frac{g}{2}(2n - 1) = \frac{g}{2}(2n - 1). \quad ...(ii)$$

Since, **Both the distances are equal**, therefore equating equations (*i*) and (*ii*)

$$4{\cdot}5\ g = \frac{g}{2}(2n-1) \qquad \text{or} \qquad 2n-1 = \frac{4{\cdot}5\ g}{g/2} = 9$$

or $$2n = 9 + 1 = 10 \qquad \text{or} \qquad n = 5\ \text{s} \quad \textbf{Ans.}$$

***Example 29.** *A train starts from rest with an acceleration a and describes distances s_1, s_2 and s_3 in the first, second and third seconds of its journey. Find the ratio of $s_1 : s_2 : s_3$.*

Solution. Given : Initial velocity of train (u) = 0 (because it starts from rest) ; Acceleration = a ; Distance described in 1st second = s_1; Distance described in 2nd second = s_2 and distance described in 3rd second = s_3.

Distance described by the train in first second,

$$s_1 = u + \frac{a}{2}(2n_1 - 1) = 0 + \frac{a}{2}[(2\times 1)-1] = \frac{a}{2} \qquad \text{...}(i)$$

Distance described in second second,

$$s_2 = u + \frac{a}{2}(2n_2 - 1) = 0 + \frac{a}{2}[(2\times 2)-1] = \frac{3a}{2} \qquad \text{...}(ii)$$

and Distance described in third second,

$$s_3 = u + \frac{a}{2}(2n_3 - 1) = 0 + \frac{a}{2}[(2\times 3)-1] = \frac{5a}{2} \qquad \text{...}(iii)$$

∴ Ratio of distances $s_1 : s_2 : s_3$

$$= \frac{a}{2} : \frac{3a}{2} : \frac{5a}{2} = 1 : 3 : 5 \quad \textbf{Ans.}$$

***Example 30.** *By what initial velocity a ball should be projected vertically upwards, so that the distance covered by it in 5th second is twice the distance it covered in its 6th second ? (Take g = 10 m/s²)*

Solution. Given : Initial no. of second (n_1) = 5 ; Final no. of second (n_2) = 6 and acceleration due to gravity (g) = 10 m/s^2.

Let u = Initial velocity of the ball.

Distance covered by the ball in the 5th second after it starts

$$s_5 = -u + \frac{g}{2}(2n_1 - 1) = -u + \frac{10}{2}[(2\times 5)-1]$$

$$= -u + 45 \qquad \text{...}(i)$$

...(Minus sign due to upward direction)

Distance covered by it in the 6th second after it starts

$$s_6 = -u + \frac{g}{2}(2n_2 - 1) = -u + \frac{10}{2}[(2\times 6)-1]$$

$$= -u + 55 \qquad \text{...}(ii)$$

...(Minus sign due to upward direction)

Since, **Distance covered by the ball in the 5th second is twice the distance covered by it in the 6th second**, therefore

$$-u + 45 = 2(-u + 55) = -2u + 110$$

or $$u = 110 - 45 = 65 \text{ m/s} \quad \textbf{Ans.}$$

6. *GRAPHICAL REPRESENTATION OF VELOCITY, TIME AND DISTANCE TRAVELLED BY A BODY

[AM-IETE]

The motion of body may also be represented by means of a graph. Such a graph may be drawn by plotting velocity as ordinate and the corresponding time as abscissa as shown in Fig. 5. (*a*) and (*b*). Here we shall discuss the following two cases :

1. ***When the body is moving with a uniform velocity***

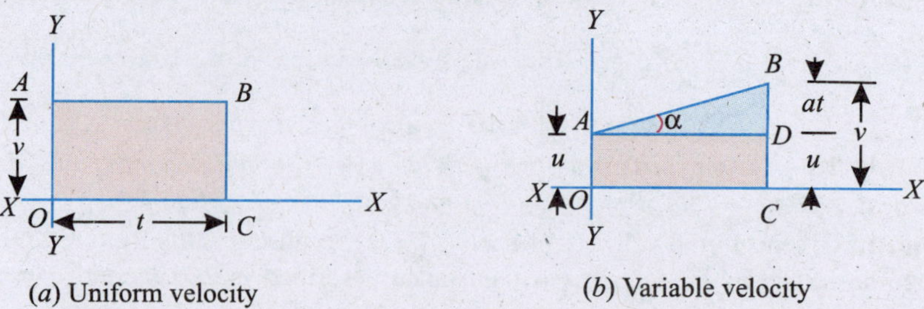

Fig. 5.

Consider the motion of a body, which is represented by the graph *OABC* as shown in Fig. 5(*a*). We know that the distance traversed by the body,

$$s = \text{Velocity} \times \text{Time}$$

Thus we see that the area of the figure *OABC* (*i.e.*, velocity × time) represents the distance traversed by the body, to some scale.

2. ***When the body is moving with a variable velocity***

We know that the distance traversed by a body,

$$s = ut + \frac{1}{2}at^2$$

From the geometry of the Fig. 5 (*b*), we know that area of the figure *OABC*

$$= \text{Area } (OADC + ABD)$$

But area of figure *OADC*, $= u \times t$

and area of figure *ABD*, $= \frac{1}{2} \times t \times at = \frac{1}{2}at^2$

$\therefore$ Total area of *OABC*, $= ut + \frac{1}{2}at^2$

Thus, we see that the area of the *OABC* represents the distance traversed by the body to some scale. From the figure it is also seen

$$\tan \alpha = \frac{at}{t} = a$$

Thus, $\tan \alpha$ represents the acceleration of the body.

***Example 31.** *A lift goes up to a height of 900 m with a constant acceleration and then the next 300 m with a constant retardation and comes to rest. Find* (*i*) *maximum velocity of the lift, if the time taken to travel is 30 seconds* ; (*ii*) *acceleration of the lift* ; *and* (*iii*) *retardation of the lift. Take acceleration of the lift as 1/3 of its retardation.*

Solution. Let *OAB* be the velocity-time graph, in which the ordinate *AL* represents the maximum velocity as shown in Fig. 6.

(*i*) ***Maximum velocity of the lift***

Let t_1 = Time of acceleration

t_2 = Time of retardation, and

v = Maximum velocity of lift.

Fig. 6.

First of all, consider motion of the lift from *O* to *A*. We know that the area of triangle *OAL* (s_1),

$$900 = \frac{1}{2} \times t_1 \times v \qquad ...(i)$$

Now consider motion of the lift from A to B. We know that area of the triangle ALB (s_2)

$$300 = \frac{1}{2} \times t_2 \times v \qquad ...(ii)$$

Dividing equation (i) by (ii),

$$\frac{900}{300} = \frac{t_1}{t_2} \qquad \text{or} \qquad t_1 = 3t_2$$

But $\quad t_1 + t_2 = 30 \quad$ or $\quad 3t_2 + t_2 = 30$

or $\quad t_2 = 7.5$ s $\quad$ and $\quad t_2 = 7.5$ s

Now substituting the value of t_1 in equation (i),

$$900 = \frac{1}{2} \times 22.5 \times v = 11.25\, v$$

$$\therefore \qquad v = \frac{900}{11.25} = 80 \text{ m/s} \quad \textbf{Ans.}$$

(*ii*) ***Acceleration of the lift***

From geometry of the figure, we find that acceleration, of the lift,

$$a_1 = \tan \alpha = \frac{AL}{OL} = \frac{80}{22.5} = 3.55 \text{ m/s}^2 \quad \textbf{Ans.}$$

(*iii*) ***Retardation of the lift***

We also know that retardation of the lift,

$$a_2 = 3a_1 = 3 \times 3.55 = 10.65 \text{ m/s}^2 \quad \textbf{Ans.}$$

***Example 32.** *A train moving with a velocity of 30 km.p.h. has to slow down to 15 km.p.h. due to repairs along the road. If the distance covered during retardation be one kilometer and that covered during acceleration be half a kilometer, find the time lost in the journey.*

Solution. Let $OABCD$ be the velocity-time graph, in which AB represents the period of retardation and BC period of acceleration as shown in Fig. 7.

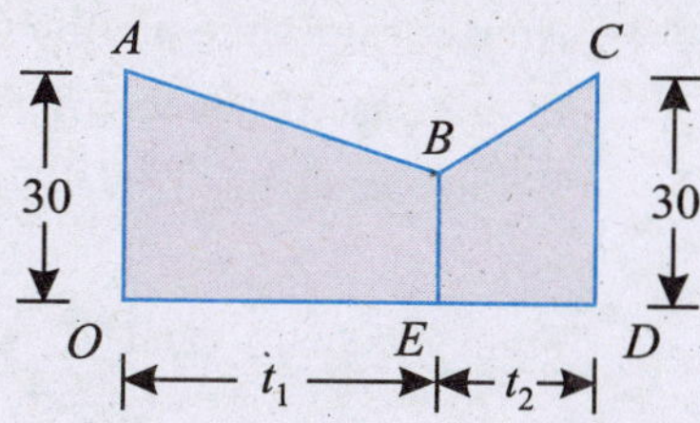

Fig. 7.

First of all, consider motion of the train from A to B. In this case, distance travelled (s_1) = 1 km; initial velocity (u_1) = 30 km. p.h. and final velocity (v_1) = 15 km.p.h.

Let $\quad t_1$ = Time taken by the train to move from A to B.

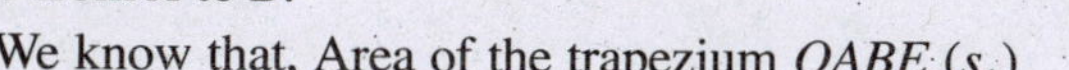

We know that, Area of the trapezium $OABE$ (s_1)

$$1 = \frac{30 + 15}{2} \times t_1 = 22.5\, t_1$$

$$\text{or} \qquad t_1 = \frac{1}{22.5} \text{ hr} = 2.67 \text{ min} \qquad ...(i)$$

Now consider motion of the train from B to C. In this case, distance travelled (s_2) = 0.5 km ; Initialvelocity (u_2) = 15 km.p.h. and final velocity (v_2) = 30 km.p.h.

Let $\quad t_2$ = Time taken by the train to move from B to C.

We also know that, Area of trapezium $BCDE$ (s_2),

$$\frac{1}{2} = \frac{15 + 30}{2} \times t = 22.5\, t_2$$

$$\text{or} \qquad t_2 = \frac{1}{45} \text{ hr} = 1.33 \text{ min} \qquad ...(ii)$$

$$\therefore \qquad \text{Total time, } t = t_1 + t_2 = 2.67 + 1.33 = 4 \text{ min}$$

If the train had moved uniformly with a velocity of 30 km/hr, then the time required to cover 1.5 km,

$$= \frac{60}{30} \times \frac{3}{2} = 3 \text{ min} \quad ...(iii)$$

$\therefore$ Time lost = 4 – 3 = 1 min **Ans.**

****Example 33.** *A cage goes down a main shaft 750 m deep, in 45 s. For the first quarter of the distance only, the speed is being uniformly accelerated and during the last quarter uniformly retarded, the acceleration and retardation being equal. Find the uniform speed of the cage, while traversing the central portion of the shaft.* **[MU]**

Solution. Let *OABC* represent the velocity-time graph in which *OA* represents the period of acceleration, *AB* the period of uniform velocity and *BC* the period of retardation as shown in Fig. 8.

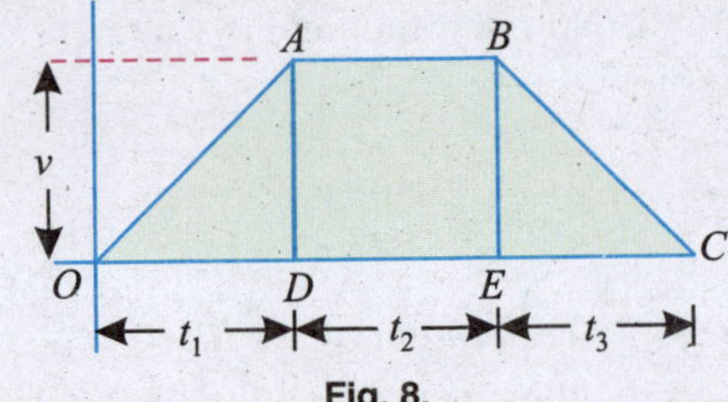

Fig. 8.

First of all consider motion of cage from *O* to *A*. In this case, initial velocity (u_1) = 0 (because it goes down from rest) and distance travelled (s_1) $= \frac{750}{4} = 187.5$ m

Let a_1 = Constant acceleration of the cage, and

v_1 = Uniform velocity of the cage (*AD* or *BE*).

We know that, Area of triangle *OAD* (s_1),

$$187.5 = \frac{1}{2} \times t_1 \times v_1$$

or $$t_1 \times v_1 = 187.5 \times 2 = 375 \quad ...(i)$$

Now consider the motion of the cage from *A* to *B*.

In this case, Distance travelled (s_2) = 750 – (2 × 187.5) = 375 m

We also know that, Area of the rectangle *ABED* (s_2),

$$375 = t_2 \times v_1 \quad ...(ii)$$

From equation (*i*) and (*ii*), we find that

$$t_1 = t_2$$

Similarly $$t_2 = t_3$$

$\therefore$ $$t_1 = t_2 = t_3$$

Since the total time taken ($t_1 + t_2 + t_3$) is 45 seconds, therefore

$$t_1 = t_2 = t_3 = 15 \text{ s}$$

Again consider the motion of the cage from *O to A*. In this case, Initial velocity (u_1) = 0 ; distance travelled (s_1) = 187.5 m and time *OABC*(t_1) = 15 s

We know that, Distance (s_1),

$$187.5 = u_1 t_1 + \frac{1}{2} a_1 t_1^2 = 0 + \frac{1}{2} a_1 (15)^2 = 112.5 a_1$$

or $$a_1 = \frac{187.5}{112.5} = 1.67 \text{ m/s}^2$$

and Speed of the cage while traversing the central portion of the shaft,

$$v_1 = u + a_1 t_1 = 0 + 1.67 \times 15 = 25 \text{ m/s} \quad \textbf{Ans.}$$

EXERCISE 3

1. A car starts from rest with an acceleration of 4 m/s^2. What is the distance travelled in 8th second ? [**Ans.** 30 m]
2. A train travels between two stations 10 kilometers apart in 15 minutes. Assuming that its motion is with uniform acceleration for the first part of the journey and with uniform retardation for the rest, show that the greatest velocity of the train is 80 kilometers per hour.
3. An electric train starts from a station and comes to another station which is 5 kilometers from the first on a straight track. For three-quarters of the distance, the train is uniformly accelerated and for the remainder uniformly retarded. If the train takes 12 minutes to cover the whole journey, find (*i*) maximum speed the train attains ; (*ii*) acceleration of the train ; and (*iii*) retardation of the train. [**Ans.** 50 km.p.h. ; 333.3 km/hr^2 ; 1000 km/hr^2]
4. An electric train, travelling between two stations 1.5 km apart is uniformly accelerated for the first 10 seconds, during which period it covers 100 m. It then runs with a constant speed, until it is finally retarded uniformly in the last 50 m. Find the maximum speed of the train and the time taken to complete the journey between the two stations.

   [**Ans.** 20 m/s ; 82.5 s]
5. A passenger train passes a certain station at 60 kilometers per hour, and covers a distance of 12 kilometers with this speed and then stops at next station 15 kilometers from the first, with uniform retardation. A local train, starting from the first station covers the same distance in double this time and stops at the next station.

   Determine the maximum speed of the local train which covers a part of the distance, with uniform acceleration and the rest with uniform retardation. [**Ans.** 50 km.p.h.]

QUESTIONS

1. How would you find out, if a particular body is at rest or in motion ?
2. Distinguish clearly between speed and velocity. Give examples.
3. What do you understand by the term 'acceleration ? Define positive acceleration and negative acceleration.
4. What is the difference between uniform acceleration and variable acceleration ?
5. How does the velocity of a moving point, that possesses a unifomly variable motion, change if acceleration is positive ? How does it change, if acceleration is negative ?
6. Prove the relationship,

$$s = ut + \frac{1}{2}at^2$$

   in a body, subjected to a uniformly accelerated motion.
7. What is the force of gravity ? How does it effect the body, when,

   (1) it is allowed to fall downwards.

   (2) it is projected upwards ?
8. Derive a relation for the distance travelled by a body in the *n*th second.
9. How would you study the motion of a body graphically ? Discuss the uses of such diagrams.

OBJECTIVE TYPE QUESTIONS

1. The relationship $s = ut + \frac{1}{2}at^2$ is applicable to bodies
   (*a*) moving with any type of motion
   (*b*) moving with uniform velocity
   (*c*) moving with uniform acceleration
   (*d*) both (*b*) and (*c*)
2. The motion under gravity is a particular case of motion under constant velocity.
   (*a*) Yes (*b*) No
3. If two bodies A and B are projected upwards such that the velocity of A is double the velocity of B, then the height to which the body A will rise will be .... the height to which the body B will rise
   (*a*) two times (*b*) four times (*c*) eight times
4. A lecturer told the class that if a body is projected upwards from any point, then the body while coming down at the same point will have the same velocity with which it was projected upwards. Is his statement correct?
   (*a*) Agree (*b*) Disagree

ANSWERS

1. (*c*) **2.** (*b*) **3.** (*b*) **4.** (*a*)

CHAPTER

16 Motion Under Variable Acceleration

Contents

1. MOTION UNDER VARIABLE ACCELERATION

In the last chapter, we have discussed the motion under constant acceleration, *i.e.*, the rate of change of velocity was constant. But in actual practice, it is seldom possible, that a body may move with a uniform velocity or uniform acceleration, at all times.

A body, which does not move with a uniform acceleration, is said to be moving with a non-uniform or variable acceleration.

In this chapter, we shall discuss the motion under variable acceleration.

2. *VELOCITY AND ACCELERATION AT ANY INSTANT

Fig 1. Motion under variable acceleration.

Consider a body moving from O in the direction OX. Let P be its position at any instant as shown in Fig 1.

Let s = Distance travelled by the body,

t = Time taken by the body, in seconds, to travel this distance

v = Velocity of the body at P,

a = Acceleration of the body.

We know that, Velocity of a body, is the rate of change of its position. Mathematically :

$$v = \frac{ds}{dt} \qquad ...(i)$$

Similarly, Acceleration of a body is the rate of change of its velocity. Mathematically :

$$a = \frac{d^2s}{dt^2} = \frac{dv}{dt} = v.\frac{dv}{ds} \qquad ...(ii)$$

3. METHODS TO FIND VELOCITY, ACCELERATION AND DISPLACEMENT FROM A MATHEMATICAL EQUATION

The velocity, acceleration and displacement of a body may be found out from a mathematical expression, by either of the following two methods :

1. By Differentiation **2. By Integration.**

4. **VELOCITY AND ACCELERATION BY DIFFERENTIATION

Sometimes, the given equation of motion is in terms of displacement (s) and time (t) *e.g.*,

$$s = 3t^3 + 2t^2 + 6t + 4 \qquad ...(i)$$

or

$$s = 6 + 5t^2 + 6t^3 \qquad ...(ii)$$

or

$$s = 2t^3 + 4t - 15 \qquad ...(iii)$$

Now differentiating both sides, of the equations, with respect to t,

$$\frac{ds}{dt} = 9t^2 + 4t + 6 \qquad ...(i)$$

$$\frac{ds}{dt} = 10t + 18t^2 \qquad ...(ii)$$

$$\frac{ds}{dt} = 6t^2 + 4 \qquad ...(iii)$$

The equations, so obtained by differentiation, give velocity of the body (as the velocity of a body is the rate of change of its position). Again differentiating, both sides of the above equations, with respect to t,

$$\frac{d^2s}{dt^2} = 18t + 4 \qquad ...(i)$$

$$\frac{d^2s}{dt^2} = 10 + 36t \qquad ...(ii)$$

$$\frac{d^2s}{dt^2} = 12t \qquad ...(iii)$$

The equations, so obtained by second differentiation, give acceleration of the body (as the acceleration of a body is the rate of change of velocity).

Example 1. *A particle, starting from rest, moves in a straight line, whose equation of motion is given by :* $s = t^3 - 2t^2 + 3$. *Find the velocity and acceleration of the particle after 5 seconds.*

Solution. Given : Equation of displacement : $s = t^3 - 2t^2 + 3$ ...(i)

Velocity after 5 seconds

Differentiating the above equation with respect to t,

$$\frac{ds}{dt} = 3t^2 - 4t \qquad ...(ii)$$

i.e., velocity, $v = 3t^2 - 4t$ ...$\left(\because \frac{ds}{dt} = \text{Velocity}\right)$

substituting t equal to 5 in the above equation,

$$v = 3\,(5)^2 - (4 \times 5) = 55 \text{ m/s} \qquad \textbf{Ans.}$$

Acceleration after 5 seconds

Again differentiating equation (ii) with respect to t,

$$\frac{d^2s}{dt^2} = 6t - 4 \qquad ...(iii)$$

i.e. acceleration, $a = 6t - 4$ ...$\left(\because \frac{d^2s}{dt^2} = \text{Acceleration}\right)$

Now substituting t equal to 5 in the above equation,

$$a = (6 \times 5) - 4 = 26 \text{ m/s}^2 \qquad \textbf{Ans.}$$

***Example 2.** *A car moves along a straight line whose equation of motion is given by* $s = 12t + 3t^2 - 2t^3$, *where (s) is in metres and (t) is in seconds. calculate*

(i) *velocity and acceleration at start, and*

(ii) *acceleration, when the velocity is zero.* **[RGPV, PU]**

Solution. Given : Equation of displacement : $s = 12t + 3t^2 - 2t^3$ ...(*i*)

Velocity at start

Differentiating the above equation with respect to t,

$$\frac{ds}{dt} = 12 + 6t - 6t^2 \quad ...(ii)$$

i.e. velocity, $v = 12 + 6t - 6t^2$ $\quad ...\left(\because \frac{ds}{dt} = v\right)$

Substituting t equal to 0 in the above equation,

$$v = 12 + 0 - 0 = 12 \text{ m/s}$$ **Ans.**

Acceleration at start

Again differentiating equation (*ii*) with respect to t ,

$$\frac{dv}{dt} = 6 - 12t \quad ...(iii)$$

i.e. accleration, $a = 6 - 12t$ $\quad ...\left(\because \frac{dv}{dt} = a\right)$

Now substituting t equal to 0 in the above equation,

$$a = 6 - 0 = 6 \text{ m/s}^2$$ **Ans.**

Acceleration, when the velocity is zero

Substituting equation (*ii*) equal to zero

or $12 + 6t - 6t^2 = 0$

or $t^2 - t - 2 = 0$ ...(Dividing by – 6)

or $t = 2$ s

It means that velocity of the car after two seconds will be zero. Now substituting the value of t equal to 2 in equation (*iii*),

$$a = 6 - (12 \times 2) = -18 \text{ m/s}^2$$ **Ans.**

***Example 3.** *The equation of motion of a particle moving in a straight line is given by*

$$s = 18t + 3t^2 - 2t^3$$

where (s) *is in metres and* (t) *in seconds. Find* (*1*) *velocity and acceleration at start,* (2) *time, when the particle reaches its maximum velocity, and* (3) *maximum velocity of the particle.*

[BPUT, WBUT, IGNOU]

Solution. Given : Equation of displacement : $s = 18t + 3t^2 - 2t^3$ ...(*i*)

(1) ***Velocity and acceleration at start***

Differentiating equation (*i*) with respect to t ,

$$\frac{ds}{dt} = 18 + 6t - 6t^2 \quad ...(ii)$$

i.e. velocity, $v = 18 + 6t - 6t^2$

Substituting, t equal to 0 in equation (*ii*),

$$v = 18 + 0 + 0 = 18 \text{ m/s}$$ **Ans.**

Again differentiating equation (*ii*) with respect to t,

$$\frac{d^2s}{dt^2} = 6 - 12t \quad ...(iii)$$

i.e. acceleration , $a = 6 - 12t$ ...(*iv*)

Substituting t equal to 0 in equation (*iv*),

$$a = 6 - 0 = 6 \text{ m/s}^2 \quad \textbf{Ans.}$$

(2) *Time, when the particle reaches its maximum velocity*

For maximum velocity, let us differentiate the equation of velocity and equate it to zero. The differentiation of the equation of velocity is given by equation (*iii*).

Therefore equating the equation (*iii*) to zero,

$$6 - 12t = 0$$

or $$t = 1/2 = 0{\cdot}5 \text{ s} \quad \textbf{Ans.}$$

(3) *Maximum velocity of the particle*

Substituting t equal to 0·5 s in equation (*ii*),

$$v = 18 + (6 \times 0{\cdot}5) - 6\,(0{\cdot}5)^2 = 19{\cdot}5 \text{ m/s} \quad \textbf{Ans.}$$

EXERCISE 1

*1. A particle, starting from rest, moves in a straight line whose equation of motion is given by :

$$s = 3t^3 - 2t$$

where (s) is in metres and (t) in seconds. Find (*i*) velocity after 3 seconds ; (*ii*) acceleration at the end of 3 seconds ; and (*iii*) average velocity in the 4th seconds.

[AUT, AM-IETE] (**Ans.** 83 m/s ; 54 m/s^2 ; 114.5 m/s)

*2. A car moves along a straight line, whose equation of motion is given by $s = 12t + 3t^2 - 2t^3$, where (s) is in metres and (t) in seconds. Calculate (*i*) velocity and acceleration at start ; (*ii*) acceleration when velocity is zero. (**Ans.** 12 m/s, 6 m/s^2 ; – 18 m/s^2)

*3. The equation of motion of an engine is given by $s = 2t^3 - 6t^2 - 5$, where (s) is in metres and (t) in seconds. Calculate (*i*) displacement and acceleration when velocity is zero ; and (*ii*) displacement and veiocity when acceleration is zero.

[WBUT] (**Ans.** – 13 m ; 12 m/s^2 ; – 9m ; – 6 m/s)

5. *VELOCITY AND DISPLACEMENT BY INTEGRATION

Sometimes, the given equation of motion is in terms of acceleration (a) and time (t) *e.g.*

$$a = 4t^3 - 3t^2 + 5t + 6 \quad \text{...}(i)$$

$$^{*}\frac{dv}{dt} = t^3 + 8t \qquad ...(ii)$$

$$^{*}v\frac{dv}{ds} = 6 + 3t \qquad ...(iii)$$

$$^{*}\frac{d^2s}{dt^2} = 4t - 8t^2 \qquad ...(iv)$$

Now integrating both sides of the above equations,

$$= \frac{t^4}{4} + \frac{8t^2}{2} + C_1 \qquad ...(i)$$

$$= 6t + \frac{3t^2}{2} + C_1 \qquad ...(ii)$$

$$= \frac{4t^2}{2} - \frac{8t^3}{3} + C_1 \qquad ...(iii)$$

where C_1 is the first constant of integration. The equations, so obtained, give the velocity of the body. Again integrating both sides of the above equations.

$$= \frac{t^5}{20} + \frac{8t^3}{6} + C_1t + C_2$$

$$= \frac{6t^2}{2} + \frac{3t^3}{6}\ C_1t + C_2$$

$$= \frac{4t^3}{6} - \frac{8t^4}{12} + C_1t + C_2$$

where C_2 is the second constant of integration. The equations, so obtained, give the displacement of the body.

It may be noticed that the method for velocity and displacement by integration is somewhat difficult, as we have to find out the values of constants of integration (*i.e.* C_1 and C_2)

---

*Note: These are the different forms of acceleration. A little consideration is very essential to use the proper form. As a thumb rule, if the variable acceleration is a function of t, then equation (*i*) or (*ii*) is used. But if it is a function of s, then equation (*iii*) or (*iv*) is used.

***Example 4.** *The motion of a particle is given by :*

$$a = t^3 - 3t^2 + 5$$

where (a) is the acceleration in m/s² and (t) is the time in seconds. The velocity of the particle at t = 1 second is 6.25 m/sec and the displacement is 8.8 metres.

Calculate the displacement and velocity at t = 2 seconds.

Solution. Given : Equation of acceleration : $a = t^3 - 3t^2 + 5$

Rewriting the given equation,

or $$\frac{dv}{dt} = t^3 - 3t^2 + 5 \qquad ...\left(\because\ a = \frac{dv}{dt}\right)$$

or $$dv = (t^3 - 3t^2 + 5)\ dt \qquad ...(i)$$

Velocity at t = 2 seconds

Integrating both sides of equation (*i*),

$$v = \frac{t^4}{4} - \frac{3t^3}{3} + 5t + C_1$$

$$= \frac{t^4}{4} - t^3 + 5t + C_1 \qquad ...(ii)$$

where C_1 is the first constant of integration. Substituting the values of $t = 1$ and $v = 6.25$ in equation (*ii*),

$$6.25 = \frac{1}{4} - 1 + 5 + C_1 = 4.25 + C_1$$

or $$C_1 = 6.25 - 4.25 = 2$$

Substituting this value of C_1 in equation (*ii*),

$$v = \frac{t^4}{4} - t^3 + 5t + 2 \quad \text{...(iii)}$$

Now for velocity of the particle, substituting the value of $t = 2$ in the above equation ,

$$v = \frac{(2)^4}{4} - (2)^3 + (5 \times 2) + 2 = 8 \text{ m/s} \quad \textbf{Ans.}$$

Displacement at t = 2 seconds

Rewriting equation (*iii*),

$$\frac{ds}{dt} = \frac{t^4}{4} - t^3 + 5t + 2 \quad \text{...}\left(\because v = \frac{ds}{dt}\right)$$

or $$ds = \left(\frac{t^4}{4} - t^3 + 5t + 2\right) dt \quad \text{...(iv)}$$

Integrating both sides of equation, (*iv*)

$$s = \frac{t^5}{20} - \frac{t^4}{4} + \frac{5t^2}{2} + 2t + C_2 \quad \text{...(v)}$$

where C_2 is the second constant of integration. Substituting the values of $t = 1$ and $s = 8.8$ in equation (*v*),

$$8.8 = \frac{1}{20} - \frac{1}{4} + \frac{5}{2} + 2 + C_2 = 4.3 + C_2$$

or $$C_2 = 8.8 - 4.3 = 4.5$$

Substituting this value of C_2 in equation (*v*) ,

$$s = \frac{t^5}{20} - \frac{t^4}{4} + \frac{5t^2}{2} + 2t + 4.5$$

Now for displacement of the particle, substituting the value of $t = 2$ in the above equation,

$$s = \frac{32}{20} - \frac{16}{4} + \frac{20}{2} + 4 + 4.5 = 16.1 \text{ m} \quad \textbf{Ans.}$$

***Example 5.** *A train, starting from rest, is uniformly accelerated. The acceleration at any instant is* $\frac{10}{v+1}$ *m/s*^{2} *, where (v) is the velocity of the train in m/s at the instant. Find the distance, in which the train will attain a velocity of 35 km.p.h.* **[AMIE]**

Solution. Given : Equation of acceleration : $a = \frac{10}{v+1}$

Rewriting the given equation,

$$v \cdot \frac{dv}{ds} = \frac{10}{v+1} \quad \text{...}\left(\because a = v \cdot \frac{dv}{ds}\right)$$

or $$v\,(v + 1)\, dv = 10\, ds$$

or $$(v^2 + v)\, dv = 10\, ds \quad \text{...(i)}$$

Integrating both sides of equation (*i*),

$$\frac{v^3}{3} + \frac{v^2}{2} = 10s + C_1 \qquad ...(ii)$$

where C_1 is the first constant of integration. Substituting the values of $s = 0$ and $v = 0$ in equation (*ii*),

$$C_1 = 0$$

Substituting this value of $C_1 = 0$ in equation (*ii*),

$$\frac{v^3}{3} + \frac{v^2}{2} = 10s$$

or $$2v^3 + 3v^2 = 60\,s \qquad ...(iii)$$

Now for distance travelled by the train, substituting $v = 36$ km.p.h. = 10 m/s in equation (*iii*),

$$2\,(10)^3 + 3\,(10)^2 = 60\,s \qquad \text{or} \qquad 2000 + 300 = 60\,s$$

or $$s = \frac{2300}{60} = 38.3 \text{ m} \quad \textbf{Ans.}$$

***Example 6.** *A particle, starting from rest, moves in a straight line, whose acceleration is given by the equation :*

$$a = 10 - 0.006\,s^2$$

where (a) is in m/s² and (s) in metres. Determine

(*i*) *velocity of the particle, when it has travelled 50 metres.*

(*ii*) *distance travelled by the particle, when it comes to rest.*

Solution. Given : Equation of acceleration : $a = 10 - 0.006\,s^2$

Rewriting the given equation,

$$v \,.\, \frac{dv}{ds} = 10 - 0.006\,s^2 \qquad ...\left(\because a = v \,.\, \frac{dv}{ds}\right)$$

$$\therefore \qquad v.dv = (10 - 0.006\,s^2)\,ds \qquad ...(i)$$

(a) Velocity of the particle, when it has travelled 50 metres

Integrating both sides, of equation (*i*),

$$\frac{v^2}{2} = 10s - \frac{0.006\,s^3}{3} + C_1 = 10s - 0.002s^3 + C_1$$

or $$v^2 = 20s - 0.004\,s^3 + 2C_1 \qquad ...(ii)$$

where C_1 is the first constant of integration. Substituting the values of $s = 0$ and $v = 0$ in equation (*ii*),

$$C_1 = 0$$

Substituting this value of C_1 in equation (*ii*),

$$v^2 = 20s - 0.004\,s^3 \qquad ...(iii)$$

Now for velocity of the particle, substituting $s = 50$ m in equation (*iii*),

$$v^2 = 20\,(50) - 0.004\,(50)^3 = 1000 - 500 = 500$$

or $$v = \sqrt{500} = 22.36 \text{ m/s} \quad \textbf{Ans.}$$

(b) Distance travelled by the particle, when it comes to rest

When the particle comes to rest, the velocity will be zero. Therefore substituting $v = 0$ in equation (*iii*),

$$20s - 0.004s^3 = 0 \qquad \text{or} \qquad s\,(20 - 0.004s^2) = 0$$

Therefore either $s = 0$ or $(20 - 0.004s^2) = 0$. A little consideration will show that when $s = 0$ the body is in its initial stage.

$$\therefore \quad 20 - 0.004s^2 = 0$$

or $$s = \sqrt{\frac{20}{0.004}} = \sqrt{5000} = 70.7 \text{ m} \quad \textbf{Ans.}$$

***Example 7.** *A body moves along a straight line and its acceleration (a) which varies with time (t) is given by a = 2 – 3t. After 5 seconds , from start of observations, its velocity is observed to be 20 m/s. After 10 seconds, from start of observation, the body was at 85 metres from the origin. Determine*

(*a*) *its acceleration and velocity at the time of start*

(*b*) *distance from the origin at the start of observations,*

(*c*) *the time after start of observation in which the velocity becomes zero.*

Solution. Given : Equation of acceleration : $a = 2 - 3t$ ...(*i*)

(a) Accelertion and velocity at the time of start

Substituting the value of t equal to 0 in the given equation (*i*),

$$a = 2 \text{ m/s}^2 \quad \textbf{Ans.}$$

Rewriting the given equation (*i*),

$$\frac{dv}{dt} = 2 - 3t \qquad ...\left(\because a = \frac{dv}{dt}\right)$$

$$\therefore \quad dv = (2 - 3t)\, dt \qquad ...(ii)$$

Integrating both sides of equation (*ii*) ,

$$v = 2t - \frac{3t^2}{2} + C_1 \qquad ...(iii)$$

where C_1 is the first constant of integration. Substituting the values of $t = 5$ and $v = 20$ in equation (*iii*),

$$20 = 2 \times 5 - \frac{3}{2}(5)^2 + C_1 = C_1 - 27.5$$

or $$C_1 = 20 + 27.5 = 47.5$$

Substituting this value of C_1 in equation (*iii*) ,

$$v = 2t - \frac{3t^2}{2} + 47.5 \qquad ...(iv)$$

Now for velocity of the body at the time of start, substituting $t = 0$ in equation (*iv*),

$$v = 47.5 \text{ m/s} \qquad \textbf{Ans.}$$

(b) Distance from the origin at the start of observation

Rewriting equation (*iv*),

$$\frac{ds}{dt} = 2t - \frac{3t^2}{2} + 47.5 \qquad ...\left(\because v = \frac{ds}{dt}\right)$$

or $$ds = \left(2t - \frac{3t^2}{2} + 47.5\right) dt$$

Integrating both sides of the above equation,

$$s = \frac{2t^2}{2} - \frac{3t^3}{6} + 47.5t + C_2 = t^2 - \frac{t^3}{2} + 47.5t + C_2 \quad ...(v)$$

where C_2 is the second constant of integration. Now substituting the values of $t = 10$ and $s = 85$ in above equation,

$$85 = (10)^2 - \frac{(10)^3}{2} + 47.5 \times 10 + C_2 = 75 + C_2$$

or $$C_2 = 85 - 75 = 10$$

Substituting this value of C_2 in equation (v),

$$s = t^2 - \frac{t^3}{2} + 4.75t + 10$$

Now for the distance from the origin at the time of start of observation , substituting t equal to 0 in the above equation,

$$s = 10 \text{ m} \quad \textbf{Ans.}$$

(c) *Time after start of observations in which the velocity becomes zero*

Substituting the value of v equal to 0 in equation (iv),

$$0 = 2t - \frac{3t^2}{2} + 47.5$$

Multiplying both sides by – 2 and rearranging

$$3t^2 - 4t - 95 = 0$$

This is a quadratic equation in t,

$$\therefore \quad t = \frac{+4 \pm \sqrt{(4)^2 + 4 \times 3 \times 95}}{2 \times 3} = 6.33 \text{ s} \quad \textbf{Ans.}$$

EXERCISE 2

1. The motion of a body is given by an equation :

$$a = t^2 - 2t + 2$$

where a is acceleration in m/s^2 and t is time in seconds.The velocity and displacement of the body after 1 second was $6\frac{1}{3}$ m/s and $14\frac{3}{4}$ m respectively. Find the velocity and displacement after 2 seconds. (**Ans.** $7\frac{2}{3}$ m/s ; $21\frac{2}{3}$ m)

2. A body starting from rest, moves along a straight line with an acceleration whose equation is given by :

$$a = 4 - \frac{t^2}{9}$$

where a is in m/s^2 and t in seconds. Find (a) velocity after 6 seconds, and (b) distance traversed in 6 seconds. **[RGPV]** (**Ans.**16 m/s ; 60 m)

3. A body starting from rest, moves in such a way that its acceleration is given by :

$$a = 3 - 0.15\,t^2$$

Find the time when the body comes to stop and distance travelled during this time.

(**Ans.** 7.75 s ; 45 m)

4. A car moving with a velocity of 10 m/s shows down in such a manner that the relation between velocity and time is given by :

$$v = 10 - t^2 - \frac{t^3}{2}$$

Find the distance travelled in two seconds, average velocity and average retardation of the car in these two seconds. (**Ans.** 16.67 m ; 8.33 m/s ; 4 m/s^2)

6. VELOCITY, ACCELERATION AND DISPLACEMENT BY PREPARING A TABLE

Sometimes, the motion of a body is given in a tabular form containing time (t) and distance (s) or time (t) and acceleration (a) *e.g.*

t	1	2	3	4	5	6
s	8	20	35	55	80	110

Or

t	0	2	4	6	8	10
a	0.5	1.0	1.5	0.9	0.6	0

In such a case, the velocity, acceleration and displacement of the body may be easily found out by preparing a table, showing the other details of the motion (*i.e.,* average acceleration, increase in velocity and final velocity etc.).

Example 8. *An electric train has velocity in m/s as shown in the following table :*

t	0	1	2	3	4	5	6
v	40	39	36	31	24	15	4

Find the distance travelled by the train in the last 3 seconds.

Solution. In the first sec, mean velocity of the train

$$= \frac{40 + 39}{2} = 39.5 \text{ m/s}$$

∴ Distance travelled in this second,

$$= 1 \times 39.5 = 39.5$$

Similarly, in the next second, mean velocity of the train

$$= \frac{39 + 36}{2} = 37.5 \text{ m/s}$$

∴ Distance travelled in this second,

$$= 1 \times 37.5 = 37.5 \text{ m}$$

and total distance travelled upto the end of 2 second,

$$= 39.5 + 37.5 = 77 \text{ m}$$

Similarly, find the distances travelled by the train at the end of each sec, and prepare the table as given below :

t	v	δt $(t_2 - t_1)$	v_{av} $\left(\frac{v_2 + v_1}{2}\right)$	*Distance travelled* in $\delta t = \delta t \times v_{av}$	*Total distance travelled* in metres
0	40				
		1	39.5	39.5	
1	39				39.5
		1	37.5	37.5	
2	36				77.0
		1	33.5	33.5	
3	31				110.5
		1	27.5	27.5	
4	24				138.0
		1	19.5	19.5	
5	15				157.5
		1	9.5	9.5	
6	4				167.0

From the last column of the table, we find that distance travelled by the train in the last 3 seconds

= Distance travelled in 6 sec – Distance travelled in the first 3 sec

= 167.0 – 110.5 m = 56.5 m **Ans.**

***Example 9.** *An automobile starting from rest, moves along a straight line. Its acceleration after every 10 m distance was observed to be as below :*

s	*0*	*10*	*20*	*30*	*40*	*50*
a	*2.2*	*2.4*	*2.8*	*2.0*	*1.6*	*1.0*

Find the velocity of the automobile at the end of *45 metres.*

Solution. In the first 10 m the mean acceleration

$$= \frac{2.2 + 2.4}{2} = 2.3 \text{ m/s}^2$$

Substituting initial velocity (u) = 0, acceleration (a) = 2.3 m/s^2 and distance travelled (s) = 10 in the equation of motion *i.e.* $v^2 = u^2 + 2as$,

$$v^2 = (0)^2 + (2 \times 2.3 \times 10) = 46 \quad \text{or} \quad v = 6.78 \text{ m/s}$$

Similarly, in the next 10 m, the mean acceleration

$$= \frac{2.4 + 2.8}{2} = 2.6 \text{ m/s}^2$$

and now substituting initial velocity (u) = 6.78 m/s, acceleration (a) = 2.6 m/s^2 and distance travelled (s) = 10 m in the equation of motion *i.e.*, $v^2 = u^2 + 2as$,

$$v^2 = (6.78)^2 + (2 \times 2.6 \times 10) = 98 \qquad \text{or} \qquad v = 9.9 \text{ m/s}$$

Similarly, calculate the mean accelerations and velocities of the automobile at the end of each 10 metres and prepare the table as shown below.

s	a	a_{av} $\frac{a_1 + a_2}{2}$	δs $s_1 - s_2$	$u^2 + 2as$	v
0	2.2				
		2.3	10	46	6.78
10	2.4				
		2.6	10	98	9.9
20	2.8				
		2.4	10	146	12.08
30	2.0				
		1.8	10	182	13.49
40	1.6				
		1.3	10	218	14.76
50	1.0				

From the above table, we find that the average velocity of the automobile between 40 and 50 or 45 m of its start is 14.76 m/s **Ans.**

Example 10. *A car, starting from rest has an acceleration a in m/s^2 after t second from its start as given in the following table*:

t	0	4	8	12	16	20
a	10	9.6	8.4	0.4	3.6	0

Find the speed of the car at the end of each 4 sec interval and the distance traversed.

Solution. In the first 4 sec the mean acceleration

$$\frac{10 + 9.6}{2} = 9.8 \text{ m/s}^2$$

∴ Increase in velocity during these 4 second

$$= 9.8 \times 4 = 39.2 \text{ m/s}$$

and velocity at the end of 4 second

$$= 0 + 39.2 = 39.2 \text{ m/s}$$

∴ Average velocity in the first 4 second

$$= \frac{0 + 39.2}{2} = 19.6 \text{ m/s}$$

and distance traversed in the first 4 second

$$= 19.6 \times 4 = 78.4 \text{ m}$$

∴ Distance traversed up to the end of 4 second.

$$= 0 + 78.4 = 78.4 \text{ m}$$

Similarly, in the next 4 second, the mean acceleration

$$= \frac{9.6 + 8.4}{2} = 9 \text{ m/s}^2$$

∴ Increase in velocity during these 4 second,

$$= 9 \times 4 = 36 \text{ m/s}$$

and velocity at the end of these 4 second,

$$= 39.2 + 36 = 75.2 \text{ m/s}$$

∴ Average velocity in these 4 second,

$$= \frac{39.2 + 75.2}{2} = 57.2 \text{ m/s}.$$

and distance traversed in these 4 second,

$$= 57.2 \times 4 = 228.8 \text{ m}$$

∴ Distance traversed up to the end of these 4 second,

$$= 78.4 + 228.8 = 307.2 \text{ m}$$

Similarly, calculate the mean accelerations, velocities and distances traversed by the car at the end of each 4 sec and prepare the table as shown in the following table.

t	a	a_{av} $\frac{a_1 + a_2}{2}$	δt $(t_2 - t_1)$	*increase in v* $(a_{av} \times t)$	v	v_{av} $\frac{v_1 + v_2}{2}$	δs $(v_{av} \times \delta t)$	s
0	10				0			0
		9.8	4	39.2		19.6	78.4	
4	9.6				39.2			78.4
		9.0	4	36.0		57.2	228.8	
8	8.4				75.2			307.2
		7.4	4	29.6		90.0	360.0	
12	6.4				104.8			667.2
		5.0	4	20.0		114.8	459.2	
16	3.6				124.8			1126.4
		1.8	4	7.2		128.4	513.6	
20	6				132.0			1640.0

In the above table is given the speed of car and distance traversed at the end of each 4 seconds. **Ans.**

***Example 11.** *A car starts from rest, and moves along a straight line. The distance covered (s) in seconds (t) from the start, were observed to be as under :*

t	*0*	*5*	*10*	*15*	*20*	*25*	*30*
s	*0*	*20*	*100*	*230*	*330*	*380*	*400*

Calculate the velocity and acceleration of the car after 10 and 20 seconds from start.

Solution. In the first 5 seconds, the distance covered = 20 m

∴ Average velocity after 2.5 seconds of start (or in all 5 seconds from 0 to 5)

$$= 20/5 = 4 \text{ m/s}$$

Similarly, in the next 5 seconds, the distance covered

$$= 100 - 20 = 80 \text{ m}$$

∴ Average velocity after 7.5 seconds of start (or in all 5 seconds from 5 to 10)

$$= 80/5 = 16 \text{ m/s}$$

and rate of increase in average velocity (or acceleration) in 5 seconds (from 2.5 to 7.5 seconds)

$$= \frac{16 - 4}{5} = 2.4 \text{ m/s}^2$$

Similarly, calculate the distances covered, average velocities and accelerations in each 5 seconds and prepare the table as shown below.

t	s	δt $(t_2 - t_1)$	δs $(s_2 - s_1)$	$v = \frac{\delta s}{\delta t}$	$v_2 - v_1$	$a = \frac{v_2 - v_1}{t_2 - t_1}$
0	0					
		5	20	4		
5	20				12	12/5 = 2.4
		5	80	16		
10	100				10	10/5 = 2.0
		5	130	26		
15	230				– 6	– 6/5 = – 1.2
		5	100	20		
20	330				– 10	– 10/5 = – 2.0
		5	50	10		
25	380				– 6	– 6/5 = – 1.2
		5	20	4		
30	400					

Velocity of the car after 10 and 20 seconds from start

From the average velocity column, we find that average velocity of the car after 7.5 seconds

$$= 16 \text{ m/s}$$

and average velocity of the car after 12.5 seconds

$$= 26 \text{ m/s}$$

∴ Average velocity of the car after 10 seconds from start

$$= \frac{16 + 26}{2} = 21 \text{ m/s} \quad \textbf{Ans.}$$

Similarly, velocity of the car after 20 seconds from start

$$= \frac{20 + 10}{2} = 15 \text{ m/s} \quad \textbf{Ans.}$$

Acceleration of the car after 10 and 20 seconds from start

From the acceleration (*i.e.* last) column, we find that acceleration of the car after 10 and 20 seconds from start is 2.0 m/s^2 and – 20 m/s^2 respectively. **Ans.**

EXERCISE 3

1. A tram car, starting with an initial velocity of 7.5 m/s moves with a variable acceleration. The time-acceleration chart is given below :

t	0	5	10	15	20	25	30	35	40
a	0.4	0.9	1.1	1.8	1.9	2.3	2.5	1.9	1.1

Find the velocity of the car after 40 seconds. **(Ans.** 73.25 m/s)

2. Abody starts moving along a straight line with an initial velocity of 8 m/s. The acceleration in m/s^2 at intervals of 5 seconds were observed to be as under :

t	0	5	10	15	20	25	30
a	0.2	0.8	1.2	1.6	2.0	1.2	0

Find the distance travelled in 30 seconds from the start. **(Ans.** 723.75)

QUESTIONS

1. How would you distinguish between the motion, when it is subjected to (*a*) constant acceleration, and (*b*) variable acceleration ?
2. Under what circumstances you would differentiate or integrate the given equation of motion of a particle to obtain, velocity, acceleration and displacement ?
3. If (s) is the distance traversed by a particle, then what does $\frac{ds}{dt}$ and $\frac{d^2s}{dt^2}$ represent ?
4. When would you prefer to prepare a table of motion of a moving body ?

OBJECTIVE TYPE QUESTIONS

1. We are given an equation of displacement (s) in terms of time (t). If we differentiate it with respect to t, the equation so obtaned will give
   (*a*) velocity (*b*) acceleration (*c*) distance traversed
2. The second differentiation, of the above equation will give
   (*a*) velocity (*b*) acceleration (*c*) distance traversed
3. If we differentiate an equation in terms of acceleration and time, it will give
   (*a*) velocity (*b*) distance traversed (*c*) none of these two
4. We are given an equation of acceleration (a) in terms of time (t). The second integration of the equation will give the velocity.
   (*a*) Yes (*b*) No
5. Which of the following statement is wrong ?
   (*a*) A body falling freely under the force of gravity is an exapmle of motion under variable acceleration.
   (*b*) A bus going down the valley may have variable acceleration.
   (*c*) A lift going down in a gold mine cannot have constant acceleration in the entire journey.
   (*d*) In a cricket match, the ball does not move with constant acceleration.

ANSWERS

1. (*a*) 2. (*b*) 3. (*c*) 4. (*b*) 5. (*a*)

CHAPTER

17 Relative Velocity

Contents

1. RELATIVE VELOCITY

It has been established since long that every motion is relative, as an absolute motion is impossible to conceive. Strictly speaking, our motion is always relative with reference to the Earth , which is supposed to be fixed or at rest. But we know that our Earth is also not at rest. It has some relative velocity with respect to the celestial bodies such as sun, moon etc. These celestial bodies, in turn, have some relative velocity with respect to the stars of the universe.

It will be interesting to know that when we say that a train is moving at 50 kilometres per hour, we simply mean that the speed of the train, which appears to an observer on the earth, is 50 kilometres per hour. Thus the relative velocity of A with respect to B, is the velocity with which A appears to move to an observer sitting on B, neglecting the motion of B relative to the Earth.

2. METHODS TO FIND RELATIVE VELOCITY

The relative velocity of two bodies may be found out either graphically or analytically. But it has been experienced that analytical method is somewhat confusing. Thus in this book, we shall follow the graphical method. The best practice is the combination of both the methods. The values should first be obtained by graphical method and then their accuracy should be checked analytically.

3. *RELATIVE VELOCITY OF RAIN AND MAN

We see in our daily walk of life, that whenever we go out in rain, we have to adjust the inclination of our umbrella (with the vertical) to protect ourselves from the rain. If the rain is falling in the opposite direction of our movement, then the inclination of umbrella is less when we are standing, than that when we are moving. This happens because the relative velocity of rain (with respect to our movement) now is inclined at a greater angle.

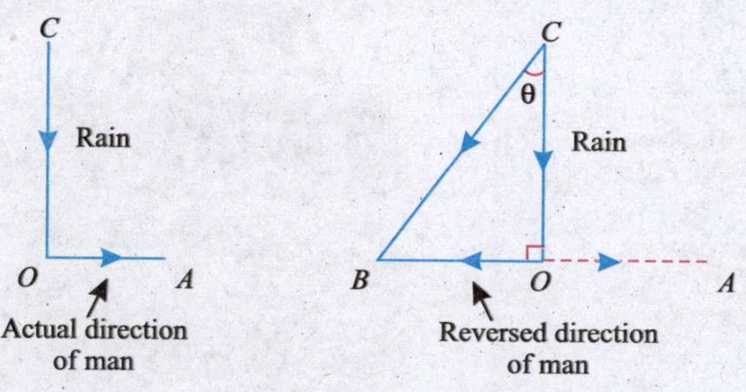

(*a*) Actual velocity diagram (*b*) Relative velocity diagram

Fig. 1.

Now consider CO and OA as actual directions of the rain and man respectively as shown in Fig. 1 (*a*). The relative velocity diagram for rain and man may be drawn as shown in Fig. 1 (*b*) and as discussed below :

1. First of all, draw a horizontal line OA to some suitable scale which represents the actual direction of motion of the man.
2. Now draw CO equal to the velocity of the rain (assumed to be vertical in this case) to the scale.
3. Let us *superimpose a velocity equal and opposite to that of the man on both the man as well as rain. It will reduce the velocity of the man to zero. And the rain will have some resultant velocity (due to its own velocity and superimposed velocity) as shown in Fig. 1 (*b*).

   *Note: If the velocity of man is superimposed on both the man and rain, their relative velocity remains unchanged.
4. Now relative velocity of the rain and man, in magnitude and direction, will be given by the resultant velocity CB, which may be found out either graphically or by Triangle Law or Forces.

Example 1. *A train is moving at 48 km. p.h. and a person sitting in it feels rain coming down at 60° to the vertical. However, a person standing in a field outside, feels the rain to be vertical. Find the actual velocity of the rain.*

Solution. Given : Velocity of train = 48 km. p.h.

Let us draw the relative velocity and actual velocity diagrams for the train and rain as shown in Fig. 2 and as discussed below :

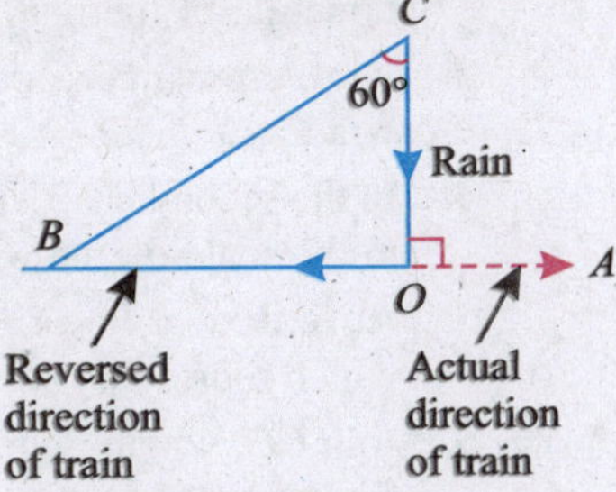

Fig. 2.

1. First of all, draw a line OA representing the actual direction of the motion of train which is moving with a velocity of 48 km. p.h.
2. Now cut off OB equal to 48 km. p.h. to some suitable scale in the opposite direction of the actual motion of the train.
3. At O, draw a perpendicular line, which represents the actual direction of the rain (or in other words, direction of the rain which is felt by a man standing in the field outside).
4. From C, draw a line CB making an angle of 60° with CO (*i.e.* vertical) which represents the relative velocity of the rain.
5. By measurement, we find that the actual velocity of the rain = CO = 27.7 km.p.h.

Mathematical check

In right angled triangle OBC,

$$\tan 60° = \frac{48}{OC}$$

or

$$OC = \frac{48}{\tan 60°} = \frac{48}{1.732} = 27.7 \text{ km.p.h.}$$

***Example 2.** *A man, running eastwards with a speed of 6 kilometres per hour, feels the wind to be blowing directly from North. On doubling his speed, he feels the wind to blow from the North-east. Find the actual direction and velocity of the wind.*

Solution. Given : Velocity of man = 6 km.p.h. (East).

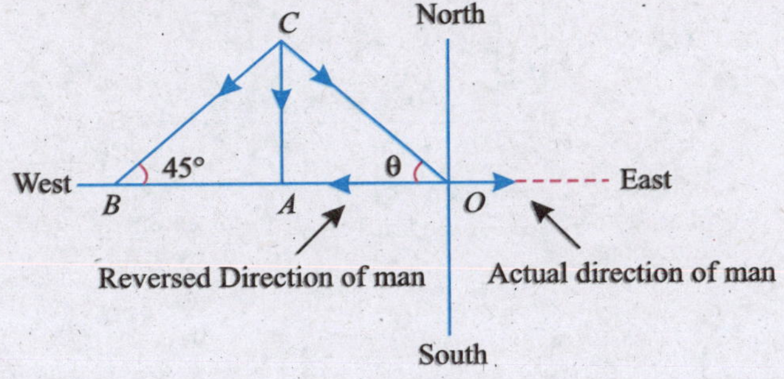

Fig. 3.

Let us draw the relative velocity diagram for the man and wind as shown in Fig. 3 and as discussed below :

1. First of all, draw East, West, North and South lines meeting at O.
2. In the first case, since the man is running eastwards at 6 km. p.h., therefore cut off OA equal to 6 km. to some suitable scale on the westwards (opposite to eastwards *i.e.* actual direction of the man).
3. At A, draw a perpendicular CA, which represents the direction of the relative velocity of wind.
4. In the second case, since the man doubles his speed, therefore cut off OB equal to 12 km. to the scale on the westward.
5. At B, draw a line CB at 45° (*i.e.* North-East) meeting the vertical line through A at C, which represents the direction of the new relative velocity of the wind.
6. Join CO, which gives the actual direction and velocity of the wind. By measurement, we find that $\angle \theta = \angle AOC = 45°$ and $CO = 8.5$ km.p.h.

Mathematical check

We know that

$$OA = AB = 6 \text{ km.} \qquad \text{...(Speed of man)}$$

and

$$CA = AB = 6 \text{ km.} \qquad \text{...}\left(\frac{CA}{AB} = \tan 45° = 1\right)$$

$$\therefore \quad OA = CA \quad \text{or} \quad \theta = 45° \textbf{ Ans.}$$

Now in triangle OAC

$$CO = \frac{CA}{\sin 45°} = \frac{6}{0.707} = 8.49 \text{ km.p.h. } \textbf{Ans.}$$

***Example 3.** *When a cyclist is riding towards West at 20 km per hour, he finds the rain meeting at an angle of 45° with the vertical. When he rides at 12 km per hour, he meets the rain at an angle of 30° with the vertical. What is the actual velocity, in magnitude and direction, of the rain ?*

Solution. Given : When velocity is 20 km. p.h., apparent direction, of the rain = 45° with vertical and when velocity is 12 km. p.h. apparent direction of rain = 30° with vertical.

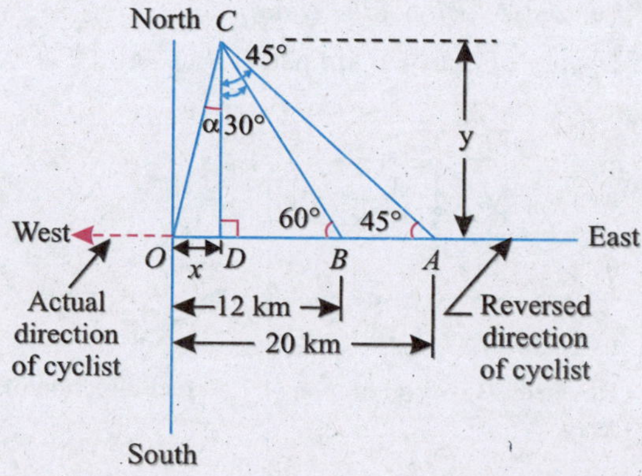

Fig. 4.

Let us draw the relative velocity diagram for both the cases as shown in Fig. 4 and as discussed below :

1. First of all, draw East, West, North, and South lines meeting at O.
2. In the first case, since the cyclist is riding towards West at 20 km. p.h., therefore cut off OA equal to 20 km. to some suitable scale towards East (opposite to West *i.e.* actual direction of the cyclist).
3. At A, draw a line at an angle of 45° with OA which represents the relative direction of the rain.
4. In the second case, since the man is riding at 12 km.p.h., therefore cut off OB equal to 12 km. to the scale towards East.
5. At B, draw a line at an angle of 60° with OA, which represents the direction of the relative velocity of the rain. This is so because the rain meets the man at angle of 30° with the vertical. Let the two lines (from A and B) meet at C.
6. Join CO, which gives the actual direction and velocity of the wind. By measurement, we find that $\angle \alpha = 3{\cdot}3°$ and $CO = 19$ km.p.h. **Ans.**

Mathematical check

From C, draw a perpendicular CD to the line OA. Let $OD = x$ and $CD = y$. From the triangle ACD,

$$\tan 45° = \frac{20 - x}{y} \quad \text{or} \quad 1 = \frac{20 - x}{y} \qquad ...(\because \tan 45° = 1)$$

or
$$y = 20 - x \qquad ..(i)$$

Similarly, in triangle BCD

$$\tan 30° = \frac{12 - x}{y} \quad \text{or} \quad 0.577 = \frac{12 - x}{y}$$

or
$$y = \frac{12 - x}{0.577} \qquad ...(ii)$$

Equating equations (*i*) and (*ii*),

$$20 - x = \frac{12 - x}{0.577}$$

or
$$11.54 - 0.577\,x = 12 - x$$

or
$$0.423\,x = 0.46$$

or
$$x = \frac{0.46}{0.423} = 1.09 \text{ km}$$

Substituting the value of x in equation (*i*),

$$y = 20 - x = 20 - 1.09 = 18.91 \text{ km}$$

$$\therefore \quad \tan \alpha = \frac{x}{y} = \frac{1.09}{18.91} = 0.0576$$

$$\therefore \quad \alpha = 3.3° \quad \textbf{Ans.}$$

Now from the triangle OCD,

$$CO = \sqrt{OD^2 + CD^2} = \sqrt{x^2 + y^2}$$

$$= \sqrt{(1.09)^2 + (18.91)^2} = 18.94 \text{ km.p.h.} \quad \textbf{Ans.}$$

EXERCISE 1

1. A passenger, travelling in a train, observed the drops of rain water to pass the railway carriage at an angle of 10° 18′ with the horizontal. If actual velocity of the water drops is 2 m/s, show that the train is moving at 39.6 km.p.h. Assume actual direction of the train to be vertical.
2. A shower of rain appears to be falling vertically downwards with a velocity of 12 km.p.h. to a man walking due to West with a velocity of 5 km.p.h. What is the actual velocity and direction of the rain ? (**Ans.** 13 km.p.h.; 23°)

4. *RELATIVE VELOCITY OF TWO BODIES MOVING ALONG INCLINED DIRECTIONS

Sometimes, two bodies are moving along two inclined directions. In such a case, the relative velocity of one, with respect to the other, may be found out by superimposing the actual velocity of any one of these two bodies, in the opposite direction, on both the bodies. It will be interesting to know that after superimposing the velocity, one of the bodies will be brought to rest. And the resultant of the two velocities, on the second, will give the required relative velocity.

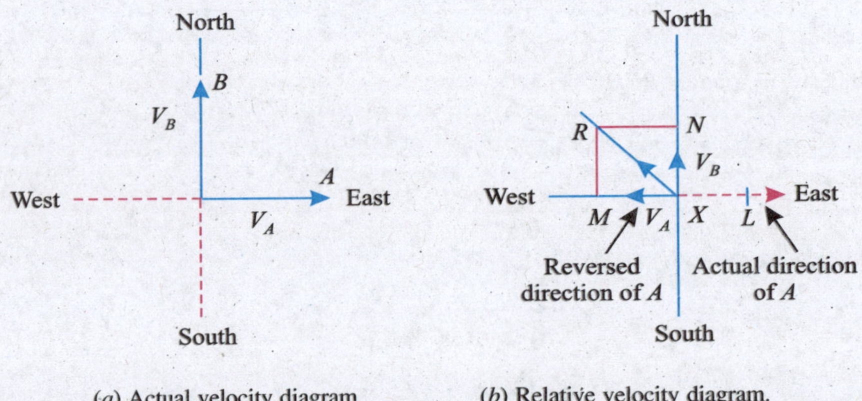

(*a*) Actual velocity diagram (*b*) Relative velocity diagram.

Fig. 5. Bodies moving along inclined directions.

Now consider two bodies *A* and *B* moving with velocities v_A and v_B respectively along East and North as shown in Fig. 5 (*a*).

Now let us draw the relative velocity diagram as shown in Fig. 5 (*b*) and as discussed below :

1. First of all, draw East, West, North and South lines meeting at *X* .
2. Since the body *A* is moving eastwards, therefore cut off *XL* equal to velocity v_A to some suitable scale towards East, representing the actual velocity of the body *A* .
3. Now cut off *XM* equal to the velocity v_A to the same scale on the opposite direction of the actual motion of the body (*i.e.* towards West).
4. Now cut off *XN* equal to the velocity v_B to the scale to represent the actual velocity of the body *B*.
5. Complete the parallelogram *XMRN* with *XM* and *XN* as adjacent sides.

6. Join the diagonal *XR*, which gives the magnitude and direction of the relative velocity of the two bodies.

Note. There is no hard and fast rule as to which velocity (in the opposite direction) is to be superimposed. Thus any one of the two velocities may be superimposed in the opposite direction.

***Example 4.** *A railway coach, having ordinary cross-seats, is travelling at 4 m/s. A person runs at 5 m/s on the platform. In what direction, he must run so that he may enter the railway coach parallel to the seats ? Also find the velocity with which he enters the coach.*

Solution. Given : Velocity of train = 4 m/s and velocity of person = 5 m/s.

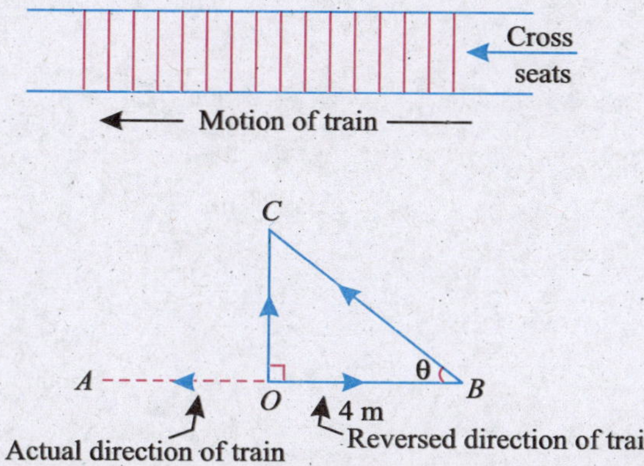

Fig. 6.

Let us draw the relative velocity diagram of the man and train as shown in Fig. 6 and as discussed below :

1. First of all, draw position of the train having cross-seats. Now draw a line *OA* representing the actual direction of motion of the train with a velocity of 4 m/s.
2. Now cut off *OB* equal to 4 m to some suitable scale on the opposite direction of the actual motion of the train.
3. At *O*, draw a perpendicular (*i.e.* parallel to the cross-seats) and cut off *BC* equal to 5 m to the scale, which represents the actual velocity of the man with which he is running on the platform.
4. Now *OC* represents the relative velocity of the man or his velocity with which he enters the train. By measurement, we find that $\angle \theta = \angle OBC = 36.8°$ and $OC = 3$ m/s **Ans.**

Mathematical check

In right angled triangle *OBC* ,

$$\cos \theta = \frac{4}{5} = 0.8 \qquad \text{or} \qquad \theta = 36.8°$$

and

$$OC = \sqrt{(5)^2 - (4)^2} = 3 \text{ m/s} \quad \textbf{Ans.}$$

***Example 5.** *Two ships leave a port at the same time. The first steams North-West at 32 kilometres per hour and the second 40° South of West at 24 kilometres per hour.*

(*a*) *What is the velocity of the second ship relative to the first in km per hour ?*

(*b*) *After what time, they will be 160 km apart ?* **[PU]**

Solution. Given : Velocity of first ship 32 km.p.h. (N-W) ; Velocity of second ship = 24 km.p.h. (40° South of West).

(a) Velocity of the second ship relative to the first

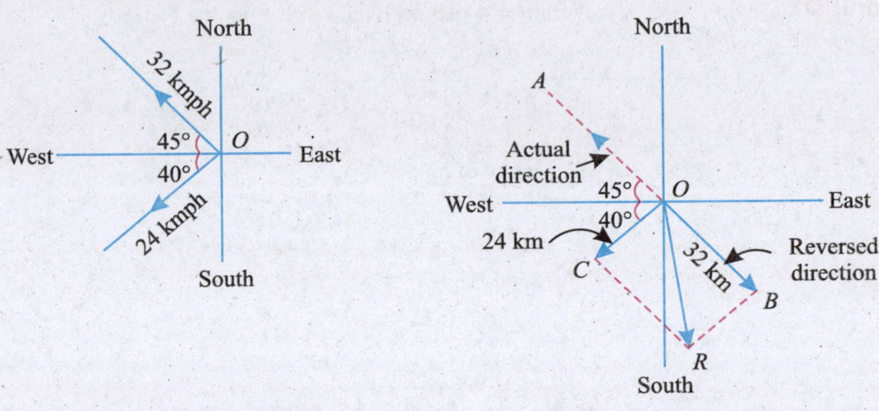

(*a*) Actual velocity diagram (*b*) Relative velocity diagram

Fig. 7.

First of all, let us draw the actual velocity diagram of the two ships as shown in Fig. 7 (*a*). Now draw the relative velocity diagram as shown in Fig. 7 (*b*) and as discussed below :

1. First of all, draw the East, West, North and South lines meeting at *O*.
2. Since the first ship steams in the North-West direction , therefore, draw a line *OA* at 45° to the North (or West) representing the actual direction of the ship at 32 km.p.h.
3. Now cut off *OB* equal to 32 km to some suitable scale on the opposite direction of the actual motion of the first ship (*i.e.* South-East).
4. Now draw a line at an angle of 40° South of West and cut off *OC* equal to 24 km to the scale to represent the actual direction of the second ship.
5. Complete the parallelogram *OBRC* with *OB* and *OC* as adjacent sides.
6. Join *OR* ,which gives the magnitude and direction of the second ship relative to the first. By measurement, we find that *OR* = 38.3 km.p.h. **Ans.**

(b) Time when the two ships will be* 160 *km apart

We know that the two ships will be 160 km apart after

$$= \frac{160}{38.3} = 4.18 \text{ hrs} \quad \textbf{Ans.}$$

Mathematical check

In parallelogram *OBRC*

$$OR = \sqrt{(32)^2 + (24)^2 + 2 \times 32 \times 24 \cos 95^\circ}$$

$$= \sqrt{1024 + 576 + 1536 \cos (90° + 5°)}$$

$$= \sqrt{1600 - 1536 \sin 5°} \qquad ...[\because \cos (90° + \theta) = -\sin\theta]$$

$$= \sqrt{1600 - (1536 \times 0.0872)} = 38.3 \text{ km.p.h.} \quad \textbf{Ans.}$$

Example 6. *A submarine, travelling on a course of bearing 80 degrees East of North with a speed of 21 knots, reports the presence of an enemy ship travelling on a course of bearing 135 degrees West of North with a speed of 15 knots with respect to the submarine . Assuming that the submarine and the enemy ship are at the same place at the time of sighting,*

(*a*) *Find the true direction and speed of the enemy ship.*

(*b*) *If a warship stationed 150 nautical miles South of the submarine immediately starts at 18 knots to intercept the enemy ship, what bearing should it take ? Find the time it will take to intercept the enemy ship.*

Solution. Given : Velocity of submarine = 21 knots (N 80° E); Velocity of enemy ship = 15 knots (N 135° W) and velocity of warship = 18 knots

(a) True direction and speed of the enemy ship

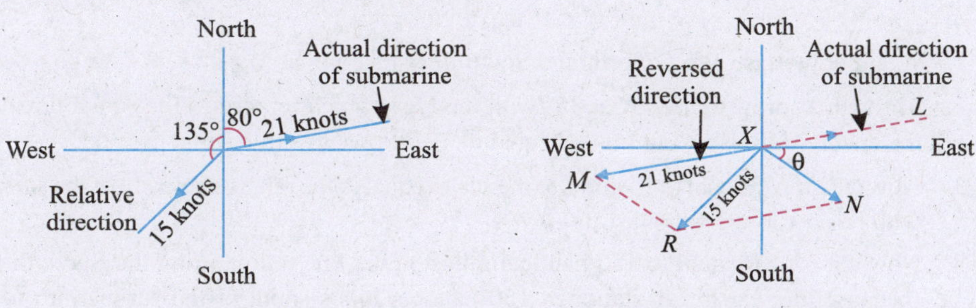

(*a*) Actual velocity diagram (*b*) Relative velocity diagram

Fig. 8.

First of all , let us draw the actual velocity diagram of the submarine and its relative velocity with enemy ship as shown in Fig. 8 (*a*). Now draw the relative velocity diagram as shown in Fig. 8 (*b*) and as discussed below :

1. First of all, draw the East, West, North and South lines meeting at *X* .
2. Since the submarine is travelling on a course of bearing 80 degrees East of North, therefore, draw a line *XL*. at 80° to the North representing the actual direction of the submarine.
3. Now cut off *XM* equal to 21 knots to some suitable scale on the opposite direction of the actual motion of the submarine.
4. Now draw a line at an angle of 135° West of North and cut off *XR* equal to 15 knots to the scale to represent the relative velocity of the enemy ship with respect to the submarine.
5. Complete the parallelogram *XMRN* with *XM* as one side and *XR* as diagonal.
6. Now the side *XN* of the diagonal gives the true direction and speed of the enemy ship. By measurement, we find that $\angle \theta = 35°$ and *XN* = 12.7 knots **Ans.**

(b) Bearing of the submarine

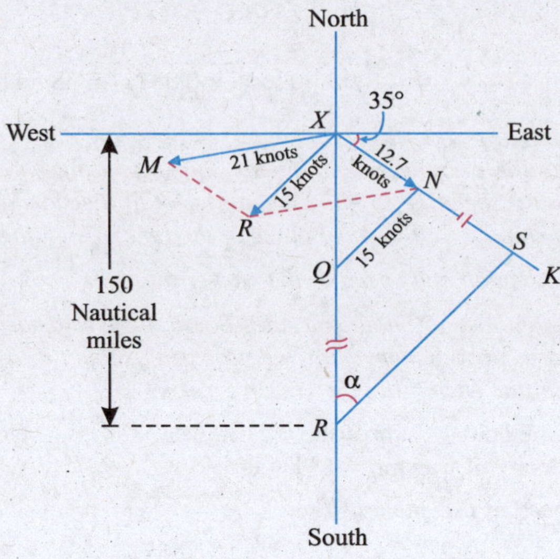

Fig. 9.

1. Again draw East, West, North and South lines meeting at X.
2. Through X. draw a line XK at 35° with the East, which represents the actual direction of the enemy ship. Now cut off XN equal to 12.7 knots to some suitable scale.
3. Now cut off NQ equal to 15 knots to the scale (equal to the relative velocity of the submarine with respect to the warship). Join NQ.
4. Now cut off XR equal to 150 nautical miles on the line representing the South direction (because the warship is stationed 150 nautical miles south of the submarine) to some other scale. Through R, draw RS parallel to the line QN meeting the line XK at S as shown in Fig. 9.
5. By measurement, we find that bearing of the submarine $\angle XRS = 43°$ with North. **Ans.**

Time taken by the warship to intercept the enemy ship

By measurement, we also find that distance $RS = 123$ km.

Therefore time taken by the warship to intercept the enemy ship

$$= \frac{\text{Distance } RS}{\text{Velocity of warship}} = \frac{123}{18} \text{ hours}$$

$$= 6.83 \text{ hours} = 6 \text{ hours } 50 \text{ min} \quad \textbf{Ans.}$$

***Example 7.** *A rifleman on a train, moving with a speed of 52 km per hour, fires an object running away from the train at right angle with a speed of 39 kilometres per hour. The line connecting the man and the object makes an angle of 30° to the train at the instant of shooting.*

At what angle to the train should he aim in order to hit the object, if the muzzle velocity is 200 metres per second ?

Solution. Given : Speed of train = 52 km.p.h. ; Speed of object = 39 km.p.h. ; Angle, which the line connecting the man and object makes with the train = 30° and muzzle velocity = 200 m/s

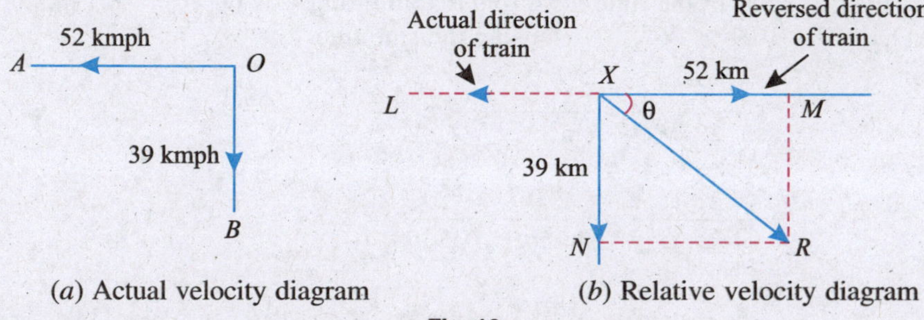

(*a*) Actual velocity diagram (*b*) Relative velocity diagram

Fig. 10.

First of all, let us draw the actual velocity diagram of the train and object as shown in Fig. 10. (*a*). Now draw the relative velocity diagram as shown in Fig. 10 (*b*) and as discussed below :

1. First of all, draw a line *XL* representing the actual direction of the motion of the train with a velocity of 52 km.p.h.
2. Now cut off *XM* equal to 52 km to some suitable scale on the opposite direction of the actual motion of the train.
3. Now cut off *XN* equal to 39 km to the scale at right angle to *LM,* which represents the actual velocity with which the object is running away.
4. Complete the parallelogram *XMRN* with *XM* and *XN* as adjacent sides.
5. Join *XR* which gives the magnitude and direction of the relative velocity. By measurement, we find that $\angle\theta = 36.9°$ and $XR = 65$ km. p.h. = 18.1 m/s

Now let us draw the muzzle velocity diagram as shown in Fig. 11. and as discussed below:

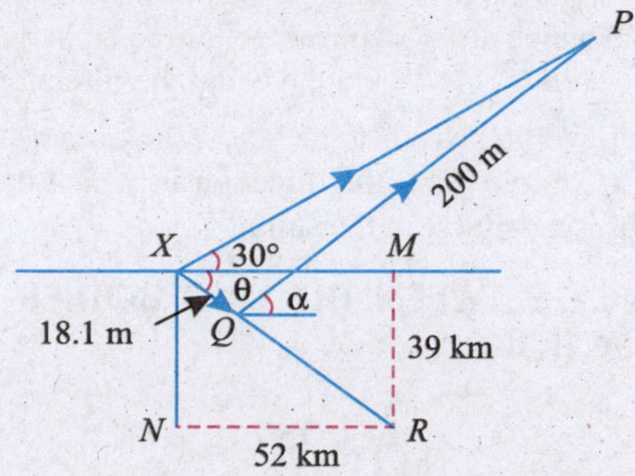

Fig. 11.

1. First of all, draw the relative velocity diagram and the parallelogram *XMRN* as discussed above.
2. Cut off *XQ* equal to 18.1 m (*i.e.* relative velocity of train and object) to some suitable scale.
3. At *X* draw a line at an angle of 30°, which represents the line connecting the man and object with the train.
4. Now cut off *QP* equal to 200 m to the scale and join *QP*. Now *P* represents the position of the object, which the rifleman wants to hit.
5. Now let the rifleman make an angle α with the train in order to hit the object as shown in Fig. 11.

From the geometry of the figure, we find that in triangle XPQ, $XQ = 18.1$ m, $QP = 200$ m and $\angle PXQ = 30° + 36.9° = 66.9°$. Now using the sine rule,

$$\frac{XQ}{\sin \angle XPQ} = \frac{QP}{\sin \angle PXQ}$$

or
$$\frac{18.1}{\sin \angle XPQ} = \frac{200}{\sin 66.9} = \frac{200}{0.9198} = 217.4$$

or
$$\sin \angle XPQ = \frac{18.1}{217.4} = 0 \cdot 0833 \quad \text{or} \quad \angle XPQ = 4 \cdot 8°$$

∴ Angle, at which rifleman should aim,

$$\alpha = 30° + 4.8° = 34.8° \quad \textbf{Ans.}$$

EXERCISE 2

1. A train moving at 30 km.p.h. is struck by a stone moving at right angles to the train with a velocity of 22.5 km.p.h. Find the velocity and direction which the stone appears to strike the train, to a person sitting in it. (**Ans.** 37.5 km.p.h. ; 53°)
2. If a ship is moving North-West at 15 knots and a second ship is moving due East at 7 knots, determine the direction and magnitude of the second ship relative to the first. (**Ans.** 31° ; 20.6 knots)
3. A man is sitting in a ship (S_1) sailing South-East with a velocity of 12 km.p.h. He notices another ship (S_2) which always appears to him to be in the East and going further away. If the speed of the ship (S_2) is 18 km.p.h., find the direction of ship (S_2). (**Ans.** 28°)
4. A steam ship is travelling North at the rate of 20 km.p.h. and there is a wind blowing from North-East at 30 km.p.h. Find the direction in which the smoke from the chimney will appear to an observer sitting in the ship. (**Ans.** 27°)

5. *LEAST DISTANCE BETWEEN THE TWO BODIES MOVING ALONG INCLINED DIRECTIONS

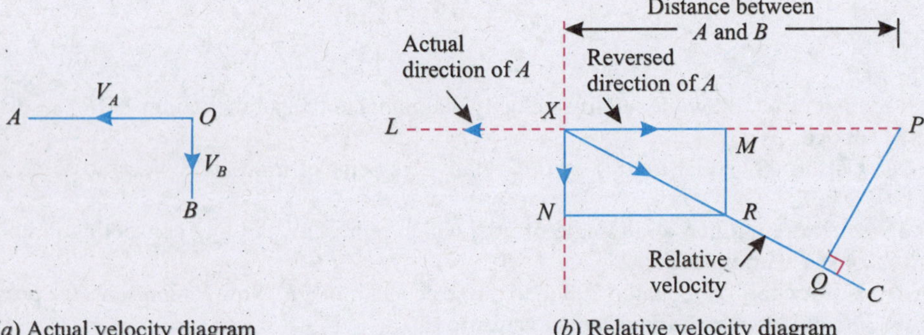

Fig. 12.

Consider two bodies A and B moving with velocities v_A and v_B respectively. Let the actual direction of v_A and v_B be as shown in Fig. 12 (a).

Now in order to find the least distance between the two bodies, first of all, we have to find out the relative velocity of one body with respect to the other. So let us draw the relative velocity diagram *XMRN* for the two bodies, as usual in which *XR* represents the relative velocity as shown in Fig. 12 (*b*).

Let us superimpose the least distance diagram on the relative velocity diagram as shown in Fig. 12 (*b*) and as discussed below :

1. First of all, extend the relative velocity line *XR* to *C*.
2. Now cut off *XP* equal to the distance between the two bodies at any given instant to the scale.
3. From *P*, draw *PQ* perpendicular to *XC*. The length *PQ* will give the least distance between the two bodies to the scale.
4. The time elapsed before the two bodies are least distance apart is given by the relation :

$$= \frac{\text{Length } PQ \text{ to the scale}}{\text{Relative velocity of two bodies}}$$

Note. A little consideration will show that the two bodies will go on coming nearer to each other till one of the body reaches *Q*. Beyond this point, the two bodies will go away from each other. This happens as the least distance between a point and a line is the perpendicular from the point to the line.

***Example 8.** *A ship sailing East with a velocity of 15 km. per hour, passes a certain point at noon ; and a second ship, sailing North with the same velocity passes the same point at 1.30 P.M.*

At what time are they closest together and what is the distance between them ?

Solution. Given. Velocity of first ship = 15 km.p.h. (East) and velocity of second ship = 15 km.p.h. (North).

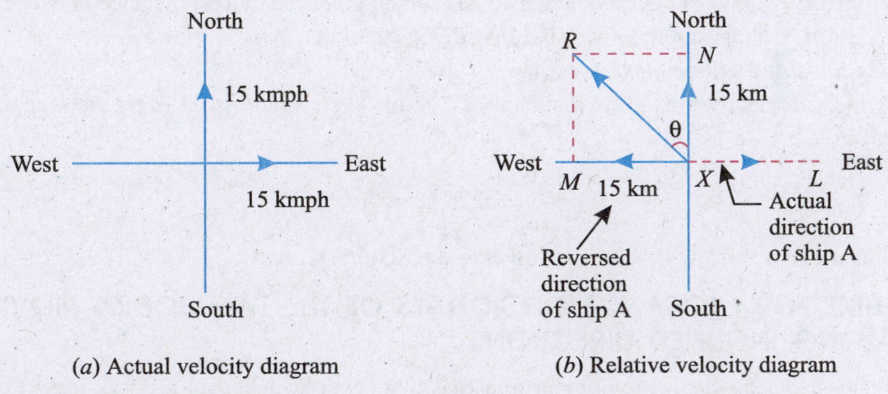

(*a*) Actual velocity diagram

(*b*) Relative velocity diagram

Fig. 13.

Closest distance between the two ships

First of all, let us draw the actual velocity diagram of the two ships as shown in Fig. 13 (*a*). Now draw the relative velocity diagram as shown in Fig. 13 (*b*) and as discussed below :

1. First of all, draw the East, West, North and South lines meeting at *X*.
2. Since the first ship is sailing East, therefore draw *XL* representing the actual direction of the first ship.
3. Now cut off *XM* equal to 15 km to some suitable scale on the opposite direction of the actual motion of the first ship (*i.e.* towards West).

4. Now cut off XN equal to 15 km to the scale on the North line to represent the actual direction of the second ship.
5. Complete the parallelogram $XMRN$ with XM and XN as adjacent sides.
6. Join XR which gives the magnitude and direction of the second ship relative to the first. By measurement, we find that XR = 21.2 km.p.h. and $\angle \theta = 45°$.

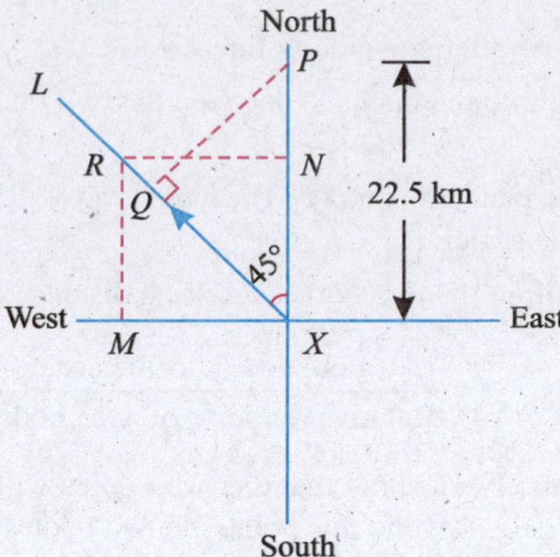

Fig. 14. Closest velocity diagram

Now let us draw the closest velocity diagram as shown in Fig. 14 and as discussed below :

1. First of all, draw the relative velocity diagram and the parallelogram $XMRN$ as discussed above. Now extend the relative velocity line XR to L.
2. At noon, let the second ship be at X. Therefore the first ship will be $15 \times 1.5 = 22.5$ km towards North or in other words, the second ship will be 22.5 km behind the first (*i.e.* towards South). So mark XP equal to 22.5 km to the scale.
3. From P, draw PQ perpendicular to XL. which gives the closest distance between the two ships. By measurement we find that PQ = 15.9 km. **Ans.**

Time when the ships will be closest together

By measurement we also find that XQ = 15.9 km. Therefore time taken by the second ship to reach Q from X

$$= \frac{15.9}{21.2} = 0.75 \text{ hr} = 45 \text{ min}$$

Thus the two ships will be closest together at 12.45 P.M. **Ans.**

6. *TIME FOR EXCHANGE OF SIGNALS OF THE TWO BODIES MOVING ALONG INCLINED DIRECTIONS

We have discussed in previous article that whenever two bodies are moving in different directions, at a certain instant they are least distance apart. It will be interesting to know that before and after this instant, the two bodies are not at the least distance apart. It is thus obvious, that the bodies start coming nearer to each other, and after coming at the least distance apart, they start going away from each other.

Sometimes, these bodies start exchanging signals, whenever they come within the vision of each other, and go on signalling so long as they remain within their vision. The moment they are beyond their vision, the bodies stop exchanging of signals. Thus the time for exchange of signals of the two moving bodies, is the time for which they remain within their vision.

Consider two bodies A and B moving with velocities v_A and v_B respectively. Let the actual directions of v_A and v_B be as shown in Fig. 15 (a).

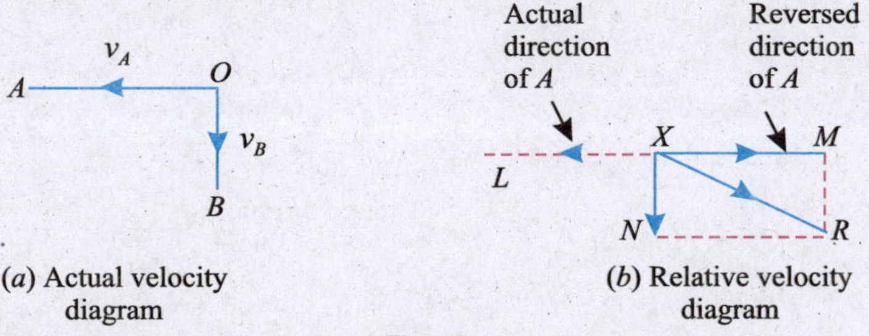

(a) Actual velocity diagram

(b) Relative velocity diagram

Fig. 15.

Now in order to find the time for exchange of signals, first of all we have to find the relative velocity of one body with respect to the other. Let us draw the relative velocity diagram $XMRN$ for the two bodies as usual, in which XR represents the relative velocity as shown in Fig. 15 (b).

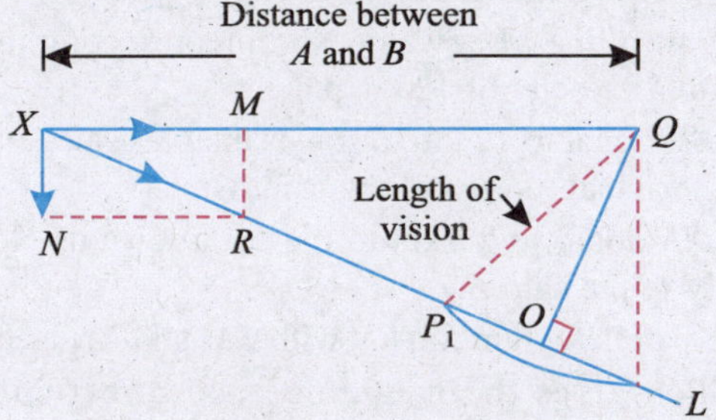

Fig. 16. Exchange of signal diagram

Now let us draw the exchange of signal diagram for the two bodies with the help of their relative velocity diagram as shown in Fig. 16 and as discussed below :

1. First of all, extend the relative velocity line XR to L.
2. Now cut off XQ equal to the distance between two bodies, at any given instant, to the scale.
3. With Q as centre and radius equal to the vision distance, draw an arc meeting the relative velocity line at P_1 and P_2.
4. The time for exchange of signals is to the time required by the body B to travel from P_1 to P_2 is given by the relation :

$$= \frac{\text{Distance } XP_2 \text{ to the scale}}{\text{Relative velocity of two bodies}} - \frac{\text{Distance } XP_1 \text{ to the scale}}{\text{Relative velocity of two bodies}}$$

Notes : 1. A little consideration will show that the two bodies will go on coming nearer to each other till the body B reaches P_1 (*i.e.* within the vision distance). At this point, the bodies will start exchanging signals.They will continue till the body B reaches the point P_2. Beyond this point, the two bodies will be beyond the vision distance.

2. Such problems should, preferably, be attempted graphically.

***Example 9.** *Two ships A and B are moving with velocities of 12 km.p.h. in North and 20 km.p.h. in North-East directions respectively. At midnight, the ship A is 40 km East of B. If they can exchange signals when they are 10 km apart, find when they will begin to signal and how long they will continue ?* **[PU]**

Solution. Given : Velocity of ship A = 12 km.p.h. (North) ; Velocity of ship B = 20 km.p.h. (North-East) ; Distance between the ships at midnight = 40 km and vision distance = 10 km

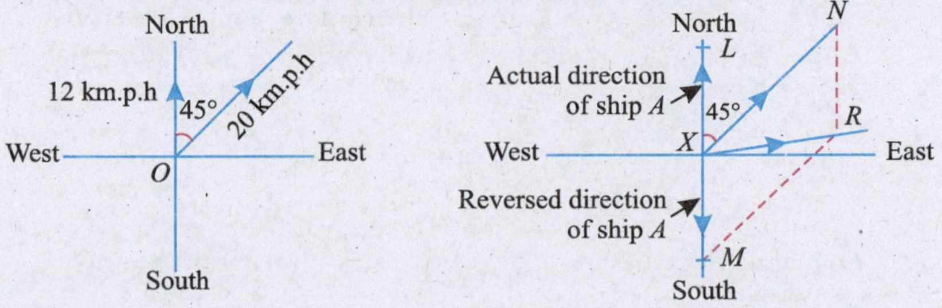

(*a*) Actual volocity diagram (*b*) Relative volocity diagram

Fig 17.

First of all, let us draw the actual velocity diagram of the two ships A and B as shown in Fig. 17 (*a*). Now draw the relative velocity diagram as shown in Fig. 17 (*b*) and as discussed below :

1. First of all, draw a line XL towards North representing the actual direction of motion of the ship A with a velocity of 12 km.p.h.
2. Now cut off XM equal to 12 km to some suitable scale towards South *i.e.* on the opposite direction of the actual motion of the ship A.
3. Now cut off XN equal to 20 km to the scale towards North-East, which represents the actual velocity of the ship B.
4. Complete the parallelogram $XMRN$ with XM and XN as adjacent sides.
5. Join XR, which gives the magnitude and direction of the relative velocity. By measurement, we find that XR = 14.3 km.p.h.

Now draw the exchange of signal diagram for the two ships with the help of their relative velocity diagram as shown in Fig. 18 and as discussed below :

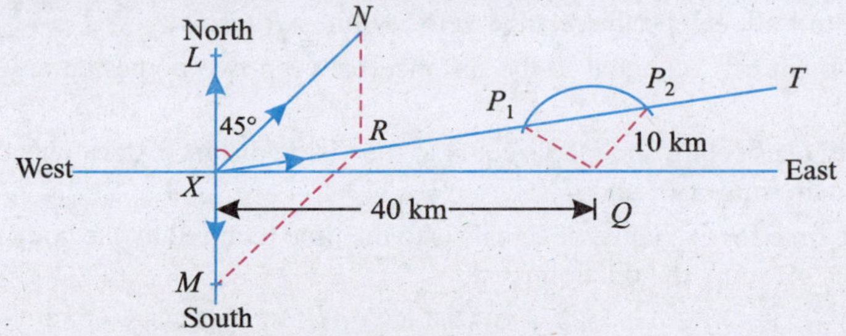

Fig. 18. Exchange of signal diagram

1. First of all, draw the relative velocity diagram and the parallelogram $XMRN$ as discussed above, and extend the line XR to T.
2. At midnight, let the ship A be at X. Therefore the ship B will be 40 km towards East. So mark XQ equal to 40 km to some scale, which represents the position of the ship B at midnight.
3. With Q as centre and radius equal to 10 km to the scale (*i.e.* vision distance) draw an arc meeting the relative velocity line at P_1 and P_2. By measurement, we find that XP_1 = 31.5 km and XP_2 = 47.5 km.
4. Thus the two ships start signalling from $\frac{31.5}{14.3} = 2.20$ hours to $\frac{47.5}{14.3} = 3.32$ hours after midnight. Therefore the two ships will continue signalling for $3.32 - 2.20 = 1.12$ hour = 1 hour 7 min 12 sec. **Ans.**

EXERCISE 3

1. Two ships A and B, at a given instant, are 5 km away from each other. The ship A is South-East of the ship B. If the ship B is travelling at 10 km.p.h. due East and the ship A is travelling at 16 km.p.h. due North, determine (*i*) velocity of the ship A relative to the ship B ; (*ii*) shortest distance between the two ships ; and (*iii*) the time when the two ships are nearest to each other. (**Ans.** 18.9 km.p.h. ; 1.1 km ; 15.5 min.)
2. At noon, a ship A is 10 kilometers due East of another ship B, and is travelling at 14 km.p.h. in a direction 30° West of North. The ship B is travelling North-East at 10 km.p.h. Find the velocity of ship A relative to B and the least distance apart and time when this occurs. (**Ans.** 15 km.p.h. ; 20° ; 37.6 min.)

*3. At midnight, two ships A and B are 80 km apart, the ship B being East of A. The ship A is steaming at a speed of 10 km.p.h. towards South, while ship B is steaming at 22 km.p.h. on South-East course. Find the velocity of ship B relative to A.

If the two ships can exchange signals when they are 30 km apart, find when the two ships can start signalling and how long they can continue to signal ?

(**Ans.** 16.5 km.p.h. ; 1 hour 36 min.)

QUESTIONS

1. What do you understand by the term 'relative velocity' ?
2. What are the different methods of finding out the relative velocity of two bodies moving along inclined directions ? Explain one of them.
3. How would you determine the time, when the two bodies moving along inclined directions, are closest to each other ? How would you find out this distance ?
4. Explain clearly, how would you find out the time for exchange of signals of the two bodies moving along inclined directions ?

OBJECTIVE TYPE QUESTIONS

1. The relative velocity of A with respect to B is the velocity with which A appears to move to an observer sitting in B when it is
   (*a*) at rest (*b*) in motion (*c*) either (*a*) or (*b*)
2. The rain is falling vertically downwards, but it appears to fall at some angle to a man walking along a road. If the man increases his speed, the inclination of the rain with the vertical will
   (*a*) increase (*b*) remain the same (*c*) decrease
3. The relative velocity of two bodies moving along inclined directions is found out by superimposing the actual velocities of both the bodies.
   (*a*) Agree (*b*) Disagree
4. When the two ships are moving along inclined directions, then the time when the two ships will be closest together depends upon
   (*a*) velocity of one of the ships
   (*b*) velocity of both the ships
   (*c*) angle between the two directions
   (*d*) all of the above

ANSWERS

1. (*c*) 2. (*a*) 3. (*b*) 4. (*d*)

CHAPTER

18 Projectile Motion

Contents

1. PROJECTILE

In the previous chapters, we have been discussing the motion of bodies, either in horizontal or vertical directions. But we see that whenever a particle is projected upwards at a certain angle (but not vertical), we find that the particle traces some path in the air and falls on the ground at a point, other than the point of projection. If we study the motion of the particle, we find that the velocity, with which the particle was projected, has two components namely vertical and horizontal.

The function of the vertical component is to project the body vertically upwards, and that of the horizontal is to move the body horizontally in its direction. The combined effect of both the components is to move the particle along a parabolic path. A particle, moving under the combined effect of vertical and horizontal forces, is called a **projectile**. It may be noted that the vertical component of the motion is always subjected to gravitational acceleration, whereas the horizontal component remains constant.

1.1. PROJECTILE MOTION AS A CURVILINEAR MOTION

Motion of an object moving in a curved path is called curvilinear motion. An object is said to be in curvilinear motion if the direction of the velocity of its movement is variable and the path of the object is a curved line. Therefore projectile motion is a type of curvilinear motion.

Examples: Motion of Stone thrown into the air at an angle. Motion of Missile fired into air at an angle.

The other types of curvilinear motion are Motion along a Circular Path and Simple Harmonic Motion.

2. IMPORTANT TERMS

The following terms, which will be frequently used in this chapter, should be clearly understood at this stage :

1. ***Trajectory***. The path, traced by a projectile in the space, is known as trajectory.
2. ***Velocity of projection***. The velocity, with which a projectile is projected, is known as the velocity of projection.
3. ***Angle of projection***. The angle, with the horizontal, at which a projectile is projected, is known as the angle of projection.
4. ***Time of flight***. The total time taken by a projectile, to reach maximum height and to return back to the ground, is known as the time of flight.
5. ***Range***. The distance, between the point of projection and the point where the projectile strikes the ground, is known as the *range*. It may be noted that the range of a projectile may be horizontal or inclined.

3. MOTION OF A BODY THROWN HORIZONTALLY INTO THE AIR

Consider a body at A thrown horizontally into the *air with a horizontal velocity (v) as shown in Fig. 1. A little consideration will show, that this body is subjected to the following two velocities :

1. Horizontal velocity (v),
2. Vertical velocity due to gravitational acceleration.

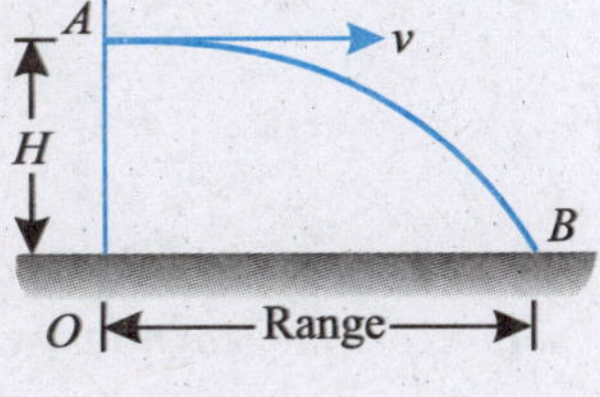

Fig. 1.

It is thus obvious, that the body will have some resultant velocity, with which it will travel into the air. We have already discussed in Article 1 of this chapter, that the vertical component of this velocity is always subjected to gravitational acceleration, whereas the horizontal component remains constant. Thus the time taken by the body to reach the ground, is calculated from the vertical component of the velocity, whereas the horizontal range is calculated from the horizontal component of the velocity. The velocity, with which the body strikes the ground at B, is the resultant of horizontal and vertical velocities.

*Note: For all types of calculation, the air resistance is generally neglected, (until specified otherwise).

***Example 1.** *An aircraft, moving horizontally at 108 km/hr at an altitude of 1000 m towards a target on the ground, releases a bomb which hits it. Estimate the horizontal distance of the aircraft from the target, when it released the bomb. Calculate also the direction and velocity with which the bomb hits the target. Neglect air friction.* **[IGNOU]**

Solution. Given : Horizontal velocity of aircraft, $(V) = 108$ km/hr = 30 m/s

Horizontal distance of the aircraft from the target when it released the bomb

First of all, consider the vertical motion of the bomb due to gravitational acceleration only. In this case, initial velocity $(u) = 0$ and distance covered $(s) = 1000$ m.

Let $\quad t$ = Time required by bomb to reach the ground.

Height of aircraft (s),

$$1000 = ut + \frac{1}{2}gt^2 = 0 + \frac{1}{2} \times 9.8t^2 = 4.9t^2$$

or $\quad t^2 = \dfrac{1000}{4.9} = 204.1 \quad$ or $\quad t = 14.3$ s

∴ Horizontal distance of the aircraft from the target when it released the bomb,

$$H = V \times t = 30 \times 14.3 = 429 \text{ m} \quad \textbf{Ans.}$$

Direction and velocity with which the bomb hits the target

Let $\quad \theta$ = Angle which the bomb makes with vertical when it hits the target.

Final velocity of the bomb in the vertical direction when it hits target (*i.e.* after 14.3 seconds),

$$v = u + gt = 0 + (9.8 \times 14.3) = 140.1 \text{ m/s}$$

$$\therefore \quad \tan\theta = \frac{30}{140.1} = 0.2141 \quad \text{or} \quad \theta = 12.1° \quad \textbf{Ans.}$$

Resultant velocity with which the bomb hits the target

$$= \sqrt{(140.1)^2 + (30)^2} = 143.3 \text{ m/s} \quad \textbf{Ans.}$$

***Example 2.** *A motor cyclist wants to jump over a ditch as shown in Fig. 2.*

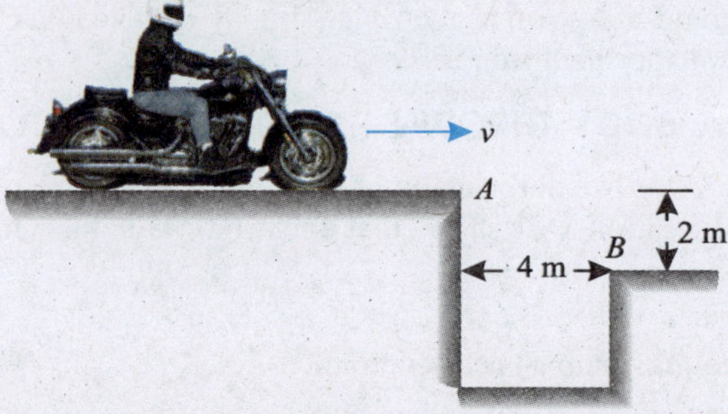

Fig. 2.

Find the necessary minimum velocity at A in km. p. hr. of the motor cycle. Also find the inclination and the magnitude of the velocity of the motor cycle just after clearing the ditch. **[IGNOU]**

Solution. Given : Width of ditch $(x) = 4$ m and vertical distance between A and B $(s) = 2$ m.

Minimum velocity of motor cycle at A

Let $\quad u$ = Minimum velocity of motor cycle at A, and

$\quad t$ = Time taken by the motor cycle to clear the ditch.

First of all, let us consider the vertical motion of the motor cycle from A to B due to gravitational acceleration only. In this case, initial velocity of motor cycle $(u) = 0$.

Vertical distance between A and B (s),

$$2 = ut + \frac{1}{2}gt^2 = 0 + \frac{1}{2} \times 9.8\,t^2 = 4.9\,t^2$$

or $$t^2 = \frac{2}{4.9} = 0.41 \quad \text{or} \quad t = 0.64 \text{ s}$$

∴ Minimum velocity of the motor cycle at A

$$= \frac{4}{0.64} = 6.25 \text{ m/s} = 22.5 \text{ km.p.h.} \quad \textbf{Ans.}$$

Inclination and magnitude of the velocity of motor cycle just after clearing the ditch (i.e. at B)

Let θ = Inclination of the velocity with the vertical.

Final velocity of the motor cycle in the vertical direction at B (*i.e.*after 0.64 second)

$$v = u + gt = 0 + (9.8 \times 0.64) = 6.27 \text{ m/s}$$

or $$\tan \theta = \frac{6.25}{6.27} = 0.9968 \quad \text{or} \quad \theta = 44.9° \quad \textbf{Ans.}$$

Magnitude of the velocity of the motor cycle just after clearing the ditch

$$= \sqrt{(6.25)^2 + (6.27)^2} = 8.85 \text{ m/s} = 31.86 \text{ km.p.h.} \quad \textbf{Ans.}$$

****Example 3.** *An aeroplane is flying on a straight level course at 200 km per hour at a height of 1000 metres above the ground. An anti-aircraft gun located on the ground fires a shell with an initial velocity of 300 m/s, at the instant when the plane is vertically above it. At what inclination, to the horizontal, should the gun be fired to hit the plane ? What time after firing, the gun shell will hit the plane ? What will then be the horizontal distance of the plane from the gun ?*

[AMIE]

Solution. Given : Aeroplane velocity = 200 km.p.h. = 55.56 m/s ; Height of plane = 1000 m and velocity of shell (u) = 300 m/s

Inclination of the gun

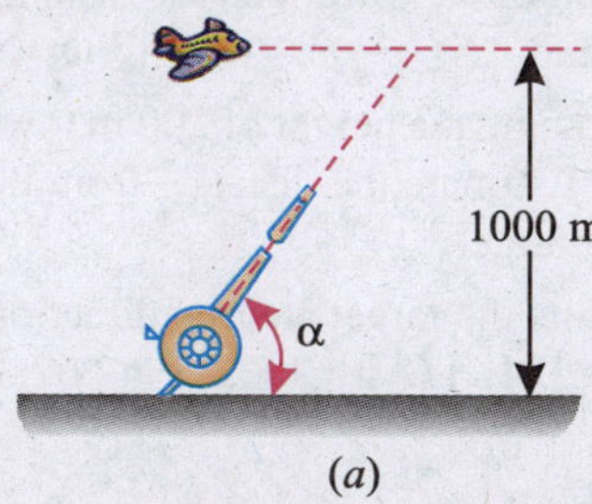

(*a*) (*b*)

Fig. 3.

The actual position of the plane and anti-aircraft gun is shown in Fig. 3 (*a*). Let the anti-aircraft gun be located at O and the plane at the time of firing the shell be at A. Now after sometime, let the plane reach at B, when it is hit by the shell as shown in Fig. 3 (*b*).

Now let α = Inclination of gun with the horizontal,

t = Time taken by the shell to hit the plane.

First of all, consider the motion of the plane. We know that the distance AB

$$= 55.56 \times t = 55.56\, t \text{ metres} \quad ...(i)$$

Now consider motion of the plane. We know that horizontal component of the shell velocity

$$u_x = u \cos \alpha = 300 \cos \alpha$$

and distance, $$AB = 300 \cos \alpha .\, t \quad ...(ii)$$

Equating equation (*i*) and (*ii*),

$$55.56\, t = 300 \cos \alpha . t$$

or $\cos \alpha = \dfrac{55.56}{300} = 0.1852$ or $\alpha = 79.3°$ **Ans.**

Time after firing the shell will hit the plane

Now consider the vertical motion of the shell. We know that,

Vertical component of the shell velocity,

$$u_y = 300 \sin 79.3° = 300 \times 0.9826 = 295 \text{ m/s}$$

Vertical distance OA (s),

$$1000 = u_y t - \frac{1}{2} gt^2 = 295t - \frac{1}{2} \times 9.8\, t^2 = 295\, t - 4.9\, t^2$$

or $4.9\, t^2 - 295\, t + 1000 = 0$

This is a quadratic equation in t,

$$\therefore \quad t = \frac{+\,295 \pm \sqrt{(295)^2 - (4 \times 4.9 \times 1000)}}{2 \times 4.9} = 3.57 \text{ s} \quad \textbf{Ans.}$$

Horizontal distance of the plane from the gun

We know that horizontal distance of the plane from the gun

$$AB = 55.56\, t = 55.56 \times 3.57 = 198.3 \text{ m} \quad \textbf{Ans.}$$

EXERCISE 1

1. A bomber, flying horizontally at a height of 500 m with a velocity of 450 km.p.h., has aimed to hit a target. Find at what distance from the target, he should release the bomb in order to hit the target. (**Ans.** 1262.5 m)
2. A shot is fired horizontally from the top of a tower with a velocity of 100 m/s. If the shot hits the ground after 2 seconds, find the height of the tower and the distance from the foot of the tower, where the shot strikes the ground. (**Ans.** 19.2 m ; 200 m)

*3. A helicopter is moving horizontally at 90 km.p.h. at a height of 200 m towards a target on the ground, which is intended to be shelled. Estimate the distance from the target, where the shell must be released in order to hit the target.

Also find the velocity with which the shell hits the target and the direction of shell at the time of hitting the target. **[MU]** (**Ans.** 173.25 m ; 67.4 m/s ; 21° 46')

4. *MOTION OF A PROJECTILE

Consider a particle projected upwards from a point O at an angle α, with the horizontal, with an initial velocity u m/sec as shown in Fig. 4.

Now resolving this velocity into its vertical and horizontal components,

$$V = u \sin \alpha \quad \text{and } H = u \cos \alpha$$

We know that the vertical component ($u \sin \alpha$) is subjected to retardation due to gravity. The particle will reach maximum height, when the vertical component becomes zero. After this the particle will come down, due to gravity, and this motion will be subjected to acceleration due to gravity.

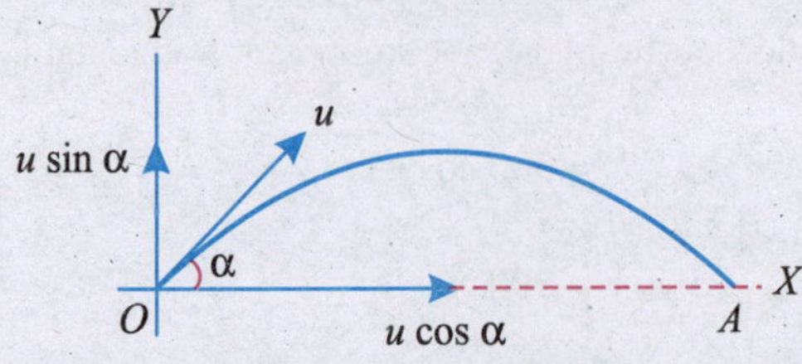

Fig. 4. Projectile on a horizontal plane.

The horizontal component ($u \cos \alpha$) will remain constant, since there is no acceleration or retardation (neglecting air resistance). The combined effect of the horizontal and the vertical components will be to move the particle, along some path in the air and then the particle falls on the ground at some point A, other than the point of projection O as shown in Fig. 4.

5. **EQUATION OF THE PATH OF A PROJECTILE

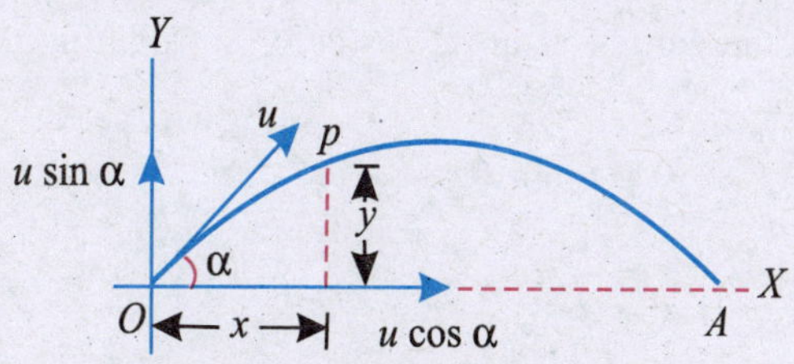

Fig. 5. Path of a projectile.

[MU, BPUT]

Consider a particle projected from a point O at a certain angle with the horizontal. As already discussed, the particle will move along certain path OPA, in the air, and will fall down at A as shown in Fig. 5.

Let u = Velocity of projection, and

α = Angle of projection with the horizontal.

Consider any point P as the position of particle, after t seconds with x and y as co-ordinates as shown in Fig. 5. We know that,

Horizontal component of the velocity of projection.

$$= u \cos \alpha$$

Vertical component $= u \sin \alpha$

$$\therefore \quad y = u \sin \alpha t - \frac{1}{2} g t^2 \quad \text{...}(i)$$

and

$$x = u \cos \alpha \, t$$

or

$$t = \frac{x}{u \cos \alpha}$$

Substituting the value of t in equation (i),

$$y = u \sin \alpha \left(\frac{x}{u \cos \alpha} \right) - \frac{1}{2} g \left(\frac{x}{u \cos \alpha} \right)^2$$

$$= x \tan \alpha - \frac{g x^2}{2 u^2 \cos^2 \alpha} \quad \text{...}(ii)$$

Since this is the equation of a parabola, therefore path of a projectile (or the equation of trajectroy) is also a parabola.

Note. It is an important equation, which helps us in obtaining the following relations of a projectile :

1. Time of flight,
2. Horizontal range, and
3. Maximum height of a projectile.

6. *TIME OF FLIGHT OF A PROJECTILE ON A HORIZONTAL PLANE

It is the time, for which the projectile has remained in the air. We have already discussed in Article 5 (of Equation of the Path of a Projectile in this chapter), that the co-ordinates of a projectile after time t.

$$y = u \sin \alpha t - \frac{1}{2} g t^2$$

We know that when the particle is at A, y is zero. Substituting this value of y in the above equation,

$$0 = u \sin \alpha t - \frac{1}{2} g t^2$$

or $$u \sin \alpha t = \frac{1}{2} g t^2$$

or $$u \sin \alpha = \frac{1}{2} g t \qquad \text{...(Dividing both sides by } t)$$

or $$t = \frac{2u \sin \alpha}{g}$$

Example 4. *A projectile is fired upwards at an angle of 30° with a velocity of 40 m/s. Calculate the time taken by the projectile to reach the ground, after the instant of firing.*

Solution. Given : Angle of projection with the horizontal (α) = 30° and velocity of projection (u) = 40 m/s.

Time taken by the projectile to reach the ground after the instant of firing,

$$t = \frac{2u \sin \alpha}{g} = \frac{2 \times 40 \sin 30^\circ}{g} = \frac{80 \times 0.5}{9.8} = 4.08 \text{ s} \quad \textbf{Ans.}$$

7. *HORIZONTAL RANGE OF A PROJECTILE

We have already discussed, that the horizontal distance between the point of projection and the point, where the projectile returns back to the earth is called horizontal range of a projectile. We have also discussed in Articles 4 and 6 of this chapter, that the

Horizontal velocity of a projectile

$$= u \cos \alpha$$

Time of flight, $$t = \frac{2u \sin \alpha}{g}$$

∴ Horizontal range $$= \text{Horizontal velocity} \times \text{Time of flight}$$

$$= u \cos \alpha \times \frac{2u \sin \alpha}{g} = \frac{2u^2 \sin \alpha \cos \alpha}{g}$$

$$R = \frac{u^2 \sin 2\alpha}{g} \qquad \text{...}(\because \sin 2\alpha = 2 \sin \alpha \cos \alpha)$$

Note. For a given velocity of projectile, the range will be maximum when $\sin 2\alpha = 1$. Therefore

$$2\alpha = 90^\circ \quad \text{or} \quad \alpha = 45^\circ$$

or $$R_{max} = \frac{u^2 \sin 90^\circ}{g} = \frac{u^2}{g} \qquad \text{...}(\because \sin 90^\circ = 1)$$

Example 5. *A ball is projected upwards with a velocity of 15 m/s at an angle of 25° with the horizontal. What is the horizontal range of the ball ?*

Solution. Given : Velocity of projection (u) = 15 m/s and angle of projection with the horizontal (α) = 25°

Horizontal range of the ball,

$$R = \frac{u^2 \sin 2\alpha}{g} = \frac{(15)^2 \times \sin(2 \times 25°)}{g} = \frac{225 \sin 50°}{g}$$

$$= \frac{225 \times 0.766}{9.8} = 17.6 \text{ m} \quad \textbf{Ans.}$$

8. *MAXIMUM HEIGHT OF A PROJECTILE ON A HORIZONTAL PLANE

[WBUT, AMIE]

We have already discussed that, the

Vertical component of the initial velocity of a projectile

$$= u \sin \alpha \quad \quad ...(i)$$

Vertical component of final velocity

$$= 0 \quad \quad ...(ii)$$

$\therefore$ Average velocity of (*i*) and (*ii*),

$$= \frac{u \sin \alpha + 0}{2} = \frac{u \sin \alpha}{2} \quad \quad ...(iii)$$

Let H be the maximum height reached by the particle and t be the time taken by the particle to reach maximum height *i.e.*,to attain zero velocity from ($u \sin \alpha$). We have also discussed that,

Time taken by the projectile to reach the maximum height,

$$= \frac{u \sin \alpha}{g}$$

$\therefore$ Maximum height of the projectile,

$$H = \text{Average vertical velocity} \times \text{Time}$$

$$= \frac{u \sin \alpha}{2} \times \frac{u \sin \alpha}{g} = \frac{u^2 \sin^2 \alpha}{2g}$$

Example 6. *A bullet is fired with a velocity of 100 m/s at an angle of 45° with the horizontal. How high the bullet will rise ?*

Solution. Given : Velocity of projection (u) = 100 m/s and angle of projection with the horizontal (α) = 45°

Maximum height to which the bullet will rise,

$$H = \frac{u^2 \sin^2 \alpha}{2g} = \frac{(100)^2 \times \sin^2 45°}{2 \times 9.8} = \frac{10000 \times (0.707)^2}{19.6} \text{ m}$$

$$= 255.1 \text{ m} \quad \textbf{Ans.}$$

Example 7. *If a particle is projected inside a horizontal tunnel which is 5 metres high with a velocity of 60 m/s, find the angle of projection and the greatest possible range.*

Solution. Given : Height of the tunnel (H) = 5 m and velocity of projection (u) = 60 m/s.

Angle of projection

Let α = Angle of projection.

Height of tunnel (H),

$$5 = \frac{u^2 \sin^2 \alpha}{2g} = \frac{(60)^2 \sin^2 \alpha}{2 \times 9.8} = 183.7 \sin^2 \alpha$$

or $$\sin^2 \alpha = \frac{5}{183.7} = 0.0272$$

or $\sin \alpha = 0.1650$ or $\alpha = 9.5°$ **Ans.**

Greatest possible range

Greatest possible range, $$R = \frac{u^2 \sin 2\alpha}{g} = \frac{(60)^2 \sin (2 \times 9.5°)}{9.8} = \frac{(60)^2 \sin 19°}{9.8} \text{ m}$$

$$= \frac{3600 \times 0.3256}{9.8} = 119.6 \text{ m}$$ **Ans.**

***Example 8.** *A body is projected at such an angle that the horizontal range is three times the greatest height. Find the angle of projection.* **[WBUT, UPTU]**

Solution. Given : Horizontal range (R) = 3 H (where H is the greatest height). ...(*i*)

Let α = Angle of projection.

Horizontal range,

$$R = \frac{u^2 \sin 2\alpha}{g}$$

Greatest height, $$H = \frac{u^2 \sin^2 \alpha}{2g}$$

Substituting these values of R and H in the given equation (*i*),

$$\frac{u^2 \sin 2\alpha}{g} = 3 \times \frac{u^2 \sin^2 \alpha}{2g}$$

or $$\frac{u^2 \times 2 \sin \alpha \cos \alpha}{g} = 3 \times \frac{u^2 \sin^2 \alpha}{2g}$$ ...($\because$ $2\alpha = 2 \sin \alpha \cos \alpha$)

or $$2 \cos \alpha = 1.5 \sin \alpha$$

or $\tan \alpha = \frac{2}{1.5} = 1.333$ or $\alpha = 53.1°$ **Ans.**

***Example 9.** *A particle is thrown with a velocity of 5 m/s at an elevation of 60° to the horizontal. Find the velocity of another particle thrown at an elevation of 45° which will have (a) equal horizontal range, (b) equal maximum height, and (c) equal time of flight.* **[IGNOU]**

Solution. Given : Velocity of projection of first particle (u_1) = 5 m/s ; Angle of projection of first particle with the horizontal (α_1) = 60° and angle of projection of second particle with the horizontal (α_2) = 45°

Let u_2 = Velocity of projection of the second particle.

(a) Velocity of the second particle for equal horizontal range

Horizontal range of a projectile,

$$R = \frac{u^2 \sin 2\alpha}{g}$$

$\therefore$ For equal horizontal range,

$$\frac{u_1^2 \sin 2\alpha_1}{g} = \frac{u_2^2 \sin 2\alpha_2}{g}$$

or $$(5)^2 \sin (2 \times 60°) = u_2^2 \sin (2 \times 45°)$$

or $$u_2^2 = 25 \times \frac{\sin 120°}{\sin 90°} = 25 \times \frac{0.866}{1.0} = 21.65$$

or $u_2 = 4.65$ m/s **Ans.**

(b) Velocity of the second particle for equal maximum height

We know that maximum height of a projectile,

$$H = \frac{u^2 \sin^2 \alpha}{2g}$$

∴ For equal maximum height,

$$\frac{u_1^2 \sin^2 \alpha_1}{2g} = \frac{u_2^2 \sin^2 \alpha_2}{2g}$$

or $$(5)^2 \sin^2 60° = {u_2}^2 \sin^2 45°$$

or $${u_2}^2 = 25 \times \frac{\sin^2 60°}{\sin^2 45°} = 25 \times \frac{(0.866)^2}{(0.707)^2} = 37.5$$

or $$u_2 = 6.12 \text{ m/s}$$ **Ans.**

(c) Velocity of the second particle for equal time of flight

Time of flight of a projectile,

$$t = \frac{2u \sin \alpha}{g}$$

∴ For equal time of flight,

$$\frac{2u_1 \sin \alpha_1}{g} = \frac{2u_2 \sin \alpha_2}{g}$$

or $$2 \times 5 \sin 60° = 2u_2 \sin 45°$$

or $$u_2 = 5 \times \frac{\sin 60°}{\sin 45°} = 5 \times \frac{0.866}{0.707} = 6.12 \text{ m/s}$$ **Ans.**

****Example 10.** *A particle is projected from the base of a hill whose shape is that of a right circular cone with axis vertical. The projectile grazes the vertex and strikes the hill again at a point on the base. If θ be the semi-vertical angle of the cone, h its height, u the initial velocity of the projectile and α the angle of projection measured from the horizontal, show that :*

$$\tan \alpha = 2 \cot \theta \quad \text{and} \quad u^2 = gh\left(2 + \frac{1}{2}\tan^2 \theta\right).$$

where g is acceleration due to gravity. **[JNTU, AMIE]**

Solution. Given : Semi-vertical angle = θ ; Initial velocity of the projectile = u and angle of projection with the horizontal = α

(i) Maximum height through which the particle will rise,

$$h = \frac{u^2 \sin^2 \alpha}{2g}$$

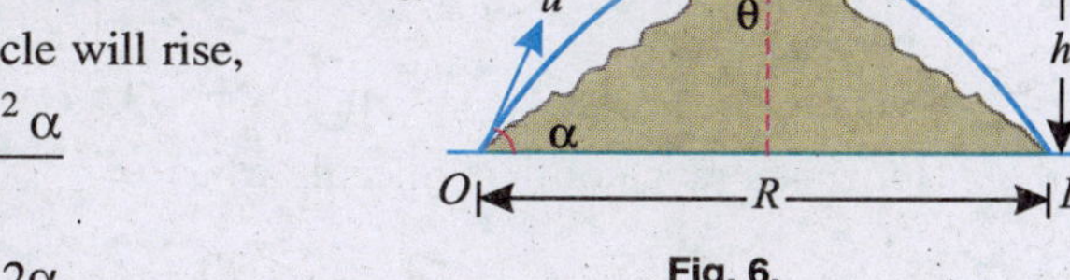

Fig. 6.

Horizontal range OB, $$R = \frac{u^2 \sin 2\alpha}{g}$$

From the geometry of the figure we find that,

$$\cot \theta = \frac{h}{\frac{R}{2}} = \frac{2h}{R} = \frac{2 \times \frac{u^2 \sin^2 \alpha}{2g}}{\frac{u^2 \sin 2\alpha}{g}} = \frac{\sin^2 \alpha}{\sin 2\alpha} = \frac{\sin^2 \alpha}{2 \sin \alpha \cos \alpha} = \frac{\tan \alpha}{2}$$

or $$\tan \alpha = 2 \cot \theta$$ **Ans.**

(*ii*) The above equation, may be written as,

$$\frac{1}{\tan\alpha} = \frac{1}{2\cot\theta} \quad \text{or} \quad \cot\alpha = \frac{1}{2}\tan\theta$$

or $$\cot^2\alpha = \frac{1}{4}\tan^2\theta \qquad \text{...(Squaring both sides)}$$

Maximum height through which the particle will rise,

$$h = \frac{u^2\sin^2\alpha}{2g}$$

or $$u^2 = \frac{2gh}{\sin^2\alpha} = 2gh\ \text{cosec}^2\,\alpha$$

$$u^2 = 2gh(1+\cot^2\alpha) = 2gh\left(1+\frac{1}{4}\tan^2\theta\right)$$

$$\text{...}\left(\text{Substituting } \cot^2\alpha = \frac{1}{4}\tan^2\theta\right)$$

$$= gh\left(2+\frac{1}{2}\tan^2\theta\right) \quad \textbf{Ans.}$$

***Example 11.** *A projectile is aimed at a mark on the horizontal plane through the point of projection. It falls 12 metres short when the angle of projection is 15° ; while it overshoots the mark by 24 metres when the same angle is 45°. Find the angle of projection to hit the mark. Assume no air resistance.* **[RGPV]**

Solution. Given : When angle of projection with the horizontal (α_1) = 15°, horizontal range (R_1) = R – 12 m and when angle of projection with the horizontal (α_2) = 45°, horizontal range (R_2) = R + 24 m (where R is the horizontal range).

Let u = Velocity of projection, and

α = Angle of projection to hit the mark.

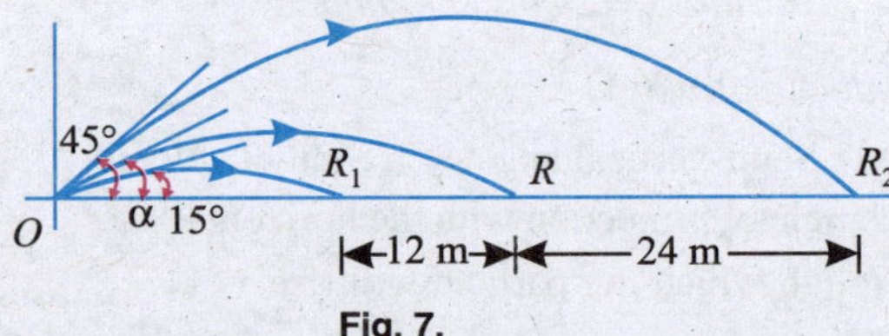

Fig. 7.

Horizontal range of the projectile when $\alpha = 15°$,

$$R_1 = \frac{u^2\sin 2\alpha_1}{g} = \frac{u^2\sin(2\times 15°)}{g}$$

or $$(R-12) = \frac{u^2\sin 30°}{g} = \frac{u^2\times 0.5}{g} \qquad \text{...(i)}$$

Similarly, $$R_2 = \frac{u^2\sin 2\alpha_2}{g} = \frac{u^2\sin(2\times 45°)}{g}$$

or $$(R+24) = \frac{u^2\sin 90°}{g} = \frac{u^2\times 1}{g} \qquad \text{...(ii)}$$

Dividing equation (*i*) by (*ii*),

$$\frac{R-12}{R+24} = \frac{0.5}{1} \quad \text{or} \quad R - 12 = 0.5R + 12$$

$$\text{or} \quad 0.5\,R = 12 + 12 = 24 \quad \text{or} \quad R = \frac{24}{0.5} = 48 \text{ m}$$

Substituting the value of R in equation (*i*),

$$48 - 12 = \frac{u^2 \times 0.5}{g} = \frac{u^2}{2g}$$

$$\text{or} \quad u^2 = 36 \times 2g = 72g$$

Horizontal distance between the point of projection and the mark (R),

$$48 = \frac{u^2 \sin 2\alpha}{g} = \frac{(72g)\sin 2\alpha}{g} = 72 \sin 2\alpha$$

$$\text{or} \quad \sin 2\alpha = \frac{48}{72} = 0.667 \quad \text{or} \quad 2\alpha = 41.8°$$

$$\text{or} \quad \alpha = 20.9° \quad \textbf{Ans.}$$

****Example 12.** *A projectile fired from the edge of a 150 m high cliff with an initial velocity of 180 m/s at an angle of elevation of 30° with the horizontal. Neglecting air resistance find :*

1. *The greatest elevation above the ground reached by the projectile ; and*
2. *Horizontal distance from the gun to the point, where the projectile strikes the ground.*

[WBUT, AUT, AMIE]

Solution. Given : Height of cliff = 150 m ; Velocity of projection (u) = 180 m/s and angle of projection with the horizontal (α) = 30°.

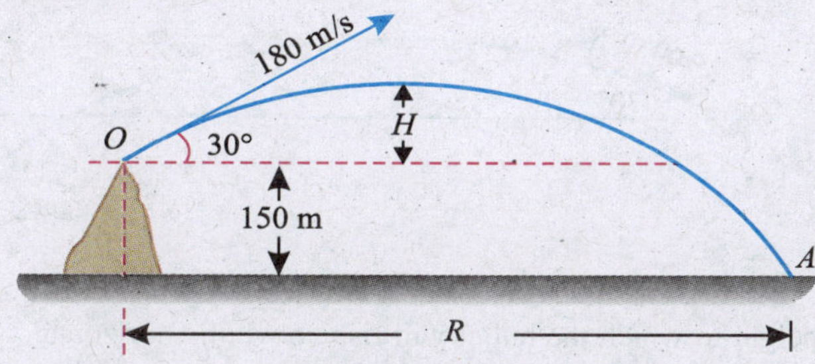

Fig. 8.

1. *The greatest elevation above the ground reached by the projectile*

Maximum height to which the projectile will rise above the edge O of the cliff,

$$H = \frac{u^2 \sin^2 \alpha}{2g} = \frac{(180)^2 \sin^2 30°}{2 \times 9.8} = \frac{(180)^2 \times (0.5)^2}{19.6} = 413.3 \text{ m}$$

∴ Greatest elevation above the ground reached by the projectile,

$$s = 413.3 + 150 = 563.3 \text{ m} \quad \textbf{Ans.}$$

2. *The horizontal distance from the gun to the point, where the projectile strikes the ground*

First of all, consider vertical motion of the projectile from the edge of the cliff to the maximum height. We know that the time taken by the projectile to reach maximum height from the edge of the cliff,

$$t_1 = \frac{u \sin \alpha}{g} = \frac{180 \sin 30°}{9.8} = \frac{180 \times 0.5}{9.8} = 9.2 \text{ s} \quad ...(i)$$

Now consider vertical motion of the projectile from the maximum height to the ground due to gravitational acceleration only. In this case, $u = 0$ and $s = 563.3$ m.

Let t_2 = Time taken by the projectile to reach the ground from the maximum height.

Vertical distance (s),

$$563.3 = u t_2 + \frac{1}{2} g t_2^2 = 0 + \frac{1}{2} \times 9.8 t_2^2 = 4.9 t_2^2$$

or $$t^2 = \frac{563.3}{4.9} = 115 \quad \text{or} \quad t_2 = 10.7 \text{ s} \quad ...(ii)$$

∴ Total time taken by the projectile to reach the ground from the edge of the cliff

$$= t_1 + t_2 = 9.2 + 10.7 = 19.9 \text{ s}$$

Horizontal distance from the gun to the point, where the projectile strikes the ground,

R = Horizontal components of velocity × Time

$= 180 \cos 30° \times 19.9 = (180 \times 0.866) \times 19.9$ m

$= 3102$ m $= 3.102$ km **Ans.**

***Example 13.** *A bullet is fired upwards at an angle of 30° to the horizontal from a point P on a hill, and it strikes a target which is 80 m lower than P. The initial velocity of bullet is 100 m/s. Calculate the actual velocity with which the bullet will strike the target.* **[GTU, WBUT]**

Solution. Given : Angle of projection with the horizontal (α) = 30° and initial velocity of projection (u) = 100 m/s

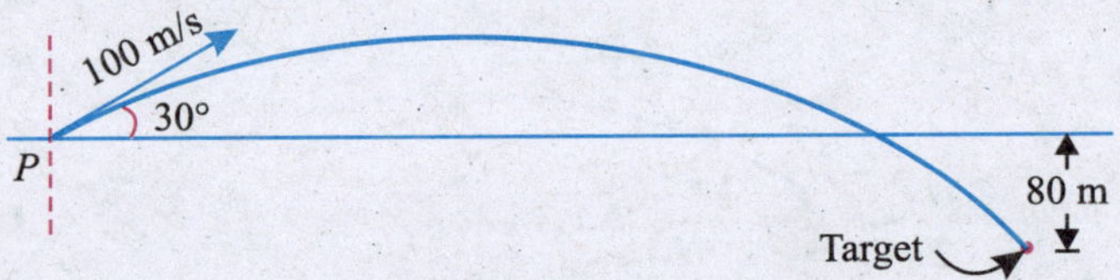

Fig. 9.

Maximum height to which the bullet will rise above the horizontal,

$$H = \frac{u^2 \sin^2 \alpha}{2g} = \frac{(100)^2 \sin^2 30°}{2 \times 9.8} = \frac{(100)^2 \times (0.5)^2}{19.6} = 127.6 \text{ m}$$

First of all, consider vertical motion of the bullet to reach maximum height. We know that time taken by the bullet to reach maximum height,

$$t_1 = \frac{u \sin \alpha}{g} = \frac{100 \sin 30°}{9.8} = \frac{100 \times 0.5}{9.8} = 5.1 \text{ s} \quad ...(i)$$

Now consider vertical motion of the bullet from the maximum height to the target due to gravitational acceleration only.

In this case, initial velocity (u) = 0 and total distance (s) = 127.6 + 80 = 207.6 m.

Let t_2 = Time taken by the bullet to reach the target from the maximum height.

Vertical distance (s),

$$207.6 = u t_2 + \frac{1}{2} g t_2^2 = (0 \times t_2) + \left(\frac{1}{2} \times 9.8 t_2^2\right) = 4.9 t_2^2$$

or $$t_2^2 = \frac{207.6}{4.9} = 42.4 \quad \text{or} \quad t_2 = 6.5 \text{ s} \quad ...(ii)$$

∴ Total time required for the flight of the bullet

$$= t_1 + t_2 = 5.1 + 6.5 = 11.6 \text{ s}$$

Final velocity of the bullet in the vertical direction, when it strikes the target (*i.e.* after 6.5 seconds),

$$v = u + gt = (0) + (9.8 \times 6.5) = 63.7 \text{ m/s}$$

Horizontal component of the initial velocity of the bullet

$$= u \cos \alpha = 100 \cos 30° = 100 \times 0.866 = 86.6 \text{ m/s}$$

∴ Actual velocity with which the bullet will strike the target

$$= \sqrt{(63.7)^2 + (86.6)^2} = 107.5 \text{ m/s} \quad \textbf{Ans.}$$

****Example 14.** *A shot is fired with a velocity if 30 m/s from a point 15 metres in front of a vertical wall 6 metres high. Find the angle of projection, to the horizontal for the shot just to clear the top of the wall.* **[AUT, JNTU, WBUT]**

Solution. Given : Initial velocity = 30 m/s; Distance of point of projection from wall (OB) = 15 m and height of the wall AB = 6 m.

Let α = Angle of projection.

∴ Vertical component of the velocity of the projection = 30 cos α

First of all, consider vertical motion of the shot. Let the bullet take t seconds to cross the wall. In order to enable the shot just to clear the top of the wall, it must rise 6 metres high in t seconds. In this case, s = 6 m and $u = 30 \sin \alpha$

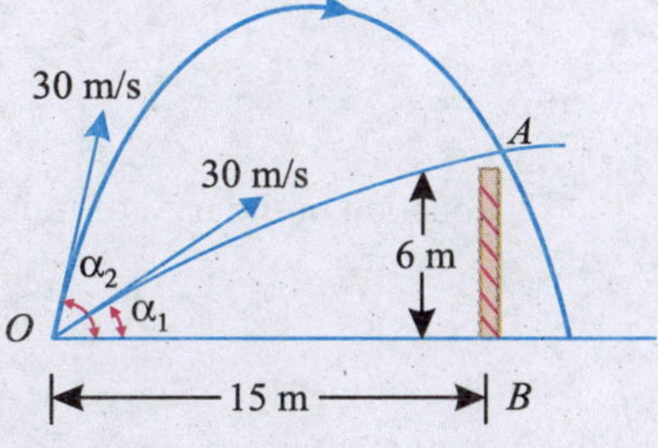

Fig. 10.

We know that vertical distance travelled by the shot (s),

$$6 = ut - \frac{1}{2} g t^2 = (30 \sin \alpha)t - \frac{1}{2} \times 9.8 t^2$$

$$= (30 \sin \alpha)\, t - 4.9\, t^2 \quad ...(i)$$

Now consider the horizontal motion of the shot. In order to enable the shot just to clear the top of the wall, it must traverse 15 m in t seconds.

∴ $$15 = \text{Horizontal velocity} \times \text{Time} = (30 \cos \alpha)\, t$$

or $$t = \frac{15}{30 \cos \alpha} = \frac{0.5}{\cos \alpha} \quad ...(ii)$$

Substituting the value of t in equation (i),

$$6 = 30 \sin \alpha \left(\frac{0.5}{\cos \alpha}\right) - 4.9 \left(\frac{0.5}{\cos \alpha}\right)^2$$

or $$6 = 15 \tan \alpha - 1.225 \sec^2 \alpha \qquad \ldots\left(\because \frac{1}{\cos^2 \alpha} = \sec^2 \alpha\right)$$

$$= 15 \tan \alpha - 1.225 (1 + \tan^2 \alpha) \qquad \ldots(\because \sec^2 \alpha = 1 + \tan^2 \alpha)$$

$$= 15 \tan \alpha - 1.225 - 1.225 \tan^2 \alpha$$

or $1.225 \tan^2 \alpha - 15 \tan \alpha + 7.225 = 0$

This is quadratic equation in ($\tan \alpha$),

$\therefore$ $$\tan \alpha = \frac{+15 \pm \sqrt{15^2 - 4 \times 1.225 \times 7.225}}{2 \times 1.225} = 11.74 \quad \text{or} \quad 0.5$$

or $$\alpha = 85.1° \quad \text{or} \quad 26.6° \quad \textbf{Ans.}$$

***Example 15.** *Find the least initial velocity which a projectile may have, so that it may clear a wall 3.6 m high and 4.8 m distant (from the point of projection) and strike the horizontal plane through the foots of the wall at a distance 3.6 m beyond the wall. The point of projection is at the same level as the foot of the wall.* **[MU]**

Solution. Given : Height of wall = 3.6 m ; Distance of the wall from the point of projection (OB) = 4.8 m and distance of strike point from the foot of the wall (BC) = 3.6 m.

Let u = Initial velocity of projection,

α = Angle of projection.

Range OC of the projectile (R),

$$4.8 + 3.6 = \frac{u^2 \sin 2\alpha}{g}$$

or $$8.4 = \frac{u^2\, 2 \sin \alpha \cos \alpha}{g} \qquad \ldots(\because \sin 2\alpha = 2 \sin \alpha \cos \alpha)$$

or $$u^2 = \frac{8.4g}{2 \sin \alpha \cos \alpha} = \frac{4.2g}{\sin \alpha \cos \alpha} \qquad \ldots(i)$$

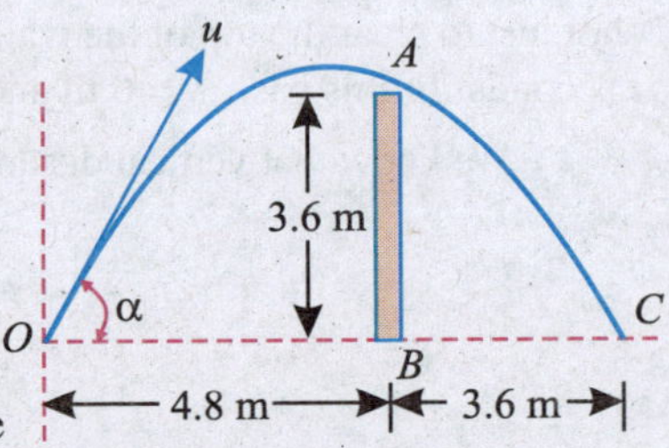

Fig. 11.

Equation of the path of trajectory,

$$y = x \tan \alpha - \frac{gx^2}{2u^2 \cos^2 \alpha}$$

Substituting the value of x = 4.8 m and y = 3.6 m in the above equation,

$$3.6 = 4.8 \tan \alpha - \frac{g\,(4.8)^2}{2u^2 \cos^2 \alpha}$$

$$= 4.8 \tan \alpha - \frac{11.52\, g}{\cos^2 \alpha} \times \frac{1}{u^2}$$

Now substituting the value of u^2 from equation (i),

$$3.6 = 4.8 \tan \alpha - \frac{11.52\, g}{\cos^2 \alpha} \times \frac{1}{\frac{4.2g}{\sin \alpha \cos \alpha}}$$

$$= 4.8 \tan \alpha - 2.74 \tan \alpha = 2.06 \tan \alpha$$

or $$\tan \alpha = \frac{3.6}{2.06} = 1.748 \quad \text{or} \quad \alpha = 60.2°$$

Now substituting the value of α in equation (*i*),

$$u^2 = \frac{4.2 \times 9.8}{\sin 60.2° \cos 60.2°} = \frac{41.16}{0.8681 \times 0.4965} = 95.5$$

or $$u = 9.77 \text{ m/s} \quad \textbf{Ans.}$$

Example 16. *Two guns are pointed at each other, one upward at an angle of 30°, and the other at the same angle of depreesion the muzzles being 30 m apart. If the guns are shot with velocities of 350 m/s upwards and 300 m/s downwards respectively, find when and where they will meet ?*

Solution. Given : Angle of projection of both the guns (α) = 30° ; Velocity of projection of first gun (v_A) = 350 m/s ; Velocity of projection of second gun (v_B) = 300 m/s and distance between the muzzles = 30 m

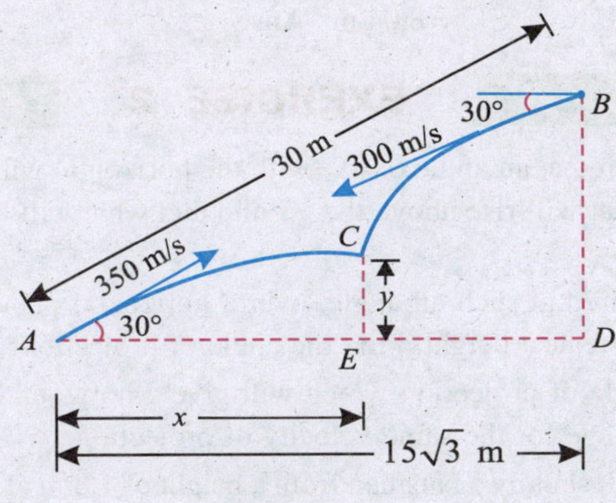

Fig. 12.

Time when the shots meet after they leave the guns

Let the two shots meet at C as shown in Fig. 12.

Now let t = Time in seconds, when the two shots meet after they leave the guns.

x = Horizontal distance between A and C

y = Vertical distance between A and C

$\therefore$ Horizontal distance between A and B (*i.e.* AD)

$$= 30 \cos 30° = \frac{30\sqrt{3}}{2} = 15\sqrt{3} \text{ m}$$

Distance covered by the shot A in t seconds,

$$x = \text{Horizontal component of } v_A \times t$$

$$= 350 \cos 30° \times t = 350 \frac{\sqrt{3}}{2} \times t = 175\sqrt{3}\, t \quad ...(i)$$

Similarly, distance covered by the shot B in t seconds

$$(15\sqrt{3} - x) = 300 \cos 30° \times t = 300 \frac{\sqrt{3}}{2} \times t = 150\sqrt{3}\, t \quad ...(ii)$$

Adding equation (*i*) and (*ii*),

$$15\sqrt{3} = 175\sqrt{3} \times t + 150\sqrt{3} \times t$$

or $$15 = 175\, t + 150\, t = 325\, t$$

or $$t = \frac{15}{325} = 0.046 \text{ s} \quad \textbf{Ans.}$$

Point where the two shots meet

Substituting this value of t in equation (i),

$$x = 175\sqrt{3} \times 0.046 = 13.94 \text{ m} \quad \textbf{Ans.}$$

Vertical component of v_A

$$= 350 \sin 30° = 300 \times 0.5 = 175 \text{ m/s}$$

∴ Vertical distance between A and C,

$$y = ut - \frac{1}{2}gt^2 = (175 \times 0.046) - \left(\frac{1}{2} \times 9.8\,(0.046)^2\right) \text{m}$$

$$= 8.04 \text{ m} \quad \textbf{Ans.}$$

EXERCISE 2

1. A bullet is fired at an angle of 45° with the horizontal with a velocity of 275 m/s. How high the bullet will rise above the ground and what will be its horizontal range ? Take $g = 9.8 \text{ m/s}^2$ (**Ans.** 1928.6 m ; 7716.8 m)
2. A bullet is fired at such an angle, over a horizontal plane, that its horizontal range is equal to its greatest height. Find the angle of projection. (**Ans.** 75° 58′)
3. Find the angle of projection which will give a horizontal range equal to 3/4 th of the maximum range for the same velocity of projection. (**Ans.** 24° 18′ ; 65° 42′)

*4. A cricket ball, shot by a batsman from a height of 1.8 m at an angle of 30° with horizontal with a velocity of 18 m/s is caught by a fields man at a height of 0.6 m from the ground. How far apart were the two players ? **[AMIE]** (**Ans.** 30.56 m)

*5. A jet of water, discharged from a nozzle, hits a screen 6 m away at a height of 4 m above the centre of a nozzle. When the screen is moved 4 m further away, the jet hits it again at the same point. Assuming the curve described by the jet to be parabolic, find the angle at which the jet is projected. (**Ans.** 46° 51′)

6. A bird is sitting on the top of a tree 10 m high. With what velocity should a person, standing at a distance of 25 m from the tree, throws a stone at an angle of 30° with the horizontal so as to hit the bird ? (**Ans.** 30.35 m/s)

*7. A projectile is fired from a point at 125 m/s so as to strike a point at a horizontal distance of 1000 m and 200 m higher than the point of firing. Neglecting air resistance, calculate (i) the angle with the horizontal, at which the projectile should be fired in order to strike the point in minimum time, and (ii) time taken by the projectile to strike the point. **[PU, IGNOU]** (**Ans.** 32° 46′ ; 9.5 s)

Fig. 13.

9. *VELOCITY AND DIRECTION OF MOTION OF A PROJECTILE, AFTER A GIVEN INTERVAL OF TIME, FROM THE INSTANT OF PROJECTION

Consider a projectile projected from O as shown in Fig. 13.

Let u = Initial velocity of projection, and

α = Angle of projection with the horizontal.

After t seconds, let the projectile reach at any point P, as shown in Fig. 13.

Now let v = Velocity of the projectile at P, and

θ = Angle, which the projectile at P makes with the horizontal.

Vertical component of the initial velocity

$$= u \sin \alpha$$

Vertical component of the final velocity after t seconds

$$= v \sin \theta$$

This change in velocity (*i.e.*, from $u \sin \alpha$ to $v \sin \theta$) is because of the retardation (g) due to gravity,

$$\therefore \quad v \sin \theta = u \sin \alpha - gt \quad \quad ...(i)$$

...(Minus sign due to upward direction)

We also know that the horizontal component of these two velocities does not change. Therefore

$$v \cos \theta = u \cos \alpha \quad \quad ...(ii)$$

Squaring the equations (*i*) and (*ii*) and adding the same,

$$v^2 \sin^2 \theta + v^2 \cos^2 \theta = u^2 \sin^2 \alpha + g^2 t^2 - 2u (\sin \alpha)\, gt + u^2 \cos^2 \alpha$$

or
$$v^2 (\sin^2 \theta + \cos^2 \theta) = u^2 (\sin^2 \alpha + \cos^2 \alpha) + g^2 t^2 - 2u (\sin \alpha)\, gt$$

or
$$= u^2 + g^2 t^2 - 2u (\sin \alpha)\, gt \quad \quad ...(\because \sin^2 \alpha + \cos^2 \alpha = 1)$$

or
$$v = \sqrt{u^2 + g^2 t^2 - 2u (\sin \alpha)\, gt}$$

The angle which the projectile makes with horizontal at P may be found out by dividing the equation (*i*) by (*ii*), *i.e.*

$$\frac{v \sin \theta}{v \cos \theta} = \frac{u \sin \alpha - gt}{u \cos \alpha}$$

or
$$\tan \theta = \frac{u \sin \alpha - gt}{u \cos \alpha} = \frac{\text{Vertical velocity after } t \text{ seconds}}{\text{Horizontal component of initial velocity}}$$

***Example 17.** *A projectile is fired with a velocity of 80 m/s at an elevation of 65°. Find its velocity and direction after 5 seconds of firing.* **[BPUT]**

Solution. Given : Initial velocity of projection (u) = 80 m/s ; Angle of projection with the horizontal (α) = 65° and time (t) = 5 s.

Velocity of the projectile

Velocity of the projectile,

$$v = \sqrt{u^2 + g^2 t^2 - 2u (\sin \alpha)\, gt.}$$

$$= \sqrt{80^2 + (9.8)^2 \times (5)^2 - 2 \times 80 \times (\sin 65°) \times 9.8 \times 5} \text{ m/s}$$

$$= \sqrt{6400 + 2401 - 160 \times 0.9063 \times 49} \text{ m/s}$$

$$= \sqrt{1696} = 41.2 \text{ m/s} \quad \textbf{Ans.}$$

Direction of the projectile

Let θ = Angle which the projectile makes with the horizontal.

We also know that, $\tan\theta = \dfrac{u\sin\alpha - gt}{u\cos\alpha} = \dfrac{80\sin 65° - 9.8\times 5}{80\cos 65°} = \dfrac{(80\times 0.9063) - 49}{80\times 0.4226}$

$= 0.6952$ or $\theta = 34.8°$ **Ans.**

Example 18. *A particle is projected upwards with a velocity of 100 m/s at an angle of 45° to the horizontal. When it reaches a certain point P, it is found to be moving at an angle of 30° to the horizontal. Find the time for the particle to reach the point P and distance OP.*

Solution. Given : Initial velocity of projectile $(u) = 100$ m/s ; Angle of projection $(\alpha) = 45°$ and angle of projection at point $P\ (\theta) = 30°$

Let t = Time for the particle to reach the point P from O.

We know that, $$\tan\theta = \frac{u\sin\alpha - gt}{u\cos\alpha}$$

or $$\tan 30° = \frac{100\sin 45° - 9.8t}{100\cos 45°}$$

or $$0.5774 = \frac{(100\times 0.707) - 9.8t}{100\times 0.707}$$

or $$40.82 = 70.7 - 9.8\,t$$

$\therefore$ $$t = \frac{70.7 - 40.82}{9.8} = 3.05 \text{ s} \quad \textbf{Ans.}$$

Fig. 14.

Distance OP

First of all, consider the horizontal motion of the particle. We know that the horizontal distance OA

$$x = \text{Horizontal component of velocity} \times \text{Time}$$
$$= 100\cos 45° \times 3.05 = 100\times 0.707\times 3.05 = 215.6 \text{ m} \quad ...(i)$$

Now consider vertical motion of the particle. We know that vertical component of the velocity.

$$u_y = 100\sin 45° = 100\times 0.707 = 70.7 \text{ m}$$

Vertical distance AP, $$y = u_y t - \frac{1}{2}.gt^2 = (70.7\times 3.05) - \frac{1}{2}\times 9.8\ (3.05)^2$$
$$= 215.6 - 45.6 = 170 \text{ m}$$

$\therefore$ Distance, $$OP = \sqrt{(215.6)^2 + (170)^2} = 274.6 \text{ m} \quad \textbf{Ans.}$$

***Example 19.** *A projectile is fired with a velocity of 500 m/s at an elevation of 35°. Neglecting air friction, find the velocity and direction of the projectile moving after 29 seconds and 30 seconds of firing.* **[JNTU, WBUT]**

Solution. Given : Velocity of projection $(u) = 500$ m/s and angle of projection $(\alpha) = 35°$

Velocity of the projectile after 29 and 30 seconds of firing

Velocity of projectile after 29 seconds

$$= \sqrt{u^2 + g^2t^2 - 2u\,(\sin\alpha)\,gt}$$

$$v_{39} = \sqrt{\left[(500)^2 + (9.8)^2\times(29)^2\right] - (2\times 500\times(\sin 35°)\times 9.8\times 29)}$$

$$= \sqrt{(250\,000 + 80770) - (1000\times 0.5736\times 284.2)} \text{ m/s}$$

$$= 409.58 \text{ m/s} \quad \textbf{Ans.}$$

Similarly,
$$v_{30} = \sqrt{\left[(500)^2 + (9.8)^2 \times (30)^2\right] - (2 \times 500 \times (\sin 35°) \times 9.8 \times 30)}$$

$$= \sqrt{[250\,000 + 86440] - [1000 \times 0.5736 \times 294]} \text{ m/s}$$

$$= 409.64 \text{ m/s} \quad \textbf{Ans.}$$

Direction of the projectile after 29 and 30 seconds

Angle, which the projectile makes with the horizontal after 29 seconds,

$$\tan \theta_{29} = \frac{u \sin \alpha - gt}{u \cos \alpha} = \frac{(500 \sin 35°) - (9.8 \times 29)}{500 \cos 35°}$$

$$= \frac{(500 \times 0.5736) - 284.2}{500 \times 0.8192} = 0.00635$$

or
$$\theta_{29} = 0.36° \quad \textbf{Ans.}$$

Similarly,
$$\tan \theta_{30} = \frac{(500 \sin 35°) - (9.8 \times 30)}{500 \cos 35°} = \frac{(500 \times 0.5736) - 294.0}{500 \times 0.8192}$$

$$= -0.0176 \quad \text{or} \quad \theta_{30} = -1°$$

Note : Minus sign means that the projectile is moving downwards after reaching the highest point.

***Example 20.** *A particle is projected upwards with a velocity of 100 m/s at an angle of 30° with the horizontal.*

Find the time, when the particle will move perpendicular to its initial direction.

Solution. Given : Initial velocity of projection (u) = 100 m/s and angle of projection with the horizontal (α) = 30°

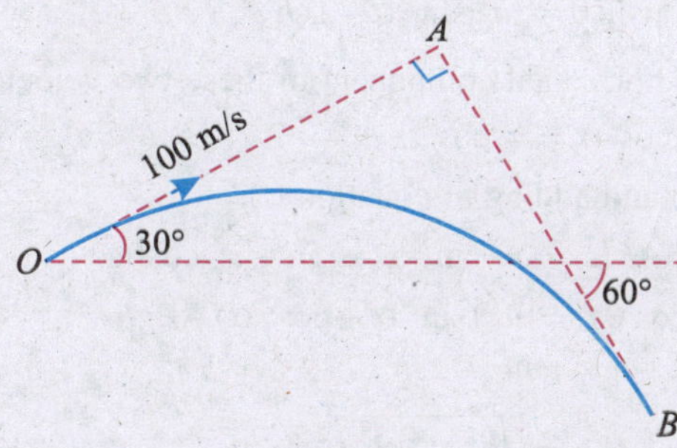

Fig. 15.

Let t = Time from the instant of projection, when the particle will move perpendicular to its initial direction.

We know that when the particle will move perpendicular to its initial direction, it will make an angle of 90° – 30° = 60° with the horizontal, but in the downward direction as shown in Fig. 14. Therefore actual angle,

$$\theta = (-60°) \qquad \text{...(Minus sign due to downward)}$$

We also know that the angle, which the particle makes with the horizontal after t seconds (θ),

$$\tan(-60°) = \frac{u \sin \alpha - gt}{u \cos \alpha} = \frac{100 \sin 30° - 9.8t}{100 \cos 30°}$$

or $$-1.732 = \frac{(100 \times 0.5) - 9.8t}{100 \times 0.866} = \frac{50 - 9.8t}{86.6}$$

or $$-150 = 50 - 9.8\,t$$

or $$t = \frac{50 + 150}{9.8} = 20.4 \text{ s} \quad \textbf{Ans.}$$

10. *VELOCITY AND DIRECTION OF MOTION OF A PROJECTILE, AT A GIVEN HEIGHT ABOVE THE POINT OF PROJECTION

Consider a projectile projected from O as shown in Fig. 16.

Let u = Initial velocity of projection, and

α = Angle of projection with the horizontal.

After reaching a height h, let the projectile reach at any point P as shown in Fig. 15.

Let v = Velocity of the projectile at P, and

θ = Angle, which the projectile at P makes with the horizontal.

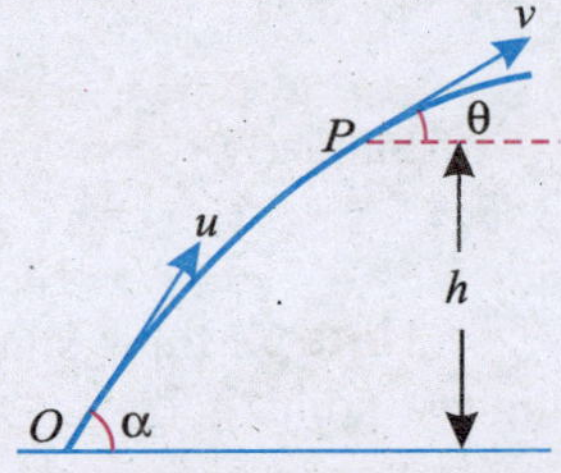

Fig. 16.

Vertical component of initial velocity

$= u \sin \alpha$

Vertical component of the final velocity after t seconds

$= v \sin \theta$

This change in velocity (*i.e.* from $u \sin \alpha$ to $v \sin \theta$) is becasue of the retardation (g) due to gravity.

$\therefore$ $$v^2 \sin^2 \theta = u^2 \sin^2 \alpha - 2gh \qquad ...(\because\ v^2 = u^2 - 2as) \ ...(i)$$

or $$v \sin \theta = \sqrt{u^2 \sin^2 \alpha - 2gh} \qquad ...(ii)$$

We also know that the horizontal component of these two velocities does not change.

$\therefore$ $$v \cos \theta = u \cos \alpha \qquad ...(iii)$$

Squaring equation (*iii*) and adding to equation (*i*),

$$v^2 \sin^2 \theta + v^2 \cos^2 \theta = u^2 \sin^2 \alpha - 2gh + u^2 \cos^2 \alpha$$

or $$v^2 (\sin^2 \theta + \cos \theta) = u^2 (\sin^2 \alpha + \cos^2 \alpha) - 2gh$$

or $$v^2 = u^2 - 2gh \qquad ...(\because\ \sin^2 \theta + \cos^2 \theta = 1)$$

or $$v = \sqrt{u^2 - 2gh}$$

This angle which the particle makes with the horizontal at P may be found out by dividing the equation (*ii*) by (*iii*) *i.e.*,

$$\frac{v \sin \theta}{v \cos \theta} = \frac{\sqrt{u^2 \sin^2 \alpha - 2gh}}{u \cos \alpha}$$

or $$\tan \theta = \frac{\text{Vertical velocity at a height } h}{\text{Horizontal component of initial velocity}}$$

***Example 21.** *A body is projected upwards with a velocity of 50 m/s at angle of 50° with the horizontal. What will be its (i) velocity and (ii) direction at a height of 30 m from the point of projection.*

Solution. Given : Initial velocity of projection (u) = 50 m/s ; Angle of projection (α) = 50° and height (h) = 30 m.

(i) Velocity of the projectile

Velocity of the projectile,

$$v = \sqrt{u^2 - 2gh} = \sqrt{(50)^2 - (2 \times 9.8 \times 30)} = 43.7 \text{ m/s} \quad \textbf{Ans.}$$

(ii) Direction of the projectile

Let θ = Angle which the projectile makes with the horizontal.

We also know that,
$$\tan\theta = \frac{\sqrt{u^2 \sin^2\alpha - 2gh}}{u\cos\alpha} = \frac{\sqrt{(50)^2 \sin^2 50° - 2 \times 9.8 \times 30}}{50 \cos 50°}$$

$$= \frac{\sqrt{2500 \times (0.766)^2 - 588}}{50 \times 0.6428} = 0.9224$$

or $\theta = 42.7°$ **Ans.**

****Example 22.** *The velocity of a particle, at its greatest height is $\sqrt{2/5}$ times of its velocity at half of its greatest height. Show that the angle of projection is 60°.*

Solution. Given : Velocity of the particle at its greatest height $= \sqrt{2/5} \times$ Velocity at half of its greatest height.

Let u = Initial velocity of projection, and

α = Angle of projection with the horizontal.

Velocity of a projectile at its greatest height

= Horizontal component of velocity of projection

$= u\cos\alpha$ ...(i)

Maximum height of projection

$$= \frac{u^2 \sin^2\alpha}{2g}$$

Half of the greatest height, $h = \frac{1}{2} \times \frac{u^2 \sin^2\alpha}{2g} = \frac{u^2 \sin^2\alpha}{4g}$

∴ Velocity of projectile at half of the greatest height,

$$v = \sqrt{u^2 - 2gh} = \sqrt{u^2 - 2g \times \frac{u^2 \sin^2\alpha}{4g}}$$

$$= \sqrt{u^2 - \frac{u^2 \sin^2\alpha}{2}} \quad \text{...(ii)}$$

Now substituting the values from equations (i) and (ii) in the given equation,

$$u\cos\alpha = \sqrt{\frac{2}{5}} \times \sqrt{u^2 - \frac{u^2 \sin^2\alpha}{2}} = \sqrt{\frac{2}{5}} \times \sqrt{\frac{2u^2 - u^2 \sin^2\alpha}{2}}$$

Squaring both sides,

$$u^2 \cos^2\alpha = \frac{2}{5} \times \frac{2u^2 - u^2 \sin^2\alpha}{2} = \frac{u^2 (2 - \sin^2\alpha)}{5}$$

or $$5\cos^2\alpha = 2 - \sin^2\alpha$$

or $$5(1 - \sin^2\alpha) = 2 - \sin^2\alpha \qquad ...(\sin^2\alpha + \cos^2\alpha = 1)$$

or $$4\sin^2\alpha = 3 \quad \text{or} \quad \sin^2\alpha = \frac{3}{4} = 0.75$$

or $$\sin\alpha = 0.866 \quad \text{or} \quad \alpha = 60° \quad \textbf{Ans.}$$

11. *TIME OF FLIGHT OF A PROJECTILE ON AN INCLINED PLANE

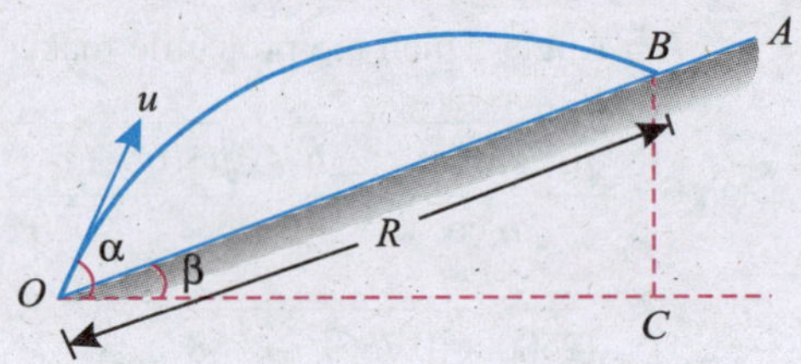

Fig. 17. Projectile on an inclined plane.

Consider a projectile projected from O on an upward inclined plane OA. Let the projectile strike B as shown in Fig. 17.

Let u = Initial velocity of projection,

α = Angle of projection with the horizontal.

β = Inclination of the plane OA with the horizontal,

R = Range of flight from O to B, and

t = Time of flight from O to B.

∴ Component of initial velocity, normal to the plane OA

$$= u\sin(\alpha - \beta) \qquad ...(i)$$

Acceleration due to gravity normal to the plane OA

$$= g\cos\beta \qquad ...(ii)$$

Acceleration due to gravity along the plane OA

$$= g\sin\beta \qquad ...(iii)$$

Now consider the motion of the projectile normal to the plane. We know that distance covered by the projectile normal to the plane OA is zero. Therefore substituting these values in the general equation of motion, *i.e.*

$$s = ut - \frac{1}{2}gt^2$$

or $$0 = u\sin(\alpha - \beta)\,t - \frac{1}{2}(g\cos\beta)t^2$$

or $$0 = u\sin(\alpha - \beta) - \frac{1}{2}g(\cos\beta)t \qquad \text{... (Dividing both sides by } t)$$

or $$t = \frac{2u\sin(\alpha - \beta)}{g\cos\beta}$$

Note : When the projectile is projected on a downward inclined plane, the time of flight may be found out by substituting $-\beta$ instead of $+\beta$ in the above equation. Therefore time of flight in this case,

$$t = \frac{2u \sin(\alpha + \beta)}{g \cos \beta} \qquad \text{... } [\because \cos(-\beta) = \cos \beta]$$

***Example 23.** *A ball is projected from a point with a velocity of 10 m/s on an inclined plane. the angle of projection and inclination of the plane are 35° and 15° respectively with the horizontal. Find the time of flight of the ball, when it is projected upwards and downwards the plane.*

Solution. Given : Velocity of projection (u) = 10 m/s ; Angle of projection with the horizontal (α) = 35° and inclination of the plane (β) = 15°

Time of flight when the ball is projected upwards

Time of flight when the ball is projected upwards,

$$t_1 = \frac{2u \sin(\alpha - \beta)}{g \cos \beta} = \frac{2 \times 10 \sin(35° - 15°)}{9.8 \cos 15°} = \frac{20 \sin 20°}{9.8 \cos 15°} \text{ s}$$

$$= \frac{20 \times 0.342}{9.8 \times 0.9659} = 0.72 \text{ s} \quad \textbf{Ans.}$$

Time of flight when the ball is projected downwards

Time of flight when the ball is projected downwards,

$$t_2 = \frac{2u \sin(\alpha + \beta)}{g \cos \beta} = \frac{2 \times 10 \sin(35° + 15°)}{9.8 \cos 15°} = \frac{20 \sin 50°}{9.8 \cos 15°}$$

$$= \frac{20 \times 0.766}{9.8 \times 0.9659} = 1.62 \text{ s} \quad \textbf{Ans.}$$

EXERCISE 3

1. A shot is fired with a velocity of 420 m/s at an elevation of 32°. Find the velocity and direction of the shot after 20 seconds of its firing. (**Ans.** 4° 16')
2. A stone is projected with a velocity 21 m/s at an angle of 30° with the horizontal. Find its velocity at a height of 5 m from the point of projection. Also find the interval of time between two points at which the stone has the same velocity of 20 m/s.

(**Ans.** 18.52 m/s ; 1.69 s)

12. *RANGE OF PROJECTILE ON AN INCLINED PLANE

We have already discussed in Article 11 of this chapter, that the

Time of Flight of a Projectile on an Inclined Plane,

$$t = \frac{2u \sin(\alpha - \beta)}{g \cos \beta}$$

Horizontal components of the range,

$$OC = \text{Horizontal component of velocity} \times \text{Time}$$

$$= u \cos \alpha \times \frac{2u \sin(\alpha - \beta)}{g \cos \beta} = \frac{2u^2 \sin(\alpha - \beta) \cos \alpha}{g \cos \beta}$$

$\therefore$ Actual range on the inclined plane,

$$R = \frac{OC}{\cos\beta} = \frac{2u^2 \sin(\alpha-\beta)\cos\alpha}{g\cos\beta \times \cos\beta} = \frac{2u^2 \sin(\alpha-\beta)\cos\alpha}{g\cos^2\beta}$$

$$= \frac{u^2}{g\cos^2\beta}[2\cos\alpha\sin(\alpha-\beta)]$$

$$= \frac{u^2}{g\cos^2\beta}[\sin(2\alpha-\beta) - \sin\beta] \quad ...(i)$$

$$...[\because 2\cos A\sin B = \sin(A+B) - \sin(A-B)]$$

From the above equation, we find that for the given values of u and β, the range will be maximum, when $\sin(2\alpha - \beta)$ is maximum (as the values of u, g and β are constant). We know that for maximum value of sine of any angle, the angle must be equal to 90° or $\pi/2$.

$$\therefore \quad (2\alpha-\beta) = \frac{\pi}{2} \quad \text{or} \quad \alpha = \left(\frac{\pi}{4}+\frac{\beta}{2}\right)$$

Or in other words, the range on the given plane is maximum, when the direction of projection bisects the angle between the vertical and inclined plane.

Now for maximum range, substituting the value of α in equation (*i*),

$$R_{max} = \frac{u^2}{g\cos^2\beta}\left\{\sin\left[2\left(\frac{\pi}{4}+\frac{\beta}{2}\right)-\beta\right] - \sin\beta\right\}$$

$$= \frac{u^2}{g\cos^2\beta}\left[\sin\left(\frac{\pi}{2}+\beta-\beta\right) - \sin\beta\right]$$

$$= \frac{u^2}{g\cos^2\beta}\left[\sin\left(\frac{\pi}{2}\right) - \sin\beta\right]$$

$$= \frac{u^2}{g\cos^2\beta}[1-\sin\beta] \quad ...\left[\because \sin\left(\frac{\pi}{2}\right) = 1\right]$$

$$= \frac{u^2(1-\sin\beta)}{g(1-\sin^2\beta)} \quad ...(\because \sin^2\beta + \cos^2\beta = 1)$$

$$= \frac{u^2}{g(1+\sin\beta)}$$

Notes : 1. When the projectile is projected on a downward inclined plane, then the range of flight will be given by substituting $-\beta$ instead of $+\beta$ in the above equation. Therefore range of flight in this case,

$$R = \frac{u^2}{g\cos^2\beta}[\sin(2\alpha+\beta) + \sin\beta]$$

2. When the projectile is projected on a downward inclined plane, the range will be maximum, when

$$\alpha = \left(\frac{\pi}{2} - \frac{\beta}{2}\right)$$

3. When the projectile is projected on a downward inclined plane,the value of maximum range will be,

$$R_{max} = \frac{u^2}{g(1-\sin\beta)}$$

***Example 24.** *A particle is projected from a point, on an inclined plane,with a velocity of 30 m/s. The angle of projection and the angle of plane are 55° and 20° to the horizontal respectively. Show that the range up the plane is maximum one for the given plane. Find the range and the time of flight of the particle.* **[WBUT]**

Solution. Given : Velocity of projection (u) = 30 m/s ; Angle of projection with the horizontal (α) = 55° and angle of plane (β) = 20°

Maximum Range

Maximum range, the angle of projection,

$$\alpha = \frac{\pi}{4} + \frac{\beta}{2} = \frac{180°}{4} + \frac{20°}{2} = 55°$$

Since the given angle of projection is 55°, therefore range up the plane is maximum one for the given plane. **Ans.**

Range of the projectile

Range of the projectile,

$$R = \frac{u^2}{g\cos^2\beta}\left[\sin(2\alpha - \beta) - \sin\beta\right] \text{ m}$$

$$= \frac{(30)^2}{9.8\cos^2 20°}\left[\sin(2\times 55° - 20°) - \sin 20°\right] \text{ m}$$

$$= \frac{900}{9.8\,(0.9397)^2}\left[\sin 90° - \sin 20°\right] \text{ m}$$

$$= 104.0\,(1 - 0.3420) = 68.43 \text{ m} \quad \textbf{Ans.}$$

Time of flight

Time of flight,

$$t = \frac{2u\sin(\alpha - \beta)}{g\cos\beta} = \frac{2\times 30\sin(55° - 20°)}{9.8\cos 20°} = \frac{60\times\sin 35°}{9.8\cos 20°} \text{ s}$$

$$= \frac{60\times 0.5736}{9.8\times 0.9397} = 3.74 \text{ s} \quad \textbf{Ans.}$$

Note : Since the angle of projection is for the maximum range, therefore the range may also be found out from the relation :

$$R = \frac{u^2}{g\,(1+\sin\beta)} = \frac{(30)^2}{9.8\,(1+\sin 20°)} = \frac{900}{9.8\,(1+0.3420)} \text{ m}$$

$$= 68.43 \text{ m} \quad \textbf{Ans.}$$

***Example 25.** *A plane has a rise of 5 in 12. A shot is projected with a velocity of 200 m/s at an elevation of 30°. Find the range of the plane, if (a) the shot is fired up the plane, (b) the shot is fired down the plane.*

Solution. Given : $\tan\beta = 5/12 = 0.4167$ or $\beta = 22.6°$; Velocity of projection with the horizontal (u) = 200 m/s and angle of projection (α) = 30°.

(a) Range of the plane, when the shot is fired up the plane

Range of the plane, when the shot is fired up the plane,

$$R_1 = \frac{u^2}{g\cos^2\beta}\left[\sin(2\alpha - \beta) - \sin\beta\right]$$

$$= \frac{(200)^2}{9.8\cos^2 22.6°}[\sin(2\times 30° - 22.6°) - \sin 22.6°]$$

$$= \frac{40\,000}{9.8\,(0.9231)^2}[\sin 37.4° - \sin 22.6°]$$

$$= 4790\,(0.6072 - 0.3846) = 1066 \text{ m} \quad \textbf{Ans.}$$

(b) Range of the plane, when the shot is fired down the plane

Range of the plane, when the shot is fired down the plane,

$$R_2 = \frac{u^2}{g\cos^2\beta}[\sin(2\alpha + \beta) + \sin\beta]$$

$$= \frac{(200)^2}{9.8\cos^2 22.6°}[\sin(2\times 30° + 22.6°) + \sin 22.6°]$$

$$= \frac{40\,000}{9.8\,(0.9231)^2}[\sin 82.6° + \sin 22.6°] \text{ m}$$

$$= 4790\,(0.9917 + 0.3846) = 6592 \text{ m} \quad \textbf{Ans.}$$

EXERCISE 4

1. A player can throw a cricket ball 100 m on a level ground. Find the distance through which he can throw the same ball from the top of hill at angle of 52° 30', if slope of the hill is 15°.

   (**Ans.** 120.5 m)

2. A shot is fired with a velocity of 100 m/s at an angle of 45° with the horizontal on a plane inclined at an angle of 30° with the horizontal. Find the maximum range of the shot.

   (**Ans.** 680.3 m)

3. A projectile is projected up a plane of inclination (β) with an initial velocity of (u) at an angle (α) to the horizontal. Show that condition for the projectile to strike the inclined plane at right angles is

$$\cot\beta = 2\tan(\alpha - \beta).$$

QUESTIONS

1. What is a projectile ? Give an example of a projectile.
2. Define the terms : velocity of projection and angle of projection.
3. Obtain an equation for the trajectory of a projectile, and show that it is a parabola.
4. Derive an expression for the maximum height and range of a projectile traversed by a stone, thrown with an initial velocity of u and an inclination of α.
5. At what angle,the projectile should be projected in order to have maximum range ? Justify your answer by calculations.
6. Derive a relation for the velocity and direction of motion of a projectile :
   (*a*) after a given interval of time t from the instant of projection.
   (*b*) at a given height h above the point of projection.

7. How would you find out (*a*) time of flight (*b*) range of a projectile, when projected upwards on an inclined plane ?

   What happens to the above equations, when the same projectile is projected on the same plane, but in a downward direction ?

OBJECTIVE TYPE QUESTIONS

1. The path of a projectile is not a parabola.

   (*a*) True (*b*) False

2. The time of flight of a projectile on a horizontal plane is

   (*a*) $\frac{2u\sin\alpha}{g}$ (*b*) $\frac{2u\cos\alpha}{g}$ (*c*) $\frac{2u\sin\alpha}{2g}$ (*d*) $\frac{u\cos 2\alpha}{2g}$

3. The horizontal range of a projectile is

   (*a*) $\frac{u\sin 2\alpha}{g}$ (*b*) $\frac{u^2\sin 2\alpha}{g}$ (*c*) $\frac{u\sin 2\alpha}{2g}$ (*d*) $\frac{u^2\sin 2\alpha}{2g}$

4. The horizontal range of a projectile is maximum when the angle of projectile is

   (*a*) 30° (*b*) 45° (*c*) 60° (*d*) 75°

5. The maximum height of a projectile on a horizontal range is

   (*a*) $\frac{u^2\sin 2\alpha}{2g}$ (*b*) $\frac{u^2\sin\alpha}{2g}$ (*c*) $\frac{u^2\sin^2 2\alpha}{2g}$ (*d*) $\frac{u^2\sin^2\alpha}{2g}$

6. The time of flight of a projectile on un upward inclined plane depends upon

   (*a*) angle of projection (*b*) angle of inclination of the plane

   (*c*) both '*a*' and '*b*' (*d*) none of the above

7. The range of projectile on a downward inclined plane is ....the range on upward inclined plane for the same velocity of projection and angle of projection.

   (*a*) less than (*b*) equal to (*c*) more than

ANSWERS

1. (*b*) 2. (*a*) 3. (*b*) 4. (*b*) 5. (*d*) 6. (*c*) 7. (*c*)

CHAPTER

19 Motion of Rotation

Contents

1. MOTION OF ROTATION

Some bodies like pulley, shafts, flywheels etc., have **motion of rotation** (*i.e.*, angular motion) which takes place about the geometric axis of the body. The angular velocity of a body is always expressed in terms of revolutions described in one minute, *e.g.*, if at an instant the angular velocity of rotating body in N r.p.m. (*i.e.* revolutions per min) the corresponding angular velocity ω (in rad) may be found out as discussed below :

$$1 \text{ revolution/min} = 2\pi \text{ rad/min}$$

$$\therefore \quad N \text{ revolutions/min} = 2\pi N \text{ rad/min}$$

and Angular Velocity,

$$\omega = 2\pi N \text{ rad/min} = \frac{2\pi N}{60} \text{ rad/sec}$$

2. IMPORTANT TERMS

The following terms, which will be frequently used in this chapter, should be clerarly understood at this stage :

1. ***Angular velocity.*** It is the rate of change of angular displacement of a body, and is expressed in r.p.m. (revolutions per minute) or in radian per second. It is, usually, denoted by ω (omega).
2. ***Angular acceleration.*** It is the rate of change of angular velocity and is expressed in radian per second per second (rad/s^2) and is usually, denoted by α. It may be constant or variable.
3. ***Angular displacement.*** It is the total angle, through which a body has rotated, and is usually denoted by θ. Mathematically, if a body is rotating with a uniform angular velocity (ω) then in t seconds, the angular displacement

$$\theta = \omega t$$

3. *MOTION OF ROTATION UNDER CONSTANT ANGULAR ACCELERATION

[PU]

Consider a particle, rotating about its axis.

Let ω_0 = Initial angular velocity,

ω = Final angular velocity,

t = Time (in seconds) taken by the particle to change its velocity from ω_0 to ω.

α = Constant angular acceleration in rad/s^2, and

θ = Total angular displacement in radians.

Since in t seconds, the angular velocity of the particle has increased steadily from ω_0 to ω at the rate of α rad/s^2, therefore

$$\omega = \omega_0 + \alpha t \qquad \text{...}(i)$$

Average angular velocity $= \dfrac{\omega_0 + \omega}{2}$

Total angular displacement,

$$\theta = \text{Average velocity} \times \text{Time} = \left(\frac{\omega_0 + \omega}{2}\right) \times t \qquad \text{...}(ii)$$

Substituting the value of ω from equation (i),

$$\theta = \frac{\omega_0 + (\omega_0 + \alpha t)}{2} \times t = \frac{2\omega_0 + \alpha t}{2} \times t = \omega_0 t + \frac{1}{2}\alpha t^2 \qquad \text{...}(iii)$$

and from equation (i), we find that

$$t = \frac{\omega - \omega_0}{\alpha}$$

Substituting this value of t in equation (ii),

$$\theta = \left(\frac{\omega_0 + \omega}{2}\right) \times \left(\frac{\omega - \omega_0}{\alpha}\right) = \frac{\omega^2 - \omega_0^2}{2\alpha}$$

$$\therefore \qquad \omega^2 = \omega_0^2 + 2\alpha\theta \qquad \text{...}(iv)$$

4. *RELATION BETWEEN LINEAR MOTION AND ANGULAR MOTION

[RTU, BPUT]

Following are the relations between the linear motion and the angular motion of a body :

S. No.	*Particulars*	*Linear motion*	*Angular motion*
1.	Initial velocity	u	ω_0
2.	Final velocity	v	ω
3.	Constant acceleration	a	α
4.	Total distance traversed	s	θ
5.	Formula for final velocity	$v = u + at$	$\omega = \omega_0 + \alpha t$
6.	Formula for distance traversed	$s = ut + \frac{1}{2}at^2$	$\theta = \omega_0 t + \frac{1}{2}\alpha t^2$
7.	Formula for final velocity	$v^2 = u^2 + 2as$	$\omega^2 = \omega_0^2 + 2\alpha\theta$
8.	Differential formula for velocity	$v = \frac{ds}{dt}$	$\omega = \frac{d\theta}{dt}$
9.	Differential formula for acceleration	$a = \frac{dv}{dt}$	$\alpha = \frac{d\omega}{dt}$

Example. 1. *A flywheel starts from rest and revolves with an acceleration of 0.5 rad/sec^2. What will be its angular velocity and angular displacement after 10 seconds.*

Solution. Given : Initial angular velocity $(\omega_0) = 0$ (becasue it starts from rest) ; Angular acceleration $(\alpha) = 0.5$ rad/sec^2 and time $(t) = 10$ sec.

Angular velocity of the flywheel

Angular velocity of the flywheel,

$$\omega = \omega_0 + \alpha t = 0 + (0.5 \times 10) = 5 \text{ rad/sec} \quad \textbf{Ans.}$$

Angular displacement of the flywheel

Angular displacement of the flywheel,

$$\theta = \omega_0 t + \frac{1}{2}\alpha t^2 = (0 \times 10) + \left[\frac{1}{2} \times 0.5 \times (10)^2\right] = 25 \text{ rad} \quad \textbf{Ans.}$$

Example 2. *A wheel increases its speed from 45 r.p.m. to 90 r.p.m. in 30 seconds. Find (a) angular acceleration of the wheel, and (b) no. of revolutions made by the wheel in these 30 seconds.*

Solution. Given : Initial angular velocity $(\omega_0) = 45$ r.p.m. $= 1.5\,\pi$ rad/sec ; Final angular velocity $(\omega) = 90$ r.p.m. $= 3\,\pi$ rad/sec and time $(t) = 30$ sec

(a) Angular acceleration of the wheel

Let α = Angular acceleration of the wheel.

Final angular velocity of the wheel (ω),

$$3\pi = \omega_0 + \alpha t = 1.5\,\pi + (\alpha \times 30) = 1.5\pi + 30\alpha$$

or

$$\alpha = \frac{3\pi - 1.5\pi}{30} = \frac{1.5\pi}{30} = 0.05\pi \text{ rad/sec}^2 \quad \textbf{Ans.}$$

(b) No. of revolutions made by the wheel in 30 seconds

Total angle turned by the wheel in 30 seconds,

$$\theta = \omega_0 t + \frac{1}{2}\alpha t^2 = [1.5\pi \times 30] + \left[\frac{1}{2} \times 0.05\ \pi(30)^2\right] = 67.5\ \pi\ \text{rad}$$

$$= \frac{67.5\pi}{2\pi} = 33.75\,\text{rev}\quad \textbf{Ans.} \qquad \text{...(1 rev} = 2\pi\ \text{rad)}$$

Example 3. *A flywheel is making 180 r.p.m. and after 20 sec it is running at 120 r.p.m. How many revolutions will it make and what time will elapse before it stops, if the retardation is uniform ?*

Solution. Given : Initial angular velocity (ω_0) = 180 r.p.m. = 6π rad/sec ; Final angular velocity (ω) = 120 r.p.m. = 4π rad/sec and time (t) = 20 sec.

Revolutions of the wheel, before it stops

Let α = Uniform angular acceleration, and

θ = Angular displacement of the flywheel before coming to rest.

First of all, consider the angular motion of the flywheel from 180 r.p.m. to 120 r.p.m. in 20 seconds. We know that final angular velocity (ω),

$$4\pi = \omega_0 + \alpha t = 6\pi + \alpha \times 20$$

or

$$\alpha = \frac{4\pi - 6\pi}{20} = -\ 0.1\ \pi\ \text{rad/sec}^2$$

...(Minus sign indicates retardation)

Now consider angular motion of the flywheel from 180 r.p.m. (or 6π rad/s) to zero r.p.m. (*i.e.*, coming to stop) or ω = 0 with a constant acceleration of – 0.1 π rad/s^2. We know that (ω^2).

$$0 = \omega^2{}_0 + 2\alpha\theta = (6\pi)^2 + 2 \times (-\ 0.1\pi)\ \theta = 36\pi^2 - 0.2\pi\theta$$

$$\therefore \qquad \theta = \frac{36\pi^2}{0.2\pi} = 180\ \pi = \frac{180\ \pi}{2\pi} = 90\ \text{rev}\quad \textbf{Ans.}$$

...(1 rev = 2π rad)

Time in which the wheel will come to rest

Let t = Time in which the wheel will come to rest.

Final velocity of flywheel (ω),

$$0 = \omega_0 + \alpha t = 6\pi - 0.1\ \pi\ t$$

or

$$t = \frac{6\pi}{0.1\pi} = 60\,\text{s} = 1\ \text{min}\quad \textbf{Ans.}$$

Example 4. *A pulley, starting from rest, is given an acceleration of 0.5 rad/s^2. What will be its speed in r.p.m. at the end of 2 minutes ? If it is uniformly retarded at the rate of 0.3 rad /s^2, in how many minutes the pulley will come to rest ?*

Solution. First of all, consider angular motion of pully from rest. In this case, initial angular velocity (ω_0) = 0 ; Acceleration (α_1) = 0.5 rad/sec^2 and time taken (t_1) = 2 minutes = 120 sec.

Angular speed of pully in r.p.m. at the end of 2 min

Final angular speed of the pulley

$$\omega = \omega_0 + \alpha t_1 = 0 + (0.5 \times 120) = 60\ \text{rad/sec}$$

$$= \frac{60}{2\pi} = 9.55\ \text{r.p.s.} = 9.55 \times 60 = 573\ \text{r.p.m.}\quad \textbf{Ans.}$$

Time in which the pulley will come to rest

Let t_2 = Time in which the pulley will come to rest.

Now consider angular motion of the pulley in coming to rest. In this case, initial angular velocity (ω_0) = 60 rad/sec ; Final angular velocity (ω) = 0 and retardation (α_2) = – 0.3 rad/sec^2 (Minus sign due to retardation).

We know that final velocity of the pulley

$$0 = \omega_0 + \alpha t_2 = 60 - 0.3\, t_2$$

or
$$t_2 = \frac{60}{0.3} = 200 \text{ sec} \quad \textbf{Ans.}$$

Example 5. *A wheel rotates for 5 seconds with a constant angular acceleration and describes during this time 100 radians. It then rotates with a constant angular velocity and during the next five seconds describes 80 radians.*

Find the initial angular velocity and the angular acceleration.

Solution. Given : Time (t) = 5 sec and angular displacement (θ) = 100 rad

Initial angular velocity

Let ω_0 = Initial angular velocity in rad/s,

α = Angular acceleration in rad/s^2, and

ω = Angular velocity after 5 s in rad/s.

First of all, consider the angular motion of the wheel with constant acceleration for 5 seconds. We know that angular displacement (θ),

$$100 = \omega_0 t + \frac{1}{2}\alpha t^2 = \omega_0 \times 5 + \frac{1}{2} \times \alpha (5)^2 = 5\omega_0 + 12.5\alpha$$

or
$$40 = 2\omega_0 + 5\alpha \quad ...(i)$$

or Final velocity,
$$\omega = \omega_0 + \alpha t = \omega_0 + \alpha \times 5 = \omega_0 + 5\alpha$$

Now consider the angular motion of the wheel with a constant angular velocityof ($\omega_0 + 5\alpha$) for 5 seconds and describe 80 radians. We know that the angular displacement,

$$80 = 5\,(\omega_0 + 5\alpha)$$

or
$$16 = \omega_0 + 5\alpha \quad ...(ii)$$

Subtracting equation (*ii*) from (*i*),

$$24 = \omega_0 \quad \text{or} \quad \omega_0 = 24 \text{ rad/s} \quad \textbf{Ans.}$$

Angular acceleration

Substituting this value of ω_0 in equation (*ii*),

$$16 = 24 + 5\alpha \quad \text{or} \quad \alpha = \frac{16 - 24}{5} = -1.6 \text{ rad/s}^2 \quad \textbf{Ans.}$$

...(Minus sign means retardation)

Example 6. *A shaft is uniformly accelerated from 10 rev/s to 18 rev/s in 4 seconds. The shaft cotinues to accelerate at this rate for the next 8 seconds. Thereafter the shaft rotates with a uniform angular speed. Find the total time to complete 400 revolutions.*

Solution. Given : Initial angular velocity (ω_0) = 10 rev/s = 20 π rad/s ; Final angular velocity (ω) = 18 rev/s = 36 π rad/s ; Time taken during constant acceleration (t_1) = 4 sec ; Time taken during uniform angular velocity (t_2) = 8 sec and total angular displacement (θ) = 400 rev = 800 π rad

Let α = Angular acceleration of the shaft.

First of all, consider the motion of the shaft in the first 4 seconds. In this case, initial angular velocity (ω_0) = 20 π rad/s ; Final angular velocity (ω) = 36 rad/s.

Final velocity of the shaft (ω),

$$36\,\pi = \omega_0 + \alpha t_1 = 20\,\pi + (\alpha \times 4) = 20\,\pi + 4\alpha$$

or $$\alpha = \frac{36\pi - 20\pi}{4} = 4\pi \text{ rad/s}^2 \qquad ...(i)$$

Angular displacement, $$\theta_1 = \omega_0 t + \frac{1}{2}\alpha t^2 = 20\pi \times 4 + \frac{1}{2} \times 4\pi\,(4)^2 = 112\pi \text{ rad} \qquad ...(ii)$$

Now consider the motion of the shaft for the next 8 seconds. In this case, initial velocity $(\omega_0) = 36\pi$ rad/s and angular acceleration $(\alpha) = 4\pi$ rad/s^2 as obtained in equation (*i*).

Angular velocity of the shaft,

$$\omega = \omega_0 + \alpha t_2 = 36\pi + (4\pi \times 8) = 68\pi \text{ rad/s} \qquad ...(iii)$$

Angular displacement, $$\theta_2 = \omega_0 t + \frac{1}{2}\alpha t^2 = 36\pi \times 8 + \frac{1}{2} \times 4\pi\,(8)^2 = 416\,\pi \text{ rad} \qquad ...(iv)$$

Now consider motion of the shaft with a constant angular velocity of 68π rad/s as obtainted in equation (*iii*). We know that angular displacement of the shaft at this speed

$$= 800\,\pi - 112\,\pi - 416\,\pi = 272\,\pi \text{ rad}$$

∴ Time taken by the shaft to complete 272 π rad,

$$t_3 = \frac{272\pi}{68\pi} = 4 \text{ sec}$$

Total time to complete 400 revolutions or 800 π rad,

$$= t_1 + t_2 + t_3 = 4 + 8 + 4 = 16 \text{ sec}$$ **Ans.**

Example 7. *A swing bridge turns through 90° in 120 seconds. The bridge is uniformly accelerated from rest for the first 40 seconds. Subsequently, it turns with a uniform angular velocity for the next 60 seconds. Now the motion of the bridge is uniformly retarded for the last 20 seconds. Find (i) angular acceleration ; (ii) maximum angular velocity ; and (iii) angular retardation of the bridge.*

Solution. Given : Angular displacement (θ) $= 90° = \frac{\pi}{2} = 0.5\pi$ rad ; Total time $(T) = 120$ s; Time for acceleration $(t_1) = 40$ sec; Time for uniform velocity $(t_2) = 60$ sec and time for retardation $(t_3) = 20$ sec

(i) Angular acceleration of the bridge

Let α_1 = Angular acceleration of the bridge, and

α_2 = Angular retardation of the bridge.

First of all, consider the motion of the bridge from rest in the first 40 sec. In this case, initial angular velocity $(\omega_1) = 0$ and time $(t_1) = 40$ s. We know that final angular velocity of bridge,

$$\omega = \omega_o + \alpha t_1 = 0 + \alpha_1 \times 40 = 40\,\alpha_1 \qquad ...(i)$$

Angular displacement, $$\theta_1 = \omega_0 t_1 + \frac{1}{2}\alpha_1 t_2^2 = 0 + \frac{1}{2}\alpha_1\,(40)^2 = 800\,\alpha_1 \qquad ...(ii)$$

Now consider the motion of the bridge in the next 60 sec. In this case, constant angular velocity of ($40\ \alpha_1$) as obtained in equation (*i*) and time (t_2) = 60 sec. Therefore angular displacement during 60 sec,

$$\theta_2 = 40\ \alpha_1 \times t_2 = 40\ \alpha_1 \times 60 = 2400\ \alpha_1 \text{ rad} \qquad ...(iii)$$

Now consider the motion of the bridge in the last 20 sec. In this case, initial angular velocity (ω_0) = $40\ \alpha_1$ as obtained in equation (*i*) ; Final angular velocity (ω) = 0 and time (t_3) = 20 sec. We know that final angular velocity of the bridge (ω)

$$0 = \omega_0 - \alpha t_3 = 40\ \alpha_1 - \alpha_2 \times 20 \quad \text{or} \quad \alpha_2 = 2\alpha_1$$

...(Minus sign due to retardation)

Angular displacement, $\theta_3 = \omega_0 t_3 - \frac{1}{2}\alpha_2 t_3^2 = (40\alpha_1 \times 20) - \frac{1}{2} \times 2\alpha_1\ (20)^2$

$$= 400\ \alpha_1 \qquad ...(iv)$$

...(Minus sign due to retardation)

Total angular displacement of the bridge (θ)

$$0.5\pi = \theta_1 + \theta_2 + \theta_3 = 800\ \alpha_1 + 2400\ \alpha_1 + 400\ \alpha_1 = 3600\ \alpha_1$$

$$\therefore \qquad \alpha_1 = \frac{0.5\pi}{3600} = 0.436 \times 10^{-3} \text{ rad/sec}^2 \quad \textbf{Ans.}$$

(*ii*) *Maximum angular velocity of the bridge*

Maximum angular velocity of the bridge,

$$\omega = 40\ \alpha_1 = 40 \times 0.436 \times 10^{-3} = 17.44 \times 10^{-3} \text{ rad/s} \quad \textbf{Ans.}$$

(*iii*) *Angular retardation of the bridge*

Angular retardation of the bridge,

$$\alpha_2 = 2\alpha_1 = 2 \times (0.436 \times 10^{-3}) = 0.872 \times 10^{-3} \text{ rad/sec}^2 \quad \textbf{Ans.}$$

Example 8. *A flywheel rotates with a constant retardation due to braking. From t = 0 to t = 10 seconds, it made 300 revolutions. At time t = 7.5 sec, its angular velocity was 40 π rad/sec. Determine (i) value of constant ratardation ; (ii) total time taken to come to rest and (iii) total revolutions made till it comes to rest.*

Solution. Given : Time interval (t) = 10 – 0 = 10 sec ; Angular displacement (θ) = 300 rev = $2\pi \times 300 = 600\ \pi$ rad ; when time (t) = 7.5 sec and angular velocity (ω) = $40\ \pi$ rad/sec

(*i*) *Value of constant retardation*

Let α = Constant retardation, and

ω_0 = Initial angular velocity.

First of all, consider motion of the flywheel from $t = 0$ to $t = 10$ seconds. We know that total angular displacement (θ)

$$600\pi = \omega_0 t - \frac{1}{2}\alpha t^2 = (\omega_0 \times 10) - \left(\frac{1}{2}\alpha\ (10)^2\right)$$

$$= 10\ \omega_0 - 50\alpha \qquad ...(i)$$

...(Minus sign due to retardation)

Now consider motion of the flywheel from $t = 0$ to $t = 7.5$ seconds. We know that final angular velocity (ω)

$$40\ \pi = \omega_0 - \alpha t = \omega_0 - \alpha \times 7.5 \qquad ...(ii)$$

...(Minus sign due to retardation)

$$\alpha = \frac{\omega_0 - 40\ \pi}{7.5} \text{ rad/sec}^2 \qquad ...(iii)$$

Substituting this value of α in equation (*i*),

$$600\pi = 10\,\omega_0 - 50\left(\frac{\omega_0 - 40\pi}{7.5}\right)$$

$$= 10\,\omega_0 - 6.67\,\omega_0 + 266.7\,\pi = 3.33\,\omega_0 + 266.7\,\pi$$

$$\therefore \qquad \omega_0 = \frac{600\pi - 266.7\pi}{3.33} = 100\pi \text{ rad/sec}$$

and now substituting the value of ω_0 in equation (*iii*),

$$\alpha = \frac{100\pi - 40\pi}{7.5} = 8\pi = 25.1 \text{ rad/sec}^2 \quad \textbf{Ans.}$$

(ii) Total time taken by the flywheel to come to rest

Now consider motion of the flywheel till it comes to rest. In this case, Initial angular velocity $(\omega_0) = 100\pi$ rad/sec ; Final angular velocity $(\omega) = 0$ (becasue it comes to rest) and angular acceleration $(\alpha) = 8\pi$ rad/sec^2

Let $\qquad t$ = Total time taken by the flywheel to come to rest.

Final angular velocity of the flywheel (ω),

$$0 = \omega_0 - \alpha t = 100\pi - 8\pi t \qquad \text{...(Minus sign due to retardation)}$$

$$\therefore \qquad t = \frac{100\pi}{8\pi} = 12.5 \text{ sec} \quad \textbf{Ans.}$$

(iii) Total revolutions made till it comes to rest

Total revolutions made by the flywheel till it comes to rest (or in other words revolutions made in 12.5 seconds),

$$\theta = \omega_0 t - \frac{1}{2}\alpha t^2 = 100\pi \times 12.5 - \frac{1}{2} \times 8\pi\,(12.5)^2$$

...(Minus sign due to retardation)

$$= 625\pi \text{ rad} = \frac{625\pi}{2\pi} = 312.5 \text{ rev} \quad \textbf{Ans.}$$

EXERCISE 1

1. A flywheel increases its speed from 30 r.p.m. to 60 r.p.m. in 10 seconds. Calculate (*i*) the angular acceleration ; and (*ii*) no. of revolutions made by the wheel in these 10 seconds.
   (**Ans.** $\pi/10$ rad/s^2 ; 7.5 rev)
2. A wheel, starting form rest, is accelerated at the rate of 5 rad/s^2 for a period of 10 seconds. It is then made to stop in the next 5 seconds by applying brakes. Find (*a*) maximum velocity attained by the wheel, and (*b*) total angle turned by the wheel.
   (**Ans.** 50 rad/sec ; 375 rad)
3. A wheel is running at a constant speed of 360 r.p.m. At what constant rate, in rad/s, its motion must be retarded to bring the wheel to rest in (*i*) 2 minutes, and (*ii*) 18 revolution.
   (**Ans.** 0.314 rad/s^2 ; 6.28 rad/s^2)

*4. A wheel rotating about a fixed axis at 24 r.p.m. is uniformly accelerated for 70 seconds, during which time it makes 50 revolution. Find (*i*) angular velocity at the end of this interval, and (*ii*) time required for the speed to reach 150 r.p.m. **[IGNOU]**
   (**Ans.** – 61.4 r.p.m. ; 3 min 55.6 s)

5. A wheel rotates for 5 seconds with a constant angular acceleration and describes 80 radians. It then rotates with a constant angular velocity in the next 5 seconds and describes 100 radians. Find the initial angular velocity and angular acceleration of the wheel.
   (**Ans.** 12 rad/s ; 1.6 rad/s^2)

5. LINEAR (OR TANGENTIAL) VELOCITY OF A ROTATING BODY

[PU, BPUT]

Consider a body rotating about its axis as shown in Fig. 1.

Let ω = Angular velocity of the body in rad/s,

r = Radius of the circular path in metres, and

v = Linear velocity of the particle on the periphery in m/s.

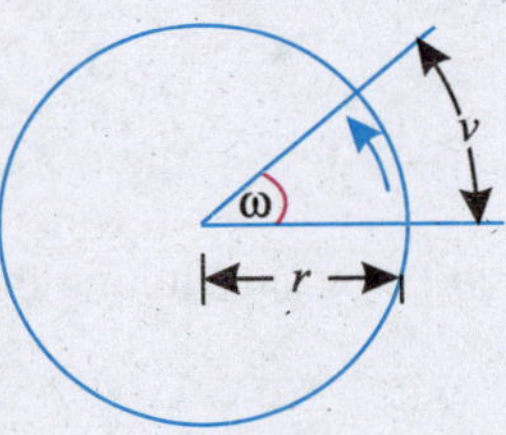

Fig. 1.

After one second, the particle will move v metres along the circular path and the angular displacement will be ω rad.

We know that length of arc = Radius of arc × Angle subtended in rad.

$$\therefore \quad v = r\,\omega$$

Example 9. *A wheel of 1.2 m diameter starts from rest and is accelerated at the rate of 0.8 rad/s². Find the linear velocity of a point on its periphery after 5 seconds.*

Solution. Given : Diameter of wheel = 1.2 m or radius (r) = 0.6 m ; Initial angular velocity (ω_0) = 0 (becasue, it starts from rest) ; Angular acceleration (α) = 0.8 rad/s^2 and time (t) = 5 s

Angular velocity of the wheel after 5 seconds,

$$\omega = \omega_0 + \alpha t = 0 + (0.8 \times 5) = 4 \text{ rad/s}$$

∴ Linear velocity of the point on the periphery of the wheel,

$$v = r\omega = 0.6 \times 4 = 2.4 \text{ m/s} \quad \textbf{Ans.}$$

Example 10. *A pulley 2 m in diameter is keyed to a shaft which makes 240 r.p.m. Find the linear velocity of a particle on the periphery of the pulley.*

Solution. Given : Diameter of pulley = 2 m or radius (r) = 1 m and angular frequency (N) = 240 r.p.m.

Angular velocity of the pulley,

$$\omega = \frac{2\pi N}{60} = \frac{2\pi \times 240}{60} = 25.1 \text{ rad/s} \quad \textbf{Ans.}$$

∴ Linear velocity of the particle on the periphery of the pulley,

$$v = r\omega = 1 \times 25.1 = 25.1 \text{ m/s} \quad \textbf{Ans.}$$

6. *LINEAR (OR TANGENTIAL) ACCELERATION OF A ROTATING BODY

Consider a body rotating about its axis with a constant angular (as well as linear) acceleration. We know that linear acceleration,

$$a = \frac{dv}{dt} = \frac{d}{dt}(v) \quad \ldots(i)$$

We also know that, For the motion of rotation, the linear velocity,

$$v = r\omega$$

Now substituting the value of v in equation (i),

$$a = \frac{d}{dt}(r\omega) = r\frac{d\omega}{dt} = r\alpha$$

Where α = Angular acceleration in rad/sec^2 and is equal to $d\omega/dt$.

Note. The above relation, in terms of angular acceleration may also be written as :

$$\alpha = \frac{a}{r} \quad \textbf{Ans.}$$

***Example 11.** *A car is moving at 72 k.m.p.h., If the wheels are 75 cm diameter, find the angular velocity of the tyre about its axis. If the car comes to rest in a distance of 20 metres, under a uniform retardation, find angular retardation of the wheels.*

Solution. Given : Linear velocity (v) = 72 k.m.p.h. = 20 m/s; Diameter of wheel (d) = 75 cm or radius (r) = 37.5 m = 0.375 m and distance travelled by the car (s) = 20 m.

Angular retardation of the wheel

Angular velocity of the wheel,

$$\omega = \frac{v}{r} = \frac{20}{0.375} = 53.3 \text{ rad/sec}$$

Let a = Linear retardation of the wheel.

We know that, $v^2 = u^2 + 2as$

or $0 = (20)^2 + 2 \times a \times 20 = 400 + 40a$

or $a = -\dfrac{400}{40} = -10 \text{ m/sec}^2$ ...(Minus sign indicates retardation)

Angular retardation of the wheel,

$$\alpha = \frac{a}{r} = \frac{-10}{0.375} = -26.7 \text{ rad/sec}^2 \quad \textbf{Ans.}$$

...(Minus sign indicates retardation)

EXERCISE 2

1. A horizontal bar 1.5 m long and of small cross-section rotates about vertical axis through one end. It accelerates uniformly from 30 r.p.m. to 45 r.p.m. for 10 seconds. What is the linear velocity at the beginning and end of this interval ? What is the tangential component of the acceleration of the mid-point of the bar after 10 seconds.
   (**Ans.** 4.71 m/s ; 7.07 m/s ; 0.118 m/s^2)
2. In a children park, a train is moving in a circular path. If the linear and angular speeds of the train are 10 m/s and 0.25 rad/s respectively, find the radius of the circular path.
   (**Ans.** 40 m)
3. A motor cycle starts form rest and moves with a constant acceleration of 2.25 m/s^2. What is its angular acceleration, if the diameter of the motor cycle wheels is 750 mm.
   (**Ans.** 6 rad/s^2)

7. *MOTION OF ROTATION OF A BODY UNDER VARIABLE ANGULAR ACCELERATION

[PU]

In the previous articles, we have discussed the cases of angular motion under constant acceleration. But sometimes the motion takes place under variable acceleration also. All the relations discussed in chapter 17 about the motion in a straight line under variable acceleration are applicable to the motion of rotation also.

Example 12. *The equation for angular displacement of a body moving on a circular path is given by :*

$$\theta = 2t^3 + 0.5$$

where θ is in rad and t in sec. Find angular velocity, displacement and acceleration after 2 sec.

Solution. Given : Equation for angular displacement $\theta = 2t^3 + 0.5$ ...(*i*)

Angular displacement after 2 seconds

Substituting $t = 2$ in equation (*i*),

$$\theta = 2\,(2)^3 + 0.5 = 16.5 \text{ rad} \quad \textbf{Ans.}$$

Angular velocity after 2 seconds

Differentiating both sides equation (*i*) with respect to *t*,

$$\frac{d\theta}{dt} = 6\,t^2 \quad \text{...(ii)}$$

or velocity, $\omega = 6\,t^2$ ...(*iii*)

Substituting $t = 2$ in equation (*iii*),

$$\omega = 6\,(2)^2 = 24 \text{ rad/sec} \quad \textbf{Ans.}$$

Angular acceleration after 2 seconds

Differentiating both sides of equation (*iii*) with respect to *t*,

$$\frac{d\omega}{dt} = 12t \text{ or Acceleration } \alpha = 12t$$

Now substituting $t = 2$ in above equation,

$$\alpha = 12 \times 2 = 24 \text{ rad/sec}^2 \quad \textbf{Ans.}$$

***Example 13.** *The equation for angular displacement of a particle, moving in a circular path (radius 200 m) is given by :*

$$\theta = 18t + 3t^2 - 2t^3$$

where θ is the angular displacement at the end of t sec. Find (i) angular velocity and acceleration at start, (ii) time when the particle reaches its maximum angular velocity; and (iii) maximum angular velocity of the particle. **[WBUT, AMIE]**

Solution. Given : Equation for angular displacement $\theta = 18t + 3t^2 - 2t^3$ ...(*i*)

(i) Angular velocity and acceleration at start

Differentiating both sides of equation (*i*) with respect to *t*,

$$\frac{d\theta}{dt} = 18 + 6t - 6t^2$$

i.e. angular velocity, $\omega = 18 + 6t - 6t^2$ ...(*ii*)

Substituting $t = 0$ in equation (*ii*),

$$\omega = 18 + 0 - 0 = 18 \text{ rad/s} \quad \textbf{Ans.}$$

Differentiating both sides of equation (*ii*) with respect to *t*,

$$\frac{d\omega}{dt} = 6 - 12t$$

i.e. angular acceleration, $\alpha = 6 - 12t$ ...(*iii*)

Now substituting $t = 0$ in equation (*iii*),

$$\alpha = 6 \text{ rad/s}^2 \quad \textbf{Ans.}$$

(ii) Time when the particle reaches maximum angular velocity

For maximum angular velocity, differentiating the equation for angular velocity (*ii*) with respect to *t i.e.* equation (*iii*) and equating it to zero.

$$6 - 12t = 0 \quad \text{or} \quad t = \frac{6}{12} = 0.5 \text{ sec} \quad \textbf{Ans.}$$

(iii) Maximum angular velocity of the particle

The maximum angular velocity of the particle may now be found out by substituting $t = 0.5$ in equation (*ii*),

$$\omega_{max} = 18 + (6 \times 0.5) - 6\,(0.5)^2 = 19.5 \text{ rad/s} \quad \textbf{Ans.}$$

EXERCISE 3

1. The angular displacement of a body is given by equation $(\theta) = a + bt + ct^2$. What is the angular acceleration of the body ? (**Ans.** 2*c*)
2. The relation between the angle of rotation (θ) in radians and time (*t*) in seconds of a rotating body is given by the equation.

   $$\theta = 2t^3 + 3t^2 + 10.$$

   Find displacement, angular velocity and angular acceleration after 4 seconds. **[RGPV]**

   (**Ans.** 186 rad; 120 rad/sec ; 54 rad/sec^2)

QUESTIONS

1. Define motion of rotation and give three examples of it.
2. What do you understand by the term 'angular velocity' and 'angular acceleration'? Do they have any relation between them ?
3. How would you find out linear velocity of a rotating body ?
4. Obtain an equation between the linear acceleration and angular acceleration of a rotating body.

OBJECTIVE TYPE QUESTIONS

1. The angular velocity of rotating body is expressed in terms of
   (*a*) revolution per minute (*b*) radians per second
   (*c*) any one of the two (*d*) none of the two
2. The linear velocity of a rotating body is given by the relation
   (*a*) $v = r.\omega$ (*b*) $v = r/\omega$
   (*c*) $v = \omega/r$ (*d*) ω^2/r
   where r = Radius of the circular path, and
   ω = Angular velocity of the body in radians/s.
3. The linear acceleration of a rotating body is given by the relation
   (*a*) $a = r.\alpha$ (*b*) $a = r/\alpha$
   (*c*) $a = \alpha/r$ (*d*) α^2/r
   where r = Radius of the circular path, and
   α = Angular acceleration of the body in radians/s^2
4. If at any given instant, we know that linear velocity and acceleration of a car, we can mathematically obtain its
   (*a*) angular velocity (*b*) angular acceleration
   (*c*) none of the two (*d*) both of the two
5. The relationship between linear velocity and angular velocity of a cycle
   (*a*) exists under all conditions
   (*b*) does not exist under all conditions
   (*c*) exists only when it does not slip
   (*d*) exists only when it moves on horizontal plane

ANSWERS

1. (*c*) 2. (*a*) 3. (*a*) 4. (*d*) 5. (*a*)

CHAPTER

20 Combined Motion of Rotation and Translation

Contents

1. INTRODUCTION

We have discussed seperately about the motion of translation (or linear motion) and the motion of rotation (or angular motion). But sometimes a body has simultaneously, the **motion of rotation as well as translation**. Examples of such motion are a wheel of a car, a sphere rolling (without slipping) on the ground etc. A little consideration will show, that such a motion will have the combined effect of rotation as well as translation. Now in this chapter, we shall discuss the combined motion of rotation and translation.

2. RIGID LINK AND ITS MOTION

Consider a rigid link AB, which moves from its initial position AB to its final position A_1B_1. It will be interesting to know that the link neither has, wholly a motion of translation nor wholly rotational, but a combination of the two. But the point B moves faster than the point A as shown in Fig. 1. (*a*).

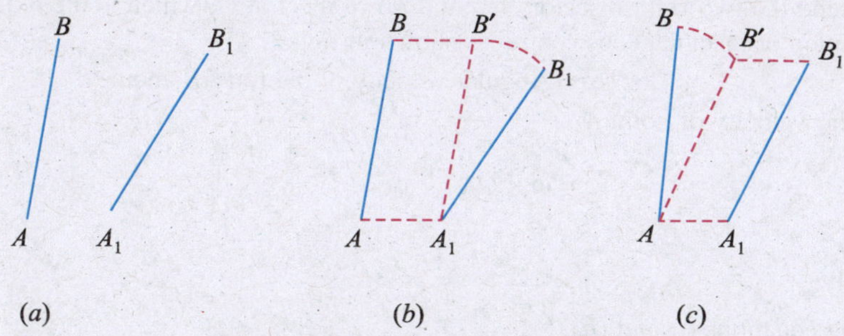

Fig. 1. Motion of a rigid link AB to A_1B_1.

If we split up the motion of the link AB, we shall find that the link has first motion of translation from AB to A_1B', and then the motion of rotation about A_1, till it occupies the final position A_1B_1 as shown in Fig. 1. (*b*).

The motion of link AB may also be considered to be first motion of rotation from AB to AB' about A, and then motion of translation from AB' to A_1B_1 as shown in Fig. 1 (*c*)

Such a motion of AB to A_1B_1 is an excellent example of combined motion of rotatiton and translation ; it being immaterial, whether the motion of rotation takes place first, or the motion of translation takes first.

3. *INSTANTANEOUS CENTRE

[MU]

We have discussed in the last article the motion of a rigid link as an example of combined motion of rotation and translation. Now consider the motion of link from AB to A_1B_1 as shown in fig. 2. In the last article, we had split up such a motion for the sake of analysis into the following two parts :

1. Motion of rotation, and
2. Motion of translation.

Both these motions were considered to take place one after the other. But in actual practice the motion of link AB is so gradual, that it is difficult to see the two separate motions. But we see a smooth motion of the link, though the point B moves faster than the point A.

This combined motion of rotation and translation, may be assumed to be a motion of pure rotation about some centre. As the position of link AB goes on changing, therefore the centre, about which the motion of rotation is assumed to take place, also goes on changing. Such a centre, which goes on changing, from one instant to another, is known as *instantaneous centre*. The locus of all such instantaneous centres, is known as centrode. The position of instantaneous

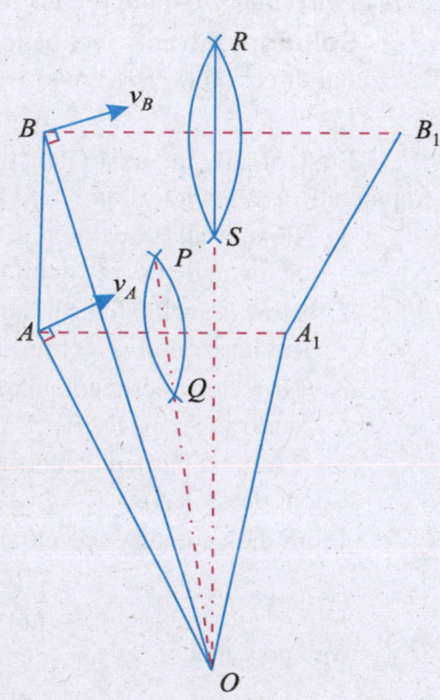

Fig. 2. Instantaneous centre.

centre may be located graphically as discussed below :

1. First of all, draw the initial and final positions of the rigid link as AB and A_1B_1 respectively.
2. Join AA_1 and BB_1.
3. Draw PQ right bisector of AA_1, and RS right bisector of BB_1.
4. Extend the two right bisectors PQ and RS to meet at O, which is the required centre of rotation or instantaneous centre as shown in Fig. 2.

Let ω = Angular velocity of the rod AB about A.

$\therefore$ Linear velocity of point A.

$$v_A = \omega \times OA \quad \text{or} \quad \omega = \frac{v_A}{OA} \qquad ...(i)$$

Similarly, $$\omega = \frac{v_B}{OB} \qquad ...(ii)$$

Equating equations (*i*) and (*ii*),

$$\frac{v_B}{v_A} = \frac{OB}{OA}$$

The direction of v_A and v_B will be at right angles to OA and OB respectively as shown in Fig. 2.

Note. Sometimes, three or more members are hinged together to form a triangle, a quadrilateral etc., In such a case , each member will have its own instantaneous centre. The instantaneous centre for each member may be found out by drawing perpendiculars at the directions of motions at the two ends of the member as usual.

***Example 1.** *A link AB is moving in a vertical plane. At a certain instant, when the link is inclined at 60° to the horizontal, the point A is moving horizontally at 2 m/s, while B is moving in a vertical direction. Find the velocity of B.* **[MU]**

Solution. Given : Inclination of the link with horizontal = 60° and velocity of point A in horizontal direction (v_A) = 2 m/s

Let v_B = Velocity of B in the vertical direction.

First of all, let us locate the position of instantaneous centre O, graphically, as shown in Fig. 3 and as discussed below :

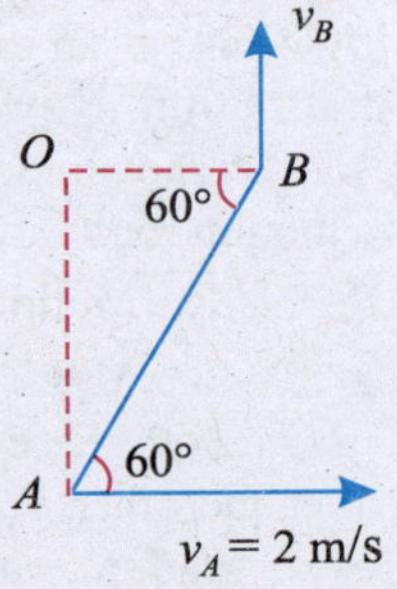

1. First of all draw the position of the link AB, such that it is inclined at an angle of 60° with the horizontal.
2. Now draw the lines indicating the directions of motions of points A (in horizontal direction) and B (in vertical direction).
3. Now draw perpendiculars at A and B on the directions of motion v_A and v_B.
4. Let these perpendiculars meet at O, which is the instantaneous centre of the link AB.

Fig. 3.

From the geometry of velocity of B, the triangle AOB, we find that

$$\frac{OB}{OA} = \cot 60° = 0.577$$

We know that,

$$\frac{v_B}{v_A} = \frac{OB}{OA} = 0.577$$

or $$v_B = v_A \times 0.577 = 2 \times 0.577 = 1.15 \text{ m/s} \quad \textbf{Ans.}$$

Example 2. *Three links are hinged together to form a triangle ABC as shown in Fig. 4.*

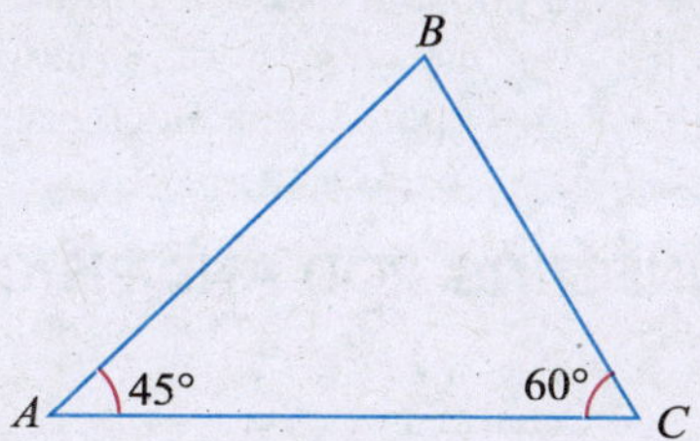

Fig. 4.

At a certain instant, the point A is moving towards the mid-point of BC with a velocity of 5 m/s, and B is moving at a perpendicular direction to AC. Find the velocity of C.

Solution. Given : Velocity of A (v_A) = 5 m/s

First of all, let us locate the position of instantaneous centre of the points A and B graphically as shown in Fig. 5 and as discussed below :

1. Draw the triangle ABC with the given data.

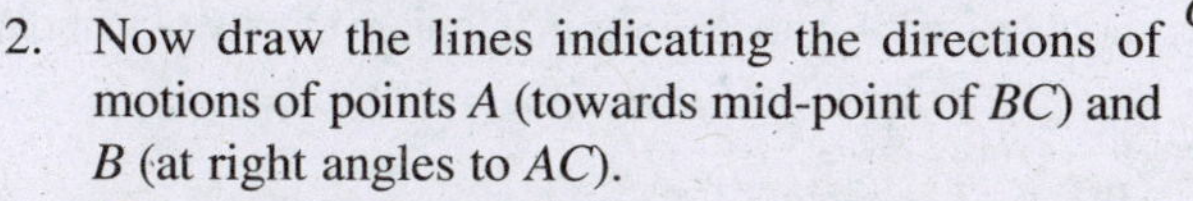

2. Now draw the lines indicating the directions of motions of points A (towards mid-point of BC) and B (at right angles to AC).
3. Now draw perpendiculars at A and B on the directions of motion of v_A and v_B.

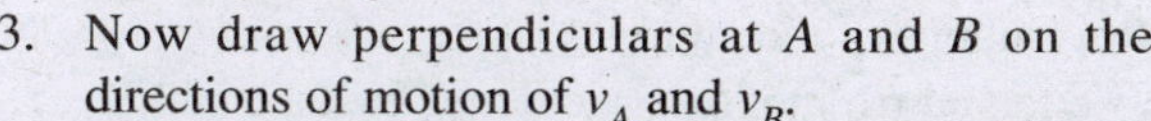

4. Let these perpendiculars meet at O, which is the instantaneous centre of the link AB and BC.
5. Now join OC and draw a line at right angle to OC indicating the direction of motion of the point C.

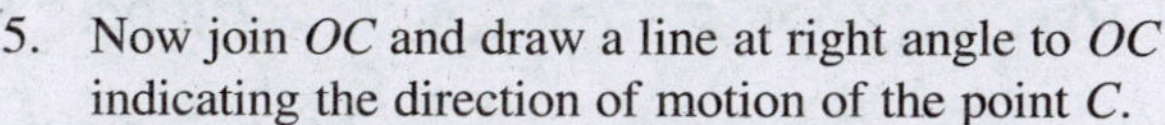

Fig. 5.

Measuring the diagram to some scale, we find that OA = 2.6 cm and OC = 5.4 cm

We know that, $$\frac{v_C}{v_A} = \frac{OC}{OA} = \frac{5.4}{2.6} = 2.08$$

∴ Velocity of C,

$$v_C = v_A \times 2.08 = 5 \times 2.08 = 10.4 \text{ m/s} \quad \textbf{Ans.}$$

EXERCISE 1

1. The ends A and B of a link 1.5 m long are constrained to move in vertical and horizontal guides as shown in Fig. 6.
   At a given instant, when A is 0.9 m above C it was moving at 3 m/s upwards. Find the velocity of B at this instant. **(Ans.** 2.25 m/s)

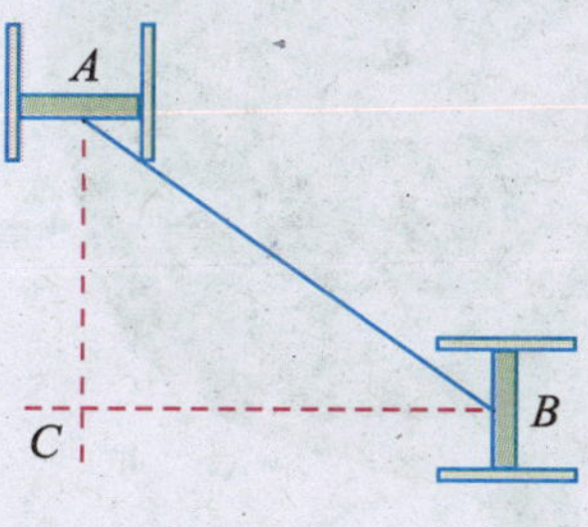

Fig. 6

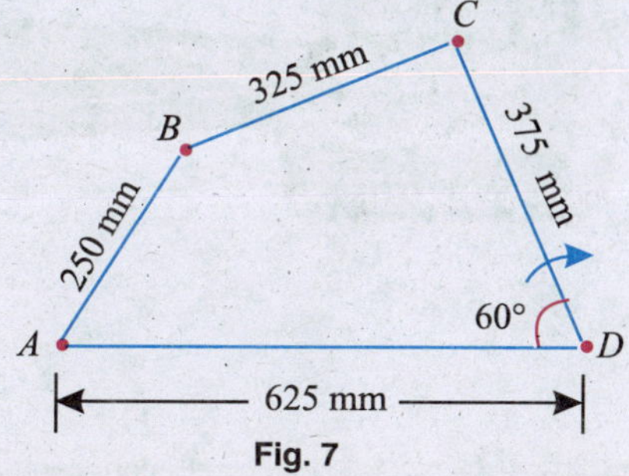

Fig. 7

2. In a four-bar mechanism *ABCD*, the two points *A* and *D* are fixed while the links *AB* and *CD* can rotate about the fixed points *A* and *D* as shown in Fig. 7
Find graphically, or otherwise, the velocity of *B* and angular velocity of the link *AB*, when the link *CD* rotates at 240 r.p.m. in the clockwise direction.
(**Ans.** 18.3 m/s ; 700 r.p.m.)

4. *MOTION OF A CONNECTING ROD AND PISTON OF A RECIPROCATING PUMP

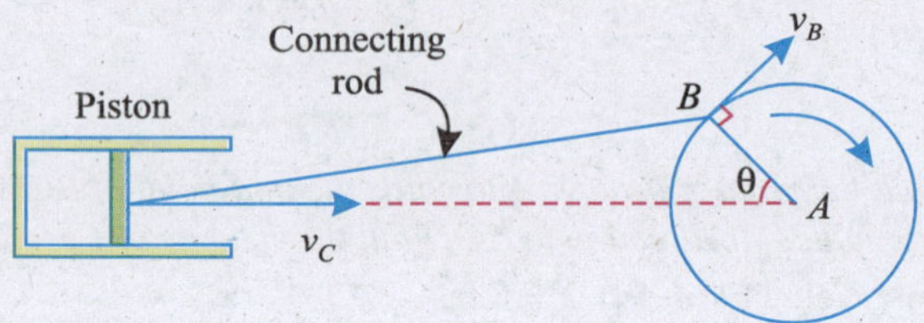

Fig. 8. Motion of connecting rod and piston of a reciprocating pump.

In Fig.8 is shown the mechanism of a crank connecting rod and piston of a reciprocating pump. The working of main parts of the pump mechanism is as follows ;

1. The crank *AB*, which rotates about its axis *A* with a uniform angular velocity of ω radians/sec, such that

$$\omega = \frac{2\pi N}{60} \text{ rad/s}$$

where N = Revolutions of the crank in r.p.m.

2. The connecting rod *BC*, whose to and fro motion is governed by the crank *B*.
3. The piston *C*, whose to and fro motion is governed by the connecting rod *BC*.

5. *METHODS TO FIND THE VELOCITY OF PISTON OF A RECIPROCATING PUMP

In the last article, we have discussed the working of the main parts of a reciprocating pump. In most of the cases, we are required to find out the velocity of piston. This may be found out by any one of the following two methods :

1. Graphical method **2. Analytical method.**

Piston, connecting rod and crank.

6. *ANALYTICAL METHOD TO FIND THE VELOCITY OF PISTON OF A RECIPROCATING PUMP

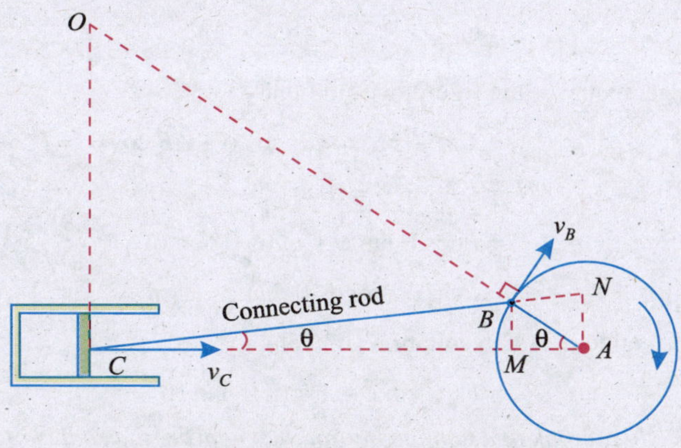

Fig. 9 Veloctiy of a piston of reciprocating pump.

Consider the mechanism of a reciprocating pump, in which *AB* be the crank, *BC* the connecting rod, and *C* the piston. Let *O* be the instantaneous centre of the connecting rod *BC*. Produce the line *CB* to meet the vertical line through *A* at *N* as shown in Fig. 9. Now through *B*, draw *BM* perpendicular to *CA*.

From the geometry of the triangle *OCB* and *ABN*, we find that these two triangles are similar. Therefore

$$\frac{OC}{OB} = \frac{AN}{AB}$$

Let ω_1 = Angular velocity of the crank *AB* in rad/s.

ω_2 = Angular velocity of the connecting rod *BC* about *O*, in rad/s.

l = Length of the connecting rod *BC*,

r = Radius of the crank *AB*,

θ = Angle traversed by the crank in degrees,

ϕ = Angle, which the connecting rod makes with *CA* (known as obliquity of the connecting rod).

We know that, Velocity of piston or *C*,

$$v_C = \omega_2 \times OC \qquad ...(i)$$

and Velocity of *B*,

$$v_B = \omega_2 \times OB \qquad ...(ii)$$

We also know that, Velocity of *B*,

$$v_B = \omega_1 \times AB \qquad ...(iii)$$

Dividing equation (*i*) by (*ii*),

$$\frac{v_C}{v_B} = \frac{\omega_2 \times OC}{\omega_2 \times OB} = \frac{OC}{OB} = \frac{AN}{AB}$$

or $$v_C = v_B \times \frac{AN}{AB} = \omega_1 \times AB \times \frac{AN}{AB}$$

$$= \omega_1 \times r \times \frac{AN}{r} = \omega_1 \times AN \qquad ...(iv)$$

Now from the geometry of the figure, we find that

$$CA = CM + MA = CB \cos \phi + AB \cos \theta = l \cos \phi + r \cos \theta$$

and $$AN = CA \tan \phi = (l \cos \phi + r \cos \theta) \tan \phi$$

$$= l \cos \phi \tan \phi + r \cos \theta \tan \phi$$

$$= l \sin \phi + r \cos \theta \tan \phi$$

Substituting this value of AN in equation (*iv*) velocity of C,

$$v_C = \omega_1 (l \sin \phi + r \cos \theta \tan \phi)$$

****Example 3.** *In a crank and connecting rod mechanism, the radius of crank and length of the connecting rod are 300 mm and 1200 mm respectively. The crank is rotating at 180 r.p.m. Find the velocity of the piston, when the crank is at an angle of 45°, with the horizontal.* **[MU]**

Solution. Given : Radius of the crank (r) = 300 mm = 0.3 m ; Length of connecting rod (l) = 1200 mm = 1.2 m ; Angular rotation of crank (N) = 180 r.p.m and angle traversed by the crank (θ) = 45°

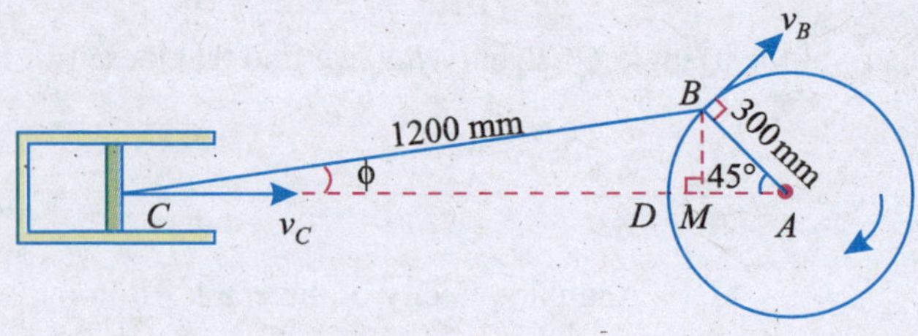

Fig. 10.

We know that, Angular velocity of the crank,

$$\omega_1 = \frac{2\pi \times 180}{60} = 6\pi \text{ rad/s}$$

and Velocity of B, $$v_B = \omega r = 6\pi \times 0.3 = 5.66 \text{ m/s}$$

From the geometry of the figure, we find that

$$\sin \phi = \frac{BM}{BC} = \frac{AB \sin 45°}{BC} = \frac{0.3 \times 0.707}{1.2}$$

$$= 0.1768 \quad \text{or} \quad \phi = 10.18°$$

We also know that, Velocity of the piston,

$$v_C = \omega_1 (l \sin \phi + r \cos \theta \tan \phi)$$

$$= 6\pi (1.2 \sin 10.18° + 0.3 \cos 45° \tan 10.18°) \text{ m/s}$$

$$= 6\pi [(1.2 \times 0.1768) + (0.3 \times 0.707 \times 0.1796)] \text{ m/s}$$

$$= 4.72 \text{ m/s} \quad \textbf{Ans.}$$

7. VELOCITY DIAGRAM METHOD TO FIND THE VELOCITY OF PISTON OF ARECIPROCATING PUMP

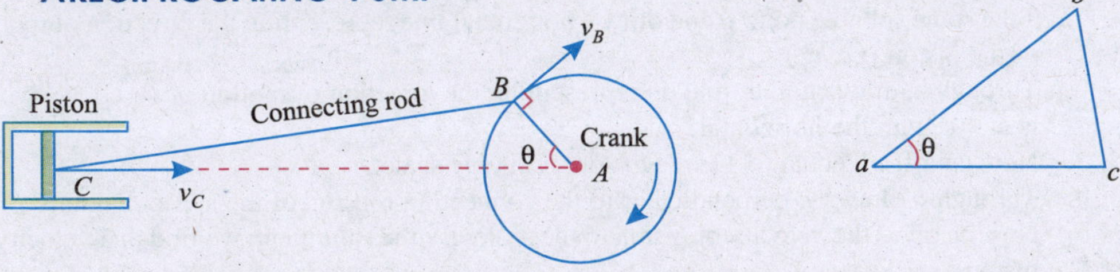

(a) Space diagram (b) Velocity diagram

Fig. 11. Velocity diagram method.

Consider the mechanism of a reciprocating pump, in which *AB* be the crank, *BC* the connecting rod and *C* the piston as shown in Fig. 11 (*a*). Now let us draw the velocity triangle as shown in Fig. 11 (*b*) and as discussed below :

1. First of all, take some suitable point *a* and draw a horizontal line *ac* representing the direction of motion of the piston. (*i.e.* v_C).;
2. Through *a* draw another line *ab* representing the direction of motion of *B* (*i.e.* v_B). It is at right angles to the crank *AB*.
3. Now cut off *ab* equal to the velocity v_B to some suitable scale (such that $v_B = \omega.r$) where ω is the angular velocity of the crank and *r* is the radius of the crank.
4. Through *b* draw a line *bc* perpendicular to the connecting rod *BC* of the space diagram.
5. Now *ac* of the velocity diagram gives the velocity of piston v_C to the scale.

Note. The length *bc* to the scale gives the velocity of the connecting rod. The angular velocity of the connecting rod may now be found out from the relation *bc/BC*, where *BC* is the length of the connecting rod.

Example 4. *In a reciprocating pump, the lengths of connecting rod and crank is 1125 mm and 250 mm respectively. The crank is rotating at 420 r.p.m. Find the velocity with which the piston will move, when the crank has turned through an angle of 40° from the inner dead centre.*

[BPUT]

Solution. Given: Length of connecting rod (*l*) = 1125 mm = 1.125 m ; Length of crank (*r*) = 250 mm = 0.25 m ; Angular rotation of crank (*N*) = 420 r.p.m and angle traversed by the crank (θ) = 40°

We know that, Angular velocity of crank,

$$\omega = \frac{2\pi N}{60} = \frac{2\pi \times 420}{60} = 44 \text{ rad/s}$$

and Velocity of *B*,

$$v_B = \omega r = 44 \times 0.25 = 11 \text{ m/s}$$

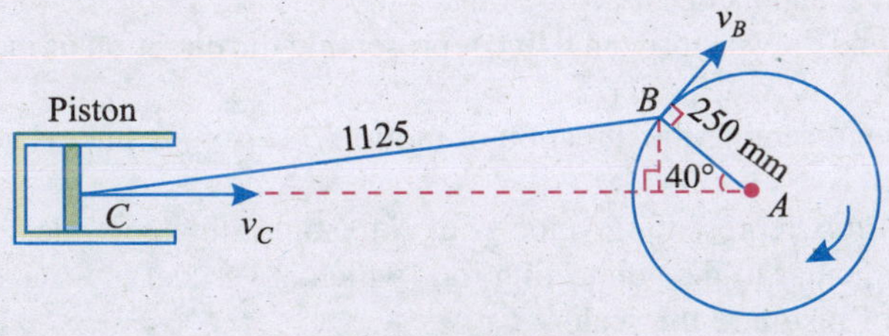

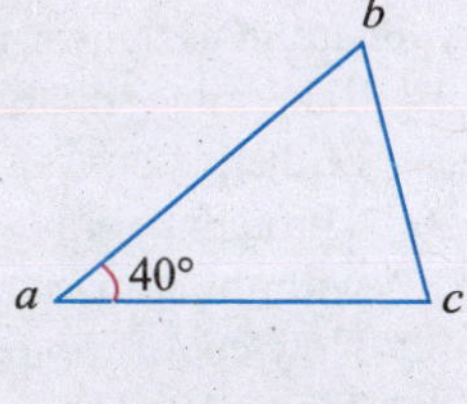

(a) Space diagram (b) Velocity diagram

Fig. 12.

First of all, draw the space diagram for the reciprocating pump machanism as shown in Fig. 12 (*a*). Now draw the velocity diagram as shown in Fig. 12 (*b*) and as discussed below :

1. Take some suitable point *a* and draw a horizontal line representing the direction of motion of the piston (*i.e.* v_C).
2. Through *a*, draw another line *ab* representing the direction of motion of *B* (*i.e.* v_B) which is at 40° with the horizontal.
3. Now cut off *ab* equal to 11 m/s to some suitable scale.
4. Through *b*, draw *bc* perpendicular to the connecting rod *BC* of the space diagram.
5. Now *ac* gives the velocity of piston to the scale. By measurement, we find that velocity of piston,

$$v_C = ac = 8.3 \text{ m/s} \quad \textbf{Ans.}$$

***Example 5.** *The crank AB in the mechanism shown in Fig. 13 rotates at 5 rev/s is 300 mm long.*

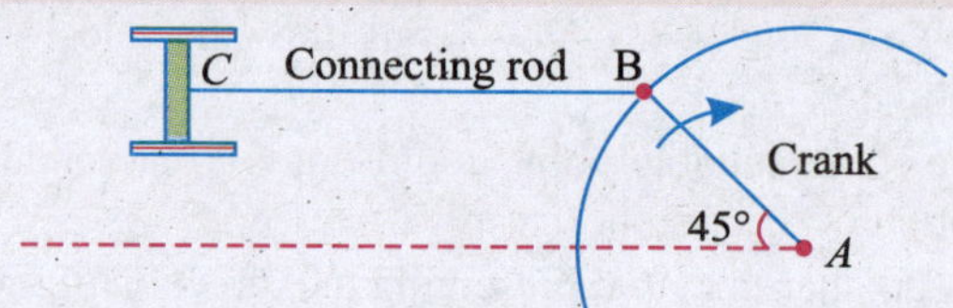

Fig. 13.

The link CB is 600 mm long, and the piston C moves in horizontal guides. Find for the position shown (i) velocity of piston C, (ii) angular velocity of the connecting rod BC and (iii) velocity of a point D at the centre AB. **[MU]**

Solution. Given : Angular rotation of crank (N) = 5 rev/s ; Length of the crank (r) = 300 mm = 0.3 m and length of the link CB (l) = 600 mm = 0.6 m

We know that, Angular velocity of crank,

$$\omega = 2\pi N = 2\pi \times 5 = 10\pi \text{ rad/s}$$

and Velocity of *B*,

$$v_B = \omega r = 10\pi \times 0.3 = 9.4 \text{ m/s}$$

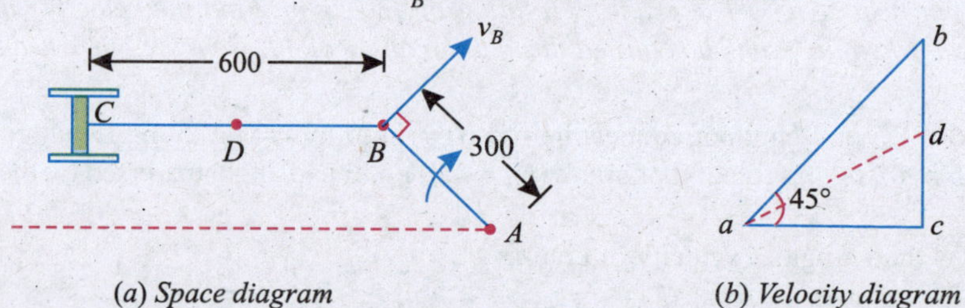

(*a*) *Space diagram* (*b*) *Velocity diagram*

Fig. 14.

First of all, draw the space diagram for the mechanism as shown in Fig. 14 (*a*). Now draw the velocity diagram as shown in Fig. 14 (*b*) and as discussed below :

1. Take some suitable point *a*, and draw a horizontal line representing the direction of motion of the piston (*i.e.* v_C).
2. Through *a*, draw a line *ab* representing the direction of motion *B* (*i.e.* v_B) which is at 45° with the horizontal.
3. Now cut off *ab* equal to 9.4 m/s representing velocity of *B* to some suitable scale.
4. Through *b*, draw *bc* perpendicular to the connecting rod (which is vertical).
5. Now *ac* gives the velocity of piston to the scale.

By measurement, we find that the velocity of piston

$$v_C = ac = 6.65 \text{ m/s} \quad \textbf{Ans.}$$

By measurement, we also find that $bc = 6.65$ m/s

∴ Angular velocity of connecting rod,

$$\omega = \frac{bc}{BC} = \frac{6.65}{0.6} = 11.1 \text{ rad/s} = \frac{11.1 \times 60}{2\pi} = 106 \text{ r.p.m.} \quad \textbf{Ans.}$$

Now locate a point d at the centre of bc. Join ad. By measurement, we also find that velocity of D,

$$v_D = ad = 7.43 \text{ m/s} \quad \textbf{Ans.}$$

8. *MOTION OF A ROLLING WHEEL WITHOUT SLIPPING

Consider a wheel rolling along a straight path, in such a way that there is no slipping of the wheel on its path. A little consideration will show, that the centre of the wheel O moves with some linear velocity. And each particle on the periphery of the wheel rotates with some angular velocity. Thus the motion of any particle on the periphery of the wheel is a combination of linear and angular velocity as shown in Fig. 15.

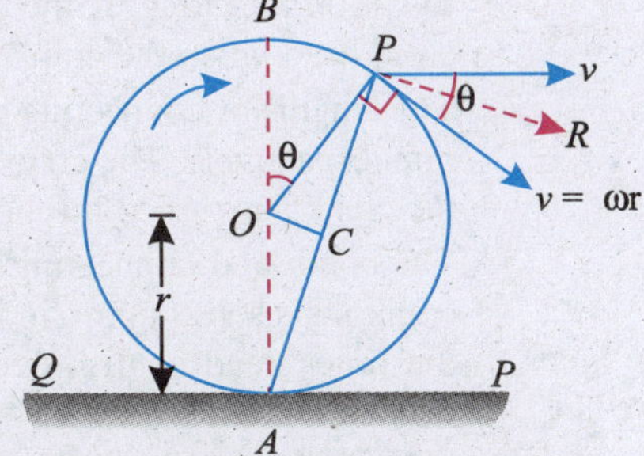

Fig. 15. Motion of a rolling wheel without slipping.

Let v = Linear velocity of the centre of the wheel

ω = Angular velocity of the wheel, and

r = Radius of the wheel

Now consider any particle P on the periphery of the wheel making an angle θ with the vertical through the centre of the wheel as shown in Fig. 15. We know that particle (P) is subjected to the following two motions simultaneously.

1. Linear velocity (v) acting in the horizontal direction.
2. Tangential velocity (equal to linear velocity, such that $v = \omega r$) acting at right angles to OP.

From the geometry of the figure, we find that the angle between these two velocities is equal to θ. The resultant of these velocities (R) will act along the bisector of the angle between the two forces, such that

$$R = 2v \cos\left(\frac{\theta}{2}\right)$$

We know that, Angular velocity of P about A

$$= \frac{v_P}{AP}$$

We know that, From the triangle AOP,

$$AP = 2AC = 2r \cos\left(\frac{\theta}{2}\right)$$

∴ Angular velocity of C about A

$$= \frac{2v \cos\left(\frac{\theta}{2}\right)}{AP} = \frac{2v \cos\left(\frac{\theta}{2}\right)}{2r \cos\left(\frac{\theta}{2}\right)} = \frac{v}{r}$$

It is thus obvious, that any point on the wheel rotates about the lowest point A (which is in touch with the ground) with the same angular velocity $(\omega = \frac{v}{r})$.

Notes. 1. The particle A is subjected to the following two velocities :

(*i*) linear velocity (v) towards P, and

(*ii*) tangential velocity ($v = \omega r$) towards Q

Since these two velocities are equal, opposite and collinear, therefore their resultant is zero. Or in other words, the particle A will be momentarily at rest.

2. Similarly, it can be proved the particle B (*i.e.* highest point of the wheel) will be subjected to a resultant velocity of $2v$ and acts perpendicular to OB.
3. The lowest point (A) is the instantaneous centre of the motion of the wheel.

EXERCISE 2

1. In a crank and connecting rod mechanism, the crank is 300 mm long and the connecting rod 1500 mm long. If the crank rotates uniformly at 300 r.p.m., find the velocity of the cross head when the crank is inclined at 30° with the inner dead centre. (**Ans.** 5.39 m/s)
2. The lengths of connecting rod and crank in a reciprocating engine are 1000 mm and 225 mm respectively. The crank is rotating at 360 r.p.m. Find the velocity of cross head, when the crank has turned through 30° with the line joining the centres. (**Ans.** 5.19 m/s)
3. The crank and connecting rod of a steam engine are 0.5 m and 2 m respectively. The crank makes 180 r.p.m. in the clockwise direction. When the crank has turned 45° from the inner dead centre, determine (*i*) velocity of piston, and (*ii*) angular velocity of connecting rod. (**Ans.** 8.15 m/s ; 32.5 r.p.m.)

QUESTIONS

1. Describe the phenomenon of combined motion of rotation and translation. Give few examples.
2. Explain the term 'instantaneous centre'.
3. How would you locate the instantaneous centre of a rigid link moving with combined motion of rotation and translation ?
4. What are the various methods for finding out the velocity of piston of a reciprocating pump ? Explain one of them.
5. Derive a relation for the velocity of piston in a crank and connecting rod mechanism.

OBJECTIVE TYPE QUESTIONS

1. In a combined motion of rotation and translation
   (*a*) the motion of roation takes place before the motion of translation
   (*b*) the motion of translation takes place before the motion of rotation
   (*c*) both the motions take place simultaneously
2. The instantaneous centre is a point which is always fixed.
   (*a*) yes (*b*) No
3. The velocity of piston in a reciprocating pump mechanism depends upon
   (*a*) angular velocity of the crank (*b*) radius of the crank
   (*c*) length of the connecting rod (*d*) both (*a*) and (*b*)
   (*e*) all the above
4. The velocity of a piston in a reciprocating pump mechanism is maximum when the obliquity is
   (*a*) zero (*b*) maximum (*c*) average (*d*) minimum

ANSWERS

1. (*c*) **2.** (*b*) **3.** (*e*) **4.** (*b*)

CHAPTER

21 Simple Harmonic Motion

Contents

1. SIMPLE HARMONIC MOTION

In the previous chapters, we have discussed the motion of particles or bodies in a straight line or in rotation. But in this chapter, we shall discuss the motion of a particle or point along a circular path and its effect on one of its diameters.

Consider a particle starting from x and moving round the circumference of circle in an anti-clockwise direction, with a constant angular velocity, as shown in Fig. 1. Let P be the position of the particle at any instant and N be the projection of P on the diameter Y-Y' of the circle.

It will be interesting to know that when the point P moves round the circumference of the circle from x to y, N moves from O to y ; when P moves from y to x', N moves from y to O. Similarly, when P moves from x' to y', N moves from O to y', and finally when P moves from y' to x, N moves from y' to O. Hence, as P completes one revolution, the point N completes one vibration about the point O. This to and fro motion of N is known as *Simple harmonic motion*, briefly written as S.H.M.

2. IMPORTANT TERMS

The following terms, which will be frequently used in this chapter, should be clearly understood at this stage :

1. ***Amplitude.*** It is the maximum displacement of a body, from its mean position. In Fig. 1, Oy and Oy' is the amplitude of the particle N. The amplitude is always equal to the radius of the circle.

2. ***Oscillation.*** It is one complete vibration of a body. In Fig. 1, when the body moves from y to y' and then back to y (or in other words from O to y, y to y' and then y' to O), it is said to have completed one oscillation.

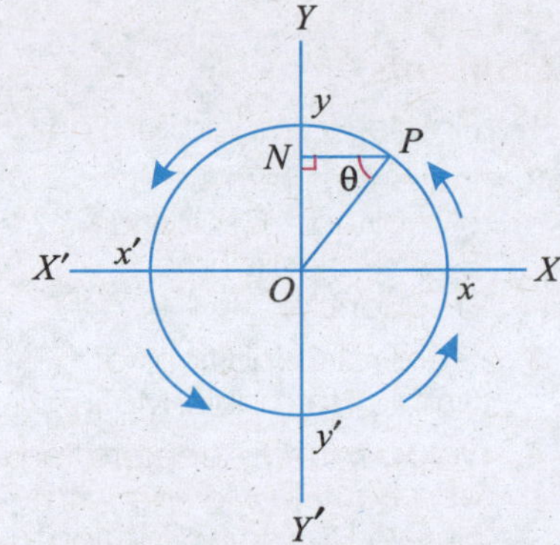

Fig. 1. S.H.M.

3. ***Beat.*** It is half of the oscillation. In Fig. 1, when the body moves from y to y' or y' to y (or in other words O to y' and then y' to O), it is said to have completed one beat.

4. ***Periodic time.*** It is the time taken by a particle for one complete oscillation. Mathematically, periodic time,

$$T = \frac{2\pi}{\omega}$$

where, ω = Angular velocity of the particle in rad/s.

It is thus obvious, that the periodic time of a S.H.M. is independent of its amplitude.

5. ***Frequency.*** It is the number of cycles per second and is equal to $\frac{1}{T}$ where T is the periodic time. Frequency is generally denoted by the letter 'n'. The unit of frequency is hertz (briefly written Hz) which means frequency of one cycle per second.

3. GENERAL CONDITIONS OF SIMPLE HARMONIC MOTION

In general, a body is said to move or vibrate, with simple harmonic motion, if it satisfies the following two conditions :

1. Its acceleration is always directed towards the centre, known as the point of reference or mean position.

2. Its acceleration is proportional to the distance from that point.

The pendulum clock keeps time as the pendulum swings back and forth.

4. *VELOCITY AND ACCELERATION OF A PARTICLE MOVING WITH SIMPLE HARMONIC MOTION

Consider a particle moving along the circumference of a circle, of radius r with a uniform angular velocity of ω radians/sec as shown in Fig. 2.

Fig. 2. Particle moving with S.H.M.

Let P be the position of the particle at some instant after t sec from X. Therefore, angle turned by the particle,

$$\theta = \omega t \quad \text{rad}$$

and displacement of the point N (*i.e.* projection of P on the vertical diameter Y-Y' of the circle),

$$y = ON = r \sin \theta$$

$$= r \sin \omega t \qquad ...(i)$$

Differentiating this equation with respect to time t,

$$\frac{dy}{dt} = r\omega \cos \omega t \qquad ...(ii)$$

or Velocity, $$v = r\omega\sqrt{1 - \sin^2 \omega t} \qquad ...(\because \sin^2 \theta + \cos^2 \theta = 1)$$

From equation (*i*), we find that

$$\sin \omega t = \frac{y}{r}$$

Substituting this value of $\sin \omega t$ in the above equation,

$$v = r\omega\sqrt{\left(1 - \frac{y^2}{r^2}\right)}$$

or Velocity, $$v = \omega\sqrt{r^2 - y^2} \qquad ...(iii)$$

Now differentiating equation (*ii*) with respect to time t,

$$\frac{d^2 y}{dt^2} = -r\omega^2 \sin \omega t$$

or acceleration, $$a = -\omega^2 y \qquad ...(\text{Substituting } y = r \sin \omega t)$$

Note. The minus sign shows that the direction of acceleration is opposite to the direction in which y increases *i.e.*, the acceleration is always directed towards the point O. But in actual practice, this relation is used as $a = \omega^2 y$

***Example 1.** *The piston of a steam engine moves with simple harmonic motion. The crank rotates at 120 r.p.m. and the stroke length is 2 metres. Find the velocity and acceleration of the piston, when it is at a distance of 0.75 metre from the centre.* **[JNTU]**

Solution. Given : Frequency of piston (N) = 120 r.p.m ; stroke length l = 2 m or radius (r) = 1 m and distance of piston from the centre (y) = 0.75 m

Velocity of piston

We know that angular velocity of piston,

$$\omega = \frac{2\pi N}{60} = \frac{2\pi \times 120}{60} = 4\pi \text{ rad/sec}$$

∴ Velocity of piston, $$v = \omega\sqrt{r^2 - y^2} = 4\pi\sqrt{(1)^2 - (0.75)^2} = 8.3 \text{ m/s} \quad \textbf{Ans.}$$

Acceleration of piston

Acceleration of piston,

$$a = \omega^2 y = (4\pi)^2 \times 0.75 = 118.4 \text{ m/s}^2 \quad \textbf{Ans.}$$

***Example 2.** *A body, moving with simple harmonic motion, has an amplitude of 1 meter and the period of complete oscillation is 2 seconds. What will be the velocity and acceleration of the body after 0.4 second from the extreme position?* **[AMIE]**

Solution. Given : Amplitude (r) = 1 m; Periodic time (T) = 2 s and time taken by the body from extreme position = 0.4 s

Velocity of the body

Now let O be the centre, Y an extremity of the motion and P the position of the body after 0.4 sec from Y as shown in Fig. 3. Therefore time required by the body to travel from Y to P

$$= 0.4 \text{ s}$$

Time required by the body to travel from O to Y

$$= \frac{1}{4} \times T = \frac{1}{4} \times 2 = 0.5 \text{ s}$$

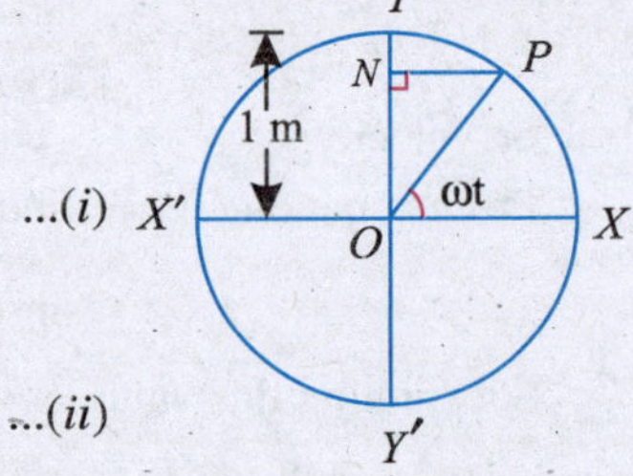

Fig. 3.

∴ Time required by the body to travel from O to N

$$t = 0.5 - 0.4 = 0.1 \text{ s} \qquad ...(i)$$

Angular velocity of the body

$$\omega = \frac{2\pi}{T} = \frac{2\pi}{2} = \pi \text{ rad/s} \qquad ...(ii)$$

∴ Displacement of the body after 0.4 sec from the extreme position (or 0.1 second from the mean position),

$$y = r \cos \omega t = 1 \cos (\pi \times 0.1) = 1 \cos 18° \text{ m}$$

$$...(\pi \times 0.1 = 180 \times 0.1 = 18°)$$

$$= 0.95 \text{ m}$$

∴ Velocity of the body,

$$v = \omega \sqrt{r^2 - y^2} = \pi \sqrt{1^2 - (0.95)^2} = 0.98 \text{ m/s} \quad \textbf{Ans.}$$

Acceleration of the body

Acceleration of the body,

$$a = \omega^2 y = (\pi)^2 \times 0.95 = 9.38 \text{ m/s}^2 \quad \textbf{Ans.}$$

***Example 3.** *Find amplitude and time period of a particle moving with simple harmonic motion, which has a velocity of 9 m/s and 4 m/s at the distance of 2 m and 3 m respectively from the centre.* **[AMIE]**

Solution. Given : When velocity (v_1) = 9 m/s, distance from centre (y_1) = 2 m and when velocity (v_2) = 4 m/s, distance from centre (y_2) = 3 m

Amplitude of the particle

Let r = Amplitude of the particle, and

ω = Angular velocity of the particle.

Velocity of the particle,

$$v = \omega \sqrt{r^2 - y^2}$$

Now $$9 = \omega\sqrt{r^2 - (2)^2} = \omega\sqrt{r^2 - 4} \quad ...(i)$$

Also $$4 = \omega\sqrt{r^2 - (3)^2} = \omega\sqrt{r^2 - 9} \quad ...(ii)$$

Dividing equation (*i*) by (*ii*) $$\frac{9}{4} = \frac{\sqrt{r^2 - 4}}{\sqrt{r^2 - 9}}$$

or $$\frac{81}{16} = \frac{r^2 - 4}{r^2 - 9} \quad ...\text{(Squaring both sides)}$$

$$81r^2 - 729 = 16r^2 - 64 \quad \text{or} \quad 65r^2 = 665$$

or $$r = \sqrt{\frac{665}{65}} = \sqrt{10.23} = 3.2 \text{ m}$$

Time-period of the particle

Substituting this value of r in equation (*i*),

$$9 = \omega\sqrt{(3.2)^2 - 4} = \omega\sqrt{6.24} = \omega \times 2.5$$

or $$\omega = \frac{9}{2.5} = 3.6 \text{ rad/s}$$

We know that time period,

$$T = \frac{2\pi}{\omega} = \frac{2\pi}{3.6} = 1.75 \text{ s} \qquad \textbf{Ans.}$$

***Example 4.** *In a system, the amplitude of the motion is 5 m and the time is 4 seconds. Find the time required by the particle in passing between points which are at distances of 4 m and 2 m from the centre of force and are on the same side of it.* **[JNTU, IGNOU]**

Solution. Given : Amplitude (r) = 5 m ; Time taken (T) = 4 sec and distances of the point (y_1) = 4 m and (y_2) = 2m

Angular velocity of the particle,

$$\omega = \frac{2\pi}{T} = \frac{2\pi}{4} = 90°/\text{s}$$

Displacement of particle (y_1),

$$4 = r \sin \omega t = 5 \sin \omega t_1$$

or $$\sin \omega t_1 = \frac{4}{5} = 0.8$$

or $$\omega t_1 = 53.1°$$

or $$t_1 = \frac{53.1}{90} = 0.59 \text{ s}$$

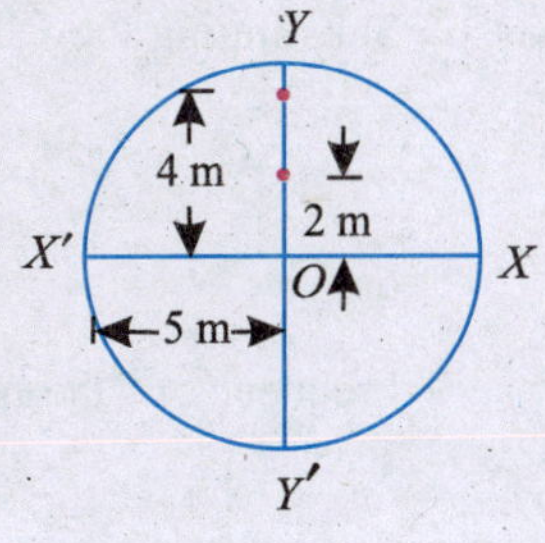

Fig. 4.

Similarly (y_2), $$2 = r \sin \omega t_2$$

or $$\sin \omega t_2 = \frac{2}{5} = 0.4$$

or $$\omega t_2 = 23.6°$$

or $$t_2 = \frac{23.6}{90} = 0.26 \text{ s}$$

Time required in passing between the two points,

$$t = t_1 - t_2 = 0.59 - 0.26 = 0.33 \text{ s} \qquad \textbf{Ans.}$$

****Example 5.** *A body performing simple harmonic motion has a velocity of 12 m/s when the displacement is 50 mm, and 3 m/s when the displacement is 100 mm, the displacement being measuerd from the mid-point. Calculate the frequency and amplitude of the motion. What is the acceleration when the displacement is 75 mm ?* **[IGNOU]**

Solution. Given : Velocity (v_1) = 12 m/s, when displacement (y_1) = 50 mm = 0.05 m ; and velocity (v_2) = 3 m/s, when displacement (y_2) = 100 mm = 0.1 m

Amplitude of the motion

Let r = Amplitude of the motion, and

ω = Angular displacement of the body,

Velocity of the body,

$$v = \omega \sqrt{r^2 - y^2}$$

$$\therefore \quad 12 = \omega \sqrt{r^2 - (0.05)^2} = \omega \sqrt{r^2 - 0.0025} \quad \text{...}(i)$$

Similarly

$$3 = \omega \sqrt{r^2 - (0.1)^2} = \omega \sqrt{r^2 - 0.01} \quad \text{...}(ii)$$

Dividing equation (*i*) and (*ii*)

$$\frac{12}{3} = \frac{\omega \sqrt{r^2 - 0.0025}}{\omega \sqrt{r^2 - 0.01}} \quad \text{or} \quad 4 = \frac{\sqrt{r^2 - 0.0025}}{\sqrt{r^2 - 0.01}}$$

or

$$16 = \frac{r^2 - 0.0025}{r^2 - 0.01} \quad \text{...(Squaring both sides)}$$

or

$$16r^2 - 0.16 = r^2 - 0.0025$$

or

$$15r^2 = -0.0025 + 0.16 = 0.1575$$

or

$$r^2 = \frac{0.1575}{15} = 0.0105$$

or

$$r = 0.1025 \text{ m} \quad \textbf{Ans.}$$

Frequency of the motion

Substituting the value of r in equation (*i*),

$$12 = \omega \sqrt{(0.1025)^2 - 0.0025} = \omega \times 0.09$$

or

$$\omega = \frac{12}{0.09} = 133.3 \text{ rad/sec}$$

Frequency of the motion,

$$N = \frac{1}{T} = \frac{133.3}{2\pi} = 21.2 \text{ Hz} \quad \textbf{Ans.}$$

Acceleration when the displacement is 75 mm

Acceleration of the body when y is 75 mm or 0.075 m,

$$a = \omega^2 y = (133.3)^2 \times 0.075 = 1332.6 \text{ m/s}^2 \quad \textbf{Ans.}$$

EXERCISE 1

1. A particle, moving with simple harmonic motion, has an acceleration of 6 m/s^2 at a distance of 1.5 m from the centre of oscillation. Find the time period of the oscillation. (**Ans.** 3.142 s)
2. A body weighing 150 N, moves with simple harmonic motion. The velocity and acceleration of the body when it is 200 mm from the centre of oscillation, are 5 m/s and 20 m/s^2 respectively. Determine (*a*) amplitude of motion and (*b*) no. of vibrations per minute. (**Ans.** 539 mm ; 95.5)
3. A particle moves with simple harmonic motion. When the particle is 0.75 m from the mid-path, its velocity is 11 m/s and when 2 m from the mid-path its velocity is 3 m/s. Find the angular velocity, periodic time and its maximum acceleration. (**Ans.** 5.7 rad/s ; 1.1 s ; 67.25 m/s^2)
4. A particle moving with simple harmonic motion, has a velocity of 20 m/s at its central position. If the particle makes two oscillations per second, find (*i*) amplitude of motion and (*ii*) velocity at 1/4th the distance of the amplitude. (**Ans.** 1.59 m ; 19.35 m/s)

5. MAXIMUM VELOCITY AND ACCELERATION OF A PARTICLE MOVING WITH SIMPLE HARMONIC MOTION

We have already discussed in previous article, that the velocity of a particle moving with simple harmonic motion,

$$v = \omega \sqrt{r^2 - y^2} \qquad ...(i)$$

A little consideration will show, that the velocity is maximum, when $y = 0$ or when N passes through O *i.e.* its mean position. Therefore, maximum velocity

$$v_{max} = \omega r \qquad ...(ii)$$

It may be noted from equation (*i*) that its velocity is zero when $y = r$, *i.e.* when N passes through Y' or Y as shown in Fig. 2. At these points, N is momentarily at rest. We have also discussed that the acceleration of a particle moving with simple harmonic motion,

$$a = \omega^2 y \qquad ...(iii)$$

A little consideration will show, that the acceleration is maximum when the value of y is maximum or $y = r$ *i.e.* when N passes through Y or Y'. Therefore maximum acceleration,

$$a_{max} = \omega^2 r \qquad ...(iv)$$

It may also be noted from equation (*iii*) that the acceleration is zero, when $y = 0$ or when N passes through O *i.e.* its mean position. It is thus obvious, that the acceleration is proportional to the distance from O, *i.e*, mean position.

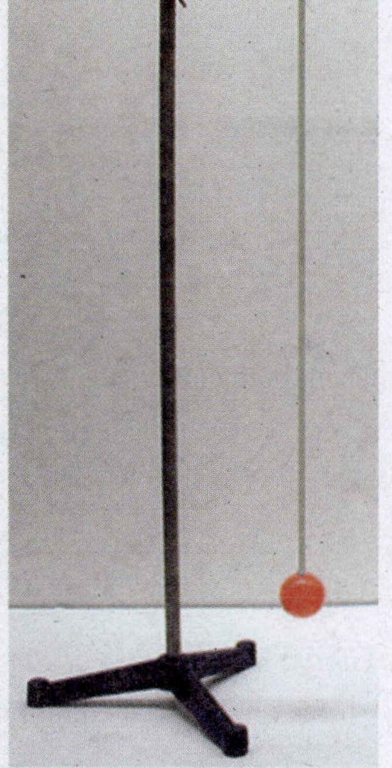

Simple pendulum is the most common example for SHM.

Example 6. *A body is vibrating with simple harmonic motion of amplitude 100 mm, and frequency 2 vibrations/sec. Calculate the maximum velocity and acceleration.*

Solution. Given : Amplitude (r) = 100 mm = 0.1 m and frequency of body (N) = 2 vib/sec.

Maximum velocity

Angular velocity of the body,

$$\omega = 2\pi N = 2\pi \times 2 = 4\pi \text{ rad/s}$$

and maximum velocity, $v_{max} = r\omega = 0.1 \times 4\pi = 1.257$ m/s **Ans.**

Maximum acceleration

We also know that maximum acceleration,

$$a_{max} = \omega^2 r = (4\pi)^2 \times 0.1 = 15.79 \text{ m/s}^2 \quad \textbf{Ans.}$$

***Example 7.** *A particle, moving with simple harmonic motion, performs 10 complete oscillation per minute and its speed, is 60% of the maximum speed when it is at a distance of 8 cm from the centre of oscillation, . Find amplitude, maximum acceleration of the particle. Also find speed of the particle, when it is 6 cm far from the centre of oscillation.* **[JNTU]**

Solution. Given : No. of oscillation/min = 10 and when displacement (y) = 8 cm velocity (v) = 60% v_{max} = 0.6 v_{max}.

Amplitude of the particle

No. of oscillations per second

$$= \frac{10}{60} = \frac{1}{6}$$

∴ Time-period of the motion

$$T = \frac{6}{1} = 6 \text{ s}$$

Angular velocity, $$\omega = \frac{2\pi}{T} = \frac{2\pi}{6} = \frac{\pi}{3} \text{ rad/s}$$

Linear velocity, $$v = \omega\sqrt{r^2 - y^2}$$

or $$0.6\, v_{max} = \omega\sqrt{r^2 - (8)^2} \quad \text{...}(\because y = 8 \text{ cm})$$

or $$0.6\, \omega r = \omega\sqrt{r^2 - 64} \quad \text{...}(v_{max} = \omega r)$$

or $$0.6\, r = \sqrt{r^2 - 64}$$

Squaring both sides,

$$0.36r^2 = r^2 - 64 \quad \text{or} \quad 0.64r^2 = 64$$

or $$r^2 = \frac{64}{0.64} = 100$$

or $$r = \sqrt{100} = 10 \text{ cm} \quad \textbf{Ans.}$$

Maximum acceleration of the particle

Maximum acceleration of the particle,

$$a_{max} = \omega r = \left(\frac{\pi}{3}\right)^2 \times 10 = 10.97 \text{ cm}^2\text{/s} \quad \textbf{Ans.}$$

Speed of the particle when it is 6 cm from the centre of oscillation

Speed of the particle when it is 6 cm from the centre of oscillation,

$$v = \omega\sqrt{r^2 - y^2} = \frac{\pi}{3}\sqrt{(10)^2 - (6)^2} = 8.38 \text{ cm/s} \quad \textbf{Ans.}$$

***Example 8.** *A prismatic bar AB of weight W is resting on rough rollers rotating with equal angular velocity in opposite directions as shown below :*

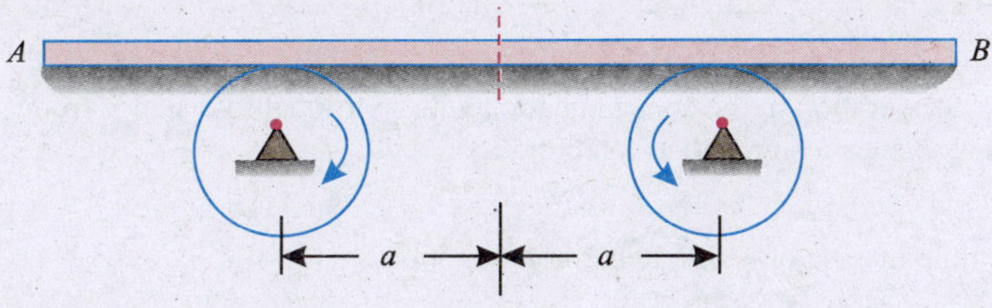

Fig. 5.

If the bar is so placed on the rollers that its c.g. is displaced from the middle plane and released, show that the bar executes a simple harmonic motion. Also find the time period, if the coefficient of friction is μ and the distance between the two rollers is 2a. **[NIT]**

Solution. Let the weight W of the bar AB acting at its centre of gravity G be at a distance x from its mid point as M as shown in Fig. 6.

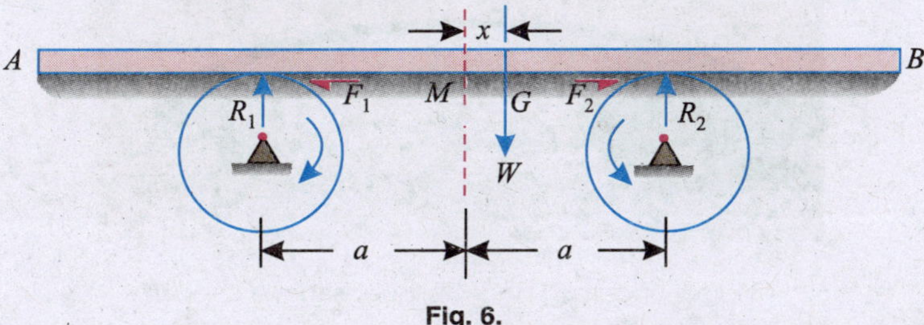

Fig. 6.

Let R_1 = Vertical reaction of the left roller,

F_1 = Force of friction between the bar and left roller, and

R_2, F_2 = Corresponding values for the right roller.

Taking moments about the left support and equating the anti-clockwise moments with clockwise moments,

$$R_2 \times 2a = W(a + x)$$

or
$$R_2 = \frac{W(a + x)}{2a}$$

Similarly,
$$R_1 = \frac{W(a - x)}{2a}$$

Force of friction at the left roller,

$$F_1 = \mu R_1 = \frac{\mu W(a - x)}{2a}$$

Similarly,
$$F_2 = \mu R_2 = \frac{\mu W(a + x)}{2a}$$

Resultant forces of friction,

$$F = F_2 - F_1 = \frac{\mu W(a + x)}{2a} - \frac{\mu W(a - x)}{2a} = \frac{\mu Wx}{a}$$

We know that, this resultant force of friction is equal to the product of mass of the bar and its acceleration.

$$\therefore \qquad \frac{\mu\, Wx}{a} = \frac{W}{g} \times a \qquad \text{or} \qquad a = \frac{\mu\, gx}{a}$$

Since the acceleration in the above equation is proportional to distance from the centre of the bar, thus the bar executes a simple harmonic motion. **Ans.**

Periodic time

Periodic time in case of a simple harmonic motion,

$$t = 2\pi\sqrt{\frac{\text{Displacement}}{\text{Acceleration}}} = 2\pi\sqrt{\frac{x}{\frac{\mu g x}{a}}} = 2\pi\sqrt{\frac{a}{\mu g}} \qquad \textbf{Ans.}$$

Bridges undergo SHM when loads move over it.

EXERCISE 2

1. A particle moving with simple harmonic motion of amplitude 150 mm is subjected to an angular velocity of 2 rad/s. What is the maximum velocity and maximum acceleration of the particle ? (**Ans.** 300 mm/s ; 600 mm/s^2)
2. The time period of a simple harmonic motion is 6 seconds, and the particle oscillates through a distance of 300 mm on each side of the mean position. Find the maximum velocity and maximum acceleration of the particle. (**Ans.** 0.315 m/s ; 0.33 m/s^2)

QUESTIONS

1. Explain the meaning of S.H.M. and give its one example.
2. Define the term amplitude as applied to S.H.M.
3. What do you understand by the terms 'periodic time' and 'frequency'? What relation do they have ?
4. Show that when a particle moves with simple harmonic motion, its time for a complete oscillation is independent of the amplitude of its motion.

OBJECTIVE TYPE QUESTIONS

1. The maximum displacement of a body moving with simple harmonic motion from its mean position is called

   (*a*) oscillation (*b*) amplitude (*c*) Beat (*d*) none of them.

2. The frequency of vibration in case of simple harmonic motion

   (*a*) means the number of cycles per second

   (*b*) represents time taken by the particle for one complete oscillation

   (*c*) depends upon its amplitude.

   (*d*) is directly proportional to its beat.

3. The periodic time of a body moving with simple harmonic motion

   (*a*) depends upon its amplitude under all conditions.

   (*b*) is independent of its amplitude

   (*c*) depends upon its amplitude under certain conditions

   (*d*) has no relation with its frequency.

4. The maximum acceleration of a particle moving with S.H.M. takes place, when

   (*a*) it passes through its extreme positions

   (*b*) it passes through its mid-point

   (*c*) it has maximum velocity

   (*d*) none of the above.

5. The velocity of a particle moving with simple harmonic motion is maximum when its acceleration is

   (*a*) zero (*b*) maximum (*c*) average (*d*) both (*a*) and (*b*).

ANSWERS

1. (*b*) **2.** (*a*) **3.** (*b*) **4.** (*a*) **5.** (*a*)

The ship undergoes SHM when in water.

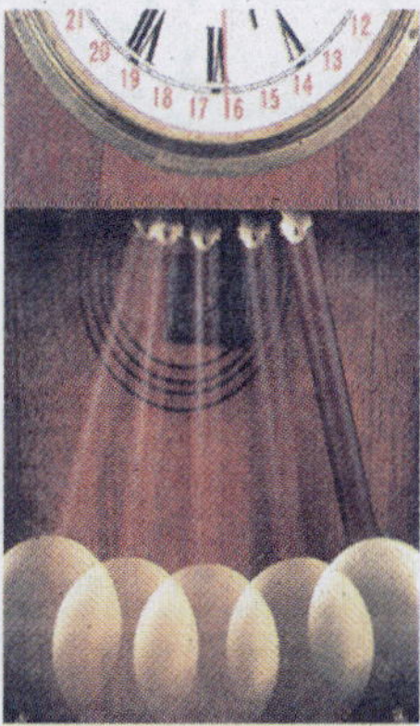

The pendulum clock keeps time as the pendulum swings back and forth.

CHAPTER

22 Laws of Motion

Contents

1. INTRODUCTION

The entire system of dynamics is based on three laws of motion, which are the fundamental laws, and were formulated by Sir Issac Newton in 1680. Like other scientific laws, these are also stated in the mathematical forms which agree with actual observations.

2. IMPORTANT TERMS

The following terms, which will be mostly used in this chapter, should be clearly understood at this stage :

1. ***Mass.*** It is the matter contained in a body. The units of mass are kilogram, tonne etc.

2. ***Weight.*** It is the force, by which the body is attracted towards the centre of the earth. The units of weight are the same as those of force *i.e.* N, kN etc.
3. ***Momentum.*** It is the quantity of motion possessed by a body. It is expressed mathematically as

$$\text{Momentum} = \text{Mass} \times \text{Velocity.}$$

   The units of momentum depend upon the units of mass and velocity. In S.I. units, the mass is measured in kg, and velocity in m/s, therefore the unit of momentum will be kg-m/s.
4. ***Force.*** It is a very important factor in the field of dynamics also, and may be defined as any cause which produces or tends to produce, stops or tends to stop motion. The units of force, like those of weight, are N, kN etc.
5. ***Inertia.*** It is an inherent property of a body, which offers resistance to the change of its state of rest or uniform motion.

3. RIGID BODY

Strictly speaking, the laws of motion, enunciated by Newton, are applicable only to the rigid bodies. Though a rigid body (or sometimes written as 'body' for the sake of simplicity) is defined in many ways by the different scientists, yet there is not much of difference between all the definitions. But the following definition of a rigid body is universally recognised.

A rigid body consists of a system of innumerable particles. If the positions of its various particles remain fixed, relative to one another (or in other words, distance between any two of its particles remain constant), it is called a solid body. It will be interesting to know that in actual practice, all the solid bodies are not perfectly rigid bodies. However, they are regarded as such, since all the solid bodies behave more or less like rigid bodies.

4. NEWTON'S LAWS OF MOTION

Following are the three laws of motion, which were enunciated by Newton, who is regarded as father of the Science.

1. **Newton's First Law of Motion states, "*Everybody continues in its state of rest or of uniform motion, in a straight line, unless it is acted upon by some external force.*"**
2. **Newton's Second Law of Motion states, "*The rate of change of momentum is directly proportional to the impressed force, and takes place in the same direction, in which the force acts.*"**
3. **Newton's Third Law of Motion states, "*To every action, there is always an equal and opposite reaction.*"**

5. NEWTON'S FIRST LAW OF MOTION

It states ***"Everybody continues in its state of rest or of uniform motion, in a straight line, unless it is acted upon by some external force."*** It is also called the law of inertia, and consists of the following two parts :

1. A body at rest continues in the same state, unless acted upon by some external force. It appears to be self-evident, as a train at rest on a level track will not move unless pulled by an engine. Similarly, a book lying on a table remains at rest, unless it is lifted or pushed.
2. A body moving with a uniform velocity continues its state of uniform motion in a straight line, unless it is compelled by some external force to change its state. It cannot be exemplified because it is, practically, impossible to get rid of the forces acting on a body.

Note. The second part of the law furnishes us with an idea about the function of a force. It also implies that a force, which is to produce a change in the rest or motion of a body must be externally impressed ; or in other words, must act from outside. A little consideration will show, that the effect of inertia is of the following two types :

1. A body at rest has a tendency to remain at rest. It is called *inertia of rest.*
2. A body in uniform motion in a straight line has a tendency to preserve its motion. It is called *inertia of motion.*

6. NEWTON'S SECOND LAW OF MOTION

It states, ***"The rate of change of momentum is directly proportional to the impressed force and takes place, in the same direction in which the force acts."*** This law enables us to measure a force, and establishes the fundamental equation of dynamics.

It is also called the Law of dynamics and consists of the following two parts :

1. A body can posses acceleration only when some force is applied on it. Or in other words, if no force is applied on the body, then there will be no acceleration, and the body will continue to move with the existing uniform velocity.
2. The force applied on a body is proportional to the product of the mass of the body and the acceleration produced in it.

It will be interesting to know that first part of the above law appears to be an extension of the First Law of Motion. However, the second part is independent of the First Law of Motion.

6.1 Force Relation Derived from Newton's Second Law of Motion

Consider a body moving in a straight line. Let its velocity be changed while moving.

Let m = Mass of a body,

u = Initial velocity of the body,

v = Final velocity of the body,

a = Constant acceleration,

t = Time, in seconds required to change the velocity from u to v, and

F = Force required to change velocity from u to v in t seconds.

Initial momentum $= mu$

Final momentum $= mv$

Rate of change of momentum

$$= \frac{mv - mu}{t} = \frac{m(v-u)}{t} = ma \qquad \ldots \left[\because \frac{v-u}{t} = a \right]$$

According to Newton's Second Law of Motion, the rate of change of momentum is directly proportional to the the impressed force.

$$\therefore \qquad F \propto ma = kma$$

where k is a constant of proportionality.

For the sake of convenience, the unit of force adopted is such that it produces unit acceleration to a unit mass.

$$\therefore \qquad F = ma \qquad \text{or} \qquad \text{Force} = \text{Mass} \times \text{Acceleration}$$

In S.I. system of units, the unit of force is called newton briefly written as N.

A Newton may be defined as the force while acting upon a mass of 1 kg, produces an acceleration of 1 m/s^2 in the direction of which it acts.

7. ABSOLUTE AND GRAVITATIONAL UNITS OF FORCE

We have already discussed, that when a body of mass 1 kg is moving with an acceleration of 1 m/s^2, the force acting on the body is 1N.

Therefore when the same body of mass 1 kg is moving with an acceleration of 9.8 m/s^2, the force acting on the body is 9.8 N. (But sometimes in engineering use, we denote 1 kg mass attracted towards the earth with an acceleration of 9.8 m/s^2 as 1 kg(wt) instead of 9.8 N).

∴ **1 kg(wt) = 9.8 N**

Similarly, 1 t(wt) = 9.8 kN

Here

(1). **kg(wt)** and **t(wt)** (also written as **kg(f)** and **t(f)** are **Gravitational or Engineer's Units of Force**

(2). **N** or **kN** are called **Absolute or Scientific Units of Force**.

It is thus obvious that,

Gravitational or Engineer's Units of Force

= '*g*' times greater the corresponding Absolute or Scientific Units of Force

It will be interesting to know that the mass of a body, in absolute units, is **numerically equal** *to* the weight of the same body in gravitational units *e.g.*, consider a body whose mass,

$$\boldsymbol{m = 100 \text{ kg}}$$

∴ The force, with which it will be attracted towards the centre of the earth,

$$\boldsymbol{F = ma = mg = 100 \times 9.8 = 980 \text{ N}}$$

Now, as per definition, we know that the weight of a body is the force by which it is attracted towards the centre of the earth. Therefore weight of the body,

$$\boldsymbol{W = 980 \text{ N} = \frac{980}{9.8} = 100 \text{ kg(wt)}} \qquad ...(\because 1 \text{ kg(wt)} = 9.8 \text{ N})$$

In engineering practice, the weight of a body is of primary importance. In order to avoid inconvenience of always multiplying the force in kg(f) by 9.8 to determine its value in newtons, the engineers use kg(f) as a unit of force. To preserve the force equation (*i.e.* $F = ma$) we take the mass of the body in metric slugs. In general, the mass of a body in kg is divided by gravitational acceleration '*g*' gives the mass in slugs. Mathematically 1 slug = kg/9.8. Here the following points should be clearly understood to avoid uncalled confusion :

1. If the weight of the body is given, it will be in gravitational units. Its numerical value is equal to its mass in absolute untis. *e.g.* consider a body of weight 100 kg. Then it may be written as,

   Weight, $\boldsymbol{w = 100 \text{ kg(wt)}}$ **...(In Gravitational Units)**

   or **Mass,** $\boldsymbol{m = 100 \text{ kg}}$ **...(In Absolute Units)**

   or **Weight,** $\boldsymbol{w = mg = 100 \text{ kg} \times 9.81 \text{ m/s}^2 = 980 \text{ N}}$ **...(In Absolute Units)**

2. Sometimes, the mass of a body is given. It is always in absolute units.
3. In the force equation (*i.e.* $F = ma$) the value of mass is taken in absolute units.

Notes. From the above discussion, it may be clearly understood that

Force = Mass × Acceleration

1. If force is in newtons, then mass is in kg (absolute units).
2. If force is in kg(f), then mass is in slugs (absolute units)
3. The value of acceleration is in m/s^2 in both the cases.

Example 1. *Determine the force, which can move a body of mass 100 kg with an acceleration of 3.5 m/s².*

Solution. Given : Mass of body (m) = 100 kg and acceleration (a) = 3.5 m/s²

Force, $F = ma = 100 \times 3.5 = 350$ N **Ans.**

***Example 2.** *A body has 50 kg mass on the earth. Find its weight (a) on the earth, where g = 9.8 m/s² ; (b) on the moon, where g = 1.7 m/s² and (c) on the sun, where g = 270 m/s².* **[JNTU]**

Solution. Given: Mass of body (m) = 50 kg ; Acceleration due to gravity on earth (g_e) = 9.8 m/s²;Acceleration due to gravity on moon (g_m) = 1.7 m/s² and acceleration due to gravity on sun (g_s) = 270 m/s².

(a) Weight of the body on the earth

Weight of the body on the earth

$$F_1 = mg_e = 50 \times 9.8 = 490 \text{ N} \quad \textbf{Ans.}$$

(b) Weight of the body on the moon

Weight of the body on the moon,

$$F_2 = mg_m = 50 \times 1.7 = 85 \text{ N} \quad \textbf{Ans.}$$

(c) Weight of the body on the sun

Weight of the body on the sun,

$$F_3 = mg_s = 50 \times 270 = 13500 \text{ N} = 13.5 \text{ kN} \quad \textbf{Ans.}$$

Example 3. *A body of mass 7.5 kg is moving with a velcoity of 1.2 m/s. If a force of 15 N is applied on the body, determine its velocity after 2 s.*

Solution. Given: Mass of body = 7.5 kg ; Velocity (u) = 1.2 m/s ; Force (F) = 15 N and time (t) = 2 s.

Acceleration of the body,

$$a = \frac{F}{m} = \frac{15}{7.5} = 2 \text{ m/s}^2$$

∴ Velocity of the body after 2 seconds

$$v = u + at = 1.2 + (2 \times 2) = 5.2 \text{ m/s} \quad \textbf{Ans.}$$

***Example 4.** *A vehicle, of mass 500 kg, is moving with a velocity of 25 m/s. A force of 200 N acts on it for 2 minutes. Find the velocity of the vehicle :*

(1) when the force acts in the direction of motion, and

(2) when the force acts in the opposite direction of the motion. **[WBUT]**

Solution. Given : Mass of vehicle (m) = 500 kg ; Initial velocity (u) = 25 m/s ; Force (F) = 200 N and time (t) = 2 min = 120 s

1. *Velocity of vehicle when the force acts in the dirction of motion*

Acceleration of the vehicle,

$$a = \frac{F}{m} = \frac{200}{500} = 0.4 \text{ m/s}^2$$

∴ Velocity of the vehicle after 120 seconds

$$v_1 = u + at = 25 + (0{\cdot}4 \times 120) = 73 \text{ m/s} \quad \textbf{Ans.}$$

2. *Velocity of the vehicle when the force acts in the opposite direction of motion*

Velcoity of the vehicle in this case after 120 seconds, (when $a = -0.4$ m/s²),

$$v_2 = u + at = 25 + (-0.4 \times 120) = -23 \text{ m/s} \quad \textbf{Ans.}$$

Minus sign means that the vehicle is moving in the reverse direction or in other words opposite to the direction in which the vehicle was moving before the force was made to act.

Example 5. *A constant retarding force of 50* N *is applied to a body of mass 20 kg moving initially with a velocity of 15 m/s. How long the body will take to stop?*

Solution. Given: Retarding force (F) = 50 N ; Mass of the body (m) = 20 kg ; Initial velocity (u) = 15 m/s and final velocity (v) = 0 (because it stops)

Let t = Time taken by the body to stop.

Retardation of the body

$$a = \frac{F}{m} = \frac{50}{20} = 2.5 \text{ m/s}^2$$

Final velocity of the body,

$$0 = u + at = 15 - 2.5\,t \qquad \text{...(Minus sign due to retardation)}$$

or

$$t = \frac{15}{2.5} = 6 \text{ s} \qquad \textbf{Ans.}$$

Example 6. *A car of mass 2.5 tonnes moves on a level road under the action of 1 kN propelling force. Find the time taken by the car to increase its velocity from 36 km. p.h. to 54 km.p.h.*

Solution. Given : Mass of the car (m) = 2.5 t ; Propelling force (F) = 1 kN ; Initial velocity (u) = 36 km.p.h. = 10 m/s and final velocity (v) = 54 km.p.h. = 15 m/s

Let t = Time taken by the car to increase its speed.

Acceleration of the car,

$$a = \frac{F}{m} = \frac{1}{2.5} = 0.4 \text{ m/s}^2$$

Final velocity of the car (v),

$$15 = u + at = 10 + 0.4\,t$$

or

$$t = \frac{15 - 10}{0.4} = \frac{5}{0.4} = 12.5 \text{ s} \qquad \textbf{Ans.}$$

Example 7. *A multiple unit electric train has 800 tonnes mass. The resistance to motion is 100 N per tonne of the train mass. If the electric motors can provide 200 kN tractive force, how long does it take to accelerate the train to a speed of 90 km/hr from rest.*

Solution. Given: Mass of electric train (m) = 800 t ; Resistance to motion = 100 N/t = 100 × 800 = 80000 N = 80 kN ; Tractive force = 200 kN ; Final velocity (v) = 90 km/hr = 25 m/s and initial velocity (u) = 0 (because it starts from rest)

Let t = Time taken by the electric train.

Net force available to move the train,

$$F = \text{Tractive force} - \text{Resistance to motion}$$

$$= 200 - 80 = 120 \text{ kN}$$

Acceleration of the train

$$a = \frac{F}{m} = \frac{120}{800} = 0.15 \text{ m/s}^2$$

Final velocity of the body (v)

$$25 = u + at = 0 + 0.15\,t$$

or

$$t = \frac{25}{0.15} = 166.7 \text{ s} \qquad \textbf{Ans.}$$

Example 8. *A man of mass 60 kg dives vertically downwards into a swimming pool from a tower of height 20 m. He was found to go down in water by 2 m and then started rising. Find the average resistance of the water. Neglect the resistance of air.*

Solution. Given : Mass (m) = 60 kg and height of tower (s) = 20 m.

First of all, consider the motion of the man from the top of the tower to the water surface. In this case, initial velocity (u) = 0 (because the man dives) and distance covered (s) = 20 m

Let v = Final velocity of the man when he reaches the water surface.

We know that, $v^2 = u^2 + 2gs = (0)^2 + 2 \times 9.8 \times 20 = 392$

$\therefore \quad v = \sqrt{392} = 19.8 \text{ m/s}$

Now consider motion of the man from the water surface up to the point in water from where he started rising. In this case, initial velocity (u) = 19.8 m/s ; final velocity (v) = 0 (because the man comes to rest) and distance covered (s) = 2 m

Let a = Retardation due to water resistance.

We know that, $v^2 = u^2 + 2as$

$$0 = (19.8)^2 - 2a \times 2 = 392 - 4a$$

...(Minus sign due to retardation)

$$\therefore \quad a = \frac{392}{4} = 98 \text{ m/s}^2$$

Average resistance of the water,

$$F = ma = 60 \times 98 = 5880 \text{ N}$$ **Ans.**

Example 9. *At a certain instant, a body of mass 10 kg, falling freely under the force of gravity, was found to be falling at the rate of 20 m/s. What force will stop the body in (i) 2 seconds and (ii) 2 metres?*

Solution. Given : Mass of the body (m) = 10 kg ; Initial velocity (u) = 20 m/s and final velocity (v) = 0 (because, it stops)

(i) Force which will stop the body in 2 seconds

Let a = Constant retardation.

Final velocity of the body (v),

$$0 = u - a_1 t = 20 - 2a_1$$ ...(Minus sign due to retardation)

$$\therefore \quad a_1 = \frac{20}{2} = 10 \text{ m/s}^2$$

A little consideration will show that an upward acceleration of 10 m/s^2 is required to stop the body in 2 seconds. But as the body is falling under the force of gravity (*i.e.* with an acceleration of 9.8 m/s^2) therefore, the applied force must be able to produce an acceleration of 10 + 9.8 = 19.8 m/s^2.

$\therefore$ Force required to stop the body,

$$F = ma = 10 \times 19.8 = 198 \text{ N}$$ **Ans.**

(ii) Force which will stop the body in 2 metres

We know that, $v^2 = u^2 - 2a_2 s$ ...(Minus sign due to retardation)

$$0 = (20)^2 - 2a_2 \times 2 = 400 - 4a_2$$

or $$a_2 = \frac{400}{4} = 100 \text{ m/s}^2$$

As discussed above, the applied force must be able to produce an acceleration of 100 + 9.8 = 109.8 m/s^2.

∴ Force required to stop the body,

$$F = ma_2 = 10 \times 109.8 = 1098 \text{ N } \textbf{Ans.}$$

Example 10. *A body of mass 10 kg is moving over a smooth surface, whose equation of motion is given by the relation.*

$$s = 5t + 2t^2$$

where (s) is in metres and (t) in seconds. Find the magnitude of force responsible for the motion.

Solution. Given : Equation of motion : $s = 5t + 2t^2$

Differentiating both sides of the above equation with respect to t,

$$\frac{ds}{dt} = 5 + 4t$$

Again differentiating both sides of the above equation with respect to t,

$$\frac{d^2s}{dt^2} = 4 \quad \text{or} \quad \text{acceleration, } a = 4 \text{ m/s}^2.$$

∴ Force responsible for the motion,

$$F = ma = 10 \times 4 = 40 \text{ N } \textbf{Ans.}$$

EXERCISE 1

1. Find the force required to give an acceleration of 1.5 m/s^2 to a body of mass 40 kg. (**Ans.** 60 N)
2. A force of 30 N acts on a body of mass 8 kg. Calculate the acceleration it can produce. (**Ans.** 3.75 m/s^2)
3. A body of mass 40 kg is moving with a constant velocity of 2.5 m/s. Now a force of 100 N is applied on the body in its direction of motion. What will be its velocity after 2 second. (**Ans.** 7.5 m/s)
4. A constant force of 100 N is applied on a body of mass 50 kg at rest. Find the distance travelled by it in 12 seconds. (**Ans.** 144 m)
5. A force of 10 N acts on a body at rest for 10 s and causes it to more 20 m during this period. What is the mass of the body? (**Ans.** 25 kg)
6. A constant force acting on a body of mass 20 kg changes its speed from 2.5 m/s to 10 m/s in 15 seconds. What is the magnitude of the force? (**Ans.** 10 N)
7. A railway coach of mass 50 tonne can exert a tractive force of 20 kN. Find the acceleration of the coach on a level track if the resistance is 150 N per tone. (**Ans.** 0.25 m/s^2)
8. An engine of mass of 25 tonnes was moving with a velocity of 72 km.p.h. The steam was shut off and brakes were applied to bring the engine to rest in 400 m. Find the uniform force exerted by the brakes. (**Ans.** 12.5 kN)

8. MOTION OF A LIFT

Consider a lift (elevator or cage etc.) carrying some mass and moving with a uniform acceleration

Let m = Mass carried by the lift,

a = Uniform acceleration of the lift, and

R = Reaction of the lift or tension in the cable, supporting the lift,

Here we shall discuss the following two cases as shown in Fig. 1 (*a*) and (*b*) :

1. When the lift is moving upwards.
2. When the lift if moving downwards.

1. *When the lift is moving upwards*

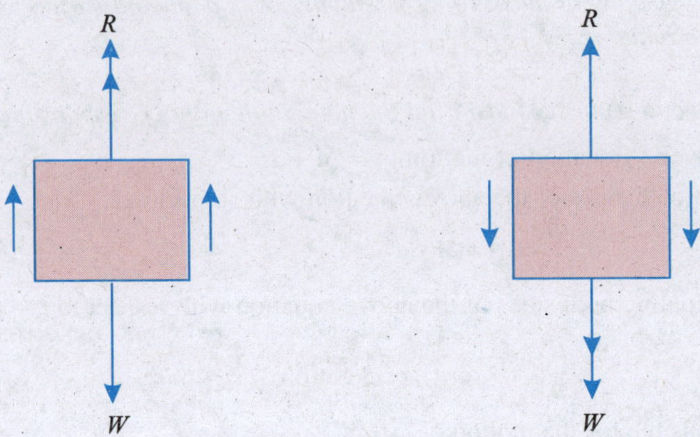

(*a*) Lift moving upwards (*b*) Lift moving downwards

Fig. 1. Motion of a lift.

Downward force due to mass of the lift

$$= mg$$

Net upward force on lift, $F = R - mg$ ...(*i*)

We also know that, this Force = Mass × Acceleration = $m.a$ ...(*ii*)

From equations (*i*) and (*ii*),

$$R - mg = ma$$

$$\therefore \quad R = ma + mg = m(a + g)$$

2. *When the lift is moving downwards*

In this case, the net downward force, which is responsible for the motion of the lift.

$$= mg - R \quad ...(iii)$$

From equations (*ii*) and (*iii*),

$$ma = mg - R$$

or $$R = mg - ma = m(g - a)$$

Note. In the above cases, we have taken mass carried by the lift only. We have assumed that it includes mass of the lift also. But sometimes the example contains mass of the lift and mass carried by the lift separately.

In such a case, the mass carried by the lift (or mass of the operator etc.) will exert a pressure on the floor of the lift. Whereas tension in the cable will be given by the algebraic sum of the masses of the lift and mass carried by the lift. Mathematically. (When the lift is moving upwards), then

Pressure exerted by the mass carried by the lift on its floor

$$= m_2 (g + a)$$

Tension in the cable $= (m_1 + m_2)(g + a)$

where, m_1 = Mass of the lift and

m_2 = Mass carried by the lift.

Example 11. *A body of mass 50 kg is being lifted by a lift in an office. Find the force exerted by the body on the lift floor, when it is moving with a uniform acceleration of 1.2 m/s².*

Solution. Given : Mass of the body (m) = 50 kg and acceleration (a) = 1.2 m/s^2

Pressure exerted by the body on the floor, when it is being lifted

$$F = m\,(g + a) = 50\,(9.8 + 1.2) = 550 \text{ N} \quad \textbf{Ans.}$$

Example 12. *In a factory, an elevator is required to carry a body of mass 100 kg. What will be the force exerted by the body on the floor of the lift, when (a) the lift is moving upwards with retardation of 0.8 m/s² ; (b) moving downwards with a retardation of 0.8 m/s².*

Solution. Given : Mass of the body (m) = 100 kg and acceleration (a) = – 0.8 m/s^2 (Minus sign due to retardation)

(a) ***When the lift is moving upwards***

Force exerted by the body on the floor of the lift

$$F_1 = m\,(g + a) = 100\,(9.8 - 0.8) = 900 \text{ N} \quad \textbf{Ans.}$$

(b) ***When the lift is moving downwards***

Force exerted by the body on the floor of the lift.

$$F_2 = m\,(g - a) = 100\,(9.8 + 0.8) = 1060 \text{ N} \quad \textbf{Ans.}$$

Example 13. *An elevator is required to lift a body of mass 65 kg. Find the acceleration of the elevator, which could cause a force of 800 N on the floor.*

Solution. Given : Mass of the body (m) = 65 kg and Force (R) = 800 N

Let a = Acceleration of the elevator.

Force caused on the floor when the elevator is going up (R),

$$800 = m\,(g + a) = 65\,(9.8 + a)$$

or

$$a = \frac{800}{65} - 9.8 = 2.5 \text{ m/s}^2 \quad \textbf{Ans.}$$

***Example 14.** *An elevator of mass 500 kg is ascending with an acceleration of 3 m/s². During this ascent, its operator whose mass is 70 kg is standing on the scale placed on the floor. What is the scale reading? What will be the total tension in the cables of the elevator during this motion?* **[DTU]**

Solution. Given : Mass of the elevator (m_1) = 500 kg ; Acceleration (a) = 3 m/s^2 and mass of operator (m_2) = 70 kg

Scale Reading

Scale reading when the elevator is ascending,

$$R_1 = m_2\,(g + a) = 70\,(9.8 + 3) = 896 \text{ N} \quad \textbf{Ans.}$$

Total tension in the cable of the elevator

Total tension in the cable of the elevator when it is ascending,

$$R_2 = (m_1 + m_2)\,(g + a) = (500 + 70)\,(9.8 + 3) \text{ N}$$
$$= 7296 \text{ N} \quad \textbf{Ans.}$$

***Example 15.** *An elevator of mass 2500 kg is moving vertically downwards with a constant acceleration. Starting from rest, it travels a distance of 35 m during an interval of 10 seconds. Find the cable tension during this time. Neglecting all other resistances to motion, what are the limits of cable tension?*

Solution. Given : Mass of elevator $(m) = 2500$ kg ; Initial velocity $(u) = 0$ (because it starts from rest) ; Distance travelled $(s) = 35$ m and time $(t) = 10$ *s*.

Cable tension

Let a = Constant acceleration of the elevator.

Distance travelled by the elevator (s)

$$35 = ut + \frac{1}{2}at^2 = (0 \times 10) + \frac{1}{2}a(10)^2 = 50a$$

or $$a = \frac{35}{50} = 0.7 \text{ m/s}^2$$

∴ Tension in the cable when the elevator is moving vertically downwards,

$$R = m(g - a) = 2500(9.8 - 0.7) = 22750 \text{ N} = 22.75 \text{ kN} \quad \textbf{Ans.}$$

Limits of cable tension

A little consideration will show, that the cable tension will have two limits *i.e.* when acceleration is zero and when acceleration is maximum (*i.e.* 9.8 m/s^2).

∴ Cable tension when the elevator is moving vertically downwards with zero acceleration,

$$R = m(g - a) = 2500(9.8 - 0) = 24\,500 \text{ N} = 24.5 \text{ kN} \quad \textbf{Ans.}$$

and cable tension when the acceleration is maximum (*i.e.* 9.8 m/s^2).

$$R = m(g - a) = 2500(9.8 - 9.8) = 0 \quad \textbf{Ans.}$$

***Example 16.** *An elevator of gross mass 500 kg starts moving upwards with a constant acceleration, and acquires a velocity of 2 m/s, after travelling a distance of 3 m. Find the pull in the cables during the accelerated motion.*

If the elevator, when stopping moves with a constant deceleration from a constant velocity of 2 m/s and comes to rest in 2 s, calculate the force transmitted by a man of mass 75 kg the floor during stopping. **[MU, BPUT, IGNOU]**

Solution. Given : Gross mass of elevator $(m_1) = 500$ kg

Pull in the cable, during accelerated motion

First of all consider motion of the elevator with a constant acceleration. In this case, initial velocity $(u) = 0$ (because it starts from rest) ; Final velocity acquired $(v) = 2$ m/s and distance travelled $(s) = 3$ m

Let a_1 = Constant acceleration.

We know that, $$v^2 = u^2 + 2a_1 s$$

or $$(2)^2 = 0 + 2a_1 \times 3 = 6a_1$$

or $$a_1 = \frac{4}{6} = 0.67 \text{ m/s}^2$$

∴ Force (or pull) required to produce this acceleration,

$$F_1 = m_1 a_1 = 500 \times 0.67 = 335 \text{ N}$$

Total pull in the cable, $$R = (500 \times 9.8) + 335 = 5235 \text{ N} \quad \textbf{Ans.}$$

Force transmitted by the man during the decelerating motion

Now consider motion of the elevator the decelaration (*i.e.* retardation). In this case, initial velocity $(u) = 2$m/s ; Final velocity $(v) = 0$ (because it comes to rest) ; time $(t) = 2$ s and mass of man $(m_2) = 75$ kg

Let a_2 = Constant deceleration *i.e.* retardation

Final velocity of the elevator (v),

$$0 = u + a_2 t = 2 + a_2 \times 2 = 2 + 2a_2$$

or $$a_2 = -1 \text{ m/s}^2 \quad \text{...(Minus sign means retardation)}$$

∴ Force transmitted by the man during decelerating motion,

$$R = m_2 (g - a) = 75 (9.8 - 1) = 660 \text{ N} \textbf{ Ans.}$$

EXERCISE 2

1. In an office, a lift is moving upwards with an acceleration of 1.5 m/s^2. Find the pressure exerted by a body of mass 30 kg on the floor of the lift. **(Ans.** 339 N)
2. An elevator of mass 2 t is to be lifted and lowered by means of a rope. Find the tension in the rope, when the elevator is moving (*i*) upward with an acceleration of 2 m/s^2 and (*ii*) downward with an acceleration of 1.5 m/s^2. **(Ans.** 23.6 kN ; 16.6 kN)
3. A lift has an upward acceleration of 1 m/s^2. Find the pressure exerted by the man of mass 62.5 kg on the floor of the lift. If the lift had a downward acceleration of 1 m/s^2, find the pressure exerted by the man. Also find an upward acceleration of the lift, which would cause the man to exert a pressure of 750 N. **[UPTU](Ans.** 675 N ; 550 N ; 2.2 m/s^2)

9. **D'ALEMBERT'S PRINCIPLE AND DYNAMIC EQUILIBRIUM*

[UPTU, AUT, WBUT, BPUT, RGPV]

It states, **"If a rigid body is acted upon by a system of forces, this system may be reduced to a single resultant force whose magnitude, direction and the line of action may be found out by the methods of graphic statics."**

We have already discussed in article of Newtons's Second Law of Motion in this chapter, that force acting on a body.

$$P = ma \quad \text{...}(i)$$

where m = mass of the body, and

a = Acceleration of the body.

The equation (*i*) may also be written as :

$$P - ma = 0 \quad \text{...}(ii)$$

It may be noted that equation (*i*) is the equation of *dynamics* whereas the equation (*ii*) is the equation of *statics*. The equation (*ii*) is also known as the **equation of dynamic equilibrium** under the action of the real force *P.* This principle is known as *D' Alembert's principle.*

D'Alembert's Principle is used for solving questions on dynamics which involve **force and acceleration**.

(*Note: It is also known as the **principle of kinostatics**.)

Example 17. *Two bodies A and B of mass 80 kg and 20 kg are connected by a thread and move along a rough horizontal plane under the action of a force 400 N applied to the first body of mass 80 kg as shown in Fig. 2.*

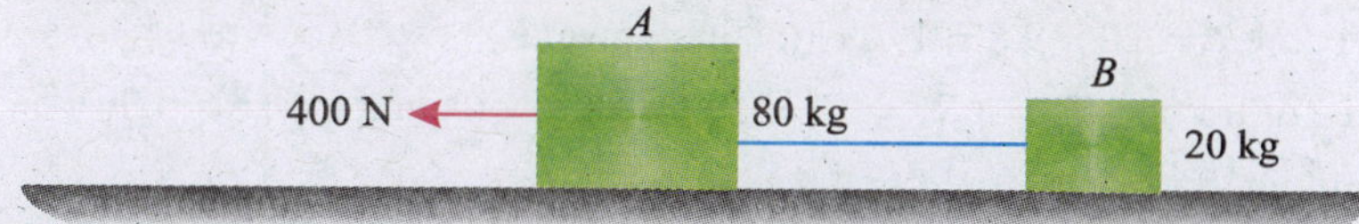

Fig. 2.

The coefficient of friction between the sliding surfaces of the bodies and the plane is 0.3. Determine the acceleration of the two bodies and the tension in the thread, using D' Alembert's principle.

Solution. Given : Mass of body A (m_1) = 80 kg ; Mass of the body B (m_2) = 20 kg; Force applied on first body (P) = 400 N and coefficient of friction (μ) = 0.3

Acceleration of the two bodies

Let a = Acceleration of the bodies, and

T = Tension in the thread.

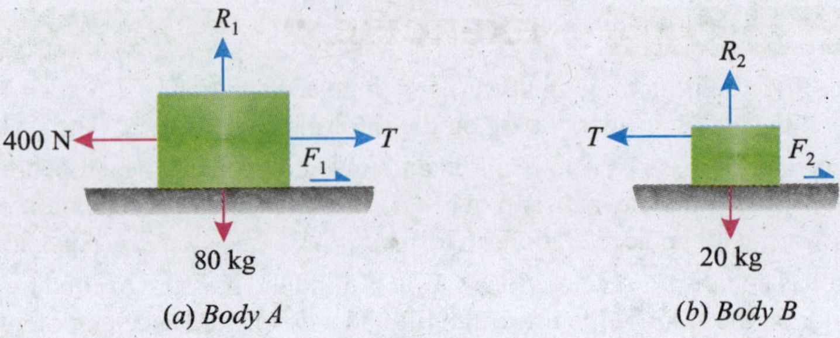

(*a*) *Body A* (*b*) *Body B*

Fig. 3.

First of all, consider the body A. The forces acting on it are :

1. 400 N force (acting towards left)
2. Mass of the body = 80 kg (acting downwards)
3. Reaction $R_1 = 80 \times 9.8 = 784$ N (acting upwards)
4. Force of friction, $F_1 = \mu R_1 = 0.3 \times 784 = 235.2$ N (acting towards right)
5. Tension in the thread = T (acting towards right).

∴ Resultant horizontal force,

$$P_1 = 400 - T - F_1 = 400 - T - 235.2$$
$$= 164.8 - T \text{ (acting towards left)}$$

We know that, Force causing acceleration to the body A

$$= m_1 a = 80\,a$$

and According to D' Alembert's principle,

$$P_1 - m_1 a = 0$$

or $$164.8 - T - 80\,a = 0$$

or $$T = 164.8 - 80a \quad ...(i)$$

Now consider the body B. The forces acting on it are :

1. Tension in the thread = T (acting towards left)
2. Mass of the body = 20 kg (acting downwards)
3. Reaction $R_2 = 20 \times 9.8 = 196$ N (acting upwards)
4. Force of friction, $F_2 = \mu R_2 = 0.3 \times 196 = 58.8$ N (acting towards right)

∴ Resulting horizontal force,

$$P_2 = T - F_2 = T - 58.8$$

We know that, Force causing acceleration to the body B

$$= m_2 a = 20\,a$$

and According to *D'* Alembert's principle ($P_2 - m_2 a = 0$)

$$(T - 58.8) - 20\,a = 0$$

or $$T = 58.8 + 20\,a \qquad ...(ii)$$

Now equating the two values of T from equation (*i*) and (*ii*),

$$164.8 - 80\,a = 58.8 + 20\,a$$

or $$100\,a = 106$$

or $$a = \frac{106}{100} = 1.06 \text{ m/s}^2 \textbf{ Ans.}$$

Tension in the thread

Substituting the value of a in equation (*ii*),

$$T = 58.8 + (20 \times 1.06) = 80 \text{ N} \textbf{ Ans.}$$

10. NEWTON'S THIRD LAW OF MOTION

It states ***"To every action, there is always an equal and opposite reaction."***

By *action* is meant the force, which a body exerts on another, and the *reaction* means the equal and opposite force, which the second body exerts on the first. This law, therefore, states that a force always occurs in pair. Each pair consisting of two equal opposite forces.

This law appears to be self-evident as when a bullet is fired from a gun, the bullet moves out with a great velocity, and the reaction of the bullet, in the opposite direction, gives an unpleasant shock to the man holding the gun. Similarly, when a swimmer tries to swim, he pushes the water backwards and the reaction of the water pushes the swimmer forward. Though the Newton's Third Law of Motion has a number of applications, yet recoil of gun and motion of a boat are important from the subject point of view.

11. Law of Conservation of Momentum

It states, "*the momentum is conserved in a system in which the external force is zero.*"
Mathematically,

Initial momentum = Final momentum

or $$MV = mv$$

Examples:

1. **Recoil of gun:** When a bullet is fired from a gun, the opposite reaction of the bullet is known as the recoil of gun.
2. **Motion of a Boat:** When a boat boy pushes the water back with the help of sticks, it in turn sets the boat in motion.

12. RECOIL OF GUN

According to Newton's Third Law of Motion, when a bullet is fired from a gun, the opposite reaction of the bullet is known as the recoil of gun.

Let M = Mass of the gun,
V = Velocity of the gun with which it recoils,
m = mass of the bullet, and
v = Velocity of the bullet after explosion.

Momentum of the bullet after explosion

$$= mv \qquad ...(i)$$

Momentum of the gun $= MV$ ...(*ii*)

Equating the equations (*i*) and (*ii*),

$$MV = mv$$

Note. This relation is Law of Conservation of Momentum.

***Example 18.** *A machine gun of mass 25 kg fires a bullet of mass 30 gram with a velocity of 250 m/s. Find the velocity with which the machine gun will recoil.* **[AM-IETE]**

Solution. Given : Mass of the machine gun (M) = 25 kg ; Mass of the bullet (m) = 30 g = 0.03 kg and velocity of firing (ν) = 250 m/s.

Let V = Velocity with which the machine gun will recoil.

We know that, $MV = mv$

or $25 \times v = 0.03 \times 250 = 7.5$

or $v = \dfrac{7.5}{25} = 0.3$ m/s **Ans.**

***Example 19.** *A bullet of mass 20 g is fired horizontally with a velocity of 300 m/s, from a gun carried in a carriage. The combined mass of gun and carraige is 100 kg. The resistance to sliding of the carriage over the ice on which it rests is 20 N. Find (a) velocity, with which the gun will recoil, (b) distance, in which it comes to rest, and (c) time taken to do so.* **[JNTU, RGPV]**

Solution. Given : Mass of the bullet (m) = 20 g = 0.02 kg ; Velocity of bullet (v) = 300 m/s; Mass of the carriage with gun (M) = 100 kg and resistance to sliding (F) = 20 N

(a) Velocity, with which the gun will recoil

Let V = velocity with which the gun will recoil.

We know that, $MV = mv$

$$100 \times V = 0.02 \times 300 = 6$$

or $V = \dfrac{6}{100} = 0.06$ m/s **Ans.**

(b) Distance, in which the gun comes to rest

Now consider motion of the gun. In this case, initial velocity (u) = 0.06 m/s and final velocity (v) = 0 (because it comes to rest)

Let a = Retardation of the gun, and

s = Distance in which the gun comes to rest.

We know that, Resisting force to sliding of carriage (F),

$$20 = Ma = 100\,a$$

$\therefore$ $a = \dfrac{20}{100} = 0.2$ m/s^2

We also know that, $v^2 = u^2 - 2as$ ...(Minus sign due to retardation)

$$0 = (0.06)^2 - 2 \times 0.2\,s = 0.0036 - 0.4\,s$$

or $s = \dfrac{0.0036}{0.4} = 0.009$ m = 9 mm **Ans.**

(c) *Time taken by the gun in coming to rest*

Let t = Time taken by the gun in coming to rest.

We know that, Final velocity of the gun (v),

$$0 = u + at = 0.06 - 0.2\,t \qquad \text{...(Minus sign due to retardation)}$$

$$t = \frac{0.06}{0.2} = 0.3 \text{ s} \quad \textbf{Ans.}$$

13. *MOTION OF A BOAT

We see that a boat boy always pushes the water back, with the help of sticks, which in turn, sets the boat in motion. It has also been experienced that if the boat is at rest and the boat boy runs on it and dives off into the water, the boat will also move backward.

The movement of the boat may be easily found out by the application of the Newton's Third Law of Motion (*i.e.,* by equating the momentum of the boat boy and the boat). Now consider a boat on which a boat boy runs and then dives off into the water.

Let M = Mass of the boat,

V = Velocity of the boat,

m = Mass of the boat boy, and

v = Velocity of the boat boy.

Momentum of the boat, after the boy jumps

$$= MV \qquad \text{...}(i)$$

Momentum of the boat boy $= mv$ ...(*ii*)

Equating equations (*i*) and (*ii*),

$$MV = mv$$

Note. This relation is Law of Conservation of Momentum.

***Example 20.** *Two men, standing on a floating boat, run in succession, along its length, with a speed of 4.2 m/sec relative to the boat and dive off from the end. The weight of each man is 80 kg and that of the boat is 400 kg. If the boat was initially at rest, find the final velocity of the boat. Neglect water friction.*

Solution. Given : $v = 4.2$ m/sec ; Weight of each man, $w = 80$ kg-wt (in gravitational units) or $m = 80$ kg (in absolute units) ; Weight of boat = 400 kg-wt (in gravitational units) or $M = 400$ kg (in absolute units).

Let V = Velocity of the boat after the second man dives off the boat.

A little consideration will show, that when the first man dives off the boat, it will give some momentum to the boat as well as the second man (who is still standing on the boat). When the second man also dives off the boat, it will also give some momentum to the boat. Therefore the total momentum gained by the boat is equal to the momentum given by the first man *plus* the momentum given by the second man to the boat.

Final momentum of the boat

$$= 400\ V \text{ kg-m/s} \quad ...(i)$$

Momentum given by the first man to the boat

$$= 80 \times 4.2 = 336 \text{ kg-m/s}$$

Similarly, Momentum given by the second man to the boat

$$= 80 \times 4.2 = 336 \text{ kg-m/s}$$

We know that, Final momentum of the boat

= Momentum given by the first man

+ Momentum given by the second man

$$= 336 + 336 = 672 \text{ kg-m/s} \quad ...(ii)$$

Now equating equations (*i*) and (*ii*),

$$400\ V = 672$$

or $$V = \frac{672}{400} = 1.68 \text{ m/s}$$ **Ans.**

EXERCISE 3

1. A bullet of 10 gm mass is fired horizontally with a velocity of 1000 m/s from a gun of mass 50 kg. Find (*a*) velocity with which the gun will recoil, and (*b*) force necessary to be ring the gun to rest in 250 mm. (**Ans.** 0.2 m/s ; 4 N)
2. A block of mass 10 kg slides over a frictionless horizontal plane with a constant velocity of 5 m/s. After some distance, the plane is inclined at an angle of 20° with the horizontal. Find the distance through which the block will slide upwards on the inclined plane before coming to rest. (**Ans** 36.5 m)

14. *MOTION ON INCLINED PLANE

[BPUT]

In the previous articles, we have been discussing the motion of bodies on level surface. But sometimes, the motion of body takes place down or up an inclined plane as shown in Fig. 4 (*a*) and (*b*).

(*a*) *Downward motion* (*b*) *Upward motion*

Fig. 4. Motion on inclined plane.

Now consider a body moving downwards on an inclined plane as shown in Fig. 4 (*a*).

Let m = Mass of the body, and

α = Inclination of the plane.

μ = Coefficient of friction between body and inclination of the plane.

Normal reaction on the inclined plane due to mass of the body.

$$R = mg \cos \alpha \qquad ...(i)$$

This component is responsible for the force of friction experienced by the body such that

$$F = \mu R$$

Component of the force along the inclined plane due to mass of the body.

$$= mg \sin \alpha \qquad ...(ii)$$

This component is responsible for sliding (or moving) the body downwards. Now we can find out any detail of the motion by subtracting the force of friction (due to normal reaction) from the component along the inclined surface ($mg \sin \alpha$).

Note. If the body is moving upwards, then the component along the inclined surface is taken as an additional resistance. *i.e.* this component is added to other types to resistances.

***Example 21.** *A vehicle of mass 2 tonnes has a frictional resistance of 50 N/tonne. As one instant, the speed of this vehicle at the top of an incline was observed to be 36 km.p.h. as shown in Fig.5.*

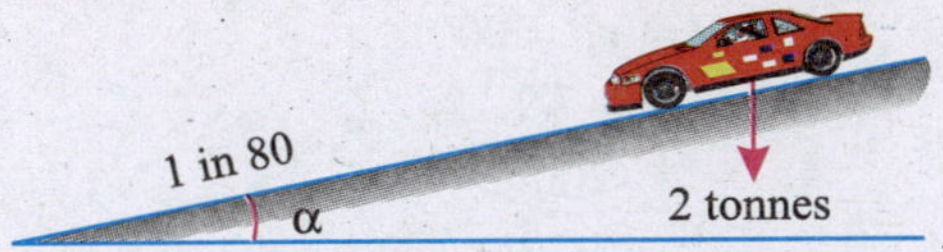

Fig. 5.

Find the speed of the vehicle after running down the incline for 100 seconds. **[IGNOU]**

Solution. Given : Mass of vehicle (m) = 2 t = 2000 kg; Frictional resistance = 50 N/t = 50 × 2 = 100 N ; Initial velocity of vehicle (u) = 36 km.p.h. = 10 m/s ; Slope of inclination ($\sin \alpha$) = $\frac{1}{80}$ = 0.0125 and time (t) = 100 s.

Let a = Acceleration of the vehicle.

Force due to inclination

$$= mg \sin \alpha = 2000 \times 9.8 \times 0.0125 = 245 \text{ N}$$

∴ Net force available to move the vehicle.

$$F = \text{Force due to inclination} - \text{Frictional resistance}$$

$$= 245 - 100 = 145 \text{ N}$$

We also know that, Available force (F)

$$145 = ma = 2000 \times a$$

$$\therefore \qquad \alpha = \frac{145}{2000} = 0.0725 \text{ m/s}^2$$

Speed of the train after running down the incline

$$v = u + at = 10 + (0.0725 \times 100) = 17.25 \text{ m/s} = 62.1 \text{ km.p.h.} \quad \textbf{Ans.}$$

***Example 22.** *A body of mass 200 kg is initially stationary on a 15° inclined plane. What distance along the incline must the body slide before it reaches a speed of 10 m/s? Take coefficient of friction between the body and the plane as 0.1.*

Solution. Given : Mass of the body (m) = 200 kg ; Initial velocity (u) = 0 (because, it is stationary) ; Inclination of the plane (α) = 15° ; Final velocity (v) = 10 m/s and coefficient of friction (μ) = 0.1

Let s = Distance through which the body will slide.

Force responsible for sliding down the body

$$= mg \sin \alpha$$
$$= 200 \times 9.8 \sin 15° \text{ N}$$
$$= 1960 \times 0.2588 = 507.2 \text{ N}$$

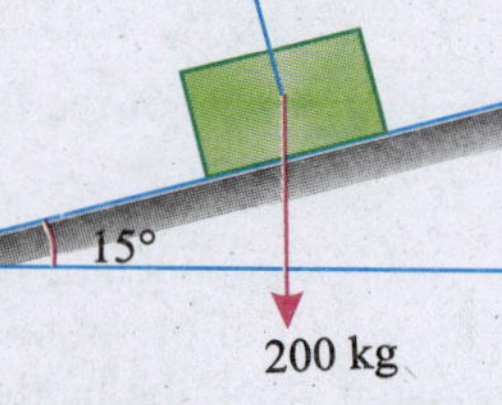

Fig. 6.

Normal reaction, $R = mg \cos \alpha$

$$= 200 \times 9.8 \cos 15° \text{ N}$$
$$= 1960 \times 0.9659 = 1893 \text{ N}$$

∴ Force of friction $F = \mu R = 0.1 \times 1893 = 189.3$ N

Net force available to move the body,

F = Force responsible for sliding – Force of friction

$$= 507.2 - 189.3 = 317.9 \text{ N}$$

We know that, Available force (F)

$$317.9 = ma = 200 \times a$$

or $$a = \frac{317.9}{200} = 1.59 \text{ m/s}^2$$

We also know that, $v^2 = u^2 + 2as$

or $$(10)^2 = (0)^2 + (2 \times 1.59 \times s) = 3.18\, s$$

or $$s = \frac{100}{3.18} = 31.4 \text{ m} \quad \textbf{Ans.}$$

***Example 23.** *A train of wagons is first pulled on a level track from A to B and then up a 5% upgrade as shown in Fig. 7.*

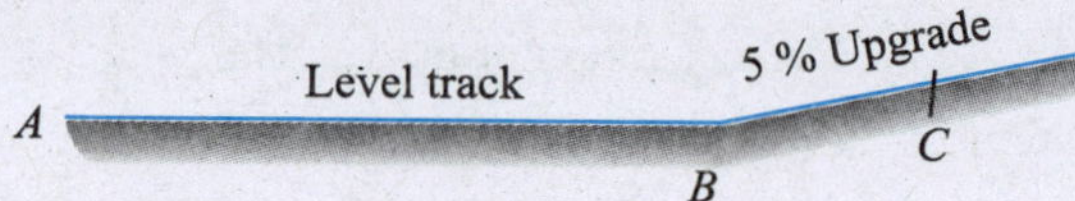

Fig. 7.

At some point C, the last wagon gets detached from the train, when it was travelling with a velocity of 36 km.p.h. If the detached wagon has a mass of 50 tonnes and the track resistance is 100 N per tonne, find the distance through which the wagon will travel before coming to rest.

Solution. Given : Grade = 5% or $\sin \alpha = 0.05$; Initial velocity (u) = 36 km.p.h. = 10 m/s ; Mass of the detached wagon (m) = 50 t ; Final velocity (v) = 0 (because it comes to rest) and track resistance = 100 N/t = 100 × 50 = 5000 N = 5 kN

Let s = Distance through which the wagon will travel before coming to rest, and

a = Retardation of the train.

Resistance to the train due to upgrade

$$= mg \sin \alpha = 50 \times 9.8 \times 0.05 = 24.5 \text{ kN}$$

∴ Total resistance to the movement of the train

F = Resistance due to upgrade + Track resistance

$$= 24.5 + 5 = 29.5 \text{ kN}$$

Total resistance (F)

$$29.5 = ma = 50 \times a$$

$$\therefore \qquad a = \frac{29.5}{50} = 0.59 \text{ m/s}^2$$

We also know that, $v^2 = u^2 + 2as = (10)^2 - 2 \times 0.59 \times s$

...(Minus sign due to retardation)

$$0 = 100 - 1.18 \text{ s}$$

or $$s = \frac{100}{1.18} = 84.7 \text{ m}$$ **Ans.**

***Example 24.** *A truck is moving down a 10° incline when the driver applies brakes, with the result that the truck decelerates at a steady rate of 1 m/s². Investigate whether a 500 kg placed on the truck will slide or remain stationary relative to the truck. Assume the coefficient of friction between the truck surface and the load as 0.4.*

What will be the factor of safety against slipping for this load?

Solution. Given : Inclination of the plane (α) = 10° ; Acceleration (a) = 1 m/s²; Mass of the body placed on the truck (m) = 500 kg and coefficient of friction (μ) = 0.4.

Stability of the load

We know that when the truck is decelerated, the body will tend to slip forward (*i.e.* downward).

$\therefore$ Force caused due to deceleration,

$$P_1 = ma = 500 \times 1 = 500 \text{ N}$$

and Component of the load along the plane

$$P_2 = mg \sin \alpha = 500 \times 9.8 \sin 10°$$
$$= 4900 \times 0.1736 \text{ N}$$
$$= 850.6 \text{ N}$$

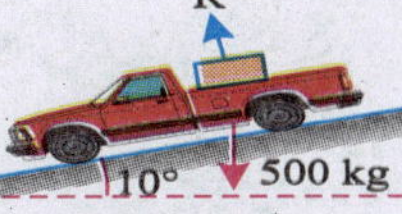

Fig. 8.

$\therefore$ Total force, which will cause slipping,

$$= P_1 + P_2 = 500 + 850.6 = 1350.6 \text{ N}$$

We know that, Normal reaction of the load,

$$R = mg \cos 10° = 500 \times 9.8 \times 0.9848 = 4825.5 \text{ N}$$

and Force of friction, $F = \mu R = 0.4 \times 4825.5 = 1930.2 \text{ N}$

Since the force of friction (1930.2 N) is more than the force which will cause slipping (1350.6 N), therefore the load will not slip. **Ans.**

Factor of safety against slipping for this load

Factor of safety against slipping for this load

$$= \frac{\text{Force of friction}}{\text{Force causing slipping}} = \frac{1930.2}{1350.6} = 1.43$$ **Ans.**

EXERCISE 4

1. A body of mass 500 kg, initially at rest at A 50 m from B on an 15% upgrade, is allowed to slide down as shown in Fig. 9.

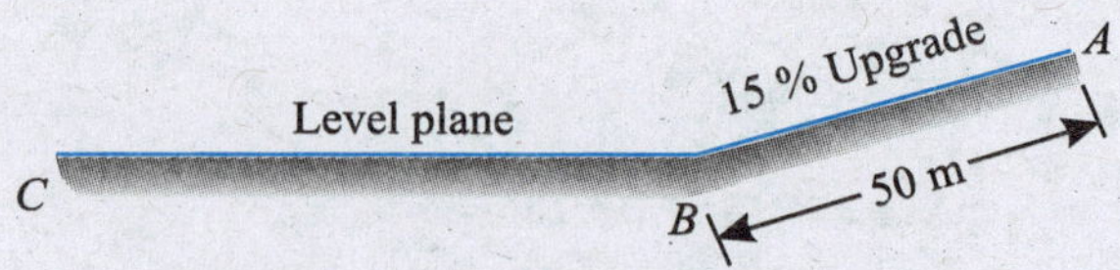

Fig. 9

If the coefficient of friction between the body and the plane is 0.1, find the velocity of the body at B. Also find the distance through which the body will travel beyond B on the level plane. (**Ans.** 8 m/s ; 32.65 m)

2. A locomotive of mass 200 tonnes draws a train of mass 450 tonnes. The frictional resistance is constant and equal to 75 N/t. Find the tractive force which will be required for the train to reach a speed of 72 km.p.h. in a distance of 2 km from the starting point (*i*) on a level track ; (*ii*) for going upward on an inclined plane of 1 in 240 and (*ii*) for going downward on the same inclined plane. (**Ans.** 113.75 kN ; 140.5 kN ; 87 kN)

15. Impulse and Impulse-Momentum Equation

[AM-IETE, BPUT]

The product of force and the time for which it is acting on a body is called the *impulse* of the force. Mathematically, Impulse = Force × Time interval

Let force P act on a body for a time t.

According to Newton's second law of motion,

$$\text{Force} = \text{Mass} \times \text{Acceleration}$$

$$P = ma$$

$$= m\left(\frac{v-u}{t}\right)$$

$$Pt = m(v-u)$$

i.e. Force × Time interval = Mass × Change in Velocity

or **Impulse = Mass × Change in Velocity**

or **Impulse = Change in Momentum**

This is known as **Impulse-Momentum equation.**

Impulse of a force is a useful concept when a force acts for a short duration of time. It tells how much the velocity of a body of a given mass will change if a given impulse is applied to it. Example: When we hit a hockey stick to a ball, we give to the ball an impulse.

The SI unit of impulse is newton-second (N-s).

Impulse-Momentum Method is used for solving questions on dynamics which involve **force, velocity** and **time.**

***Example 25.** *A pile hammer (driver) of mass 300 kg falls on a pile. The hammer falls freely from a height of 6 metres and the hammer comes to rest in 1/80 second. Find the average impulsive force of blow.*

Solution. Mass of pile hammer $= 300$ kg ; Initial velocity of hammer $u = 0$

We know that, $v^2 - u^2 = 2gh$, when $u = 0$, then $v^2 = 2gh$ or $v = \sqrt{2gh}$

Therefore, Final velocity of hammer just before striking the pile

$$v = \sqrt{2gh} = \sqrt{2 \times 9.81 \times 6} = 10.85 \text{ m/s}$$

We know form the **Impulse-Momentum equation** that,

Impulse = Change in Momentum or $P \times t = mv - mu$

Therefore, Average Force, $P = m\left(\frac{v-u}{t}\right) = 300\left(\frac{10.85-0}{1/80}\right) = 260400$ N **Ans.**

***Example 26.** *A bullet of mass 0.045 kg leaves the barrel of a gun with a velocity of 400 m/s. Suppose the force lasted for 0.0025 seconds. Find the*

(i) Impulse (ii) Average Force. Take $g = 9.80$ m/s^2.

Solution. Mass of the bullet = 0.045 kg ; Velocity of bullet = 400m/s ; Time of application of force = 0.0025 s.

Impulse

We know from the **Impulse-Momentum equation** that,

Impulse = Change in Momentum or $P \times t = mv - mu$

Now, Impulse $= mv - mu = m(v-u) = 0.045\,(400-0) = 18.0$ kg-m/s **Ans.**

Average Force

Average Force, $P = m\left(\frac{v-u}{t}\right) = 0.045 \times \left(\frac{400-0}{0.0025}\right) = 7200$ N **Ans.**

***Example 27.** *A glass marble, whose mass is 25 g, falls from a height of 15 m and rebounds to a height of 10 m. Find the impulse and the average force between the marble and the floor if the time during which they are in contact be one-tenth of a second.*

Solution. Given : Mass of the glass marble = 25 g ; Height of fall = 15 m ; Rebound height = 10 m ; Time of application of force = 1/10 s.

Impulse

Velocity aquired by the glass marble in falling through 15 m

$$= \sqrt{2gh} = \sqrt{2 \times 9.81 \times 15} = 17.15 \text{ m/s}$$

Velocity of the rebound = Velocity necessary to carry the marble to a height of 10 m

$$= \sqrt{2gh} = \sqrt{2 \times 9.81 \times 10} = 14 \text{ m/s}$$

Momentum before impact $= mv = 0.025 \times 17.15 = 0.428$ kg-m/s

Momentum after impact $= mu = 0.025 \times (-14) = 0.35$ kg-m/s

...*(Since the velocity of rebound is in the opposite direction to the velocity before impact with the ground, Hence the negitive sign.)*

We know from the **Impulse-Momentum equation** that,

Impulse = Change in Momentum or $P \times t = mv - mu$

Now, Impulse $= mv - mu = m(v-u) = 0.428 - (-0.35) = 0.778$ kg-m/s.

Average Force

Let the average force between the marble and floor = P newtons.

Gravitational pull of the ground = 0.025×9.81 N.

Hence, Net force acting on the marble = $(P - 0.025 \times 9.81)$.

We know that, Impule = Force × Time

or $$0.778 = (P - 0.025 \times 9.81) \times \frac{1}{10}$$

or $$P = 8.025 \text{ N} \quad \textbf{Ans.}$$

QUESTIONS

1. State the Laws of Motion. Discuss the First Law in the light of Second Law.
2. Distinguish clearly between 'mass' and 'weight'.
3. Derive an expression for the tension in the cable supporting a lift when (*i*) it is going up, and (*ii*) it is coming down.
4. Explain the dynamic equilibrium of a rigid body in plane motion.
5. Explain clearly the term 'recoil of gun'. How will you find the velocity of the bullet?
6. How will you apply the Third Law of Motion in the case of horse pulling a cart?

OBJECTIVE TYPE QUESTIONS

1. The units of weight are the same as that of force.

   (*a*) Agree (*b*) Disagree

2. The Newton's Second Law of Motion gives a relation between force, mass and

   (*a*) Velocity (*b*) Acceleration (*c*) None of the two

3. Which of the following statement is wrong?

   (*a*) The matter contained in a body is called mass.

   (*b*) The force with which a body is attracted towards the centre of the earth is called weight.

   (*c*) The total motion possessed by a moving body is called impulsive force

   (*d*) none of them

4. If a lift is moving downwards with some acceleration, then tension in the cable supporting the lift is...proportional to the acceleration.

   (*a*) Directly (*b*) Indirectly

5. A science teacher told to his students that the Newton's Third Law of Motion is involved while studying the motion of rockets. Is his statement correct?

   (*a*) Yes (*b*) No

ANSWERS

1. (*a*) 2. (*b*) 3. (*d*) 4. (*b*) 5. (*a*)

CHAPTER

23

Motion of Connected Bodies

Contents

1. INTRODUCTION

In the previous chapter, we discussed the motion of a body, when it is subjected to some external force. But in this chapter, we shall discuss the motion of a body, when it is connected by a string to another body, which is subjected to some external force. A little consideration will show, that in this case the velocity and acceleration of both the bodies will be the same. Now in this chapter, we shall discuss the following cases :

1. Two bodies connected by a string and passing over a smooth pulley.
2. Two bodies connected by a string one of which is hanging free and the other lying on a horizontal plane.

3. Two bodies connected by a string one of which is hanging free and the other lying on an inclined plane.
4. Two bodies connected by a string and lying on two inclined planes.

2. MOTION OF TWO BODIES CONNECTED BY A STRING AND PASSING OVER A SMOOTH PULLEY

Consider two bodies of masses m_1 and m_2 kg respectively connected by an inextensible light string (*i.e.,* its weight is neglected) and passing over a small smooth fixed pulley as shown in Fig. 1.

It may be noted that if the string is light (*i.e.* its weight is neglected) the tension will be the same throughout its length. But if the string is heavy (*i.e.,* its weight is considered) the tension will vary, depending upon the weight per unit length. Moreover, if the string is extensible the tension will also vary with the extension. It may also be noted that if the string passes over a smooth pulley, the tension will be the same on both sides. But if the string does not pass over a smooth pulley, the tension, in the two strings will also vary.

For simplicity, we shall consider light inextensible string passing over a smooth pulley, so that the tension in both the strings may be the same.

T T
m_2
m_1

Fig. 1.

Let m_1 be greater than m_2. A little consideration will show, that the greater mass m_1 will move downwards, whereas the smaller one will move upwards. Since the string is inextensible, the upward acceleration of the mass m_2 will be equal to the downward acceleration of the mass m_1.

Let a = Acceleration of the bodies and

T = Tension in both the strings.

First of all, consider the motion of body 1 of mass m_1 kg, which is coming downwards. We know that the forces acting on it are $m_1 . g$ newtons (downwards) and T newtons (upwards). As the body is moving downwards, therefore resultant force

$$= m_1 g - T \text{ (downwards)} \qquad ...(i)$$

Since this body is moving downwards with an acceleration (a), therefore force acting on this body

$$= m_1 a \qquad ...(ii)$$

Equating equations (i) and (ii),

$$m_1 g - T = m_1 a \qquad ...(iii)$$

Now consider the motion of body 2, of mass m_2 kg, which is moving upwards. We know that the forces acting on it are $m_2\, g$ newtons (downwards) and T newtons (upwards). As the body is moving upwards, therefore resultant force

$$= T - m_2\, g \text{ (upwards)} \qquad ...(iv)$$

Since this body is moving upwards with an acceleration (a), therefore, force acting on this body

$$= m_2\, a \qquad ...(v)$$

Equating the equations (iv) and (v),

$$T - m_2\, g = m_2\, a \qquad ...(vi)$$

Now adding equations (iv) and (v),

$$m_1\, g - m_2\, g = m_1\, a + m_2\, a$$

$$g\,(m_1 - m_2) = a\,(m_1 + m_2)$$

$$\therefore \qquad a = \frac{g\left(m_1 - m_2\right)}{m_1 + m_2}$$

From equation (*vi*) we have

$$T = m_2\,a + m_2\,g = m_2\,(a + g)$$

$$= m_2\left[\frac{g\left(m_1 - m_2\right)}{m_1 + m_2} + g\right]$$

$$= m_2\left[\frac{g\left(m_1 - m_2\right) + g\left(m_1 + m_2\right)}{m_1 + m_2}\right]$$

$$= \frac{m_2 . g}{m_1 + m_2}\left(m_1 - m_2 + m_1 + m_2\right)$$

$$= \frac{2\,m_1\,m_2\,g}{m_1 + m_2}$$

Example 1. *Two bodies of masses 45 and 30 kg are hung to the ends of a rope, passing over a frictionless pulley. With what acceleration the heavier mass comes down? What is the tension in the string?*

Solution. Given : Mass of first body $(m_1) =$ 45 kg and mass of the second body $(m_2) = 30$ kg

Acceleration of the heavier mass

We know that acceleration of the heavier mass,

$$a = \frac{g\left(m_1 - m_2\right)}{m_1 + m_2} = \frac{9.8\left(45 - 30\right)}{45 + 30} = 1.96 \text{ m/s}^2 \qquad \textbf{Ans.}$$

Tension in the string

We also know that tension in the string,

$$T = \frac{2m_1 m_2 g}{m_1 + m_2} = \frac{2 \times 45 \times 30 \times 9.8}{45 + 30} = 352.8 \text{ N} \qquad \textbf{Ans.}$$

Example 2. *Determine the tension in the strings and accelerations of two blocks of mass 150 kg and 50 kg connected by a string and a frictionless and weightless pulley as shown in Fig. 2.*

Solution. Given : Mass of first block $(m_1) = 150$ kg and mass of second block $(m_2) = 50$ kg

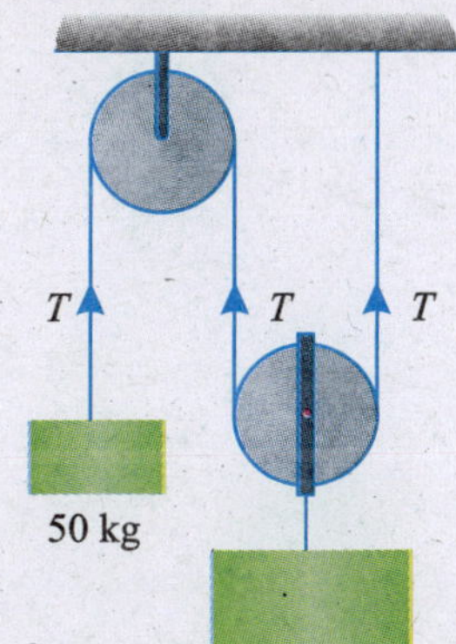

Fig. 2.

Acceleration of the blocks

Let $\qquad a =$ Acceleration of the blocks

$\qquad T =$ Tension in the string in N.

First of all, consider the motion of the 150 kg block, which is coming downwards. We know that forces acting on it are $m_1.g = 150\,g$ newtons (downwards) and $2T$ newtons (upwards).

Therefore resultant force $= 150\,g - 2T$ (downwards) ...(*i*)

Since the block is moving downwards with an acceleration (a), therefore force acting on this block

$$= 150\,a \qquad ...(ii)$$

Equating equations (*i*) and (*ii*),

$$150\,g - 2\,T = 150\,a \qquad ...(iii)$$

Now consider the motion of 50 kg block, which is going upwards. A little consideration will show that its acceleration will be (2*a*). We know that forces acting on it are $m_2.g = 50\,g$ newtons (downwards) and *T* newtons upwards. Therefore resultant force

$$= T - 50\,g \text{ (upwards)} \qquad ...(iv)$$

Since the block is moving upwards with an acceleration of (2*a*) therefore force acting on this block

$$= 50 \times 2a = 100\,a \qquad ..(v)$$

Equating equations (*iv*) and (*v*),

$$T - 50\,g = 100\,a$$

Multiplying the above equation by 2,

$$2T - 100\,g = 200\,a \qquad ...(vi)$$

Adding equations (*iii*) and (*vi*),

$$50\,g = 350\,a$$

$$\therefore \qquad a = \frac{50g}{350} = \frac{50 \times 9.8}{350} = 1.4 \text{ m/s}^2 \qquad \textbf{Ans.}$$

and acceleration of the block *B*

$$= 2a = 2 \times 1.4 = 2.8 \text{ m/s}^2 \qquad \textbf{Ans.}$$

Tension in the strings

Substituting the value of *a* in equation (*iii*),

$$150g - 2T = 150 \times 1.4 = 210$$

$$\therefore \qquad 2T = 150g - 210 = 150 \times 9.8 - 210 = 1260$$

$$\text{or} \qquad T = \frac{1260}{2} = 630 \text{ N} \qquad \textbf{Ans.}$$

Example 3. *A system of masses connected by string, passing over pulleys A and B is shown in Fig. 3 below :*

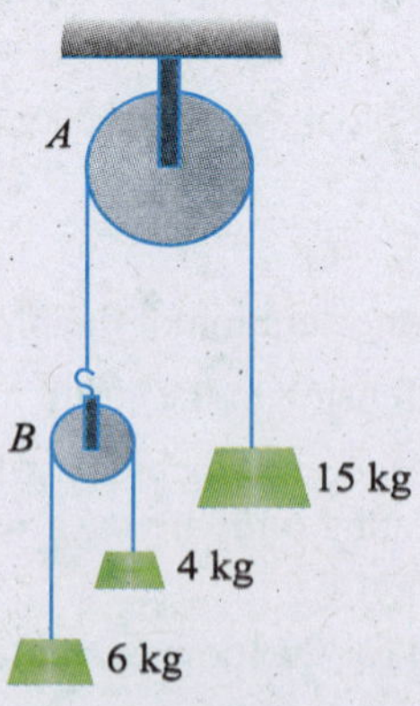

Fig. 3.

Find the acceleration of the three masses, assuming weightless strings and ideal conditions for pulleys.

Solution. Given : First mass $(m_1) = 15$ kg ; Second mass $(m_2) = 6$ kg and third mass $(m_3) = 4$ kg

From the system of pulleys and masses, we find that at pulley *A*, the 15 kg mass will come down with some acceleration as the total mass on the other side of the string is less than 15 kg. Similarly, at pulley *B*, the 6 kg mass will come down with some acceleration.

Let a_{15} = Acceleration of 15 kg mass

a_6 = Acceleration of 6 kg mass, and

a_4 = Acceleration of 4 kg mass.

We know that acceleration of the 15 kg mass,

$$a_{15} = \frac{g(m_1 - m_2)}{m_1 + m_2} = \frac{9.8[15-(6+4)]}{15+(6+4)} = 1.96 \text{ m/s}^2$$

Similarly, $$a_6 = a_4 = \frac{9.8(6-4)}{6+4} = 1.96 \text{ m/s}^2$$

Now if we look at all the three masses to move simultaneously at pulleys *A* and *B*, we find that

1. The mass 15 kg will come downwards with an acceleration of 1.96 m/s^2. **Ans.**
2. The pulley *B* will go up with an acceleration of 1.96 m/s^2.
3. The 6 and 4 kg masses will go up (because the pulley *B* is going up) with an acceleration of 1.96 m/s^2. Moreover, at pulley *B*, the 6 kg mass will come down (because 4 kg mass will go up) with an acceleration of 1.96 m/s^2. Thus the net acceleration of the 6 kg mass will be zero. **Ans.**
4. At pulley *B* the 4 kg mass will go up (because the 6 kg mass will come down) with an acceleration of 1.96 m/s^2. Thus the net acceleration of the 4 kg mass will be 1.96 + 1.96 = 3.92 m/s^2. **Ans.**

3. MOTION OF TWO BODIES CONNECTED BY A STRING, ONE OF WHICH IS HANGING FREE AND THE OTHER LYING ON A SMOOTH HORIZONTAL PLANE

Consider two bodies of masses m_1 and m_2 kg respectively connected by a light inextensible string as shown in Fig. 4.

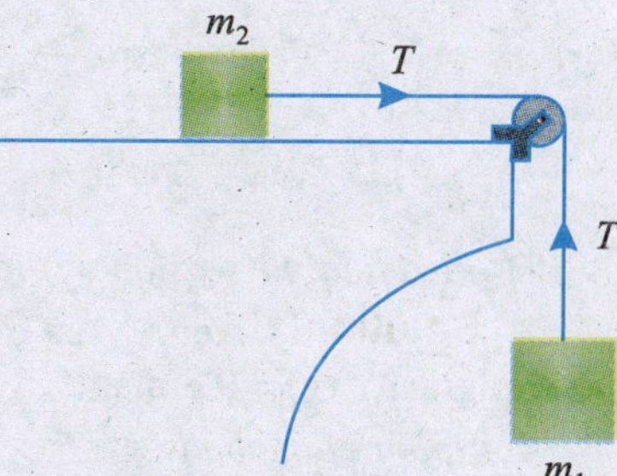

Fig. 4 Body lying over smooth horizontal plane.

Let the body of mass m_1 hang free and the body of mass m_2 be placed on a smooth (*i.e.*, friction between the body of mass m_2 and the horizontal plane is neglected) horizontal plane. It may be noted that the body of mass m_1 will move downwards and the body of mass m_2 along the surface of the plane.

We know that velocity and acceleration of the body of mass m_1 will be the same as that of the body of mass m_2. Since the string is inextensible, therefore, tension in both the strings will also be equal,

Let a = Acceleration of the system and

T = Tension in the strings

First of all, consider the motion of body 1 of mass m_1 downwards. We know that forces acting on it are $m_1.g$ (downwards) and T (upwards). As the body is moving downwards, therefore resultant force

$$= m_1 g - T \text{ (downwards)} \quad ...(i)$$

Since this body is moving downwards with an accleration (a) therefore force acting on the body

$$= m_1\, a \qquad ...(ii)$$

Equating the equations (i) and (ii),

$$m_1\, g - T = m_1\, a \qquad ...(iii)$$

Now consider the motion of body 2 of mass m_2, which is moving horizontally. We know that only force acting on it

$$= T \text{ (horizontal)} \qquad ...(iv)$$

Since this body is moving horizontally with an acceleration (a), therefore force acting on this body

$$= m_2\, a \qquad ...(v)$$

Equating the equations (iv) and (v),

$$T = m_2\, a \qquad ...(vi)$$

Adding equation (iii) and (vi),

$$m_1\, g = m_1\, a + m_2\, a = a\,(m_1 + m_2)$$

$$\therefore \quad a = \frac{m_1\, g}{m_1 + m_2}$$

Substituting this value of a in equation (vi),

$$T = m_2 \times \frac{m_1 g}{m_1 + m_2} = \frac{m_1 m_2 g}{m_1 + m_2}$$

Example 4. *Find the acceleration of a solid body A of mass 10 kg, when it is being pulled by another body B of mass 5 kg along a smooth horizontal plane as shown in Fig. 5.*

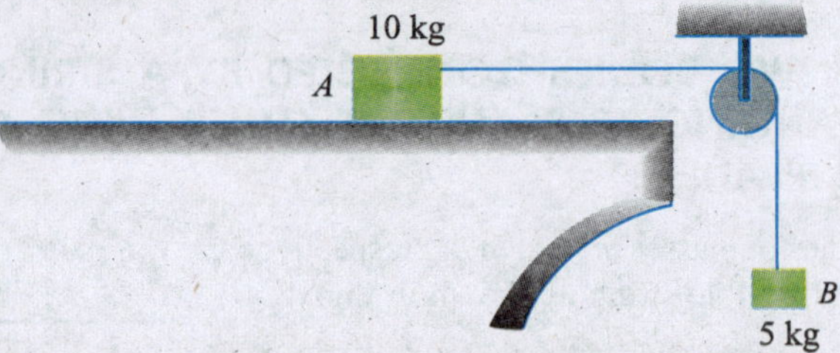

Fig. 5.

Also find the tension in the string, assuming the string to be inextensible. Take g = 9.8 m/s².

Solution. Given : Mass of body $A(m_2)$ = 10 kg : mass of body $B(m_1)$ = 5 kg and acceleration due to gravity (g) = 9.8 m/s^2.

Acceleration of the body A

We know that the acceleration of the body A,

$$a = \frac{m_1\, g}{m_1 + m_2} = \frac{5 \times 9.8}{5 + 10} = 3.27 \text{ m/s}^2 \quad \textbf{Ans.}$$

Tension in the string

We know that tension in the string,

$$T = \frac{m_1 m_2 g}{m_1 + m_2} = \frac{5 \times 10 \times 9.8}{5 + 10} = 32.7 \text{ N} \quad \textbf{Ans.}$$

4. MOTION OF TWO BODIES CONNECTED BY A STRING, ONE OF WHICH IS HANGING FREE AND THE OTHER LYING ON A ROUGH HORIZONTAL PLANE

Consider two bodies of masses m_1 and m_2 kg respectively, connected by a light inextensible string as shown in Fig. 6.

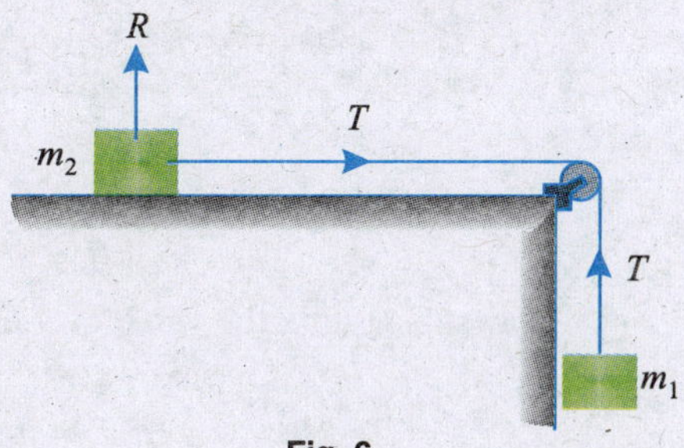

Fig. 6.

Let the body of mass m_1 hang free, and the body of mass m_2 be placed on a rough horizontal plane. Let the body of mass m_1 move downwards and the body of mass m_2 move along the surface of the plane.

We know that velocity and acceleration of the body of mass m_1 will be the same as that of mass m_2. Since the string is inextensible, therefore tensions in both the strings will also be equal.

Let a = Acceleration of the system,

T = Tension in the string and

μ = Coefficient of friction.

We know that the normal reaction on the horizontal surface due to body of mass m_2 kg (as shown in Fig. 6)

$$R = m_2\, g$$

$\therefore$ Frictional force $= \mu R = \mu\, m_2\, g$

This frictional force will act in the opposite direction to the motion of the body of mass 2.

First of all, consider the motion of the body of mass m_1 kg, which is coming downwards. We know that forces acting on it are $m_1.g$ (downwards) and T upwards. As the body is moving downwards, therefore, resultant force

$$= m_1\, g - T \text{ (downwards)} \qquad ...(i)$$

Since this body is moving downwards with an acceleration (a), therefore force acting on this body

$$= m_1\, a \qquad ...(ii)$$

Equating equations (i) and (ii),

$$m_1\, g - T = m_1\, a \qquad ...(iii)$$

Now consider the motion of body 2 of mass m_2 kg, which is moving horizontally. We know that the forces acting on it are T towards right and frictional force $\mu m_2 g$ towards left. As the body is moving towards right, therefore, resultant force

$$= T - \mu\, m_2\, g \qquad ...(iv)$$

Since this body is moving horizontally with an acceleration (a) therfore force acting on this body

$$= m_2\, a \qquad ...(v)$$

Equating the equations (iv) and (v),

$$T - \mu\, m_2\, g = m_2\, a \qquad ...(vi)$$

Adding equation (*iii*) and (*vi*),

$$m_1 g - \mu m_2 g = m_1 a + m_2 a$$

$$g(m_1 - \mu m_2) = a(m_1 + m_2)$$

$$\therefore \quad a = \frac{g(m_1 - \mu m_2)}{m_1 + m_2} \text{ m/s}^2$$

From equation (*vi*) we find that

$$T = m_2 a + \mu m_2 g = m_2 (a + \mu g)$$

Now substituting the value of a in the above equation,

$$T = m_2 \left[\frac{g(m_1 - \mu m_2)}{m_1 + m_2} + \mu g \right]$$

$$= m_2 g \left[\frac{m_1 - \mu m_2 + \mu m_1 + \mu m_2}{m_1 + m_2} \right]$$

$$= \frac{m_1 m_2 g (1 + \mu)}{m_1 + m_2}$$

Note. For smooth surface, if we substitute the value of $\mu = 0$ in the above equations for a and T, the relations obtained will be the same as we derived in the last article 3 of this chapter.

Example 5. *Two blocks shown in Fig. 7, have masses A = 20 kg and B = 10 kg and the coefficient of friction between the block A and the horizontal plane, μ = 0.25.*

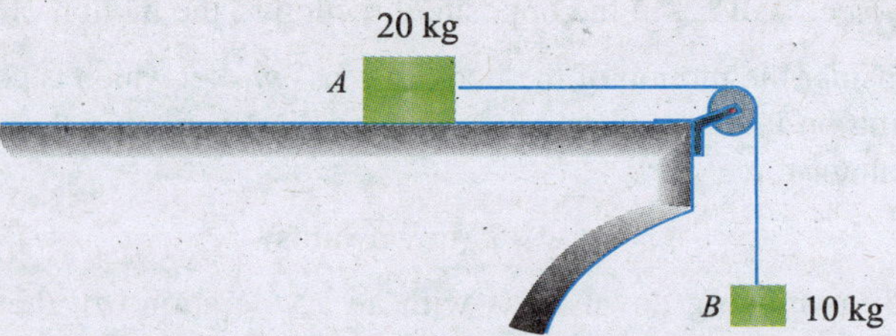

Fig. 7.

If the system is released, from rest, and the block B falls through a vertical distance of 1m, what is the velocity acquired by it? Neglect the friction in the pulley and the extension of the string.

Solution. Given : Mass of block A $(m_2) = 20$ kg ; Mass of block B $(m_1) = 10$ kg ; Coefficient of friction between block A and horizontal plane $(\mu) = 0.25$; Initial velocity $(u) = 0$ (because the system is released from rest) and vertical distance $(s) = 1$ m

Let $\quad v$ = Final velocity of the block A.

We know that acceleration of the block A,

$$a = \frac{g(m_1 - \mu m_2)}{m_1 + m_2} = \frac{9.8(10 - 0.25 \times 20)}{10 + 20} = 1.63 \text{ m/s}^2$$

and $\quad v^2 = u^2 + 2as = 0 + 2 \times 1.63 \times 1 = 3.26$

$\therefore \quad v = 1.81$ m/s **Ans.**

Example 6. *A block of wood A of mass 10 kg is held on a rough horizontal table. An elastic string connected to the block passes over a smooth pulley at the end of the table and then under a second smooth pulley carrying a body B of mass 5 kg as shown in Fig. 8.*

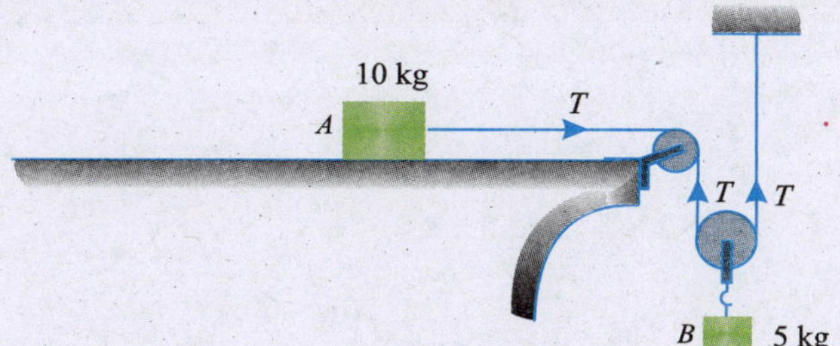

Fig. 8.

The other end of the string is fixed to a point above the second pulley. When the 10 kg block is released, it moves with an acceleration of g/9. Determine the value of coefficient of friction between the block and the table.

Solution. Given : Mass of block A (m_2) = 10 kg ; Mass of body B (m_1) = 5 kg and accelaration of block $A = \frac{g}{9}$.

Let T = Tension in the string in N, and

μ = Coefficient of friction between block and table.

We know that the normal reaction on the horizontal surface due to body of mass 10 kg

$$R = 10\ g$$

∴ Frictional force $= \mu R = \mu \times 10\ g = 10\ \mu g$

First of all consider the motion of block A, which is moving horizontally. We know that the forces acting on it are T (towards right) and frictional force 10 μg (towards left). As the block is moving towards right, therefore resultant force

$$= T - 10\ \mu g \quad ...(i)$$

Since the block is moving with an acceleration of (g/9) therefore force acting on it

$$= 10 \times \frac{g}{9} = \frac{10g}{9} \quad ...(ii)$$

Equating the equations (*i*) and (*ii*)

$$T - 10\mu g = \frac{10g}{9}$$

Multiplying both sides by 2,

or
$$2T - 20\mu g = \frac{20g}{9} \quad ...(iii)$$

Now consider the motion of the block B, which is coming downwards. A little consideration will show the acceleration of this block will be half of that of the block A *i.e.* $g/18$. We know that the forces acting on it are $mg = 5\ g$ (downwards) and $2T$ (upwards). Therefore resultant force

$$= 5g - 2T \quad ...(iv)$$

Since the block is moving with an acceleration of ($g/18$) therefore force acting in it

$$= ma = 5 \times \frac{g}{18} = \frac{5g}{18} \quad ...(v)$$

Equating equations (*iv*) and (*v*),

$$5g - 2T = \frac{5g}{18} \qquad ...(vi)$$

Adding equations (*iii*) and (*vi*),

$$5g - 20\mu g = \frac{20g}{9} + \frac{5g}{18} = \frac{45g}{18} = 2.5\ g$$

$\therefore$ $$20\mu = 5 - 2.5 = 2.5$$

or $$\mu = \frac{2.5}{20} = 0.125 \quad \textbf{Ans.}$$

EXERCISE 1

1. A mass of 9 kg, while descending vertically down, drags up a mass of 6 kg by means of a string passing over a smooth pulley. Find the acceleration of the system and tension in the string. (**Ans.** 1.96 m/s^2 ; 70.6 N)
2. Two bodies of mass 3 kg and 2.5 kg are hung to the ends of a string passing over a smooth pulley. At the end of 5 seconds, the string breaks. How much higher the 2 kg mass will go? (**Ans.** 11.1 m)
3. A body of mass 4.5 kg is placed on a smooth table at a distance of 2 m from the edge. The body is connected by a light string passing over a smooth pulley. The other end of the string is connected with a body of mass 2.5 kg. Find (*i*) acceleration of the system ; and (*ii*) time that elapses before the body reaches edge of the table. (**Ans.** 2.8 m/s^2 ; 1.2 s)
4. As body of mass 4 kg lying on a rough horizontal plane is attached to one end of a string. The string passes over a smooth pulley and carries, at its other end, a body of mass 10 kg which hangs freely vertical down. If the system starts from rest, and attains an acceleration of 6 m/s^2, find the coefficient of friction. (**Ans.** 0.35)
5. Two blocks of mass 50 kg and 40 kg are connected by a light inextensible string as shown in Fig. 9.

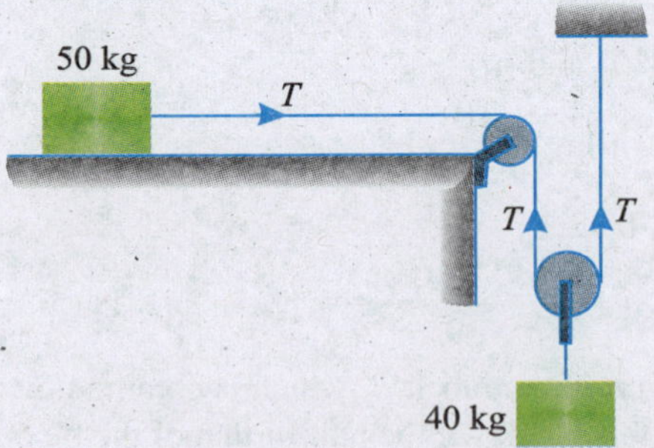

Fig. 9.

Find, from first principles, the acceleration of the system and tensions in the cable. Take $\mu = 0.3$. (**Ans.** 0.82 m/s^2 ; 188 N)

5. MOTION OF TWO BODIES CONNECTED BY A STRING, ONE OF WHICH IS HANGING FREE AND OTHER LYING ON A SMOOTH INCLINED PLANE

Consider two bodies of masses m_1 and m_2 kg respectively connected by a light inextensible string as shown in Fig. 10.

Let the body of mass m_1 hang free and the body of mass m_2 be placed on an inclined smooth plane (*i.e.*, the friction between the mass m_2 and the plane is neglected).

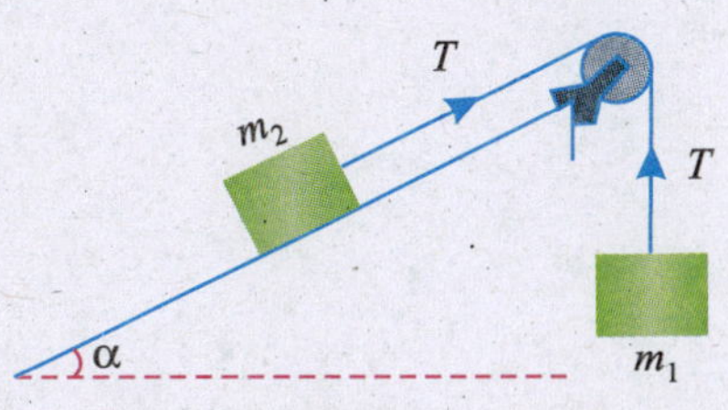

Fig. 10.

It may be noted that the body of mass m_1 will move downwards and the body of mass m_2 will move upwards along the inclined surface. A little consideration will show that the velocity and acceleration of the body of mass m_1 will be the same as that of the body of mass m_2. Since the string is inextensible, therefore tension in both the strings will also be equal.

Let a = Acceleration of the system

T = Tension in the strings and

α = Inclination of the plane.

First of all, consider the motion of the body 1 of mass m_1 kg, which is coming downwards. We know that the forces acting, on it are m_1g (downwards) and T (upwards). As the body is moving downwards, therefore resultant force

$$= m_1\,g - T \qquad ...(i)$$

Since this body is moving downwards with an acceleration (a), therefore force acting on this body

$$= m_1\,a \qquad ...(ii)$$

Equating equations (*i*) and (*ii*),

$$m_1\,g - T = m_1\,a \qquad ...(iii)$$

Now consider the motion of body 2 of mass m_2, which is moving upward along the inclined surface. We know that forces acting on it, along the plane, are T (upwards) and $m_2.g \sin \alpha$ (downwards). As the body is moving upwards, therefore resultant force

$$= T - m_2\,g \sin \alpha \qquad ...(iv)$$

Since this body is moving upwards along the inclined surface with an acceleration (a), therefore force acting on this body

$$= m_2\,a \qquad ..(v)$$

Equating the equations (*iv*) and (*v*),

$$T - m_2\,g \sin \alpha = m_2\,a \qquad ...(vi)$$

Adding equations (*iii*) and (*vi*),

$$m_1\,g - m_2\,g \sin \alpha = m_1\,a + m_2\,a$$

$$g\,(m_1 - m_2 \sin \alpha) = a\,(m_1 + m_2)$$

$$a = \frac{g\left(m_1 - m_2 \sin \alpha\right)}{m_1 + m_2}\ \text{m/s}^2$$

From equation (*iii*) we find that

$$T = m_1 g - m_1 a = m_1 (g - a)$$

Substituting the value of a in the above equation,

$$T = m_1 \left[g - \frac{g(m_1 - m_2 \sin\alpha)}{m_2 + m_2} \right]$$

$$= m_1 g \left[\frac{m_1 + m_2 - m_1 + m_2 \sin\alpha}{m_1 + m_2} \right]$$

$$= \frac{m_1 m_2 g (1 + \sin\alpha)}{m_1 + m_2}$$

Note. For horizontal surface, if we substitute value of $\alpha = 0$ in the above equations for a and T, the relations obtained will be the same as derived in Artilce 4 of this chapter.

Example 7. *A body of mass 30 kg, lying on a smooth plane inclined at 15° to the horizontal, is being pulled by a body of mass 20 kg. The 20 kg body is connected to the first body by a light inextensible string and hangs freely beyond the frictionless pulley.*

Find the acceleration, with which the body will come down.

Solution. Given : Mass of the body lying on smooth plane (m_2) = 30 kg ; Inclination of the plane with horizontal (α) = 15° and mass of the body which hangs freely beyond the pully (m_1) = 20 kg

We know that the acceleration with which the body will come down,

$$a = \frac{g(m_1 - m_2 \sin\alpha)}{m_1 + m_2} = \frac{9.8(20 - 30 \sin 15°)}{20 + 30} \text{ m/s}^2$$

$$= \frac{9.8(20 - 30 \times 0.2588)}{50} = 2.4 \text{ m/s}^2 \quad \textbf{Ans.}$$

6. MOTION OF TWO BODIES CONNECTED BY A STRING, ONE OF WHICH IS HANGING FREE AND THE OTHER LYING ON A ROUGH INCLINED PLANE

Consider two bodies of masses m_1 and m_2 respectively, connected by a light inextensible string as shown in Fig. 11.

Let the body of mass m_1 hang free and the body of mass m_2 be placed on an inclined *rough* surface. Let the body of mass m_1 move downwards and the body of mass m_2 move upwards along the inclined surface.

We know that velocity and acceleration of the body of mass m_1 will be the same, as that of the body of mass m_2. Since the string is inextensible, therefore tension in both the string will also be equal.

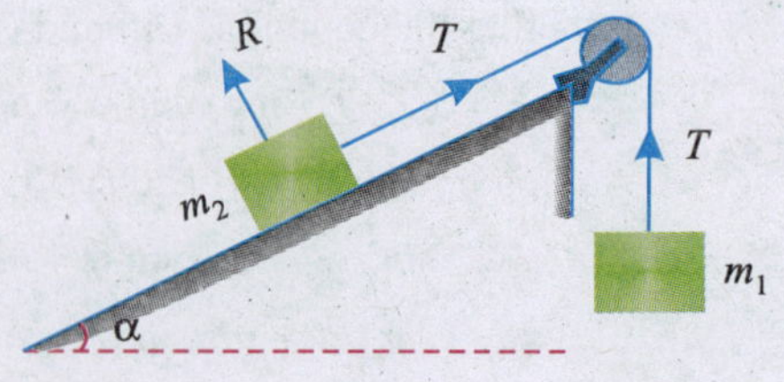

Fig. 11.

Let a = Acceleration of the system in m/s²

T = Tension in the string in N,

μ = Coefficient of friction, and

α = Inclination of the plane.

We know that the normal reaction on the inclined surface due to body of mass m_2 (as shown in Fig. 11).

$$= m_2 g \cos \alpha$$

∴ Frictional force $= \mu m_2 g \cos \alpha$

This frictional force will act in the opposite direction to the motion of the body of mass 2.

First of all, consider the motion of body 1 of mass m_1, which is coming down. We know that the forces acting on it are $m_1.g$ (downwards) and T (upwards). As the body is moving downwards, therefore, resultant force

$$= m_1 g - T \quad ...(i)$$

Since the body is moving downwards with an acceleration (a), therefore force acting on this body

$$= m_1 a \quad ...(ii)$$

Equating equations (i) and (ii),

$$m_1 g - T = m_1 a \quad ...(iii)$$

Now consider the motion of the body 2 of mass m_2, which is moving upwards on inclined surface. We know that the forces acting on it, along the plane, are T (upwards), $m_2.g \sin \alpha$ (downwards) and force of friction $\mu.m_2.g$ (downwards). As the body is moving upwards, therefore resultant force

$$= T - m_2 g \sin \alpha - \mu m_2 g \cos \alpha \quad ...(iv)$$

Since this body is moving upwards along the inclined surface with an acceleration (a) therefore force acting on this body

$$= m_2 a \quad ...(v)$$

Equating the equations (iv) and (v),

$$T - m_2 g \sin \alpha - \mu m_2 g \cos \alpha = m_2 a \quad ...(vi)$$

Adding equations (iii) and (vi),

$$m_1 g - m_2 g \sin \alpha - \mu m_2 g \cos \alpha = m_1 a + m_2 a$$

$$g (m_1 - m_2 \sin \alpha - \mu m_2 \cos \alpha) = a (m_1 + m_2)$$

$$a = \frac{g \left(m_1 - m_2 \sin \alpha - \mu m_2 \cos \alpha\right)}{m_1 + m_2}$$

From equation (iii) we find that

$$T = m_1 g - m_1 a = m_1 (g - a)$$

Substituting the value of a in the above equation,

$$T = m_1 \left[g - \frac{g \left(m_1 - m_2 \sin \alpha - \mu m_2 \cos \alpha\right)}{m_1 + m_2} \right]$$

$$= m_1 g \left[1 - \frac{m_1 - m_2 \sin \alpha - \mu m_2 \cos \alpha}{m_1 + m_2} \right]$$

$$= m_1 g \left[\frac{m_1 + m_2 - m_1 + m_2 \sin \alpha + \mu m_2 \cos \alpha}{m_1 + m_2} \right]$$

$$= \frac{m_1 m_2 g \left(1 + \sin \alpha + \mu \cos \alpha\right)}{m_1 + m_2}$$

Notes. 1. For horizontal surface, if we substitute the value of $\alpha = 0$ in the above equations for a and T, the rotations obtained will be the same as derived in Artice 4 of this chapter.

2. For smooth surface, if we substitue the value of $\mu = 0$ in the above equations for a and T, the relations obtained will be the same as derived in the last article 5 of this chapter.

Example 8. *A body of mass 150 kg, rests on a rough plane inclined at 10º to the horizontal. It is pulled up the plane, from rest, by means of a light flexible rope running parallel to the plane. The portion of the rope, beyond the pulley hangs vertically down and carries a man of 80 kg at the end. If the coefficient of friction for the plane and the body is 0.2, find*

(i) *the tension in the rope,*

(ii) *the acceleration in m/s², with which the body moves up the plane, and*

(iii) *the distance in metres moved by the body in 4 seconds starting from rest.*

Solution. Given : Mass of the body (m_2) = 150 kg ; Inclination of plane (α) = 10º ; Mass of the man (m_1) = 80 kg and coefficient of friction (μ) = 0.2

(i) Tension in the rope

We know that tension in the rope,

$$T = \frac{m_1\, m_2\, g\,(1 + \sin\alpha + \mu\cos\alpha)}{m_1 + m_2}$$

$$= \frac{80 \times 150 \times 9.8\,(1 + \sin 10^\circ + 0.2 \cos 10^\circ)}{80 + 150}\ \text{N}$$

$$= \frac{117\,600\,[(1 + 0.1736) + (0.2 \times 0.9848)]}{230}\ \text{N}$$

$$= 700\ \text{N}\ \textbf{Ans.}$$

(ii) Acceleration, with which the body moves up the plane

We also know that the acceleration with which the body moves up the plane,

$$a = \frac{g\,(m_1 - m_2 \sin\alpha - \mu m_2 \cos\alpha)}{m_1 + m_2}$$

$$= \frac{9.8\,(80 - 150 \sin 10^\circ - 0.2 \times 150 \cos 10^\circ)}{80 + 150}\ \text{m/s}^2$$

$$= \frac{9.8\,(80 - 150 \times 0.1736 - 0.2 \times 150 \times 0.9848)}{230}\ \text{m/s}^2$$

$$= 1.04\ \text{m/s}^2\ \textbf{Ans.}$$

(iii) Distance moved by the body in 4 sec starting from rest

In this case, initial velocity (u) = 0 (because starting from rest) ; Time (t) = 4 sec and acceleration (a) = 1.04 m/s²

∴ Distance moved by the body,

$$s = ut + \frac{1}{2}at^2 = 0 + \frac{1}{2} \times 1.04\,(4)^2 = 8.32\ \text{m}\ \textbf{Ans.}$$

Example 9. *Determine the resulting motion of the body A assuming the pulleys to be smooth and weightless as shown in Fig. 12.*

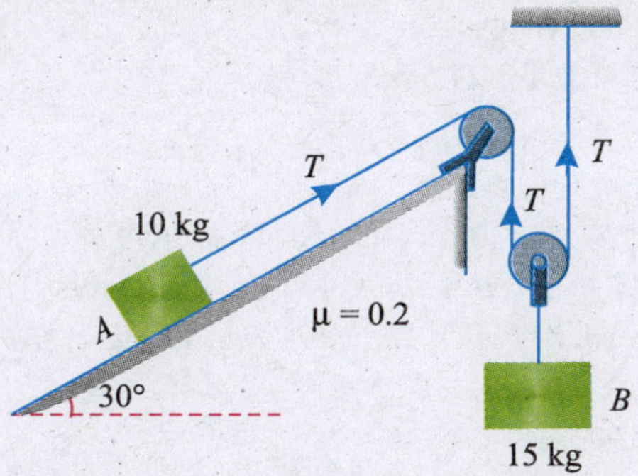

Fig. 12.

If the system starts from rest, determine the velocity of the body A after 10 seconds.

Solution. Given : Mass of body B (m_1) = 15 kg ; Mass of body A (m_2) = 10 kg; Inclination of plane (α) = 30° and coefficient of friction (μ) = 0.2

Let T = Tension in the string, and

a = Acceleration of the block A.

We know that normal reaction on the inclined surface due to body A of mass 10 kg.

$$R = m_2\, g \cos \alpha = 10 \times 9.8 \times \cos 10° = 98 \times 0.9848 = 96.5 \text{ N}$$

∴ Frictional force $= \mu\, R \times 96.5 = 0.2 \times 96.5 = 19.3 \text{ N}$

First of all, consider the motion of the block A, which is moving upwards. We know that the forces acting on it, along the plane, are T newtons (upwards), $m_2.g \sin \alpha$ newtons (downwards), and frictional force equal to 19.3 newtons (downwards, as the body is moving upwards). Therefore resultant force

$$= T - m_2\, g \sin \alpha - 19.3$$
$$= T - 10 \times 9.8 \sin 30° - 19.3$$
$$= T - 98 \times 0.5 - 19.3 = T - 68.3 \quad ...(i)$$

Since the body is moving with an acceleration (a) therefore force acting on it

$$= m_2\, a = 10\, a \quad ...(ii)$$

Equating equations (i) and (ii),

$$T - 68.3 = 10\, a$$

Multiplying both sides by 2,

$$2T - 136.6 = 20\, a \quad ...(iii)$$

Now consider motion of the body B, which is coming downwards. A little consideration will show that acceleration of the body B will be half the acceleration of the block A (*i.e.* $a/2$).

We know that the forces acting on it are $m_1\, g = 15 \times 9.8 = 147$ N (downwards) and $2T$ newtons (upwards). As the body is moving downwards, therefore resultant force

$$= 147 - 2T \quad ...(iv)$$

Since the body is moving with an acceleration of 0.5 a therefore force acting on it

$$= 15 \times 0.5\, a = 7.5\, a \quad ...(v)$$

Equating equations (*iv*) and (*v*),

$$147 - 2T = 7.5\,a \qquad \text{...}(vi)$$

Adding equations (*iii*) and (*vi*),

$$10.4 = 27.5\,a$$

or
$$a = \frac{10.4}{27.5} = 0.4 \text{ m/s}^2$$

∴ Velocity of the body *A* after 10 seconds, if the system starts from rest.

$$v = u + at = 0 + 0.4 \times 10 = 4 \text{ m/s} \quad \textbf{Ans.}$$

Example 10. *The system of bodies shown in Fig. 13 starts from rest.*

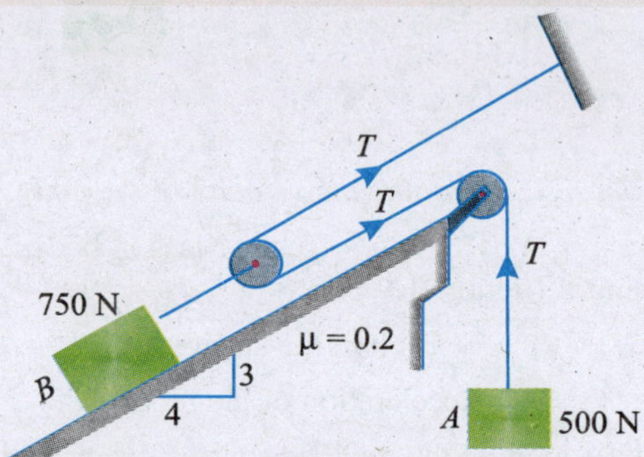

Fig. 13.

Determine the acceleration of body B and the tension in the string supporting body A.

Solution. Given : Weight of body *A* (W_1) = 500 N ; Weight of body *B* (W_2) = 750 N and coefficient of friction (μ) = 0.2

Acceleration of the body B

T = Tension in the strings, and

a = Acceleration of the body *B*.

From the slope of the surface, we find that

$$\tan\alpha = \frac{3}{4} = 0.75 \text{ or } \sin\alpha = 0.6 \text{ and } \cos\alpha = 0.8$$

We know that normal reaction on the inclined surface due to body of weight 750 N

$$= W_2 \cos\alpha = 750 \times 0.8 = 600 \text{ N}$$

∴ Frictional force
$$= \mu R = \mu \times 600 = 0.2 \times 600 = 120 \text{ N}$$

First of all, consider the motion of the body *B*, which is moving, upwards. We know that the forces acting on it, along the plane, are $2T$ newtons (upwards), $W_2 \sin\alpha$ newtons (downwards) and force of friction equal to 120 newtons (downwards as the body is moving upwards). Therefore resultant force

$$= 2T - W_2 \sin\alpha - 120 = 2T - 750 \times 0.6 - 120 \text{ N}$$

$$= 2T - 570 \text{ N} \qquad \text{...}(i)$$

Since the body is moving with an acceleration (*a*), therefore force acting on it

$$m_2\, g = \frac{W_2}{g} \times a = \frac{750}{9.8} \times a = 76.5\,a \qquad \text{...}(ii)$$

Equating equations (*i*) and (*ii*),

$$2T - 570 = 76.5\,a \qquad \text{...}(iii)$$

Now consider the motion of the body A, which is coming downwards. A little consideration will show that the acceleration of the body A will be double the acceleration of the block A (*i.e.* $2a$). We know that the force acting on it 500 N (downwards) and T (upwards). As the body is moving downwards, therefore resultant force

$$= 500 - T \qquad \ldots(iv)$$

Since the body is moving with an acceleration $2a$, therefore force acting on it

$$m_1\, 2a = \frac{W_1}{g} \times 2a = \frac{500}{9.8} \times 2a = 102a \qquad \ldots(v)$$

Equating equations (iv) and (v),

$$500 - T = 102\, a$$

Multiplying both sides by 2,

$$1000 - 2T = 204\, a \qquad \ldots(vi)$$

Adding equations (iii) and (vi),

$$430 = 280.5\, a$$

or

$$a = \frac{430}{280.5} = 1.5 \text{ m/s}^2$$

Tension in the string supporting body A

Substituting the value of a in equation (vi),

$$1000 - 2T = 204a = 204 \times 1.5 = 306$$

$$T = \frac{1000 - 306}{2} = 347 \text{ N} \qquad \textbf{Ans.}$$

7. MOTION OF TWO BODIES, CONNECTED BY A STRING AND LYING ON SMOOTH INCLINED PLANES

Consider two bodies of masses m_1 and m_2 kg respectively connected by a light inextensible string on two smooth surfaces (*i.e.* friction between the masses and the surfaces is neglected) as shown in Fig. 14.

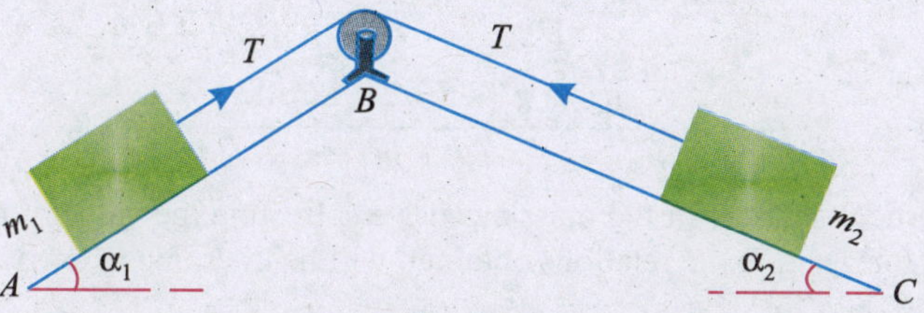

Fig. 14.

Let the body of mass m_1 move downwards along the inclined plane surface AB and the body of mass m_2 move upwards along the inclined surface BC. A little consideration will show, that the velocity and acceleration of the body of mass m_1 will be the same as that of the body of mass m_2. Since the string is inextensible, therefore tension in both the strings will also be equal.

Let

a = Acceleration of the system

T = Tension in the string and

α_1 and α_2 = Inclination of surfaces AB and BC

First of all, consider the motion of body 1 of mass m_1 kg which is coming down along the inclined plane AB. We know that forces acting on it, along the plane, are $m_1.g \sin \alpha_1$ newtons (downwards) and T newtons (upwards). As the body is moving downwards, therefore resultant force

$$= m_1\, g \sin \alpha_1 - T \qquad \ldots(i)$$

Since the body is moving downwards with an acceleration (a), therefore force acting on this body

$$= m_1 a \qquad ...(ii)$$

Equating equations (i) and (ii),

$$m_1 g \sin \alpha_1 - T = m_1 a \qquad ...(iii)$$

Now consider the motion of the body 2 of mass m_2 which is moving upwards along the inclined plane BC. We know that the forces acting on it, along the plane, are T (upwards) and $m_2.g \sin \alpha_2$ (downwards). As the body is moving upwards, therefore resultant force

$$= T - m_2 g \sin \alpha_2 \qquad ...(iv)$$

Since the body is moving upwards with an acceleration (a) therefore force acting on this body

$$= m_2 a \qquad ...(v)$$

Equating equations (iv) and (v),

$$T - m_2 g \sin \alpha_2 = m_2 a \qquad ...(vi)$$

Adding equations (iii) and (vi)

$$m_1 g \sin \alpha_1 - m_2 g \sin \alpha_2 = m_1 a + m_2 a$$

$$g (m_1 \sin \alpha_1 - m_2 \sin \alpha_2) = a (m_1 + m_2)$$

$$\therefore \qquad a = \frac{g\left(m_1 \sin \alpha_1 - m_2 \sin \alpha_2\right)}{m_1 + m_2}$$

From equation (iii), we have

$$T = m_1 g \sin \alpha_1 - m_1 a = m_1 (g \sin \alpha_1 - a)$$

Substituting the value of a in above equation,

$$T = m_1\left[g \sin \alpha_1 - \frac{g\left(m_1 \sin \alpha_1 - m_2 \sin \alpha_2\right)}{m_1 + m_2}\right]$$

$$= m_1 g\left[\frac{m_1 \sin \alpha_1 + m_2 \sin \alpha_1 - m_1 \sin \alpha_1 + m_2 \sin \alpha_2}{m_1 + m_2}\right]$$

$$= \frac{m_1 m_2 g\left(\sin \alpha_1 + \sin \alpha_2\right)}{m_1 + m_2}$$

Note. For vertical surface for the mass m_1, if we substitute the value of $\alpha_1 = 0$ and $\alpha_2 = \alpha$ in the above equations for a and T, the relations obtained will be the same as derived in Article 5 of this chapter.

Example 11. *Two smooth inclined planes whose inclinations with the horizontal are 30° and 20° are placed back to back. Two bodies of mass 10 kg and 6 kg are placed on them and are connected by a light inextensible string passing over a smooth pulley as shown in Fig 15.*

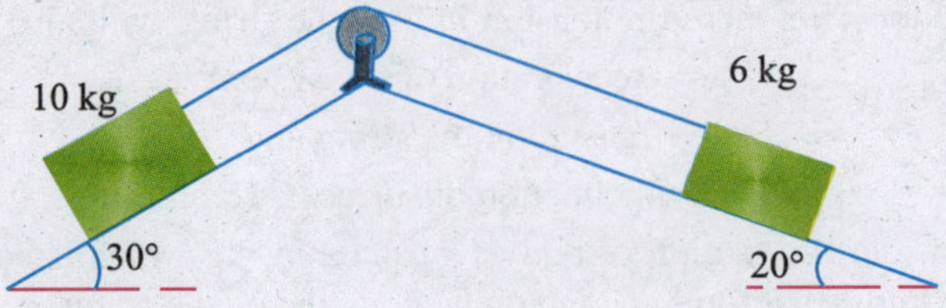

Fig. 15.

Find the tension in the string. Take $g = 9.8$ m/s^2.

Solution. Given : Inclination of first plane (α_1) = 30° ; Inclination of second plane (α_2) = 20° ; Mass of first body (m_1) = 10 kg and mass of second body (m_2) = 6 kg

We know that tension in the string,

$$T = \frac{m_1 m_2 g(\sin \alpha_1 + \sin \alpha_2)}{m_1 + m_2}$$

$$= \frac{10 \times 6 \times 9.8(\sin 30° + \sin 20°)}{10 + 6} \text{ N}$$

$$= \frac{60 \times 9.8(0.5 + 0.3420)}{16} = 31 \text{ N} \quad \textbf{Ans.}$$

Example 12. *Two bodies A and B are connected by a light inextensible cord as shown in Fig. 16.*

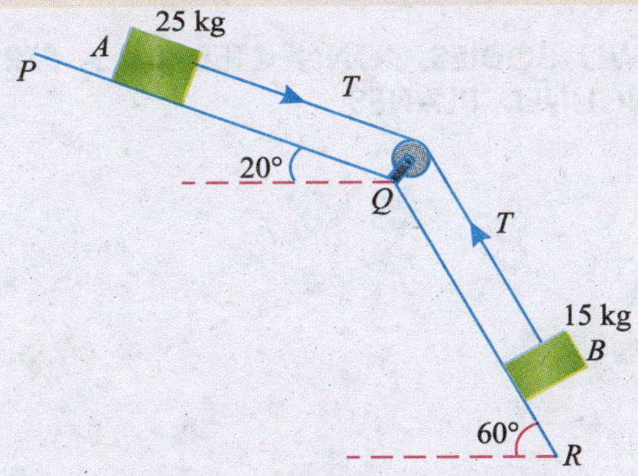

Fig. 16.

If both the bodies are released simultaneously, what distance do they move in 3 seconds? Neglect friction between the two bodies and the inclined surfaces.

Solution. Given : Mass of body B (m_1) = 15 kg ; Mass of body A (m_2) = 25 kg ; Inclination of plane PQ with horizontal (α_2) = 20° ; Inclination of plane QR with horizontal (α_1) = 60° ; Initial velocity (u) = 0 (because the bodies are released) and time (t) = 3 s.

Let a = Acceleration of the system, and

T = Tension in the cord.

First of all, consider the motion of body B of mass 15 kg which is coming down along the inclined surface QR. We know that forces acting on it, along the plane, QR are $m_1.g \sin \alpha_1$ newtons (downwards) and T newtons (upwards). As the body is moving downwards, therefore resultant force

$$= m_1 g \sin \alpha_1 - T = 15 \times 9{\cdot}8 \sin 60° - T$$

$$= 147 \times 0{\cdot}866 - T = 127{\cdot}3 - T \quad ...(i)$$

Since the body is moving downwards with an acceleration (a), therefore force acting on it.

$$= m_1 a = 15\, a \quad ...(ii)$$

Equating equations (i) and (ii)

$$127{\cdot}3 - T = 15\, a \quad ...(iii)$$

Now consider the motion of body A of mass 25 kg which is also coming down along the inclined surface PQ. We know that the forces acting on it along the plane PQ, are $m_2.g \sin \alpha_2$ (downwards) and T (again downwards). Therefore resultant force

$$= m_2 g \sin \alpha_2 + T = (25 \times 9.8 \sin 20°) + T$$

$$= 245 \times 0.342 + T = 83.8 + T \quad ...(iv)$$

Since the body is moving downwards with an acceleration (a), therefore force acting on it.

$$= m_2 \, a = 25 \, a \qquad \text{...}(v)$$

Equating equations (iv) and (v)

$$83.8 + T = 25 \, a \qquad \text{...}(vi)$$

Adding equations (iii) and (vi),

$$211.1 = 40 \, a$$

or

$$a = \frac{211.1}{40} = 5.3 \text{ m/s}^2$$

∴ Distance moved by the bodies in 3 seconds,

$$s = ut + \frac{1}{2}at^2 = 0 + \frac{1}{2} \times 5.3(3)^2 = 23.85 \text{ m} \quad \textbf{Ans.}$$

8. MOTION OF TWO BODIES CONNECTED BY A STRING AND LYING ON ROUGH INCLINED PLANES

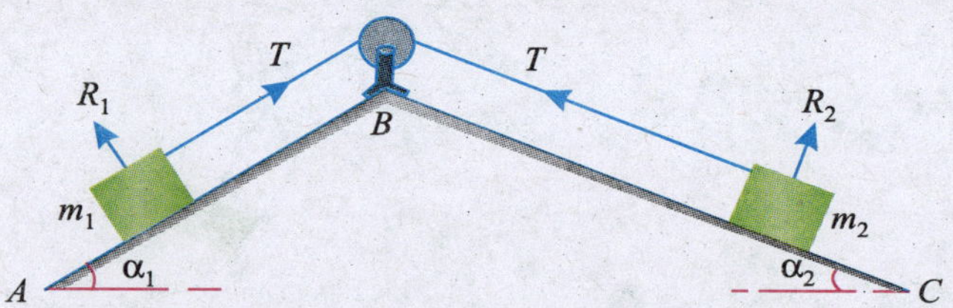

Fig. 17.

In this case, the friction between the bodies of masses m_1 and m_2 and the surfaces AB and BC will also be considered.

It may be noted that the force of the friction always acts in the opposite direction to the motion of a mass.

If the mass m_1 comes down and the mass m_2 goes up, the frictional force at m_1 will act upwards and the frictional force at m_2 will act downwards. Thus, while calculating the forces on any mass, the frictional force, acting in the opposite direction, should also be considered.

Example 13. *Two rough planes inclined at 30° and 15° to the horizontal and of the same height are placed back to back. Two bodies of masses of 15 kg and 5 kg are placed on the faces and connected by a string over the top of the planes If the coefficient of friction be 0.3 find from fundamentals the resulting acceleration.*

Solution. Inclination of first plane (α_1) = 30° ; Inclination of second plane (α_2) = 15° ; Mass of first body (m_1) = 15 kg ; Mass of second body (m_2) = 5 kg and coefficient of friction (μ) = 0.3.

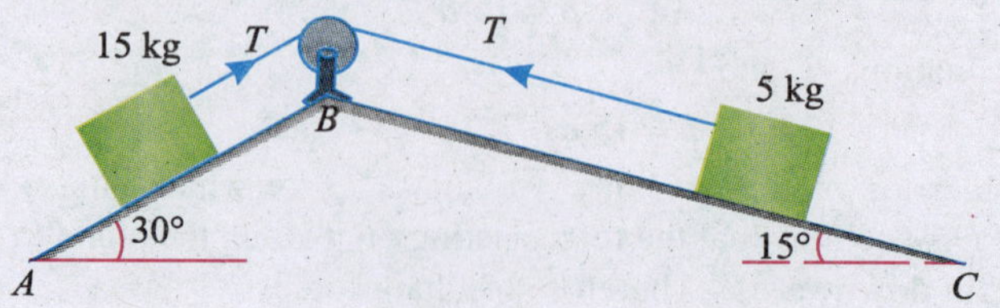

Fig. 18.

Let a = Acceleration of the system and

T = Tension in the string.

We know that normal reaction on the plane AB due to body of mass 15 kg.

$$R_1 = 15\,g \cos \alpha_1 = 15 \times 9.8 \cos 30^\circ$$
$$= 15 \times 9.8 \times 0.866 = 127.3 \text{ N}$$

∴ Frictional force $= \mu\, R_1 = 0.3 \times 127.3 = 38.2 \text{ N}$

Similarly, normal reaction on the plane BC due to the body of mass 5 kg.

$$R_2 = 5 \text{ g} \cos \alpha_2 = 5 \times 9.8 \cos 15^\circ$$
$$= 5 \times 9.8 \times 0.9659 = 48 \text{ N}$$

∴ Frictional force $= \mu R_2 = 0.3 \times 4.8 = 14.4 \text{ N}$

The frictional forces will act in the opposite directions to the motions of the two bodies.

First of all consider the motion of body 1 of mass 15 kg, which is coming down along the inclined plane AB. We know that the forces acting on it, along the plane, are $m_1.g \sin \alpha_1$ newtons (downwards), T newtons (upwards) and force of friction equal to 38.2 newton (upwards as the body is moving downwards). Therefore resultant force

$$= m_1\, g \sin \alpha_1 - T - 38.2 = (15 \times 9.8 \sin 30^\circ) - T - 38.2$$
$$= 147 \times 0.5 - T - 38.2 = 35.3 - T \qquad ...(i)$$

Since the body is moving downwards with an acceleration (a), therefore force acting on it

$$= m_1\, a = 15\, a \qquad ...(ii)$$

Equating the equations (i) and (ii),

$$35.3 - T = 15\, a \qquad ...(iii)$$

Now consider the motion of body 2 of mass 5 kg, which is moving upwards along the inclined plane BC. We know that the forces acting on it, along the plane, are T (upwards), $m_2.g \sin \alpha_2$ (downwards) and force of friction equal to 14.4 N (downwards as the body is moving upwards). Therefore resultant force

$$= T - m_2\, g \sin \alpha_2 - 14.4 = T - (5 \times 9.8 \sin 15^\circ) - 14.4$$
$$= T - 5 \times 9.8 \times 0.2588 - 14.4 = T - 27 \qquad ...(iv)$$

Since the body is moving upwards with an acceleration (a), therefore force acting on this body

$$= m_2\, a = 5\, a \qquad ..(v)$$

Equating equations (iv) and (v),

$$T - 27 = 5\, a \qquad ...(vi)$$

Adding equations (iii) and (vi),

$$35.3 - 27 = 15\, a + 5\, a = 20\, a$$

$$a = \frac{35.3 - 27}{20} = 0.42 \text{ m/s}^2$$ **Ans.**

Example 14. *A system of bodies A,B and C in Fig. 19 is released from rest.*

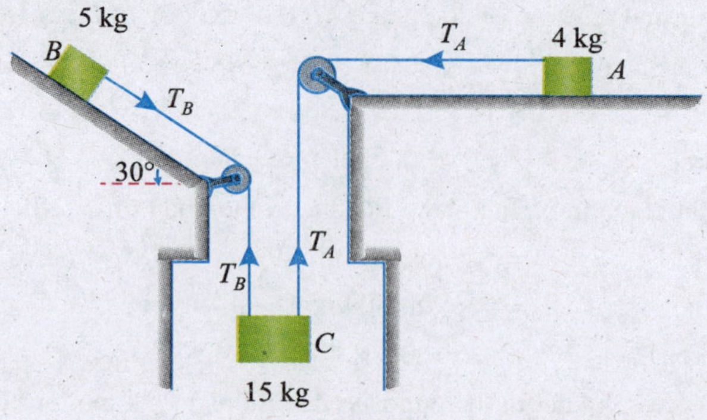

Fig. 19.

Find (i) acceleration of the masses and (ii) tension in the two strings. Take coefficient of friction for the contact surfaces of bodies A and B as 0.4.

Solution. Given : Mass of body A (m_A) = 4 kg ; Mass of body B (m_B) = 5 kg ; Mass of body C (m_C) = 15 kg and Coefficient of friction (μ) = 0.4

(i) Acceleration of the bodies

Let a = Acceleration of the bodies

T_A = Tension in the string connected with body A

and T_B = Tension in the string connected with body B.

We know that normal reaction on the horizontal surface due to body A,

$$R_A = m_A\, g = 4 \times 9.8 = 39.2$$

∴ Frictional force, $F_A = \mu\, R_A = 0.4 \times 39.2 = 15.68$ N

Similarly, normal reaction on the inclined surface due to body B,

$$R_B = m_B\, g \cos \alpha = 5 \times 9.8 \cos 30° = 49 \times 0.866 = 42.43$$

∴ Frictional force $F_B = \mu\, R_B = 0.4 \times 42.43 = 16.97$ N

First of all, consider the motion of body A which is moving horizontally. We know that forces acting on it are T_A (towards left) and frictional force of 15.68 N (towards right). As the body is moving towards left, therefore resultant force

$$= T_A - 15.68 \qquad ...(i)$$

Since the body A is moving with an acceleration of (a) therefore force acting on it

$$= 4\, a \qquad ...(ii)$$

Equating equations (i) and (ii)

$$T_A - 15.68 = 4\, a \qquad ...(iii)$$

or

$$T_A = 4\, a + 15.68 \qquad ...(iv)$$

Now consider the motion of the body B, which is moving downwards on inclined surface. We know that the forces acting on it, along the plane, are T_B (downwards), $m_B.g \sin \alpha$ (again downwards), frictional force equal to 16.97 N (upwards, as the block is moving downwards). Therefore resultant force

$$= T_B + m_B\, g \sin \alpha - 16.97$$

$$= T_B + 5 \times 9.8 \sin 30^\circ - 16.97$$

$$= T_B + 49 \times 0.5 - 16.97 = T_B + 7.53 \qquad ...(v)$$

Since the body is moving with an acceleration of (a), therefore force acting on it

$$= 5a \qquad ...(vi)$$

Equating equations (v) and (vi),

$$T_B + 7.53 = 5\,a$$

or

$$T_B = 5\,a - 7.53 \qquad ...(vii)$$

Now consider the motion of the body C, which is coming down. We know that forces acting on it are $m_C.g = 15 \times 9.8 = 147$ N (downwards) and $(T_A + T_B)$ upwards. As the body is moving downwards, therefore resultant force

$$= 147 - (T_A + T_B) \qquad ...(viii)$$

Since the body is moving with an acceleration (a), therefore force acting on it

$$= 15\,a \qquad ...(ix)$$

Equating equations ($viii$) and (ix),

$$147 - (T_A + T_B) = 15\,a$$

Substituting the values of T_A and T_B from equations (iv) and (vii)

$$147 - [(4a + 15.68) + (5a - 7.53)] = 15a$$

$$147 - 4a - 15.68 - 5a + 7.53 = 15a$$

$$\therefore \quad 24\,a = 138.85$$

$$a = \frac{138.85}{24} = 5.8 \text{ m/s}^2 \qquad \textbf{Ans.}$$

(ii) Tension in the two strings

Substituting the value of (a) in equation (iv),

$$T_A = 4a + 15.68 = (4 \times 5.8) + 15.68 = 38.88 \text{ N} \quad \textbf{Ans.}$$

Again substituting the value of (a) in equation (vii),

$$T_B = 5a - 7.53 = (5 \times 5.8) - 7.53 = 21.47 \text{ N} \quad \textbf{Ans.}$$

EXERCISE 2

1. Fig. 20 shows two masses connected by a light inextensible string passing over a smooth pulley.

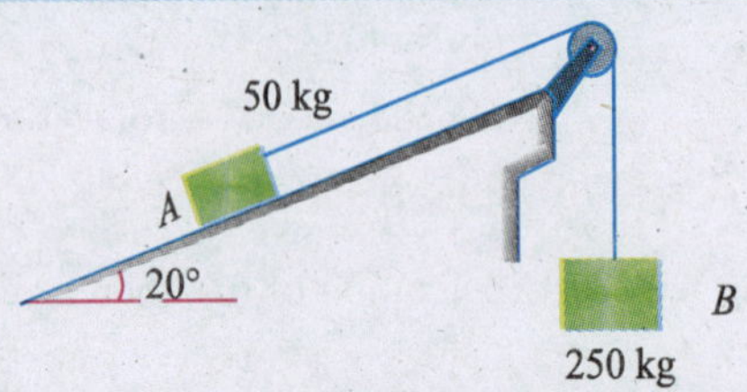

Fig. 20.

If the system is released from rest, with what velocity 50 kg mass will touch the floor, if initially it was 1.6 m higher than the floor level. Take coefficient of friction between 50 kg mass and inclined surface as 0.2. (**Ans.** 4.83 m/s)

2. Two bodies *A* and *B* of mass 8 kg and 10 kg are placed on two smooth inclined planes as shown in Fig. 21.

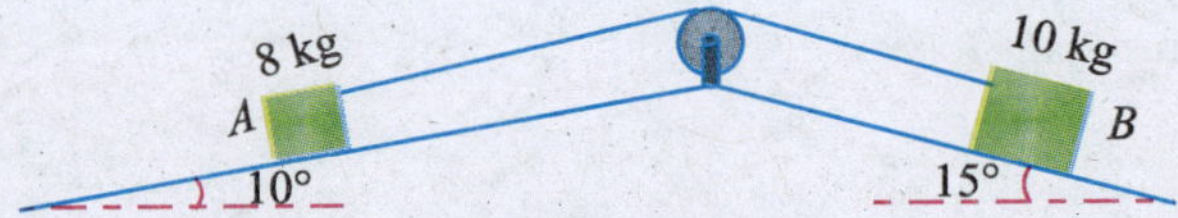

Fig. 21.

Find, the acceleration of the body of mass 8 kg. (**Ans.** 0.47 m/s^2)

3. Two bodies *P* and *Q* are connected by a light inextensible string as shown in Fig. 22.

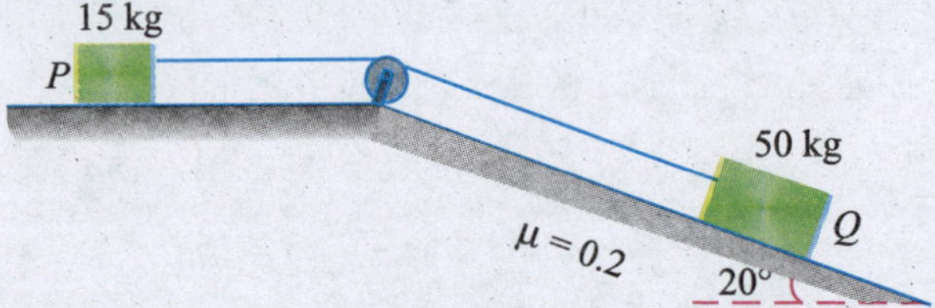

Fig. 22.

Find acceleration of two bodies and tension in the string. (**Ans.** 0.94 m/s^2 : 28.8)

QUESTIONS

1. Explain the reason for the tension in both the strings to be equal, when two masses are attached to its ends, and the inextensible string is made to pass over a smooth pulley.
2. Obtain a relation for the acceleration of two bodies connected by a string, when one body is hanging free and the other is lying on a smooth horizontal plane.
3. Derive an equation for the tension in the string, when one body is free and the other is lying on a rough horizontal plane.
4. Explain the procedure adopted for obtaining the relations for acceleration of two bodies and tension in the string when one of the bodies is hanging free and the other is lying on (*a*) smooth inclined plane ; and (*b*) rough inclined plane.

5. If two bodies connected by an inextensible string are lying on two rough inclined planes, then how will you judge, as to which of the two bodies will come down?

OBJECTIVE TYPE QUESTIONS

1. Two bodies of masses 10 kg and 15 kg are hung from the ends of an extensible rope passing over a frictionless pulley. If masses of both the bodies are doubled the acceleration of the system will also be doubled.

   (*a*) Yes (*b*) No

2. Two masses of 10 kg and 5 kg are connected to two ends of a rope which is passing over a smooth pulley. The 10 kg mass is lying on a smooth horizontal plane and 5 kg mass is hanging. If the position of the two masses is interchanged, its acceleration will also change.

   (*a*) Agree (*b*) Disagree

3. If two masses are connected to the two ends of an inextensible string, passing over a pulley. One of the mass is lying on a rough horizontal plane and the other is hanging free. If the value of coefficient of friction is increased, it will increase its

   (*a*) acceleration (*b*) tension (*c*) both of them

4. Two masses of 10 kg and 15 kg are connected to two ends of an inextensible rope and passing over a smooth pulley. The 10 kg mass is lying over a rough plane, which is inclined at an angle of 25° with the horizontal. If this angle is made 30°, then

   (*a*) tension in the string will increase

   (*b*) tension in the string will decrease

   (*c*) acceleration of the system will remain the same.

ANSWERS

1. (*b*) **2.** (*a*) **3.** (*b*) **4.** (*a*)

CHAPTER

24

Helical Springs and Pendulums

Contents

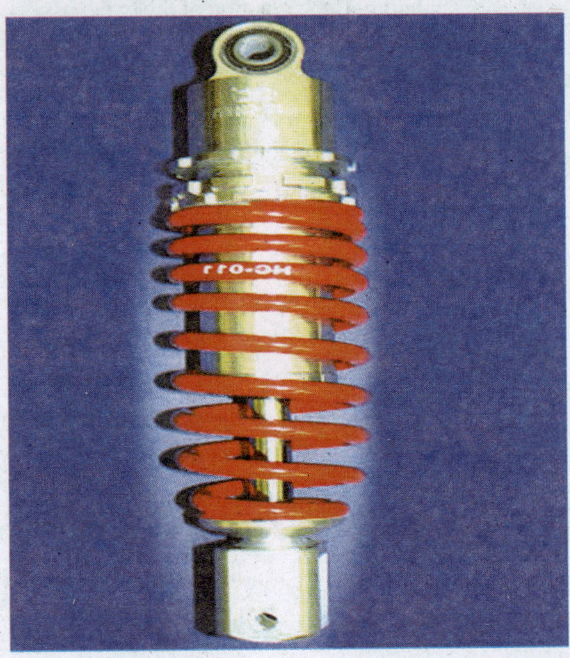

1. INTRODUCTION

We have studied the simple harmonic motion of bodies without any reference to the force causing such a motion. But in this chapter, we shall discuss the forces causing simple harmonic motion of the following bodies :

1. Helical springs,
2. Simple pendulum,
3. Compound pendulum, and
4. Conical pendulum.

2. HELICAL SPRINGS

[BPUT]

Consider a closely-coiled helical springs whose upper end is fixed as shown in Fig 1.

Let a body be attached to the lower end. Let A-A be the equilibrium position of the spring, after the body is attached. If the spring is stretched up to B-B and then released, the body will move up and down with a simple harmonic motion.

Let m = Mass of the body in kg (such that its weight (W) is mg newtons)

s = Stiffness of the spring in N/m

x = Displacement of the load below the equilibrium position in metres.

a = Acceleration of the body in m/s^2

g = Gravitational acceleration, and

t = Periodic time.

Fig. 1 Helical spring.

We know that the deflection of spring,

$$\delta = \frac{W}{s} = \frac{mg}{s}$$

Then disturbing force $=$ Mass $\times$ Acceleration $= m.a$ ...(i)

and restoring force $= sx$ ...(ii)

Equating equations (i) and (ii),

$$ma = sx \quad \text{or} \quad \frac{m}{s} = \frac{x}{a} \qquad ...(iii)$$

We know that in simple harmonic motion, time period,

$$t = 2\pi\sqrt{\frac{\text{Displacement}}{\text{Acceleration}}} = 2\pi\sqrt{\frac{x}{a}} = 2\pi\sqrt{\frac{m}{s}}$$

$$\frac{2\pi}{\omega} = 2\pi\sqrt{\frac{\delta}{g}} \qquad ...\left(\because t = \frac{2\pi}{\omega} \text{ and } \frac{mg}{s} = \delta\right)$$

$$\therefore \qquad \omega = \sqrt{\frac{g}{\delta}}$$

where ω is the angular velocity in rad/ sec.

Notes. 1. Frequency of motion,

$$n = \frac{1}{t} = \frac{1}{2\pi}\sqrt{\frac{s}{m}} = \frac{1}{2\pi}\sqrt{\frac{g}{\delta}}$$

2. From equation (iii) we find that

$$x = \frac{ma}{s}$$

Thus we see that x is directly proportional to m/s.

***Example 1.** *A 4 kg mass hung at one end of a helical spring and is set vibrating vertically. The mass makes 2 vibrations per second. Determine the stiffness of the spring.*

Solution. Given : Mass (m) = 4 kg and frequency (n) = 2 vib/s = 2 Hz

Let s = Stiffness of the spring.

We know that periodic time,

$$t = \frac{1}{n} = \frac{1}{2} = 0.5 \text{ s}$$

We also know that periodic time,

$$0.5 = 2\pi\sqrt{\frac{m}{s}} = 2\pi\sqrt{\frac{4}{s}}$$

Squaring both sides,

$$0.25 = (2\pi)^2 \times \frac{4}{s} = \frac{157.9}{s}$$

$$s = \frac{157.9}{0.25} = 631.6 \text{ N/m} \qquad \textbf{Ans.}$$

Example 2. *A spiral spring hung up at one end, and carrying a mass of 7 kg at the other is made to vibrate. Find the period of oscillation, if the spring is found to extend 10 mm for each 0.5 kg of mass.*

Solution. Given: Mass = 7kg and when mass = 0.5 kg, then deflection (δ) = 10 mm = 0.01 m.

We know that when mass is equal to 7 kg, then deflection of the spring

$$\delta = \frac{0.01}{0.5} \times 7 = 0.14 \text{ m}$$

and period of oscillation,

$$t = 2\pi\sqrt{\frac{\delta}{g}} = 2\pi\sqrt{\frac{0{\cdot}14}{9{\cdot}8}} = 0{\cdot}75 \text{ s} \qquad \textbf{Ans.}$$

***Example 3.** *A mass supported by a spring has a static deflection of 0.5 mm. Determine its natural frequency of oscillation.* **[AMIE]**

Solution. Given: Deflection (δ) = 0.5 mm = 0.0005 m

We know that natural frequency of oscillation,

$$n = \frac{1}{2\pi}\sqrt{\frac{g}{\delta}} = \frac{1}{2\pi}\sqrt{\frac{9{\cdot}8}{0{\cdot}0005}} = 22{\cdot}3 \text{ vib/s} = 22{\cdot}3 \text{ Hz} \qquad \textbf{Ans.}$$

***Example 4.** *A body of mass 3 kg, suspended from a vertically mounted spring, deflects it by 12 mm. Determine the no. of oscillations of the body.*

Also determine the maximum force in the spring, when it is displaced through a distance of 25 mm from its rest position and then released. **[JNTU]**

Solution. Given : Mass of the body (m) = 3 kg ; Deflection (δ) = 12 mm = 0.012 m and displacement (x) = 25 mm = 0.025 m

No. of oscillations of the body

We know that the no. of oscillations of the body,

$$n = \frac{1}{2\pi}\sqrt{\frac{g}{\delta}} = \frac{1}{2\pi}\sqrt{\frac{9{\cdot}8}{0.012}} = 4{\cdot}55 \text{ vib/s} = 4{\cdot}55 \text{ Hz} \qquad \textbf{Ans.}$$

Maximum force in the spring

We know that the angular velocity of the body,

$$\omega = \sqrt{\frac{g}{\delta}} = \sqrt{\frac{9{\cdot}8}{0{\cdot}012}} = 28{\cdot}6 \text{ rad/s}$$

and maximum acceleration, $a_{max} = \omega^2 x = (28{\cdot}6)^2 \times 0{\cdot}025 = 20{\cdot}4 \text{ m/s}^2$

∴ Maximum inertia force

$$= \text{Mass} \times \text{Acceleration} = 3 \times 20{\cdot}4 = 61{\cdot}2 \text{ N}$$

We also know that maximum force in the spring occurs when the mass is at its lowest position. In this position, the force in the spring is equal to the sum of weight of the body and the inertia force. Therefore maximum force in the spring

$$= (3 \times 9{\cdot}8) + 61{\cdot}2 = 90{\cdot}6 \text{ N} \quad \textbf{Ans.}$$

***Example 5.** *The weight of an empty railway wagon is 240 kN. On loading it with goods weighing 320 kN, its spring gets compressed by 80 mm.*

(*a*) *Calculate its natural period of vibrations when the railway wagon is* (*i*) *empty and* (*ii*) *loaded as above.*

(*b*) *It is set into natural vibrations with an amplitude of 100 mm when empty. Calculate the velocity of the railway wagon when its displacement is 40 mm from statical equilibrium position.*

[AMIE]

Solution. Given : Weight of empty wagon = 240 kN and when the load (W) is 320 kN, then deflection (δ) = 80 mm = 0.08 m

(*i*) *Period of vibrations when the wagon is empty*

We know that deflection of the spring, when wagon is empty,

$$\delta_1 = \frac{0{\cdot}08}{320} \times 240 = 0{\cdot}06 \text{ m}$$

and period of vibrations, $t_1 = 2\pi \sqrt{\frac{\delta_1}{g}} = 2\pi \sqrt{\frac{0{\cdot}06}{9{\cdot}8}} = 0{\cdot}49 \text{ s}$ **Ans.**

(*ii*) *Period of vibrations when the wagon is loaded*

We know that total load on the springs when the wagon is loaded

$$= 240 + 320 = 560 \text{ kN}$$

∴ Deflection of the spring when the wagon is loaded

$$\delta_2 = \frac{0{\cdot}08}{320} \times 560 = 0.14 \text{ m}$$

and period of vibrations, $t_2 = 2\pi \sqrt{\frac{\delta_2}{g}} = 2\pi \sqrt{\frac{0{\cdot}14}{9{\cdot}8}} = 0{\cdot}75 \text{ s}$

(*b*) *Velocity of the railway wagon when it is empty*

When the wagon is empty, amplitude (r) = 100 mm = 0.1 m and displacement (y) = 40 mm = 0.04 m

We know that angular velocity of the wagon,

$$\omega = \frac{2\pi}{t_1} = \frac{2\pi}{0{\cdot}49} = 12{\cdot}82 \text{ rad/s}$$

and velocity, $v = \omega \sqrt{r^2 - y^2} = 12{\cdot}82 \sqrt{(0{\cdot}1)^2 - (0{\cdot}04)^2} = 1{\cdot}175 \text{ m/s}$ **Ans.**

EXERCISE 1

1. A helical spring has a stiffness of 1 N/mm. What mass should be hung on it, so that it may oscillate with a periodic time of 1·5 second ? (**Ans.** 56·96 kg)
2. A helical spring, of negligible mass is found to extend 0·25 mm under a mass of 1·5 kg, is made to support a mass of 60 kg. The spring and the mass system is displaced vertically through 12·5 mm and then released. Determine the frequency of natural vibration of the system. Take g as 9·81 m/s^2.

   Find also the velocity of the mass, when it is 5 mm below its rest position.

   (**Ans.** 4·98 Hz ; 358·6 mm/s)
3. A mass of 1·8 kg suspended from a spring of stiffness 45 N/mm is set in oscillation. What length of simple pendulum will have the same frequency of oscillation ? What is the frequency of oscillation ? (**Ans.** 394 mm ; 0·79 Hz)

3. *HELICAL SPRINGS IN SERIES AND PARALLEL

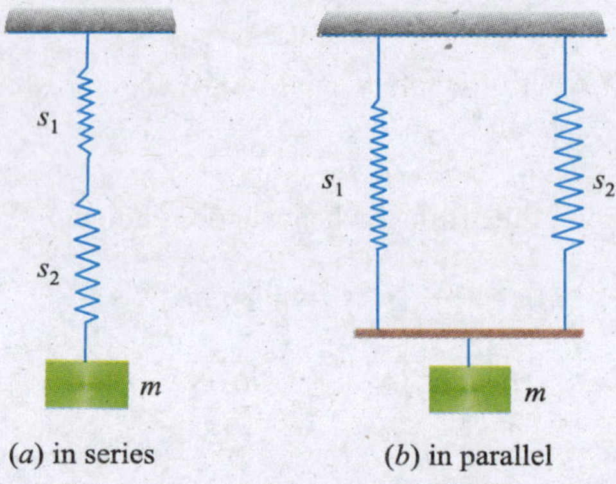

(*a*) in series (*b*) in parallel

Fig. 2.

In the previous article, we have discussed the arrangement of one helical spring only. But sometimes two or more helical springs are used at one place in the system. It will be interesting to know that the arrangement of helical springs may be in series or parallel as shown in Fig. 2 (*a*) & (*b*).

Now consider two helical springs, in series, as shown in Fig. 3 (*a*). We know that in this case, both the springs will be subjected to the same load.

Let s_1 = Stiffness of the spring 1, and

s_2 = Stiffness of the spring 2.

Now both the springs may be assumed to be replaced by an equivalent spring of stiffness (s) such that

$$\frac{1}{s} = \frac{1}{s_1} + \frac{1}{s_2} = \frac{s_1 + s_2}{s_1 . s_2}$$

Similarly, when the two springs are arranged in parallel, then they will share the given load. And the springs may also be assumed to be replaced by an equivalent springs of stiffness (s) such that,

$$s = s_1 + s_2$$

Notes. 1. We have considered only two springs for the sake of simplicity. But this principle may be extended for any no. of springs.

2. All other relations regarding periodic time, frequency etc. are also applicable in this case.

***Example 6.** *A block of mass 50 kg supported by two springs connected in series hangs from the ceiling. It can move between smooth vertical guides. The spring constants are 4 kN /m and 6 kN /m as shown in Fig. 3.*

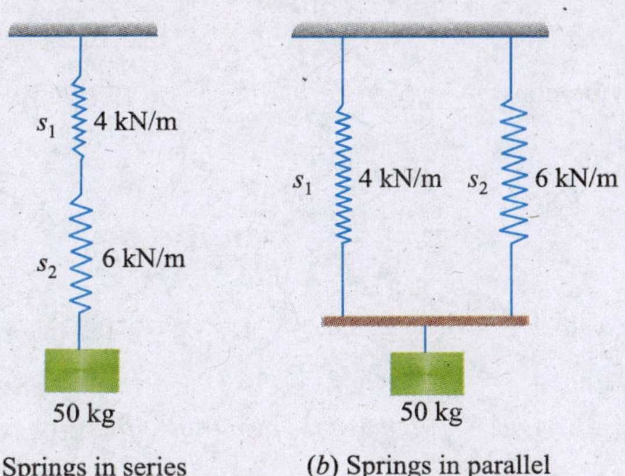

(*a*) Springs in series (*b*) Springs in parallel

Fig. 3.

The block is pulled 40 mm down from its position of equilibrium and then released. Determine

(a) period of vibrations, maximum velocity and acceleration of the block.

(b) quantities in (a) above, when the block is supported by the springs connected in parallel.

[AMIE, IGNOU]

Solution. Given : Mass (m) = 50 kg = 0.005 t, Stiffness of first spring (s_1) = 4 kN/m ; Stiffness of second spring (s_2) = 6 kN /m and displacement (r) = 40 mm = 0·04 m.

(a) When the springs are connected in series

We know that spring constant of an equivalent spring,

$$\frac{1}{s} = \frac{1}{s_1} + \frac{1}{s_2} = \frac{1}{4} + \frac{1}{6} = \frac{10}{24} = \frac{1}{2{\cdot}4}$$

or $$s = 2{\cdot}4 \text{ kN/m}$$

and deflection of the spring $$\delta = \frac{mg}{s} = \frac{0.05 \times 9.8}{2{\cdot}4} = 0{\cdot}204 \text{ m}$$

∴ Period of vibrations, $$t = 2\pi\sqrt{\frac{\delta}{g}} = 2\pi\sqrt{\frac{0{\cdot}204}{9{\cdot}8}} = 0{\cdot}91 \text{ s} \quad \textbf{Ans.}$$

We know that the angular velocity of the block,

$$\omega = \frac{2\pi}{t} = \frac{2\pi}{0{\cdot}91} = 6{\cdot}9 \text{ rad/s}$$

∴ Maximum velocity,

$$v_{max} = \omega r = 6{\cdot}9 \times 0{\cdot}04 = 0{\cdot}276 \text{ m/s} \quad \textbf{Ans.}$$

and maximum acceleration, $$a_{max} = \omega^2 r = (6{\cdot}9)^2 \times 0{\cdot}04 = 1{\cdot}9 \text{ m/s}^2 \quad \textbf{Ans.}$$

(b) When the springs are connected in parallel

We know that spring constant of an equivalent spring,

$$s = s_1 + s_2 = 4 + 6 = 10 \text{ kN/m}$$

and deflection of the spring due to block of weight 0·49 kN

$$\delta = \frac{0.05 \times 9.8}{10} = 0{\cdot}049 \text{ m}$$

∴ Period of vibrations $\quad t = 2\pi \sqrt{\frac{\delta}{g}} = 2\pi \times \sqrt{\frac{0{\cdot}049}{9{\cdot}8}} = 0.44 \text{ s}$ **Ans.**

We know that angular velocity of the block,

$$\omega = \frac{2\pi}{t} = \frac{2\pi}{0{\cdot}44} = 14{\cdot}28 \text{ rad/s}$$

∴ Maximum velocity, $\quad v_{max} = \omega r = 14{\cdot}28 \times 0{\cdot}04 = 0{\cdot}57 \text{ m/s}$ **Ans.**

and maximum acceleration, $\quad = \omega^2 r = (14{\cdot}28)^2 \times 0{\cdot}04 = 8{\cdot}16 \text{ m/s}^2$ **Ans.**

***Example 7.** *A weight P is attached to springs of stiffness C_1 and C_2 in two different cases as shown in Fig. 4.*

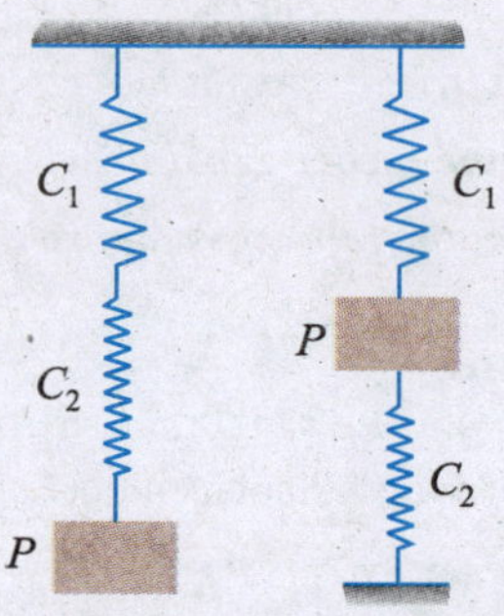

Fig. 4.

Determine the period of vibrations in both the cases.

Solution. Given : Weight $= P$

Period of vibrations in the first case

We know that in this case both the springs will be subjected to the weight P. Therefore total displacement of the spring

$$= \frac{P}{C_1} + \frac{P}{C_2} = P\left(\frac{1}{C_1} + \frac{1}{C_2}\right) = \frac{P(C_1 + C_2)}{C_1 C_2}$$

and period of vibration $\quad = 2\pi\sqrt{\frac{\text{Displacement}}{\text{Acceleration}}} = 2\pi\sqrt{\frac{P(C_1 + C_2)}{g\, C_1 C_2}}$ **Ans.**

Period of vibrations in the second case

We know that in this case, the upper spring will be subjected to tension, whereas the lower one will be subjected to compression.

Let P_1 = Weight shared by the upper spring, and

P_2 = Weight shared by the lower spring.

∴ Elongation of the upper spring,

$$\delta_1 = \frac{P_1}{C_1} \quad \text{...(i)}$$

and shortening of the lower spring,

$$\delta_2 = \frac{P_2}{C_2} \quad \text{...(ii)}$$

Since elongation of the upper spring is equal to shortening of the lower spring, therefore equating the values of δ_1 and δ_2 from the above equations.

$$\frac{P_1}{C_1} = \frac{P_2}{C_2}$$

$$\therefore \quad P_1\, C_2 = P_2\, C_1$$

$$(P - P_2)\, C_2 = P_2\, C_1 \quad \text{...}(\because P_1 + P_2 = P)$$

$$P\, C_2 - P_2\, C_2 = P_2 C_1$$

$$P_2\,(C_1 + C_2) = P\, C_2$$

or

$$\frac{P_2}{C_2} = \frac{P}{C_1 + C_2}$$

$$\therefore \text{ Displacement } = \frac{P}{C_1 + C_2} \quad \text{...}\left(\because \frac{P_2}{C_2} = \delta_2\right)$$

We know that the period of vibration,

$$= 2\pi\sqrt{\frac{\text{Displacement}}{\text{Acceleration}}} = 2\pi\sqrt{\frac{P}{g\,(C_1 + C_2)}} \quad \textbf{Ans.}$$

4. *SIMPLE PENDULUM

[AMIETE]

A simple pendulum, in its simplest form, consists of a heavy bob suspended at the end of a light inextensible, flexible string and the other end of the string is fixed at O as shown in Fig. 5.

Let l = Length of the string in metres, and

m = Mass of the bob in kg (such that its weight is $m.g$ newtons).

We know that the pendulum is in equilibrium, when the bob is at A. If the bob is brought to B or C and released, it will start vibrating between the positions B and C with A as the mean position.

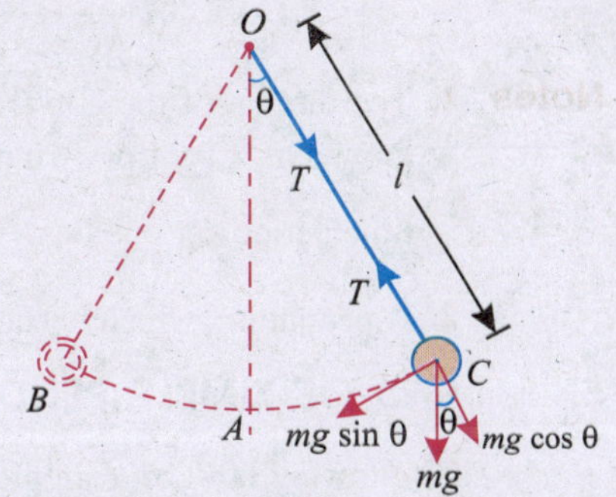

Fig. 5.

It has been observed that if the angle $\angle AOC$ is very small (*i.e.* less than 4°), the bob will move with simple harmonic motion.

Consider the equilibrium of the system at C. The weight mg of the bob can be resolved into two components *i.e.*, $mg \cos \theta$ and $mg \sin \theta$ at right angles to each other.

The component $mg \cos \theta$ will act along the thread. This will balance the tension in the string as shown in Fig. 5. The other component $mg \sin \theta$, being unbalanced, will give rise to an acceleration in the direction CA.

Let a = Acceleration caused by the component.

∴ Force responsible for this acceleration

$$= mg \sin \theta \qquad ...(i)$$

We also know that the force

$$= \text{Mass} \times \text{Acceleration} = m.a \qquad ...(ii)$$

Equating equations (*i*) and (*ii*),

$$ma = mg \sin \theta$$

$$\therefore \quad a = g \sin \theta$$

Since the angle θ is very small, therefore substituting $\sin \theta = \theta$ in the above equation,

$$a = g\,\theta \qquad ...(iii)$$

From the geometry of the figure, we know that

$$\theta = \frac{\text{Length of the arc}}{\text{Radius}} = \frac{AC}{l}$$

$$\therefore \quad a = g\,\theta = g \times \frac{AC}{l}$$

$$\text{or} \quad \frac{AC}{a} = \frac{l}{g} \qquad ...(\text{where } AC = \text{Displacement of the body})$$

We know that in a simple harmonic motion, the time period,

$$t = 2\pi \sqrt{\frac{\text{Displacement}}{\text{Acceleration}}} = 2\pi \sqrt{\frac{AC}{a}} = 2\pi \sqrt{\frac{l}{g}} \qquad ...\left(\because \frac{AC}{a} = \frac{l}{g}\right)$$

Notes 1. The motion of the bob from one extremity to the other (*i.e.* from B to C or C to B) is known as a beat or swing. It is thus obvious, that one beat $= \frac{1}{2}$ oscillation. Therefore time period for one beat $= \pi \sqrt{\frac{l}{g}}$

2. A pendulum, which executes one beat per second, is known as a *second's pendulum.*

5. LAWS OF SIMPLE PENDULUM

The following laws of a simple pendulum are important from the subject point of view :

1. ***Law of isochronism.*** It states, "*The time period* (t) *of simple pendulum does not depend on its amplitude of vibrations, and remain the same provided the angular amplitude* (θ) *does not exceed 4°.*"

2. **Law of mass.** It states, "*The time period (t) of a simple pendulum does not depend upon the mass of the body suspended at the free end of the string.*"
3. **Law of Length.** It states, "*The time period (t) of a simple pendulum is proportional to $\sqrt{l}$, where l is the length of the string.*"
4. **Law of gravity.** It states, "*The time period (t) of a simple pendulum is inversely proportional to $\sqrt{g}$. where g is the acceleration due to gravity.*"

Notes. The above laws of a simple pendulum are true from the equation of the time period

i.e.
$$t = 2\pi\sqrt{\frac{l}{g}}.$$

***Example 8.** *A simple pendulum of amplitude 4° performs 24 oscillations in one minute. Find (a) length of the pendulum (b) maximum acceleration of the bob, (c) maximum linear velocity of the bob; and (d) maximum angular velocity of the bob.*

Solution. Given : Angular amplitude $(\theta) = 4° = \frac{4\pi}{180}$ rad Frequency (n) = 24 Hz and time (t) = 1 min = 60 s.

(a) Length of the pendulum

Let l = Length of the pendulum

We know that time period for one oscillation,

$$t = \frac{60}{24} = 2{\cdot}5\,\text{s}$$

$$\therefore \quad 2{\cdot}5 = 2\pi\sqrt{\frac{l}{g}} = 2\pi\sqrt{\frac{l}{9{\cdot}8}}$$

Squaring both sides of the equation

$$6{\cdot}25 = (2\pi)^2\,\frac{l}{9{\cdot}8}$$

or
$$l = \frac{6{\cdot}25 \times 9{\cdot}8}{(2\pi)^2} = 1{\cdot}55 \text{ m} \quad \textbf{Ans.}$$

(b) Maximum linear acceleration of the bob

We know that maximum linear acceleration of the bob takes place, when it is at its extreme position (or in other words, the displacement is maximum). We also know that displacement of the bob (as per Fig. 6).

$$= \text{Arc } AC = OC \times \theta \text{ (in radians)} = 1.55 \times \frac{4\pi}{180} = 0{\cdot}108 \text{ m}$$

and angular velocity,
$$\omega = \frac{2\pi}{t} = \frac{2\pi}{2{\cdot}5} = 2{\cdot}51 \text{ rads/s}$$

∴ Maximum linear acceleration of the bob

$$a_{max} = \omega^2 \times AC = (2{\cdot}51)^2 \times 0{\cdot}108 = 0{\cdot}68 \text{ m/s}^2 \quad \textbf{Ans.}$$

(c) Maximum linear velocity of the bob

We know that maximum linear velocity of the bob,

$$v_{max} = \omega \times AC = 2{\cdot}51 \times 0{\cdot}108 = 0{\cdot}27 \text{ m/s} \quad \textbf{Ans.}$$

(*d*) *Maximum angular velocity of the bob*

We also know that maximum angular velocity of the bob,

$$= \frac{v_{max}}{l} = \frac{0{\cdot}27}{1{\cdot}55} = 0{\cdot}174 \text{ rad/s} \quad \textbf{Ans.}$$

***Example 9.** *A simple pendulum consists of a 600 mm long cord and a bob of mass 2 kg. Find the no. of oscillations made by the bob per second.*

If the same pendulum is suspended inside a train, accelerating smoothly on a level track at the rate of 3 m/s², find the angle which the cord will make with the vertical. Also find the tension in the cord.

Solution. Given: Length of cord (l) = 600 mm = 0·6 m ; Mass of bob (m) = 2 kg and acceleration (a) = 3 m/s²

Number of oscillations made by the bob per second

We know that time-period for one oscillation,

$$t = 2\pi\sqrt{\frac{l}{g}} = 2\pi\sqrt{\frac{0{\cdot}6}{9{\cdot}8}} = 1{\cdot}55 \text{ s}$$

and no. of oscillations made by the bob per second,

$$n = \frac{l}{t} = \frac{1}{1{\cdot}55} = 0{\cdot}65 \text{ Hz} \qquad \textbf{Ans.}$$

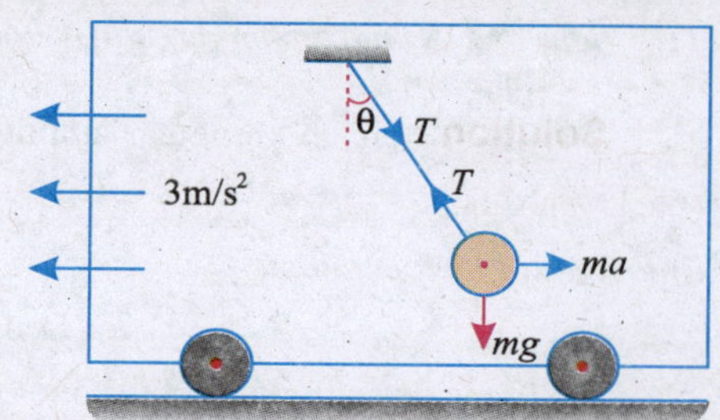

Fig. 6.

Angle, which the cord will make with the vertical

Let θ = Angle, which the cord will make with the vertical.

We know that weight of the bob (acting downwards)

$$W = mg = 2 \times 9{\cdot}8$$
$$= 19{\cdot}6 \text{ N}$$

and inertia force acting on the bob (opposite to the acceleration of the train)

$$F = ma = 2 \times 3$$
$$= 6 \text{ N}$$

$$\therefore \quad \tan\theta = \frac{F}{W} = \frac{6}{19{\cdot}6} = 0{\cdot}3061 \quad \text{or} \quad \theta = 17°$$

Tension in the cord

We know that tension in the cord,

$$T = \sqrt{(19{\cdot}6)^2 + (6)^2} = 20{\cdot}5 \text{ N} \quad \textbf{Ans.}$$

6. *GAIN OR LOSS IN THE NO. OF OSCILLATIONS DUE TO CHANGE IN THE LENGTH OF STRING OR ACCELERATION DUE TO GRAVITY OF A SIMPLE PENDULUM

Consider a simple pendulum, oscillating with a simple harmonic motion.

Let l = Length of the string,

t = Time period for one oscillation and

n = No. of beats or swings in time t.

We have seen in Article 4 of this chapter, that the time period for one beat.

$$= \pi \sqrt{\frac{l}{g}}$$

$\therefore$ Time-period for n beats $\quad t = n\pi \sqrt{\frac{l}{g}}$

or
$$n = \frac{t}{\pi \sqrt{\frac{l}{g}}} = \frac{t}{\pi} \sqrt{\frac{g}{l}} \qquad \text{...(i)}$$

Taking logs of both sides,

$$\log n = \log t - \log \pi + \frac{1}{2} (\log g - \log l)$$

Taking differential coefficients of variable (considering t and π as constants),

$$\frac{dn}{n} = \frac{1}{2}\left(\frac{dg}{g} - \frac{dl}{l}\right) = \frac{dg}{2g} - \frac{dl}{2l} \qquad \text{...(ii)}$$

where dn is the change in time in n seconds.

Now, if the gravity changes (keeping length of pendulum constant), then

$$\frac{dn}{n} = \frac{dg}{2g} \qquad \text{...}\left(\because \frac{dl}{l} = 0\right)$$

and if the length of pendulum changes (keeping g constant) then

$$\frac{dn}{n} = -\frac{dl}{2l} \qquad \text{...}\left(\because \frac{dg}{g} = 0\right)$$

Note. The minus sign indicates that as the value of dn/n increases, the value of dl/l decreases, and vice versa. Or in other words, if dn increases, dl decreases (keeping n and l constant) *i.e.*, if the no. of beats in a clock increases, it is due to decrease in length of the pendulum. Similarly, if the no. of beats decreases it means that it is due to increase in length of the pendulum.

Thus in order to set the pendulum correct, we have to reduce its length. Similarly, if the no. of beats increase (or the pendulum gains time), it is due to decrease in the length of the pendulum. Thus in order to set the pendulum correct, we have to increase its length.

***Example 10.** *The gravity at the poles exceeds the gravity at the equator in the ratio of 301 : 300. If a pendulum regulated at the poles is taken to the equator, find how many seconds a day will it lose ?*

Solution. Given : Gravity at pole (g_p) = 301 and gravity at the equator (g_e) = 300

$\therefore$ Change in gravity, when the pendulum is taken from the poles to the equator,

$$dg = 300 - 301 = -1$$

Let $\quad dn$ = No. of seconds the pendulum will lose in one day.

We know that no. of seconds in one day or 24 hours

$$n = 24 \times 60 \times 60 = 86\,400$$

We also know that $\frac{dn}{n} = \frac{dg}{2g}$

$$\frac{dn}{86\,400} = \frac{-1}{2 \times 301}$$

$$\therefore \qquad dn = -\frac{86\,400}{2 \times 301} = -143{\cdot}5 \text{ s} \quad \textbf{Ans.}$$

Minus sign means that the pendulum will lose 143·5 seconds per day.

***Example 11.** *Find the length of a pendulum, which will have one beat per second. If such a pendulum loses 5 seconds a day, by how much length must it be shortened to keep the correct time?*

Solution. Given : Time (t) = 1 s and no. of second the pendulum loses in one day (dn) = – 5 (Minus sign due is loss of seconds.)

Length of the pendulum which will have one beat per second

Let l = Length of the pendulum.

We know that time for one beat (t).

$$1 = \pi\sqrt{\frac{l}{g}} = \pi\sqrt{\frac{l}{9{\cdot}8}}$$

Squaring both sides, $1 = \pi^2 \times \frac{l}{9{\cdot}8}$

$$\therefore \qquad l = \frac{9{\cdot}8}{\pi^2} = 0{\cdot}993 \text{ m} = 993 \text{ mm} \quad \textbf{Ans.}$$

Length, by which the pendulum should by shortened

Let dl = Length in mm by which the pendulum should be shortened, to keep the correct time.

We know that no. of seconds in one day or 24 hours,

$$n = 24 \times 60 \times 60 = 86\,400$$

We also know that $\frac{dn}{n} = -\frac{dl}{2l}$

$$\frac{-5}{86\,400} = -\frac{dl}{2 \times 993}$$

$$\therefore \qquad dl = \frac{5 \times 2 \times 993}{86\,400} = 0{\cdot}115 \text{ mm} \quad \textbf{Ans.}$$

It means that the length of the pendulum is 0·115 mm more than the correct length. Thus, for correct time, the length of the pendulum should be shortened by 0·115 mm **Ans.**

7. *GAIN OR LOSS IN THE NO. OF OSCILLATIONS DUE TO CHANGE IN THE POSITION OF A SIMPLE PENDULUM

Consider a simple pendulum oscillating with a simple harmonic motion.

Let l = Length of the string,

t = Time period,

n = No. of beats in time t, and

r = Radius of the earth

We know that the value of g (*i.e.*, acceleration due to gravity) varies inversely as r^2. Mathematically :

$$g \propto \frac{1}{r^2}$$

$$\therefore \quad g = k \times \frac{1}{r^2} = \frac{k}{r^2} \quad \text{...(where } k \text{ is constant)}$$

Taking logs of both sides,

$$\log g = \log k - 2 \log r$$

Now taking differentials,

$$\frac{dg}{g} = -2\frac{dr}{r}$$

or
$$\frac{dg}{2g} = -\frac{dr}{r}$$

Now if h is height of the point where pendulum is placed above the earth's surface, then $dr = h$. Therefore

$$\frac{dg}{2g} = -\frac{h}{r}$$

We have studied in Article 6 of this chapter. that

$$\frac{dg}{2g} = \frac{dn}{n}$$

$$\therefore \quad \frac{dn}{n} = -\frac{h}{r}$$

Notes : 1. Minus sign indicates that as $\frac{dn}{n}$ increases, the value of $\frac{h}{r}$ decreases and *vice versa*.

2. Similarly, it can be proved that if the value of h is taken as negative (*i.e.*, the pendulum is taken below the surface of the earth as in the case of mines etc.) the value of dn/n will increase.

Example 12. *Find the approximate height of a mountain, at the top of which a pendulum, which beats seconds at sea level, loses 20 seconds a day. Take radius of the earth as 6400 km and the acceleration due to gravity varies as the square of the distance from the centre of the earth.*

Solution. Given : No. of seconds the pendulum loses in one day (dn) = – 20 (Minus sign due to loss of seconds) and radius of the earth (r) = 6400 km.

Let h = Height of mountain in km.

We know that no. of seconds in one day or 24 hours

$$n = 24 \times 60 \times 60 = 86\,400$$

We also know that
$$\frac{dn}{n} = -\frac{h}{r}$$

$$\frac{-20}{86\,400} = -\frac{h}{6400}$$

$$\therefore \qquad h = \frac{20 \times 6400}{86\,400} = 1{\cdot}48 \text{ km} \qquad \textbf{Ans.}$$

EXERCISE 2

1. A weight of 100 N is attached to two springs of stiffness 400 N/m and 500 N/m connected together vertically. Determine the period of oscillation of the weight. (**Ans.** 1.35 s)
2. A weight of 200 N is suspended from two springs arranged in parallel. Determine the periodic time of the weight, if the spring constants are 800 N/m and 1000 N/m respectively. Also determine angular velocity of the weight. (**Ans.** 0·67 s ; 9·38 rad/s)
3. A pendulum having 500 mm long string is carrying a bob of mass 100 gm. Find the time period of the pendulum. (**Ans.** 1·42 s)
4. In a laboratory, a spiral spring of stiffness 1·5 N/mm is available. Find the magnitude of the weight, which should be hung from the spring, so that it oscillates with a periodic time of 1 second. (**Ans.** 407 N)
5. Find the number of seconds a clock would gain per day, if the acceleration due to gravity is increased in the ratio of 900 : 901. (**Ans.** 48)
6. Calculate the number of beats lost per day by a seconds's pendulum, when its string is increased by 1/10 000th of its length. (**Ans.** – 4·32)
7. A simple pendulum gains 4 seconds per day. Determine the change in length of the pendulum in order to correct the time. (**Ans.** – 0·092 mm)
8. A second's pendulum is taken on a mountain 1200 metres high. Find the number of beats it will lose or gain per day on the mountain. Take radius of the earth as 6400 km. (**Ans.** – 16·2 beats)
9. A second's pendulum loses 10 seconds per day at the bottom of a mine. Find the depth of the mine. Take radius of the earth as 6400 km. (**Ans.** 740 m)

8. *COMPOUND PENDULUM

A compound pendulum, in its simplest form, consists of a rigid body suspended vertically at O and oscillating with a small amplitude under the action of the force of gravity. At some instant, let the position of the rigid body be at its extreme position as shown in Fig. 7.

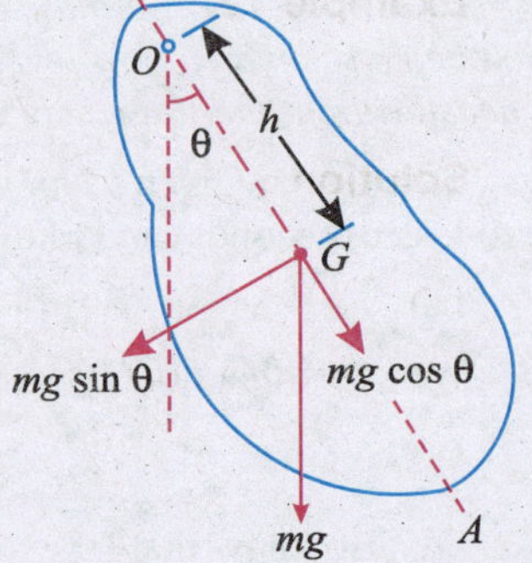

Fig. 7. Compound pendulum

Let $\quad m$ = mass of the pendulum

h = Distance between point of suspension (O) and the centre of gravity (G) of the body.

A little consideration will show, that if the pendulum is given a small angular displacement θ, then the moment of the couple tending to restore the pendulum in the equilibrium position OA

$$= mgh \sin\theta = mgh\theta \qquad ...(i)$$

...(Since θ is very small, therefore substituting $\sin\theta = \theta$)

Now let I_0 = Mass moment of inertia of the body about O

α = Angular acceleration of the body.

∴ Disturbing moment due to angular displacement

$$= I_0\,\alpha \qquad \text{...}(ii)$$

Equating two moments *i.e.* equations (*i*) and (*ii*),

$$mgh\,\theta = I_0\,\alpha$$

∴ $$\alpha = \frac{mgh\,\theta}{I_0}$$

We know that periodic time,

$$t = 2\pi\sqrt{\frac{\text{Displacement}}{\text{Acceleration}}} = 2\pi\sqrt{\frac{\theta}{\alpha}} = 2\pi\sqrt{\frac{\theta}{\dfrac{mgh\,\theta}{I_0}}} = 2\pi\sqrt{\frac{I_0}{mgh}} \qquad \text{...}(iii)$$

and frequency of motion, $$n = \frac{1}{t} = \frac{1}{2\pi}\sqrt{\frac{mgh}{I_0}} \qquad \text{...}(iv)$$

Note. The above formula for periodic time (or frequency of motion may also be expressed in terms of radius of gyration (k) as discussed below :

We know from the Theorem of Parallel Axis (studied in chapter of Moment of Inertia) that the mass moment of inertia of the pendulum about O,

$$I_0 = I_g + mh^2 = mk^2 + mh^2 = m\,(k^2 + h^2)$$

Now substituting the value of I_0 in equation (*iii*),

$$t = 2\pi\sqrt{\frac{m(k^2 + h^2)}{mgh}} = 2\pi\sqrt{\frac{k^2 + h^2}{gh}}$$

Similarly $$n = \frac{1}{t} = \frac{1}{2\pi}\sqrt{\frac{gh}{k^2 + h^2}}$$

Example 13. *A uniform straight rod of length 600 mm and mass 250 g is smoothly pivoted about a point, which is 40 mm from one end. Find the period of small oscillation about the pivot, if the rod can turn freely in the vertical plane.*

Solution. Given : Length of rod (l) = 600 mm = 0.6 m ; Mass of rod (m) =250 g = 0·25 kg. and distance between the point of suspension and the centre of gravity of the body (h) = 300 – 40 = 260 mm = 0·26 m

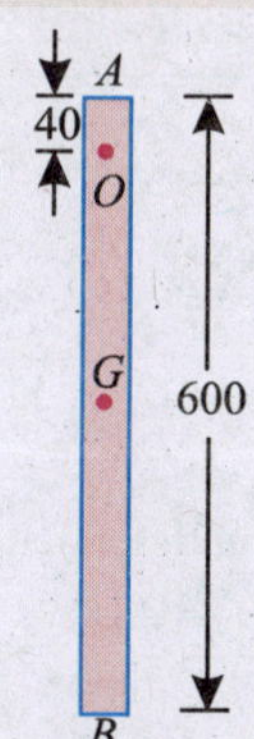

Fig. 8.

We know that mass moment of inertia of the rod about the pivot G,

$$I_0 = \frac{0{\cdot}25\,(0{\cdot}3)^2}{3} + 0{\cdot}25\,(0{\cdot}26)^2 = 0{\cdot}0244 \text{ kg-m}^2$$

and period of small oscillation, $$t = 2\pi\sqrt{\frac{I_0}{mgh}} = 2\pi\sqrt{\frac{0{\cdot}0244}{0{\cdot}25 \times 9{\cdot}8 \times 0{\cdot}26}}\ \text{s}$$

$$= 1{\cdot}23 \text{ s} \quad \textbf{Ans.}$$

***Example 14.** *A block whose weight = W is suspended by means of a spring whose spring constant = k_2 from the end of a rigid weightless beam, which is of length l and hinged to a wall by a frictionless connection at its left. It is held in a horizontal position by a spring whose spring constant = k_1. attached to it at a distance b from the hinge end to the ceiling as shown in Fig. 9.*

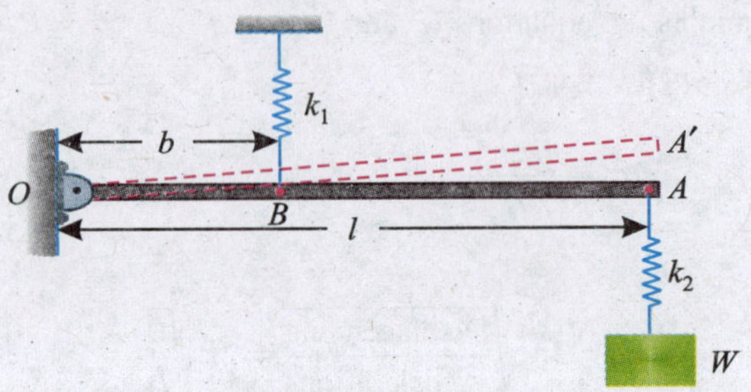

Fig. 9.

Prove that (n), the natural frequency of vibration of the system in Hz is given by :

$$n = \frac{1}{2\pi}\sqrt{\frac{gk_1k_2}{W[k_1 + (l/b)^2 k_2]}}$$

Solution. Fig. 9 shows *OA* the cantilever beam in the horizontal position with springs of constant k_1 and k_2 and weight *W* attached at *A*. A little consideration will show, that if the weight *W* is removed, the cantilever will spring upwards and *A* will occupy a new position *A′* as shown in the figure.

Let T = Tension in the spring k_1.

Taking moments about *O* and equating the same,

$$T \times b = W \times l$$

or

$$T = \frac{Wl}{b} \qquad ...(i)$$

∴ Deflection of spring k_1 (at *B*) due to tension *T*

$$= \frac{T}{k_1} = \frac{Wl}{k_1 b} \qquad ...(ii)$$

From the geometry of the figure, we find that

$$\frac{\text{Deflection at } B}{b} = \frac{\text{Deflection at } A}{l}$$

∴ Deflection at *A*

$$= \frac{l \times \text{Deflection at } B}{b} = \frac{\frac{lWl}{k_1 b}}{b} = \frac{W}{k_1}\left(\frac{l}{b}\right)^2$$

and deflection of spring k_2 (at *A*) due to weight *W*

$$= \frac{W}{k_2}$$

∴ Total deflection of cantilever at A,

$$\delta = \frac{W}{k_1}\left(\frac{l}{b}\right)^2 + \frac{W}{k_2} = W\left[\frac{1}{k_1}\left(\frac{l}{b}\right)^2 + \frac{1}{k_2}\right] = W\left[\frac{k_2\left(\frac{l}{b}\right)^2 + k_1}{k_1 k_2}\right]$$

and natural frequency of vibration of the system,

$$n = \frac{1}{2\pi}\sqrt{\frac{g}{\delta}} = \frac{1}{2\pi}\sqrt{\frac{g k_1 k_2}{W\left[k_1 + \left(\frac{l}{b}\right)^2 k_2\right]}} \text{ Hz} \quad \textbf{Ans.}$$

***Example 15.** *A uniform thin rod as shown in Fig. 10 has a mass of 1 kg and carries a concentrated mass of 2·5 kg at B. The rod is hinged at A, and is maintained in the horizontal position by a spring of stiffness 1·8 kN/m at C.*

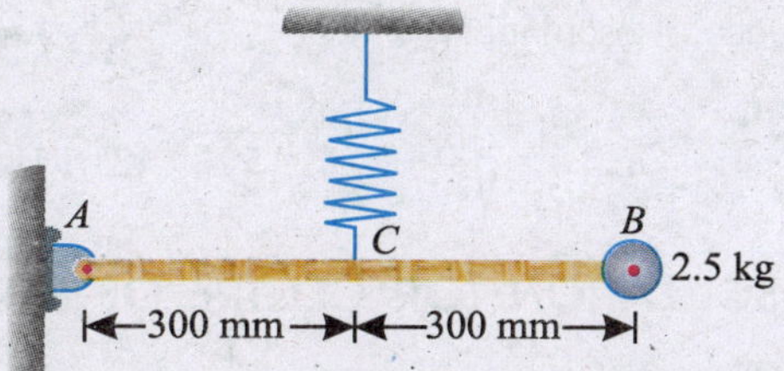

Fig. 10.

Find the frequency of oscillation. Neglect the effect of the mass of the spring.

Solution. Given : Mass of rod = 1 kg ; Mass at B = 2·5 kg ; Length of rod AB = 300 + 300 = 600 mm = 0·6 m ; Stiffness of spring (s) = 1·8 kN/ m = 1800 N /m.

Let n = Frequency of oscillation.

θ = Small angular displacement of the rod, and

α = Angular acceleration of the rod AB.

We know that mass moment of inertia of the system about A,

$$I_A = \text{M.I. of 1 kg about } A + \text{M.I. of 2·5 kg about } A$$

$$= \frac{1\,(0·6)^2}{3} + 2·5\,(0·6)^2 = 1·02 \text{ kg-m}^2$$

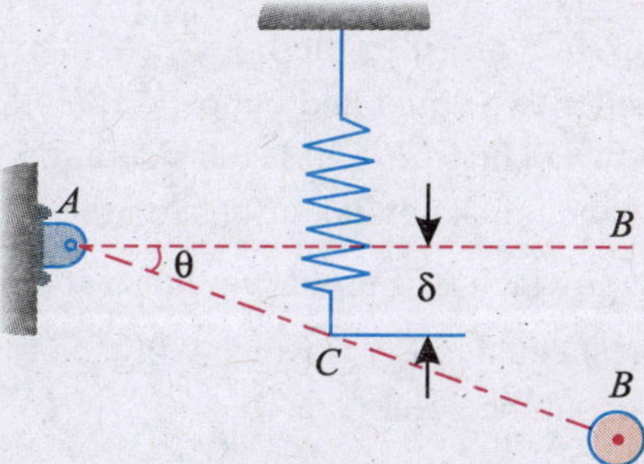

Fig. 11.

If the rod AB is given a small angular displacement as shown in Fig. 11 and then released, it will start oscillating about A with simple harmonic motion.

We know that extension of the spring,

$$\delta = 0{\cdot}3 \sin \theta = 0{\cdot}3 \times \theta \text{ m}$$

...($\because$ θ is very small, therefore substituting $\sin \theta = \theta$)

and restoring force $= s\,\delta = 1800 \times 0{\cdot}3\,\theta = 540\,\theta$ N

Therefore restoring moment about A

$$= 540\,\theta \times 0{\cdot}3 = 162\,\theta \text{ N-m} \quad ...(i)$$

and disturbing moment about A $= I_A\,\alpha = 1{\cdot}02\,\alpha$ N-m ...(*ii*)

Equating equations (*i*) and (*ii*)

$$1{\cdot}02\,\alpha = 162\,\theta$$

$$\therefore \quad \frac{\alpha}{\theta} = \frac{162}{1{\cdot}02} = 158{\cdot}8$$

We know that frequency of oscillation,

$$n = \frac{1}{2\pi}\sqrt{\frac{\alpha}{\theta}} = \frac{1}{2\pi}\sqrt{158{\cdot}8} = 2{\cdot}01 \text{ Hz} \quad \textbf{Ans.}$$

9. *CENTRE OF OSCILLATION (OR CENTRE OF PERCUSSION)

It has been experimentally found that whenever a suspended body is given a blow, it causes:

1. the body to oscillate about the point of suspension O.
2. an impulse on the body.

However, it has been experimentally found that there is always a point in the body, such that if a blow is given at that point, it will not produce any impulse on the body. Such a point is called *centre of oscillation* or *centre of percussion e.g.* in cricket, the batsman always intends to hit the ball at the centre of oscillation. But, if the ball is hit at some point, other than the centre of oscillation, the bat transmits a blow to the hands.

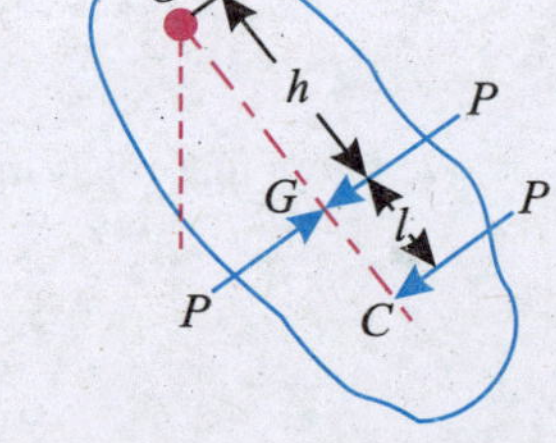

Fig. 12. Centre of oscillation.

The centre of oscillation (C) may also be defined as a point on the line joining the axis of suspension O and the centre of gravity (G) through which the resultant of the effective forces act as shown in Fig. 12.

Consider a force (*i.e.* blow) P given to the body at C *i.e.*, centre of oscillation. Now consider two equal and opposite forces equal to P to be acting at G as shown in Fig. 12. A little consideration will show, that the forces will have the following effects on the body :

1. A force (P) acting at G which will produce a linear motion with an acceleration a.
2. A couple (with moment equal to $P \times l$) which will tend to produce a motion of rotation in clockwise direction about the point G.

Now the force (P) acting at G will produce an acceleration (a) such that,

$$P = ma \qquad \text{or} \qquad a = \frac{P}{m} \qquad \ldots(i)$$

where m is the mass of the body.

Similarly, the turning moment ($P \times l$) will produce an angular acceleration (α) such that,

$$\alpha = \frac{P \times l}{I_G}$$

where I_G is the moment of inertia of the body about the axis parallel to the axis of rotation and passing through G. Now the corresponding linear acceleration of O,

$$a_1 = \alpha\, h = \frac{P\, l\, h}{I_G} \qquad \ldots\left(\because \alpha = \frac{Pl}{I_G}\right)$$

$$a_1 = \frac{P\, l\, h}{m\, k_G^{\,2}} \qquad \ldots(ii)$$

Now, if the axis through O is not to have any impulse, then a should be equal to a_1. Therefore equating (i) and (ii),

$$\frac{P}{m} = \frac{P\, l\, h}{m\, k_G^{\,2}}$$

$$\therefore \qquad k_G^{\,2} = l\, h \qquad \ldots(iii)$$

We know that

$$I_0 = I_G + m\, h^2$$

$$\therefore \qquad m\, k_0^{\,2} = m\, k_G^{\,2} + m\, h^2$$

$$\text{or} \qquad k_0^{\,2} = k_G^{\,2} + h^2$$

$$\therefore \qquad k_G^{\,2} = k_0^{\,2} - h^2 \qquad \ldots(iv)$$

Equating equations (iii) and (iv),

$$k_0^{\,2} - h^2 = l\, h$$

$$\therefore \qquad k_0^{\,2} = l\, h + h^2 = h\,(l + h) = OG \times OC$$

$$\text{or} \qquad OG = \frac{k_0^{\,2}}{OC}$$

It is thus obvious, that centres of suspension (O) and oscillation (C) are interchangeable. The distance between these two centres (*i.e.*, OC) known as the **length of equivalent simple pendulum**.

$$\therefore \qquad L = h + l = h + \frac{k_G^{\,2}}{h} = \frac{h^2 + k_G^{\,2}}{h}$$

Now time period in terms of the length of equivalent simple pendulum (L),

$$t = 2\pi\sqrt{\frac{L}{g}} = 2\pi\sqrt{\frac{h^2 + k_G^{\,2}}{gh}}$$

∴ Frequency of motion, $n = \frac{1}{t} = \frac{1}{2\pi}\sqrt{\frac{g}{L}} = \frac{1}{2\pi}\sqrt{\frac{gh}{h^2 + k_G^2}}$

***Example 16.** *A body of mass 0·5 kg oscillates about an axis at a distance 300 mm from the centre of gravity. If the mass moment of inertia about the centroidal axis, parallel to the axis of rotation, be 0·125 kg-m², find the length of the equivalent simple pendulum.*

Solution. Given : Mass of the body (m) = 0·5 kg ; Distance of Centre of oscillation from the centre of gravity (h) = 300 mm = 0·3 m and moment of inertia about centroidal axis (I_G) = 0·125 kg-m²

Let k_G = Radius of gyration about the centroidal axis.

We know that mass moment of inertia about the centroidal axis (I_G)

$$0{\cdot}125 = m\,k_G^2 = 0{\cdot}5\,k_G^2$$

$$\therefore \quad k_G^2 = \frac{0{\cdot}125}{0{\cdot}5} = 0{\cdot}25$$

We also know that the length of equivalent simple pendulum,

$$L = h + \frac{k_G^2}{h} = 0{\cdot}3 + \frac{0{\cdot}25}{0{\cdot}3} = 1{\cdot}133 \text{ m} \quad \textbf{Ans.}$$

***Example 17.** *The pendulum AB of an Izod impact testing machine makes 40 oscillations per minute about a knife edge through the hole A. The distance of centre of gravity from the centre of oscillation 450 mm. Find the radius of gyration of the pendulum about an axis through its centre of gravity and parallel to the knife edge.*

Also find the number of oscillations per minute, which the pendulum will make, if supported on a knife edge through another hole at B such that AB = 1·5 m. Take g as 9·81 m/s².

Solution. Given : No. of oscillations/ min = 40 ; Distance of centre of gravity from the centre of oscillation (l) = 450 mm = 0·45 m ; Length AB = 1·5 m and centre of gravity (g) = 9·81 m/s².

Radius of gyration

Let k_C = Radius of gyration about an axis through centre of gravity and parallel to the knife edge, and

L = Length of the equivalent simple pendulum.

We know that frequency of pendulum

$$n = \frac{40}{60} = \frac{2}{3}$$

$$\therefore \quad \text{Time period}\,(T) = \frac{3}{2} = 1{\cdot}5 \text{ s}$$

We also know that time period (T).

$$1{\cdot}5 = 2\pi\sqrt{\frac{L}{g}} = 2\pi \times \sqrt{\frac{L}{9{\cdot}81}}$$

Squaring both sides, $2{\cdot}25 = 4\pi^2 \times \frac{L}{9{\cdot}81} = 4{\cdot}024\,L$

$$\therefore \quad L = \frac{2{\cdot}25}{4{\cdot}024} = 0{\cdot}559 \text{ m}$$

and length of equivalent simple pendulum (L),

$$0{\cdot}559 = h + \frac{k_G^{\,2}}{h} = 0{\cdot}45 + \frac{k_G^{\,2}}{0{\cdot}45}$$

$\therefore$ $$k_G^{\,2} = (0{\cdot}559 - 0{\cdot}45) \times 0{\cdot}45 = 0{\cdot}049$$

or $$k_G = 0{\cdot}22 \text{ m} = 220 \text{ mm} \quad \textbf{Ans.}$$

No. of oscillations per minute

Given. Distance between A and B = 1·5 m

or $$h = AB - l = 1{\cdot}5 - 0{\cdot}45 = 1{\cdot}05 \text{ m}$$

We know that length of the equivalent simple pendulum,

$$L = h + \frac{k_G^{\,2}}{h} = 1{\cdot}05 + \frac{(0{\cdot}22)^2}{1{\cdot}05} = 1{\cdot}1 \text{ m}$$

and no. of oscillations per second of the pendulum when supported at B.

$$n_1 = \frac{1}{2\pi}\sqrt{\frac{g}{L}} = \frac{1}{2\pi}\sqrt{\frac{9{\cdot}81}{1{\cdot}1}} = 0{\cdot}475$$

$\therefore$ No. of oscillations per minute

$$= 0{\cdot}475 \times 60 = 28{\cdot}5 \quad \textbf{Ans.}$$

10. *CONICAL PENDULUM

A conical pendulum, in its simplest form, consists of a heavy bob suspended at the end of a light inextensible flexible string ; and the other end of the string is fixed at O as shown in Fig. 13. The pendulum bob rotates about the vertical axis with a uniform angular velocity. A little consideration will show, that the bob moves in a horizontal plane and describes a circle.

Let l = Length of the string,

ω = Angular velocity of the bob,

r = Radius of the horizontal circle described by the bob,

h = Vertical distance of the bob from O,

θ = Inclination of the string with the vertical

m = Mass of the bob in kg (such that its weight is $m.g.$ newtons).

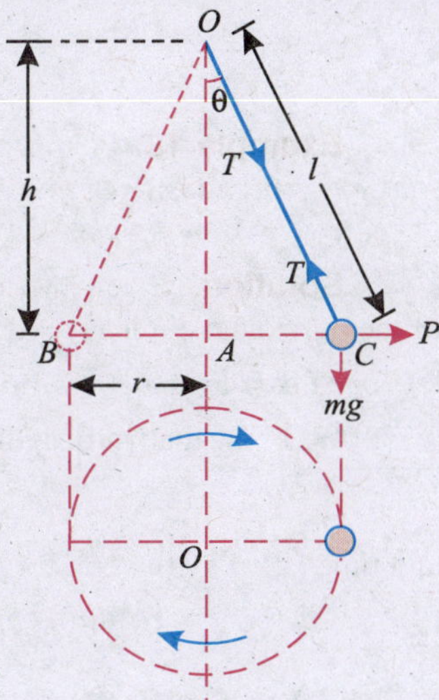

Fig. 13. Conical pendulum.

We know that the pendulum is in equilibrium, when the bob is at A. But when the bob is made to rotate, it will move outwards due to centrifugal force. Now consider equilibrium of the bob at C. We know that at this point, the forces acting on it are :

1. Weight (equal to mg) of the bob acting downwards in newtons.
2. Tension in the string (equal to T) in newtons.
3. Centrifugal force (P equal to $m\omega^2 r$) in newtons.

Resolving the forces horizontally and equating the same,

$$T \sin \theta = m\omega^2 r \qquad ...(i)$$

or
$$T = \frac{m\omega^2 r}{\sin \theta} = \frac{m\omega^2 r}{\left(\frac{r}{l}\right)} = m\omega^2 l \qquad ...(ii)$$

Now taking moments about O, and equating the same,

$$(m\omega^2 r)\, h = mgr$$

$$\therefore \qquad \omega^2 = \frac{g}{h}$$

We know that the periodic time (*i.e.* time taken by the bob for one revolution),

$$t = \frac{2\pi}{\omega} = 2\pi\sqrt{\frac{h}{g}}$$

and now resolving the forces vertically and equating the clockwise moments with anti-clockwise moments,

$$T \cos \theta = mg \qquad ...(iii)$$

or
$$T = \frac{mg}{\cos \theta} = \frac{mg}{\frac{h}{l}} = \frac{mgl}{h} \qquad ...(iv)$$

Now dividing equation (*i*) by (*iii*),

$$\frac{T \sin \theta}{T \cos \theta} = \frac{m\omega^2 r}{mg}$$

$$\therefore \qquad \tan \theta = \frac{\omega^2 r}{g} \qquad ...(v)$$

Example 18. *A sphere of 2 kg mass is attached to an inextensible string of length 1·3 m, whose upper end is fixed to the ceiling. The sphere is made to describe a horizontal circle of radius 0·5 m (i) Calculate the time taken by the bob for one revolution (ii) What is the tension in the string ?*

Solution. Given : Mass of the sphere (m) = 2 kg ; Length of the string (l) = 1·3 m and radius of the horizontal circle (r) = 0·5 m

(*i*) *Time taken by the bob for one revloution*

We know that vertical distance between the bob and O (*i.e* AO),

$$h = \sqrt{(1{\cdot}3)^2 - (0{\cdot}5)^2} = 1{\cdot}2 \text{ m}$$

and time taken by the bob for one revolution.

$$t = 2\pi\sqrt{\frac{h}{g}} = 2\pi\sqrt{\frac{1{\cdot}2}{9{\cdot}8}} \text{ s}$$

$$= 2{\cdot}2 \text{ s} \quad \textbf{Ans.}$$

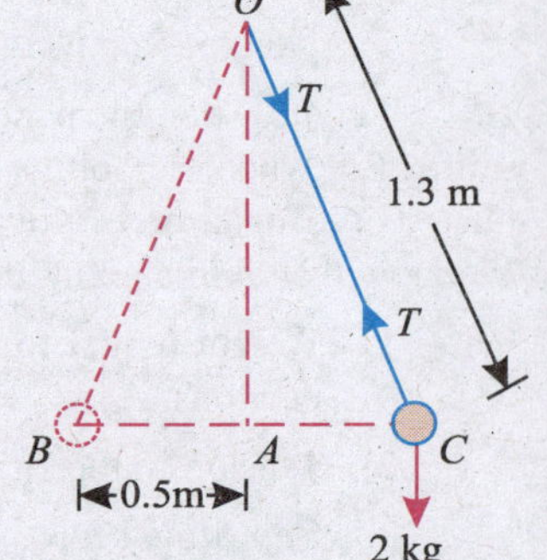

Fig. 14.

(*ii*) *Tension in the string*

We also know that tension in the string,

$$T = \frac{mgl}{h} = \frac{2 \times 9{\cdot}8 \times 1{\cdot}3}{1{\cdot}2} = 21.2 \text{ N} \quad \textbf{Ans.}$$

Example 19. *A conical pendulum 1·5 m long is revolving at 30 revolutions per minute. Find the angle which the string will make with the vertical, if the bob describes a circle of 500 mm radius.*

Solution. Given : Length of pendulum (l) = 1·5 m ; Angular speed of the pendulum (N) = 30 r.p.m and radius of the circle (r) = 500 mm = 0·5 m

Let θ = Angle which the string will make with the vertical.

We know that angular velocity of the bob,

$$\omega = \frac{2\pi N}{60} = \frac{2\pi \times 30}{60} = \pi \text{ rad/s}$$

$$\therefore \quad \tan\theta = \frac{\omega^2 r}{g} = \frac{\pi^2 \times 0{\cdot}5}{9{\cdot}8} = 0{\cdot}5036$$

or $\theta = 26.7°$ **Ans.**

EXERCISE 3

1. A uniform rod of mass 1 kg is 1 m long. The rod is pivoted about a point 100 mm from one end. Find the frequency of the rod about the pivot, when it turns freely in the vertical plane. (**Ans.** 0·64 Hz)
2. A body of mass 2·5 kg oscillates about an axis at a distance of 500 mm from its centre of gravity. Find the length of the equivalent simple pendulum, if the mass moment of inertia parallel to the axis rotation and about the centroidal axis is 0·4 kg-m^2. (**Ans.** 0·82 m)
3. A body of mass 1·5 kg is oscillating about an axis, which is at a distance of 0·5 m from the centre of gravity of the body. If the radius of gyration of the body about its centroidal axis is 0·8 m, find (*i*) mass moment of inertia of the body, (*ii*) length of the equivalent simple pendulum, and (*iii*) time period. (**Ans.** 0·96 kg-m^2 ; 1·78 m ; 2·68 s)
4. A conical pendulum consists of a 500 gm bob and 1 m long string. If the bob describes a horizontal circle of radius 250 mm, find the time taken by the bob for one revolution. (**Ans.** 1·97 s)
5. A conical pendulum 2 m long has a bob of 1 kg mass. If the bob describes a circle of 1 m radius at 45 r.p.m, find (*i*) the angle which the string will make with the vertical ; and (*ii*) tension in the string. (**Ans.** 68.2° ; 11.32 N)

QUESTIONS

1. Derive an expression for the period of oscillation of a weight, when attached to the helical spring.
2. What is a simple pendulum ? Under what conditions its motion is regarded as simple harmonic.
3. Differentiate the equation for the stiffness of two springs, when they are arranged in series and parallel.
4. How does the change in length or gravity affects the time period of a simple pendulum. Justify your answers with the help of derivations.
5. Obtain a relation for the change of no. of oscillations due to change in the position of a pendulum.
6. What is a compound pendulum ? Derive an expression for the time period of a compound pendulum.
7. What is meant by centre of oscillation ? Describe its importance.

8. Define the term 'length of equivalent simple pendulum'. Obtain an expression for the same.

9. What is a conical pendulum ? How does it differ from a simple pendulum ?

OBJECTIVE TYPE QUESTIONS

1. One end of a helical spring is fixed while the other carries a load W. The weight moves with simple harmonic motion, whose frequency of motion is

(*a*) $2\pi\sqrt{\dfrac{g}{\delta}}$ (*b*) $2\pi\sqrt{\dfrac{\delta}{g}}$ (*c*) $\dfrac{1}{2\pi}\sqrt{\dfrac{g}{\delta}}$ (*d*) $\dfrac{1}{2\pi}\sqrt{\dfrac{\delta}{g}}$

2. The periodic time of one oscillation for a simple pendulum is

(*a*) $\dfrac{1}{2\pi}\sqrt{\dfrac{l}{g}}$ (*b*) $\dfrac{1}{2\pi}\sqrt{\dfrac{g}{l}}$ (*c*) $2\pi\sqrt{\dfrac{l}{g}}$ (*d*) $2\pi\sqrt{\dfrac{g}{l}}$

3. In order to double the period of simple pendulum, the length of the string should be

(*a*) halved (*b*) double (*c*) quadrupled

4. Which of the following statement is wrong ?

(*a*) If two springs of stiffness s_1 and s_2 are arranged in series, then stiffness of the equiva lent spring is $s_1 + s_2$.

(*b*) The motion of a body from one extremity to another is known as beat.

(*c*) A pendulum, which executes one beat per second is known as second's pendulum.

(*d*) none of them.

5. If a simple pendulum is taken 1 km below the earth's surface in a mine, it will lose in time.

(*a*) Yes (*b*) No

6. The centre of percussion is below the centre of gravity of the body is at a distance equal to

(*a*) $\dfrac{k_G}{h^2}$ (*b*) $\dfrac{{k_G}^2}{h}$ (*c*) $\dfrac{h}{k_G}$ (*d*) $\dfrac{h^2}{k_G}$

7. The periodic time of a conical pendulum is

(*a*) $2\pi\sqrt{\dfrac{h}{g}}$ (*b*) $2\pi\sqrt{\dfrac{g}{h}}$ (*c*) $\dfrac{1}{2\pi}\sqrt{\dfrac{h}{g}}$ (*d*) $\dfrac{1}{2\pi}\sqrt{\dfrac{g}{h}}$

ANSWERS

1. (*c*) **2.** (*c*) **3.** (*c*) **4.** (*a*) **5.** (*b*) **6.** (*b*) **7.** (*a*)

CHAPTER

25 Collision of Elastic Bodies

Contents

1. ELASTICITY AND ELASTIC BODIES

We daily see children playing with balls of glass, rubber and other elastic materials. If we see them carefully, we will find that if a glass ball is allowed to fall on a marble floor, from a certain height, it rebounds (*i.e.* rises up, after striking the floor) to a certain height. This height is, generally, less than the height from which the ball is allowed to fall. If the same ball is allowed to fall on a wooden floor, then it will rebound to a lesser height. We will also see that if the balls of different materials are allowed to fall on a marble floor, they will rebound to different heights.

This property of bodies, by virtue of which, they rebound, after impact, is called **elasticity**. It may be noted that a body, which rebounds to a greater height is said to be more elastic, than that which rebounds to a lesser height. But, if a body does not rebound at all, after its impact, it is called an *inelastic* body.

2. COLLISION PHENOMENON

Whenever two elastic bodies collide with each other, the phenomenon of collision takes place as given below :

1. The bodies, immediately after collision, come momentarily to rest.
2. The two bodies tend to compress each other, so long as they are compressed to the maximum value.
3. The two bodies attempt to regain its original shape due to their elasticity. This process of regaining the original shape is called restitution.

The time taken by the two bodies in compression, after the instant of collision, is called the time of compression and time for which restitution takes place is called the time of restitution. The sum of the two times of collision and restitution is called time of collision, period of collision, or period of impact.

3. *LAW OF CONSERVATION OF MOMENTUM FOR COLLISION

It* states, "*The total momentum of two bodies remains constant after their collision or any other mutual action.*" Mathematically

$$m_1u_1 + m_2u_2 = m_1v_1 + m_2v_2$$

where m_1 = Mass of the first body,

u_1 = Initial velocity of the first body,

v_1 = Final velocity of the first body, and

m_2, u_2, v_2 = Corresponding values for the second body.

*Note: Mathematically this law may be explained as stated below :

Initial momentum of the first body

$= \text{Mass} \times \text{Velocity} = m_1u_1$

Similarly, initial momentum of the second body

$= m_2u_2$

∴ Total initial momentum $= m_1u_1 + m_2u_2$

Similarly, total final momentum

$= m_1v_1 + m_2v_2$

∴ According to the law of conservation of momentum

$$m_1u_1 + m_2u_2 = m_1v_1 + m_2v$$

4. NEWTON'S LAW OF COLLISION OF ELASTIC BODIES

It states, "*When two moving bodies collide with each other, their velocity of separation bears a constant ratio to their velocity of approach.*" Mathematically,

$$(v_2 - v_1) = e\,(u_1 - u_2)$$

where v_1 = Final velocity of the first body,

u_1 = Initial velocity of the first body,

v_2, u_2 = Corresponding values for the second body, and

e = Constant of proportionality.

5. COEFFICIENT OF RESTITUTION

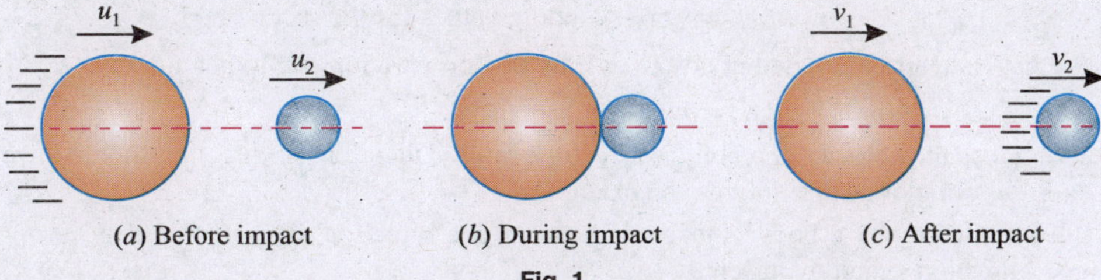

Fig. 1.

Consider two bodies A and B having a direct impact as shown in Fig. 1 (a).

Let u_1 = Initial velocity of the first body,

v_1 = Final velocity of the first body, and

u_2, v_2 = Corresponding values for the second body.

A little consideration will show, that the impact will take place only if u_1 is greater than u_2. Therefore, the velocity of approach will be equal to $(u_1 - u_2)$. After impact, the separation of the two bodies will take place, only if v_2 is greater than v_1. Therefore the velocity of separation will be equal to $(v_2 - v_1)$. Now as per Newton's Law of Collision of Elastic Bodies :

Velocity of separation = $e \times$ Velocity of approach

$$(v_2 - v_1) = e\,(u_1 - u_2)$$

where e is a constant of proportionality, and is called the *coefficient of restitution.* Its value lies between 0 and 1. It may be noted that if $e = 0$, the two bodies are inelastic. But if $e = 1$, the two bodies are perfectly elastic.

Notes : **1.** If the two bodies are moving in the same direction, before or after impact, then the velocity of approach or separation is the difference of their velocities. But if the two bodies are moving in the opposite directions, then the velocity of approach or separation is the algebraic sum of their velocities.

2. The above formula holds good under the assumed conditions (*i.e.* $u_1 > u_2$ and $v_2 > v_1$). But if the above assumptions do not hold good, in an example, then the formula may be adjusted accordingly, to keep both the sides of the equation as positive.

6. *TYPES OF COLLISIONS

When two bodies collide with one another, they are said to have an impact. Following are the two types of impacts.

1. Direct impact, and 2. Indirect (or oblique) impact.

7. DIRECT COLLISION OF TWO BODIES

The line of impact, of the two colliding bodies, is the lien joining the centres of these bodies and passes through the point of contact or point of collision as shown in Fig. 2.

If the two bodies, before impact, are moving along the line of impact, the collision is called as *direct impact* as shown in Fig. 2.

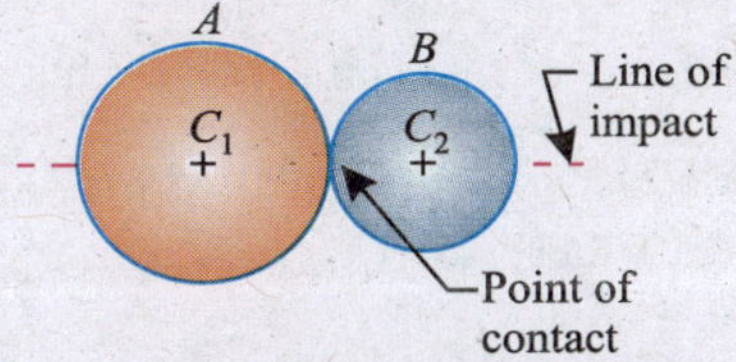

Fig. 2. Direct impact

Now consider the two bodies A and B having a direct impact as shown in Fig. 2.

Let m_1 = Mass of the first body,

u_1 = Initial velocity of the first body,

v_1 = Final velocity of the first body and

m_2, u_2, v_2 = Corresponding values for the second body.

We have already discussed in article of Law of Conservation of Momentum that

$$m_1u_1 + m_2u_2 = m_1v_1 + m_2v_2$$

Notes. 1. Since the velocity of a body is a vector quantity, therefore its direction should always be kept in view while solving the examples.

2. If velocity of a body is taken as + ve in one direction, then the velocity in opposite direction should be taken as – ve.

3. If one of the body is initially at rest, then such a collision is also called impact.

***Example 1.** *A ball of mass 1 kg moving with a velocity of 2 m/s impinges directly on a ball of mass 2 kg at rest. The first ball, after impinging, comes to rest. Find the velocity of the second ball after the impact and the coefficient of restitution.* **[IGNOU]**

Solution. Given : Mass of first ball (m_1) = 1 kg ; Initial velocity of first ball (u_1) = 2 m/s ; Mass of second ball (m_2) = 2 kg ; Initial velocity of second ball (u_2) = 0 (because it is at rest) and final velocity of first ball after impact (v_1) = 0 (because, it comes to rest)

Velocity of the second ball after impact.

Let v_2 = Velocity of the second ball after impact.

We know from the law of conservation of momentum that

$$m_1u_1 + m_2u_2 = m_1v_1 + m_2v_2$$

$$(1 \times 2) + (2 \times 0) = (1 \times 0) + (2 \times v_2)$$

$$\therefore \quad 2 = 2v_2$$

or $$v_2 = 1 \text{ m/s} \quad \textbf{Ans.}$$

Coefficient of restitution

Let e = Coefficient of restitution.

We also know from the law of collision of elastic bodies that

$$(v_2 - v_1) = e\,(u_1 - u_2)$$

$$(1 - 0) = e\,(2 - 0)$$

or $$e = \frac{1}{2} = 0.5 \quad \textbf{Ans.}$$

***Example 2.** *A ball overtakes another ball of twice its own mass and moving with 1/7 of its own velocity. If coefficient of restitution between the two balls is 0.75, show that the first ball will come to rest after impact.*

Solution. Given : Mass of first ball (m_1) = M kg ; Mass of second ball (m_2) = 2 M ; Initial velocity of first ball (u_1) = U ; Initial velocity of second ball (u_2) = $\frac{U}{7}$ and coefficient of restitution (e) = 0.75

Let v_1 = Velocity of the first ball after impact, and

v_2 = Velocity of the second ball after impact.

We know from the law of conservation of momentum that

$$m_1u_1 + m_2u_2 = m_1v_1 + m_2v_2$$

$$MU + \frac{2MU}{7} = Mv_1 + 2Mv_2$$

$$\frac{9MU}{7} = Mv_1 + 2Mv_2 \quad \text{or} \quad \frac{9U}{7} = v_1 + 2v_2 \qquad ...(i)$$

We also know from the law of collision of elastic bodies that

$$(v_2 - v_1) = e\,(u_1 - u_2) = 0.75\left(U - \frac{U}{7}\right) = \frac{9U}{14}$$

or $$v_2 = \frac{9U}{14} + v_1$$

Substituting this value of v_2 in equation (*i*),

$$\frac{9U}{7} = v_1 + 2\left(\frac{9U}{14} + v_1\right) = 3v_1 + \frac{9U}{7} \quad \text{or} \quad v_1 = 0$$

Thus the first ball will come to rest after impact. **Ans.**

***Example 3.** *The masses of two balls are in the ratio of 2 : 1 and their velocities are in the ratio of 1 : 2, but in the opposite direction before impact. If the coefficient of restitution be 5/6, prove that after the impact, each ball will move back with 5/6th of its original velocity.* **[AMIE]**

Solution. Given : Mass of first ball $(m_1) = 2\,M$; Mass of second ball $(M_2) = M$; Initial velocity of first ball $(u_1) = U$; Initial velocity of second ball $(u_2) = -\,2U$ (Minus sign due to opposite direction) and coefficient of restitution $(e) = \frac{5}{6}$

Let v_1 = Final velocity of the first ball, and

v_2 = Final velocity of the second ball.

We know from the law of conservation of momentum that

$$m_1u_1 + m_2u_2 = m_1v_1 + m_2v_2$$

$$2M \times U + M\,(-\,2U) = 2Mv_1 + Mv_2$$

or $$0 = 2Mv_1 + Mv_2$$

∴ $$v_2 = -\,2v_1 \qquad ...(i)$$

We also know from the law of collision of elastic bodies that

$$(v_2 - v_1) = e\,(u_1 - u_2) = \frac{5}{6}\left[U - (-2U)\right] = \frac{5U}{2} \qquad ...(ii)$$

Substituting the value of v_2 from equation (*i*)

$$\left[-2v_1 - (v_1)\right] = \frac{5U}{2} \quad \text{or} \quad v_1 = -\frac{5}{6} \times U$$

Minus sign indicates that the direction of v_1 is opposite to that of U. Thus the first ball will move back with $\frac{5}{6}$th of its original velocity. **Ans.**

Now substituting the value of v_1 in equation (*i*),

$$v_2 = -\,2\left(-\frac{5}{6} \times U\right) = +\frac{5}{6} \times 2U$$

Plus sign indicates that the direction of v_2 is the same as that of v_1 or opposite to that of u_2. Thus the second ball will also move back with $\frac{5}{6}$th of its original velocity. **Ans.**

***Example 4.** *Three perfectly elastic balls A, B and C of masses 2 kg, 4 kg and 8 kg move in the same direction with velocities of 4 m/s, 1m/s and 0.75 m/s respectively. If the ball A impinges with the ball B, which in turn, impinges with the ball C, prove that the balls A and B will be brought to rest by the impacts.* **[RTU]**

Solution. Given : Coefficient of restitution (e) = 1(because the balls are perfectly elastic) ; Mass of ball A (m_1) = 2 kg ; mass of ball B (m_2) = 4 kg ; Mass of ball C (m_3) = 8 kg ; Initial velocity of ball A (u_1) = 4 m/s ; Initial velocity of ball B (u_2) = 1 m/s and initial velocity of ball C (u_3) = 0.75 m/s

Final velocity of the first ball after impact

First of all, consider the impact of the first and second ball.

Let v_1 = Final velocity of the first ball after impact, and

v_2 = Final velocity of the second ball after impact.

We know from the law of conservation of momentum that

$$m_1u_1 + m_2u_2 = m_1v_1 + m_2v_2$$

$$(2 \times 4) + (4 \times 1) = 2 \times v_1 + 4 \times v_2$$

$$2v_1 + 4v_2 = 12 \quad \text{or} \quad v_1 + 2v_2 = 6 \quad \text{...(i)}$$

We also know from the law of collision of elastic bodies that

$$(v_2 - v_1) = e\,(u_1 - u_2) = 1\,(4 - 1) = 3 \quad \text{...(ii)}$$

Adding equations (*ii*) and (*i*),

$$3v_2 = 9 \quad \text{or} \quad v_2 = 3 \text{ m/s}$$

Substituting the value of v_2 in equation (*ii*),

$$3 - v_1 = 3 \quad \text{or} \quad v_1 = 0$$

Thus the first ball will be brought to rest by the impact of first and second ball. **Ans.**

Final velocity of the second ball

Now consider the impact of second and third ball. In this case $u_2 = v_2 = 3$ m/s

Let v_2 = Final velocity of the second ball, after the impact of second and third ball, and

v_3 = Final velocity of the third ball after impact.

We know from the law of conservation of momentum that

$$m_2u_2 + m_3u_3 = m_2v_2 + m_3v_3$$

$$(4 \times 3) + (8 \times 0.75) = 4 \times v_2 + 8 \times v_3$$

$$4v_2 + 8v_3 = 18 \quad \text{or} \quad 2v_2 + 4v_3 = 9 \quad \text{...(iii)}$$

We also know from the law of collision of elastic bodies that

$$(v_3 - v_2) = e\,(u_2 - u_3) = 1\,(3 - 0.75) = 2.25$$

Multiplying the above equation by 4,

$$\therefore \quad 4v_3 - 4v_2 = 9 \quad \text{...(iv)}$$

Subtracting equation (*iv*) from (*iii*),

$$6v_2 = 0 \quad \text{or} \quad v_2 = 0$$

Hence the second ball will also be brought to rest by the impact of second and third ball. **Ans.**

8. *LOSS OF KINETIC ENERGY DURING COLLISION

The kinetic energy may be broadly defined as the energy possessed by a body by virtue of its mass and velocity. Mathematically kinetic energy,

$$E = \frac{1}{2} mv^2$$

where m = Mass of the body, and

v = Velocity of the body,

The loss of kinetic energy, during impact, may be obtained by finding out the kinetic energy of the two bodies before and after the impact. The difference between the kinetic energies of the system, gives the required loss of kinetic energy during impact. Consider two bodies A and B having a direct impact.

Let m_1 = Mass of the first body,

u_1 = Initial velocity of the first body,

v_1 = Final velocity of the first body,

m_2, u_2, v_2 = Corresponding values for the second body,

e = Coefficient of restitution.

We know that kinetic energy of the first body, before impact

$$= \frac{1}{2} m_1 u_1^2$$

and kinetic energy of the second body, before impact,

$$= \frac{1}{2} m_2 u_2^2$$

∴ Total kinetic energy of the two bodies, before impact,

$$E_1 = \frac{1}{2} m_1 u_1^2 + \frac{1}{2} m_2 u_2^2 = \frac{1}{2}\left(m_1 u_1^2 + m_2 u_2^2\right) \qquad ...(i)$$

Similarly, total kinetic energy of two bodies, after impact

$$E_2 = \frac{1}{2} m_1 v_1^2 + \frac{1}{2} m_2 v_2^2 = \frac{1}{2}\left(m_1 v_1^2 + m_2 v_2^2\right) \qquad ...(ii)$$

∴ Loss of kinetic energy, during impact

$$E_L = E_1 - E_2 = \frac{1}{2}\left[\left(m_1u_1^2 + m_2u_2^2\right) - \left(m_1v_1^2 + m_2v_2^2\right)\right]$$

Multiplying the numerator and denominator of the right hand side by $(m_1 + m_2)$,

$$E_L = \frac{1}{2(m_1 + m_2)}\left[(m_1 + m_2)\left(m_1u_1^2 + m_2u_2^2\right) - (m_1 + m_2)\left(m_1v_1^2 + m_2v_2^2\right)\right]$$

$$= \frac{1}{2(m_1 + m_2)}\left[(m_1^2u_1^2 + m_1m_2u_2^2 + m_1m_2u_1^2 + m_2^2u_2^2) - (m_1^2v_1^2 + m_1m_2v_2^2 + m_2m_1v_1^2 + m_2^2v_2^2)\right]$$

$$= \frac{1}{2(m_1 + m_2)}\left[\left\{(m_1^2u_1^2 + m_2^2u_2^2 + m_1m_2(u_1^2 + u_2^2)\right\} - \left\{(m_1^2v_1^2 + m_2^2v_2^2 + m_1m_2(v_1^2 + v_2^2)\right\}\right]$$

$$= \frac{1}{2(m_1 + m_2)}\left[\left\{(m_1u_1 + m_2u_2)^2 - (2m_1m_2u_1u_2) + m_1m_2(u_1 - u_2)^2 + (2m_1m_2u_1u_2)\right\} - \left\{(m_1v_1 + m_2v_2)^2 - (2m_1m_2v_1v_2) + m_1m_2(v_1 - v_2)^2 + (2m_1m_2v_1v_2)\right\}\right]$$

$$= \frac{1}{2(m_1 + m_2)}\left[\left\{(m_1u_1 + m_2u_2)^2 + m_1m_2(u_1 - u_2)^2\right\} - \{(m_1v_1 + m_2v_2)^2 + m_1m_2(v_1 - v_2)^2\}\right]$$

We know that in a direct impact,

Initial momentum = Final momentum

or $\qquad m_1u_1 + m_2u_2 = m_1v_1 + m_2v_2$

or $\qquad (m_1u_1 + m_2u_2)^2 = (m_1v_1 + m_2v_2)^2 \qquad$ ...(Squaring both sides)

Therefore loss of kinetic energy due to impact,

$$E_L = \frac{1}{2(m_1 + m_2)}\left[m_1m_2(u_1 - u_2)^2 - m_1m_2(v_1 - v_2)^2\right]$$

Now substituting $(v_1 - v_2) = e(u_1 - u_2)$ in the above equation,

$$E_L = \frac{1}{2(m_1 + m_2)}\left[m_1m_2(u_1 - u_2)^2 - m_1m_2e^2(u_1 - u_2)^2\right]$$

$$= \frac{m_1m_2}{2(m_1 + m_2)}(u_1 - u_2)^2\left(1 - e^2\right)$$

Note. The loss of kinetic energy may also be found out by calculating the kinetic energy of the system before impact, and then by subtracting from it the kinetic energy of the system after impact.

***Example 5.** *A ball impinges directly on a similar ball at rest. The first ball is reduced to rest by the impact. Find the coefficient of restitution, if half of the initial kinetic energy is lost by impact.*

Solution. Given : Initial velocity of second body $(u_2) = 0$ (because it is at rest) and final velocity of the first body $(v_1) = 0$ (because it comes to rest by the impact)

Let m_1 = Mass of the first body,

$m_2 = m_1$ = Mass of the second body,

...($\because$ both the balls are similar)

u_1 = Initial velocity of the first body,

v_2 = Final velocity of the second body, and

e = Coefficient of restitution.

We know that kinetic energy of the system before impact,

$$E_1 = \frac{1}{2} m_1 u_1^2 + \frac{1}{2} m_2 u_2^2 = \frac{1}{2} m_1 u_1^2 \qquad ...(\because u_2 = 0)$$

and kinetic energy of the system after impact,

$$E_2 = \frac{1}{2} m_1 v_1^2 + \frac{1}{2} m_2 v_2^2 = \frac{1}{2} m_1 v_2^2 \qquad ...(\because v_1 = 0)$$

$\therefore$ Loss of kinetic energy during impact

$$E_L = E_1 - E_2 = \left(\frac{1}{2} m_1 u_1^2\right) - \left(\frac{1}{2} m_1 v_2^2\right)$$

Since half of the initial K.E. is equal to loss of K.E. by impact, therefore

$$\therefore \quad \frac{1}{2}\left(\frac{1}{2} m_1 u_1^2\right) = \left(\frac{1}{2} m_1 u_1^2\right) - \left(\frac{1}{2} m_1 v_2^2\right)$$

$$\frac{m_1 u_1^2}{2} = m_1 u_1^2 - m_1 v_2^2$$

$$\therefore \quad m_1 u_1^2 = 2\, m_1 v_2^2$$

or
$$u_1^2 = 2\, v_2^2 \qquad ...(i)$$

We know from the law of conservation of elastic bodies that

$$(v_2 - v_1) = e\,(u_1 - u_2)$$

$$v_2 - 0 = e\,(u_1 - 0) \qquad ...(\because v_1 = 0 \text{ and } u_2 = 0)$$

$$\therefore \quad v_2 = e u_1 \qquad ...(ii)$$

Substituting the value of v_2 in equation (*i*),

$$u_1^2 = 2\left(e u_1\right)^2 = 2e^2 u_1^2$$

or
$$e^2 = \tfrac{1}{2} \quad \text{or} \quad e = 0.707 \quad \textbf{Ans.}$$

***Example 6.** *A sphere of mass 1 kg, moving at 3 m/s, overtakes another sphere of mass 5 kg moving in the same line at 60 cm/s. Find the loss of kinetic energy during impact, and show that the direction of motion of the first sphere is reversed. Take coefficient of restitution as 0.75.*

[AM-IETE, JNTU]

Solution. Given : Mass of first sphere $(m_1) = 1$ kg; Initial velocity of first sphere $(u_1) = 3$ m/s; Mass of second sphere $(m_2) = 5$ kg ; Initial velocity of second sphere $(u_2) = 60$ cm/s $= 0.6$ m/s and coefficient of restitution $(e) = 0.75$.

Loss of kinetic energy during impact

We know that loss of kinetic energy during impact,

$$E_L = \frac{m_1 m_2}{2(m_1 + m_2)} (u_1 - u_2)^2 (1 - e^2)$$

$$= \frac{1 \times 5}{2(1 + 5)} (3 - 0.6)^2 (1 - 0.75^2) \text{ N-m}$$

$$= \frac{5}{12} (2.4)^2 \times 0.4375 = 1.05 \text{ kg-m} = 1.05 \text{ J} \quad \textbf{Ans.}$$

Final velocity of the first sphere0

Let v_1 = Final velocity of the first sphere, and

v_2 = Final velocity of the second sphere.

We know from the law of conservation of momentum that

$$m_1 u_1 + m_2 u_2 = m_1 v_1 + m_2 v_2$$

$$(1 \times 3) + (5 \times 0.6) = 1 \times v_1 + 5 \times v_2$$

$$\therefore \quad v_1 + 5v_2 = 6 \quad \text{...}(i)$$

We also know from the law of collision of elastic bodies that

$$(v_2 - v_1) = e\,(u_1 - u_2) = 0.75\,(3 - 0.6) = 0.75 \times 2.4$$

or $$v_2 - v_1 = 1.8 \quad \text{...}(ii)$$

Adding equations (*i*) and (*ii*),

$$6v_2 = 7.8$$

$$\therefore \quad v_2 = \frac{7.8}{6} = 1.3 \text{ m/s}$$

Substituting this value of v_2 in equation (*i*),

$$v_1 + (5 \times 1.3) = 6$$

or $$v_1 = 6 - (5 \times 1.3) = -0.5 \text{ m/s}$$

Minus sign indicates, that direction of motion of the first body is reversed after impact. **Ans.**

EXERCISE 1

1. A ball of mass 2 kg impinges directly with a ball of mass 1 kg, which is at rest. If the velocity of the smaller mass after impact, be the same as that of the first ball before impact, find the coefficient of restitution. (**Ans.** 0.5)
2. Two balls of masses 2 kg and 3 kg are moving with velocities 2 m/s and 3 m/s towards each other. If the coefficient of restitution is 0.5, find the velocity of the two balls after impact. (**Ans.** 2.5 m/s ; 0)
3. Three spheres of masses 2 kg, 6 kg and 12 kg are moving with velocities of 12 m/s, 4 m/s and 2 m/s respectively in a straight line. Show that after impact of first and second as well as second and third, the first two spheres will be brought to rest. Take $e = 1.0$.
4. A bullet of mass 50 gm is fired into a freely suspended target of 2.5 kg. On impact, the target moves with a velocity of 2.5 m/s. Find the velocity of the bullet and the loss of kinetic energy, if the impact is perfectly inelastic. (**Ans.** 127.5 m/s ; 398.4 kg-m)

   Hint. After impact, the bullet and target will move with the same velocity.

9. *INDIRECT IMPACT OF TWO BODIES

If the two bodies, before impact, are *not* moving along the line of impact, the collision is called an *indirect* (or *oblique*) impact as shown in Fig. 3.

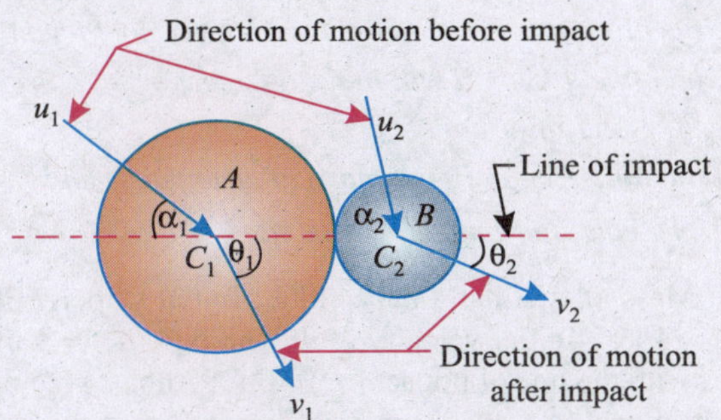

Fig. 3. Indirect impact.

Consider two bodies A and B having an indirect impact as shown in Fig. 3.

Let m_1 = Mass of the first body,

u_1 = Initial velocity of the first body,

α_1 = Angle which the initial velocity of the first body, makes with the line of impact,

v_1 = Final velocity of the first body,

θ_1 = Angle, which the final velocity of the first body, makes with the line of impact, and

$m_2, u_2, \alpha_2, v_2, \theta_2$ = Corresponding values for the second body,

A little consideration will show, that the components of u_1, u_2, v_1 and v_2 along the line of impact, will cause the direct impact and all the relations for direct impact, will hold good for these components. The other components of u_1, u_2, v_1 and v_2, (*i.e.*, along the perpendicular to the line of impact) will not effect the phenomenon of impact in any way, and the component of initial velocity will be equal to the component of final velocity of body.

Now $u_1 \cos \alpha_1$ = Component of the initial velocity of the first body along the line of impact.

Similarly, $u_2 \cos \alpha_2$ = Component of initial velocity of the second body along the line of impact.

$v_1 \cos \theta_1$ = Component of the final velocity of the first body along the line of impact.

and $v_2 \cos \theta_2$ = Component of the final velocity of the second body along the line of impact.

Now the law of conservation of momentum may be applied in the amended form in this case also. *i.e.*,

Total initial momentum along the line of impact

= Total final momentum along the line of impact

$$m_1 u_1 \cos \alpha_1 + m_2 u_2 \cos \alpha_2 = m_1 v_1 \cos \theta_1 + m_2 v_2 \cos \theta_2$$

Newton's Law of Collision of Elastic Bodies, also holds good for indirect impact. *i.e.*,

$$(v_2 \cos \theta_2 - v_1 \cos \theta) = e\,(u_1 \cos \alpha_1 - u_2 \cos \alpha_2)$$

***Example 7.** *A ball of mass 2 kg, moving with a velocity of 3 m/sec, impinges on a ball of mass 4 kg moving with a velocity of 1 m/sec. The velocities of two balls are parallel and inclined at 30° to the line joining their centres at the instant of impact. If the coefficient of restitution be 0.5, find*

(*a*) *direction, in which the 4 kg ball will move after impact* ;

(*b*) *velocity of the 4 kg ball after impact* ;

(*c*) *direction, in which the 2 kg ball will move after impact* ; *and*

(*d*) *velocity of the 2 kg ball after impact.* **[AUT]**

Solution. Given : Mass of first ball $(m_1) = 2$ kg ; Initial velocity of first ball $(u_1) = 3$ m/s ; Mass of second ball $(m_2) = 4$ kg ; Initial velocity of second ball $(u_2) = 1$ m/s ; Angle, which initial velocity of first ball makes with the line of impact $(\alpha_1) = 30°$; Angle, which initial velocity of second ball makes with the line of impact $(\alpha_2) = 30°$ and coefficient of restitution $(e) = 0.5$

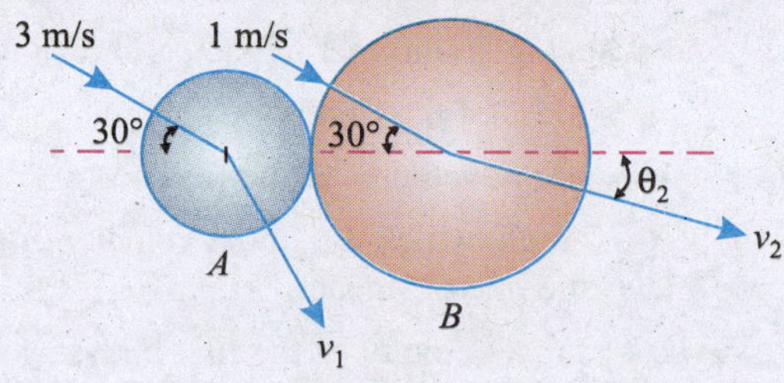

Fig. 4.

(*a*) *Direction, in which the 4 kg ball will move after the impact*

Let θ_1 = Angle, which the 2 kg ball makes with the line of impact,

θ_2 = Angle, which the 4 kg ball makes with the line of impact,

v_1 = Velocity of the 2 kg ball after impact, and

v_2 = Velocity of the 4 kg ball after impact,

We know that the components of velocities, perpendicular to the line of impact, remain unchanged before and after impact.

$\therefore \quad v_1 \sin \theta_1 = u_1 \sin \alpha_1 = 3 \sin 30° = 3 \times 0.5$

or $\quad v_1 \sin \theta_1 = 1.5 \quad$ ...(*i*)

Similarly $\quad v_2 \sin \theta_2 = u_2 \sin \alpha_2 = 1 \sin 30° = 1 \times 0.5$

or $\quad v_2 \sin \theta_2 = 0.5 \quad$ ...(*ii*)

We also know from the law of conservation of momentum

$$m_1 u_1 \cos \alpha_1 + m_2 u_2 \cos \alpha_2 = m_1 v_1 \cos \theta_1 + m_2 v_2 \cos \theta_2$$

$$(2 \times 3 \cos 30°) + (4 \times 1 \times \cos 30°) = 2v_1 \cos \theta_1 + 4v_2 \cos \theta_2$$

$$(6 \times 0.866) + (4 \times 0.866) = 2v_1 \cos \theta_1 + 4v_2 \cos \theta_2$$

$$8.66 = 2v_1 \cos \theta_1 + 4v_2 \cos \theta_2$$

$\therefore \quad v_1 \cos \theta_1 + 2v_2 \cos \theta_2 = 4.33 \quad$ ...(*iii*)

We know from the law of collision of elastic bodies that

$$v_2 \cos \theta_2 - v_1 \cos \theta_1 = e\,(u_1 \cos \alpha_1 - u_2 \cos \alpha_2) = 0.5\,(3 \cos 30° - 1 \cos 30°)$$

$$= 0.5\,(3 \times 0.866 - 1 \times 0.866)$$

$$v_2 \cos \theta_2 - v_1 \cos \theta_1 = 0.866 \qquad ...(iv)$$

Adding equations (*iii*) and (*iv*),

$$3v_2 \cos \theta_2 = 5.196$$

or

$$v_2 \cos \theta_2 = 1.732 \qquad ...(v)$$

Dividing equation (*ii*) by (*v*),

$$\tan \theta_2 = \frac{0.5}{1.732} = 0.2887 \quad \text{or} \quad \theta_2 = 16.1° \quad \textbf{Ans.}$$

(*b*) *Velocity of the 4 kg ball after impact*

Substituting the value of θ_2 in equation (*ii*),

$$v_2 \sin 16.1° = 0.5$$

$$\therefore \quad v_2 = \frac{0.5}{\sin 16.1°} = \frac{0.5}{0.2773} = 1.803 \text{ m/s} \quad \textbf{Ans.}$$

(*c*) *Direction, in which the 2 kg ball will move after impact*

Substituting the values of θ_2 and v_2 in equation (*iv*),

$$1.803 \cos 16.1° - v_1 \cos \theta_1 = 0.866$$

or

$$v_1 \cos \theta_1 = 1.803 \cos 16.1° - 0.866$$

$$= (1.803 \times 0.9608) - 0.866 = 0.866 \qquad ...(vi)$$

Dividing equation (*i*) by (*vi*)

$$\tan \theta_1 = \frac{1.5}{0.866} = 1.732 \quad \text{or} \quad \theta_1 = 60° \quad \textbf{Ans.}$$

(*d*) *Velocity of 2 kg ball after impact*

Now substituting the value of θ_1 in equation (*i*),

$$v_1 \sin 60° = 1.5$$

$$\therefore \quad v_1 = \frac{1.5}{\sin 60°} = \frac{1.5}{0.866} = 1.732 \text{ m/s} \quad \textbf{Ans.}$$

10. *DIRECT IMPACT OF A BODY WITH A FIXED PLANE

In the previous articles, we have been discussing the impacts of two bodies. Both these bodies had some initial velocities, and after impact they had some final velocities (in particular cases, some of these velocities were zero also). But in the following articles, we shall discuss the impact of a body with a fixed plane.

Now consider a body having a direct impact on a fixed plane.

Let u = Initial velocity of the body,

v = Final velocity of the body, and

e = Coefficient of restitution.

We know that the fixed plane will not move even after impact. Thus the velocity of approach is equal to (u) and velocity of separation is equal to (v).

Newton's Law of Collision of Elastic Bodies also holds good for this type of impact. *i.e.*

$$v = eu$$

Notes. 1. In such cases, we do not apply the principle of momentum (*i.e.*, equating the initial momentum and the final momentum), since the fixed plane has infinite mass.

2. If a body is allowed to fall from some height on a floor, then the velocity, with which the body impinges on the floor, should be calculated by the relations of plane motion as discussed below :

Let H = Height from which the body is allowed to fall.

∴ Velocity with which the body impinges on the floor,

$$u = \sqrt{2gH}$$

3. If a body is first projected upwards from the ground with some initial velocity, it will reach the greatest height and will return to the ground with the same velocity, with which it was projected upwards.

***Example 8.** *From a point, on a smooth floor of a room, a toy ball is shot to hit a wall. The ball then returns back to the point of projection. If the time taken by the ball in returning is twice the time taken in reaching the wall, find the coefficient of restitution between the ball and the wall.*

Solution.

Let s = Distance between the point of shot and the wall, and

t = Time taken by the ball in reaching the wall.

∴ Time taken by the ball in returning to the point of shot

$$= 2t \qquad \text{...(given)}$$

Let e = Coefficient of restitution between the ball and the wall.

Since the ball is rolling on a smooth floor, therefore its velocity will remain constant.

∴ Velocity, with which the ball will hit the wall,

$$u = \frac{\text{Distance}}{\text{Time}} = \frac{s}{t} \qquad \text{...(i)}$$

and velocity with which the ball will rebound after hitting,

$$v = \frac{\text{Distance}}{\text{Time}} = \frac{s}{2t} \qquad \text{...(ii)}$$

We know that the velocity after hitting (v),

$$\frac{s}{2t} = eu = e \times \frac{s}{t}$$

$$\therefore \qquad e = \frac{1}{2} = 0.5 \qquad \textbf{Ans.}$$

***Example 9.** *A ball is dropped from a height $h_0 = 1$ m on a smooth floor. Knowing that the height of the first bounce is $h_1 = 81$cm, determine*

(*a*) *coefficient of restitution, and*

(*b*) *expected height h_2 after the second bounce.* **[RGPV]**

Solution. Given : Height from which the ball is dropped (h_0) = 1m and height to which the ball rose after first bounce (h_1) = 81 cm. = 0.81 m.

(*a*) *Coefficient of restitution*

Let e = Coefficient of restitution.

We know that the velocity with which the ball impinges on the floor,

$$u = \sqrt{2gh_0} = \sqrt{2g \times 1} = \sqrt{2g} \text{ m/s} \qquad \text{...(i)}$$

and velocity with which the ball rebounds,

$$v = \sqrt{2gh_1} = \sqrt{2g \times 0.81} = 0.9\sqrt{2g} \text{ m/s} \quad ...(ii)$$

We also know that the velocity with which the ball rebounds (v)

$$0.9\sqrt{2g} = e\sqrt{2g}$$

$\therefore$ $e = 0.9$ **Ans.**

(*b*) *Expected height after the second bounce*

Let h_2 = Expected height after the second bounce.

We know that *velocity, with which the ball impinges second time,

$$u = 0.9\sqrt{2g} \text{ m/s}$$

and velocity, with which the ball rebounds,

$$v = \sqrt{2gh_2}$$

We also know that the velocity with which the ball rebounds second time (v)

$$\sqrt{2gh_2} = eu = 0.9 \times 0.9\sqrt{2g} = 0.81\sqrt{2g}$$

$\therefore$ $2gh_2 = (0.81)^2\, 2g = 0.656 \times 2g$

or $h_2 = 0.656$ m **Ans.**

*Note: The velocity, with which the ball impinges second time is the same with which the ball rebounded first time.

***Example 10.** *From what height, must a heavy elastic ball be dropped on a floor, so that after rebounding thrice it will reach a height of 16 metres ? Take $e = (0.5)^{1/3}$.*

Solution. The system of rebounding is shown in Fig. 5

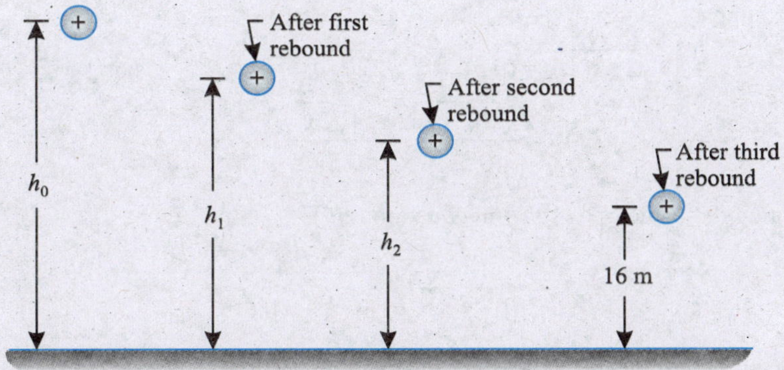

Fig. 5.

Let h_0 = Height from which the ball is dropped,

h_1 = Height after first rebound, and

h_2 = Height after second rebound

We know that the velocity with which the ball impinges on the floor,

$$u = \sqrt{2gh_0} \qquad ...(i)$$

and the velocity with which the ball rebounds first time

$$v = \sqrt{2gh_1} \qquad ...(ii)$$

Similarly, velocity with which the ball impinges after first rebound,

$$u_1 = v = \sqrt{2gh_1} \qquad ...(iii)$$

and velocity with which the ball rebound second time

$$v_1 = \sqrt{2gh_2}$$

Similarly, velocity with which the ball impinges after second rebound,

$$u_2 = v_1 = \sqrt{2gh_2} \qquad ...(iv)$$

and velocity with which the ball rebounds third time

$$u_3 = \sqrt{2g \times 16} = 4\sqrt{2g} \text{ m/s} \qquad ...(v)$$

We know that during first impact,

$$v_1 = eu_1 \qquad ...(vi)$$

or

$$\sqrt{2gh_1} = (0.5)^{1/3}\sqrt{2gh_0} \qquad ...(vii)$$

Similarly, during second impact,

$$\sqrt{2gh_1} = \frac{\sqrt{2gh_2}}{(0.5)^{1/3}} \qquad ...(viii)$$

$$\sqrt{2gh_2} = (0.5)^{1/3}\sqrt{2gh_1} \qquad ...(ix)$$

and during third impact

$$\sqrt{2g \times 16} = (0.5)^{1/3} \times \sqrt{2gh_2}$$

or

$$\sqrt{2gh_2} = \frac{\sqrt{2g \times 16}}{(0.5)^{1/3}} = \frac{4\sqrt{2g}}{(0.5)^{1/3}}$$

Substituting the value of $\sqrt{2gh_2}$ in equation (*viii*),

$$\sqrt{2gh_1} = \frac{4\sqrt{2g}}{(0.5)^{1/3} \times (0.5)^{1/3}} = \frac{4\sqrt{(2g)}}{(0.5)^{2/3}}$$

Now substituting the value of $\sqrt{2gh_1}$ in equation (*vii*)

$$\frac{4\sqrt{(2g)}}{(0.5)^{2/3}} = (0.5)^{1/3}\sqrt{2gh_0}$$

$$\sqrt{h_0} = \frac{4}{(0.5)} = 8$$

$$h_0 = 64 \text{ m} \quad \textbf{Ans.}$$

11. *INDIRECT IMPACT OF A BODY WITH A FIXED PLANE

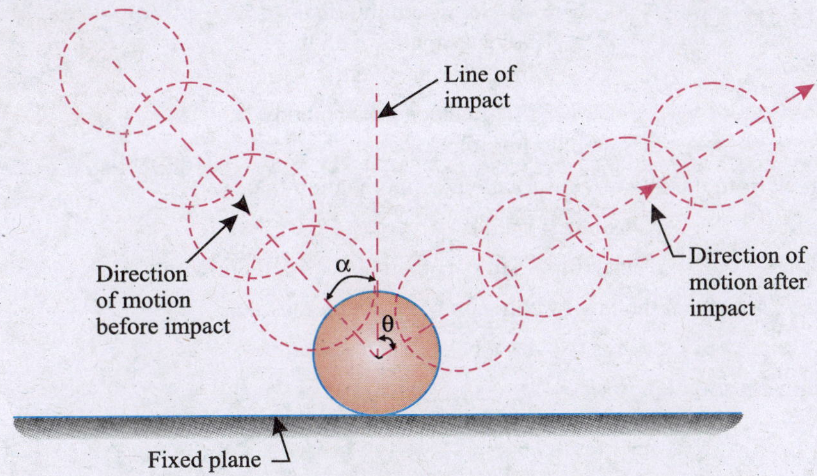

Fig. 6.

Consider a body having an indirect impact on a fixed plane as shown in Fig. 6.

Let u = Initial velocity of the body,

v = Final velocity of the body,

α= Angle, which the initial velocity of the body makes with the line of impact,

θ = Angle which the final velocity of the body makes with the line of impact, and

e = Coefficient of restitution.

A little consideration will show, that the component of u, along the line of impact will cause the direct 'impact' of the body with the fixed plane. The other component of u (*i.e.* along the perpendicular to the line of impact) will not affect the phenomenon of impact and will be equal to the other component of v (*i.e.*, along the perpendicular to the line of impact).

We know that velocity of approach

$$= u \cos \alpha$$

and velocity of separation $= v \cos \theta$

Newton's Law of Collision of Elastic Bodies also holds good for this impact. *i.e.*,

$$v \cos \theta = eu \cos \alpha$$

Notes : 1. In this impact also, we do not apply the principle of momentum (*i.e.* equating the initial momentum and the final momentum) since the fixed plane has infinite mass.

2. The components of initial and final velocities at right angles to the line of impact are same *i.e.*

$$u \sin \alpha = v \sin \theta$$

***Example 11.** *A ball, moving with a velocity of 4 m/s, impinges on a fixed plane at an angle of 30°. If the coefficient of restitution is 0.5, find,*

(*a*) *direction of the body after impact, and*

(*b*) *velocity of the body after impact.*

Solution. Given : Initial velocity of the body (u) = 4 m/s ; Angle, which the initial velocity of the body makes with the line of impact (α) = 90° – 30° = 60° and coefficient of restitution (e) = 0.5.

(*a*) *Direction of the body after impact*

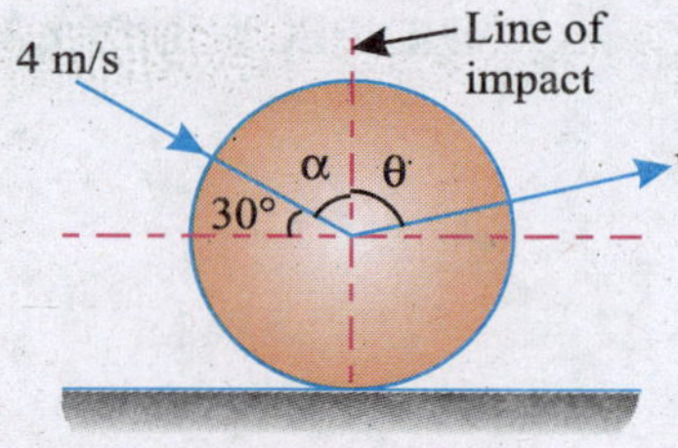

Fig. 7.

Let θ = Angle, which the final velocity makes with the line of impact, and

v = Final velocity of the body after impact.

We know from the law of conservation of momentum that

$$u \sin \alpha = v \sin \theta$$

$$\therefore \quad v \sin \theta = u \sin \alpha = 4 \sin 60° = 4 \times 0.866 = 3.464 \quad \text{...(i)}$$

We also know from the law of collision of elastic bodies that

$$v \cos \theta = e \times u \cos \alpha = 0.5 \times 4 \cos 60° = 2 \times 0.5 = 1 \quad \text{...(ii)}$$

Dividing equation (*i*) by (*ii*),

$$\frac{v \sin \theta}{v \cos \theta} = \frac{3.464}{1}$$

$$\therefore \quad \tan \theta = 3.464 \quad \text{or} \quad \theta = 73.9° \quad \textbf{Ans.}$$

(*b*) *velocity of the body after impact*

Substituting the value of θ in equation (*ii*),

$$v \cos 73.9° = 1$$

$$\text{or} \quad v = \frac{1}{\cos 73.9°} = \frac{1}{0.2773} = 3.6 \text{ m/s} \quad \textbf{Ans.}$$

EXERCISE 2

1. A ball is dropped from a height of 25 metres upon a horizontal floor. Find the coefficient of restitution between the floor and the ball, if it rebounds to a height of 16 metres. (**Ans.** 0.8)

2. A 1 kg ball traverses a frictionless tube as shown in Fig. 8.

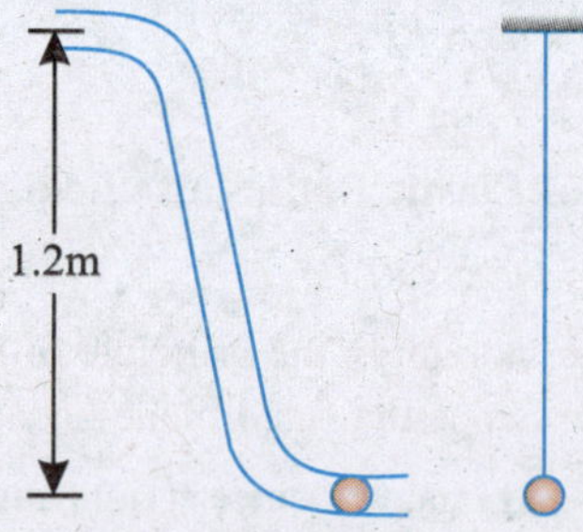

Fig. 8.

The ball, after falling through a height of 1.2 metres, strikes a 1.5 kg ball hung on a rope. Find the velocities of the two balls, if the collision is perfectly elastic.

(**Ans.** v_1 = – 0.97 m/s ; v_2 = 3.88 m/s)

3. A heavy elastic ball drops from the ceiling of a room, and after rebounding twice from the floor reaches a height of equal to one-half of the ceiling. Show that the coefficient of restitution is $(0.5)^{1/4}$.

4. A ball moving with a velocity of 2 m/s, hits a smooth fixed plane at an angle of 30°. If the coefficient of restitution is 0.8, find the angle which the ball makes with the line of impact after hitting . Also find the velocity of the ball after hitting. **[BPUT]** (**Ans.** 65.2°; 1.91 m/s)

QUESTIONS

1. Define the terms 'collision of elastic bodies'. Discuss the phenomenon of collision.
2. What are various types of impacts ? Discuss any one of them.
3. Define the coefficient of restitution.
4. If m_1 and m_2 are the masses of two bodies, u_1 and u_2, v_1 and v_2 are the velocities of the bodies before and after the impact respectively and e the coefficient of restitution, show that the loss of a kinetic energy due to direct impact is given by :

$$0.5 \frac{m_1 m_2}{m_1 + m_2} (u_1 + u_2)^2 (1 - e^2)$$

5. What is the difference between the impact of two bodies and the impact of a body on a fixed plane ?

OBJECTIVE TYPE QUESTIONS

1. The total momentum of two bodies remains constant after collision or any other mutual action. This is known as
   (*a*) Law of Conservation of Momentum
   (*b*) Newton's Law of Collision of Elastic Bodies
   (*c*) both (*a*) and (*b*)
   (*d*) none of them
2. The coefficient of restitution for inelastic bodies is
   (*a*) zero (*b*) between zero and one (*c*) one
3. The bodies which rebound after impact are called
   (*a*) inelastic bodies (*b*) elastic bodies
   (*c*) neither elastic or inelastic bodies
4. The loss of kinetic energy during elastic impact is zero.
   (*a*) Yes (*b*) No
5. The loss of kinetic energy due to direct impact of two bodies depends on
   (*a*) the mass of two bodies
   (*b*) the initial velocities of two bodies
   (*c*) the final velocities of two bodies
   (*d*) Both (*a*) and (*b*)
6. If a lead ball with a certain velocity is made to strike a wall, it does not rebound. But if a rubber ball of same mass and velocity strikes the same wall, it rebounds. Select correct reason from the followig :
   (*a*) the change in momentum suffered by the rubber ball is more than that of the lead ball.
   (*b*) the change in momentum suffered by the lead ball is more than that of the rubber ball.
   (*c*) both (*a*) and (*b*).
   (*d*) none of the above.

ANSWERS

1. (*a*) **2.** (*a*) **3.** (*b*) **4.** (*a*) **5.** (*d*) **6.** (*a*)

CHAPTER

26 Motion Along a Circular Path

Contents

1. MOTION ALONG A CIRCULAR PATH AS A CURVILINEAR MOTION

Motion of an object moving in a curved path is called curvilinear motion. An object is said to be in curvilinear motion if the direction of the velocity of its movement is variable and the path of the object is a curved line. Therefore motion along a circular path is a type of curvilinear motion.

The other types of curvilinear motion are Projectile Motion and Simple Harmonic Motion.

1.1 MOTION ALONG A CIRCULAR PATH

In this chapter we shall study *motion along a circular path.*

We see that whenever a person drives around a curve or a circular track on a bicycle or a motor cycle, he has to lean inward in order to maintain a perfect equilibrium.

As a matter of fact, the angle at which the man leans with the vertical, is more, if he is driving at a faster speed, than when he is driving at a lower speed.

2. CENTRIPETAL ACCELERATION

Consider a particle moving along the circumference of a circle with O as centre as shown in Fig. 1 (a)

Let r = Radius of the circular path in m, and

v = Linear velocity of the particle in m/s.

Now consider an instant when A be the position of the particle. After a small interval of time (δt), let its position be changed from A to B. When the body is at A its linear velocity (v) will be along the dotted line AD *i.e.*, tangent at A. Similarly, when the body is at B, its linear velocity (again equal to v) will be along the dotted line BE *i.e*, tangent at B as shown in Fig. 1 (a).

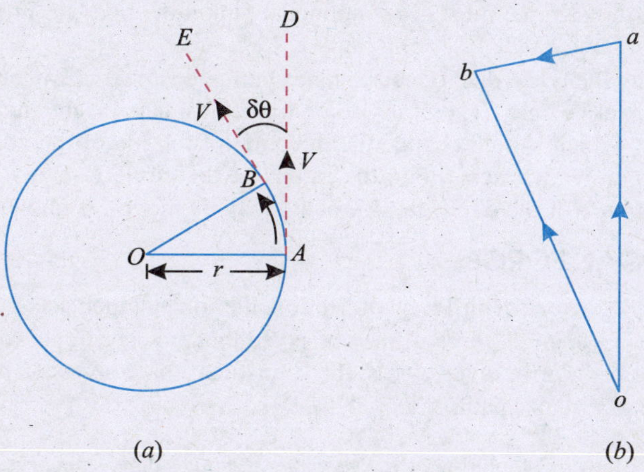

Fig. 1. Centrifugal acceleration.

First of all, take any suitable point o, draw oa parallel to AD and let it represent the velocity v at A, in direction and magnitude to some scale. Similarly, draw ob parallel to BE and let it represent the velocity v at B, in direction and magnitude to the scale. Join ab as shown in Fig. 1 (b).

Now in triangle aob, oa represents the initial velocity ; ob represents the final velocity and ab represents the change in velocity.

We know that the acceleration,

$$a = \frac{\text{Change in velocity}}{\text{Time}} = \frac{ab}{\delta t} \qquad ...(i)$$

Since the interval of time (δt) is considered to be very small, therefore chord AB may be considered to be equal to arc AB. Now consider two triangles AOB and aob, which are similar.

$$\therefore \qquad \frac{ab}{AB} = \frac{oa}{OA}$$

Substituting the values of $AB = v \times \delta t$, $oa = v$ and $OA = r$, in the above equation,

$$\frac{ab}{v \times \delta t} = \frac{v}{r}$$

or $$\frac{ab}{\delta t} = \frac{v^2}{r}$$

Therefore acceleration, $$a = \frac{ab}{\delta t} = \frac{v^2}{r} \quad ...(ii)$$

Now substituting the value $v = \omega r$ in equation (*ii*) where ω is the angular velocity,

$$a = \frac{\omega^2 r^2}{r} = \omega^2 r$$

Note. The same formula holds good for centrifugal acceleration also.

3. *CENTRIPETAL FORCE

A body, moving in a circle or along a circular path, with a constant velocity, suffers a continuous change in its direction at every point of its motion ; though the magnitude of its speed remains the same. Since the velocity involves both magnitude as well as direction, and the velocity of the body is continuously changing due to change in direction ; therefore according to *Newton's First Law of Motion, an external force must act continuously upon the body, to produce a change in the direction of the moving body.

Strictly speaking the body, due to inertia, tends to move along the tangent at every point of its motion, with the constant velocity. Therefore, some force must act at right angles to the direction of motion at every point, which should change the direction of motion of the body, leaving the speed uniform. Thus the force, which acts along the radius of the circle at every point, and is always directed towards the centre of the circle along which the body moves, is known as *centripetal force*.

4. *CENTRIFUGAL FORCE

According to **Newton's Third Law of Motion, the force, which acts opposite to the centripetal force, is known as *centrifugal force*. It may be noted that the centrifugal force always acts away from the centre of the path, or in other words, the centrifugal force always tends to throw the body away from the centre of circular path.

Note: * Newton's First Law of Motion states : "*Everybody continues in its state of rest or of uniform motion, in a straight line, unless it is acted upon by some external force*".

** Newton's Third Law of Motion states, "*To every action, there is always an equal and opposite reaction*".

Example 1. *A body of mass 5 kg is moving in a circle of radius of 1·5 m with an angular velocity of 2 rad/s. Find the centrifugal force acting on the body.*

Solution. Given : Mass of body (m) = 5 kg ; Radius of circle (r) = 1·5 m and angular velocity of the body (ω) = 2 rad/s.

We know that centrifugal force acting on the body,

$$F = m\,\omega^2 r = 5 \times (2)^2 \times 1{\cdot}5 = 30 \text{ N} \qquad \textbf{Ans.}$$

Example 2. *A stone of mass 1 kg is revolving in a circle of radius 1 m with a linear velocity of 10 m/s. What is the value of centrifugal force acting on the stone.*

Solution. Given : Mass of stone (m) = 1 kg ; Radius of circle (r) = 1 m and linear velocity of the stone (v) = 10 m/s.

We know that centrifugal force acting on the stone,

$$F = \frac{mv^2}{r} = \frac{1 \times (10)^2}{1} = 100 \text{ N} \qquad \textbf{Ans.}$$

***Example 3.** *A ball of mass 0·25 kg is attached to the end of a 2 m long string. The string will break, if tension in the string is more than 25 N. What is the maximum angular velocity at which the ball can be rotated ?*

Solution. Given : Mass of ball (m) = 0·25 kg ; Length of string or radius of circle (r) = 2 m and maximum tension in the string (F) = 25 N.

Let ω = Maximum angular velocity at which the ball can be rotated.

We know that tension in the string or centrifugal force,

$$F = m\,\omega^2\,r = 0{\cdot}25 \times \omega^2 \times 2 = 0{\cdot}5\,\omega^2$$

or
$$\omega^2 = \frac{F}{0{\cdot}5} = \frac{25}{0{\cdot}5} = 50$$

∴
$$\omega = \sqrt{50} = 7{\cdot}07 \text{ rad/s} \qquad \textbf{Ans.}$$

***Example 4.** *A body of mass 0·5 kg tied to string is whirled in a vertical circle making 2 rev/s. If radius of the circle is 1·2 m, then find tensions in the string when the body is at top of the circle, and (ii) at the bottom of the circle.*

Solution. Given : Mass of body (m) = 0·5 kg ; Angular rotation of the body (N) = 2 rev/s and radius of circle (r) = 1·2 m.

(*i*) *Tension in the string when the body is at the top of the circle*

We know that angular velocity of the body,

$$\omega = 2\pi N = 2\pi \times 2 = 4\pi \text{ rad/s}$$

∴ Tension in the string when the body is at the top of the circle

$$T_1 = m\,\omega^2\,r - mg = [0{\cdot}5 \times (4\pi)^2 \times 1{\cdot}2] - (0{\cdot}5 \times 9{\cdot}8) \text{ N}$$
$$= 94.7 - 4.9 = 89.8 \text{ N} \qquad \textbf{Ans.}$$

(*ii*) *Tension in the string when the body is at the bottom of the circle*

We know that tension in the string when the body is at the bottom of the circle

$$T_2 = m\,\omega^2\,r + mg = [0{\cdot}5 \times (4\pi)^2 \times 1{\cdot}2] + (0{\cdot}5 \times 9{\cdot}8) \text{ N}$$
$$= 94{\cdot}7 + 4{\cdot}9 = 99{\cdot}6 \text{ N} \qquad \textbf{Ans.}$$

***Example 5.** *In a circus show, a motor cyclist is moving in a spherical cage of radius 3 m. The motor cycle and the rider together has mass of 750 kg. Find the least velocity, with which the motor cyclist must pass the highest-point on the cage, without losing contact inside the cage.* **[PU]**

Solution. Given : Radius of spherical cage (r) = 3 m and mass of motor cycle and rider (m) = 750 kg.

Let v = Least velocity of the motor cyclist.

We know that centrifugal force,

$$F = \frac{mv^2}{r} = \frac{750\,v^2}{3} = 250\,v^2$$

In order to maintain the contact with the highest point of the cage, the centrifugal force must be equal to the weight of the motor cycle and the rider. Therefore force (F),

$$250\,v^2 = mg = 750 \times 9{\cdot}8 = 7350$$

∴
$$v^2 = \frac{7350}{250} = 29{\cdot}4$$

or
$$v = \sqrt{29{\cdot}4} = 5{\cdot}42 \text{ m/s} \qquad \textbf{Ans.}$$

EXERCISE 1

1. A body of mass 2 kg ; revolving in a circle of radius 1·5 m. Find the centripetal force acting on the body, if the angular velocity of the body is 3 rad/s. **(Ans.** 27 N)
2. A body of mass 2·5 kg is moving in a circle of radius 1·5 m with a velocity of 6 m/s. Calculate the centrifugal force acting on the body. **(Ans.** 60 N)
3. A ball of mass 1·5 kg is being rotated with the help of a 0·8 m long string. If velocity of the ball is 2·4 m/s, find the tension in the string. **(Ans.** 10·8 N)
4. A stone of mass 0·5 kg is tied to a string of 1 m length, which is whirled in a vertical circle with a velocity of 5 m/s. What are the magnitudes of the tensions, when the stone is at the top and bottom of the circle ? **(Ans.** 7·6 N ; 17·4 N)

5 *CENTRIFUGAL FORCE ACTING ON A BODY MOVING ALONG A CIRCULAR PATH

Consider a body moving along a circular path with a constant velocity.

Let m = Mass of the body,

r = Radius of the circular path, and

v = Linear velocity of the body.

We know that the centrifugal acceleration of the body,

$$a = \frac{v^2}{r}$$

and centrifugal force, P_c = Mass × Centrifugal acceleration

$$= m \times \frac{v^2}{r} = \frac{mv^2}{r} \qquad \text{...(when } v \text{ is given)}$$

$$= m\,\omega^2 r \qquad \text{...(when } \omega \text{ is given)}$$

Example 6. *A railway engine of mass 60 tonnes, is moving in a circular track of radius 200 metres with a velocity of 36 km.p.h. Find the force exerted on the rails towards the centre of the circle.*

Solution. Given : Mass of railway engine (m) = 60 t ; Radius of circular path (r) = 200 m and velocity of engine (v) = 36 km.p.h. = 10 m/s.

We know that the force exerted on the rails,

$$P_c = \frac{mv^2}{r} = \frac{60\,(10)^2}{200} = 30 \text{ kN} \quad \textbf{Ans.}$$

***Example 7.** *An automobile of mass 1·5 t travelling at of 54 km.p.h.traverses a sag in the road. The sag is a part of a circle of radius 25 m. Find the reaction between the automobile and road while travelling at the lowest part of the sag.*

Solution. Given : Mass of automobile (m) = 1·5 t; Velocity of automobile (v) = 54 km.p.h = 15 m/s and radius of the sag (r) = 25 m

We know that reaction between the automobile and the load while travelling at the lowest part of the sag

$$= \frac{mv^2}{r} + mg = \frac{1{\cdot}5\,(15)^2}{25} + 1.5 \times 9.8 \text{ kN}$$

$$= 13{\cdot}5 + 14{\cdot}7 = 28{\cdot}2 \text{ kN} \quad \textbf{Ans.}$$

5.1. APPLICATIONS OF BALANCING CENTRIPETAL AND CENTRIFUGAL FORCES

Applications of balancing of centripetal force and centrifugal force can be seen in following cases:

1. Motion of Cyclist Around a Curve.
2. Motion of Aeroplane Around a Curve.
3. Motion of Vehicle on Level Circular Path.
4. Super-elevation or Banking of Curves for Roadways.
5. Super-elevation of Curves for Railways.

6. SUPERELEVATION

In our day to day life, we see that whenever a roadway (or railway) is laid on a curve or curved path, then both the edges are not at the same level. In such a case, the outer edge is always higher than the inner one. This is done to keep the vehicle in equilibrium while going on the curved path.

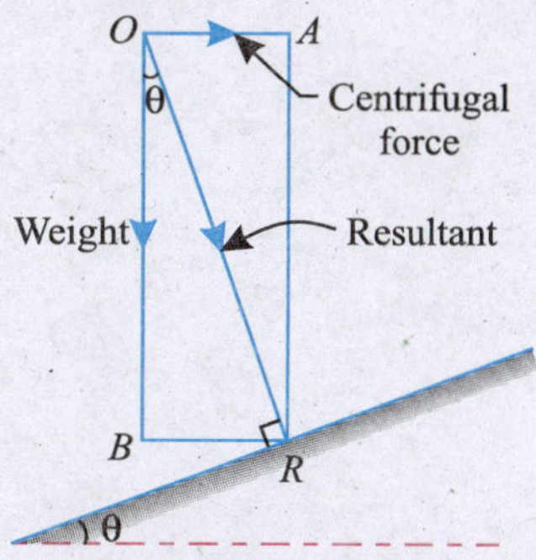

Fig. 2. Superelevation.

We know that the body, moving along a curved path, is subjected to the following forces :

1. Its own weight, and
2. Centrifugal force.

The resultant of these forces is inclined with the vertical as shown in Fig 2. It may be noted, that if curved path be made level, then the resultant force will be inclined at some angle, with the vertical, and thus the reactions on both the supports, of a vehicle, will not be equal, which will effect the equilibrium of the vehicle.

To counterbalance this effect and maintain equilibrium of the vehicle, the surface of the path is made perpendicular to the resultant by keeping the inner edge level and raising the *outer edge of the roadway or (outer rail of the railway). The amount by which the outer edge of rail is raised is known as cant or *superelevation.*

(*Note: In some countries the outer edge is raised from the centre line by half of the superelevation, and the inner edge is also lowered from the centre line by half of the superelevation.)

To counteract the centrifugal force roads are built slant-inwards.

7. EFFECT OF SUPERELEVATION IN ROADWAYS

In case of roadways, the outer edge is raised with respect to the inner edge of the road surface. The amount, by which the outer edge is raised is known as **cant or superelevation**. The process of providing superelevation is known as **banking of the road**. The general practice, to define the superelevation, is to mention the angle of inclination of the road surface as shown in Fig. 3. Now consider a vehicle moving on a roadway and along a curved path with a uniform velocity as shown in Fig. 3.

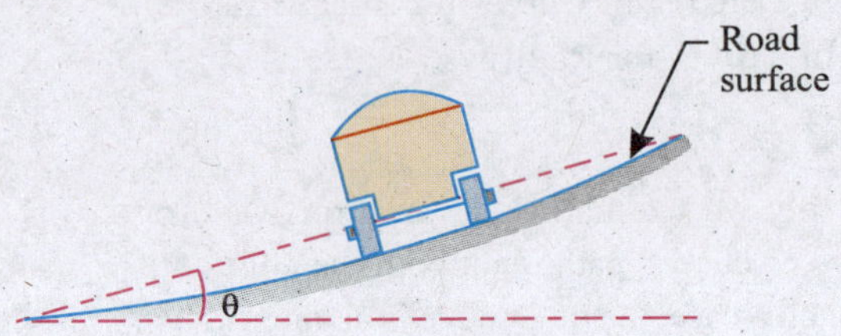

Fig. 3. Superelevation in roadways.

Let m = Mass of the body in tonnes,

r = Radius of circular path in m,

v = Velocity of the body in m/s, and

θ = Angle of the bank.

We have seen in the previous articles, that whenever a body is moving along circumference of a circle, it is subjected to the following forces :

1. Its own weight $= mg$

2. Centrifugal force $= \dfrac{mv^2}{r}$

From the geometry of the figure, we find that

$$\tan\theta = \frac{\text{Centrifugal force}}{\text{Weight of the vehicle}} = \frac{\left(\dfrac{mv^2}{r}\right)}{mg} = \frac{v^2}{gr}$$

It may be noted from the above expression that the superelevation is independent of the mass of the body.

Example 8. *A circular automobile test track has a radius of 200 m. The track is so designed that when a car travels at a speed of 90 kilometres per hour, the force between the automobile and the track is normal to the surface of the track. Find the angle of the bank.*

Solution. Given : Radius of the track (r) = 200 m and speed of the car (v) = 90 km.p.h. = 25 m/s.

Let θ = Angle of the bank.

We know that $\tan\theta = \dfrac{v^2}{gr} = \dfrac{(25)^2}{9{\cdot}8 \times 200} = 0.3189$

$\theta = 17{\cdot}7°$ **Ans.**

8. EFFECT OF SUPERELEVATION IN RAILWAYS

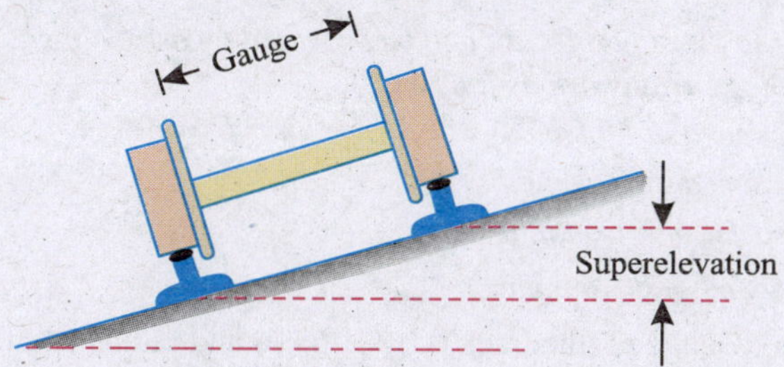

Fig. 4. Superelevation in railways.

In case of railways, the outer rail is raised with respect to the inner rail of the track. The amount by which the *outer rail is raised is known as superelevation.

(*Note: In some countries, the outer rail is raised from the centre line by half the superelevation ; and the inner rail is also lowered from the centre line by half of the superelevation.)

The general practice, to define the superelevation, is to mention the difference of levels between the two rails as shown in Fig. 4.

In such a case, the superelevation is given by the relation,

$$S = \frac{Gv^2}{gr}$$

where G is the gauge of the track.

Example 9. *The distance between the rails of the track is 1·67 m. How much the outer rail be elevated for a curve of 500 m radius, in order that the resultant force may be normal at a speed of 45 km. p.h.*

Solution. Given : Gauge of the track (G) = 1·67 m ; Radius of the curve (r) = 500 m and speed (v) = 45 km.p.h. = 12·5 m/s

We know that the superelevation,

$$S = \frac{Gv^2}{gr} = \frac{1 \cdot 67 \times (12 \cdot 5)^2}{9 \cdot 8 \times 500} = 0 \cdot 0533 \text{ m} = 53 \cdot 3 \text{ mm} \quad \textbf{Ans.}$$

9. EQUILIBRIUM SPEED FOR SUPERELEVATION

In last article, we have obtained a relation for the superelevation in railways. The superelevation obtained, by this relation, is popularly known as **equilibrium superelevation**. It has been experienced, all over the world, that different trains pass over the curve with different speeds. It is therefore obvious that superelevation provided for a particular speed would not suit any other speed. A little consideration will show, that as the superelevation increases with the square of the speed, therefore at higher speeds there is a tendency for the train to overturn. Similarly, at lower speeds there is a tendency for the train to derail.

It will be interesting to know, that there are so many theories prevalent for the calculation of superelevation all over the world. In most of the countries (including India) superelevation is provided in such a way that the faster trains may run safely without the danger of overturning or discomfort to the passengers due to insufficient superelevation. And at the same time, the slower trains may also travel safely without the danger of derailment due to excessive superelevation. These days, the superelevation is usually provided for equilibrium speed or weighted average speed under average conditions. This point is illustrated in the followng example.

Example 10. *Find the superelevation to be provided on a 1·67 m gauge curved track of 1000 m radius, if the speeds of the trains are as follows :*

(*a*) *15 trains at a speed of 50 km.p.h.*

(*b*) *10 trains at a speed of 60 km.p.h.*

(*c*) *5 trains at a speed of 70 km.p.h.*

(*d*) *2 trains at a speed of 80 km.p.h.*

Solution. Given : Gauge of the track (G) = 1·67 m and radius of curved track (r) = 1000 m

We know that the equilibrium speed,

$$v = \frac{(15 \times 50) + (10 \times 60) + (5 \times 70) + (2 \times 80)}{15 + 10 + 5 + 2} = \frac{1860}{32} \text{ km.p.h.}$$

$$= 58 \cdot 125 \text{ km.p.h.} = 16 \cdot 15 \text{ m/s}$$

and superelevation,

$$S = \frac{Gv^2}{gr} = \frac{1 \cdot 67\ (16 \cdot 15)^2}{9 \cdot 8 \times 1000} = 0 \cdot 044 \text{ m} = 44 \text{ mm} \qquad \textbf{Ans.}$$

10. REACTIONS OF A VEHICLE MOVING ALONG A LEVEL CIRCULAR PATH

In the last articles, we have discussed that whenever a vehicle is moving along circular path, it is subjected to some centrifugal force. In order to avoid any unpleasant effect of the centrifugal force, on the moving vehicle or passengers sitting in it, we provide some superelevation in the roadways and railways. Now we shall discuss the effects of not providing any superelevation *i.e.* providing a level circular path.

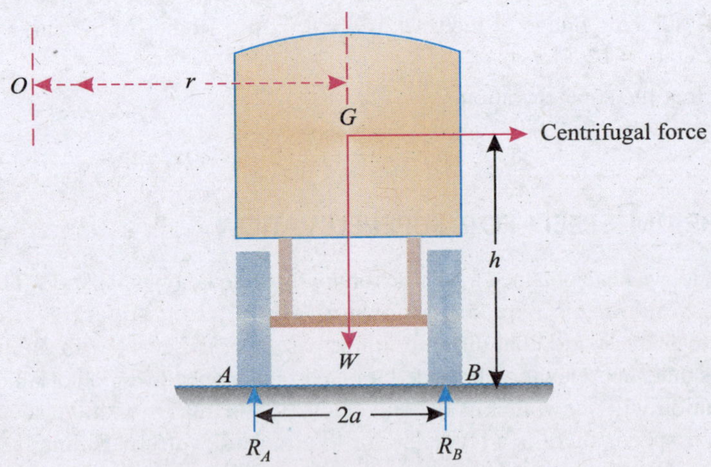

Fig. 5. Reactions of a vehicle.

Consider a vehicle moving on a level circular path, with O as centres as shown in Fig. 5.

Let m = Mass of the vehicle,

v = Velocity of the vehicle,

r = Radius of the circular path,

h = Height of the centre of gravity of the vehicle from the ground level,

R_A = Reaction at A (*i.e.* inner wheel)

R_B = Reaction at B, (*i.e.* outer wheel) and

$2a$ = Distance between the reactions at A and B

We know that the centrifugal force, acting through the centre of gravity of the vehicle,

$$P_c = \frac{mv^2}{r}$$

We also know that the weight of the vehicle (equal to mg) acting downwards through its centre of gravity will exert a reaction equal to $\frac{mg}{2}$ on A and B (because of symmetry).

Taking moments about A and equating the same,

$$R_B \times 2a = (mg \times a) + \left(\frac{mv^2}{r} \times h\right) = mag\left(1 + \frac{v^2 h}{gra}\right)$$

$$\therefore \qquad R_B = \frac{mg}{2}\left(1 + \frac{v^2 h}{gra}\right)$$

Similarly, by taking moments about B and equating the same, we get

$$R_A = \frac{mg}{2}\left(1 - \frac{v^2 h}{gra}\right)$$

Example 11. *A vehicle of mass 1200 kg. is to turn a level circular curve of radius 100 metres with a velocity of 30 km.p.h. The height of its c.g. above the road level is 1 metre and the distance between the centre lines of the wheels is 1·5 metre. Find the reactions of the wheels.*

Solution. Given : Mass of the vehicle (m) = 1200 kg. = 1·2 t ; Radius of the curve (r) = 100 m ; Velocity of the vehicle (v) = 30 km.p.h. = 8·33 m/s ; Height of the c.g. of the vehicle from the road level (h) = 1 m and distance between the centre lines of the wheel ($2a$) = 1·5 m or a = 0·75 m

Reaction at the inner wheel

We know that reaction at the inner wheel,

$$R_A = \frac{mg}{2}\left(1 - \frac{v^2 h}{gra}\right) = \frac{1·2 \times 9·8}{2}\left(1 - \frac{(8·33)^2 \times 1}{9·8 \times 100 \times 0·75}\right) \text{ kN}$$

$$= 5.325 \text{ kN} \qquad \textbf{Ans.}$$

Reaction at the outer wheel

We also know that reaction at the outer wheel,

$$R_B = \frac{mg}{2}\left(1 + \frac{v^2 h}{gra}\right) = \frac{1·2 \times 9·8}{2}\left(1 + \frac{(8·33)^2 \times 1}{9·8 \times 100 \times 0·75}\right) \text{ kN}$$

$$= 6·435 \text{ kN} \qquad \textbf{Ans.}$$

11. EQUILIBRIUM OF A VEHICLE MOVING ALONG A LEVEL CIRCULAR PATH

We have discussed in the previous articles, that whenever the alignment of a road takes a circular turn, the outer edge of the road is always raised. This is known as banking of the road.

Similarly, whenever the alignment of a railway takes a circular turn, the outer rail is raised, by an amount equal to superelevation, than the inner rail. The main idea, of banking the road or providing the superelevation in the railway lines, is to distribute the load of the vehicle equally on both the wheels. But, if the roads are not banked, then a vehicle may also have to face the following mishappenings. This will also happen, when the vehicle moves with a velocity more than the permissible velocity.

1. The vehicle may overturn, or
2. The vehicle may skid away.

Now we shall discuss the maximum velocity of a vehicle, so that it may remain in an equilibrium state on a circular track. Or in other words, we shall discuss the maximum velocity to avoid both the above mishappenings, one by one.

12. MAXIMUM VELOCITY TO AVOID OVERTURNING OF A VEHICLE MOVING ALONG A LEVEL CIRCULAR PATH

Note : The maximum velocity of the vehicle to avoid overturning is independent of its mass.

Example 12. *A vehicle, weighing 1 tonnes is to turn on a circular curve of 40 m radius. The height of its centre of gravity above the road level is 75 cm and the distance between the centre lines of the wheel is 120 cm. Find the speed, at which the vehicle should be run, in order to avoid overturning.*

Solution. Given : Weight of vehicle = 1 t ; Radius of the curve (r) = 40 m ; Height of the centre of gravity of the vehicle from road level (h) = 75 cm = 0·75 m and distance between centre lines of the wheels ($2a$) = 120 cm = 1·2 m or a = 0·6 m

We know that maximum speed at which the vehicle should run, in order to avoid overturning,

$$v_{max} = \sqrt{\frac{gra}{h}} = \sqrt{\frac{9{\cdot}8 \times 40 \times 0{\cdot}6}{0{\cdot}75}} = 17{\cdot}7 \text{ m/s}$$

$$= 63{\cdot}72 \text{ km.p.h.} \quad \textbf{Ans.}$$

13. MAXIMUM VELOCITY TO AVOID SKIDDING AWAY OF A VEHICLE MOVING ALONG A LEVEL CIRCULAR PATH

Note. The maximum velocity of the vehicle to avoid skidding is independent of its mass.

Example 13. *A car is travelling on a level track of radius 50 m. Find the maximum speed, at which he can travel on the curved track, if the coefficient of friction between the tyres and track is 0.45. Take g = 9·8 m/s^2.*

Solution. Given : Radius of level track (r) = 50 m ; Coefficient of friction (μ) = 0·45 and g = 9·8 m/s^2.

We know that maximum speed at which the car can travel,

$$v_{max} = \sqrt{\mu g r} = \sqrt{0{\cdot}45 \times 9{\cdot}8 \times 50} = 14{\cdot}85 \text{ m/s}$$

$$= 53.5 \text{ km.p.h.} \quad \textbf{Ans.}$$

Example 14. *A cyclist, riding at 5 m/s has to turn a corner. What is least radius of the curve, he has to describe, if the coefficient of friction between the tyres and the road be 0·25 ?*

Solution. Given : Velocity of the cycle (v) = 5 m/s and coefficient of friction (μ) = 0·25

Let r = Radius of the curve in metres.

We know that velocity of the cyclist (v),

$$5 = \sqrt{\mu g r} = \sqrt{0{\cdot}25 \times 9{\cdot}8 \times r} = \sqrt{2{\cdot}45\, r}$$

$$25 = 2{\cdot}45\, r \qquad \text{...(Squaring both sides)}$$

$$\therefore \qquad r = \frac{25}{2{\cdot}45} = 10{\cdot}2 \text{ m} \quad \textbf{Ans.}$$

EXERCISE 2

1. A motor car of mass 1000 kg is travelling round a circular track with a velocity of 15 m/s. If the radius of the track is 400 m, find the horizontal thrust exerted on the wheels. (**Ans.** 562·5 N)
2. An automobile of mass 1500 kg, moving at a speed of 54 km.p.h. traverses a sag in the road. The sag is part of a circle of radius 20 m. Find the reaction between the car and road, while travelling at the lowest part of the sag. (**Ans.** 31·575 kN)
3. An aeroplane is travelling with a velocity of 450 km.p.h. in a curve of radius 1000 m. Find the angle at which it must incline with the horizontal. Take $g = 9{\cdot}8$ m/s^2. (**Ans.** 32.1°)
4. A metre gauge railway track is to be taken along a circular path of radius 400 m. Find the minimum superelevation required for a train travelling at 36 km.p.h. (**Ans.** 25·5 mm)
5. A loaded truck is transmitting 10 kN pressure on each wheel. The centre of gravity of the loaded truck is 1·2 m above the road level and the centre to centre distance of wheels is 2 m. Find the pressure exerted by the inner and outer wheels, when the truck is going round a curve of 440 m radius at 48 km.p.h. (**Ans.** 9·95 kN ; 10·05 kN)
6. A racing motor car is negotiating a curve of radius 50 m. The distance between the centre to centre of the wheels is 1·5 m and height of the c.g. of the car is 90 cm above the ground level. Find the maximum speed of the motor car to avoid overturning. (**Ans.** 72·75 km. p.h.)
7. A motor cyclist riding at 20 km.p.h. has to turn a corner. Find the least radius of the curve, he should follow for safe travelling, if the coefficient of friction between the tyres and the road is 0·2. (**Ans.** 15·8 m)

QUESTIONS

1. What is centripetal force ? Explain clearly when it comes into play ?
2. Derive an expression for the centrifugal force of a body moving along a circular path.
3. What do you understand by the term 'superelevation' ? Discuss the necessity of providing a superelevation on railways.
4. Explain, why a cyclist has to lean inwards while negotiating a curve.
5. Obtain an expression for the reactions of a vehicle, moving on a level circular path.
6. Derive an equation for the maximum velocity of a vehicle moving on a level circular path.

   (*a*) to avoid its overturning path :

   (*b*) to avoid its skidding away.

OBJECTIVE TYPE QUESTIONS

1. Which of the following statements is correct ?

   (*a*) When a body moves along a circular path with a uniform velocity, there will be a tangential acceleration.

(*b*) When a body moves along a circular path with a uniform acceleration, there will be a tangential acceleration.

(*c*) When a body moves along a circular path with a uniform velocity, there will be no tangential acceleration.

(*d*) When a body moves along a circular path with a uniform acceleration, there will be no tangential acceleration.

2. The slope on the road surface generally provided on the curves is known as

(*a*) angle of friction (*b*) angle of repose

(*c*) angle of banking (*d*) none of the above

3. On a curved railway track, the amount by which the outer rail is raised, is known as superelevation.

(*a*) Yes (*b*) No

4. The superelevation is given by the relation

(*a*) $\frac{Gv}{gr}$ (*b*) $\frac{Gv^2}{gr}$ (*c*) $\frac{Gv}{gr^2}$ (*d*) $\frac{Gv^2}{gr^2}$

where G = Gauge of the track

v = Velocity of the vehicle, and

r = Radius of the circular path.

5. When a body is moving along a circular path, the centrifugal force tends to overturn the body. The chances of overturning can be decreased by decreasing the

(*a*) weight of the vehicle

(*b*) speed of the vehicle

(*c*) height of c.g.of the vehicle from the road level.

(*d*) all of the above.

6. The maximum velocity of a vehicle, in order to avoid skidding on a level circular path is

(*a*) μgr (*b*) $\frac{1}{2}\mu gr$ (*c*) $\sqrt{\mu gr}$ (*d*) $\frac{1}{2}\sqrt{\mu gr}$

ANSWERS

1. (*c*) **2.** (*c*) **3.** (*a*) **4.** (*b*) **5.** (*b*) **6.** (*c*)

CHAPTER

27

Work, Power and Energy

Contents

1. WORK

Whenever a force acts on a body, and the body undergoes some displacement, then work is said to be done. *e.g.*, if a force *P*, acting on a body, causes it to move through a distance *s* as shown in Fig. 1 (*a*). Then,

Work done (by the force *P*)

$$= \textbf{Force} \times \textbf{Distance}$$

$$= P \times s$$

Sometimes, the force *P* does not act in the direction of motion of the body, or in other words, the body does not move in the direction of the force as shown in Fig. 1 (*b*).

Then work done by the force P

$$= \text{Component of the force in the direction of motion} \times \text{Distance}$$

$$= P \cos \theta \times s$$

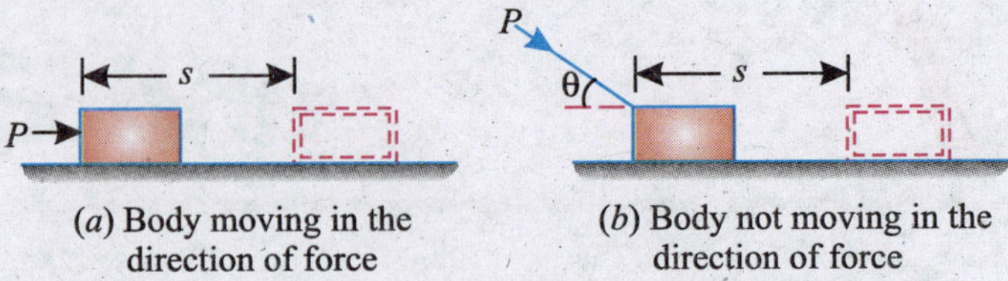

(*a*) Body moving in the direction of force

(*b*) Body not moving in the direction of force

Fig. 1. Work done by a force

2. UNITS OF WORK

We have already discussed that the work is the product of force and distance, through which the body moves due to action of the force. Thus the units of work depend upon the units of the force and distance. The units of work (or work done) are :

1. **One N-m.** It is the work done by a force of 1 N, when it displaces the body through 1 m. It is called joule (briefly written as J), Mathematically.

$$1 \text{ joule} = 1 \text{ N-m}$$

2. **One kN-m.** It is the work done by a force of 1 kN, when it displaces the body through 1 m. It is also called kilojoule (briefly written as kJ). Mathematically.

$$1 \text{ kilo-joule} = 1 \text{ kN-m}$$

Note. Sometimes, the force stretches or compresses a spring or penetrates into a body. In such a case, the average force is taken as half of the force for the purpose of calculating the work done.

3. GRAPHICAL REPRESENTATION OF WORK

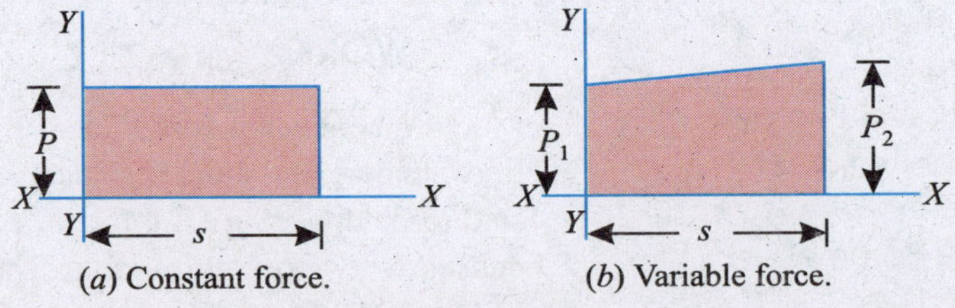

(*a*) Constant force.

(*b*) Variable force.

Fig. 2. Graphical representation of work.

The work done, during any operation, may also be represented by a graph, by plotting distance (s) along X-X axis and the force (P) along Y-Y axis as shown in Fig. 2 (*a*) and (*b*).

Since the work done is equal to the product of force and distance, therefore area of the figure enclosed, represents the work done to some scale. Such diagrams are called force-distance diagrams. If the force is not constant, but varies uniformly with the distance P_1 to P_2, the force distance diagram is not a rectangle; but a trapezium as shown in Fig. 2 (*b*).

Thus,

$$\text{Work done} = \frac{P_1 + P_2}{2} \times s$$

Example 1. *A horse pulling a cart exerts a steady horizontal pull of 300 N and walks at the rate of 4·5 km.p.h. How much work is done by the horse in 5 minutes ?*

Solution. Given : Pull (*i.e.* force) = 300 N ; Velocity (v) = 4·5 km.p.h. = 75 m/ min and time (t) = 5 min.

Distance travelled in 5 minutes,

$$s = 75 \times 5 = 375 \text{ m}$$

Work done by the horse, $W = \text{Force} \times \text{Distance} = 300 \times 375 = 112\ 500 \text{ N-m}$

$$= 112.5 \text{ kN-m} = 112.5 \text{ kJ} \quad \textbf{Ans.}$$

Example 2. *A spring is stretched by 50 mm by the application of a force. Find the work done, if the force required to stretch 1 mm of the spring is 10 N.*

Solution. Given : Spring stretched by the application of force (s) = 50 mm ; Stretching of spring = 1 mm and force = 10 N.

Force required to stretch the spring by 50 mm = 10 × 50 = 500 N

$$\therefore \quad \text{Average force} = \frac{500}{2} = 250 \text{ N}$$

and $\text{Work done} = \text{Average force} \times \text{Distance} = 250 \times 50 = 12\ 500 \text{ N-mm}$

$$= 12.5 \text{ N-m} = 12.5 \text{ J} \quad \textbf{Ans.}$$

4. *POWER

The power may be defined as the rate of doing work. It is thus the measure of performance of engines. *e.g.* an engine doing a certain amount of work, in one second, will be twice as powerful as an engine doing the same amount of work in two seconds.

5. UNITS OF POWER

In S.I. units, the unit of *power is watt (briefly written as W) which is equal to 1 N-m/s or 1 J/s. Generally, a bigger unit of power (kW) is used, which is equal to 10^3 W. Sometimes, a still bigger unit of power (MW) is also used, which is equal to 10^6 W.

*Note: First of all, the term horse power was introduced by James Watt, during his experiments on his engine. He chose a normal horse and found that it could do a work of 33 000 ft-lb (converted to 4 500 kg-m in metric units) in one minute. He thus, adopted this measure for comparing the performance of his engines.

6. TYPES OF ENGINE POWERS

In the case of engines, the following two terms are commonly used for power.

1. **Indicated power.** 2. **Brake power.**

7. *INDICATED POWER

The actual power generated in the engine cylinder is called indicated power (briefly written as I.P.). Sometimes, the indicated power is also defined as the power, which is fed into the engine in the form of steam or calorific value of the fuel.

8. *BRAKE POWER

It has been observed that the entire power, generated by the engine cylinder, is not available for useful work. This happens because a part of it is always utilized in overcoming internal friction of the moving parts of the engine. The net output of the engine (*i.e.* I.P. – Losses) is called brake power (briefly written as B.P).

9. *EFFICIENCY OF AN ENGINE

The efficiency of an engine is defined as the ratio of brake power to the indicated power. Mathematically, efficiency,

$$\eta = \frac{\text{B.P.}}{\text{I.P.}}$$

It is also called mechanical efficiency of an engine and is expressed in percentage.

As a matter of fact, the procedure for calculating the power (or brake power) of an engine, is as given below :

1. First of all, we obtain the force and the distance travelled (in one second).
2. Then we find out the work done by multiplying the force and distance (in one second).
3. Now the power is obtained from the work done.

10. MEASUREMENT OF BRAKE POWER

The brake power of an engine is measured by an apparatus called dynamometers (or brake). Following are the two types of dynamometers :

1. Absorption type, and 2. Transmission type.

In the absorption type dynamometers, the entire power, available from the engine, is wasted in friction of the brakes during the process of measurement. But in the transmission type dynamometers, the entire power available from the engine is transmitted to some other shaft, where it is suitably measured. Following are the common types of dynamometers :

1. Rope Brake Dynamometer (Absorption Type Dynamometer).
2. Proney Brake Dynamometer (Absorption Type Dynamometer).
3. Froude and Thornycraft Transmission Dynamometer (Transmission Type Dynamometer).

Example 3. *Find the power of an engine, which can do a work of 1200 joules in 8 seconds.*

Solution. Work = 1200 J and time (t) = 8 s

Work done by the engine in one second

$$= \frac{1200}{8} = 150 \text{ J/s}$$

$\therefore$ Power = 150 W **Ans.**

Example 4. *A motor boat is moving with a steady speed of 10 m/s. If the water resistance to the motion of the boat is 600 N, determine the power of the boat engine.*

Solution. Given : Speed of motor boat = 10 m/s and resistance = 600 N

Work done by the boat engine in one second

$$= \text{Resistance} \times \text{Distance} = 600 \times 10 = 6000 \text{ N-m/s}$$

$$= 6 \text{ kN-m/s} = 6 \text{ kJ/s}$$

$\therefore$ Power = 6 kW **Ans.**

Example 5. *A railway engine of mass 20 tonnes is moving on a level track with a constant speed of 45 km.p.h. Find the power of the engine, if the frictional resistance is 80 N/t. Take efficiency of the engine as 80 %.*

Solution. Given : Mass of railway engine (m) = 20 tonnes ; Velocity (v) = 45 km.p.h. = 12·5 m/s ; Frictional resistance = 80 N/t = 80 × 20 = 1600 N = 1·6 kN and efficiency of the engine (η) = 80% = 0·8.

Work done by the railway engine in one second

$$= \text{Resistance} \times \text{Distance} = 1{\cdot}6 \times 12{\cdot}5 \text{ kN-m/s}$$

$$= 20 \text{ kN-m/s} = 20 \text{ kJ/s}$$

$\therefore$ Power = 20 kW

Since efficiency of the engine is 0·8, therefore, actual power of the engine,

$$P = \frac{20}{0{\cdot}8} = 25 \text{ kW} \quad \textbf{Ans.}$$

***Example 6.** *A train of weight 1000 kN is pulled by an engine on a level track at a constant speed of 45 km.p.h. The resistance due to friction is 1% of the weight of the train. Find the power of the engine.* **[RGPV]**

Solution. Given : Weight of the train = 1000 kN ; Speed of the train (v) = 45 km.p.h. = 12·5 m/s and resistance due to friction = 1% of the weight of train.

Frictional force (or resistance)

$$= 0{\cdot}01 \times 1000 = 10 \text{ kN}$$

Work done in one second = Resistance × Distance = 10 × 125 = 125 kN-m/s = 125 kJ/s

$\therefore$ Power = 125 kW **Ans.**

***Example 7.** *A locomotive draws a train of mass 400 tonnes, including its own mass, on a level ground with a uniform acceleration, until it acquires a velocity of 54 km.p.h in 5 minutes.*

If the frictional resistance is 40 newtons per tonne of mass and the air resistance varies with the square of the velocity, find the power of the engine. Take air resistance as 500 newtons at 18 km.p.h.

Solution. Given : Mass of the locomotive *i.e.* mass of the train + own mass (m) = 400 t ; Velocity acquired (v) = 54 km.p.h. = 15 m/s ; Time (t) = 5 min = 300 s and frictional resistance = 40 N/t = 40 × 400 = 16 000 N = 16 kN

Let a = Acceleration of the locomotive train.

Final velocity of the locomotive after 300 seconds (v)

$$15 = 0 + a \times 300 = 300\,a \qquad \text{...}(\because v = u + at)$$

$$\therefore \quad a = \frac{15}{300} = 0{\cdot}05 \text{ m/s}^2$$

Force required for this acceleration

$$= ma = 400 \times 0{\cdot}05 = 20 \text{ kN} \qquad \text{...}(i)$$

As the air resistance varies with the square of the velocity, therefore air resistance at 54 km.p.h.

$$= 500\left(\frac{54}{18}\right)^2 = 4500 \text{ N} = 4{\cdot}5 \text{ kN} \qquad \text{...}(ii)$$

$\therefore$ Total resistance $= 16 + 20 + 4{\cdot}5 = 40{\cdot}5$ kN

Work done in one second = Total resistance × Distance = 40·5 × 15

$$= 607{\cdot}5 \text{ kN-m/s} = 607{\cdot}5 \text{ kJ/s}$$

$\therefore$ Power = 607·5 kW **Ans.**

EXERCISE 1

1. A trolley of mass 200 kg moves on a level track for a distance of 500 metres. If the resistance of the track is 100 N, find the work done in moving the trolley. (**Ans.** 50 kJ)
2. What is the power of an engine, which can do a work of 5 kJ in 10 s ? (**Ans.** 500 W)
3. An army truck of mass 8 tonnes has a resistance of 75 N/t. Find the power of the truck for moving with a constant speed of 45 km.p.h. (**Ans.** 7.5 kW)
4. A train of mass 200 tonnes moves on a level track having a track resistance of 85 newtons per tonne. Find the maximum speed of the engine, when the power developed is 320 kW.

   (**Ans.** 67.75 km.p.h.)
5. A train of mass 150 tonnes moves on a level track with a speed of 20 m/s. The tractive resistance is 100 newtons per tonne. Determine the power of the engine to maintain this speed.

   Also determine the power of the engine, when the train is to move with an acceleration of $0 \cdot 3$ m/s^2 on a level track. (**Ans.** 300 kW ; 1200 kW)

11. *MOTION ON INCLINED PLANE

In the previous articles, we have been discussing the motion of bodies on level surface. But sometimes, the motion of a body takes place up or down an inclined plane as shown in Fig. 3 (*a*) and (*b*).

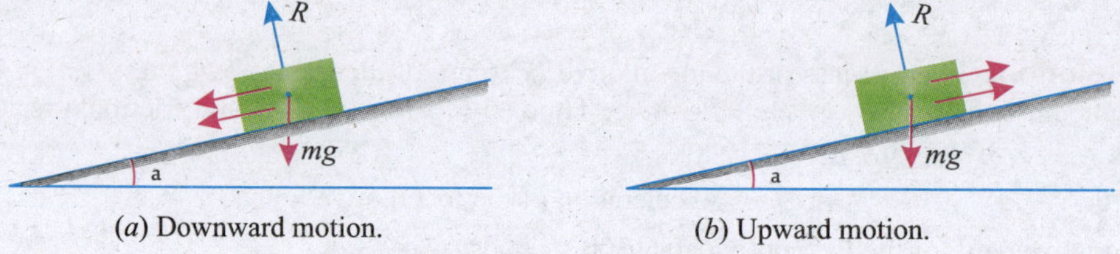

(*a*) Downward motion. (*b*) Upward motion.

Fig. 3. Motion on inclined plane.

Now consider a body moving downwards on an inclined plane as shown in Fig. 3 (*a*).

Let m = Mass of the body, and

α = Inclination of the plane.

Normal reaction on the inclined plane due to weight of the body.

$$R = mg \cos \alpha \qquad ...(i)$$

This component is responsible for the force of friction experienced by the body.

Component of the force along the inclined plane due to weight of the body

$$= mg \sin \alpha \qquad ..(ii)$$

This component, responsible for sliding (or moving) the body downwards, is known as gravitational pull. Now we can find out work done in moving the body or power required for the motion by subtracting the force of friction (due to normal reaction) from the component along the inclined surface ($mg \sin \alpha$).

Note. If the body is moving upwards, then the component along the inclined surface is taken as an additional resistance *i.e.* this component is added to other types of resistances.

Example 8. *Calculate the work done in pulling up a block of mass 200 kg for 10 m on a smooth plane inclined at an angle of 15° with the horizontal.*

Solution. Given : Mass of the block (m) = 200 kg ; Distance (s) = 10 m and inclination of plane (α) = 15°

Resistance due to inclination $= mg \sin \alpha = 200 \times 9{\cdot}8 \sin 15° = 1960 \times 0.2588 = 507.2$ N

Work done $=$ Resisting force $\times$ Distance $= 507{\cdot}2 \times 10 = 5072$ N-m/s

$= 5.072$ kN-m/s $= 5.072$ kJ **Ans.**

***Example 9.** *A locomotive and train together has a mass of 200 t and tractive resistance 100 N per tonne. If the train can move up a grade of 1 in 125 with a constant speed of 28·8 km.p.h. find the power of the locomotive.*

Also find the speed, which the train can attain, on a level track, with the same tractive resistance and power of the locomotive.

Solution. Given : Mass of the body *i.e.* locomotive and train together (m) = 200 t ; Tractive resistance = 100 N/t = 100 × 200 = 20 000 N = 20 kN ; Grade $\sin \alpha = \frac{1}{125} = 0{\cdot}008$ and speed of the train (v) = 28·8 km/h = 8 m/s

Power of the locomotive

Resistance due to inclination

$= mg \sin \alpha = 200 \times 9.8 \times 0.008 = 15.7$ kN

∴ Total resisting force = Tractive resistance + Resistance due to inclination

$= 20 + 15.7 = 35.7$ kN

Work done in one second = Total resisting force × Distance

$= 35.7 \times 8 = 285.6$ kN-m/s $= 285.6$ kJ/s

∴ Power = 285.6 kW **Ans.**

Speed which the train can attain on a level track

Since the train is to move on a level track, therefore it has no resistance due to inclination. Thus it has to overcome tractive resistance only.

Power of the locomotive,

285.6 = Tractive resistance × Speed of the train

= 20 × Speed of the train

∴ Speed of the train $= \frac{285.6}{20} = 14.3$ m/s $= 51.5$ km.p.h. **Ans.**

***Example 10.** *An engine of mass 50 tonnes pulls a train of mass of 250 tonnes up a gradient of 1 in 125 with a uniform speed of 36 km. p.h. Find the power transmitted by the engine, if the tractive resistance is 60 newtons per tonnes.*

Also find the power transmitted by the engine, if the acceleration of the engine is 0.2 m/s^2 up the gradient.

Solution. Given : Mass of the body i.e. mass of the engine + mass of the train (m) = 50 + 250 = 300 t ; Gradient ($\sin \alpha$) $= \frac{1}{125} = 0.008$; Speed of the train (v) = 36 km.p.h. = 10 m/s ; Tractive resistance = 60 N/t = 60 × 300 = 18 000 N = 18 kN and acceleration of the engine (a) = 0.2 m/s^2.

Power transmitted by the engine when the train moves with a uniform speed

Resistance due to inclination,

$$mg \sin \alpha = 300 \times 9.8 \times 0.008 = 23.5 \text{ kN}$$

$\therefore$ Total resisting force = Resistance due to inclination + Tractive resistance

$$= 23.5 + 18 = 41.5 \text{ kN}$$

Work done in one second = Total resisting force × Distance

$$= 41.5 \times 10 = 415 \text{ kN-m/s} = 415 \text{ kJ/s}$$

$\therefore$ Power = 415 kW **Ans.**

Power transmitted by the engine when the train is moving up with a uniform acceleration

Force required to accelerate the engine and train

$$= ma = 300 \times 0.2 = 60 \text{ kN}$$

$\therefore$ Total resisting force = Resistance due to inclination + Tractive resistance + Force required for acceleration

$$= 23.5 + 18 + 60 = 101.5 \text{ kN}$$

Work done in one second = Total resisting force × Distance

$$= 101.5 \times 10 = 1015 \text{ kN-m/s} = 1015 \text{ kJ/s}$$

$\therefore$ Power = 1015 kW **Ans.**

***Example 11.** *A truck of mass 5 tonnes just moves freely without working the engine at 18 km.p.h. down a slope of 1 in 50. The road resistance at this speed is just sufficient to prevent any acceleration.*

Determine the track resistance in newtons per tonne mass of the engine. What power will it have to exert to run up the same slope at double the speed, when the track resistance remains the same ?

Solution. Given : Mass of the truck (m) = 5t = 5000 kg ; Velocity of the truck v = 18 km.p.h. = 5 m/s and slope $(\sin \alpha) = \frac{1}{50} = 0.02$.

Track resistance per tonne mass of truck

Resistance due to inclination

$$= mg \sin \alpha = 5000 \times 9.8 \times 0.02 = 980 \text{ N}$$

Since the engine is not working and the truck is moving without any acceleration, therefore no external force is acting on it. Or in other words, the track resistance is equal to the resistance due to inclination (*i.e.* 980 N). Therefore track resistance per tonne mass of the truck

$$\frac{980}{5} = 196 \text{ N} \quad \textbf{Ans.}$$

Power to be exerted by the engine for moving the truck upwards

Since the truck is moving up the slope, therefore it has to overcome tractive resistance (980 N) ***plus*** **resistance due to inclination (980 N).**

$\therefore$ Total resisting force = Tractive resistance + Resistance due to inclination

$$= 980 + 980 = 1960 \text{ N} = 1.96 \text{ kN}$$

Work done in one second = Total resisting force × Distance

$= 1.96 \times (2 \times 5) = 19.6$ kN-m/s $= 19.6$ kJ/s

∴ Power = 19.6 kW **Ans.**

***Example 12.** *An army truck of mass 5 tonnes has tractive resistance of 150 N/t. Find the power required to propel the truck at a uniform speed of 36 km.p.h. (a) up an incline of 1 in 100 ; (b) on a level track ; and (c) down an incline of 1 in 100.*

Solution. Given : Mass of the truck $(m) = 5$ t $= 5000$ kg ; Tractive resistance $= 150$ N/t $= 150 \times 5 = 750$ N ; Speed of the truck $(u) = 36$ km.p.h $= 10$ m/s and slope $(\sin \alpha) = \dfrac{1}{100} = 0.01$.

(a) Power required to propel the truck up the incline

Since the **truck is propelled up the incline, therefore it has to overcome tractive resistance *plus* resistance due to inclination.**

Resistance due to inclination $= mg \sin \alpha = 5000 \times 9.8 \times 0.01 = 490$ N

∴ Total resisting force = Tractive resistance + Resistance due to inclination

$= 750 + 490 = 1240$ N

Work done in one second = Total resisting force × Distance $= 1240 \times 10$ N-m/s

$= 12\,400$ N-m/s $= 12.4$ kN-m/s $= 12.4$ kJ/s

∴ Power = 12.4 kW **Ans.**

(b) Power required to propel the truck on a level track

Since the truck is to move on a level track, therefore it has no resistance due to inclination. Thus it has to overcome tractive resistance only.

∴ Work done in one second = Tractive resistance × Distance $= 750 \times 10$ N- m/s

$= 7500$ N-m/s $= 7.5$ kN-m/s $= 7.5$ kJ/s

∴ Power = 7.5 kW **Ans.**

(c) Power required to propel the truck down the incline

Since the truck is propelled down the incline, therefore, it has to overcome tractive resistance *minus* gravitational pull *i.e.* resistance due to inclination.

∴ Net resisting force = Tractive resistance – Resistance due to inclination

$= 750 - 490 = 260$ N

Work done in one second = Net resisting force × Distance $= 260 \times 10$ N-m/s

$= 2600$ N-m/s $= 2.6$ kN-m/s $= 2.6$ kJ/s

∴ Power = 2.6 kW **Ans.**

***Example 13.** *An engine of mass 50 tonnes pulls a train of mass 300 tonnes up an incline of 1 in 100. The train starts from rest and moves with a constant acceleration against a total resistance of 50 newtons per tonnes. If the train attains a speed of 36 km.p.h. in a distance of 1 kilometre, find power of the engine. Also find tension in the coupling between the engine and train.*

Solution. Given : Mass of the engine $(m_1) = 50$ t ; Mass of the train $(m_2) = 300$ t or total mass $(m) = 50 + 300 = 350$ t ; Slope $(\sin \theta) = \dfrac{1}{100} = 0.01$; Initial velocity $(u) = 0$ (because, it starts from rest); Tractive resistance $= 50$ N/t $= 50 \times 350 = 17\,500$ N $= 17.5$ kN ; Final velocity $(v) = 36$ km.p.h. $= 10$ m/s and distance $(s) = 1$ km $= 1000$ m

Fig. 4.

Power of the engine

Let a = Acceleration of the train.

Resistance due to inclination $= mg \sin \alpha = 350 \times 9.8 \times 0.01 = 34.3$ kN

Relation for acceleration,

$$v^2 = u^2 + 2\,as$$

$$(10)^2 = (0)^2 + 2a \times 1000 = 2000\,a$$

or

$$a = \frac{100}{2000} = 0.05 \text{ m/s}^2$$

Force required to accelerate the engine and train

$= ma = 350 \times 0.05 = 17.5$ kN

∴ Total resisting force = Tractive resistance + Resistance due to inclination + Force required for acceleration

$= 17.5 + 34.3 + 17.5 = 69.3$ kN

Work done in one second = Total Resisting force × Distance

$= 69.3 \times 10 = 693$ kN-m/s $= 693$ kJ/s

∴ Power = 693 kW **Ans.**

Tension in the coupling

It is the tension which is responsible for driving the train only.

Tractive resistance (for train only)

$= 50 \times 300 = 15\,000$ N $= 15$ kN

Similarly, Resistance due to inclination (for train only)

$= m_2 g \sin \alpha = 300 \times 9.8 \times 0.01 = 29.4$ kN

Force required for the acceleration (for train only)

$= 300 \times 0.05 = 15$ kN

∴ Tension in the coupling $= 15 + 29.4 + 15 = 59.4$ kN **Ans.**

EXERCISE 2

1. A mass of 50 tonnes is to be pulled up a smooth plane having an inclination of 1 in 50. Find the work done in pulling up the mass for a distance of 2 metres. (**Ans.** 19.6 kJ)
2. A locomotive pulls a train of mass 125 tonnes including its own mass up a slope of 1 in 100 with a velocity of 18 km.p.h. Find the power of the engine, if the frictional resistance is 50 newtons per tonne. (**Ans.** 92.5 kW)

*3. A train of mass 300 tonnes moves down a slope of 1 in 200 at 45 km.p.h. with the engine developing 75 kW. Find the power required to pull the train up the slope with the same velocity. **[AM-IETE]** (**Ans.** 332.5 kW)

4. A train of mass 200 tonnes is ascending a track, which has an inclination of 1 in 100, the resistance being 75 N per tonne. What is the acceleration of the train when its speed has reached 18 km.p.h. if the power developed by the engine is 450 kW ? (**Ans.** 0.28 m/s^2)

12. *ENERGY

The energy may be defined as the capacity to do work. It exists in many forms *i.e.*, mechanical, electrical chemical, heat, light etc. But in this subject, we shall deal in mechanical energy only.

13. UNITS OF ENERGY

We have discussed in the previous article, that the energy is the capacity to do work. Since the energy of a body is measured by the work it can do, therefore the units of energy will be the same as those of the work (N-m or J)

14. MECHANICAL ENERGY

Though there are many types of mechanical energies, yet the following two types are important from the subject point of view :

1. Potential energy. **2. Kinetic energy.**

15. POTENTIAL ENERGY

It is the energy possessed by a body, for doing work, by virtue of its position. *e.g.*,

1. A body, raised to some height above the ground level, possesses some potential energy, because it can do some work by falling on the earth's surface.
2. Compressed air also possesses potential energy, because it can do some work in expanding, to the volume it would occupy at atmospheric pressure.
3. A compressed spring also possesses potential energy, because it can do some work in recovering to its original shape.

Now consider a body of mass (m) raised through a height (h) above the datum level. We know that,

Work done in raising the body = Weight × Distance = (mg) h = mgh

This work (equal to $m.g.h$) is stored in the body as potential energy. A little consideration will show, that body, while coming down to its original level , is capable of doing work equal to ($m.g.h$).

***Example 14.** *A man of mass 60 kg dives vertically downwards into a swimming pool from a tower of height 20 m. He was found to go down in water by 2 m and then started rising. Find the average resistance of the water. Neglect the air resistance.*

Solution. Given : Mass of the man (m) = 60 kg and height of the tower (h) = 20 m

Let P = Average resistance of the water

Potential energy of the man before jumping

$$= mgh = 60 \times 9.8 \times 20 = 11\ 760 \text{ N-m} \quad ...(i)$$

Work done by the average resistance of water

= Average resistance of water × Depth of water

$$= P \times 2 = 2P \text{ N-m} \quad ...(ii)$$

Since the **total potential energy of the man is used in the work done by the water,** therefore equating equations (*i*) and (*ii*),

$$11\,760 = 2\,P$$

or $$P = \frac{11\,760}{2} = 5880 \text{ N} \qquad \textbf{Ans.}$$

16. *KINETIC ENERGY

It is the energy, possessed by a body, for doing work by virtue of its mass and velocity of motion. Now consider a body, which has been brought to rest by a uniform retardation due to the applied force.

Let m = Mass of the body

u = Initial velocity of the body

P = Force applied on the body to bring it to rest,

a = Constant retardation, and

s = Distance travelled by the body before coming to rest.

Since the body is brought to rest, therefore its final velocity,

$$v = 0$$

Work done, $$W = \text{Force} \times \text{Distance} = P \times s \qquad ...(i)$$

Now substituting value of ($P = m.a$) in equation (*i*),

$$W = ma \times s = mas \qquad ...(ii)$$

We know that, $$v^2 = u^2 - 2\,as \qquad \text{...(Minus sign due to retardation)}$$

or $$2as = u^2 \qquad ...(\because v = 0)$$

or $$as = \frac{u^2}{2}$$

Now substituting the value of ($a.s$) in equation (*ii*) and replacing work done with kinetic energy,

$$\text{KE} = \frac{mu^2}{2}$$

Cor. In most of the cases, the initial velocity is taken as v (instead of u), therefore kinetic energy,

$$\text{KE} = \frac{mv^2}{2}$$

***Example 15.** *A truck of mass 15 tonnes travelling at 1.6 m/s impacts with a buffer spring, which compresses 1.25 mm per kN. Find the maximum compression of the spring.* **[RGPV, IGNOU]**

Solution. Given : Mass of the truck (m) = 15 t ; Velocity of the truck (v) = 1.6 m/s and buffer spring constant (k) = 1.25 mm/ kN

Let x = Maximum compression of the spring in mm.

Kinetic energy of the truck $$= \frac{mv^2}{2} = \frac{15\,(1.6)^2}{2} = 19.2 = 19\,200 \text{ kN-mm} \qquad ...(i)$$

Compressive load $$= \frac{x}{1.25} = 0.8\,x \text{ kN}$$

∴ Work done in compressing the spring

= Average compressive load × Displacement

$$= \frac{0.8x}{2} \times x = 0.4\,x^2 \text{ kN-mm} \qquad ...(ii)$$

Since the **entire kinetic energy of the truck is used to compress the spring,** therefore equating equations (*i*) and (*ii*),

$$19\,200 = 0.4\,x^2$$

or $$x^2 = \frac{19200}{0.4} = 48000$$

or $$x = 219 \text{ mm} \qquad \textbf{Ans.}$$

***Example 16.** *A wagon of mass 50 tonnes, starts from rest and travels 30 metres down a 1% grade and strikes a post with bumper spring as shown in Fig. 5.*

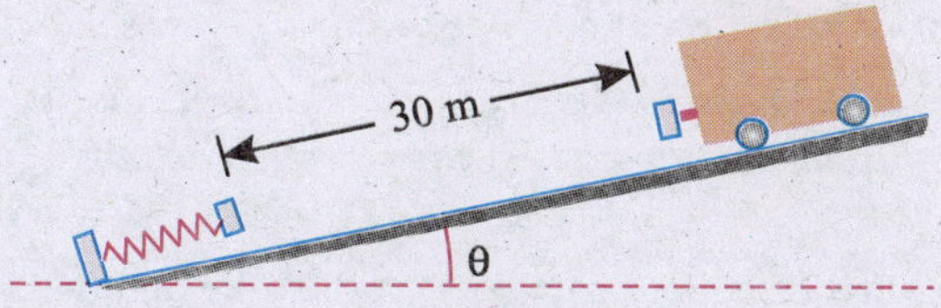

Fig. 5.

If the rolling resistance of the track is 50 N/t, find the velocity with which the wagon strikes the post. Also find the amount by which the spring will be compressed, if the bumper spring compresses 1 mm per 20 kN force. **[AUT, DTU]**

Solution. Given : Mass of the wagon (m) = 50 t ; Initial velocity (u) = 0 (because it starts from rest) ; Distance (s) = 30 m ; Gradient (sin θ) = 1% = 0.01 ; Track resistance = 50 N/t = 50 × 50 = 2500 N = 2.5 kN and bumper spring constant (k) = 1 mm/ 20 kN = 0.05 mm/ kN

Velocity with which the wagon strikes the post

Let v = Velocity with which the wagon strikes the post.

Since the wagon is travelling down the grade, therefore gravitational pull

$$= mg \sin\theta = 50 \times 9.8 \times 0.01 = 4.9 \text{ kN}$$

∴ Net force responsible for moving the wagon

= Gravitational pull – Tractive resistance

$$= 4.9 - 2.5 = 2.4 \text{ kN}$$

Net force responsible for moving the wagon,

$$2.4 = ma = 50\,a$$

∴ $$a = \frac{2.4}{50} = 0.048 \text{ m/s}^2$$

Relation for the velocity of the engine,

$$v^2 = u^2 + 2\,as = (0)^2 + 2 \times 0.048 \times 30 = 2.88$$

or $$v = 1.7 \text{ m/s} \quad \textbf{Ans.}$$

Amount by which the spring will be compressed

Let x = Amount by which the spring will be compressed in mm

Kinetic energy of the wagon

$$= \frac{mv^2}{2} = \frac{50\,(1.7)^2}{2} = 72.25 \text{ kN-m} = 72\,250 \text{ kN-mm} \qquad ...(i)$$

Compressive load $= \dfrac{x}{0.05} = 20x$

∴ Work done in compressing the spring

$$= \text{Average load} \times \text{Displacement}$$

$$= \frac{20x}{2} \times x = 10x^2 \;\text{ kN-mm} \qquad ...(ii)$$

Since the **entire kinetic energy of wagon is used to compress the spring,** therefore equating equations (*i*) and (*ii*),

$$72\,250 = 10\,x^2$$

∴ $$x^2 = \frac{72\,250}{10} = 7225$$

or $x = 85$ mm **Ans.**

***Example 17.** *A bullet of mass 30 g is fired into a body of mass 10 kg, which is suspended by a string 0.8 m long. Due to this impact, the body swings through an angle 30°. Find the velocity of the bullet.* **[AM-IETE, RGPV, BPUT]**

Solution. Given : Mass of bullet $(m) = 30$ g $= 0.03$ kg and mass of body $(M) = 10$ kg.

Let u = Initial velocity of the bullet, and

v = Velocity of the body after impact.

From the geometry of the figure, we find that when the body swings through 30° *i.e.* from *A* to *B*, it has gone up by a distance,

$$h = 0{\cdot}8 - 0{\cdot}8 \cos 30°$$
$$= 0{\cdot}8 - (0{\cdot}8 \times 0{\cdot}866) \text{ m}$$
$$= 0{\cdot}1072 \text{ m.}$$

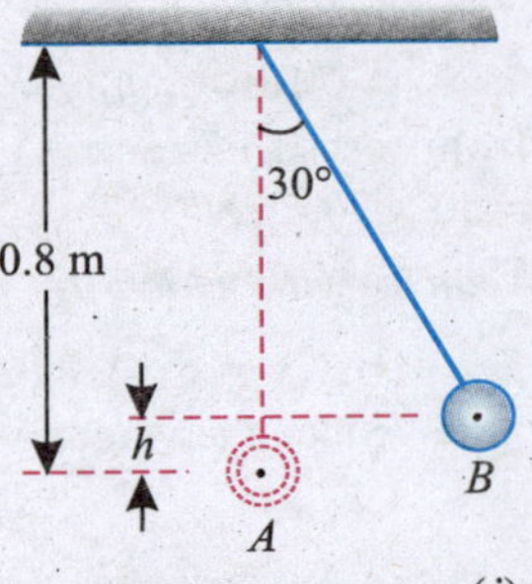

Fig. 6.

Kinetic energy of the body and bullet after impact at *A*.

$$= \frac{(m+M)v^2}{2} = \frac{(0{\cdot}03 + 10) \times v^2}{2} \text{ N-m}$$

$$= 5{\cdot}015\, v^2 \text{ N-m} \qquad ...(i)$$

Potential energy of the body at *B*

$$= (m + M)\,gh = (10 + 0{\cdot}03)\, 9{\cdot}8 \times 0{\cdot}1072 \text{ N-m}$$
$$= 10{\cdot}54 \text{ N-m} \qquad ...(ii)$$

Since the **entire kinetic energy of the body and bullet is used in raising the body (from *A* to *B*)**, therefore equating equations (*i*) and (*ii*),

$$5{\cdot}015\, v^2 = 10{\cdot}54 \qquad \text{or} \qquad v = 1{\cdot}45 \text{ m/s}$$

Momentum of the body and bullet just after impact

$$= (10 + 0{\cdot}03)\, 1{\cdot}45 = 14{\cdot}54 \text{ kg-m/s} \qquad ...(iii)$$

Momentum of the bullet just before impact

$$= 0{\cdot}03\, u \text{ kg-m/s} \qquad ...(iv)$$

Now equating equations (*iii*) and (*iv*),

$$14{\cdot}54 = 0{\cdot}03\, u$$

or $$u = \frac{14{\cdot}54}{0{\cdot}03} = 484{\cdot}7 \text{ m/s}$$ **Ans.**

17. *TRANSFORMATION OF ENERGY

In the previous articles we have discussed potential energy and kinetic energy. Now we shall discuss the transformation of potential energy into kinetic energy. Consider a body just dropped on the ground from A as shown in Fig. 7. Let us consider the ground level as the datum or reference level.

Let m = Mass of the body, and

h = Height from which the body is dropped.

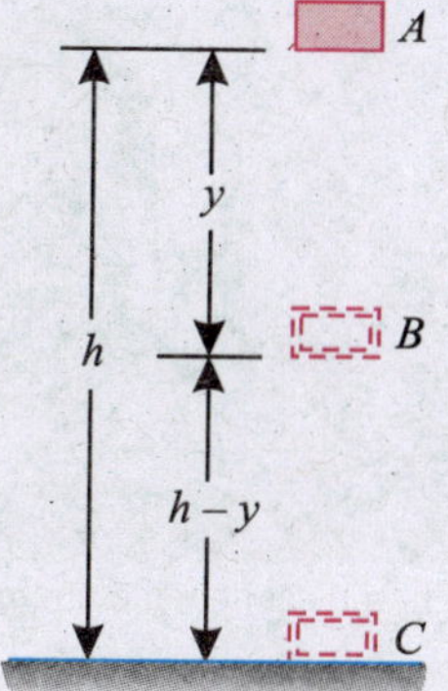

Fig. 7.

Now consider other positions B and C of the same body at various times of the fall. Now we shall find total energy of the body at these positions.

Energy at A

Since the body at A has no velocity, therefore

Kinetic energy at A $= 0$

Potential energy at A $= mgh$

$\therefore$ Total energy at A $= mgh$ ...(*i*)

Energy at B

We know that at B, the body has fallen through a distance (y). Therefore,

Velocity of the body at B $= \sqrt{2gy}$

$\therefore$ Kinetic energy at B $= \dfrac{mv^2}{2} = \dfrac{m(\sqrt{2gy})^2}{2} = mgy$

Potential energy at B $= mg(h - y)$

$\therefore$ Total energy at B $= mgy + mg(h - y) = mgh$ ...(*ii*)

Energy at C

We know that at C, the body has fallen through a distance (h). Therefore,

Velocity of the body at C $= \sqrt{2gh}$

$\therefore$ Kinetic energy at C $= \dfrac{mv^2}{2} = \dfrac{m(\sqrt{2gh})^2}{2} = mgh$

Potential energy at C $= 0$

$\therefore$ Total energy at C $= mgh$ ...(*iii*)

Thus we see that in all positions, the sum of kinetic and potential energies is the same. This proof of the transformation of energy has paved the way for the Law of Conservation of Energy.

18. *LAW OF CONSERVATION OF ENERGY

It states "***The energy can neither be created nor destroyed, though it can be transformed from one form into any of the forms, in which the energy can exist.***"

From the above statement, it is clear, that no machine can either create or destroy energy, though it can only transform from one form into another. We know that the output of a machine is always less than the input of the machine. This is due to the reason that a part of the input is utilised in overcoming friction of the machine. This does not mean that this part of energy, which is used in overcoming the friction, has been destroyed. But it reappears in the form of heat energy at the bear-

In a diesel engine chemical energy of the diesel is converted into heat energy, which is then converted into mechanical energy.

ings and other rubbing surfaces of the machine, though it is not available to us for useful work. The above statement may be exemplified as below :

1. In an electrical heater, the electrical energy is converted into heat energy.
2. In an electric bulb, the electrical energy is converted into light energy.
3. In a dynamo, the mechanical energy is converted into electrical energy.

19. *PILE AND PILE HAMMER

A pile, in its simplest form, is a body of conical shape which is driven into the ground by the impact of a pile hammer. The pile hammer is a body which is released from some height over the pile head as shown in Fig 8.

Pile hammer
h
A
x
B

Fig. 8.

The pile hammer has potential energy before it is released. After release, the potential energy is converted into kinetic energy, with which the pile hammer strikes the pile head. After impact, both the pile and pile hammer move together with a common velocity. This movement is retarded by the resistance of the soil, into which the pile is driven.

Let m = Mass of the pile hammer

M = Mass of the pile

h = Height through which the pile hammer falls before striking the pile,

x = Distance through which the pile is driven into the ground

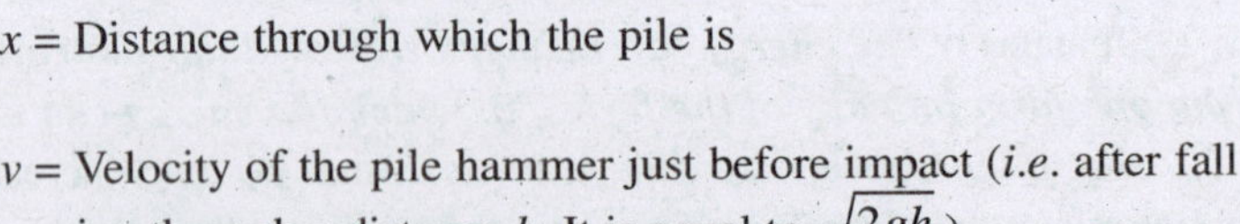

v = Velocity of the pile hammer just before impact (*i.e.* after falling through a distance h. It is equal to $\sqrt{2gh}$).

V = Common velocity of the pile and pile hammer after impact (which is reduced to zero in a distance x), and

R = Average resistance of the soil.

Momentum of the pile hammer and pile just before impact

$$= mv \quad \text{...}(i)$$

Momentum just after impact $= (m + M)\, V$ ...(*ii*)

Equating equations (*i*) and (*ii*),

$$mv = (m + M)\, V$$

$$\therefore \quad V = \frac{mv}{m + M} \quad \text{...}(iii)$$

Now consider B as the datum level. We know that kinetic energy of pile and pile hammer immediately after impact

$$= \frac{(m + M)V^2}{2}$$

Potential energy of pile and pile hammer immediately after impact

$$= (m + M)\, gx$$

$$\therefore \quad \text{Total energy,} \quad E = \frac{(m + M)V^2}{2} + (m + M)\, gx$$

Substituting the value of V from equation (*iii*),

$$E = \frac{(m + M)\left(\dfrac{mv}{m + M}\right)^2}{2} + (m + M)\, gx$$

$$= \frac{m^2 v^2}{2(m + M)} + (m + M)\, gx$$

Now substituting the value of ($v^2 = 2gh$) in the above equation,

$$E = \frac{m^2 \times 2gh}{2(m + M)} + (m + M)\, gx = \frac{m^2 gh}{(m + M)} + (m + M)\, gx \quad \text{...}(iv)$$

Work done by the soil resistance

$$= Rx \quad \text{...}(v)$$

Since the **total energy of the hammer and pile is used in the work done by the soil resistance**, therefore equating equations (*iv*) and (*v*),

$$= \frac{m^2 gh}{(m + M)} + (m + M)\, gx = Rx$$

$$\therefore \quad R = \frac{m^2 gh}{x(m + M)} + (m + M)\, g$$

Note. Sometimes the pile is of negligible mass. In such cases, the soil resistance,

$$R = \frac{m^2 gh}{xm} + mg = mg\left(\frac{h}{x} + 1\right)$$

Example. 18. *A pile of negligible mass is driven by a hammer of mass 200 kg. If the pile is driven 500 mm into the ground, when the hammer falls from a height of 4 metres, find the average force of resistance of the ground.*

Solution. Given : Mass of hammer (m) = 200 kg ; Distance through which the pile is driven into ground (x) = 500 mm = 0.5 m and the height through which hammer falls (h) = 4 m

Average force of resistance of the ground,

$$R = mg\left(\frac{h}{x}+1\right) = 200 \times 9.8\left(\frac{4}{0.5}+1\right) = 1960 \times 9 \text{ N}$$

$$= 17\,640 \text{ N} = 17.64 \text{ kN} \quad \textbf{Ans.}$$

Example 19. *A hammer of mass 0.5 kg hits a nail of 25 g with a velocity of 5 m/s and drives it into a fixed wooden block by 25 mm. Find the resistance offered by the wooden block.* **[MU]**

Solution. Given : Mass of hammer (m) = 0.5 kg ; Mass of nail (M) = 25 g = 0.025 kg ; Velocity of hammer (v) = 5 m/s and distance through which nail is driven into wooden block (x) = 25 mm = 0.025 m.

Let h = Height through which the pile hammer fell before striking the pile.

Velocity of hammer (v),

$$5 = \sqrt{2gh} = \sqrt{2 \times 9.8h} = \sqrt{19.6h}$$

$$\therefore \quad h = \frac{(5)^2}{19.6} = \frac{25}{19.6} = 1.28 \text{ m}$$

Fig. 9.

Resistance offered by the wooden block,

$$R = \frac{m^2gh}{x(m+M)} + (m+M)\,g$$

$$= \frac{(0.5)^2 \times 9.8 \times 1.28}{0.025\,(0.5+0.025)} + (0.5+0.025)9.8 \text{ N}$$

$$= 238.9 + 5.1 = 244 \text{ N} \quad \textbf{Ans.}$$

EXERCISE 3

1. A truck of mass 1.5 t is running at a speed of 54 km.p.h. Find its kinetic energy. If the resistance to the motion is 100 newtons, find how far the truck will run before it stops. (**Ans.** 168.75 kJ ; 1125 m)
2. A bullet moving at the rate of 300 m/s is fired into a thick target and penetrates up to 500 mm. If it is fired into a 250 mm thick target, find the velocity of emergence. Take resistance to be uniform in both cases. (**Ans.** 212.1 m/s)
3. A body of mass 2 kg is thrown vertically upwards with an initial velocity of 9.8 m/s. Find its kinetic energy (*i*) at the moment of its propulsion ; (*ii*) after half a second ; and (*iii*) after one second. (**Ans.** 96.04 N-m ; 2401 N-m ; 0)
4. A railway wagon of mass 20 tonnes runs into a buffer stop having two buffer springs each of 10 kN/mm stiffness. Find the maximum compression of the springs, if the wagon is travelling at 18 km.p.h. (**Ans.** 158 mm)

*5. A block of mass 5 kg is released from rest on an inclined plane as shown in Fig. 10.

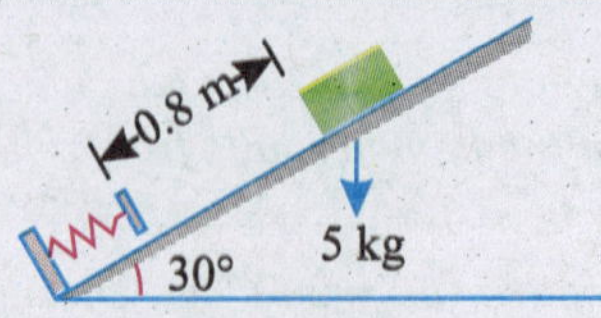

Fig. 10

Find the maximum compression of the spring, if the spring constant is 1 N/mm and the coefficient of friction between the block and the inclined plane is 0.2. (**Ans.** 160 mm)

6. A hammer of mass 400 kg falls through a height of 3 m on a pile of negligible mass. If it drives the pile 1 m into the ground, find the average resistance of the ground for penetration. (**Ans.** 15.7 kN)

*7. A pile hammer of mass 1500 kg drops from a height of 600 mm on a pile of mass 750 kg. The pile penetrates 50 mm per blow. Assuming that the motion of the pile is resisted by a constant force, find the resistance to penetration of the ground. (**Ans.** 139.7 kN)

QUESTIONS

1. Explain the term 'work'. When the work is said to be done ?
2. What are the units of work done ? What is the relation between work done and power ?
3. What do you understand by the term 'energy' ? Explain the various forms of mechanical energies.
4. Explain the term 'conservation of energy'.

OBJECTIVE TYPE QUESTIONS

1. The unit of work done in S.I. system is

   (*a*) Newton (*b*) Joule (*c*) Watt

2. The work done is said to be zero, when

   (*a*) some force acts on a body, but displacement is zero.

   (*b*) no force acts on the body, but displacement takes place,

   (*c*) either (*a*) or (*b*)

3. The rate of doing some work is power.

   (*a*) True (*b*) False

4. One watt is equal to

   (*a*) 0.1 J/s (*b*) 1 J/s

   (*c*) 10 J/s (*d*) 100 J/s

5. The units of energy and work done are the same.

   (*a*) Agree (*b*) Disagree

6. The kinetic energy of a body of mass (m) and velocity (v) is equal to

   (*a*) mv (*b*) $\frac{mv}{2}$ (*c*) $\frac{m^2 v}{2}$ (*d*) $\frac{mv^2}{2}$

7. The potential energy of a mass (m) kg raised through a height (h) metres is

   (*a*) mh newtons (*b*) gh newtons

   (*c*) mgh newtons (*d*) none of these

ANSWERS

1. (*b*) 2. (*c*) 3. (*a*) 4. (*b*) 5. (*a*) 6. (*d*) 7. (*c*)

CHAPTER

28 Mass Moment of Inertia

Contents

1. *MASS MOMENT OF INERTIA

We know about the moment of inertia of **plane figures**. But in this chapter, we shall discuss the moment of inertia of **solid bodies** or mass moment of inertia. The **mass moment of inertia of a solid body about a line is equal to the product of mass of the body and square of the distance from that line**. Mathematically :

$$I = m_1 r_1^2 + m_2 r_2^2 + m_3 r_3^2 + ...$$
$$= \Sigma mr^2$$

where m_1, m_2, m_3 ...... are the masses of the various elements of a body and r_1, r_2, r_3 ...... are distances of the various elements from the fixed line about which the moment of inertia is required to be found out. In the following pages, we shall discuss the mass moment of inertia of the bodies, which are important from the subject point of view.

2. *MASS MOMENT OF INERTIA OF A UNIFORM THIN ROD ABOUT THE MIDDLE AXIS PERPENDICULAR TO THE LENGTH

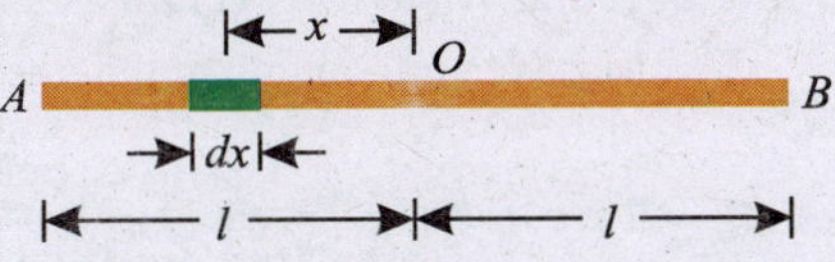

Fig. 1. Uniform rod.

Consider a uniform thin rod AB of length $2l$ with O as its mid-point as shown in Fig. 1.

Let m = Mass per unit length of the rod.

∴ Total mass of the rod,

$$M = 2ml$$

Now consider a small strip of length dx at a distance x from the mid-point O. We know that the mass moment of inertia of the strip about O

$$= (m\,dx)\,x^2 = mx^2\,dx \qquad ...(i)$$

The mass moment of inertia of the whole rod may be found out by integrating the above equation for the whole length of the rod *i.e.* from $-l$ to $+l$. Therefore

$$I = \int_{-l}^{+l} mx^2\,dx = m\left[\frac{x^3}{3}\right]_{-l}^{+l} = m\left[\frac{(+l)^3}{3} - \frac{(-l)^3}{3}\right]$$

$$= \frac{2ml^3}{3} = \frac{Ml^2}{3} \qquad ...(\because M = 2ml)$$

3. *MASS MOMENT OF INERTIA OF A UNIFORM THIN ROD ABOUT ONE OF THE ENDS PERPENDICULAR TO THE LENGTH

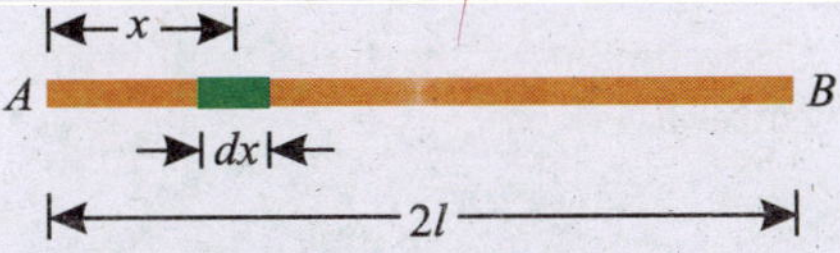

Fig. 2. Uniform rod.

Consider a uniform thin rod AB of length $2l$ as shown in Fig. 2.

Let m = Mass per unit length of the rod.

∴ Total mass of the rod

$$M = 2\,ml$$

Now consider small strip of length dx at a distance x from one of the ends (say A) as shown in Fig. 3. We know that the mass moment of inertia of the strip about A

$$= (m\,dx)\,x^2 = mx^2\,dx \qquad ...(i)$$

The mass moment of inertia of the whole rod may be found out by integrating the above equation for the whole length of the rod *i.e.* from 0 to $2l$. Therefore

$$I = \int_{0}^{2l} mx^2\,dx = m\left[\frac{x^3}{3}\right]_{0}^{2l} = m\left[\frac{(2l)^3}{3} - 0\right]$$

$$= \frac{8\,ml^3}{3} = \frac{4\,Ml^2}{3} \qquad ...(\because M = 2ml)$$

4. *MASS MOMENT OF INERTIA OF A THIN CIRCULAR RING

[JNTU]

Consider a thin circular ring of radius r and O as centre as shown in Fig. 3.

Let $\quad m$ = Mass per unit length of the ring.

∴ Total mass of the ring

$$M = 2\pi rm$$

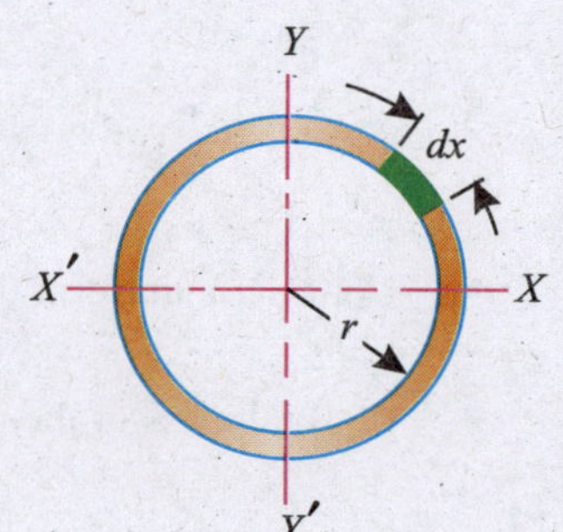

Fig. 3. Circular ring.

Now consider a small element of length dx as shown in Fig. 4. We know that mass moment of inertia of the strip about the central axis

$$= (m\,dx)\,r^2$$

The mass moment of inertia of the whole ring about central axis (*i.e.* at right angles to the plane)

$$I_{ZZ} = (2\pi rm)\,r^2 = Mr^2 \qquad ...(\because M = 2\pi rm)$$

Now as per theorem of perpendicular axis, the mass moment of inertia about X-X or Y-Y axis

$$I_{XX} = I_{YY} = \frac{I_{ZZ}}{2} = \frac{Mr^2}{2} = 0.5\,Mr^2$$

5. *MASS MOMENT OF INERTIA OF A CIRCULAR LAMINA

Consider a circular lamina of radius r with O as centre.

Let $\quad m$ = Mass per unit area of the lamina.

∴ $\quad$ Total mass of the lamina $\quad M = \pi r^2 m$

Now consider an elementary ring of thickness dx at a radius x as shown in Fig. 4. We know that mass of this thin ring

$$= m2\pi x\,dx$$

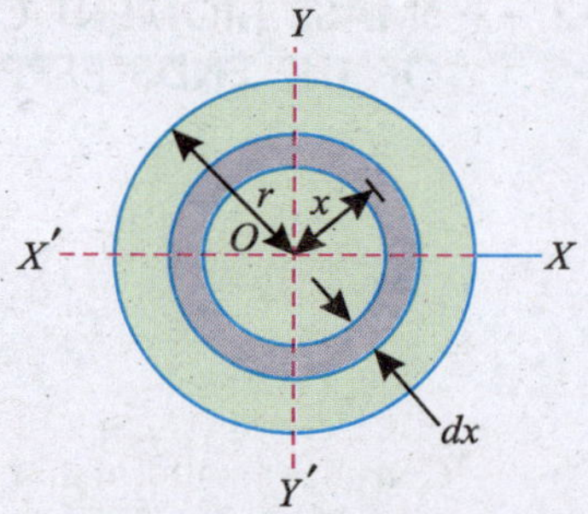

Fig. 4. Circular lamina.

We know that the mass moment of inertia of a thin ring (of mass $m.2\pi x.dx$) about central axis

$$= (m2\pi x\,dx)\,x^2$$
$$= m2\pi x^3\,dx$$

The mass moment of inertia of the whole section about the central axis, may be found out by integrating the above equation for the whole radius of the circle, *i.e.* from *0* to r. Therefore

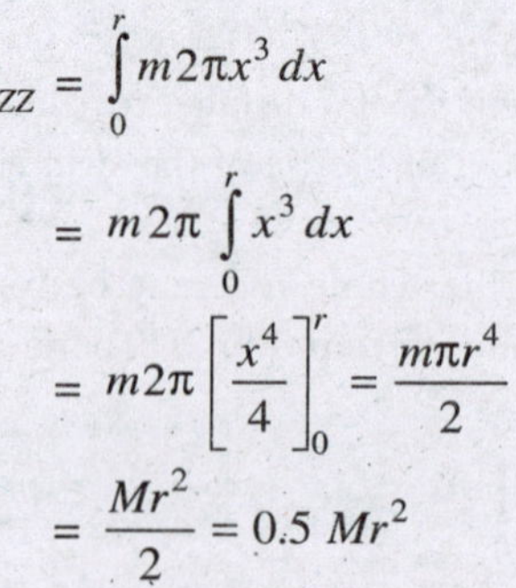

$$I_{ZZ} = \int_0^r m2\pi x^3\,dx$$

$$= m2\pi \int_0^r x^3\,dx$$

$$= m2\pi \left[\frac{x^4}{4}\right]_0^r = \frac{m\pi r^4}{2}$$

$$= \frac{Mr^2}{2} = 0.5\,Mr^2 \qquad ...(\because M = \pi r^2 m)$$

Now as per theorem of perpendicular axis, the mass moment of inertia about X-X or Y-Y axis

$$I_{XX} = I_{YY} = \frac{I_{ZZ}}{2} = \frac{0.5\,Mr^2}{2} = 0.25\,Mr^2$$

Note: The above formula holds good for the mass, moment of inertia of a solid cylinder also. In this case, M is taken as the mass of the solid cylinder.

6. *MASS MOMENT OF INERTIA OF A SOLID SPHERE

Consider a solid sphere of radius r with O as centre.

Let m = Mass per unit volume of the sphere

∴ Total mass of the sphere

$$M = \frac{m\;4\pi r^3}{3}$$

Now consider an elementary plate PQ of thickness dx and at a distance x from O as shown in Fig. 5.

We know that the radius of this plate

$$y = \sqrt{r^2 - x^2}$$

∴ Mass of this plate $= m\pi y^2\,dx = m\pi\,(r^2 - x^2)\,dx$

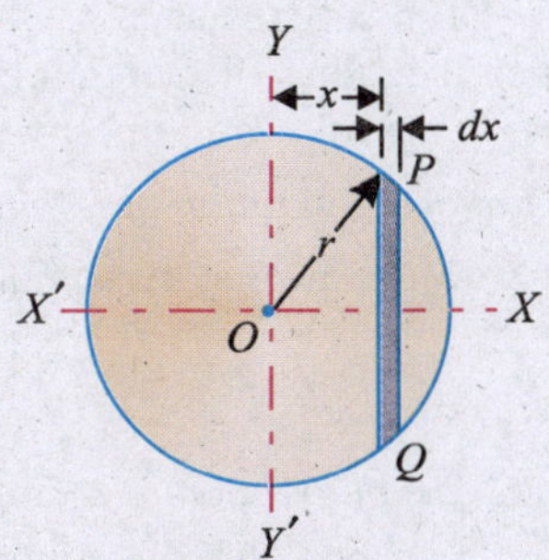

Fig. 5. Sphere.

and moment of inertia of this plate about X-X axis

$$= \text{Mass} \times \frac{(\text{Radius})^2}{2}$$

$$= m\pi\,(r^2 - x^2)\,dx \times \frac{(r^2 - x^2)}{2}$$

$$= \frac{m\pi}{2}\,(r^2 - x^2)^2\,dx$$

$$= \frac{m\pi}{2}\,(r^4 + x^4 - 2r^2x^2)\,dx$$

The mass moment of inertia of the whole sphere may now be found out by integrating the above equation from $-r$ to $+r$. Therefore

$$I = \int_{-r}^{+r} \frac{m\pi}{2}\,(r^4 + x^4 - 2r^2x^2)\,dx$$

$$= \frac{m\pi}{2} \int_{-r}^{+r} (r^4 + x^4 - 2r^2x^2)\,dx$$

$$= \frac{m\pi}{2}\left[r^4x + \frac{x^5}{5} - \frac{2r^2x^3}{3}\right]_{-r}^{+r}$$

$$= \frac{8m\pi r^5}{15} = \frac{2\,Mr^2}{5} = 0.4\,Mr^2 \qquad \ldots\left(M = \frac{m\,4\pi r^3}{3}\right)$$

Note: Since the sphere is symmetrical, therefore the mass moment of inertia of a sphere about any axis is the same.

7. UNITS OF MASS MOMENT OF INERTIA

We have already discussed that the mass moment of inertia of a body is numerically equal to its mass and the square of distance between the centre of gravity of the mass and the point about which the mass moment of inertia is required to be found out. Therefore units of mass moment of inertia depend upon the mass of the body and distance. If mass is in kg and the distance in metres, then the units of mass moment of inertia will be kg-m^2. Similarly, it may be kg-mm^2 etc.

8. *RADIUS OF GYRATION

If the entire mass of a given body be assumed to be concentrated at a certain point, at a distance k from the given axis, such that

$$Mk^2 = I \qquad \text{...(where } I \text{ is the mass moment of inertia of the body)}$$

$$\therefore \qquad k = \sqrt{\frac{I}{m}}$$

The distance k is called *radius of gyration*. Thus the radius of gyration of a body may be defined as the distance from the axis of reference where the whole mass (or area) of a body is assumed to be concentrated.

The suffixes such as X-X or Y-Y are, usually attached to k, which indicate the axis about which the radius of gyration is evaluated. Thus k_{XX} will indicate the radius of gyration about X-X axis. In the following lines, we shall discuss the radius of gyration of some important sections :

1. Radius of gyration of a thin circular ring

We know that mass moment of inertia of a thin circular ring about the central axis,

$$I_{ZZ} = Mr^2 \qquad \text{or} \qquad Mk^2 = Mr^2$$

$$\therefore \qquad k = r$$

2. Radius of gyration of a circular lamina

We know that mass moment of inertia of a circular lamina about the central axis,

$$I_{ZZ} = \frac{Mr^2}{2} \qquad \text{or} \qquad M\,k^2 = \frac{Mr^2}{2}$$

$$\therefore \qquad k = \frac{r}{\sqrt{2}} \qquad \text{or} \qquad k^2 = \frac{r^2}{2} = 0.5\,r^2$$

3. Radius of gyration of a solid sphere

We know that the mass moment of inertia of a solid sphere about any axis,

$$I = 0.4\,Mr^2 \qquad \text{or} \qquad Mk^2 = 0.4\,Mr^2$$

$$\therefore \qquad k = r \times \sqrt{0.4} \qquad \text{or} \qquad k^2 = 0.4\,r^2$$

CHAPTER

29

Kinetics of Motion of Rotation

Contents

1. KINETICS OF MOTION OF ROTATION

We have already discussed the **Kinematics of Motion of Rotation** and its applications. In that chapter, we had discussed the geometry of motion without taking into consideration the forces causing motion of rotation.

But in this chapter, we shall also discuss **Kinetics of Motion of Rotation** taking into consideration the forces, which influence the motion of rotation.

2. *TORQUE

It is the turning moment of a force on the body on which it acts. The torque is equal to the product of the force and the perpendicular distance from any point O to the line of action of the force. Mathematically, torque,

$$T = F \times l$$

where F = Force acting on the body, and

l = Perpendicular distance between the point O and line of action of the force (known as arm or leverage).

Notes:1. The units of torque depend upon the units of force and leverage. If the force is in N and leverage in mm, then the unit of torque will be N-mm. Similarly, if the force is in kN and leverage in m, then the unit of torque will be kN-m.

2. The magnitude of the moment of a force is numerically equal to that of the torque, if the force and the arm is the same. The term torque is used for the moment of a force in the motion of rotation.

3. *WORK DONE BY A TORQUE

Consider a body pivoted at O. Let a tangential force P be applied at a distance r from the pivot as shown in Fig. 1. As a result of the force, let the body rotate through a small angle (θ) in radians.

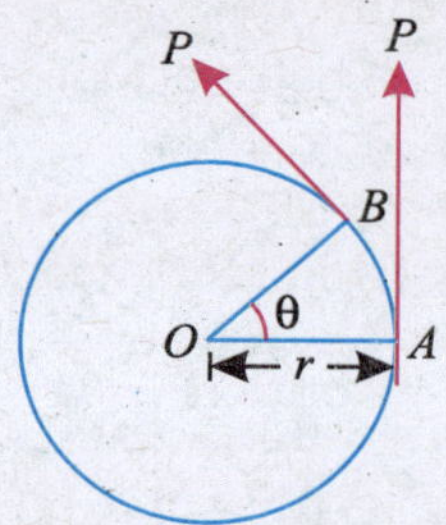

Fig. 1. Work done by a torque.

We know that the length of the arc AB

$$= r\theta$$

and work done by the force P in rotating the body from A to B

$$= P\,(AB) = P\,(r\theta) = Pr\,(\theta)$$

But ($P \times r$) is equal to the torque (T). Therefore work done by a torque is equal to the torque (T) multiplied by the angular displacement (θ) in radians.

4. *ANGULAR MOMENTUM

It is the total motion possessed by a rotating body and is expressed mathematically as :

$$\text{Momentum} = \text{Mass moment of inertia} \times \text{Angular velocity}$$
$$= I\omega$$

5. *NEWTON'S LAWS OF MOTION OF ROTATION

Following are the three Newton's Laws of Motion of Rotation :

1. Newton's First Law of Motion of Rotation states, "*Every body continues in its state of rest or of uniform motion of rotation about an axis, unless it is acted upon by some external torque*".
2. Newton's Second Law of Motion of Rotation states, "*The rate of change of angular momentum of a body is directly proportional to the impressed torque, and takes place in the same direction in which the torque acts*".
3. Newton's Third Law of Motion of Rotation states, "*To every torque, there is always an equal and opposite torque.*"

6. *KINETIC ENERGY OF ROTATION

We know that energy is the capacity to do work. Though the energy exists in many forms, yet the kinetic energy is important from the subject point of view. The kinetic energy of rotation, possesed by a body for doing work by virture of its motion of rotation. Now consider a rotating body, which is brought to rest by a uniform angular retardation due to some torque.

Let I = Mass moment of inertia of the body, and

ω = Angular velocity of the body.

The *kinetic energy of rotation,

$$E = \frac{I\omega^2}{2}$$

*Note: This equation is analogous to the equation for the kinetic energy of translation *i.e.*

$$E = \frac{mv^2}{2} \text{ N-m}$$

***Example 1.** *A circular wheel of mass 50 kg and radius 200 mm is rotating at 300 r.p.m. Find its kinetic energy.*

Solution. Given: Mass of the wheel (M) = 50 kg; Radius of the wheel (r) = 200 mm = 0.2 m and angular velocity (ω) = 300 r.p.m.

We know that mass moment of inertia of the circular wheel,

$$I = 0.5\,Mr^2 = 0.5 \times 50\,(0.2)^2 = 1 \text{ kg-m}^2$$

and angular velocity of the wheel,

$$\omega = 300 \text{ r.p.m.} = 5 \text{ r.p.s.} = 5 \times 2\pi = 10\,\pi \text{ rad/s}$$

∴ Kinetic energy of the rotating wheel,

$$E = \frac{I\omega^2}{2} = \frac{1\,(10\pi)^2}{2} = 493.5 \text{ N-m} = 493{\cdot}5 \text{ J} \quad \textbf{Ans.}$$

7. *TORQUE AND ANGULAR ACCELERATION

Consider a body rotating about its axis.

Let M = Mass of the body

ω = Angular velocity of the body

α = Angular acceleration of the body, and

T = Torque acting on the body

Now split up the body into a number of small particles of mass m_1, m_2, m_3, ...at distance of r_1, r_2, r_3, ...from the axis about which the body is rotating.

We know that the linear acceleration of the particle of mass m_1

$$= r_1\alpha$$

∴ Force acting on this particle

$$= \text{Mass} \times \text{Acceleration} = m_1(r_1\alpha)$$

and moment of this force about the axis

$$= m_1\,(r_1\alpha)\,r_1 = m_1 r_1^2\alpha$$

Similarly, moment of the force for particles of mass m_2

$$= m_2 r_2^2\alpha$$

∴ Total moment for all the particles (*i.e.* torque)

$$T = \Sigma\, m_1 r_1^2\alpha = \alpha\,\Sigma\, m_1 r_1^2$$

But $\Sigma m_1 r_1^2$ is the* mass moment of inertia (if r is taken as the radius of gyration). Therefore

$$T = I\alpha$$

8. *RELATION BETWEEN KINETICS OF LINEAR MOTION AND KINETICS OF MOTION OF ROTATION

Following are the relations between the kinetics of linear motion and kinetics of motion of rotation :

S.No.	*Linear motion*	*Motion of relation*
1.	Mass (M)	Moment of inertia (I)
2.	Force (P)	Torque (T)
3.	Force equation ($P = Ma$)	Torque equation ($T = I\alpha$)
4.	Linear velocity (v)	Angular velocity (ω)
5.	Linear momentum (Mv)	Angular momentum ($I\omega$)
6.	Linear kinetic energy $\left(E = \frac{Mv^2}{2}\right)$	Rotational kinetic energy $\left(E = \frac{I\omega^2}{2}\right)$
7.	Distance traversed (s)	Angular displacement (θ)
8.	Work done (W) = Ps	Work done (W) = $T \times \theta$

9. *FLYWHEEL

It is a circular heavy wheel, generally, fitted to a rotating engine shaft to control variation in its speed during each cycle. Strictly speaking, it serves as a reservoir to store and restore energy by virtue of its inertia. Or in other words, it shares its energy during the period, when the supply of energy is more than the requirement; and releases it during the period when the supply of energy is less than the requirement.

It will be interesting to know that in all types of steam engines and I.C. engines, the power generated and supplied to the shaft is variable. As a matter of fact, the flywheel speed increases when it absorbs energy and decreases when it releases energy. Thus a flywheel is designed to keep the engine speed within the prescribed limits during each cycle.

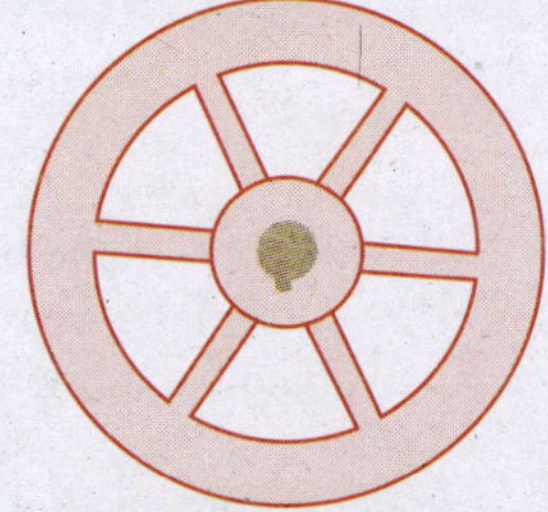

Fig. 2. Flywheel.

The difference between the maximum and minimum speeds of an engine during a cycle is called the fluctuation of speeds. And the ratio of fluctuation of speed, to the mean engine speed, is called *coefficient of fluctuation* of speed.

Let M = Mass of the flywheel,

I = Mass moment of inertia of the flywheel,

N_1 = Maximum speed of flywheel during a cycle,

N_2 = Minimum speed of flywheel during a cycle,

ω_1 = Maximum angular velocity of flywheel during a cycle, and

ω_2 = Minimum angular velocity of flywheel during a cycle.

We know that average speed of the flywheel,

$$N = \frac{1}{2}(N_1 + N_2) \quad \text{in r.p.m.}$$

$$\omega = \frac{1}{2}(\omega_1 + \omega_2) = 0.5\,(\omega_1 + \omega_2) \quad \text{in rad/s}$$

and average kinetic energy of the flywheel,

$$E = \frac{I\omega^2}{2}$$

∴ Fluctuation of energy,

$$E = \text{Maximum K.E.} - \text{Minimum K.E.} = \frac{I\omega_1^2}{2} - \frac{I\omega_2^2}{2} = \frac{I}{2}(\omega_1^2 - \omega_2^2)$$

$$= \frac{I}{2}(\omega_1 + \omega_2)(\omega_1 - \omega_2) = I\omega\,(\omega_1 - \omega_2) \qquad ...[\because \omega = 0.5\,(\omega_1 + \omega_2)]$$

$$= \frac{I \times 2\pi N}{60}\left(\frac{2\pi N_1}{60} - \frac{2\pi N_2}{60}\right) = \frac{4\pi^2}{3600} \times IN\,(N_1 - N_2)$$

$$= \frac{\pi^2}{900} \times IN\,(N_1 - N_2)\,.$$

***Example 2.** *A flywheel of an engine has a mass of 6.5 tonnes and radius of gyration 1.8 metres. If the maximum and minimum speeds of the flywheel are 120 r.p.m. and 118 r.p.m. respectively, find the fluctuation of energy.*

Solution. Given: Mass of flywheel (M) = 6.5 t = 6500 kg; Radius of gyration (k) = 1.8 m: Maximum speed of flywheel (N_1) = 120 r.p.m. and minimum speed of flywheel (N_2) = 118 r.p.m.

We know that average speed of the flywheel,

$$N = \frac{1}{2}(N_1 + N_2) = \frac{1}{2}(120 + 118) = 119 \text{ r.p.m.}$$

and mass moment of inertia,

$$I = Mk^2 = 6500\,(1.8)^2 = 21\,060 \text{ kg-m}^2$$

∴ Fluctuation of energy,

$$E = \frac{\pi^2}{900} \times IN\,(N_1 - N_2) = \frac{\pi^2}{900} \times 21\,060 \times 119\,(120 - 118) \text{ N-m}$$

$$= 54\,970 \text{ N-m} = 54.97 \text{ kN-m} = 54{\cdot}97 \text{ kJ} \qquad \textbf{Ans.}$$

***Example 3.** *A flywheel with a radius of gyration 0.9 m is fitted to a multicylinder engine, which runs at a mean speed of 360 r.p.m. If the speed varies from 2% above the mean to 2% below it and the fluctuation energy is 30 kN-m, find (i) moment of inertia of the wheel and (ii) mass of the flywheel.*

Solution. Given: Radius of gyration of flywheel (k) = 0.9 m; Mean speed of the flywheel (N) = 360 r.p.m. and fluctuation energy (E) = 30 kN-m = 30 000 N-m.

(i) Moment of inertia of the wheel

Let I = Moment of inertia of the wheel.

We know that maximum speed of the wheel

$$N_1 = 360 + (0.02 \times 360) = 367.2 \text{ r.p.m.}$$

and minimum speed of the wheel,

$$N_2 = 360 - (0.02 \times 360) = 352.8 \text{ r.p.m.}$$

∴ Fluctuation of energy (E),

$$30\ 000 = \frac{\pi^2}{900} \times IN\ (N_1 - N_2)$$

$$= \frac{\pi^2}{900} \times I \times 360\ (367.2 - 352.8) = 56.85\ I$$

$$\therefore \quad I = \frac{30\ 000}{56.85} = 527.7 \text{ kg-m}^2 \quad \textbf{Ans.}$$

(*ii*) *Mass of the wheel*

We know that mass of the wheel,

$$M = \frac{I}{k^2} = \frac{527.7}{(0.9)^2} = 651.5 \text{ kg} \quad \textbf{Ans.}$$

***Example 4.** *A flywheel of mass 8 tonnes starts from rest, and gets up a speed of 180 r.p.m. in 3 minutes. Find the average torque exerted on it, if the radius of gyration of the flywheel is 60 cm.*

[UPTU, JNTU, AMIE]

Solution. Given: Mass of the flywheel (M) = 8 t = 8000 kg; Initial angular speed (ω_0) = 0 (because it starts from rest); Final angular speed (ω) = 180 r.p.m. = $\frac{180 \times 2\pi}{60}$ = 6π rad/s.; Time (t) = 3 min = 180 s and radius of the gyration of the flywheel (k) = 60 cm = 0.6 m.

Let α = Constant angular acceleration of the flywheel

We know that the mass moment of inertia of the flywheel,

$$I = Mk^2 = 8000 \times (0.6)^2 = 2880 \text{ kg-m}^2$$

and final angular velocity of the flywheel (ω),

$$6\,\pi = \omega_0 + \alpha t = 0 + \alpha \times 180$$

or

$$\alpha = \frac{6\pi}{180} = 0.105 \text{ rad/s}^2$$

∴ Average torque exerted by the flywheel.

$$T = I\alpha = 2880 \times 0.105 = 302.4 \text{ N-m} \quad \textbf{Ans.}$$

***Example 5.** *A flywheel is made up of steel ring 40 mm thick and 200 mm wide plate with mean diameter of 2 metres. If initially the flywheel is rotating at 300 r.p.m., find the time taken by the wheel in coming to rest due to frictional couple of 100 N-m.*

Take mass density of the steel as 7900 kg/m³. Neglect the effect of the spokes. **[AMIE]**

Solution. Given: Thickness of flywheel = 40 mm = 0.04 m; Width of flywheel = 200 mm = 0.2 m; Mean diameter of flywheel = 2 m or mean radius (r) = 1 m; Initial angular speed (ω_0) = 300 r.p.m. = 5 r.p.s. = 10π rad/s; Frictional couple = 100 N-m and density of steel = 7900 kg/m³.

Let α = Constant angular acceleration of flywheel and

t = Time taken by the flywheel in coming to rest.

We know that volume of flywheel,

$$= \pi \times 2 \times 0.2 \times 0.04 = 0.05 \text{ m}^3$$

∴ Mass of the flywheel,

$$M = 0.05 \times 7900 = 395 \text{ kg}$$

and mass moment of inertia, $I = Mr^2 = 395 \times (1)^2 = 395$ kg-m²

We know that frictional couple (or torque T)

$$100 = I\alpha = 395\ \alpha$$

$$\therefore \qquad \alpha = \frac{100}{395} = 0.253 \text{ rad/s}^2$$

and final angular velocity of the flywheel,

$$0 = \omega_0 - \alpha t = 10\pi - 0.253\,t \qquad \text{...(Minus sign due to retardation)}$$

$$\therefore \qquad t = \frac{10\pi}{0.253} = 124.2 \text{ s} \qquad \textbf{Ans.}$$

EXERCISE 1

1. A wheel has a string of length 4 m wrapped round its shaft. The string is pulled with a constant force of 150 N. It is observed that when the string leaves the axle, the wheel is rotating at 3 revolutions in a second. Find the moment of inertia of the wheel. [**Ans.** 3.38 kg-m^2]
2. The flywheel of a steam engine of mass 1000 kg has radius of gyration as 1 m. If the maximum and minimum speed of the flywheel is 80 r.p.m. and 78 r.p.m. respectively, find the fluctuation of energy. [**Ans.** 1732.7 N-m]
3. A flywheel of an engine has a mass of 1250 kg and radius of gyration 600 mm. Find the angular acceleration of the wheel, when it is subjected to a torque of 12 500 N-m. [**Ans.** 27.78 rad/s^2]
4. A constant torque of 2 kN-m is exerted on a crankshaft to start the engine. The flywheel has a mass of 1800 kg and radius of gyration 1 m. If there is a resisting torque of 1 kN-m, find the speed of the engine after 1 minute. [**Ans.** 320.9 r.p.m.]

   [**Hint.** Effective torque = 2 – 1 = 1 kN-m]
5. A flywheel of mass 400 kg and radius of gyration 1 m losses its speed from 300 r.p.m. to 240 r.p.m. in 120 seconds. Determine the retarding torque acting on it.[**Ans.** 20.8 N-m]
6. A retarding torque of 600 N-m is applied to a flywheel rotating at 240 r.p.m. Find the moment of inertia of the flywheel, if it comes to rest in 100 seconds.[**Ans.** 2390 kg-m^2]

10. *MOTION OF A BODY TIED TO A STRING AND PASSING OVER A PULLEY

Consider a body tied to a string and passing over a pulley as shown in Fig. 3. Let the body descend under the force of gravity.

Let m = Mass of the body

M = Mass of the pulley

I = Moment of inertia of the pulley,

r = Radius of the pulley,

k = Radius of gyration of the pulley.

a = Linear acceleration of the body,

α = Angular acceleration of the pulley, and

P = Pull in the string

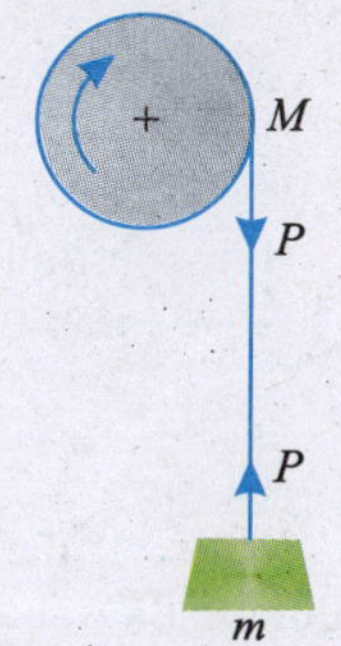

Fig. 3. Motion of one body.

First of all, consider the motion of the body, which is coming down. We know that the forces acting on it are $m.g.$ (downwards) and P (upwards). As the body is moving downwards, therefore resultant force acting on it

$$= mg - P \quad \text{...(i)}$$

Since the body is moving downwards, with an acceleration (a), therefore force acting on it.

$$= ma \quad \text{...(ii)}$$

Equating equations (i) and (ii),

$$mg - P = ma \quad \text{..(iii)}$$

Now consider motion of the pulley, which is rotating about its axis due to downward motion of the body tied to the string. We know that linear acceleration of the body is equal to the angular acceleration of the pulley.

$$\therefore \quad a = r\alpha \quad \text{...(iv)}$$

and torque, T = Tension in the string × Radius of the pulley

$$= P \times r \quad \text{...(v)}$$

We also know that, Mass Moment of Inertia of the pulley, $I = Mk^2$

and Torque on the pulley, $T = I\alpha$ ...(vi)

Equating equations (v) and (vi),

$$P \times r = I\alpha$$

$$Pr^2 = I\alpha r \quad \text{...(Multiplying both sides by } r\text{)}$$

$$= Ia \quad \text{...(}\because a = r\alpha\text{)}$$

$$\therefore \quad P = \frac{Ia}{r^2} \quad \text{...(vii)}$$

Substituting the value of P in equation (iii),

$$mg - \frac{Ia}{r^2} = ma$$

$$ma + \frac{Ia}{r^2} = mg$$

$$a\left(m + \frac{I}{r^2}\right) = mg$$

$$\therefore \quad a = \frac{mg}{\left(m + \frac{I}{r^2}\right)} = \frac{mg}{\left(m + \frac{Mk^2}{r^2}\right)} \quad \text{...(viii)}$$

Now substituting the value of a in equation (vii),

$$P = \frac{I}{r^2} \times \frac{mg}{\left(m + \frac{I}{r^2}\right)} = \frac{Img}{mr^2 + I} \quad \text{...(ix)}$$

Note: If the pulley is a solid disc, then its mass moment of inertia (I) is 0.5 Mr^2. Now substituting the value of I in equation ($viii$) and (ix).

$$a = \frac{mg}{m + \frac{0.5Mr^2}{r^2}} = \frac{mg}{m + 0.5\,M} = \frac{2mg}{2m + M}$$

and

$$P = \frac{Img}{mr^2 + I} = \frac{0.5\,Mr^2\,mg}{mr^2 + 0.5Mr^2} = \frac{M\,mg}{2m + M}$$

***Example 6.** *A homogeneous solid cylinder of mass 100 kg and 1 m diameter, whose axis is horizontal, rotates about its axis, in frictionless bearings under the action of a falling block of mass 10 kg, which is carried by a thin rope wrapped around the cylinder.* **[AMIE]**

What will be the angular velocity of the cylinder two seconds after the motion? Neglect the weight of the rope.

Solution. Given: Mass of cylinder (M) = 100 kg; Diameter of the cylinder (D) = 1 m or radius (r) = 0.5 m; Mass of the block (m) = 10 kg and time (t) = 2 s.

We know that linear acceleration of the solid cylinder,

$$a = \frac{2mg}{2m + M} = \frac{2 \times 10 \times 9.8}{(2 \times 10) + 100} = 1.63 \text{ m/s}^2$$

and angular acceleration,

$$\alpha = \frac{a}{r} = \frac{1.63}{0.5} = 3.26 \text{ rad/s}^2$$

∴ Angular velocity of the cylinder 2 seconds after the motion.

$$\omega = \omega_0 + \alpha t = 0 + (3.26 \times 2) = 6.52 \text{ rad/s} \quad \textbf{Ans.}$$

***Example 7.** *A body of mass 6 kg is suspended by a light rope wound round a solid disc of 60 kg and diameter 50 cm, the other end of the rope being fixed to the periphery of the pulley.*

Find (i) acceleration of the descending mass, (ii) pull in the rope, and (iii) velocity after the mass has descended 15 m. Take g as 9.8 m/s².

Solution. Given: Mass of the body (m) = 6 kg; Mass of the solid disc (M) = 60 kg; Diameter of the disc (D) = 50 cm and distance (s) = 15 m.

(*i*) *Acceleration of the descending mass*

We know that acceleration of the descending mass,

$$a = \frac{2mg}{2m + M} = \frac{2 \times 6 \times 9.8}{(2 \times 6) + 60} = 1.63 \text{ m/s}^2 \quad \textbf{Ans.}$$

(*ii*) *Pull in the rope*

We know that pull in the rope,

$$P = \frac{Mmg}{2m + M} = \frac{60 \times 6 \times 9.8}{(2 \times 6) + 60} = 49 \text{ N} \quad \textbf{Ans.}$$

(*iii*) *Velocity after the mass has descended 15 m*

Let v = Velocity after the mass has descended 15 m.

We know that $v^2 = u^2 + 2\,a.s = 0 + (2 \times 1.63 \times 15) = 48.9$

∴ $v = 7.0$ m/s **Ans.**

Example 8. *A wheel has a 5.4 m long string wrapped round its shaft. The string is pulled with a constant force of 98 N and it is observed that the wheel is rotating at 3 revolutions per second, when the string leaves the axle. Find the moment of inertia of the wheel about its axis.*

Solution. Given: Length of the string (l) = 5.4 m; Force (P) = 98 N and angular velocity (ω) = 3 r.p.s. = 6π rad/s.

Let I = Moment of inertia of the wheel about its axis

We know that work done in pulling the string

$$= \text{Force} \times \text{Distance} = 98 \times 5.4 = 529.2 \text{ N-m} \quad \ldots(i)$$

and kinetic energy of the wheel,

$$E = \frac{I\omega^2}{2} = \frac{I\,(6\pi)^2}{2} = 177.7\ I \text{ N-m} \qquad ...(ii)$$

Now equating the work done and kinetic energy,

$$177.7\ I = 529.2$$

$$\therefore \quad I = \frac{529.2}{177.7} = 2.98 \text{ kg-m}^2 \quad \textbf{Ans.}$$

***Example 9.** *A solid cylindrical pulley of mass 800 kg, having 0.8 m , radius of gyration and 2 m diameter, is rotated by an electric motor, which exerts a uniform torque of 60 kN-m. A body of mass 3 t is to lifted by a wire wrapped round the pulley.*

Find (i) acceleration of the body; and (ii) tension in the rope.

Solution. Given: Mass of pulley (M) = 800 kg; Radius of gyration (k) = 0.8 m; Diameter of pulley (d) = 2 m or radius (r) = 1 m; Torque (T) = 60 kN-m = 60000 N-m and mass of the body to be lifted (m) = 3 t = 3000 kg.

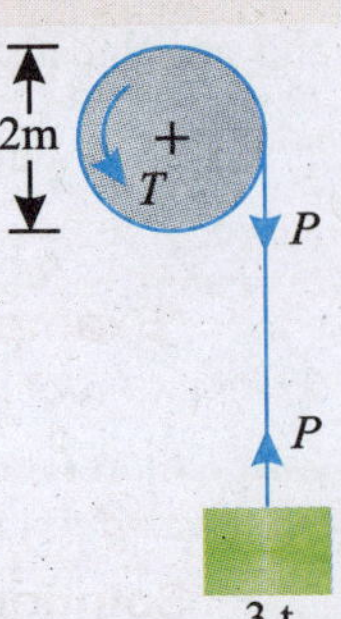

Fig. 4.

(i) Acceleration of the body

Let a = Acceleration of the body,

α = Angular acceleration of the pulley,

P = Tension in rope.

First of all, consider the motion of the hanging body, which is going upwards due to the torque. We know that forces acting on it are $mg = 3000 \times 9.8 = 29\,400$ N (downwards) and P (upwards). As the body is going upwards, therefore the resultant force

$$= P - 29\,400 \qquad ...(i)$$

Since the body is moving with an acceleration (a), therefore force acting on the body,

$$= ma = 3000\ a \qquad ...(ii)$$

Equating equations (i) and (ii),

$$P - 29\,400 = 3000\ a \qquad ...(iii)$$

Now consider the motion of the pulley, which is rotating about its axis due to the torque. We know that moment of inertia of the pulley

$$I = Mk^2 = 800\ (0.8)^2 = 512 \text{ kg-m}^2$$

$\therefore$ Accelerating torque

$$T_1 = I\alpha = 512\ \alpha = 512\ a \qquad ...(\because a = \alpha)$$

and torque due to tension in the rope

$$T_2 = P \times r = P \times 1 = P$$

Total torque $= T_1 + T_2 = 512\ a + P$

This total torque is equal to the torque exerted by the electric motor

$$\therefore \quad 60\,000 = 512\ a + P$$

or $$P = 60\,000 - 512\ a$$

Substituting the value of P in equation (iii),

$$60\,000 - 512\ a - 29\,400 = 3000\ a$$

$$3512\ a = 60\,000 - 29\,400 = 30\,600$$

$$\therefore \quad a = \frac{30\,600}{3512} = 8.71 \text{ m/s}^2 \quad \textbf{Ans.}$$

(*ii*) *Tension of the rope*

Substituting the value of a in equation (*iii*),

$$P - 29\,400 = 3000 \times 8.71 = 26\,130$$

$$P = 26\,130 + 29\,400 = 55\,530 \text{ N} = 55.53 \text{ kN} \qquad \textbf{Ans.}$$

11. *MOTION OF TWO BODIES CONNECTED BY A STRING AND PASSING OVER A PULLEY

Consider two bodies connected by an inextensible light string and passing over a simple pulley as shown in Fig. 5.

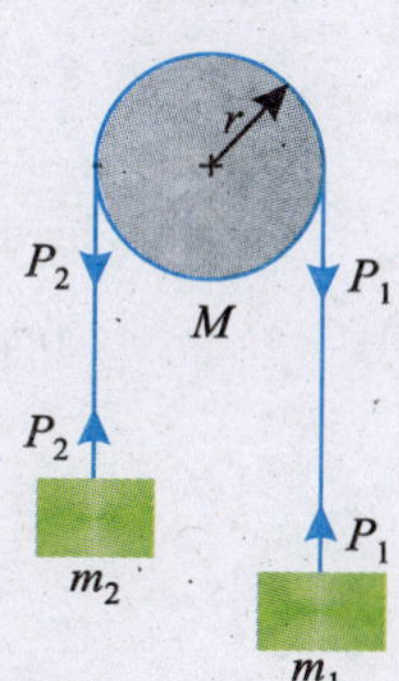

Fig. 5. Motion of two bodies

Let

m_1 and m_2 = Masses of the two bodies,

M = Mass of the pulley,

I = Moment of inertia of the pulley,

r = Radius of the pulley,

k = Radius of gyration of the pulley,

a = Acceleration of the two bodies,

α = Angular acceleration of the pulley,

and

P_1 and P_2 = Pulls in the strings.

First of all, consider the motion of body 1 of mass m_1, which is coming down. We know that the forces acting on it are $m_1 g$ (downwards) and P_1 (upwards). As the body is moving downwards, therefore, resultant force

$$= m_1 g - P_1 \qquad ...(i)$$

Since the body is moving downwards with an acceleration (a), therefore force acting on it

$$= m_1 a \qquad ...(ii)$$

Equating equations (*i*) and (*ii*),

$$m_1 g - P_1 = m_1 a \qquad ...(iii)$$

Now consider the motion of body 2 of mass m_2, which is going upwards. We know that the forces acting on it are $m_2 g$ (downwards) and P_2 (upwards). As the body in moving upwards, therefore, resultant force

$$= P_2 - m_2 g \qquad ...(iv)$$

Since the body is moving upwards with an acceleration (a) therefore force acting on it

$$= m_2 a \qquad ...(v)$$

Equating equations (*iv*) and (*v*),

$$P_2 - m_2 g = m_2 a \qquad ...(vi)$$

Now consider motion of the pulley, which is rotating about its axis due to downward motion of the body of mass m_1 tied to the string. We know that linear acceleration of the body 1 is equal to the angular acceleration of the pulley.

$$\therefore \quad a = r\alpha \qquad ...(vii)$$

and torque, T = Net tension in the string × Radius of pulley

$$= (P_1 - P_2)\, r \qquad ...(viii)$$

We also know that, Mass Moment of Inertia of the pulley, $I = Mk^2$

and Torque on the pulley,

$$T = I\alpha \qquad ...(ix)$$

Equating equations (*viii*) and (*ix*),

$$(P_1 - P_2)\, r = I\alpha$$

$$(P_1 - P_2)\, r^2 = I\alpha r \quad \text{...(Multiplying both sides by } r)$$

$$= Ia \quad \text{...}(\because a = r\alpha)$$

$$\therefore \quad (P_1 - P_2) = \frac{Ia}{r^2}$$

Adding equations (*iii*) and (*vi*)

$$m_1 g - P_1 + P_2 - m_2 g = m_1 a + m_2 a$$

$$g\,(m_1 - m_2) - (P_1 - P_2) = a\,(m_1 + m_2)$$

Substituting the value of $(P_1 - P_2)$ in the above equation,

$$g\,(m_1 - m_2) - \frac{I.a}{r^2} = a\,(m_1 + m_2)$$

$$g\,(m_1 - m_2) = a\left[\frac{I}{r^2} + (m_1 + m_2)\right]$$

$$\therefore \quad a = \frac{g\,(m_1 - m_2)}{\left[\dfrac{I}{r^2} + (m_1 + m_2)\right]}$$

Notes: 1. The values of pulls (or tensions) in the two strings (P_1 and P_2) may now be found out from the equations (*iii*) and (*vi*) *i.e.*

$$P_1 = m_1 g - m_1 a = m_1\,(g - a)$$

and

$$P_2 = m_2 a + m_2 g = m_2\,(a + g)$$

2. There will be some torque acting on the pulley due to the difference of tension in the two strings, such that

$$T_1 = (P_1 - P_2)\, r$$

where r = Radius of the pulley.

3. The net torque acting on the pulley,

$$T = T_2 - T_1$$

where T_2 is the torque obtained from the relation $(T = I\alpha)$

***Example 10.** *Two bodies of masses 15 kg and 5 kg are attached to the two ends of a flexible rope, which is passed over a pulley of mean radius 200 mm having a mass of 10 kg and radius of gyration 150 mm. Find the acceleration of the masses and pulls on either side of the rope.*

Solution. Given: Mass of the first body (m_1) = 15 kg; Mass of the second body (m_2) = 5 kg; Radius of the pulley (r) = 200 mm = 0.2 m; Mass of the pulley (M) = 10 kg and radius of gyration (k) = 150 mm = 0.15 m.

Acceleration of the masses

We know that the mass moment of inertia of the pulley,

$$I = Mk^2 = 10\,(0.15)^2 = 0.225 \text{ kg-m}^2$$

and acceleration of the masses,

$$a = \frac{g\,(m_1 - m_2)}{\left[\dfrac{I}{r^2} + (m_1 + m_2)\right]} = \frac{9.8(15 - 5)}{\left[\dfrac{0.225}{(0.2)^2} + (15 + 5)\right]} \text{ m/s}^2$$

$$= \frac{98}{5{\cdot}625 + 20} = 3.82 \text{ m/s}^2 \qquad \textbf{Ans.}$$

Pulls on either side of the rope

We know that pull in the rope with 15 kg mass,

$$P_1 = m_1 (g - a) = 15 (9.8 - 3.82) = 89.7 \text{ N} \qquad \textbf{Ans.}$$

and pull in the rope with 5 kg mass,

$$P_2 = m_2 (a + g) = 5 (3.82 + 9.8) = 68.1 \text{ N} \qquad \textbf{Ans.}$$

***Example 11.** *Two bodies A and B of masses 800 kg and 600 kg are attached at the ends of a flexible rope. The rope passes over a pulley of 800 mm diameter. The pulley has a mass of 100 kg with a radius of gyration as 400 mm about its axis of rotation.*

Find the torque, which must be applied to the pulley to raise the 800 kg body with an acceleration of 1 m/s². Neglect friction of the spindle.

Solution. Given: Mass of the body A (m_1) = 800 kg; Mass of the body B (m_2) = 600 kg; Diameter of pulley = 800 mm = 0.8 m or radius (r) = 0.4 m; Mass of the pulley (M) = 100 kg; Radius of gyration (k) = 400 mm = 0.4 m and *acceleration (a) = – 1.0 m/s².

We know that pull in the rope carrying 800 kg mass,

$$P_1 = m_1 (g - a) = 800 [9.8 - (-1)] = 800 \times 10.8 = 8640 \text{ N}$$

and pull in the rope carrying 600 kg mass,

$$P_2 = m_2 (a + g) = 600 [(-1) + 9.8] = 600 \times 8.8 = 5280 \text{ N}$$

We also know that moment of inertia of the pulley,

$$I = Mk^2 = 100 (0.4)^2 = 16 \text{ kg-m}^2$$

and angular acceleration of the pulley,

$$\alpha = \frac{a}{r} = \frac{-1}{0.4} = -2.5 \text{ rad/s}^2$$

∴ Torque due to rotation of the pulley,

$$T_1 = I\alpha = 16 (-2.5) = -40 \text{ N-m}$$

and torque acting on the pulley due to difference of pulls in the two ropes,

$$T_2 = (P_1 - P_2)\, r = (8640 - 5280)\, 0.4 = 1344 \text{ N-m}$$

∴ Torque which must be applied to the pulley

$$= T_2 - T_1 = 1344 - (-40) = 1384 \text{ N-m} \qquad \textbf{Ans.}$$

---

*Note: Under normal circumstances, the body of mass 800 kg is supposed to come down (as it is heavier than the other body). Therefore its downward motion is taken as positive. In this case, the body of mass 800 kg is being raised. Therefore its acceleration is taken as negative.

***Example 12.** *A train loaded with cars has a total mass of 6 t is hauled up an incline of 1 in 30 by a rope coiled on a winding drum at the top of the incline. The drum is 1.0 m in diameter, and has a mass of 1 ton and has a radius of gyration of 0.45 m. The drum axle friction is negligible.*

If a torque of 3 kN-m is applied to the drum from a driving motor, determine the tension in the rope. **[AMIE]**

Solution. Given: Mass of train and cars (M) = 6 t = 6000 kg; Inclination $(\sin \theta)$ = 1/30; Diameter of drum (d) = 1 m or radius (r) = 0.5 m; Mass of drawn (m) = 1 t = 1000 kg, Radius of gyration (k) = 0.45 m and torque applied (T) = 3 kN-m.

Let P = Tension in the rope

a = Linear acceleration of the train, and

α = Angular acceleration of the drum.

First of all, consider the motion of the train, which is being pulled by the rope due to the torque on the drum.

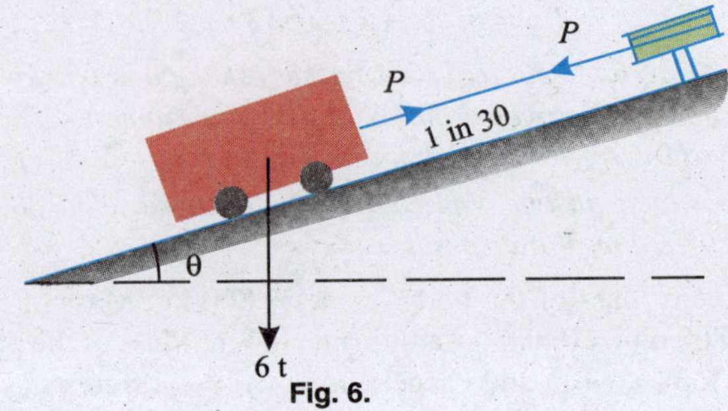

Fig. 6.

We know that component of the weight of the train along the plane

$$= Mg \sin \theta = 6000 \times 9.8 \times \frac{1}{30} = 1960 \text{ N} = 1.96 \text{ kN}$$

As the train loaded with cars is moving upwards, therefore resultant force along the plane

$$= P - 1.96 \qquad ...(i)$$

Since the cage is moving with an acceleration of (a), therefore force acting on it

$$= Ma = 6\,a \qquad ...(ii)$$

Equating equations (i) and (ii),

$$P - 1.96 = 6\,a \qquad ...(iii)$$

Now consider motion of the drum, which is rotating about its axis due to torque. We know that moment of inertia of the drum,

$$I = mk^2 = 1000 \times (0.45)^2 = 0.2 \text{ kg-m}^2$$

We know that linear acceleration of the train is equal to the angular acceleration of the drum.

$$\therefore \quad a = r\alpha = 0.5\,\alpha \quad \text{or} \quad \alpha = \frac{a}{0{\cdot}5} = 2a$$

Now accelerating torque,

$$T_1 = I\alpha = 0.2 \times 2\,a = 0.4\,a \text{ kN-m}$$

and torque due to tension in the rope,

$$T_2 = P \times r = P \times 0.5 = 0.5\,P$$

$\therefore$ Total torque, $T = T_1 + T_2 = 0.4\,a + 0.5\,P$

We know that this total torque is equal to the torque applied to the drum.

$$3 = 0.4\,a + 0.5\,P$$

or
$$P = \frac{3 - 0.4\,a}{0.5} = 6 - 0.8\,a$$

Substituting this value of P in equation (iii),

$$6 - 0.8\,a - 1.96 = 6\,a \quad \text{or} \quad 6.8\,a = 4.04$$

$$\therefore \quad a = \frac{4.04}{6.8} = 0.59 \text{ m/s}^2$$

Now substituting this value of a in equation (iii),

$$P - 1.96 = 6 \times 0.59 = 3.54$$

or
$$P = 3.54 + 1.96 = 5.5 \text{ kN}$$

***Example 13.** *Two bodies A and B of masses 30 kg and 10 kg are tied to the two ends of a light string passing over a composite pulley of radius of gyration as 70 mm and mass 4 kg as shown in Fig. 7.*

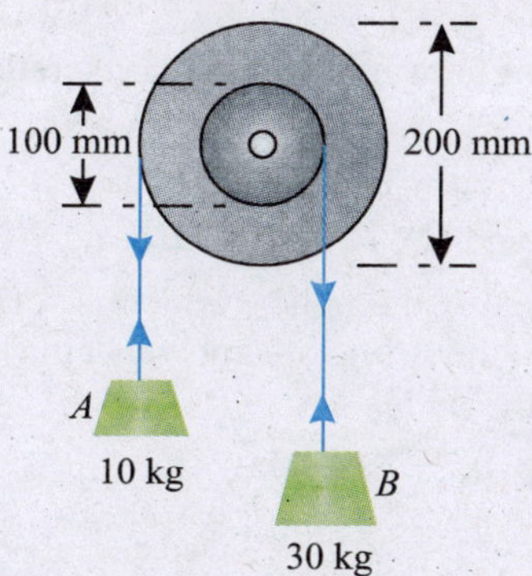

Fig. 7.

Find the pulls in the two parts of the string and the angular acceleration of the pulley. **[AUT]**

Solution. Given: Mass of the body A (m_1) = 30 kg; Mass of the body B (m_2) = 10 kg; Radius of gyration of the pulley (k) = 70 mm = 0.07 m; Mass of the pulley (M) = 4 kg; Internal diameter of the pulley (d_1) = 100 mm = 0.1 m or radius (r_1) = 0.05 m and external diameter of pulley (d_2) = 200 mm = 0.2 m or radius (r_2) = 0.1 m.

Angular acceleration of the pulley

Let P_1 = Pull in the string carrying 30 kg mass,

P_2 = Pull in the string carrying 10 kg mass, and

α = Angular acceleration of the body.

From the geometry of the masses, we find that turning moment of the mass 30 kg (*i.e.*, $30 \times 0.05 = 1.5$ kg-m) is more than that of the mass 10 kg (*i.e.*, $10 \times 0.1 = 1$ kg-m). It is thus obvious, that the 30 kg mass will come downwards and the 10 kg mass will go upwards, when the system is released.

Let a_1 = Acceleration of the mass 30 kg, and

a_2 = Acceleration of the mass 10 kg,

We know that the mass moment of inertia of the pulley,

$$I = Mk^2 = 4\,(0.07)^2 = 0.02 \text{ kg-m}^2$$

First of all consider the motion of mass 30 kg, which is coming down. We know that the forces acting on it are $m_1 g = 30 \times 9.8 = 294$ newtons (downwards) and P_1 newtons (upwards). As the mass is moving downwards therefore, resultant force

$$= 294 - P_1 \qquad ...(i)$$

Since the mass is moving downwards with an acceleration (a_1), therefore force acting on the body

$$= 30\,a_1 \qquad ...(ii)$$

Equating equations (*i*) and (*ii*),

$$294 - P_1 = 30\,a_1 \qquad ...(iii)$$

Now consider the mass 10 kg, which is going upwards. We know that the forces acting on it are $m_2 g = 10 \times 9.8 = 98$ newtons (downwards) and P_2 newtons (upwards). As the body is moving upwards, therefore, resultant force

$$= P_2 - 98 \quad ...(iv)$$

Since the mass is moving upwards with an acceleration (a_2), therefore force acting on the body

$$= 10\,a_2 \quad ...(v)$$

Equating equations (*iv*) and (*v*), $P_2 - 98 = 10\,a_2$ ...(*vi*)

Now consider the motion of the pulley, which is rotating about its axis due to downward motion of the 30 kg mass tied to the string. We know that the linear acceleration of the body is equal to the angular acceleration of the pulley.

$\therefore \quad a_1 = r_1\alpha = 0.05\,\alpha$

Similarly $\quad a_2 = r_2\alpha = 0.1\,\alpha$

and torque, $\quad T = P_1 r_1 - P_2 r_2 = P_1 \times 0.05 - P_2 \times 0.1$ ...(*vii*)

We also know that torque on the pulley, $T = I\alpha = 0.02\,\alpha$ ...(*viii*)

Equating equations (*vii*) and (*viii*),

$$0.05\,P_1 - 0.1\,P_2 = 0.02\,\alpha$$

$$P_1 - 2\,P_2 = 0.4\,\alpha \quad \text{...(Multiplying by 20)}$$

$$P_1 = 0.4\,\alpha + 2P_2 \quad ...(ix)$$

Substituting the value of P_1 in equation (*iii*),

$$294 - (0.4\,\alpha + 2P_2) = 30\,a_1 = 30 \times 0.05\,\alpha = 1.5\,\alpha \quad ...(\because a_1 = r_1\alpha = 0.05\,\alpha)$$

$$294 - 0.4\,\alpha - 2P_2 = 1.5\,\alpha$$

$$294 - 2P_2 = 1.5\,\alpha + 0.4\,\alpha = 1.9\,\alpha$$

Dividing both sides by 2,

$$147 - P_2 = 0.95\,\alpha \quad ...(x)$$

From equation (*vi*) we find that

$$P_2 - 98 = 10\,a_2 = 10 \times 0.1\,\alpha = \alpha \quad ...(xi)$$

$$...(\because a_2 = r_2\alpha = 0.1\,\alpha)$$

Adding equations (*x*) and (*xi*),

$$49 = 1.95\,\alpha$$

$$\therefore \quad \alpha = \frac{49}{1.95} = 25.1 \text{ rad/s}^2 \quad \textbf{Ans.}$$

Pulls in the two parts of the string

Substituting the value of α in equation (*x*),

$$147 - P_2 = 0.95\,\alpha = 0.95 \times 25.1 = 23.8$$

$$\therefore \quad P_2 = 147 - 23.8 = 123.2 \text{ N} \quad \textbf{Ans.}$$

Now substituting the value of α and P_2 in equation (*ix*),

$$P_1 = (0.4 \times 25.1) + 2 \times 123.2 = 256.44 \text{ N} \quad \textbf{Ans.}$$

***Example 14.** *Two bodies A and B, of mass 150 kg and 75 kg respectively are supported by a string of negligible mass and pass over a composite pulley. The bodies rest on two smooth inclined planes as shown in Fig. 8.*

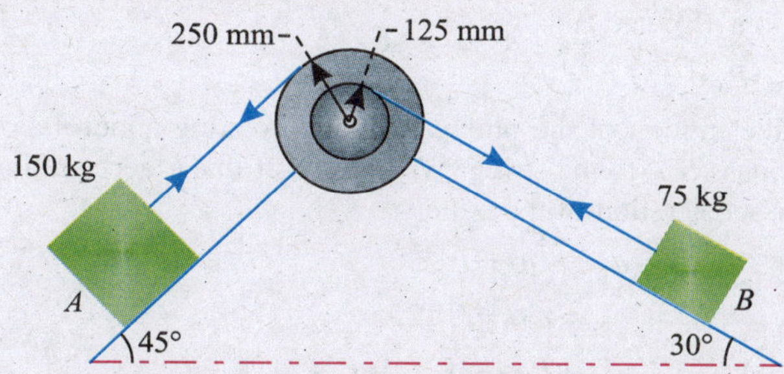

Fig. 8.

If the pulley has a mass of 75 kg and radius of gyration of 100 mm, find the accelerations of the masses A and B and pulls in the string. Neglect friction in the bearings.

Solution. Given: Mass of the body A $(m_1) = 150$ kg; Mass of the body B $(m_2) = 75$ kg; Mass of the pulley $(M) = 75$ kg; Radius of gyration $(k) = 100$ mm $= 0.1$ m; External radius of the pully $(r_1) = 250$ mm $= 0.25$ m and internal radius of the pulley $(r_2) = 125$ mm $= 0.125$ m.

Pulls in the strings

Let P_1 = Pull in the string carrying 150 kg mass, and

P_2 = Pull in the string carrying 75 kg mass.

From the geometry of the masses, we find that turning moment of mass 150 kg (*i.e.*, 150 sin 45° × 0.25 = 150 × 0.707 × 0.25 = 26.5 kg-m) is more than that of the mass 75 kg (*i.e.*, 75 sin 30° × 0.125 = 75 × 0.5 × 0.125 = 4.7 kg-m). It is thus obvious that the 150 kg mass will come downwards and the 75 kg mass will go upwards, when the system is released.

Let a_1 = Acceleration of the 150 kg mass,

a_2 = Acceleration of the 75 kg mass, and

α = Angular acceleration of the pulley.

We know that mass moment of inertia of the pulley,

$$I = Mk^2 = 75\,(0.1)^2 = 0.75 \text{ kg-m}^2$$

First of all, consider the motion of 150 kg mass, which is coming down. We know that the force acting on it, along the plane, $= m_1 g \sin \theta_1 = 150 \times 9.8 \sin 45° = 150 \times 9.8 \times 0.707 = 1039$ N. (downwards) and P_1 (upwards). As the mass is moving downwards, therefore resultant force

$$= 1039 - P_1 \quad \text{...}(i)$$

Since the mass is moving downwards with an acceleration (a_1), therefore force acting on the body

$$= 150\, a_1 \quad \text{...}(ii)$$

Equating equations (*i*) and (*ii*),

$$1039 - P_1 = 150\, a_1 \quad \text{...}(iii)$$

Now consider the motion of 75 kg mass, which is going upwards. We know that the force acting on it, along the plane, $= m_2 g \sin \theta_2 = 75 \times 9.8 \sin 30° = 75 \times 9.8 \times 0.5 = 367.5$ N (downwards) and P_2 (upwards). As the mass is moving upwards, therefore resultant force

$$= P_2 - 367.5 \quad \text{...}(iv)$$

Since the mass is moving upwards with an acceleration (a_2), therefore force acting on the body

$$= 75\ a_2 \qquad ...(v)$$

Equating equations (*iv*) and (*v*),

$$P_2 - 367.5 = 75\ a_2 \qquad ...(vi)$$

Now consider the motion of the pulley, which is rotating about its axis due to downward motion of the 150 kg mass tied to the string. We know that linear acceleration of the 150 kg mass is equal to the angular acceleration of the pulley.

$$\therefore \quad a_1 = r_1\alpha = 0.25\ \alpha$$

Similarly $\quad a_2 = r_2\alpha = 0.125\ \alpha$

and torque, $\quad T = P_1 r_1 - P_2 r_2 = P_1 \times 0.25 - P_2 \times 0.125 \qquad ...(vii)$

We also know that torque on the pulley,

$$T = I\alpha = 0.75\ \alpha \qquad ...(viii)$$

Equating equations (*vii*) and (*viii*),

$$0.25\ P_1 - 0.125\ P_2 = 0.75\ \alpha$$

or $\quad P_1 - 0.5\ P_2 = 3\ \alpha \qquad$ ...(Multiplying by 4)

$$\therefore \quad P_1 = 0.5\ P_2 + 3\ \alpha \qquad ...(ix)$$

Substituting the value of P_1 in equation (*iii*),

$$1039 - (0.5\ P_2 + 3\ \alpha) = 150\ a_1 = 150 \times 0.25\ \alpha$$

$$1039 - 0.5\ P_2 - 3\ \alpha = 37.5\ \alpha$$

$$\therefore \quad 1039 - 0.5\ P_2 = 37.5\ \alpha + 3\alpha = 40.5\ \alpha$$

Multiplying both sides by 2

$$2078 - P_2 = 81\ \alpha \qquad ...(x)$$

From equation (*vi*), we find that

$$P_2 - 367.5 = 75\ a_2 = 75 \times 0.125\ \alpha = 9.4\ \alpha \qquad ...(xi)$$

Adding equations (*x*) and (*xi*),

$$1710.5 = 90.4\ \alpha$$

$$\therefore \quad \alpha = \frac{1710.5}{90.4} = 18.9\ \text{rad/s}^2$$

Now substituting the value of α in equation (*x*),

$$2078 - P_2 = 81\ \alpha = 81 \times 18.9 = 1531$$

$$\therefore \quad P_2 = 2078 - 1531 = 547\ \text{N} \qquad \textbf{Ans.}$$

Again substituting the value of α and P_2 in equation (*ix*),

$$P_1 = (0.5 \times 547) + (3 \times 18.9) = 330\ \text{N} \qquad \textbf{Ans.}$$

Acceleration of the masses A and B

We know that the acceleration of mass *A* (*i.e.,* 150 kg),

$$a_1 = r_1\alpha = 0.25 \times 18.9 = 4.72\ \text{m/s}^2 \qquad \textbf{Ans.}$$

Similarly $\quad a_2 = r_2\alpha = 0.125 \times 18.9 = 2.36\ \text{m/s}^2 \qquad$ **Ans.**

EXERCISE 2

1. A body of mass 20 kg is suspended by a light string wound round a pulley of mass 50 kg. Find acceleration of the body and pull in the string. **[Ans.** 4.36 m/s^2; 108.9 N]
2. A body of mass 22.5 kg is hanging by a light cord from a frictionless wheel and axle of mass 45 kg as shown in Fig. 9. After 16 seconds of the release of the hanging mass from rest, the wheel was found to be revolving with an angular velocity of 144 r.p.m.

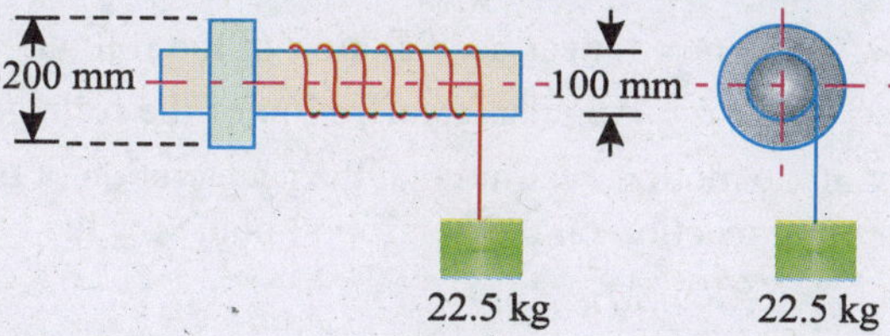

Fig. 9.

From first principles, find the moment of inertia of the wheel and axle.

[Ans. 11.64 kg-m^2]

3. A light rope passing round a pulley of mass 60 kg, radius 300 mm and radius of gyration 200 mm, has two masses 8 kg and 6 kg attached to its ends. If the rope does not slip as the pulley rotates, determine the acceleration of the two masses and the pulls in the two ropes. **[Ans.** 0.48 m/s^2; 74.6 N; 61.7 N]
4. Two masses of 100 kg and 40 kg are supported by a rope of negligible mass passing over a solid disc pulley. If the mass of the pulley is 50 kg, find the acceleration of the 100 kg mass and the pulls in the two strings. **[Ans.** 3.56 m/s^2; 624 N; 534.4 N]

Hint. For a solid disc pulley of radius r,

$$k^2 = 0.5\, r^2$$

∴ Mass moment of inertia,

$$I = Mk^2 = 50 \times 0.5\, r^2 = 25\, r^2$$

12. *MOTION OF A BODY ROLLING ON A ROUGH HORIZONTAL PLANE WITHOUT SLIPPING

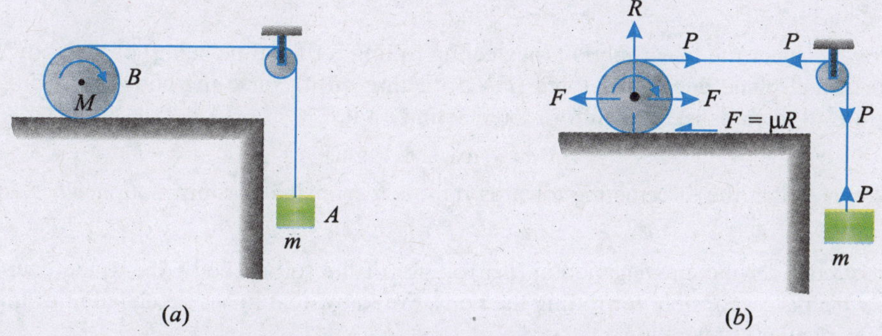

Fig. 10. Body rolling on a rough horizontal plane without slipping.

Consider a body A tied to a string passing over a smooth pulley and pulling another circular body B, which rolls on a rough horizontal plane without slipping as shown in Fig. 10 (a).

Let m = Mass of the hanging body A,

P = Tension in the string,

M = Mass of the rolling body B,

I = Moment of inertia of the rolling body B,

k = Radius of gyration of the rolling body B,

r = Radius of the rolling body B,

μ = Coefficient of friction between the plane and the body,

a = Linear acceleration of the rolling body B, and

α = Angular acceleration of the rolling body B.

From the geometry of the motion, we find that the acceleration of the hanging body A will be $(2a)$. We know that the normal reaction on the horizontal plane for the body B,

$$R = Mg$$

and force of friction,

$$F = \mu R = \mu Mg$$

Since the rolling body tends to roll towards right, therefore force of friction will act towards left as shown in Fig. 10 (*b*). Let us introduce two equal and opposite forces (each equal to the force of friction, F) through the centre of the rolling body as shown in Fig. 10. (*b*).

A little consideration will show that, these two forces will not effect the motion of the system. Now the rolling body is subjected to the following forces :

1. A force equal to $P - F$ (acting towards right)
2. A couple whose moment is equal to $F \times r$ (responsible for rolling the body).

First of all, consider motion of the hanging body A, which is coming down. We know that the forces acting on it are mg (downwards) and P (upwards). As the body is moving downwards, therefore resultant force

$$= mg - P \qquad ...(i)$$

Since the body A is moving downwards with an acceleration of $(2a)$, therefore force acting on this body

$$= m \times 2\,a = 2\,ma \qquad ...(ii)$$

Equating equations (*i*) and (*ii*),

$$mg - P = 2\,ma$$

or

$$P = mg - 2\,ma \qquad ...(iii)$$

Now consider the linear motion (neglecting rolling for the time being) of the body B, on the rough horizontal plane due to the force $(P - F)$ acting on it. Since the body is moving with an acceleration (a) towards right, therefore force acting on it.

$$= Ma \quad \text{(towards right)}$$

We know that the force acting towards right is responsible for this motion. Therefore

$$P - F = Ma \qquad ...(iv)$$

Now consider the circular motion (*i.e.* rolling) of the rolling body due to the couple (equal to $P \times r$) which is responsible for rolling the body. We know that linear acceleration of the body is equal to its angular acceleration.

$$\therefore \quad a = r\alpha$$

and torque acting on the body,

$$T = I\alpha$$

Now equating the couple (responsible for rolling) and torque on the body,

$$F \times r = I\alpha$$

$$F \times r^2 = I\alpha r \qquad \text{...(Multiplying both sides by } r)$$

$$= Ia \qquad \text{...}(\because a = r\alpha)$$

$$\therefore \quad F = \frac{Ia}{r^2} = \frac{Mk^2 a}{r^2} \qquad \text{...}(\because I = Mk^2)$$

Now substituting the value of F in equation (*iv*),

$$P - \frac{Mk^2 a}{r^2} = Ma$$

and now substituting the value of P from equation (*iii*) in the above equation,

$$(mg - 2ma) - \frac{Mk^2 a}{r^2} = Ma$$

or

$$mg = Ma + 2ma + \frac{Mk^2 a}{r^2} = a\left(M + 2m + \frac{Mk^2}{r^2}\right)$$

$$\therefore \quad a = \frac{mg}{\left(M + 2m + \dfrac{Mk^2}{r^2}\right)}$$

The above expression shows that the acceleration of the rolling body is independent of the value of coefficient of friction.

***Example 15.** *Find the acceleration (a) on a solid right circular roller A of mass 20 kg when it is being pulled by another body B of mass 10 kg along a horizontal plane as shown in Fig. 11.*

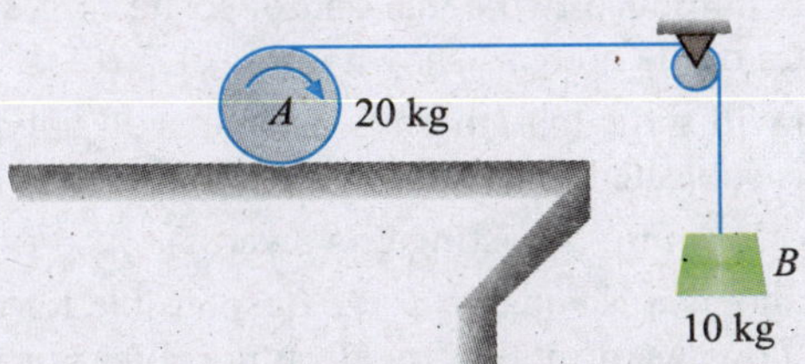

Fig. 11.

The mass B is attached to the end of a string wound round the circumference of the roller. Assume that there is no slipping of the roller and the string is inextensible.

Solution. Given: Mass of roller (M) = 20 kg and mass of hanging body B (m) = 10 kg

We know that for a solid right circular roller of radius r, $\quad 3\ k^2 = 0.5\ r^2$

and acceleration of the roller on the horizontal plane,

$$a = \frac{mg}{M + 2m + \dfrac{Mk^2}{r^2}} = \frac{10 \times 9.8}{\left(20 + (2 \times 10) + \dfrac{20 \times 0.5\ r^2}{r^2}\right)} \text{ m/s}^2$$

$$= 1.96 \text{ m/s}^2 \qquad \textbf{Ans.}$$

13. *MOTION OF A BODY ROLLING DOWN A ROUGH INCLINED PLANE WITHOUT SLIPPING

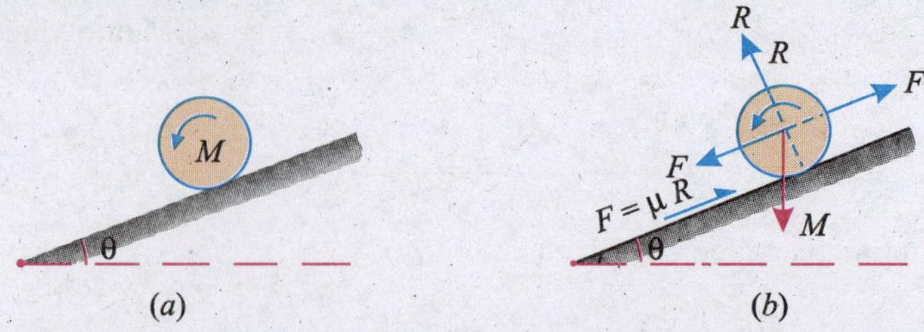

Fig. 12. Body rolling down on a rough inclined plane without slipping.

Consider a circular body rolling down an inclined rough plane without slipping as shown in Fig. 12 (*a*).

Let M = Mass of the body,

I = Moment of inertia of the body,

k = Radius of gyration of the body,

r = Radius of the body,

θ = Inclination of the plane with the horizontal,

μ = Coefficient of friction between the plane and the body,

a = Linear acceleration of the body, and

α = Angular acceleration of the body.

We know that the normal reaction,

$$R = Mg \cos \theta$$

force of friction, $F = \mu R = \mu Mg \cos \theta$

Since the body tends to roll downwards, therefore the force of friction will act upwards as shown in Fig. 16 (*b*). Let us introduce two equal and opposite forces (each equal to the force of friction F), through the centre of the body as shown in Fig. 12 (*b*).

A little consideration will show that, these two forces will not affect the motion of the system. Now the rolling body is subjected to the following two forces :

1. A force equal to $Mg \sin \theta - F$ (acting downwards)
2. A couple whose moment is equal to $F \times r$ (responsible for rolling down the body).

First of all, consider downward motion (neglecting rolling for the time being) of the body due to force ($Mg \sin \theta - F$) acting on it. Since the body is moving with an acceleration (a), therefore force acting on it

$$= Ma$$

We know that the force acting on it is responsible for this motion. Therefore

$$Mg \sin \theta - F = Ma \qquad ...(i)$$

Now consider the circular motion (*i.e.*, rolling) of the body due to the couple (equal to $P \times r$) which is responsible for the rolling of the body. We know that linear acceleration of the body is equal to its angular acceleration.

$$\therefore \qquad a = r\alpha$$

and torque on the body, $T = I\alpha$

Now equating the couple (responsible for rolling) and torque acting on the body

$$F \times r = I\alpha$$

$$F \times r^2 = I\alpha r \qquad ...(\text{Multiplying both sides by } r) \; (\because a = r\alpha)$$

$$= Ia$$

$$\therefore \qquad F = \frac{Ia}{r^2} = \frac{Mk^2 a}{r^2} \qquad \ldots (\because I = mk^2) \qquad \ldots (ii)$$

and now substituting the value of F in equation (i),

$$Mg \sin\theta - \frac{Mk^2 a}{r^2} = Ma$$

$$Mg \sin\theta = Ma + \frac{Mk^2 a}{r^2} = Ma\left(1 + \frac{k^2}{r^2}\right)$$

$$\therefore \qquad a = \frac{g \sin\theta}{1 + \dfrac{k^2}{r^2}} = \frac{g \sin\theta}{\dfrac{r^2 + k^2}{r^2}}$$

The above expression shows that the acceleration of the rolling body is independent of its mass. Now substituting the value of (a) in equation (ii),

$$F = \frac{Mk^2}{r^2} \times \frac{g \sin\theta}{\dfrac{r^2 + k^2}{r^2}} = \frac{Mk^2 \; g \sin\theta}{k^2 + r^2}$$

Now for the body (or wheel) to roll down without slipping, the applied force must be less than (or equal to) the available force of friction, in order to fulfil the condition of rolling without slipping. Or in other words

$$F \le \mu Mg \cos\theta$$

$$\frac{Mk^2 \; g \sin\theta}{k^2 + r^2} \le \mu Mg \cos\theta$$

$$\tan\theta \le \mu\left(\frac{k^2 + r^2}{k^2}\right)$$

The above equation may also be written as :

$$\mu \ge \frac{\tan\theta}{\left(\dfrac{k^2 + r^2}{k^2}\right)}$$

Note: The above expression shows that the value of (μ) should be more or equal to the value obtained from the right hand side. Thus for the minimum value of (μ), the above expression may be written as :

$$\mu = \frac{\tan\theta}{\left(\dfrac{k^2 + r^2}{k^2}\right)}$$

***Example 16.** *A sphere rolls down a plane inclined at 30° to the horizontal. Find the minimum value of the coefficient of friction between the sphere and the plane, so that the sphere may roll without slipping.*

Solution. Given: Inclination of plane $\theta = 30°$

We know that for a sphere of radius r,

$$k^2 = 0.4\, r^2$$

and minimum value of coefficient of friction,

$$\mu = \frac{\tan\theta}{\dfrac{k^2 + r^2}{k^2}} = \frac{\tan 30°}{\dfrac{0.4r^2 + r^2}{0.4r^2}} = \frac{0.5774}{3.5} = 0.165 \qquad \textbf{Ans.}$$

***Example 17.** *A solid uniformly thick wheel of radius 1 m and mass 40 kg is released with no initial velocity at the top of an inclined plane, which makes an angle of 30° with the horizontal. It rolls down without slipping. Determine (i) the minimum value of coefficient of friction, (ii) the velocity of the centre of the wheel after it has travelled a distance 4 m down the inclined plane.*

Solution. Given: Radius of the wheel (r) = 1 m; *Mass of the wheel (m) = 40 kg and inclination of plane $(\theta) = 30°$

(*i*) *Minimum value of coefficient of friction*

We know that for a uniformly thick wheel,

$$k^2 = 0.5\, r^2$$

and minimum value of coefficient of friction,

$$\mu = \frac{\tan\theta}{\dfrac{k^2 + r^2}{k^2}} = \frac{\tan 30°}{\dfrac{0.5r^2 + r^2}{0.5r^2}} = \frac{0.5774}{3} = 0.192 \quad \textbf{Ans.}$$

(*ii*) *Velocity of the centre of the wheel after it has travelled a distance of 4 m*

Let v = Velocity of the centre of the wheel.

We know that acceleration of the wheel when it rolls down the plane,

$$a = \frac{g\sin\theta}{\dfrac{k^2 + r^2}{r^2}} = \frac{9.8\sin 30°}{\dfrac{0.5r^2 + r^2}{r^2}} = \frac{9.8 \times 0.5}{1.5} = 3.27 \text{ m/s}^2$$

and $$v^2 = u^2 + 2as = 0 + 2 \times 3.27 \times 4 = 26.16$$

$$v = 5.1 \text{ m/s} \quad \textbf{Ans.}$$

EXERCISE 3

1. A solid right circular roller of mass 15 kg is being pulled by another body of mass 15 kg along a horizontal plane as shown in Fig. 13.

Fig. 13.

Find the acceleration of the roller, assuming that there is no slipping of the roller and string is inextensible. **[Ans.** 2.8 m/s^2]

2. A solid sphere of mass (m) and radius (r) rolls down a plane inclined at an angle (θ) with the horizontal. Find the acceleration of the sphere. **[Ans.** 7 sin θ]

3. A solid cylinder is placed on a plane inclined at 13° 18 with the horizontal and allowed to roll down without slipping and with its axis horizontal. Find the acceleration of the cylinder. **[Ans.** 1.5 m/s^2]

4. Find the time taken by a solid cylinder of radius 600 mm and initially at rest to roll down a distance 19.2 m on a plane inclined at 30° with the horizontal. Take $g = 9.81$ m/s^2. **[Ans.** 3.5 s]

QUESTIONS

1. What is torque ? Give an example of torque.
2. State the Newton's Laws of Motion of Rotation.
3. Define mass moment of inertia and kinetic energy of rotation.
4. Derive a relation between torque and angular acceleration of a body rotating about an axis.
5. Prove that the pull in a string having a body tied to it and passing over a pulley,

$$P = \frac{Img}{(mr^2 + I)}$$

where I = Moment of inertia of the pulley,
m = Mass of the body, and
r = Radius of the pulley

6. A uniform disc of radius r is allowed to roll down a rough inclined plane whose angle of inclination with the horizontal is θ. Prove that the linear acceleration of the disc is given by:

$$a = \frac{g \sin \theta}{\frac{r^2 + k^2}{r^2}}$$

where k is the radius of gyration.

OBJECTIVE TYPE QUESTIONS

1. The units of torque, work and energy are same.
   (*a*) Yes (*b*) No
2. Which of the following statement is correct ?
   (*a*) The moment of inertia of a thin circular ring of mass (M) and mean radius (r) is 0.25 mr^2.
   (*b*) The flywheel is a wheel which is used in flying machines.
   (*c*) The units of kinetic energy of rotation and kinetic energy of translation are different.
   (*d*) All of them.
   (*e*) None of them.
3. The torque acting on a body of moment of inertia (I) and angular acceleration (α) is
   (*a*) $I\alpha$ (*b*) $I\alpha^2$
   (*c*) 0.5 $I\alpha$ (*d*) 0.5 $I\alpha^2$
4. If the moment of inertia (I) of a pulley is doubled, then the acceleration of a body tied to a string and passing over it is
   (*a*) Halved (*b*) Remains the same
   (*c*) Doubled (*d*) None of these
5. In order to increase the acceleration of a mass rolling down on a rough inclined plane (without slipping), we have to
   (*a*) Increase the mass of the rolling body
   (*b*) Increase the inclination of the plane
   (*c*) Both (*a*) and (*b*)

ANSWERS

1. (*a*) 2. (*e*) 3. (*a*) 4. (*d*) 5. (*b*)

CHAPTER 30

Transmission of Power by Belts and Ropes

Contents

1. BELT AND BELT DRIVE

In factories, the power or rotary motion, from one shaft to another at a considerable distance is, usually, transmitted by means of flat belts, vee belts or ropes, running over the pulleys.

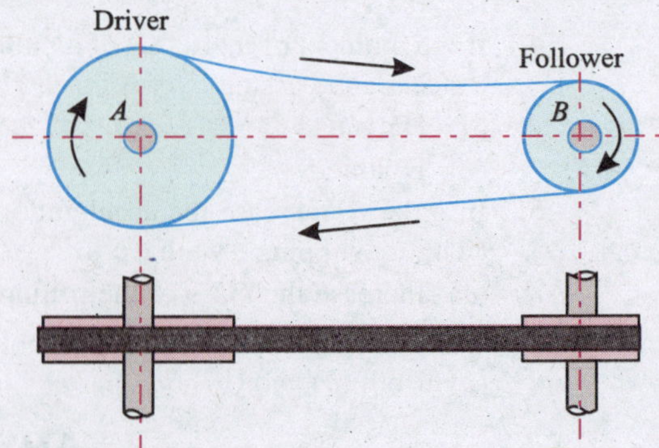

Fig. 1. Transmission of power from *A* to *B*.

In Fig. 1 is shown an open belt drive consisting of pulleys A and B. The pulley A is keyed to a rotating shaft, and is known as driver (since it drives the other pulley). The pulley B is keyed to a shaft, intended to be rotated, and is known as follower or driven (since it is driven by pulley A).

When the driver rotates, it carries the belt due to grip between its surface and the belt. The belt, in turn, carries the driven pulley which starts rotating. The grip between the pulley and the belt is obtained by friction, which arises from the pressure between the belt and the pulleys. The friction grip, if required, is increased by tightening the belt.

2. *TYPES OF BELTS

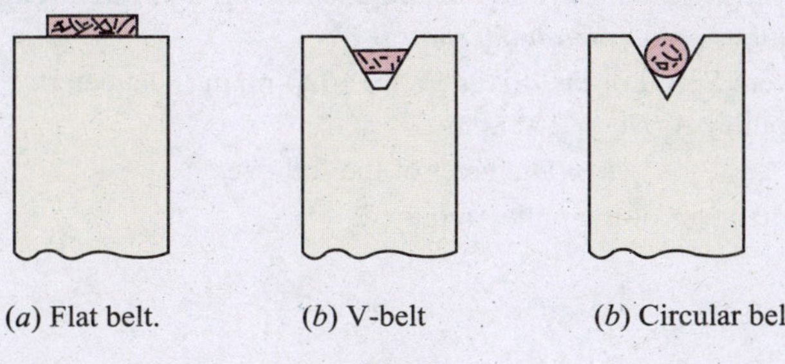

(*a*) Flat belt. (*b*) V-belt (*b*) Circular belt.

Fig. 2.

[JNTU]

Though there are many types of belts used these days, yet the following are important from the subject point of view :

1. ***Flat belt.*** The flat belt is mostly used in the factories and workshops, where a moderate amount of power is to be transmitted, from one pulley to another, when the two pulleys are not more than 10 m apart.
2. ***V-belt.*** The V-belt is mostly used in the factories and workshops where a great amount of power is to be transmitted, from One pulley to another, when the two pulleys are very near to each other.
3. ***Circular belt or rope.*** The circular belt or rope is mostly used in the factories and workshops, where a great amount of power is to be transmitted, from one pulley to another, when the two pulleys are more than 5 m apart.

If a huge amount of power is to be transmitted, then a single belt may not be sufficient. In such a case, wide pulleys (for V-belts or circular belts) with a number of grooves are used. Then one belt in each groove is provided to transmit the required amount of power from one pulley to another.

3. *VELOCITY RATIO OF A BELT DRIVE

The velocity ratio of a belt drive may be broadly defined as the ratio of the velocities of the driver and the follower or driven. Now we shall discuss the velocity ratios of :

1. Simple belt drive and 2. Compound belt drive.

4. VELOCITY RATIO OF A SIMPLE BELT DRIVE

Consider a simple belt drive (*i.e.* one driver and one follower) as shown in Fig. 1.

Let d_1 = Diameter of the driver,

N_1 = Speed of the driver in r.p.m., and

d_2, N_2 = Corresponding values for the follower.

$\therefore$ Length of the belt, that passes over the driver, in one minute $= \pi d_1 N_1$ ...(*i*)

Similarly, length of the belt that passes over the follower, in one minute $= \pi d_2 N_2$ ...(*ii*)

Since the length of belt, that passes over the driver in one minute, is equal to the length of belt that passes over the follower in one minute, therefore

$$\pi d_1 N_1 = \pi d_2 N_2$$

or

$$\frac{N_2}{N_1} = \frac{d_1}{d_2}$$

Note: The term N_2/N_1 is popularly knwon as **velocity ratio.**

Example 1. *It is required to drive a shaft at 620 revolutions per minute, by means of a belt from a parallel shaft, having a pulley A 300 mm diameter on it and running at 240 revolutions per minute. What sized pulley is required on the shaft B ?*

Solution. Given: Speed of the driver (N_2) = 620 r.p.m.; Diameter of pulley A (d_1) = 300 mm and speed of the pulley A (N_1) = 240 r.p.m.

Let d_2 = Diameter of the follower

We know that diameter of the pulley required,

$$d_2 = d_1 \times \frac{N_1}{N_2} = 300 \times \frac{240}{620} \qquad \ldots\left(\because \frac{N_2}{N_1} = \frac{d_1}{d_2}\right)$$

$$= 116{\cdot}1 \text{ mm} \qquad \textbf{Ans.}$$

5. *VELOCITY RATIO OF A COMPOUND BELT DRIVE

Compound belt drive

Sometimes, the power is transmitted, from one shaft to another, through a number of pulleys as shown in Fig. 3. Such an arrangement is known as **compound belt drive**.

In the figure is shown pulley 1, which drives the pulley 2. Since the pulleys 2 and 3 are keyed to the same shaft, therefore the pulley 1 also drives the pulley 3 which, in turn, drives the pulley 4.

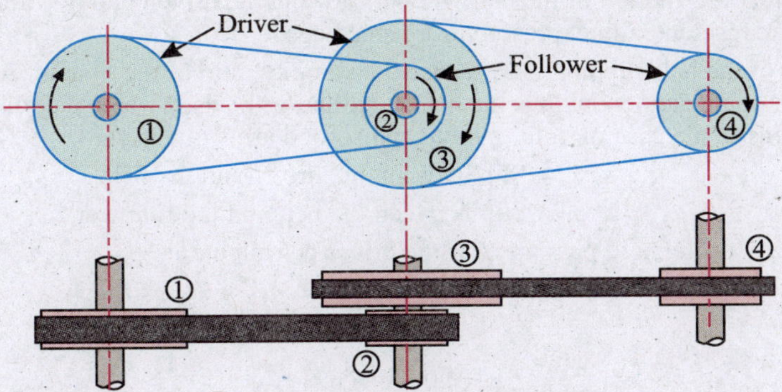

Fig. 3. Compound belt drive.

Let d_1 = Diameter of pulley 1,

N_1 = Speed of the pulley 1 in r.p.m. and

$d_2, d_3, d_4, N_2, N_3, N_4$ = Corresponding values for pulleys 2, 3 and 4.

We know that the velocity ratio of the pulleys 1 and 2,

$$\frac{N_2}{N_1} = \frac{d_1}{d_2} \quad ...(i)$$

Similarly, velocity ratio of the pulleys 3 and 4

$$\frac{N_4}{N_2} = \frac{d_3}{d_4} \quad ...(ii)$$

Multiplying equation (*i*) by equation (*ii*),

$$\frac{N_2}{N_1} \times \frac{N_4}{N_3} = \frac{d_1}{d_2} \times \frac{d_3}{d_4}$$

or

$$\frac{N_4}{N_1} = \frac{d_1 \times d_3}{d_2 \times d_4}$$

...($\because N_2 = N_3$, being keyed to the same shaft)

A little consideration will show, that if there are six pulleys, then the velocity ratio,

$$\frac{N_6}{N_1} = \frac{d_1 \times d_3 \times d_5}{d_2 \times d_4 \times d_6} = \frac{\text{Product of diameters of drivers}}{\text{Product of diameters of followers}}$$

Example 2. *In a workshop, an engine drives a shaft by a belt. The diameters of the engine pulley and the shaft pulley are 500 mm and 250 mm respectively. Another pulley of 700 mm diameter on the same shaft drives a pulley 280 mm in diameter of the follower. If the engine runs at 180 revolutions per minute, find the speed of the follower.*

Solution. Given: Diameter of the engine pulley (d_1) = 500 mm; Diameter of the shaft pulley (d_2) = 250 mm; Diameter of pulley 2 (d_3) = 700 mm, Diameter of the follower (d_4) = 280 mm and speed of the engine (N_1) = 180 r.p.m.

We know that speed of the shaft,

$$N_4 = N_1 \times \frac{d_1 \times d_3}{d_2 \times d_4} = 180 \times \frac{500 \times 700}{250 \times 280} = 900 \text{ r.p.m.} \quad \textbf{Ans.}$$

6. *SLIP OF THE BELT

In the previous articles, we have discussed the motion of belts and shafts assuming a firm frictional grip between the belts and the shafts. But sometimes the frictional grip becomes somewhat loose. This may cause some forward motion of the driver without carrying the belt with it. This may also cause some forward motion of the belt without carrying the driven pulley with it. This is called **slip of the belt**, and is generally expressed as a percentage.

The result of the belt slipping is to reduce the velocity ratio of the system. As the slipping of the belt is a common phenomenon, thus the belt should never be used where a definite velocity ratio is of importance (as in the case of hour, minute and second arms in a watch).

Let $s_1\%$ = Slip between the driver and the belt,

$s_2\%$ = Slip between the belt and the follower,

Now velocity of the belt, passing over the driver per minute,

$$v = \pi d_1 N_1 - \pi d_1 N_1 \times \frac{s_1}{100}$$

$$= \pi d_1 N_1 \left(1 - \frac{s_1}{100}\right) \qquad ...(i)$$

Similarly,
$$\pi d_2 N_2 = v - v \times \frac{s_2}{100} = v\left(1 - \frac{s_2}{100}\right)$$

Substituting the value of v from equation (i),

$$\pi d_2 N_2 = \pi d_1 N_1 \left(1 - \frac{s_1}{100}\right) \times \left(1 - \frac{s_2}{100}\right)$$

or
$$\frac{N_2}{N_1} = \frac{d_1}{d_2}\left(1 - \frac{s_1}{100} - \frac{s_2}{100}\right) \qquad ...\left(\text{Neglecting } \frac{s_1 + s_2}{100 \times 100}\right)$$

$$= \frac{d_1}{d_2}\left[1 - \left(\frac{s_1 + s_2}{100}\right)\right] = \frac{d_1}{d_2}\left(1 - \frac{s}{100}\right)$$

...(where $s = s_1 + s_2$ *i.e.*, total percentage of slip)

Note: If thickness of the belt is considered, then

$$\frac{N_2}{N_1} = \frac{d_1 + t}{d_2 + t}\left(1 - \frac{s}{100}\right) \qquad ...(\text{where } t \text{ is thickness of the belt})$$

***Example 3.** *An engine shaft running at 120 r.p.m. is required to drive a machine shaft by means of a belt. The pulley on the engine shaft is of 2 meters diameter and that of the machine shaft is of 1 meter diameter. If the belt thickness is 5 mm, find the speed of the machine shaft when (i) there is no slip, and (ii) there is a slip of 3%.*

Solution. Given: Speed of the engine shaft (N_1) = 120 r.p.m; Diameter of the pulley on the engine shaft (d_1) = 2m; Diameter of the machine shaft (d_2) = 1m; Thickness of the belt (t) = 5 mm = 0·005 m and slip (s) = 3%.

(i) Speed of the machine shaft when there is no slip

We know that speed of the machine shaft,

$$N_2 = N_1 \times \frac{d_1 + t}{d_2 + t} = 120 \times \frac{2 + 0{\cdot}005}{1 + 0{\cdot}005} = 239{\cdot}4 \text{ r.p.m.} \qquad \textbf{Ans.}$$

(ii) Speed of the machine shaft when there is a slip of 3%

We know that speed of the machine shaft,

$$N_2 = N_1 \times \frac{d_1 + t}{d_2 + t}\left(1 - \frac{s}{100}\right) = 120 \times \frac{2 + 0{\cdot}005}{1 + 0{\cdot}005}\left(1 - \frac{3}{100}\right)$$

$$= 232 \text{ r.p.m.} \qquad \textbf{Ans.}$$

7. *TYPES OF BELT DRIVES

The power, from one pulley to another, may be transmitted by any one of the following two types of belt drives :

1. **Open belt drive**
2. **Cross belt drive.**

8. OPEN BELT DRIVE

In an open belt drive, the two wheels move in the same direction as shown in Fig. 4.

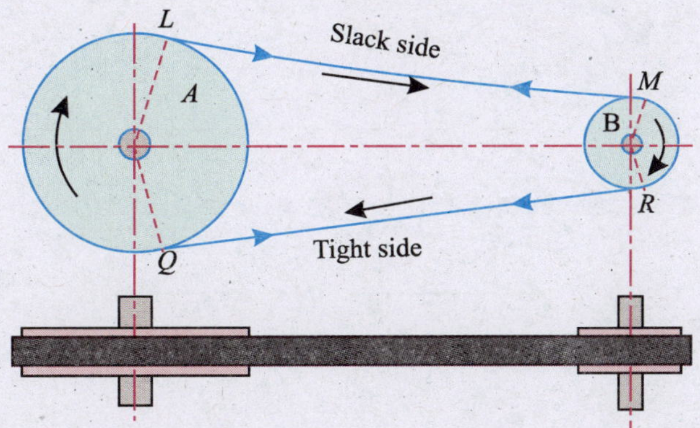

Fig. 4. Open belt drive.

The driver *A* pulls the belt from one side (*RQ* in this case) and delivers it to the other side (*LM* in this case). Thus the tension in the lower side belt will be more than that in the upper side. The lower side belt (because of more tension) is known as *tight side*; whereas the upper side belt (because of less tension) is known as *slack side*.

9. CROSS BELT DRIVE

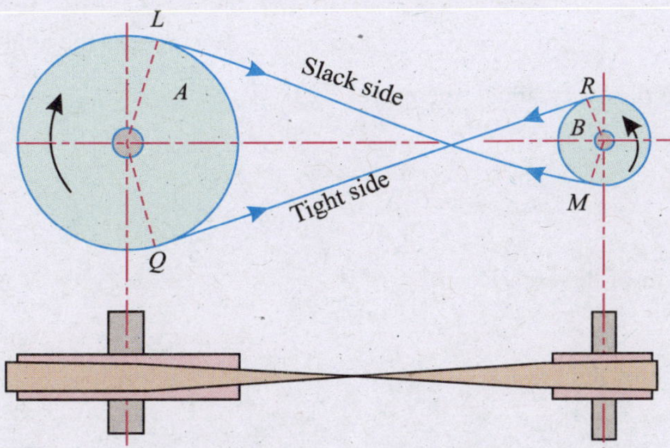

Fig. 5. Cross belt drive.

In a cross belt drive, the wheels move in the opposite directions as shown in Fig. 5.

The driver *A* pulls the belt from one side (*RQ* in this case) and delivers it to the other (*LM* in this case). Thus the tension in the belt *RQ* will be more than that in the belt *LM*. The belt *RQ* (because of more tension) is known as *tight side*; whereas the belt *LM* (because of less tension) is known as *slack side*.

10. LENGTH OF THE BELT

It means the total length of the belt, required to connect a driver and a follower. Here we shall discuss the lengths of the following two types of belt drives :

1. Open belt drive. **2.** Cross belt drive.

11. *LENGTH OF AN OPEN BELT DRIVE

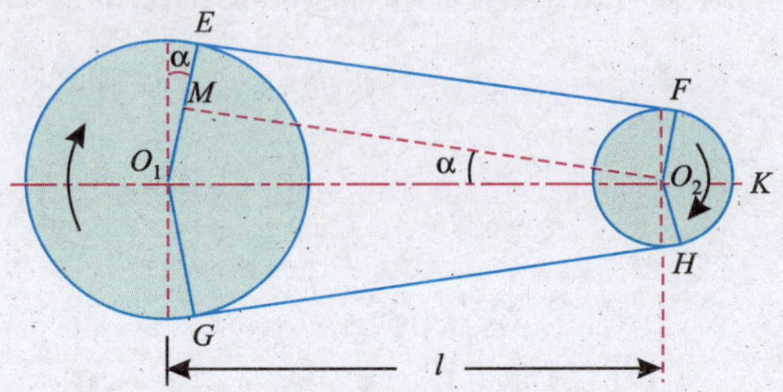

Fig. 6. Length of an open belt drive.

[PU]

In this case, both the pulleys rotate in the *same* direction as shown in Fig. 6.

Let O_1 and O_2 = Centres of the two pulleys,

r_1 and r_2 = Radii of the larger and smaller pulleys,

l = Distance between O_1 and O_2.

Let the belt leave the larger pulley at E and G and the smaller pulley at F and H as shown in Fig. 6. Through O_2 draw $O_2\ M$ parallel to EF. From the geometry of the figure, we find that $O_2\ M$ will be perpendicular to O_1E.

$$\therefore \qquad \sin\alpha = \frac{r_1 - r_2}{l} \qquad \text{... (where } \alpha \text{ is the angle } MO_2O_1 \text{ in radians.)}$$

Since the angle α is very small, therefore

$$\sin\alpha = \alpha = \frac{r_1 - r_2}{l} \qquad ...(i)$$

$$\therefore \qquad \text{Length of the arc } JE = r_1\left(\frac{\pi}{2} + \alpha\right) \qquad ...(ii)$$

$$\text{Similarly length of the arc } FK = r_2\left(\frac{\pi}{2} - \alpha\right) \qquad ...(iii)$$

and

$$EF = MO_2 = \sqrt{l^2 - (r_1 - r_2)^2} = l\sqrt{1 - \left(\frac{r_1 - r_2}{l}\right)^2}$$

Expanding this equation by Binomial theorem :

$$EF = l\left[1 - \frac{1}{2}\left(\frac{r_1 - r_2}{l}\right)^2 + ...\right] = l - \frac{(r_1 - r_2)^2}{2l} \qquad ...(iv)$$

We know that the length of the belt,

$$L = \text{Length of arc } GJE + EF + \text{Length of arc } FKH + HG$$
$$= 2\,(\text{Length of arc } JE + EF + \text{Length of arc } FK)$$

Substituting the values of length of arc *JE* from equation (*ii*), length of arc *FK* from equation (*iii*) and *EF* from equation (*iv*) in this equation,

$$L = 2\left[r_1\left(\frac{\pi}{2}+\alpha\right)+l-\frac{(r_1-r_2)^2}{2l}+r_2\left(\frac{\pi}{2}-\alpha\right)\right]$$

$$= 2\left[r_1\frac{\pi}{2}+r_1\alpha+l-\frac{(r_1-r_2)^2}{2l}+r_2\frac{\pi}{2}-r_2\alpha\right]$$

$$= 2\left[\frac{\pi}{2}(r_1+r_2)+\alpha(r_1-r_2)+l-\frac{(r_1-r_2)^2}{2l}\right]$$

$$= \pi(r_1+r_2)+2\alpha(r_1-r_2)+2l-\frac{(r_1-r_2)^2}{l}$$

Substituting the value of $\alpha = \dfrac{(r_1-r_2)}{l}$ from equation (*i*),

$$L = \pi(r_1+r_2)+2\times\frac{(r_1-r_2)}{l}(r_1-r_2)+2l-\frac{(r_1-r_2)^2}{l}$$

$$= \pi(r_1+r_2)+\frac{2(r_1-r_2)^2}{l}+2l-\frac{(r_1-r_2)^2}{l}$$

$$= \pi(r_1+r_2)+2l+\frac{(r_1-r_2)^2}{l}$$

12. *LENGTH OF A CROSS-BELT DRIVE

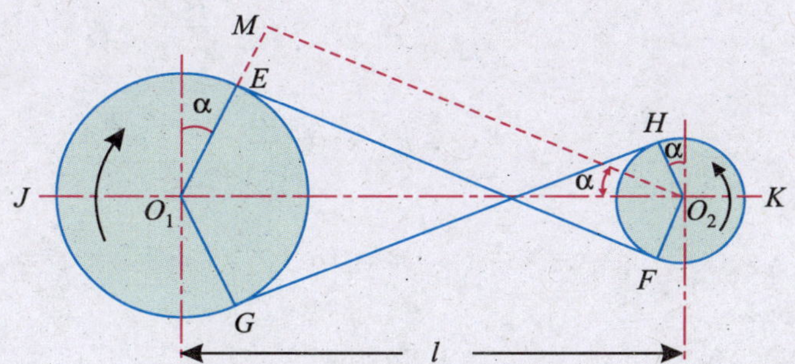

Fig. 7. Length of a cross belt drive.

In this case, both the pulleys rotate in the *opposite* directions as shown in Fig. 7.

Let O_1 and O_2 = Centre of the two pulleys,

r_1 and r_2 = Radii of the larger and smaller pulleys,

l = Distance between O_1 and O_2.

Let the belt leave the larger pulley at E and G and the smaller pulley at F and H as shown in Fig. 7. Through O_2 draw O_2M parallel to FE. From the geometry of the figure, we find that O_2M will be perpendicular to O_1E.

and $$\sin \alpha = \frac{r_1 + r_2}{l} \quad \text{...(where } \alpha \text{ is the angle } MO_2O_1 \text{ in radians.)}$$

Since the angle α is very small, therefore

$$\sin \alpha = \alpha = \frac{r_1 + r_2}{l} \quad \text{...}(i)$$

$\therefore$ $$\text{Length of arc } JE = r_1\left(\frac{\pi}{2} + \alpha\right) \quad \text{...}(ii)$$

Similarly, length of arc FK $$= r_2\left(\frac{\pi}{2} + \alpha\right) \quad \text{...}(iii)$$

and $$EF = MO_2 = \sqrt{l^2 - (r_1 + r_2)^2} = l\sqrt{1 - \left(\frac{r_1 + r_2}{l}\right)^2}$$

Expanding this equation by binomial theorem,

$$EF = l\left[1 - \frac{1}{2}\left(\frac{r_1 + r_2}{l}\right)^2 + ...\right]$$

$$= l - \frac{(r_1 + r_2)^2}{2l} \quad \text{...}(iv)$$

We know that the length of the belt,

$$L = \text{Length of arc } GJE + EF + \text{Length of arc } FKH + HG$$
$$= 2\,(\text{Length of arc } JE + EF + \text{Length of arc } FK) \quad \text{...}(v)$$

Substituting the values of length of arc JE from equation (*ii*), length of arc FK from equation (*iii*) and EF from equation (*iv*) in this equation,

$$L = 2\left[r_1\left(\frac{\pi}{2} + \alpha\right) + l - \frac{(r_1 + r_2)^2}{2l} + r_2\left(\frac{\pi}{2} + \alpha\right)\right]$$

$$= 2\left[r_1\frac{\pi}{2} + r_1\alpha + l - \frac{(r_1 + r_2)^2}{2l} + r_2\frac{\pi}{2} + r_2 x\right]$$

$$= 2\left[\frac{\pi}{2}(r_1 + r_2) + \alpha(r_1 + r_2) + l - \frac{(r_1 + r_2)^2}{2l}\right]$$

$$= \pi(r_1 + r_2) + 2\alpha(r_1 + r_2) + 2l - \frac{(r_1 + r_2)^2}{l}$$

Substituting the values of $\alpha = \frac{r_1 + r_2}{l}$ from equation (*i*),

$$L = \pi(r_1 + r_2) + \frac{2(r_1 + r_2)}{l}(r_1 + r_2) + 2l - \frac{(r_1 + r_2)^2}{l}$$

$$= \pi(r_1 + r_2) + \frac{2(r_1 + r_2)^2}{l} + 2l - \frac{(r_1 + r_2)^2}{l}$$

$$= \pi(r_1 + r_2) + 2l + \frac{(r_1 + r_2)^2}{l}$$

Note: 1. We see that the above expression is a function of $(r_1 + r_2)$. It is thus obvious, that if sum of the radii of the two pulleys be constant, length of the belt required will also remain constant ; provided the distance between the centres of the pulleys remain unchanged.

2. If width of the belt is given then the effective corresponding diameters (and then radii) will be given by the sum of actual diameter of the pulleys and width of the belt.

***Example 4.** *Find the length of belt necessary to drive a pulley of 500 mm diameter running parallel at a distance of 12 meters from the driving pulley of diameter 1600 mm.*

Solution. Given: Diameter of the driven pulley $(d_2) = 500$ mm $= 0{\cdot}5$ m or radius $(r_2) = 0{\cdot}25$ m; Distance between the centres of the two pulleys $(l) = 12$ m and diameter of the driving pulley $(d_1) = 1600$ mm $= 1{\cdot}6$ m or radius $(r_1) = 0{\cdot}8$ m.

In this example, no mention has been made whether the belt is open or crossed. Therefore we shall find out the value of length of the belt in both the cases.

Length of the belt if it is open

We know that the length of the belt if it is open,

$$L = \pi(r_1 + r_2) + 2l + \frac{(r_1 - r_2)^2}{l}$$

$$= \pi(0{\cdot}8 + 0{\cdot}25) + (2 \times 12) + \frac{(0{\cdot}8 - 0{\cdot}25)^2}{12}\ m$$

$$= 27{\cdot}32 \text{ m} \quad \textbf{Ans.}$$

Length of the belt if it is crossed

We know that length of the belt if it is crossed,

$$L = \pi(r_1 + r_2) + 2l + \frac{(r_1 + r_2)^2}{l}$$

$$= \pi(0{\cdot}8 + 0{\cdot}25) + (2 \times 12) + \frac{(0{\cdot}8 + 0{\cdot}25)^2}{12}$$

$$= 27{\cdot}39 \text{ m} \quad \textbf{Ans.}$$

Example 5. *Find the length of the belt required for driving two pulleys in a cross belt drive of 600 mm and 300 mm diameter when 3·5 m apart. Take thickness of the belt as 5 mm.*

Solution. Given: Diameter of driving pulley $(d_1) = 600 + 5 = 605$ mm $= 0{\cdot}605$ m or radius $(r_1) = 0{\cdot}3025$m; Diameter of driven pulley $(d_2) = 300 + 5 = 305$ mm $= 0{\cdot}305$ m or radius $(r_2) = 0{\cdot}1525$ m and distance between the centres of the pulleys $(l) = 3{\cdot}5$ m

We know that length of the cross belt drive

$$L = \pi(r_1 + r_2) + 2l + \frac{(r_1 + r_2)^2}{l}$$

$$= \pi(0{\cdot}3025 + 0{\cdot}1525) + (2 \times 3{\cdot}5) + \frac{(0{\cdot}3025 + 0{\cdot}1525)^2}{3{\cdot}5}\text{ m}$$

$$= 8.488 \text{ m} \quad \textbf{Ans.}$$

EXERCISE 1

1. A diesel engine shaft having a speed of 180 r.p.m. is required to drive a machine shaft with the help of a belt. Find the speed of the machine shaft, if the diameters of the engine shaft and machine shaft are 300 mm and 200 mm respectively. **[Ans.** 270 r.p.m.]
2. In a workshop, a machine shaft is driven by an electric motor with the help of belts across a main shaft and a counter-shaft. The diameters of the driving pulleys are 500 mm, 400 mm and 300 mm respectively whereas the diameters of the driven pulleys are 250 mm, 200 mm and 150 mm respectively. Find the speed of the machine shaft when the electric motor runs at 150 r.p.m. **[Ans.** 1200 r.p.m.]

3. An engine, running at 150 r.p.m. drives a line shaft by means of a belt. The engine pulley is 750 mm diameter and the pulley on the line shaft being 450 mm diameter. The 900 mm pulley on the line shaft drives a 150 mm diameter pulley keyed to a dynamo shaft. Find the speed of the dynamo shaft, when (*i*) there is no slip; and (*ii*) there is a slip of 2% at each drive.

[**Ans.** 1500 r.p.m.; 1380 r.p.m.]

4. Two parallel shafts 6 m apart are provided with 900 mm and 300 mm diameter pulleys and are connected by means of a cross belt. The direction of rotation of the follower pulley is to be reversed by changing over to an open belt drive. How much length of the belt has to be reduced?

[**Ans.** 40 mm]

13. *POWER TRANSMITTED BY A BELT

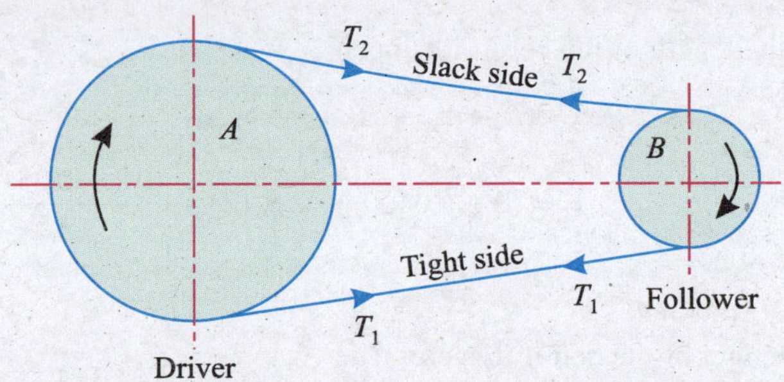

Fig. 8. Power transmitted by a belt.

[JNTU]

In Fig. 8, is shown the driving pulley (*i.e.* driver) *A* and the follower *B*. We know that the driving pulley pulls the belt from one side, and delivers the same to the other. It is thus obvious, that the tension in the former side (*i.e.*, tight side) will be more than that in the latter side (*i.e.*, slack side) as shown in Fig. 8.

Let T_1 = Tension in the tight side

T_2 = Tension in the slack side and

v = Velocity of the belt

We know that the effective turning (*i.e.* driving) force at the circumference of the follower is the difference between the two tensions (*i.e.*, $T_1 - T_2$).

$\therefore$ Work done = Force × Distance = $(T_1 - T_2) \times v$

$\therefore$ Power = $(T_1 - T_2)v$ J/s = $(T_1 - T_2)v$

Notes:1. The torque exerted on the driving pulley $(T_1 - T_2)\ r_1$

2. Similarly, the torque exerted on the follower = $(T_1 - T_2)\ r_2$ where r_1 and r_2 are in metres.

***Example 6.** *The tensions in the two sides of the belt are 1000 and 800 newtons respectively. If the speed of the belt is 75 metres per second, find the power transmitted by the belt.*

Solution. Given: Tension in the tight side (T_1) = 1000 N; Tension in the slack side (T_2) = 800 N and speed of the belt (v) = 75 m/s

We know that power transmitted by the belt,

$$P = (T_1 - T_2)\ v = (1000 - 800) \times 75 = 15\ 000 \text{ N-m/s}$$

$$= 15\ 000 \text{ W} = 15 \text{ kW} \qquad \textbf{Ans.}$$

Example 7. *Find the necessary difference in tensions in N in the two sides of a belt drive, when transmitting 120 W. at 30 m/sec.*

Solution. Given: Transmitting power $(p) = 120$ W and speed of the belt $(v) = 30$ m/s

Let $(T_1 - T_2)$ = Necessary difference in tensions in the two sides of the belt.

We know that power transmitted by the belt (P)

$$120 = \frac{(T_1 - T_2)v}{4\text{N}} = \frac{(T_1 - T_2)30}{30} = 30(T_1 - T_2)$$

$$\therefore \quad (T_1 - T_2) = \frac{120}{30} = 4 \text{ N} \quad \textbf{Ans.}$$

14. **RATIO OF TENSIONS

[AUT, JNTU]

Consider a follower (*i.e.* driven) pulley rotating in the clockwise direction as shown in Fig. 9.

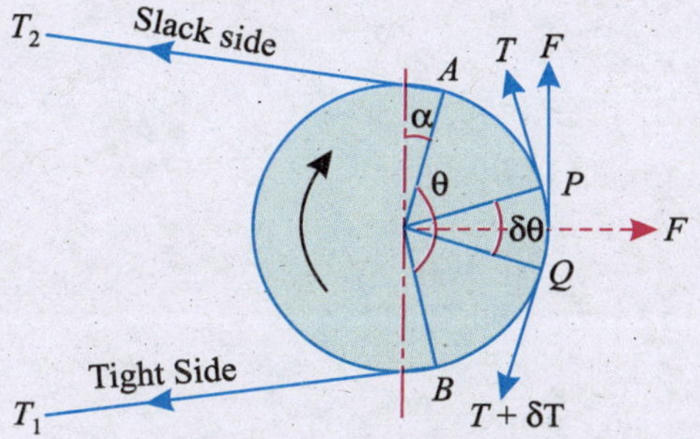

Fig. 9. Ratio of tensions.

Let T_1 = Tension in the belt on the tight side,

T_2 = Tension in the belt on the slack side, and

θ = Angle of contact in radians *i.e.*, angle subtended by the arc AB, along which the belt touches the pulley, at the centre.

Now consider a small portion of the belt PQ, subtending an angle $\delta\theta$ at the centre of the pulley as shown in Fig. 9. The belt PQ is in equilibrium under the following forces :

1. Tension T in the belt at P,
2. Tension $T + \delta T$ in the belt at Q,
3. Normal reaction R, and
4. Frictional force $F = \mu \times R$.

where μ is coefficient of friction between belt and pulley.

Resolving all the forces horizontally and equating the same,

$$R = (T + \delta T)\sin\left(\frac{\delta\theta}{2}\right) + T\sin\left(\frac{\delta\theta}{2}\right) \quad ...(i)$$

Since $\delta\theta$ is very small, therefore substituting $\sin\left(\frac{\delta\theta}{2}\right) = \frac{\delta\theta}{2}$ in equation (*i*),

$$R = (T + \delta T)\frac{\delta\theta}{2} + T\frac{\delta\theta}{2}$$

$$R = \frac{T\,\delta\theta}{2} + \frac{\delta T\,\delta\theta}{2} + \frac{T\,\delta\theta}{2} = T\,\delta\theta \qquad \text{...(ii)}$$

$$\text{...}\left(\text{Neglecting } \frac{\delta T\,\delta\theta}{2}\right)$$

Now resolving the forces vertically,

$$\mu \times R = (T + \delta T)\cos\left(\frac{\delta\theta}{2}\right) - T\cos\left(\frac{\delta\theta}{2}\right) \qquad \text{...(iii)}$$

Since the angle $\delta\theta$ is very small, therefore substituting $\cos\left(\frac{\delta\theta}{2}\right) = 1$ in equation (*iii*),

$$\mu \times R = T + \delta T - T = \delta T$$

$$\therefore \qquad R = \frac{\delta T}{\mu} \qquad \text{...(iv)}$$

Equating the values of R from equations (*ii*) and (*iv*),

$$T\,\delta\theta = \frac{\delta T}{\mu}$$

or

$$\frac{\delta T}{T} = \mu\,\delta\theta$$

Integrating both sides from A to B,

i.e.,

$$\int_{T_2}^{T_1} \frac{\delta T}{T} = \mu \int_0^{\theta} \delta\theta$$

or

$$\log_e\left(\frac{T_1}{T_2}\right) = \mu\,\theta \qquad \text{...(v)}$$

$$\therefore \qquad \left(\frac{T_1}{T_2}\right) = e^{\mu\theta}$$

The equation (v) may also be expressed in terms of corresponding logarithm to the base 10 *i.e.*

$$2{\cdot}3 \log\left(\frac{T_1}{T_2}\right) = \mu\,\theta$$

The above expression gives the relation between the tight side and slack side tensions, in terms of coefficient of friction and the angle of contact.

Notes: **1.** In the above expression (θ) is the angle of contact at the smaller pulley.

2. In an open belt drive, the angle of contact,

$$\theta = (180° - 2\alpha)$$

3. In a cross-belt drive, the angle of contact,

$$\theta = (180° + 2\alpha)$$

***Example 8.** *Find the power transmitted by a belt running over a pulley of 600 mm diameter at 200 r.p.m. The coefficient of friction between the belt and pulley is 0·25, angle of lap 160° and maximum tension in the belt is 2·5 kN.* **[UPTU, JNTU]**

Solution. Given: Diameter of pulley (d) = 600 mm = 0·6 m ; Speed of the pulley (N) = 200 r.p.m.; Coefficient of friction (μ) = 0·25; Angle of lap (θ) = 160° $= 160° \times \frac{\pi}{180°} = 2{\cdot}79$ rad and maximum tension (T_1) = 2·5 kN.

Let T_2 = Tension in the belt in slack side

We know that speed of the belt,

$$v = \frac{\pi d N}{60} = \frac{\pi \times 0{\cdot}6 \times 200}{60} = 2\pi \text{ rad/s}$$

We know that, $2{\cdot}3 \log\left(\frac{T_1}{T_2}\right) = \mu\theta = 0{\cdot}25 \times 2{\cdot}79 = 0{\cdot}6975$

$$\log\left(\frac{T_1}{T_2}\right) = \frac{0{\cdot}6975}{2{\cdot}3} = 0{\cdot}3033$$

or $$\frac{2{\cdot}5}{T_2} = 2{\cdot}01 \qquad \text{...(Taking antilog of 0·3033)}$$

$$\therefore \quad T_2 = \frac{2{\cdot}5}{2{\cdot}01} = 1{\cdot}24 \text{ kN}$$

We know that power transmitted by the belt,

$$P = (T_1 - T_2)\, v = (2{\cdot}5 - 1{\cdot}24)\, 2\pi = 7{\cdot}92 \text{ kW} \qquad \textbf{Ans.}$$

***Example 9.** *Two pulleys, one 450 mm diameter and the other 200 mm diameter are on parallel shafts 1·95 m apart and are connected by a crossed belt. Find the length of the belt required and the angle of contact between the belt and each pulley.*

What power can be transmitted by the belt, when the larger pulley rotates at 200 rev/min, if the maximum permissible tension in the belt is 1 kN, and the coefficient of friction between the belt and pulley is 0·25 ?

Solution. Given: Diameter of larger pulley (d_1) = 450 mm = 0·45 m or radius (r_1) = 0·225 m; Diameter of smaller pulley (d_2) = 200 mm = 0·2 m or radius (r_2) = 0·1 m; Distance between the centres of the pulleys (l) = 1·95 m; Speed of the larger pulley (N) = 200 r.p.m; Maximum tension in the belt (T_1) = 1 kN = 1000 N and coefficient of friction (μ) = 0·25.

Let T_2 = Tension in the belt in slack side

Length of belt

We know that length of the belt,

$$L = \pi(r_1 + r_2) + 2l + \frac{(r_1 + r_2)^2}{l}$$

$$= \pi(0{\cdot}225 + 0{\cdot}1) + 2 \times 1{\cdot}95 + \frac{(0{\cdot}225 + 0{\cdot}1)^2}{1{\cdot}95} \text{ m}$$

$$= 4{\cdot}975 \text{ m} \qquad \textbf{Ans.}$$

Angle of contact between the belt and each pulley

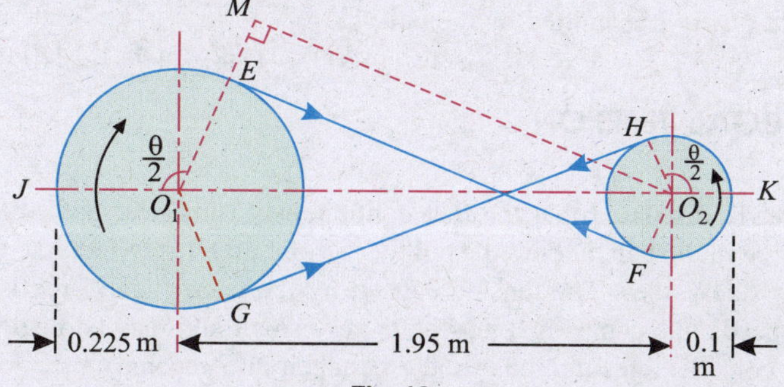

Fig. 10.

Let θ = Angle of contact between the belt and each pulley.

From the geometry of Fig. 10, we find that

$$\angle HO_2K = \angle JO_1E = \frac{\theta}{2}$$

$$\therefore \quad \angle MO_1O_2 = 180° - \frac{\theta}{2}$$

and $$\angle O_1MO_2 = 90°$$

$$\therefore \quad \cos \angle MO_1O_2 = \frac{O_1M}{O_1O_2} = \frac{O_1E + EM}{O_1O_2} = \frac{r_1 + r_2}{l}$$

$$= \frac{0{\cdot}225 + 0{\cdot}1}{1{\cdot}95} = 0{\cdot}1667$$

$$\therefore \quad \angle MO_1O_2 = 80{\cdot}4°$$

i.e. $$180° - \frac{\theta}{2} = 80{\cdot}4°$$

or $$\frac{\theta}{2} = 180° - 80{\cdot}4° = 99{\cdot}6°$$

$$\therefore \quad \theta = 2 \times 99{\cdot}6° = 199{\cdot}2° \quad \textbf{Ans.}$$

Power transmitted by the belt

We know that speed of the belt,

$$v = \frac{\pi d_1 N}{60} = \frac{\pi \times 0{\cdot}45 \times 200}{60} = 4{\cdot}71 \text{ m/s}$$

and $$\theta = 199{\cdot}2° = 199{\cdot}2° \times \frac{\pi}{180} = 3{\cdot}477 \text{ rad}$$

We know that $$2{\cdot}3 \log\left(\frac{T_1}{T_2}\right) = \mu\theta = 0{\cdot}25 \times 3{\cdot}477 = 0{\cdot}8693$$

$$\log\left(\frac{T_1}{T_2}\right) = \frac{0{\cdot}8693}{2{\cdot}3} = 0{\cdot}3780$$

or $$\left(\frac{1000}{T_2}\right) = 2{\cdot}388$$ ...(Taking antilog of 0·3780)

$$\therefore \quad T_2 = \frac{1000}{2{\cdot}388} = 418{\cdot}8 \text{ N}$$

We know that power transmitted by the belt,

$$P = (T_1 - T_2)v = (1000 - 418{\cdot}8) \times 4{\cdot}71 = 2740 \text{ W} = 2.74 \text{ kW} \quad \textbf{Ans.}$$

15. *CENTRIFUGAL TENSION

[JNTU, MU]

We have already discussed that the belt continuously runs over both the pulleys. It carries some centrifugal force in the belt, at both the pulleys, whose effect is to increase the tension on both, tight as well as the slack sides. The tension caused by the centrifugal force is called *centrifugal tension.* At lower speeds, the centrifugal tension is very small and may be neglected. But at higher speeds, its effect is considerable, and thus should be taken into account.

Consider a small portion AB of the belt as shown in Fig. 11.

Let m = Mass of the belt per unit length,

v = Linear velocity of the belt,

r = Radius of the pulley over which the belt runs,

T_C = Centrifugal tension acting tangentially at P and Q and

$d\theta$ = Angle subtended by the belt AB at the centre of the pulley.

$\therefore$ Length of belt $AB = r\ d\theta$

and total mass of the belt $M = mr\ d\theta$

We know that centrifugal force of the belt AB,

$$P_C = \frac{Mv^2}{r} = \frac{(mr\ d\theta)v^2}{r} = m\ d\theta\ v^2$$

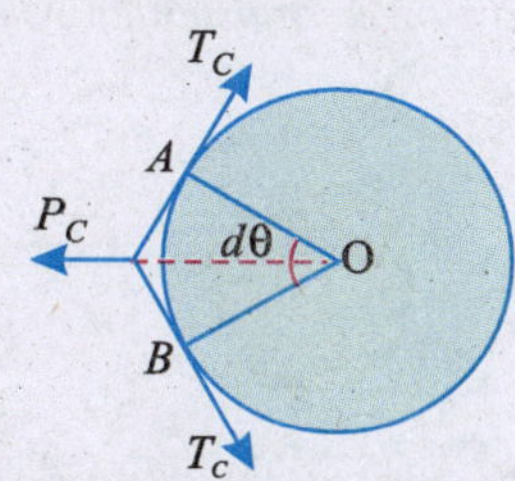

Fig. 11. Centrifugal tension.

Now resolving the forces (*i.e.*, centrifugal force and centrifugal tension) horizontally and equating the same,

$$2T_C \sin\left(\frac{d\theta}{2}\right) = m\ d\theta\ v^2$$

Since dθ is very small, therefore substituting $\sin\left(\frac{d\theta}{2}\right) = \frac{d\theta}{2}$ in the above equation,

$$2\ T_C\left(\frac{d\theta}{2}\right) = m\ d\theta\ v^2 \qquad \text{or} \qquad T_C = mv^2$$

Notes: **1.** When the centrifugal tension is taken into account, the total tension in the tight side

$= T_1 + T_C$

and total tension in the slack side

$= T_2 + T_C$

2. The centrifugal tension on the belt has no effect on the power transmitted by it. The reason for the same is that while calculating the power transmitted, we have to use the values :

= Total tension in tight side – Total tension in the slack side

$= (T_1 + T_C) - (T_2 + T_C) = (T_1 - T_2)$.

16. *MAXIMUM TENSION IN THE BELT

Consider a belt transmitting power from the driver to the follower.

Let σ = Maximum safe stress in the belt,

b = Width of the belt in mm, and

t = Thickness of the belt in mm.

We know that maximum tension in the belt,

T = Maximum stress × Cross-sectional area of belt

$= \sigma bt$

When centrifugal tension is neglected, then maximum tension,

$T = T_1$

and when centrifugal tension is considered, then maximum tension,

$T = T_1 + T_C$

***Example 10.** *A laminated belt 8 mm thick and 150 mm wide drives a pulley of 1·2 m diameter at 180 r.p.m. The angle of lap is 190° and mass of the belt material is 1000 kg/m³. If the stress in the belt is not to exceed 1·5 N/mm² and the coefficient of friction between the belt and the pulley is 0·3, determine the power transmitted when the centrifugal tension is (i) considered, and (ii) neglected.*

Solution. Given: Thickness of belt (t) = 8 mm = 0·008 m; Width of belt (b) = 150 mm = 0·15 m; Diameter of pulley (d) = 1·2 m; Speed of the pulley (N) = 180 r.p.m; Angle of lap (θ) = 190° $= 190 \times \frac{\pi}{180}$ = 3·316 rad ; Mass density of belt material (ρ) = 1000 kg/m^3; Permissible stress in the belt (σ) = 1·5 N/mm^2 = 1.5×10^6 N/m^2 and coefficient of friction (μ) = 0·3

(i) Power transmitted when the centrifugal tension is considered

Let T_1 = Tension in the tight side of the belt, and

T_2 = Tension in the slack side of the belt.

We know that speed of the belt,

$$v = \frac{\pi d N}{60} = \frac{\pi \times 1.2 \times 180}{60} = 11{\cdot}31 \text{ m/s}$$

∴ Maximum tension in the belt,

$$T = \sigma b t = (1{\cdot}5 \times 10^6) \times 0.15 \times 0.008 = 1800 \text{ N}$$

and mass of the belt per metre length,

$$m = \text{Area} \times \text{Length} \times \text{Density}$$
$$= (0{\cdot}008 \times 0{\cdot}15) \times 1 \times 1000 = 1{\cdot}2 \text{ kg}$$

∴ Centrifugal tension, $T_C = mv^2 = 1{\cdot}2 \times (11{\cdot}31)^2 = 153{\cdot}5$ N

and tension in the tight side $T_1 = T - T_C = 1800 - 153{\cdot}5 = 1646{\cdot}5$ N

We also know that,

$$2{\cdot}3 \log\left(\frac{T_1}{T_2}\right) = \mu\theta = 0{\cdot}3 \times 3{\cdot}316 = 0{\cdot}9948$$

$$\log\left(\frac{T_1}{T_2}\right) = \frac{0{\cdot}9948}{2{\cdot}3} = 0{\cdot}4325$$

$$\therefore \quad \frac{1646{\cdot}5}{T_2} = 2{\cdot}707 \qquad \text{...(Taking antilog of 0·4325)}$$

or $$T_2 = \frac{1646{\cdot}5}{2{\cdot}707} = 608{\cdot}2 \text{ N}$$

and power transmitted,

$$P = (T_1 - T_2)v = (1646{\cdot}5 - 608{\cdot}2)\ 11{\cdot}31 \text{ N-m/s}$$
$$= 11\,740 \text{ W} = 11{\cdot}74 \text{ kW} \quad \textbf{Ans.}$$

(ii) Power transmitted when the centrifugal tension is neglected

We know that tension in the tight side (without centrifugal tension),

$$T_1 = 1800 \text{ N}$$

$$\therefore \quad T_2 = \frac{1800}{2{\cdot}707} = 665 \text{ N}$$

and power transmitted,

$$P = (T_1 - T_2)v = (1800 - 665)\ 11{\cdot}31 \text{ N-m/s}$$
$$= 12\,840 \text{ W} = 12{\cdot}84 \text{ kW} \quad \textbf{Ans.}$$

Example 11. *A flat belt is required to transmit 35 kW from a pulley of 1·5 m effective diameter running at 300 r.p.m. The angle of contact is spread over $\frac{11}{24}$ of the circumference and coefficient of friction between the belt and pulley surface is 0·3. Taking centrifugal tension into account, determine the width of the belt. Take belt thickness as 9·5 mm, density as 1·1 Mg/m^3 and permissible stress as 2·5 N/mm^2.*

Solution. Given: Power to be transmitted (P) = 35 kW; Effective diameter of pulley (d) = 1·5 m; Speed of pulley (N) = 300 r.p.m; Angle of contact $(\theta) = 2\pi \times \frac{11}{24} = 2{\cdot}88$ rad ; Coefficient of friction (μ) = 0·3; Thickness of belt (t) = 9·5 mm = 0·0095 m; Mass density of the belt material (ρ) = 1·1 Mg/m^3 = 1100 kg/m^3 and permissible stress (σ) = 2·5 N/mm^2 = 2.5 × 10^6 N/m^2

Let b = Width of the belt,

T_1 = Tension on the tight side of the belt, and

T_2 = Tension on the slack side of the belt.

We know that velocity of the belt,

$$v = \frac{\pi dN}{60} = \frac{\pi \times 1.5 \times 300}{60} = 23{\cdot}56 \text{ m/s}$$

and power transmitted (P),

$$35 = (T_1 - T_2)v = (T_1 - T_2)\, 23{\cdot}56$$

$$\therefore \quad (T_1 - T_2) = \frac{35}{23{\cdot}56} = 1{\cdot}486 \text{ kN} = 1486 \text{ N} \qquad \ldots(i)$$

We also know that,

$$2{\cdot}3 \log\left(\frac{T_1}{T_2}\right) = \mu\,\theta = 0{\cdot}3 \times 2{\cdot}88 = 0{\cdot}864$$

$$\log\left(\frac{T_1}{T_2}\right) = \frac{0{\cdot}864}{2{\cdot}3} = 0{\cdot}3757$$

$$\therefore \quad \frac{T_1}{T_2} = 2{\cdot}375 \qquad \ldots\text{(Taking antilog of } 0{\cdot}3757)$$

or

$$T_1 = 2{\cdot}375\, T_2$$

Substituting the value of T_1 in equation (i),

$$2{\cdot}375\, T_2 - T_2 = 1486$$

$$\therefore \quad T_2 = \frac{1486}{1{\cdot}375} = 1081 \text{ N}$$

and

$$T_1 = 2{\cdot}375 \times 1081 = 2567 \text{ N}$$

We know that maximum tension in the belt,

$$T = \sigma b t = (2{\cdot}5 \times 10^6) \times b \times 0.0095 = 23750\, b \text{ N}$$

and mass of the belt per metre length,

$$m = \text{Area} \times \text{Length} \times \text{Density}$$

$$= (b \times 0{\cdot}0095) \times 1 \times 1100 = 10.45\, b \text{ kg}$$

$\therefore$ Centrifugal tension, $T_C = mv^2 = 10.45\, b \times (23{\cdot}56)^2 = 5800\, b$ N

and tension on the tight side of the belt (T_1)

$$2567 = T - T_C = 23750\, b - 5800\, b = 17950\, b$$

$$\therefore \quad b = \frac{2567}{17950} = 0.143 \text{ m} = 143 \text{ mm say } 150 \text{ mm}$$ **Ans.**

17. *CONDITION FOR TRANSMISSION OF MAXIMUM POWER

We know from Article 13 of this chapter, that the power transmitted by a belt,

$$P = (T_1 - T_2)\, v \qquad ...(i)$$

where T_1 = Tension on the tight side,

T_2 = Tension on the slack side, and

v = Velocity of the belt.

We have also discussed in Article. 33·14 of this chapter, the ratio of tensions,

$$\frac{T_1}{T_2} = e^{\mu\theta} \quad \text{or} \quad T_2 = \frac{T_1}{e^{\mu\theta}} \qquad ...(ii)$$

Substituting the value of T_2 in equation (*i*),

$$P = \left(T_1 - \frac{T_1}{e^{\mu\theta}}\right) \times v = T_1\left(1 - \frac{1}{e^{\mu\theta}}\right) \times v = T_1 \times v \times C \qquad ...(iii)$$

where $C = \left(1 - \frac{1}{e^{\mu\theta}}\right)$

We know that tension in the tight side,

$$T_1 = T - T_C$$

where T = Maximum tension in the belt in newtons, and

T_C = Centrifugal tension in newtons.

Substituting the value of T_1 in equation (*iii*),

$$P = (T - T_C)\, v\, C = (T - mv^2)\, v\, C = (Tv - mv^2)\, C$$

We know that for maximum power, differentiating the above equation and equating the same to zero,

$$T - 3\,mv^2 = 0 \qquad ...(iv)$$

$$T - 3T_C = 0 \qquad ...(\text{Substituting } mv^2 = T_C)$$

or $T = 3T_C$

It shows that when the power transmitted is maximum $\frac{1}{3}$rd of the maximum tension is absorbed as centrifugal tension.

18. *BELT SPEED FOR MAXIMUM POWER

We have already discussed in Article 17 of this chapter, that for maximum power transmission

$$T - 3\,mv^2 = 0 \quad \text{or} \quad 3\,mv^2 = T$$

$$\therefore \quad v = \sqrt{\frac{T}{3m}}$$

where v = Speed of the belt for maximum transmission of power,

T = *Maximum tension in the belt, and

m = Mass of the belt for unit length.

Note: The power transmitted when 1/3 of the maximum tension is absorbed as centrifugal tension (condition of last article) at belt speed for maximum power (condition of the above article) is known as *absolute maximum power or* in other words, maximum power which can be transmitted under any conditions.

*Note: Maximum tension in the belt is equal to sum of tensions in tight side (T_1) and centrifugal tension (T_C).

***Example 12.** *A belt 100 mm × 10 mm thick is transmitting power at 1200 m/min. The net driving tension is 1·8 times the tension on the slack side. If the safe stress on the belt section is 1·8 N/mm², calculate the power that can be transmitted at this speed. Assume mass density of the leather as 1 t/m³.*

Also calculate the absolute maximum power that can be transmitted by this belt and the speed at which this can be transmitted. **[MU]**

Solution. Given: Width of belt (b) = 100 mm ; = 0.1 m ; Thickness of belt (t) = 10 mm; = 0.01 m ; Velocity of belt (v) = 1200 m/min = 20 m/s; Net driving tension $(T_1 - T_2) = 1{\cdot}8\ T_2$ (where T_2 is tension in slack side); Safe stress (σ) = 1·8 N/mm² = 1.8×10^6 N/m² and mass density of leather (ρ) = 1 t/m³ = 1000 kg/m³

Power transmitted by the belt

We know that maximum tension in the belt,

$$T = \sigma b t = (1{\cdot}8 \times 10^6) \times 0.1 \times 0.01 = 1800 \text{ N}$$

and mass of the belt per metre length,

$$m = \text{Area} \times \text{Length} \times \text{Density}$$

$$= (0{\cdot}1 \times 0{\cdot}01) \times 1 \times 1000 = 1 \text{ kg}$$

∴ Centrifugal tension, $T_C = mv^2 = 1 \times 20^2 = 400$ N

and tension in the tight side, $T_1 = T - T_C = 1800 - 400 = 1400$ N

Now $1400 - T_2 = 1{\cdot}8\ T_2$ ...(Given $T_1 - T_2 = 1{\cdot}8\ T_2$)

$$T_2 = \frac{T_1}{2{\cdot}8} = \frac{1400}{2{\cdot}8} = 500 \text{ N}$$

and power transmitted by the belt,

$$P = (T_1 - T_2)\, v = (1400 - 500)\ 20 \text{ N-m/s}$$

$$= 18\ 000 \text{ W} = 18 \text{ kW}$$ **Ans.**

Speed at which absolute maximum power can be transmitted

We know that speed of the belt, at which maximum power can be transmitted,

$$v = \sqrt{\frac{T}{3m}} = \sqrt{\frac{1800}{3 \times 1}} = 24{\cdot}5 \text{ m/s}$$ **Ans.**

Absolute maximum power that can be transmitted by the belt

We know that for maximum power, the centrifugal tension,

$$T_C = \frac{T}{3} = \frac{1800}{3} = 600 \text{ N}$$

∴ Tension on the tight side, of the belt,

$$T_1 = T - T_C = 1800 - 600 = 1200 \text{ N}$$

and tension on the slack side of the belt,

$$T_2 = \frac{T_1}{2{\cdot}8} = \frac{1200}{2{\cdot}8} = 428{\cdot}6 \text{ N}$$

∴ Power transmitted by the belt,

$$P = (T_1 - T_2)\, v = (1200 - 428{\cdot}6)\ 24{\cdot}5 \text{ N-m/s}$$

$$= 18\ 900 \text{ W} = 18{\cdot}9 \text{ kW}$$ **Ans.**

19. INITIAL TENSION IN THE BELT

When a belt is wound round the two pulleys (*i.e.*, driver and follower), its two ends are jointed together; so that the belt may continuously move over the pulleys. Since the motion of the belt (from the driver) and the follower (from the belt) is governed by a firm grip due to friction between the belt and the pulleys, therefore the belt is tightened up, in order to keep a proper grip of the belt over the pulleys. Initially, even when the pulleys are stationary the belt is subject to some tension, called *initial tension.*

Let T_0 = Initial tension in the belt,

T_1 = Tension in the tight side of the belt,

T_2 = Tension in the slack side of the belt, and

μ = Coefficient of increase of the belt length per unit force.

A little consideration will show, that increase of tension in the tight side

$$= T_1 - T_0$$

and increase in the length of the belt on the tight side

$$= \mu\,(T_1 - T_0) \qquad ...(i)$$

Similarly, decrease in tension in the slack side

$$= T_0 - T_2$$

and decrease in the length of the belt on the slack side

$$= \mu\,(T_0 - T_2) \qquad ...(ii)$$

Assuming the length of the belt to be constant, when it is at rest or in motion, therefore increase in length on the tight side is equal to decrease in the length on the slack side. Therefore, equating (*i*) and (*ii*),

$$\mu\,(T_1 - T_0) = \mu\,(T_0 - T_2) \quad \text{or} \quad T_1 - T_0 = T_0 - T_2$$

$$\therefore \quad T_0 = \frac{T_1 + T_2}{2}$$

Note: If centrifugal tension is taken into consideration, then

$$T_0 = \frac{T_1 + T_2 + 2T_C}{2} = \frac{T_1 + T_2}{2} + T_C$$

***Example 13.** *Two parallel shafts whose centre lines are 4·8 m apart are connected by an open belt drive. The diameter of the larger pulley is 1·5 m and that of the smaller pulley is 1m. The initial tension in the belt, when stationary, is 3·0 kN. The mass of the material is 1·5 kg/m length and the coefficient of friction between the belt and the pulley is 0·3. Calculate the power transmitted, when the smaller pulley rotates at 400 r.p.m.* **[RTU]**

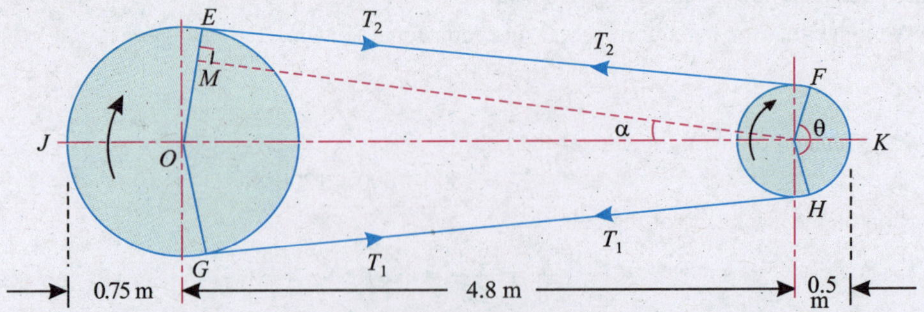

Fig. 12.

Solution. Given: Distance between the centres of shafts (l) = 4·8 m; Diameter of the larger pulley (d_1) = 1·5 m or radius (r_1) = 0·75 m; Diameter of the smaller pulley (d_2) = 1 m or radius $(r_2) = 0{\cdot}5$ m; Initial tension in the belt $(T_0) = 3$ kN; Mass of the material $(m) = 1{\cdot}5$ kg/m; Coefficient of friction $(\mu) = 0{\cdot}3$ and speed of the smaller pulley $(N_2) = 400$ r.p.m.

Let T_1 = Tension in the tight side of the belt, and

T_2 = Tension in the slack side of the belt.

We know that velocity of the belt,

$$v = \frac{\pi d_2 N}{60} = \frac{\pi \times 1 \times 400}{60} = 20{\cdot}94 \text{ m/s}$$

and initial tension in the belt when stationary (T_0)

$$3 = \frac{T_1 + T_2}{2} \quad \text{or} \quad T_1 + T_2 = 6 \text{ kN} \qquad ...(i)$$

Now for an open belt drive,

$$\sin\alpha = \frac{r_1 - r_2}{l} = \frac{0{\cdot}75 - 0{\cdot}5}{4{\cdot}8} = 0{\cdot}052$$

or $\alpha = 3°$

∴ Angle of lap for the smaller pulley,

$$\theta = 180° - 2\alpha = 180° - (2 \times 3) = 174° = \pi \times \frac{174}{180} = 3{\cdot}04 \text{ rad}$$

We also know that

$$2{\cdot}3 \log\left(\frac{T_1}{T_2}\right) = \mu\theta = 0{\cdot}3 \times 3{\cdot}04 = 0{\cdot}912$$

$$\therefore \quad \log\left(\frac{T_1}{T_2}\right) = \frac{0{\cdot}912}{2{\cdot}3} = 0{\cdot}3965$$

or $$\frac{T_1}{T_2} = 2{\cdot}49 \quad \text{or} \quad T_1 = 2{\cdot}49\, T_2 \quad \text{...(Taking antilog of } 0{\cdot}3965)$$

Substituting the value of T_1 in equation (*i*),

$$2{\cdot}49\, T_2 + T_2 = 6$$

$$\therefore \quad T_2 = \frac{6}{3{\cdot}49} = 1{\cdot}72 \text{ kN}$$

and $$T_1 = 2{\cdot}49\, T_2 = 2{\cdot}49 \times 1{\cdot}72 = 4{\cdot}28 \text{ kN}$$

∴ Power transmitted by the belt,

$$P = (T_1 - T_2)\, v = (4{\cdot}28 - 1{\cdot}72)\, 20{\cdot}94 \text{ kN-m/s} = 53{\cdot}6 \text{ kW}$$ **Ans.**

EXERCISE 2

1. Two pulleys of diameters 500 mm and 300 mm connected by a belt have tensions of 3 kN and 2.5 kN in the two sides of the belt connecting them. If the power transmitted is 25 kW, find the speed of the belt. **[Ans.** 50 m/s]

*2. A pulley is driven by a flat belt running at a speed of 600 m/min. The coefficient of friction between the pulley and the belt is 0·3 and the angle of lap is 160°. If the maximum tension in the belt is 700 N, find the power transmitted by the belt. **[JNTU]** **[Ans.** 3·97 kW]

*3. An open belt drive connects two pulleys of 1·2 m and 0·5 diameter on parallel shafts 3·6 m apart. The belt has a mass of 0·9 kg/m length and the maximum tension in it is not to exceed 2 kN. The larger pulley runs at 200 rev/min. Calculate the torque on each of the two shafts and the power transmitted $\mu = 0{\cdot}3$. **[Ans.** 654·4 N-m ; 272·7 N-m; 13·74 kW]

*4. A belt connects two pulleys *A* and *B* 4 m apart. The pulley *A* is 1 m diameter, whereas the pulley *B* is 50 cm diameter. If the coefficient of friction between the belt and the pulleys is 0·32, find the ratio of the tensions, when the drive is (*i*) an open belt drive ; and (*ii*) cross belt drive. **[JNTU]** [**Ans.** 2·623 ; 3·084]

5. The initial tension in a flat belt drive is 1800 N. The angle of lap on the smaller diameter is 170°. The coefficient of friction between the belt and pulley surface is 0·25. The pulley has a diameter of 0·9 m and it runs at 540 r.p.m.

   Determine the power that can be transmitted at the above speed. Neglect the centrifugal tension. [**Ans.** 32·5 kW]

6. A flat belt 7·5 mm thick and 100 mm wide transmits power between two pulleys, running at 1600 r.p.m. The mass of the belt is 0·9 kg/m length. The angle of lap on the smaller pulley is 165° and the coefficient of friction between the belt and pulleys is 0·3. If the maximum permissible stress in the belt is 2 MN/m^2, find (*i*) maximum power transmitted ; and (*ii*) initial tensions in the belt. [**Ans.** 15·44 kW ; 710·5 N]

20. ROPE DRIVE

Sometimes, the power, from one shaft to another, is transmitted by rope drive, instead of belt drive. This method is specially used when the shafts are a long distance apart or a considerable power is to be transmitted. The only difference between the belt drive and rope drive is that in the case of belt drive, the belt runs over the pulleys, whereas in the case of rope drive, the rim of the pulley is grooved, in which the rope runs.

The effect of the groove is to increase the frictional grip of the rope on the pulley, and thus to reduce the tendency of slipping. To have a good grip on the pulleys, the rope is in contact with side faces of the groove and not at the bottom approximately *V*-shaped as shown in Fig. 13.

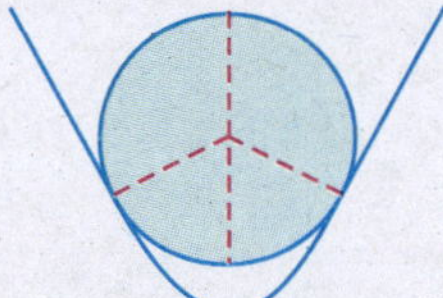

Fig. 13. Rope drive.

21. ADVANTAGES OF ROPE DRIVE

Following are the advantages of a rope drive over belt drive:

1. The rope drive is particularly suitable, when the distance between the shafts is large.
2. The frictional grip in case of rope drive is more than that in the belt drive.
3. The net driving tension (*i.e.*, difference between the two tensions) in the case of rope drive is more than the belt drive (because the ratio of tensions in the case of rope drive is cosec α times more than that in the belt drive.)

22. RATIO OF TENSIONS IN ROPE DRIVE

We have already discussed in Article 14 of this chapter, the ratio of tensions in a belt drive. The same principle may be used for the determination of ratio of tensions in case of a rope drive. Now consider a rope running in a groove as shown in Fig. 14.

Let R_1 = Normal reactions between rope and sides of the groove.

R = Total reaction in the plane of the groove.

2α = Angle of the groove.

μ = Coefficient of friction between rope and sides of the groove.

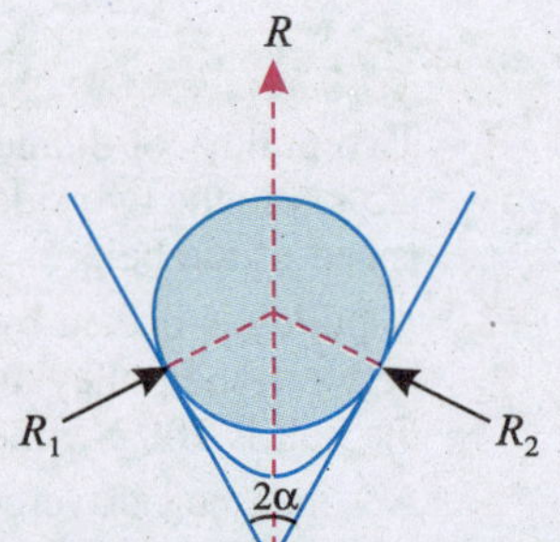

Fig. 14. Ratio of tensions in rope drive.

Resolving the reactions vertically to the groove.

$$R = R_1 \sin\alpha + R_1 \sin\alpha = 2R_1 \sin\alpha$$

or $$R_1 = \frac{R}{2\sin\alpha}$$

We know that the frictional force

$$= 2\mu R_1 = 2\mu \times \frac{R}{2\sin\alpha} = \frac{\mu R}{\sin\alpha} = \mu R \operatorname{cosec}\alpha$$

Now consider a small portion of the belt, subtending an angle $\delta\theta$ at the centre (as discussed in Article 33·14 of Ratio of Tensions in this chapter). The tension on one side will be T and on the other side $T + \delta T$. Now proceeding, we get the frictional resistance equal to (μR cosec α) against μR. Thus the relation between T_1 and T_2 for the rope drive will be:

$$2{\cdot}3 \log\left(\frac{T_1}{T_2}\right) = \mu\theta \operatorname{cosec}\alpha.$$

Example 14. *Find the power transmitted by a rope drive, from the following data:*

Angle of contact = 180°
Pulley groove angle = 60°
Coefficient of friction = 0·2
Mass of rope = 0·4 kg/metre length
Permissible tension = 1·5 kN
Velocity of rope = 15 m/s

Solution. Given: Angle of contact (θ) = 180° = 3·142 rad; Pulley groove angle (2α) = 60° or α = 30°; Coefficient of friction (μ) = 0·2; Mass of rope (m) = 0·4 kg/m; Permissible tension (T) = 1·5 kN and velocity of rope = 15 m/s.

We know that the centrifugal tension,

$$T_C = mv^2 = 0{\cdot}4(15)^2 = 90 \text{ N}$$

$\therefore$ $$T_1 = T - T_C = 1500 - 90 = 1410 \text{ N}$$

and $$2{\cdot}3 \log\left(\frac{T_1}{T_2}\right) = \mu\theta \operatorname{cosec}\alpha = 0{\cdot}2 \times 3{\cdot}142 \operatorname{cosec} 30°$$

$$= 0{\cdot}2 \times 3{\cdot}142 \times 2{\cdot}0 = 1{\cdot}257$$

$\therefore$ $$\log\left(\frac{T_1}{T_2}\right) = \frac{1{\cdot}257}{2{\cdot}3} = 0{\cdot}5465$$

or $$\frac{1410}{T_2} = 3{\cdot}52$$ ...(Taking antilog of 0·5465)

$\therefore$ $$T_2 = \frac{1410}{3{\cdot}52} = 400 \text{ N}$$

We know that power transmitted by the rope drive,

$$P = (T_1 - T_2)\, v = (1410 - 400) \times 15 \text{ N-m/s}$$

$$= 15\ 150 \text{ W} = 15{\cdot}15 \text{ kW}$$ **Ans.**

***Example 15.** *A rope drive is required to transmit 1 MW from a pulley of 1 meter diameter running at 450 r.p.m. The safe pull in each rope is 2·25 kN and the rope has mass of 1 kg per meter. The angle of lap and the groove angle is 150° and 45° respectively.*

Find the number of ropes required for the drive, if the coefficient of friction between the rope and the pulley is 0·3.

Soluton. Given: Power to be transmit (P) = 1 MW = 1000 kW; Diameter of the pulley (d) = 1 m; Speed of the pulley (N) = 450 r.p.m; Safe pull (T) = 2·25 kN = 2250 N; Mass of the rope (m) = 1 kg/m; Angle of lap (θ) = $150° = 150 \times \dfrac{\pi}{180} = 2{\cdot}62$ rad ; Groove angle (2α) = 45° or α = 22·5° and coefficient of friction (μ) = 0·3

We know that velocity of the ropes,

$$v = \frac{\pi\, d\, N}{60} = \frac{\pi \times 1 \times 450}{60} = 23{\cdot}56 \text{ m/s}$$

∴ Centrifugal tension,

$$T_C = mv^2 = 1(23{\cdot}56)^2 = 555 \text{ N}$$

∴ We know that $T_1 = T - T_C = 2250 - 555 = 1695$ N

and

$$2{\cdot}3 \log\left(\frac{T_1}{T_2}\right) = \mu\theta \operatorname{cosec} \alpha = 0{\cdot}3 \times 2{\cdot}62 \operatorname{cosec} 22{\cdot}5°$$

$$= 0{\cdot}3 \times 2{\cdot}62 \times 2{\cdot}613 = 2{\cdot}054$$

$$\log\left(\frac{T_1}{T_2}\right) = \frac{2{\cdot}054}{2{\cdot}3} = 0{\cdot}8930$$

$$\frac{1695}{T_2} = 7{\cdot}816 \qquad \text{...(Taking antilog of 0·8930)}$$

$$\therefore \quad T_2 = \frac{1695}{7{\cdot}816} = 217 \text{ N}$$

and power transmitted by one rope

$$= (T_1 - T_2)\, v = (1695 - 217)\, 23{\cdot}56 \text{ N-m/s}$$

$$= 34820 \text{ W} = 34{\cdot}82 \text{ kW}$$

$$\therefore \quad \text{No. of ropes} = \frac{\text{Total power to be transmitted}}{\text{Power transmitted by one rope}} = \frac{1000}{34{\cdot}82}$$

$$= 28{\cdot}7 \text{ say } 30 \qquad \textbf{Ans.}$$

EXERCISE 3

1. A rope drive consists of two V-belts in parallel on grooved pulleys of the same size. The angle of the groove is 30°. The cross-sectional area of each belt is 750 mm^2 and μ = 0·12. The density of the belt material is 1·2 Mg/m^3 and the maximum safe stress in the materials is 7 MN/m^2. Calculate the power that can be transmitted between the pulleys 300 mm diameter rotating at 1500 rev/min. Find also the shaft speed in rev/min at which the power transmitted would be a maximum. [**Ans.** 172 kW; 2807 r.p.m.]

QUESTIONS

1. Discuss briefly the various types of belts used for the transmission of power.
2. How does the velocity ratio of a belt drive effect, when some slip is taking place between the belt and the two pulleys?
3. Distinguish clearly the difference between an open belt drive and cross belt drive.
4. Obtain an equation for the length of a belt in :
   (*i*) an open belt drive. (*ii*) a cross belt drive.

5. Deduce the relation between the tensions on the tight side and slack side of a belt connecting two pulleys and transmitting power. Neglect the centrifugal effect of the belt mass.
6. When a flat belt is about to slip over a pulley, show that the ratio of tension in the tight and slack side depends upon friction coefficient and angle of contact.
7. Define the term 'initial tension' in a belt. How would you find out the initial tensions in a belt?
8. What are the advantages of a rope drive ?

OBJECTIVE TYPE QUESTIONS

1. The length of belt used in a cross belt drive is ..... than that used in open belt drive, provided distance between the two pulleys and their diameters remains the same.
   (*a*) less than (*b*) equal to (*c*) more than
2. The power transmitted by a belt depends upon
   (*a*) sum of the tensions in the tight side and slack side
   (*b*) difference of tension in the tight side and slack side
   (*c*) none of them
3. The relation between the tension in tight side (T_1) and slack side (T_2) of a belt drive is
   (*a*) $\frac{T_2}{T_1} = e^{\mu\theta}$ (*b*) $2{\cdot}3 \log\left(\frac{T_2}{T_1}\right) = \mu\theta$
   (*c*) both of them (*d*) none of them
   where μ = Coefficient of friction, and
   θ = Angle of lap.
4. Which of the following statement is correct?
   (*a*) The effect of centrifugal tension developed in a belt drive has no effect on the power transmitted by it.
   (*b*) The power developed is maximum when 1/3 of the maximum tension is absorbed as centrifugal tension.
   (*c*) The maximum tension in a belt drive is equal to the sum of tension in the tight side and centrifugal tension.
   (*d*) Both (*b*) and (*c*)
   (*e*) All (*a*), (*b*) and (*c*)
5. In a rope drive, the relation between the tension in the tight side and slack side is

$$2{\cdot}3\log\left(\frac{T_1}{T_2}\right) = \mu\theta \sec\alpha$$

   where 2α = Angle of groove.
   (*a*) True (*b*) False

ANSWERS

1. (*c*) **2.** (*b*) **3.** (*d*) **4.** (*e*) **5.** (*b*)

CHAPTER

31 Transmission of Power by Gear Trains

Contents

1. GEAR AND GEAR DRIVE

We have discussed (in Article of Slip of the Belt in chapter of Transmission of Power by Belts and Ropes), that the slipping in a belt or rope is a common phenomenon, in the transmission of power. The effect of slipping is to reduce the velocity ratio of the system. In **precision machines**, in which a definite velocity ratio is of importance (as in the case of hour, minute and second arms of a watch) the only positive drive is by **gears** or **toothed wheels**. A **gear drive** is provided, when the distance between the driver and the follower is very small.

A **gear** may be defined as a pulley or wheel having projections on its rim known as teeth or cogs. Sometimes, a pulley is casted with teeth on its rim. But, sometimes the teeth are cut on the rim of the pulley.

2. FRICTION WHEELS

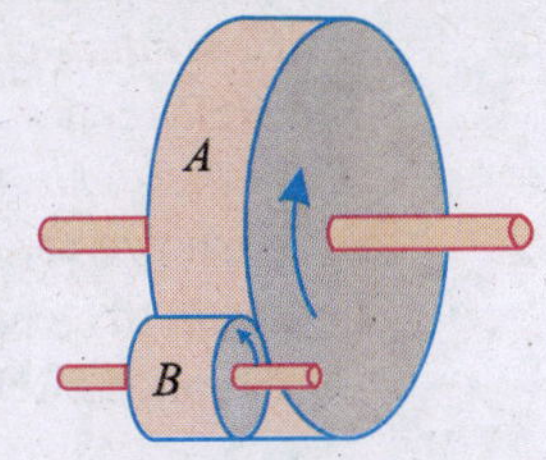

Fig. 1. Friction wheels.

Consider two plain circular wheels A and B having sufficient rough surfaces and pressed against each other as shown in Fig. 1.

Let the wheel A be keyed to a rotating shaft, and the wheel B to the shaft, to be rotated. A little consideration will show, that when the wheel A is rotated by a rotating shaft, it will rotate the wheel B in the opposite direction as shown in Fig. 1.

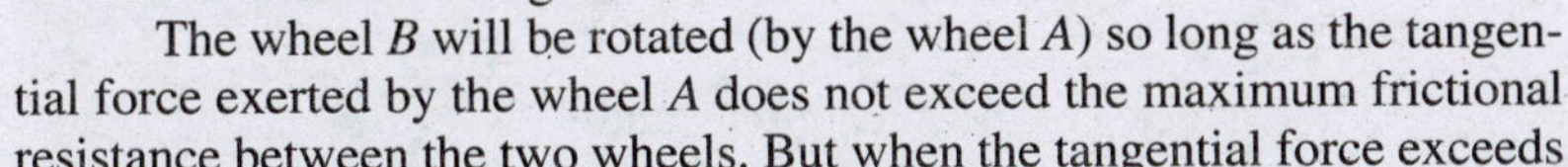

The wheel B will be rotated (by the wheel A) so long as the tangential force exerted by the wheel A does not exceed the maximum frictional resistance between the two wheels. But when the tangential force exceeds the frictional resistance, slipping will take place between the two wheels. In order to avoid the slipping, a number of projections (called teeth) are provided on the periphery of the wheel A, which will fit into the corresponding recesses on the periphery of the wheel B.

3. ADVANTAGES AND DISADVANTAGES OF A GEAR DRIVE

A gear drive has the following advantages and disadvantages :

(*a*) *Advantages*

1. It transmits exact velocity ratio.
2. It has a high efficiency.
3. It has a compact lay out.
4. It can transmit a large power.
5. It has a reliable service.

(*b*) *Disadvantages*

1. The manufacture of toothed wheels requires a special equipment and tools.
2. Any error in teeth machinery causes vibrations and noise during operation.
3. Any defect in one wheel damages the whole set up.

4. IMPORTANT TERMS

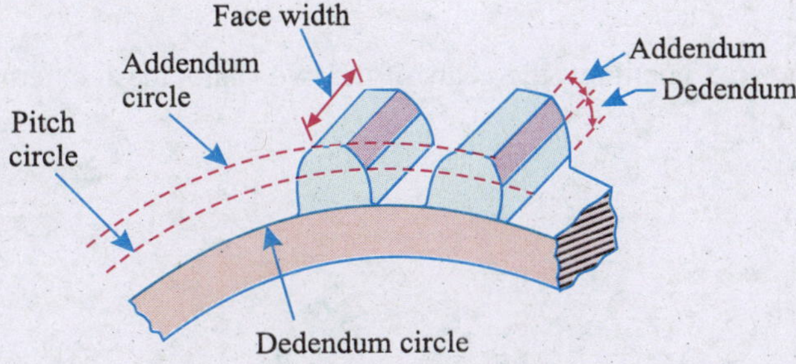

Fig. 2. Terms of gears.

The following terms, which will be mostly used in this chapter, should be clearly understood at this stage :

1. ***Pitch circle*.** An imaginary circle, which would transmit the same motion as the actual gear, by pure rolling action, is called *pitch circle* as shown in Fig. 2. The diameter of the pitch circle is known as pitch circle diameter.
2. ***Addendum circle*.** The part of a gear outside the pitch circle is called addendum, and the circle (concentric with the pitch circle) drawn through the top of the teeth is known as *addendum circle* as shown in Fig. 2. The diameter of the *addendum circle* is known as addendum circle diameter.

3. ***Dedendum circle.*** The part of a gear inside the pitch circle is called dedendum, and the circle (concentric with the pitch circle) drawn through the bottom of the teeth is known as *dedendum circle* or root circle shown in Fig. 2. The diameter of the *dedendum circle* is known as dedendum circle diameter.
4. ***Pitch.*** The centre to centre distance between any two teeth, measured along the arc of the pitch circle, is called *pitch* of the toothed wheel. Mathematically, pitch,

$$p = \frac{\pi d}{T}$$

where d = Diameter of the pitch circle, and

T = No. of teeth on the wheels.

A little consideration will show, that the two wheels will only gear together smoothly, if the two wheels have the same pitch.

5. ***Clearance.*** When the two gears mesh together, the addendum of one gear projects inside the dedendum of the other. For smooth working, a small space is left between the addendum circle of one gear and the dedendum circle of the other. This space is known as *clearance*.
6. ***Depth of the tooth.*** The radial distance between the addendum circle and dedendum circle of a gear is known as *depth of the tooth*.
7. ***Face of the tooth.*** The surface of a tooth along its width, and above the pitch circle is known as *face of the tooth* as shown in Fig. 2.
8. ***Flank of the tooth.*** The surface of a tooth along its width, and below the pitch circle is known as *flank of tooth* as shown in Fig. 2.
9. ***Face width of the tooth.*** The width of a tooth, measured parallel to its axis, is known as *face width* of the tooth as shown in Fig. 2.

5. TYPES OF GEARS

Though there are many types of gears, yet the following are important from the subject point of view :

1. ***External gears.*** Sometimes, the gears of the two shafts mesh externally with each other as shown in Fig. 3.

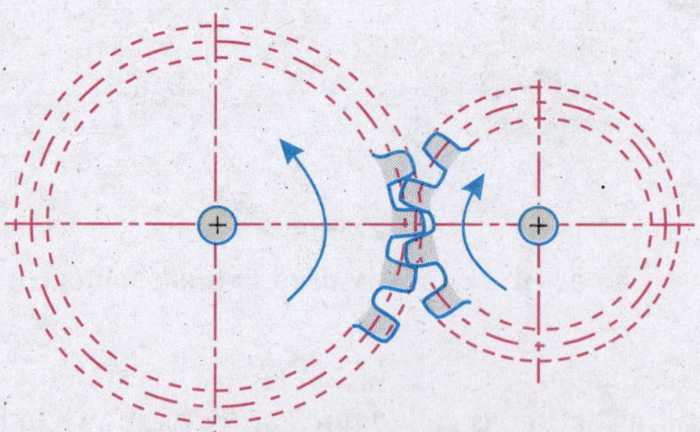

Fig. 3. External gearing.

Such a type of gear is called *external gearing*. The larger of these two wheels is called **spur wheel** and the smaller one is called **pinion**. In an external gearing, the motion of the two wheels is always unlike, as shown in Fig. 3.

2. ***Internal gears***. Sometimes, the gear of the two shafts mesh *internally* with each other as shown in Fig. 4. Such a type of gear is called *internal gearing*. The larger of these two wheels is called annula wheel and the smaller one is called pinion.

   In an internal gearing, the motion of the two wheels is always like as shown in Fig. 4.

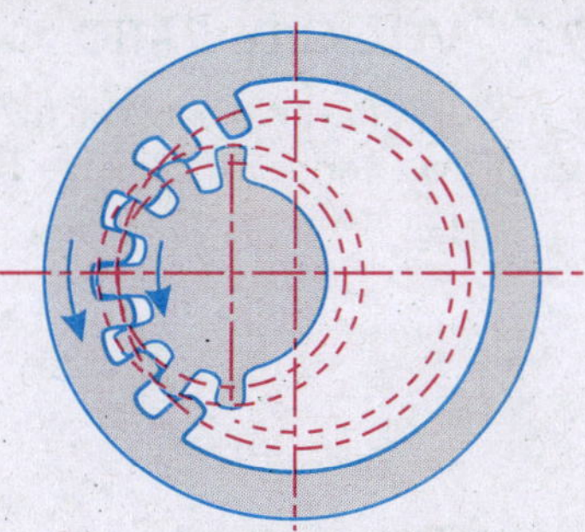

Fig. 4. Internal gearing.

3. ***Rack and pinion.*** Sometimes, the gear of a shaft meshes externally and internally with the gears in a straight line* as shown in Fig. 5.

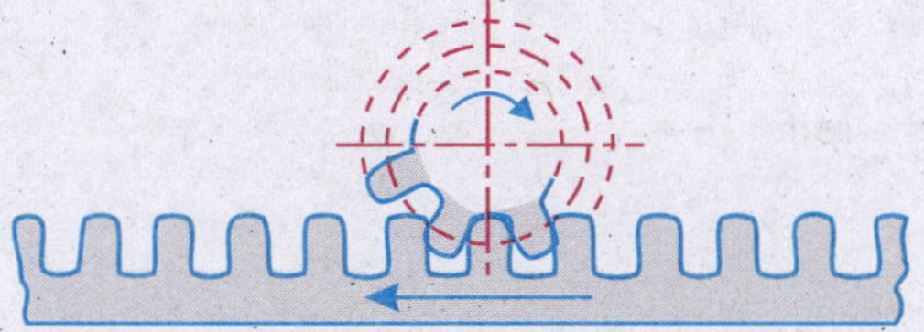

Fig. 5. Rack and pinion.

Such a type of gear is called *rack and pinion*. The straight line gear is called rack and the circular wheel is called pinion. A little consideration will show that with the help of a rack and pinion gear, we can convert linear motion into a rotary motion and *vice versa* as shown in Fig. 5.

[*Note: A straight line may also be defined as a wheel of infinite radius.]

6. SIMPLE GEAR DRIVE

A simple gear drive consists of two shafts, containing wheels, with similar teeth, geared together as shown in Fig. 6.

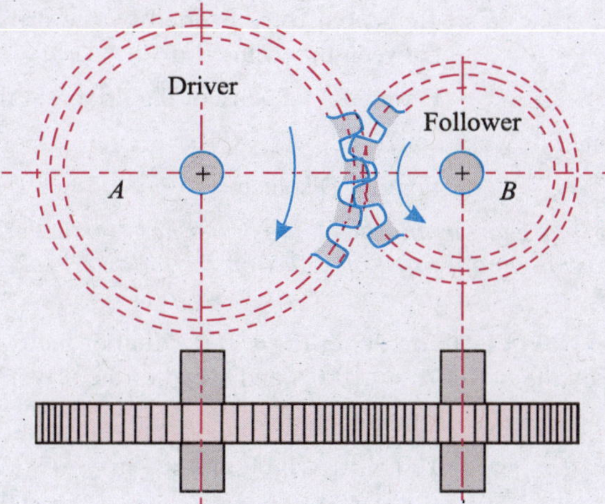

Fig. 6. Simple gear drive.

The wheel *A* is keyed to the rotating shaft, and is known as driver (since it drives the other wheel). The wheel *B* is keyed to shaft, intended to be rotated and is known as follower or driven (since it is driven by the wheel *A*). When the driver rotates, its teeth drive the teeth of the follower, which rotate it in the opposite direction of its motion as shown in Fig. 6.

7. VELOCITY RATIO OF A SIMPLE GEAR DRIVE

It is the ratio between the velocities of the driver and the follower or driven. Now consider a simple gear drive as shown in Fig. 6.

Let N_1 = Speed of the driver,

T_1 = No. of teeth on the driver,

d_1 = Diameter of the pitch circle of the driver,

N_2, T_2, d_2 = Corresponding values for the follower, and

p = Pitch of the two wheels.

We know that the pitch of the driver

$$p = \frac{\pi d_1}{T_1} \quad \text{...}(i)$$

Similarly, pitch of the follower

$$p = \frac{\pi d_2}{T_2} \quad \text{...}(ii)$$

Since the pitch of both the wheels is the same, therefore equating (*i*) and (*ii*),

$$\frac{\pi d_1}{T_1} = \frac{\pi d_2}{T_2}$$

or

$$\frac{d_1}{d_2} = \frac{T_1}{T_2}$$

∴ Velocity ratio

$$\frac{N_2}{N_1} = \frac{d_1}{d_2} = \frac{T_1}{T_2}.$$

8. POWER TRANSMITTED BY A SIMPLE GEAR

Consider a simple gear drive transmitting power, from one shaft to another.

Let F = Tangential force exerted by the driver (also called pressure between the teeth) and

v = Peripheral velocity of the driver, at the pitch point,

∴ Power transmitted or work done

$$= \text{Force} \times \text{Distance} = Fv$$

Example 1. *In a spur gear arrangement, the driver has 100 teeth of 40 mm pitch. Find the power, if it can transmit a tangential force of 100 N on the follower. Take speed of driver as 225 r.p.m.*

Solution. Given: No. of teeth on driver (T) = 100; Pitch of the two wheels (p) = 40 mm; Tangential force exerted by the driver (F) = 100 N and speed of the driver (N) = 225 r.p.m.

We know that circumference of the pitch circle

$$= 100 \times 40 = 4000 \text{ mm} = 4 \text{ m}$$

∴ Velocity of driver at the pitch point

$$v = \frac{4 \times 225}{60} = 15 \text{ m/s}$$

and power transmitted by the gear,

$$P = F \times v = 100 \times 15 = 1500 \text{ N-m/s}$$

$$= 1500 \text{ W} = 1{\cdot}5 \text{ kW} \quad \textbf{Ans.}$$

9. TRAIN OF WHEELS

Sometimes, two or more gears are made to mesh with each other, so as to operate as a single system, to transmit power from one shaft to another. Such a combination is called gear train or *train of wheels*. Following are the two types of train of wheels depending upon the arrangement of wheels :

1. Simple train of wheels, and
2. Compound train of wheels.

10. SIMPLE TRAIN OF WHEELS

Sometimes the distance between the two wheels is great. The motion from one wheel to another, in such a case, may be transmitted by either of the following two methods :

1. By providing a large sized wheel, or
2. By providing intermediate wheels,

A little consideration will show, that the former (*i.e.*, providing large sized wheel) is very inconvenient and uneconomical; whereas the latter (*i.e.*, providing intermediate wheels) is very convenient and economical.

It may be noted that when the number of intermediate wheels is *odd*, the motion of both the wheels (*i.e.*, driver and follower) is like as shown in Fig. 7. (*a*). But, if the number of intermediate wheels is *even*, the motion of the follower is the opposite direction of the driver as shown in Fig. 7. (*b*).

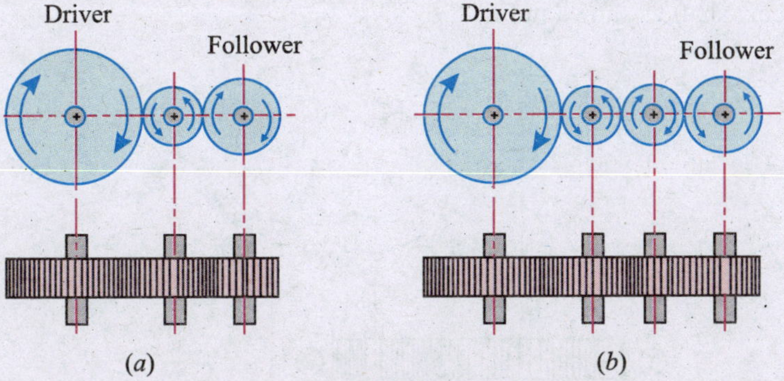

Fig. 7. Simple train of wheels.

Now consider a simple train of wheels with one intermediate wheel as shown in Fig. 7 (*a*).

Let N_1 = Speed of the driver

T_1 = No. of teeth on the driver,

N_2, T_2 = Corresponding values for the intermediate wheel, and

N_3, T_3 = Corresponding values for the follower.

Since the driver gears with the intermediate wheel, therefore

$$\frac{N_2}{N_1} = \frac{T_1}{T_2} \qquad ...(i)$$

Similarly, as the intermediate wheel gears with the follower, therefore

$$\frac{N_3}{N_2} = \frac{T_2}{T_3} \qquad ...(ii)$$

Multiplying equation (*ii*) by (*i*),

$$\frac{N_3}{N_2} \times \frac{N_2}{N_1} = \frac{T_2}{T_3} \times \frac{T_1}{T_2}$$

or

$$\frac{N_3}{N_1} = \frac{T_1}{T_3}$$

$$\therefore \quad \frac{\text{Speed of the follower}}{\text{Speed of the driver}} = \frac{\text{No. of teeth on the driver}}{\text{No. of teeth on the follower}}$$

Similarly, it can be proved that the above equation also holds good, even if there are any number of intermediate wheels. It is thus obvious, that the velocity ratio, in a simple train of wheels, is independent of the intermediate wheels. These intermediate wheels are also called idle wheels, as they do not effect the velocity ratio of the system.

11. COMPOUND TRAIN OF WHEELS

We have seen in previous article of Simple Train of Wheels, that the idle wheels do not affect the velocity ratio of the system. But these wheels are useful in bridging over the space between the driver and the follower.

But whenever the distance between the driver and follower has to be bridged over by intermediate wheels and at the same time a great (or much less) velocity ratio is required then the advantage of intermediate wheels is intensified by providing compound wheels on intermediate shafts. In this case, each intermediate shaft has two wheels rigidly fixed to it, so that they may have the same speed. One of these two wheels gears with the driver and the other with the follower attached to the next shaft as shown in Fig. 8.

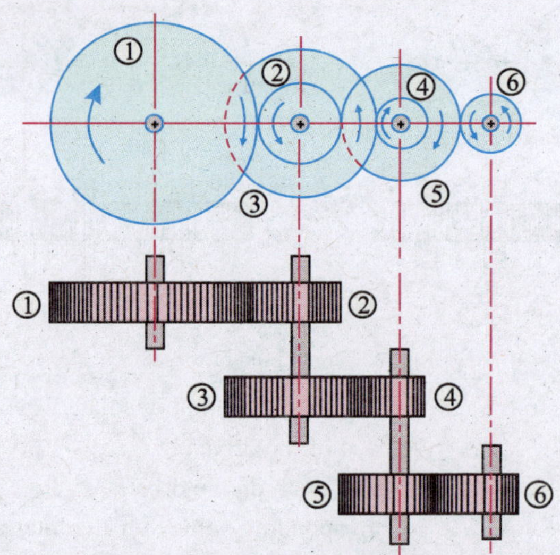

Fig. 8. Compound train of wheels.

Let N_1 = Speed of the driver 1

T_1 = No. of teeth on the driver 1,

Similarly $N_2, N_3, ...N_6$ = Speed of the respective wheels

$T_2, T_3, ...T_6$ = No. of teeth on the respective wheels.

Since the wheel 1 gears with the wheel 2, therefore

$$\frac{N_2}{N_1} = \frac{T_1}{T_2} \qquad ...(i)$$

Similarly $$\frac{N_4}{N_3} = \frac{T_3}{T_4} \qquad ...(ii)$$

and $$\frac{N_6}{N_5} = \frac{T_5}{T_6} \qquad ...(iii)$$

Multiplying equations (*i*), (*ii*) and (*iii*),

$$\frac{N_2}{N_1} \times \frac{N_4}{N_3} \times \frac{N_6}{N_5} = \frac{T_1}{T_2} \times \frac{T_3}{T_4} \times \frac{T_5}{T_6}$$

$$\therefore \quad \frac{N_6}{N_1} = \frac{T_1 \times T_3 \times T_5}{T_2 \times T_4 \times T_6} \qquad (\because \; N_2 = N_3 \text{ and } N_4 = N_5)$$

$$= \frac{\text{Product of the teeth on the drivers}}{\text{Product of the teeth on the followers}}$$

Example 2. *The gearing of a machine tools is shown in Fig. 9.*

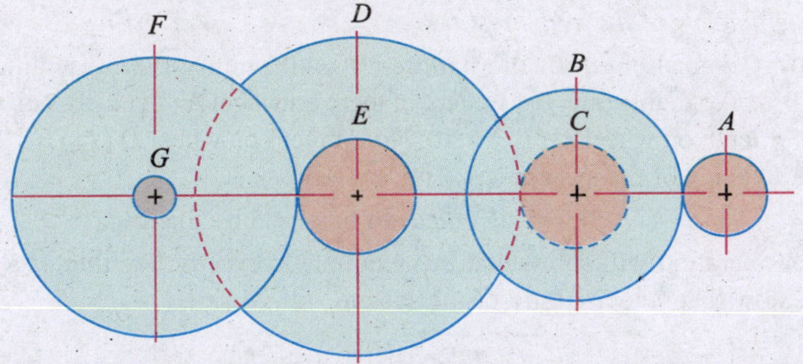

Fig. 9.

The motor shaft is connected to A and rotates at 975 r.p.m. The gear wheels B, C, D and E are fixed to parallel shafts rotating together. The final gear F is fixed on the output shaft G. What is the speed of F ? The number of teeth on each wheel is as given below :

Gear	*A*	*B*	*C*	*D*	*E*	*F*
No. of teeth	*20*	*50*	*25*	*75*	*26*	*65*

Solution. Given: Speed of the gear wheel A (N_A) = 975 r.p.m.; No. of teeth on wheel A (T_A) = 20; No. of teeth on wheel B (T_B) = 50; No. of teeth on wheel C (T_C) = 25; No. of teeth on wheel D (T_D) = 75; No. of teeth on wheel E (T_E) = 26 and no. of teeth on wheel F (T_F) = 65.

Let N_F = Speed of the shaft F.

We know that $$\frac{N_F}{N_A} = \frac{T_A \times T_C \times T_E}{T_B \times T_D \times T_F}$$

$$\therefore \quad \frac{N_F}{975} = \frac{20 \times 25 \times 26}{50 \times 75 \times 65} = \frac{4}{75}$$

or $$N_F = 975 \times \frac{4}{75} = 52 \text{ r.p.m.}$$ **Ans.**

Example 3. *A handle H drives a pinion A, which drives a drum E through gear wheels B, C and D as shown in Fig. 10.*

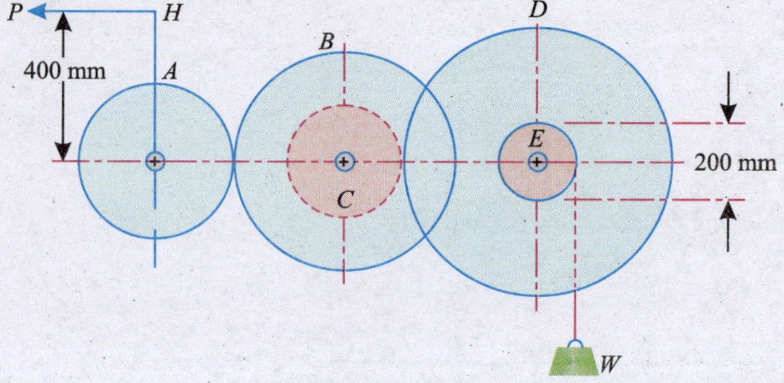

Fig. 10.

The length of handle is 400 mm, and diameter of the drum is 200 mm. The wheel A has 20 teeth, which gears with wheel B of 80 teeth. The wheel C has 20 teeth which gears with wheel D of 100 teeth.

Find the load (W) that can be raised by the drum, if an effort of 10 N is applied at the end of the handle. Take efficiency of the system as 60%.

Solution. Given: Length of the handle (l) = 400 mm = 0·4 m; Diameter of the drum (d) = 200 mm = 0·2 m or radius (r) = 0·1 m; No. of teeth on wheel A (T_A) = 20; No. of teeth on wheel B (T_B) = 80; No. of teeth on wheel C (T_C) = 20; No. of teeth on wheel D (T_D) = 100; Effort applied (P) = 10 N and efficiency of the system (η) = 60% = 0·6.

Let $\quad W$ = Load that can be raised by the drum.

A little consideration will show, that this example is exactly like that of a double purchase crap winch. We know that velocity ratio of the system,

$$\text{V. R.} = \frac{l}{r}\left(\frac{T_B \times T_D}{T_A \times T_C}\right) = \frac{0{\cdot}4}{0{\cdot}1}\left(\frac{80 \times 100}{20 \times 20}\right) = 80$$

$$\text{M.A.} = \frac{W}{P} = \frac{W}{10}$$

and efficiency,

$$0{\cdot}6 = \frac{\text{M.A.}}{\text{V.R.}} = \frac{\frac{W}{10}}{80} = \frac{W}{800}$$

$$\therefore \quad W = 0{\cdot}6 \times 800 = 480 \text{ N} \quad \textbf{Ans.}$$

EXERCISE 1

1. Two spur wheels, in gear are transmitting 40 kW. The follower has 100 teeth of 2 cm pitch. Find the pressure between the teeth when the follower runs at 300 r.p.m. [**Ans.** 4 kN]
2. In a compound train of wheels, the number of teeth on the wheels 1, 2, 3, 4, 5 and 6 are 80, 40, 50, 25, 30, and 12 respectively. Find the speed of the wheel 6, when the wheel 1 is running at 20 r.p.m. [**Ans.** 200 r.p.m.]

12. DESIGN OF SPUR WHEELS

Sometimes, the spur wheels (*i.e.* driver and follower) are to be designed for the given velocity ratio and distance between centres of their shafts.

Let l = Distance between the centres of the two shafts,

T_1 = No. of teeth on the driver,

d_1 = Diameter of the pitch circle of the driver,

T_2, d_2 = Corresponding values for the follower, and

p = Pitch of the teeth.

A little consideration will show, that the distance between the centres of two shafts,

$$l = \frac{1}{2}(d_1 + d_2) \quad \text{...}(i)$$

and

$$\text{V. R.} = \frac{d_1}{d_2} = \frac{T_1}{T_2} \quad \text{...}(ii)$$

From the above equations, we can conveniently find out the values of d_1 and d_2 [or T_1 and T_2 and pitch (p)]. The values of T_1 and T_2 as obtained above, may or may not be whole numbers. But in a wheel, since the no. of its teeth is always a whole number, therefore a slight alteration must be made in the values of l, d_1 and d_2, so that the number of teeth in the two wheels may be a complete number.

Example 4. *Two parallel shafts, about 600 mm apart are to be connected by spur wheels. One shaft is to run at 360 r.p.m. and the other at 120 r.p.m. Design the wheels, if the pitch of the teeth is to be 25 mm.*

Solution. Given: Distance between the centres of the two shafts (l) = 600 mm; Speed of the first shaft (N_1) = 120 and pitch of the wheel (p) = 25 mm.

Let d_1 = Diameter of the first wheel,

T_1 = No. of teeth on the first wheel, and

d_2, T_2 = Corresponding values for the follower,

We know that the distance between the shafts (l),

$$600 = \frac{1}{2}(d_1 + d_2)$$

$$\therefore \quad d_1 + d_2 = 600 \times 2 = 1200 \quad \text{...}(i)$$

and

$$\frac{d_1}{d_2} = \frac{N_2}{N_1} = \frac{360}{120} = 3$$

or

$$d_1 = 3d_2 \quad \text{...}(ii)$$

From equations (i) and (ii), we find that

$$d_1 = 900 \text{ mm and } d_2 = 300 \text{ mm}$$

$\therefore$ No. of teeth on the first wheel,

$$T_1 = \frac{\pi d_1}{p} = \frac{\pi \times 900}{25} = 113{\cdot}1$$

Similarly

$$T_2 = \frac{\pi d_2}{p} = \frac{\pi \times 300}{25} = 37{\cdot}7$$

Since the no. of teeth on both the wheels are to be in complete numbers, therefore let us make the no. of teeth on the second wheel as 38 instead of 37.7. Therefore for a velocity ratio of 3 [as obtained in equation (ii)] the no. of teeth on the first wheel should be 38 × 3 = 114.

Now the exact diameter of first wheel,

$$d_1' = \frac{T_1 \times p}{\pi} = \frac{114 \times 25}{\pi} = 907{\cdot}2 \text{ mm}$$

and $$d_2' = \frac{T_2 \times p}{\pi} = \frac{38 \times 25}{\pi} = 302{\cdot}4 \text{ mm}$$

Now the exact distance between the two shafts,

$$l' = \frac{1}{2}(d_1' + d_2') = \frac{1}{2}(907{\cdot}2 + 302{\cdot}4) = 604{\cdot}8 \quad \text{mm}$$

Hence the no. of teeth on the first and second wheels must be 114 and 38; and their diameters must be 907.2 mm and 302.4 mm respectively. The exact distance between the two shafts must be 604.8 mm **Ans.**

13 TRAIN OF WHEELS FOR THE HOUR AND MINUTE HANDS OF A 12-HOUR CLOCK

The train* of wheels for the hour and minute hands of a 12 hour clock, in its simplest form, consists of solid shaft containing minute hand and another hollow shaft (concentric with the shaft containing minute hand) containing hour hand. Since the motion of both the hands is to be like, with a velocity ratio of 1 : 12, therefore a compound wheel is provided as shown in Fig. 11.

*Note: When the axes of the first and last wheels (1 and 4 in this case) are co-axial, it is called as reverted gear train. In a reverted gear train, the motion of the first and last wheel is always *like*.]

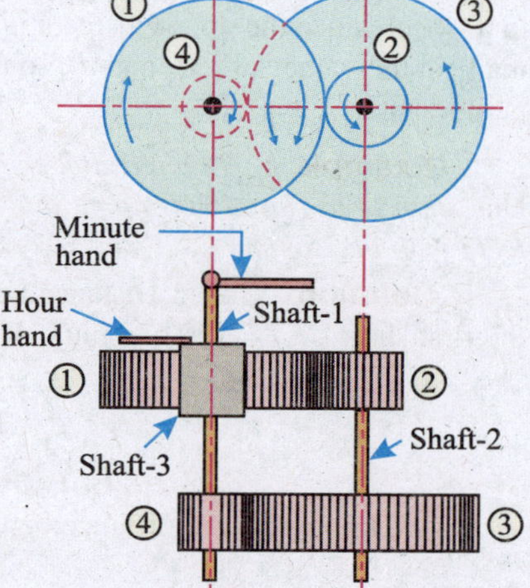

Fig. 11. Train of wheels for clock.

Let T_1 = No. of teeth on the wheel 1,

r_1 = Radius of the wheel 1,

N_1 = Speed of the wheel 1 in r.p.m.

T_2, T_3, T_4 = No. of teeth on the respective wheels,

r_2, r_3, r_4 = Radii of the respective wheels,

N_2, N_3, N_4 = Speed of respective wheels.

Since the distance between the centres of the shafts 1 and 2 as well as 3 and 4 is the same, therefore

$$r_1 + r_2 = r_3 + r_4$$

Moreover, as the pitch of all the wheels is also the same, therefore the no. of teeth on each wheel is proportional to its circumference or radius.

$$\therefore \quad T_1 + T_2 = T_3 + T_4 \qquad \text{...}(i)$$

We know that the velocity ratio of the minute hand to the hour hand is equal to 12.

Therefore $$\frac{N_4}{N_1} = \frac{T_1 \times T_3}{T_2 \times T_4} = 12$$

or $$T_1 \times T_3 = 12\, T_2 \times T_4 \qquad \text{...}(ii)$$

Now we have only two above equations for evaluating the four unknowns (*i.e.* T_1, T_2, T_3 and T_4). Strictly speaking, these two equations are insufficient for evaluating these four unknowns.

Therefore, for the purpose of evaluating these four unknowns, we should have four equations. The same can also be done either by assuming some data or superimposing two more equations.

From practical point of view, none of these wheels should have less than 12 teeth. Therefore let us assume.

$$T_4 = 12$$

and $$T_3 = 4\ T_4 = 4 \times 12 = 48$$

Substituting the value of $T_3 = 48$ and $T_4 = 12$ in equation (*i*),

$$T_1 + T_2 = 48 + 12 = 60$$

$$\therefore \quad T_1 = 60 - T_2 \qquad ...(iii)$$

and substituting these values of T_3 and T_4 in equation (*ii*),

$$T_1 \times 48 = 12 \times T_2 \times 12 = 144\ T_2 \quad \text{or} \quad T_1 = 3\ T_2$$

Now substituting the value of T_1 in equation (*iii*),

$$3\ T_2 = 60 - T_2 \quad \text{or} \quad 4\ T_2 = 60$$

$$\therefore \quad T_2 = 15 \qquad ...(iv)$$

Substituting this value of T_2 in equation (*iii*),

$$T_1 = 60 - 15 = 45$$

Hence the most suitable combination for the wheels is

$$T_1 = 45;\ T_2 = 15;\ T_3 = 48 \text{ and } T_4 = 12.$$

Example 5. *A watch is wound up regularly, at the same time everyday, and the main spring spindle receives 4·5 complete turns during the winding. State the velocity ratio of the train of wheels connecting the main spindle with the hour spindle, also the velocity ratio of the train of wheels connecting the minute hand spindle with hour hand spindle.*

Give a suitable train of wheels for the latter, with no wheel having more than 40 teeth.

Solution. Given: Turns taken by the spring spindle for one day = 4·5.

Velocity ratio of main spindle to the hour hand spindle

We know that the hour hand spindle turns through 2 revolutions in one day (*i.e.*, 24 hours). Therefore velocity ratio of the main spindle with the hour hand spindle

$$= \frac{4{\cdot}5}{2} = 2{\cdot}25 \quad \textbf{Ans.}$$

Velocity ratio of the minute hand spindle to the hour spindle

We also know that when the minute hand turns through 12 revolutions, the hour hand turns through 1 revolution in the same interval of time. Therefore velocity ratio of the minute hand spindle to the hour spindle

$$= \frac{12}{1} = 12 \quad \textbf{Ans.}$$

Train of wheels

Let T_1 = No. of teeth on the hour hand spindle,

T_2 = No. of teeth on the wheel 2,

T_3 = No. of teeth on the wheel 3, and

T_4 = No. of teeth on the minute hand spindle.

We know that $$T_1 + T_2 = T_3 + T_4 \qquad ...(i)$$

and $$T_1 \times T_3 = 12\ T_2 \times T_4 \qquad ...(ii)$$

Now assume* $T_4 = 8$ and $T_3 = 4\,T_4 = 4 \times 8 = 32$

Substituting the values of T_3 and T_4 in equation (*i*),

$$T_1 + T_2 = 32 + 8 = 40$$

$$\therefore \quad T_1 = 40 - T_2 \qquad \text{...(iii)}$$

Now substituting the value of T_3 and T_4 in equation (*ii*),

$$T_1 \times 32 = 12 \times T_2 \times 8 = 96\,T_2 \qquad \text{...(iv)}$$

$$\therefore \quad T_1 = 3\,T_2$$

Substituting this value of T_1 in equation (*iii*), $3T_2 = 40 - T_2$

$$\therefore \quad T_2 = \frac{40}{4} = 10$$

Now substituting this value of T_2 in equation (*iii*),

$$T_1 = 40 - 10 = 30$$

$$\therefore \quad T_1 = 30;\ T_2 = 10;\ T_3 = 32;\ T_4 = 8 \qquad \textbf{Ans.}$$

Since the number of teeth are within the given limit (of 40), therefore the arrangement of the teeth is correct.

[*Note: This assumption is by trial and error. If we assume the value of T_4 as 1, 2 and 3, the no. of teeth on the other wheels obtained are not complete integers. If we assume T_4 as 4, we get $T_1 = 15$, $T_2 = 5$, $T_3 = 16$, and $T_4 = 4$. The no. of teeth on each wheel is too less for practical purposes. Similarly, if we assume the values of T_4 as 5, 6 and 7, the no. of teeth on the other wheels obtained are not complete integers.

It will be also seen that if we assume the values of T_4 as 9, 10 and 11, the no. of teeth on the other wheels are not complete integers. If we assume T_4 as 12, we get $T_1 = 45$, $T_2 = 15$, $T_3 = 48$, $T_4 = 12$. This is contrary to the given condition (*i.e.*, no wheel should have more than 40 teeth).]

14. EPICYCLIC GEAR TRAIN

In the previous articles, we have discussed the cases of gear trains in which each wheel is free to revolve about it own axis. But sometimes, one of the wheel is fixed and the other makes revolution about the fixed wheel with the help of an arm. Such a system is called an *epicyclic gear train.*

In Fig. 12 is shown wheel A and an arm C having a common axis at O_1, about which they can rotate. The wheel B meshes with the wheel A and has its own axis on the arm at O_2. Now, if the arm C is considered to be fixed, the wheels A and B form a simple gear train. But, if the wheel A is fixed and the arm is rotated, the system forms an epicyclic gear train.

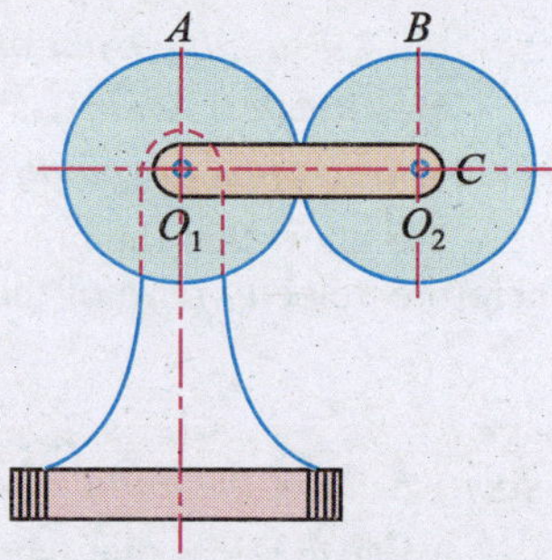

Fig. 12. Epicyclic gear train.

EXERCISE 2

1. Design the spur wheels, having a velocity ratio of 4 and the pitch of teeth as 7·5 mm. The approximate distance between the two shafts is 300 mm.

[**Ans.** $T_1 = 200$; $T_2 = 50$; $d_1 = 477{\cdot}3$ mm; $d_2 = 119{\cdot}3$ mm; $l = 298{\cdot}3$ mm]

QUESTIONS

1. Explain the advantages of gear drive over the belt and rope drives.
2. Distinguish between the terms pitch circle, addendum circle and dedendum circle.
3. What are the various types of gears ? Explain each one of them with sketches.
4. Explain clearly the use of providing intermediate wheels in :
   (*a*) a simple train of wheels, and
   (*b*) a compound train of wheels.
5. Explain the procedure adopted for designing the spur wheels.
6. Discuss the working of train of wheels for the hour and minute hands of a 12-hour clock.

OBJECTIVE TYPE QUESTIONS

1. Which of the following circle is an imaginary circle in the study of toothed gears
   (*a*) Pitch circle (*b*) Addendum circle
   (*c*) Dedendum circle (*d*) All of them
2. The pitch of a toothed wheel is
   (*a*) $\frac{\pi d}{T}$ (*b*) $\frac{\pi T}{d}$
   (*c*) $\frac{\pi}{T d}$ (*d*) $\frac{T}{\pi d}$
   where d = Diameter of the pitch circle, and
   T = No. of teeth on the wheel.
3. Which of the following statement is correct ?
   (*a*) The gears are used for transmission of power.
   (*b*) The gears are used where exact velocity ratio is required.
   (*c*) In a single train of wheels, motion of both the driver and follower is like.
   (*d*) Both (*a*) and (*b*).
   (*e*) All (*a*), (*b*) and (*c*).
4. In a compound train of wheels, we provide idle wheels, which do not affect the velocity ratio of the system.
   (*a*) Agreez (*b*) Disagree
5. In an epicyclic gear train, one of the wheel is fixed and the other makes revolution about the fixed wheel.
   (*a*) True (*b*) False

ANSWERS

1. (*a*) 2. (*a*) 3. (*e*) 4. (*b*) 5. (*a*)

CHAPTER

32

Work-Energy Method

Contents

1. INTRODUCTION

The questions on dynamics can be solved with the help of following methods:

1. **D'Alembert's Principle** is used for solving questions on dynamics which involve **force and acceleration**.
2. **Impulse-Momentum Method** is used for solving questions on dynamics which involve **force**, **velocity** and **time**.
3. **Work-Energy Principle** or **Method** is used for solving questions on dynamics which involve **force, velocity and displacement**.

2. WORK DONE

(a) Work Done for Rectilinear Motion: Whenever a force acts on a body and the body undergoes some displacement, then work is said to be done. Mathematically, if a force P acting in the direction of motion of a body which causes it to move through a distance s in the direction of motion, then work done by the force P on the body is given by,

$$\text{Work Done} = \text{Force} \times \text{Distance}$$

or

$$W = F \times s \qquad \ldots(i)$$

Sometimes, the force P does not act in the direction of motion of the body, then work done by the force P is given by

$$\text{Work done} = \text{Component of the force in the direction of motion} \times \text{Distance}$$

or

$$W = F \cos\theta \times s \qquad \ldots(ii)$$

Where θ is the angle made by the force in the direction of motion.

The work or work done is a scalar quantity. The S.I. unit of work is joule (or N-m).

(b) Work Done for Rotational Motion: Whenever a torque T acts on a circular body at a certain distance and the body undergoes some angular displacement (θ) about a fixed point, then work is said to be done. Mathematically, Work done (W) by a torque (T) in rotating a circular body about a fixed axis is given by

$$\text{Work Done} = \text{Torque} \times \text{Angular displacement}$$

or

$$W = T \times \theta \qquad \ldots(iii)$$

where θ = Angular displacement of the body in radians.

T = Torque acting on the body

(c) Gravitational work done when a body moving along a curved path or inclined plane: When a body moving along a curved path or inclined plane upward or downward and undergoes vertical displacement h_1 to h_2, then Gravitational work done in moving the vertical displacement from h_1 to h_2 is given by

$$W_G = mg(h_1 - h_2) \qquad \ldots(iv)$$

where m = Mass of the body,

h_1 = Initial vertical displacement of the body and

h_2 = Final vertical displacement of the body.

(d) Work Done by a Spring: Work done (W) by a spring on a body during the displacement from x_1 to x_2 is given by

$$W = \frac{1}{2}k(x_1^2 - x_2^2) \qquad \ldots(v)$$

where k = Spring constant or Stiffness of spring,

x_1 = Initial displacement of the body and

x_2 = Final displacement of the body.

3. WORK-ENERGY PRINCIPLE OR METHOD

Statement: The work-energy principle states, **"The total work done by a system of forces acting on a body is equal to the change in kinetic energy of the body."**

Proof: Consider a system of forces F_1, F_2, ... acting on a body which may displaced it from one position to another position.

Let m = Mass of a body,

u = Initial velocity of the body at initial position when distance is s_1,

v = Final velocity of the body at final position when distance is s_2,

$F = \sum F_i$ = Resultant of all forces acting on the body and

a = Constant acceleration in the direction of resultant force.

We know from **Newton's Second Law of Motion** that the rate of change of momentum is directly proportional to the impressed force and takes places in the direction of the force. Mathematically,

$$F = ma = v\frac{dv}{dt}$$

or
$$F = m\frac{dv}{ds}\cdot\frac{ds}{dt} = mv\frac{dv}{ds}$$

or
$$F \cdot ds = mv\ dv$$

Since $F \cdot ds$ is the work done by the force F in displacing the body a small distance ds, therefore the total work done by the force in displacing the body from s_1 to s_2 is calculated by integrating the above equation.

Integrating the above equation both sides, we get

or
$$F \cdot ds = mv\ dv$$

$$\int_{s_1}^{s_2} Fds = m\int_u^v vdv = m\left[\frac{v^2}{2}\right]_u^v = \frac{1}{2}mv^2 - \frac{1}{2}mu^2$$

or
$$F \times (s_2 - s_1) = \frac{1}{2}mv^2 - \frac{1}{2}mu^2$$

or
$$\textbf{(W.D.)}_{1-2} = \textbf{(K.E.)}_2 - \textbf{(K.E.)}_1$$

or
$$\textbf{(W.D.)}_{1-2} = \textbf{[Change in Kinetic Energy]}_{1-2} \qquad ...(i)$$

This equation is known as **work-energy equation**. This can be also written as

$$\textbf{(K.E.)}_1 \pm \textbf{(W.D.)}_{1-2} = \textbf{(K.E.)}_2 \qquad ...(ii)$$

The method of solving dynamic problems using Work-Energy Equations is known as **Work-Energy Method**.

4. APPLICATIONS OF WORK-ENERGY PRINCIPLE

Though there are various applications of Work-Energy principle, but following are the important from the subject point of view:

1. Work-Energy Principle in Case of Linear, Inclined and Circular Motions
2. Work-Energy Principle in Case of Spring
3. Work-Energy Principle in Case of Connected bodies
4. Work-Energy Principle in Case of Rotational Motion
5. Work-Energy Principle in Case of Pile and Pile Hammer

We shall discuss these applications of Work-Energy principle one by one.

WORK-ENERGY PRINCIPLE IN CASE OF LINEAR, INCLINED AND CIRCULAR MOTIONS

Example 1. *A crate of mass 30 kg, slides down on the curved path from A to B in the vertical plane, as shown in Fig. 1. If the crate has a velocity of 2.4 m/s down the incline at point A, and 8 m/s at B, then find the work done by the force of friction during the motion from A to B. Use Work-Energy Principle.*

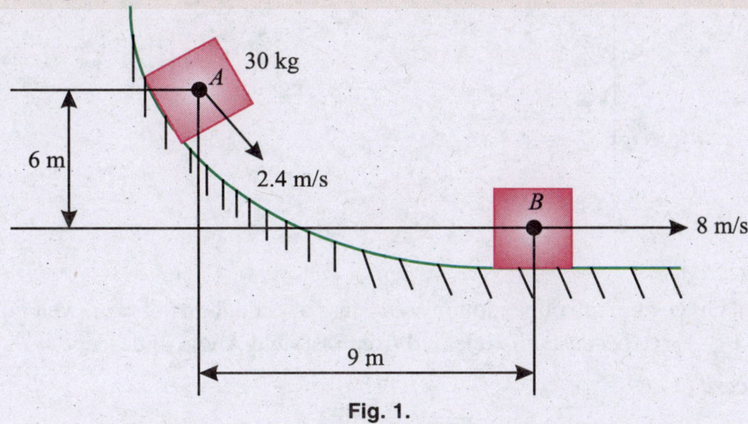

Fig. 1.

Solution. Given: Mass of crate $(m) = 30$ kg ; Velocity of crate at point A $(v_A) = 2.4$ m/s ; Velocity of crate at point B $(v_B) = 8$ m/s ; Vertical distance between point A and B $(h) = 6$ m and Horizontal distance between A and B $(d) = 9$ m.

Work done by force of friction during the motion from A to B

Let W_F = Work done by force of friction during the motion from A to B

We know that, Gravitational work done by the crate when it moved downward by 6 m from point A to B,

$$W_G = mgh = 30 \times 9.81 \times 6 = +1765.8 \text{ J}$$

We also know that, Initial kinetic energy of crate at point A,

$$(\text{K.E.})_A = \frac{1}{2} m v_A^2 = \frac{1}{2} \times 30 \times 2.4^2 = 86.4 \text{ J}$$

Similarly, Final kinetic energy of crate at point B,

$$(\text{K.E.})_B = \frac{1}{2} m v_B^2 = \frac{1}{2} \times 30 \times 8^2 = 960 \text{ J}$$

Now, Work done (W_F) by force of friction during the motion from A to B can be calculated by applying **Work-Energy Principle** between A and B,

$$(\text{W.D.})_{A-B} = (\text{K.E.})_B - (\text{K.E.})_A$$

or $$W_G + W_F = (\text{K.E.})_B - (\text{K.E.})_A$$

or $$1765.8 + W_F = 960 - 86.4$$

or $$1765.8 + W_F = 873.6$$

or $$W_F = 873.6 - 1765.8 = -892.2 \text{ J}$$ **Ans.**

...(Negative sign is due to opposite direction of force of friction)

Example 2. *The pendulum bob has a mass M and is released from rest when θ = 0° as shown in the Fig. 2. Determine the tension in the cord as a function of angle of descent θ. Neglect the size of the bob.* **[MU-Mum-Mah]**

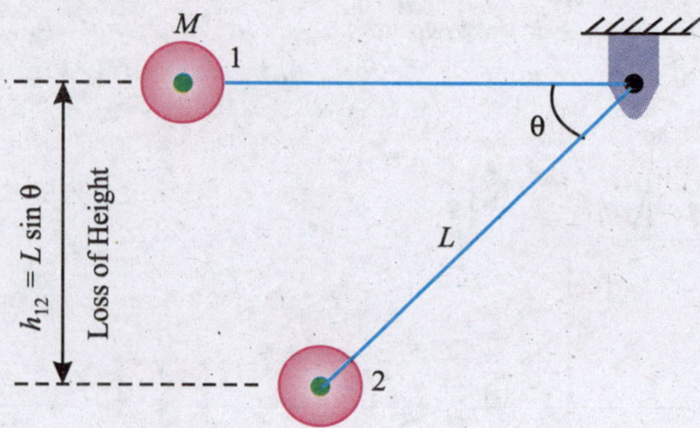

Fig. 2.

Solution. Given : Length of pendulum = L ; Mass of pendulum bob = m ; Velocity energy of the bob at position 1 (v_1) = 0 (because it is released from rest) and Angle of descent = θ.

Tension in the cord (T)

Let T = Tension in the cord

We know that, Kinetic energy of the bob at position 1,

$$(\text{K.E.})_1 = \frac{1}{2}Mv_1^2 = \frac{1}{2}\times M \times 0^2 = 0$$

Similarly, Kinetic energy of the bob at position 2,

$$(\text{K.E.})_2 = \frac{1}{2}Mv_2^2$$

And Positive gravitational work done by the bob in going from position 1 to 2,

$$W_G = Mgh_{12} = Mg(L\sin\theta)$$

Now, Velocity (v_2) of the bob at position 2 can be calculated by applying **Work-Energy Principle**,

$$(\text{W.D.})_{1-2} = (\text{K.E.})_2 - (\text{K.E.})_1$$

or $$W_G = (\text{K.E.})_2 - (\text{K.E.})_1$$

or $$Mg(L\sin\theta) = \frac{1}{2}Mv_2^2 - 0$$

or $$v_2^2 = 2gL\sin\theta \qquad (i)$$

Now, draw the free body diagram of bob at position 2 as shown in the Fig. 3 and forces acting on the bob are:

1. Tension, T in the direction along cord
2. Weight, $W = Mg$ acting vertically downward
3. Inertial force acting, $F = Ma_2$ along the path in opposite direction
4. Centrifugal force, $F_C = \dfrac{Mv_2^2}{L}$ acting outward direction

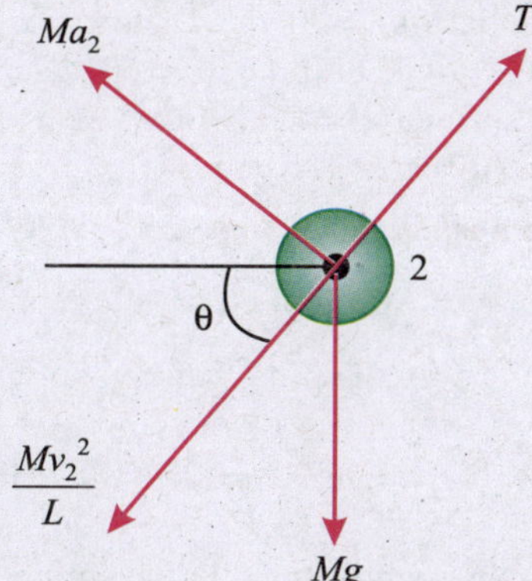

Fig. 3. Forces acting on the Bob at position 2.

We also know that, Tension (T) in the cord can be calculated by resolving forces parallel to the cord and equating the forces acting in right direction with the forces acting in left direction,

$$T = F_C + Mg\cos(90° - \theta)$$

or $$= \frac{Mv_2^2}{L} + Mg\sin\theta$$

or $$= \frac{M\left(2gL\sin\theta\right)}{L} + Mg\sin\theta \qquad \text{(Using relation } (i))$$

or $$= 2Mg\sin\theta + Mg\sin\theta = 3\,Mg\sin\theta \qquad \textbf{Ans.}$$

WORK-ENERGY PRINCIPLE IN CASE OF SPRING

Example 3. *A collar of mass 20 kg, which is supported on the smooth rod as shown in Fig. 4. The attached springs are undeformed when d = 0.5 m. Determine the speed of the collar after the applied force of 500 N causes it to displace so that d = 0.3 m. The collar is at rest when d = 0.5 m. Use Work Energy Principle or otherwise,* **[MU-Mum-Mah]**

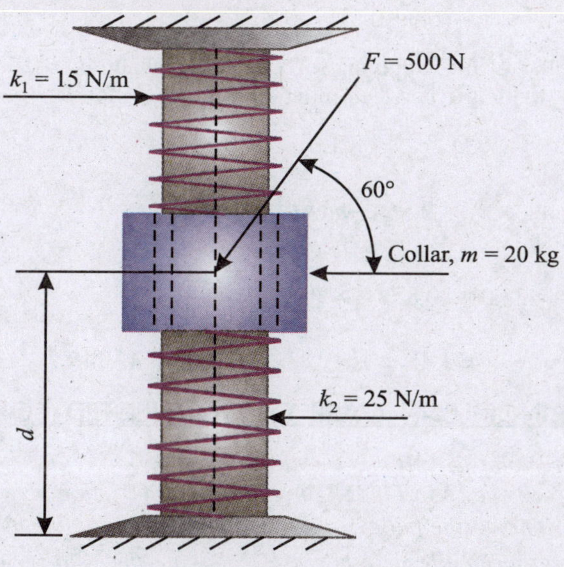

Fig. 4.

Solution. Given: Mass of collar (m) = 20 kg ; Distance before applying force (d_1) = 0.5 m ; Force applied on spring (F) = 500 N ; Angle of force (θ) = 60° ; Spring constant of upper spring (k_1) = 15 N/m ; Spring constant of lower spring (k_2) = 25 N/m ; Distance after applying force (d_2) = 0.3 m and Initial speed of collar (v_1) = 0 ...(because it is at rest).

Speed of collar after applying force when d = 0.3 m

Let v_2 = Speed of collar after applying force of 500 N

We know that, Initial kinetic energy of spring before applying force,

$$(\text{K.E.})_1 = \frac{1}{2}mv_1^2 = \frac{1}{2}\times 20\times(0)^2 = 0$$

And, Initial work done or strain energy stored in the spring before applying force,

$$W_{\text{Spring}} = 0$$

And Displacement of the collar after applying force,

$$x = d_1 - d_2 = 0.5 - 0.3 = 0.2 \text{ m}$$

Now, Positive gravitational work done by the force,

$$W_G = mgh = 20 \times 9.81 \times 0.2 = +\ 39.24 \text{ J}$$

And, Positive work done on applying force,

$$W_1 = F \times \sin\theta = 500 \times \sin 60° = 433.01 \text{ J}$$

We know that, Negative spring work or strain energy stored in the two springs,

$$W_2 = -\left(\frac{1}{2}k_1x^2 + \frac{1}{2}k_2x^2\right) = -\left[\left(\frac{1}{2}\times 15\times 0.2^2\right)+\left(\frac{1}{2}\times 25\times 0.2^2\right)\right]$$

or $$= -(0.3+0.5) = -0.8 \text{ J}$$

Now, Speed of collar v_2 after applying force of 500 N when d = 0.3 m can be calculated by applying **Work-Energy Principle** between initial and final positions,

$$(\text{W.D.})_{1-2} = (\text{K.E.})_2 - (\text{K.E.})_1$$

or $$W_{\text{Spring}} + W_G + W_1 + W_2 = \frac{1}{2}mv_2^2 - 0$$

or $$0 + 39.24 + 433.01 - 0.8 = \frac{1}{2}\times 20\times v_2^2 - 0$$

or $$471.45 = 10\,v_2^2 \quad \text{or} \quad v_2^2 = 47.145 \quad \text{or} \quad v_2 = 6.87 \text{ m/s}$$ **Ans.**

WORK-ENERGY PRINCIPLE IN CASE OF CONNECTED BODIES

Example 4. *Three blocks A, B and C, having mass of 30 kg, 20 kg and 80 kg, respectively, are connected by strings as shown in Fig. 5. The coefficient of friction between the block B and the table surface is 0.4. Assume all the pulleys to be smooth with negligible mass.*

Using work-energy principle or otherwise, calculate the velocity of block C, after it has moved through a distance on 1 m starting from rest.

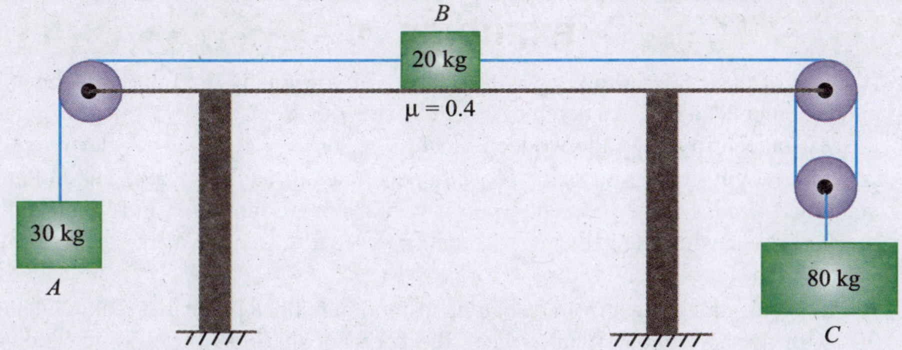

Fig. 5.

Solution. Given: Mass of the block A $(m_A) = 30$ kg ; Mass of the block B $(m_B) = 20$ kg ; Mass of the block C $(m_C) = 80$ kg ; Coefficient of friction between the block B and the table $(\mu) = 0.4$ and Vertical distance moved by the mass C $(h_C) = 1$ m.

Since the system starts from rest, therefore initial kinetic energy of the system of blocks,

$$(\text{K.E.})_1 = 0.$$

Since, $m_C > 2m_A$, therefore mass C will move downward.

We know according to **String rule** that, as mass C attached to two strings, has moved downward by 1 m, the mass B will move rightward by 2 m and mass A will move upward by 2 m. Thus, $s_B = h_A = 2$ m.

Therefore, Gravitational work done by mass C,

$$(W_G)_C = m_C\, g\, h_C = 80 \times 9.81 \times 1 = +\,784.8 \text{ J}$$

And Gravitational work done on mass A,

$$(W_G)_A = -\,m_A\, g\, h_A = -\,30 \times 9.81 \times 2$$

$$= -\,588.6 \text{ J} \qquad \text{(Minus sign due to mass has moved up)}$$

Work done on the surface of the table due to friction,

$$W_F = -\,\mu . m_B\, g\, s_B = -\,0.4 \times 20 \times 9.81 \times 2 = -\,156.96 \text{ J}$$

Since block C is attached to two springs, therefore velocity of blocks A, B and C are given as

$$v_A = v_B = 2v_c$$

We also know that, Final kinetic energy of the system of the blocks,

$$(\text{K.E.})_2 = \frac{1}{2} m_A v_A^2 + \frac{1}{2} m_B v_B^2 + \frac{1}{2} m_C v_C^2$$

$$= \left[\frac{1}{2} \times 30 \times (2v_C)^2\right] + \left[\frac{1}{2} \times 20 \times (2v_C)^2\right] + \left[\frac{1}{2} \times 80 \times (v_C^2)\right]$$

$$= 140\, v_C^2$$

Now, velocity (v_C) of the block C after it has moved through a distance 1 m starting from rest can be calculated by applying **Work-Energy Principle** between positions 1 and 2, we get

$$(\text{W.D.})_{1-2} = (\text{K.E.})_2 - (\text{K.E.})_1$$

or $$(W_G)_C + (W_G)_A + W_F = (\text{K.E.})_2 - (\text{K.E.})_1$$

or $$784.8 - 588.6 - 156.96 = 140\, v_C^2 - 0$$

or $$v_C^2 = 0.28$$

or $$v_C = 0.529 \text{ m/s} = 529 \text{ mm/s} \qquad \textbf{Ans.}$$

EXERCISE 1

1. A bullet of mass 80 gm moving with velocity of 250 m/s is fired into a wooden block and get embedded in it to a depth of 20 cm. Using the Work-Energy Principle, determine the resistance offered by the wooden block. (**Ans.** 12.5 kN)
2. A bullet moving with a velocity 165 m/s penetrates through a plank. The bullet has a velocity 143 m/s at the moment when it comes out of the plank. Find the number of planks through which it penetrates before it comes to rest. Use the Work-Energy Principle. (**Ans.** 2)
3. A 4 kg block slides down from rest along a smooth inclined plane that makes an angle of 30° with the horizontal. What will be the speed of the block at a point which lies a distance 2.5 mm from the initial point? Use Work-Energy Principle. (**Ans.** 4.95 m/s)
4. Two weights 300 N and 450 N are hung to the two ends of a light inextensible string over an ideal/smooth pulley. Using the Work-Energy Principle, determine
   (***a***) The acceleration of the system and
   (***b***) Tension in the string (**Ans.** 1.962 m/s^2; 360 N)
5. A vehicle of weight 500 kN starts from rest and moves 30 m down 1% grade and hits the bumper post in which bumper spring compresses 1 mm for every 15 kN. The rolling resistance of the track is 5N/kN. Using Work-Energy Principle, determine
   (***a***) The velocity of the vehicle when it hits the bumper post.
   (***b***) How much bumper spring get compressed ? (**Ans.** 1.716 m/s; 10.02 cm)
6. A thin circular disc, of radius R and mass M starts rotating about fixed support from the initial position as vertical. Calculate angular velocity of the circular disc when it has rotated through 30° from vertical in anti-clockwise direction. (**Ans.** $\frac{1.75}{\sqrt{R}}$)

QUESTIONS

1. Explain the term "Work". When the work is said to be done ?
2. What are the units of work done ?
3. State and prove the "Work-Energy Principle".

OBJECTIVE TYPE QUESTIONS

1. In solving kinetic/dynamic problems, which law/principle/method is used for determining force, velocity and displacement ?
   (*a*) D'Alembert's Principle (*b*) Work-Energy Principle
   (*c*) Impulse-Momentum Method (*d*) Conservation of momentum
2. The Work-Energy Principle is more useful than ...., when the dynamic/kinetic problems involves velocity and distance rather than acceleration.
   (*a*) D'Alembert's Principle (*b*) Work-Energy Principle
   (*c*) Impulse-Momentum Method (*d*) Conservation of momentum
3. According to Work-Energy Principle, the work done on a body equal to the change in
   (*a*) momentum (*b*) impulse
   (*c*) potential energy (*d*) kinetic energy

ANSWERS

1. (*b*) 2. (*a*) 3. (*d*)

CHAPTER

33 Forces in Space (In Vector Form)

Contents

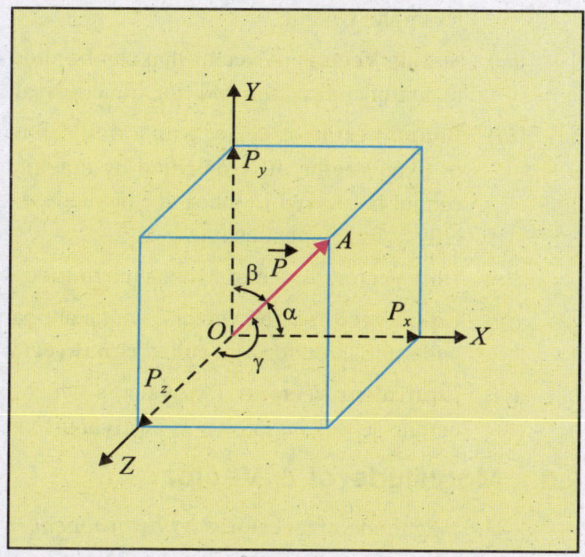

1. INTRODUCTION

In many engineering problems, the bodies or particles are subjected to external forces which lie in space (or in different planes).

Generally, the vector approach is followed for forces in space because writing a force in vector form/ notation is convenient and efficient for solving the problems of forces in space.

This chapter deals with the determination of *Resultant force* and *Equilibrium conditions* of particles subjected to non-coplanar concurrent forces in space (or in three dimensions)

2. VECTORS

Any quantity which is specified by magnitude and direction both, are called **vector** or **vector quantity**. For example, displacement, velocity, acceleration, force, momentum, moment, couple, impulse, torque etc. are all vector quantities.

2.1. Representation of a Vector

A vector quantity is represented by a bold capital letter or by putting an arrow over it. For example, the force *P* which is a vector quantity, will be represented by **P** or by putting an arrow as $\vec{P}$.

2.2. Types of Vectors

The vectors may be of the following types:

(*i*) **Free Vector:** A vector that can be placed anywhere in the space and moved parallel to itself at will, keeping its magnitude and direction same, is called a **free vector**. Example: Couple.

(*ii*) **Sliding Vector:** A vector that can be slide or moved along its line of action, keeping its magnitude, direction and line of action same, is called a **sliding vector**.

(*iii*) **Bound Vector:** A vector with a well defined point of application, is called **bound vector** or **fixed vector**. It is specified by magnitude, direction and its points of application. It cannot be moved in the space or plane. Example: moment of a force, which is applied only at its moment centre.

(*iv*) **Unit Vector:** A vector, whose magnitude is unity, is called a **unit vector**.

(*v*) **Equal Vectors:** The vectors, which are parallel to each other and have equal magnitude and same direction, are called **equal vectors**.

(*vi*) **Equivalent Vectors:** Two vectors which produce the same effect on a rigid body, in a certain sense, are known as **equivalent vectors.**

2.3. Magnitude of a Vector

The magnitude of a vector is a *positive number* corresponding to its *length*. Hence, the magnitude of a vector $\vec{A}$ is represented as:

$$\text{Magnitude of vector } \vec{A} = |\vec{A}|$$

where $|\vec{A}|$ is a *positive* scalar quantity. It is also represented by *A* (*i.e.,* capital letter *A*)

$$\therefore \quad \text{Magnitude of vector } \vec{A} = |\vec{A}| = A$$

2.4. Multiplication of a Vector by a Scalar

Let '*m*' be a given scalar and $\vec{A}$ be any given vector. Then the product $m\vec{A}$ is called the multiplication of a vector by a scalar. Mathematically,

$$\text{Vector } \vec{A} \text{ multiplied by scalar '}m\text{'} = m\vec{A}$$

The product $m\vec{A}$ represents a vector having same direction as $\vec{A}$ and its magnitude equal to *m* times the magnitude of $\vec{A}$.

3. COMPONENTS OF A VECTOR

A set of rectangular axes *X*, *Y* and *Z* is shown in the Fig. 1. The vector $\vec{A}$ is represented by line segment $\overrightarrow{OP}$ and to be resolved into three components A_x, A_y and A_z along the rectangular axes.

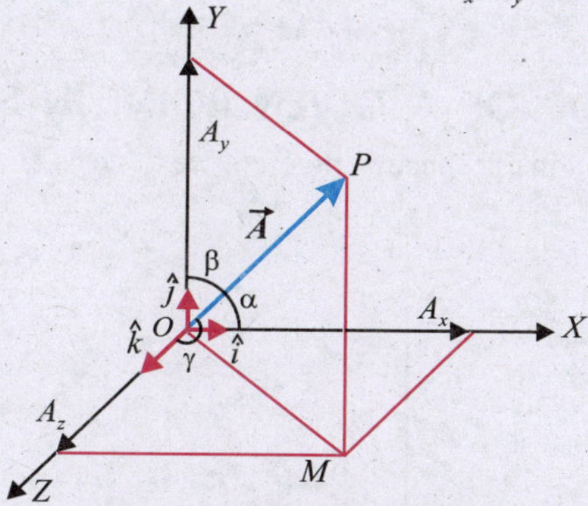

Fig. 1. Components of a vector

Let $\hat{i}$ = Unit vector in the positive *X*-direction

$\hat{j}$ = Unit vector in the positive *Y*-direction

$\hat{k}$ = Unit vector in the positive *Z*-direction

The vector $\vec{A}$ in its components form can be written as:

$$\vec{A} = A_x\hat{i} + A_y\hat{j} + A_z\hat{k} \qquad \text{...(1)}$$

where A_x, A_y and A_z are called the *scalar components* of the vector $\vec{A}$ in the directions of *X*, *Y* and *Z* respectively. The above equation is also called a **vector form** of a vector $\vec{A}$.

Magnitude of the vector $\vec{A}$ is given by:

$$A = |\vec{A}| = \sqrt{A_x^2 + A_y^2 + A_z^2} \qquad \text{...(2)}$$

Let α = Angle made by the vector $\vec{A}$ with *X*-axis,

β = Angle made by the vector $\vec{A}$ with *Y*-axis,

γ = Angle made by the vector $\vec{A}$ with *Z*-axis, and

A = Magnitude of vector $\vec{A}$

Then, scalar components of vector $\vec{A}$ are given as:

$$A_x = A\cos\alpha,\ A_y = A\cos\beta \text{ and } A_z = A\cos\gamma \qquad \text{...(3)}$$

Substituting the values of A_x, A_y and A_z in relation (1), the vector $\vec{A}$ can be written as:

$$\vec{A} = (A\cos\alpha)\hat{i} + (A\cos\beta)\hat{j} + (A\cos\gamma)\hat{k} \qquad \text{...(4)}$$

3.1. Unit Vector in the Direction of $\vec{A}$

A vector having a unit magnitude or unit length, is called **unit vector**. The unit vector in the direction of a vector $\vec{A}$ is denoted by $\vec{A}$. It is read as "*A* cap". Thus, $|\vec{A}| = 1$.

Unit vector in the direction of a vector $\vec{A}$,

$$\hat{A} = \frac{\vec{A}}{A} = \frac{\vec{A}}{\sqrt{A_x^2 + A_y^2 + A_z^2}} \quad ...(5)$$

4. POSITION VECTOR OF A GIVEN POINT IN SPACE

Let $A(x, y, z)$ be the point in the space with reference to OX, OY and OZ as the coordinate axes as shown in the Fig. 2. Here $A_x = x$, $A_y = y$ and $A_z = z$.

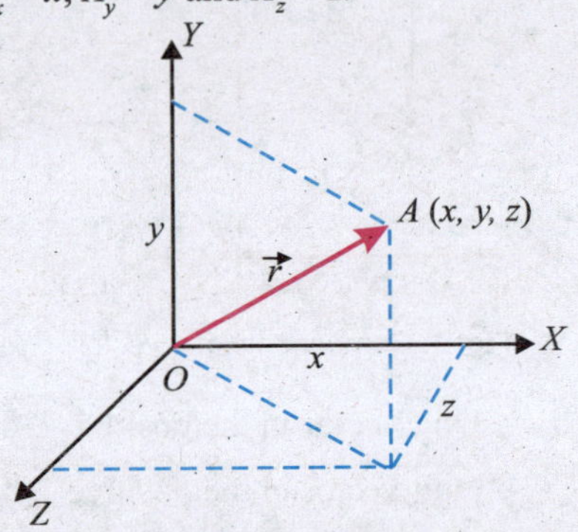

Fig. 2. Position vector of a given point in space

The position vector of a point A with respect to a reference point O is the vector $\overrightarrow{OA}$. It is denoted by $\vec{r}$ and is used to specify the position of a point A relative to O as:

$$\vec{r} = x\hat{i} + y\hat{j} + z\hat{k} \quad ...(6)$$

The magnitude of position vector of a point $A(x, y, z)$ in the space is given by:

$$|\vec{r}| = r = \sqrt{x^2 + y^2 + z^2} \quad ...(7)$$

5. VECTOR OPERATIONS

The following are the important vector operations:

1. Vector addition
2. Dot or Scalar product
3. Cross or Vector product.

5.1. Vector Addition

The sum of two or more vectors can be obtained by adding their corresponding components.

Let
$$\vec{A} = A_x\hat{i} + A_y\hat{j} + A_z\hat{k} \text{ and}$$
$$\vec{B} = B_x\hat{i} + B_y\hat{j} + B_z\hat{k}$$

be two vectors in the space, then addition of two vectors $\vec{A}$ and $\vec{B}$ is define as:

$$\vec{A} + \vec{B} = (A_x + B_x)\hat{i} + (A_y + B_y)\hat{j} + (A_z + B_z)\hat{k} \quad ...(8)$$

Note: The vector obtained after adding two or more than two vectors is called resultant (or resultant vector). It is usually denoted by $\vec{R}$ and its magnitude by R.

5.1.1. Properties of Vector Addition

Let $\vec{V}$ be the set of vectors. Then

(a) $\vec{A}, \vec{B} \in \vec{V}, \vec{A}+\vec{B} \in \vec{V}$ (Closure law of addition)

(b) $\vec{A}+\vec{B} = \vec{B}+\vec{A}$ (Cumulative law of addition)

(c) $\vec{A}+(\vec{B}+\vec{C}) = (\vec{A}+\vec{B})+\vec{C}$ (Associative law of addition)

(d) $\vec{A}+\vec{0} = \vec{A}$ (Additive Identity law of addition)

(e) $\vec{A}+(-\vec{A}) = \vec{0}$ (Additive Inverse law of addition)

Example 1. *Find a unit vector in the direction of the resultant of vectors:* $\vec{A} = 8\hat{i}+5\hat{j}+3\hat{k}$, $\vec{B} = 3\hat{i}-3\hat{j}+5\hat{k}$ *and* $\vec{C} = 7\hat{i}-6\hat{j}+4\hat{k}$.

Solution. Given: Three Vectors: $\vec{A} = 8\hat{i}+5\hat{j}+3\hat{k}$, $\vec{B} = 3\hat{i}-3\hat{j}+5\hat{k}$ and $\vec{C} = 7\hat{i}-6\hat{j}+4\hat{k}$

We know that, Resultant of three vectors,

$$\begin{aligned}\vec{R} &= \vec{A}+\vec{B}+\vec{C}\\ &= (8\hat{i}+5\hat{j}+3\hat{k}) + (3\hat{i}-3\hat{j}+5\hat{k}) + (7\hat{i}-6\hat{j}+4\hat{k})\\ &= (8+3+7)\hat{i} + (5-3-6)\hat{j} + (3+5+4)\hat{k}\\ &= 18\hat{i} - 4\hat{j} + 12\hat{k}\end{aligned}$$

∴ Magnitude of the resultant,

$$|\vec{R}| = \sqrt{18^2+(-4)^2+12^2} = 22$$

We also know that, Unit vector in the direction of the resultant $\vec{R}$,

$$\begin{aligned}\hat{R} &= \frac{\vec{R}}{|\vec{R}|} = \frac{18\hat{i}-4\hat{j}+12\hat{k}}{22} = \frac{18}{22}\hat{i}-\frac{4}{22}\hat{j}+\frac{12}{22}\hat{k}\\ &= \frac{9}{11}\hat{i}-\frac{2}{11}\hat{j}+\frac{6}{11}\hat{k} \quad \textbf{Ans.}\end{aligned}$$

5.2. Dot or Scalar Product

The dot or scalar product of two non-zero vectors $\vec{A}$ and $\vec{B}$ is denoted by $\vec{A}\cdot\vec{B}$ (*read as A dot B*) and is defined as a scalar $|\vec{A}||\vec{B}|\cos\theta$. Mathematically,

$$\vec{A}\cdot\vec{B} = |\vec{A}||\vec{B}|\cos\theta = AB\cos\theta \quad ...(9)$$

where θ is the angle between the vectors $\vec{A}$ and $\vec{B}$.

Clearly, the scalar product of two vectors is a scalar quantity due to which it is called **scalar product**. Since, there is a dot between $\vec{A}$ and $\vec{B}$, therefore it is also called **dot product**.

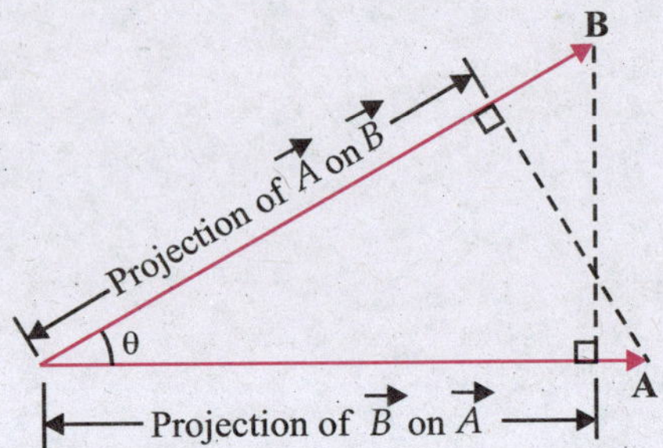

Fig. 3. Dot or Scalar product of Two vectors

Important Results:

1. When θ = 0 *i.e.,* when the vectors $\vec{A}$ and $\vec{B}$ have the *same direction*, then

$$\vec{A} \cdot \vec{B} = |\vec{A}||\vec{B}| = AB$$

2. When θ = 180° or π *i.e.,* when the vectors $\vec{A}$ and $\vec{B}$ have the *opposite directions*, then

$$\vec{A} \cdot \vec{B} = -|\vec{A}||\vec{B}| = -AB$$

3. When θ = 90° *i.e.,* when the vectors $\vec{A}$ and $\vec{B}$ are *perpendicular* or *at right angle*, then

$$\vec{A} \cdot \vec{B} = 0$$

4. Geometrical Interpretation:

$$\vec{A} \cdot \vec{B} = |\vec{A}| \text{ times (projection of } \vec{B} \text{ on } \vec{A})$$

$$\vec{A} \cdot \vec{B} = |\vec{B}| \text{ times (projection of } \vec{A} \text{ on } \vec{B})$$

5.2.1. Scalar Product in Terms of Scalar Components

Let $\vec{A} = A_x\hat{i} + A_y\hat{j} + A_z\hat{k}$ and

$$\vec{B} = B_x\hat{i} + B_y\hat{j} + B_z\hat{k}$$

The scalar product in terms of components,

$$\vec{A} \cdot \vec{B} = (A_x\hat{i} + A_y\hat{j} + A_z\hat{k}) \cdot (B_x\hat{i} + B_y\hat{j} + B_z\hat{k})$$

or $$\vec{A} \cdot \vec{B} = A_x B_x + A_y B_y + A_z B_z \quad ...(10)$$

Thus, the scalar product of two vectors is equal to the sum of the products of their corresponding components.

5.2.2. Angle Between Two Vectors in Terms of Scalar Product

The angle (θ) between vectors $\vec{A}$ and $\vec{B}$ can be calculated by the relation of dot product:

$$\cos\theta = \frac{\vec{A}\cdot\vec{B}}{|\vec{A}||\vec{B}|} = \frac{\vec{A}\cdot\vec{B}}{AB} \qquad \text{...(11a)}$$

or
$$\theta = \cos^{-1}\left(\frac{\vec{A}\cdot\vec{B}}{|\vec{A}||\vec{B}|}\right) = \cos^{-1}\left(\frac{\vec{A}\cdot\vec{B}}{AB}\right) \qquad \text{...(11b)}$$

5.2.3. Properties of Scalar Product

(*a*) $\vec{A}\cdot\vec{B} = \vec{B}\cdot\vec{A}$ (Cumulative law)

(*b*) $m\vec{A}\cdot n\vec{B} = mn(\vec{A}\cdot\vec{B}) = (mn\vec{A})\cdot\vec{B} = \vec{A}\cdot(mn\vec{B})$ ...(where *m* and *n* are scalars)

(*c*) $\vec{A}\cdot(\vec{B}+\vec{C}) = \vec{A}\cdot\vec{B}+\vec{A}\cdot\vec{C}$ (Distributive over addition–Left distributivity)

$(\vec{B}+\vec{C})\cdot\vec{A} = \vec{B}\cdot\vec{A}+\vec{C}\cdot\vec{A}$ (Distributive over addition–Right distributivity)

(*d*) $|\vec{A}+\vec{B}|^2 = |\vec{A}|^2 + |\vec{B}|^2 + 2\vec{A}\cdot\vec{B}$

(*e*) $|\vec{A}-\vec{B}|^2 = |\vec{A}|^2 + |\vec{B}|^2 - 2\vec{A}\cdot\vec{B}$

Example 2. *Find the dot product of the two vectors* $\vec{A}$ *and* $\vec{B}$ *given by:*

$$\vec{A} = 5\hat{i} + 10\hat{j} - 6\hat{k} \text{ and } \vec{B} = 8\hat{i} - 7\hat{j} - 4\hat{k}$$

Also, find the angle between them.

Solution. Given: Two Vectors: $\vec{A} = A_x\hat{i} + A_y\hat{j} + A_z\hat{k} = 5\hat{i} + 10\hat{j} - 6\hat{k}$ and $\vec{B} = B_x\hat{i} + B_y\hat{j} + B_z\hat{k} = 8\hat{i} - 7\hat{j} - 4\hat{k}$.

Dot product of two vectors

We know that, the dot product of two vectors,

$$\begin{aligned}\vec{A}\bullet\vec{B} &= A_x B_x + A_y B_y + A_z B_z \\ &= [5 \times 8] + [10 \times (-7)] + [(-6) \times (-4)] \\ &= 40 - 70 + 24 = -6 \qquad \textbf{Ans.}\end{aligned}$$

Angle between the vectors (θ)

We also know that, Angle (θ) between two vectors can be calculated from the relation:

$$\cos\theta = \frac{\vec{A}\cdot\vec{B}}{|\vec{A}||\vec{B}|}$$

$$= \frac{-6}{\sqrt{(5)^2 + (10)^2 + (-6)^2}\sqrt{(8)^2 + (-7)^2 + (-4)^2}}$$

$$= \frac{-6}{12.69 \times 11.36} = -0.0416$$

or $$\theta = \cos^{-1}(-0.0416) = \mathbf{92.38^\circ} \qquad \textbf{Ans.}$$

5.3. Cross or Vector Product

The cross or vector product of two non-zero vectors $\vec{A}$ and $\vec{B}$ denoted by $\vec{A} \times \vec{B}$ (read as A cross B) and is defined as vector $|\vec{A}||\vec{B}|\sin\theta\,\hat{n}$, where θ is the angle between $\vec{A}$ and $\vec{B}$ and whose direction is perpendicular to the plane containing the vectors $\vec{A}$ and $\vec{B}$. Mathematically,

$$\vec{A} \times \vec{B} = |\vec{A}||\vec{B}|\sin\theta\ \hat{n} = AB\sin\theta\ \hat{n} \qquad ...(12)$$

where $\hat{n}$ is the unit vector perpendicular to the plane containing the vectors $\vec{A}$ and $\vec{B}$ and $\vec{A}, \vec{B}, \hat{n}$ form a right handed system.

The magnitude of cross product of two vectors $\vec{A}$ and $\vec{B}$ is given by,

$$|\vec{A} \times \vec{B}| = |\vec{A}||\vec{B}|\sin\theta = AB\sin\theta \qquad ...(13)$$

Important Results:

1. When $\theta = 0$ or π *i.e.*, when the vectors $\vec{A}$ and $\vec{B}$ have the *same or opposite directions*, then

$$\vec{A} \times \vec{B} = \vec{0}$$

2. When $\theta = 90^\circ$ *i.e.*, when the vectors $\vec{A}$ and $\vec{B}$ are *perpendicular* or *at right angle*, then

$$\vec{A} \times \vec{B} = |\vec{A}||\vec{B}|\ \hat{n}$$

or $$|\vec{A} \times \vec{B}| = |\vec{A}||\vec{B}| = AB \qquad ...(\text{where } |\hat{n}| = 1)$$

3. Geometrical Interpretation:

$$\vec{A} \times \vec{B} = \text{Vector area of parallelogram having adjacent sides } \vec{A} \text{ and } \vec{B}$$

4. $\hat{i} \times \hat{i} = 0, \hat{j} \times \hat{j} = 0, \hat{k} \times \hat{k} = 0$

$\hat{i} \times \hat{j} = \hat{k}, \hat{j} \times \hat{k} = \hat{i}, \hat{k} \times \hat{i} = \hat{j}$

$\hat{j} \times \hat{i} = -\hat{k}, \hat{k} \times \hat{j} = -\hat{i}, \hat{i} \times \hat{k} = -\hat{j}$

5.3.1. Cross Product in Terms of Scalar Components

Let $$\vec{A} = A_x\hat{i} + A_y\hat{j} + A_z\hat{k} \text{ and}$$

$$\vec{B} = B_x\hat{i} + B_y\hat{j} + B_z\hat{k}$$

The vector product in terms of scalar components,

$$\vec{A} \times \vec{B} = (A_x\hat{i} + A_y\hat{j} + A_z\hat{k}) \times (B_x\hat{i} + B_y\hat{j} + B_z\hat{k})$$

$$= \begin{vmatrix} \hat{i} & \hat{j} & \hat{k} \\ A_x & A_y & A_z \\ B_x & B_y & B_z \end{vmatrix}$$

or $$\vec{A} \times \vec{B} = (A_y B_z - A_z B_y)\hat{i} + (A_z B_x - A_x B_z)\hat{j} + (A_x B_y - A_y B_x)\hat{k} \quad ...(14)$$

5.3.2. Angle Between Two Vectors in Terms of Cross Product

The angle between two vectors $\vec{A}$ and $\vec{B}$ can be calculated by the relation for cross product as:

$$\sin\theta = \frac{|\vec{A} \times \vec{B}|}{|\vec{A}||\vec{B}|} = \frac{|\vec{A} \times \vec{B}|}{AB} \quad ...(15a)$$

or $$\theta = \sin^{-1}\left(\frac{|\vec{A} \times \vec{B}|}{|\vec{A}||\vec{B}|}\right) = \sin^{-1}\left(\frac{|\vec{A} \times \vec{B}|}{AB}\right) \quad ...(15b)$$

5.3.3. Properties of Cross Product

(a) $\vec{A} \times \vec{A} = \vec{0}$

(b) $\vec{A} \times \vec{B} \neq \vec{B} \times \vec{A}$ (Non-cumulative)

(c) $\vec{A} \times \vec{B} = -(\vec{B} \times \vec{A})$

(d) $m\vec{A} \times n\vec{B} = mn(\vec{A} \times \vec{B})$ ...(where *m* and *n* are scalars)

(e) $\vec{A} \times (\vec{B} + \vec{C}) = (\vec{A} \times \vec{B}) + (\vec{A} \times \vec{C})$

$(\vec{B} + \vec{C}) \times \vec{A} = (\vec{B} \times \vec{A}) + (\vec{C} \times \vec{A})$

(f) $|\vec{A} \times \vec{B}|^2 = |\vec{A}|^2 |\vec{B}|^2 - (\vec{A} \cdot \vec{B})^2$ ...(Lagrange's Identity)

Example 3. *Two non-zero vectors* $\vec{A}$ *and* $\vec{B}$ *are given by:* $\vec{A} = 30\hat{i} + 20\hat{j} - 3\hat{k}$ *and* $\vec{B} = 50\hat{i} - 20\hat{j} - 3\hat{k}$. *Determine:*

(a) Corss product of the vectors,

(b) Angle between the vectors and

(c) Unit vector normal to both the vectors.

Solution. Given: Two Vectors: $\vec{A} = 30\hat{i} + 20\hat{j} - 3\hat{k}$ and $\vec{B} = 50\hat{i} - 20\hat{j} - 3\hat{k}$.

(a) Cross product of two vectors

We know that, the cross product of two vectors,

$$\vec{A} \times \vec{B} = \begin{vmatrix} \hat{i} & \hat{j} & \hat{k} \\ 30 & 20 & -3 \\ 50 & -20 & -3 \end{vmatrix}$$

$$= [(20)(-3) - (-20)(-3)]\ \hat{i} + [(-3)(50) - (30)(-3)]\ \hat{j} + [(30)(-20) - (50)(20)]\ \hat{k}$$

$$= [-60 - 60]\ \hat{i} + [-150 + 90]\ \hat{j} + [-600 - 1000]\ \hat{k}$$

or $$\vec{A} \times \vec{B} = -120\hat{i} - 60\hat{j} - 1600\hat{k} \quad \textbf{Ans.}$$

(*b*) Angle between the vectors (θ)

We also know that, Magnitude of vector $\vec{A}$,

$$|\vec{A}| = \sqrt{(30)^2 + (20)^2 + (-3)^2} = 36.18$$

and Magnitude of vector $\vec{B}$,

$$|\vec{B}| = \sqrt{(50)^2 + (-20)^2 + (-3)^2} = 53.94$$

Similarly, Magnitude of $\vec{A} \times \vec{B}$,

$$|\vec{A} \times \vec{B}| = \sqrt{(-120)^2 + (-60)^2 + (-1600)^2} = 1605.62$$

We also know that, Angle (θ) between the vectors $\vec{A}$ and $\vec{B}$ can be calculated from the relation,

$$\sin\theta = \frac{|\vec{A} \times \vec{B}|}{|\vec{A}||\vec{B}|} = \frac{1605.62}{36.18 \times 53.94} = 0.8227$$

or $$\theta = \sin^{-1}(0.8227) = \mathbf{53.36°} \quad \textbf{Ans.}$$

(*c*) Unit vector normal to the both vectors

We know that, Unit vector normal to the vectors $\vec{A}$ and $\vec{B}$,

$$\hat{n} = \frac{\vec{A} \times \vec{B}}{|\vec{A} \times \vec{B}|} = \frac{-120\hat{i} - 60\hat{j} - 1600\hat{k}}{1605.62}$$

$$= -0.075\,\hat{i} - 0.037\,\hat{j} - 0.996\,\hat{k} \quad \textbf{Ans.}$$

6. COMPONENTS OF A FORCE IN SPACE

Consider a force $\vec{P}$ acting in space, at the origin *O* of the rectangular coordinates system and represented by the vector $\overrightarrow{OP}$ as shown in the Fig. 4. To determine components of force, through point *A*, draw planes parallel to the coordinate planes. These planes along with the coordinate planes form a rectangular parallele piped or a box in which the force *P* is represented by its diagonal *OP* and its three scalar components P_x, P_y and P_z by its edges *OA*, *OB* and *OC* respectively.

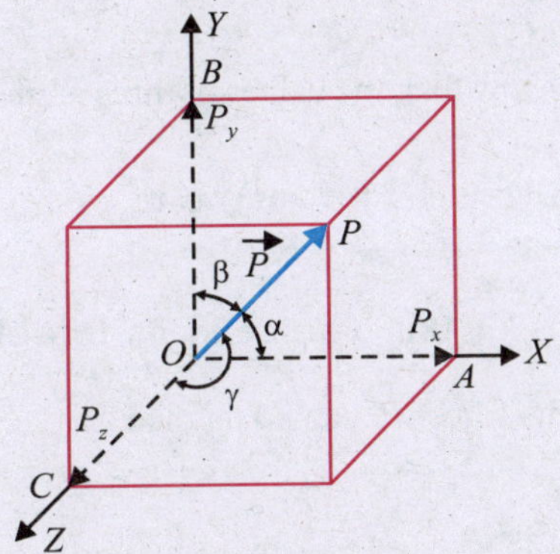

Fig. 4. Components of a force

∴ The force $\vec{P}$ in terms of its components along X, Y and Z axes can be written as:

$$\vec{P} = \vec{P}_x + \vec{P}_y + \vec{P}_z \qquad ...(16)$$

where $\vec{P}_x$, $\vec{P}_y$ and $\vec{P}_z$ are the **vector components** of the force vector $\vec{P}$ along X, Y and Z axes respectively.

Let α, β and γ be the angle made by the force $\vec{P}$ with X, Y and Z axes respectively. The scalar components of force $\vec{P}$ along X, Y and Z axes are:

$$P_x = P\cos\alpha,\ P_y = P\cos\beta,\ P_z = P\cos\gamma \qquad ...(17)$$

where P = Magnitude of force $\vec{P}$

To get the vector components of force $\vec{P}$ along the coordinates axes, multiply the scalar components by the respective unit vectors as:

$$\vec{P}_x = P_x\hat{i} = P\cos\alpha\ \hat{i}$$

$$\vec{P}_y = P_y\hat{j} = P\cos\beta\ \hat{j} \qquad ...(18)$$

and $$\vec{P}_z = P_z\hat{k} = P\cos\gamma\ \hat{k}$$

Substituting the above values in equation (16), the force vector $\vec{P}$ in space can be written as:

$$\vec{P} = P_x\hat{i} + P_y\hat{j} + P_z\hat{k} \qquad ...(19a)$$

$$= (P\cos\alpha)\hat{i} + (P\cos\beta)\hat{j} + (P\cos\gamma)\hat{k} \qquad ...(19b)$$

where P_x, P_y and P_z = Scalar components of the force P along X, Y and Z axes, and
$\hat{i}$, $\hat{j}$ and $\hat{k}$ = Unit vectors in the positive X, Y and Z directions.

And Magnitude of the force vector $\vec{P}$ is given by,

$$P = |\vec{P}| = \sqrt{P_x^2 + P_y^2 + P_z^2} \qquad ...(20)$$

6.1. Direction Cosines

The cosines of angles α, β and γ are known as the **direction cosines** of the force $\vec{P}$ and are denoted as:

$$l = \cos\alpha;\; m = \cos\beta;\; n = \cos\gamma \quad \text{...(21)}$$

A little consideration will show that, the values of three angles α, β and γ are *not independent* and are related as follows:

$$\cos^2\alpha + \cos^2\beta + \cos^2\gamma = 1 \quad \text{...(22a)}$$

or

$$l^2 + m^2 + n^2 = 1 \quad \text{...(22b)}$$

6.2. Force Vector in Terms of Unit vector in its Direction

The unit vector $\hat{\lambda}$ along the force $\vec{P}$ can be defined as:

$$\hat{\lambda} = l\hat{i} + m\hat{j} + n\hat{k} \quad \text{...(23a)}$$

$$= \cos\alpha\,\hat{i} + \cos\beta\,\hat{j} + \cos\gamma\,\hat{k} \quad \text{...(23b)}$$

∴ Force vector $\vec{P}$ in terms of unit vector $\hat{\lambda}$ in its direction is given by,

$$\vec{P} = \text{Magnitude of } \vec{P} \times \text{Unit vector in the direction of } \vec{P}$$

$$= P\,\hat{\lambda} \quad \text{...(24a)}$$

$$= P(l\hat{i} + m\hat{j} + n\hat{k}) \quad \text{...(24b)}$$

$$= P(\cos\alpha\,\hat{i} + \cos\beta\,\hat{j} + \cos\gamma\,\hat{k}) \quad \text{...(24c)}$$

Example 4. *A force of 250 N makes angles of 45°, 60° and 120° respectively with X, Y and Z axes. Express the force in vector form.*

Solution. Given: Magnitude of force $(P) = 250$ N; $\alpha = 45°$, $\beta = 60°$ and $\gamma = 120°$.

We know that scalar components of force along *X*, *Y* and *Z* axes are given as:

$$P_x = P\cos\alpha = 250\cos 45° = 176.78 \text{ N}$$

$$P_y = P\cos\beta = 250\cos 60° = 125 \text{ N}$$

and

$$P_z = P\cos\gamma = 250\cos 120° = -125 \text{ N}$$

∴ The force in vector form is given as:

$$\vec{P} = P_x\,\hat{i} + P_y\,\hat{j} + P_z\,\hat{k}$$

$$= 176.78\,\hat{i} + 125\,\hat{j} - 125\,\hat{k} \qquad \textbf{Ans.}$$

7. FORCE DEFINED BY ITS MAGNITUDE AND TWO POINTS ON ITS LINE OF ACTION

Generally, **a force in space is described by its magnitude and two points on its line of action,** instead of the angles it makes with the coordinates axes. Knowing the magnitude of force and coordinates of two points on its line of action, the force can be spilt into three components and can be written in a vectorial form as discussed below.

Consider a force $\vec{P}$ which is defined by $\overrightarrow{AB}$, joining two points $A\,(x_1, y_1, z_1)$ and $B\,(x_2, y_2, z_2)$ as shown in the Fig 5.

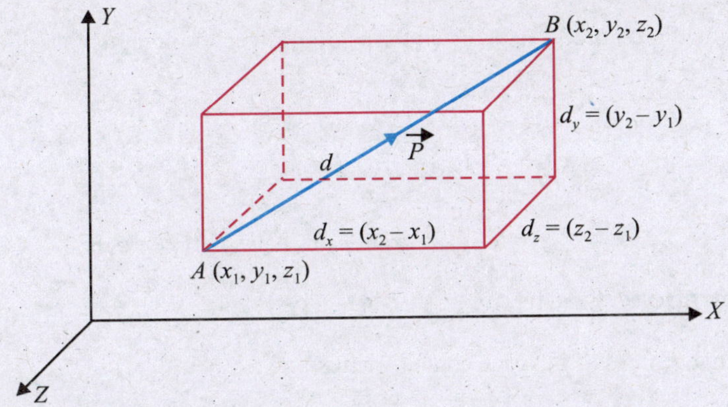

Fig. 5. A Force defined by its Magnitude and Two Points on Its Line of Action

Let $\qquad d$ = Distance between A and $B = |\overrightarrow{AB}| = AB$

α, β and γ = Angles made by the force $\vec{P}$ with the X, Y and Z axes .

Now, this distance AB (or d) can be resolved into three components along three coordinates axes as d_x, d_y and d_z as shown in Fig. 5.

From the geometry of the figure, the components of distance AB along X, Y and Z axes in terms of coordinates are:

$$d_x = (x_2 - x_1),\ d_y = (y_2 - y_1) \text{ and } d_z = (z_2 - z_1) \qquad ...(25)$$

Also, the components of distance AB along X, Y and Z axes in terms of distance d and direction cosines are:

$$d_x = d\cos\alpha,\ d_y = d\cos\beta \text{ and } d_z = d\cos\gamma \qquad ...(26)$$

$\therefore$ Vector $\overrightarrow{AB}$ can be written as:

$$\overrightarrow{AB} = d_x\,\hat{i} + d_y\,\hat{j} + d_z\,\hat{k} \qquad ...(27a)$$

$$= (x_2 - x_1)\,\hat{i} + (y_2 - y_1)\,\hat{j} + (z_2 - z_1)\,\hat{k} \qquad ...(27b)$$

$$= (d\cos\alpha)\,\hat{i} + (d\cos\beta)\,\hat{j} + (d\cos\gamma)\,\hat{k} \qquad ...(27c)$$

Hence, distance AB (or d) in terms of coordinates,

$$d = |\overrightarrow{AB}| = \sqrt{d_x^2 + d_y^2 + d_z^2} \qquad ...(28a)$$

$$= \sqrt{(x_2 - x_1)^2 + (y_2 - y_1)^2 + (z_2 - z_1)^2} \qquad ...(28b)$$

Let $\qquad \hat{\lambda}_{AB}$ = Unit vector along the line of action of $\overrightarrow{AB}$

We know that, Unit vector $\vec{\lambda}_{AB}$ along the line of action of $\overrightarrow{AB}$ is equal to the vector $\overrightarrow{AB}$ divided by its magnitude AB (*i.e.*, d).

$$\therefore \qquad \hat{\lambda}_{AB} = \frac{\overrightarrow{AB}}{d} = \frac{d_x\hat{i} + d_y\hat{j} + d_z\hat{k}}{d}$$

We also know that, Force vector $\vec{P}$ in terms of unit vector along its line of action,

$$\vec{P} = P\,\hat{\lambda}_{AB} \quad \text{..(29a)}$$

$$= \frac{P}{d}(d_x\hat{i} + d_y\hat{j} + d_z\hat{k}) \quad \text{...(29b)}$$

$$= \frac{P}{d}(x_2 - x_1)\hat{i} + (y_2 - y_1)\hat{j} + (z_2 - z_1)\hat{k} \quad \text{...(29c)}$$

7.1. Scalar Components of Force Vector

We know that, Force $\vec{P}$ in space can be written as:

$$\vec{P} = P_x\hat{i} + P_y\hat{j} + P_z\hat{k} \quad \text{...(30)}$$

From eqns. (29*a*–29*c*) and eqn. (30), we get

$$P_x = \frac{P}{d}d_x = \frac{P}{d}(x_2 - x_1) \quad \text{...(31a)}$$

$$P_y = \frac{P}{d}d_y = \frac{P}{d}(y_2 - y_1) \quad \text{...(31b)}$$

and $$P_z = \frac{P}{d}d_z = -(z_2 - z_1) \quad \text{...(31c)}$$

Equations (31a) – (31c) represent scalar components of force along coordinate axes *X*, *Y* and *Z* respectively.

The equations (31a) to (31c) can be written as:

$$\frac{P_x}{d_x} = \frac{P}{d}, \quad \frac{P_y}{d_y} = \frac{P}{d} \quad \text{and} \quad \frac{P_z}{d_z} = \frac{P}{d}$$

$$\therefore \quad \frac{P_x}{d_x} = \frac{P_y}{d_y} = \frac{P_z}{d_z} = \frac{P}{d} \quad (= P_m, \text{ called } \textbf{Force Multiplier}) \quad \text{...(32)}$$

or $$\frac{P_x}{(x_2 - x_1)} = \frac{P_y}{(y_2 - y_1)} = \frac{P_z}{(z_2 - z_1)} = \frac{P}{d} \quad \text{...(33)}$$

7.2. Force Vector in Terms of Force Multiplier (P_m)

If the force vector $\vec{P}$ is defined by $\overrightarrow{AB}$, joining two points $A\,(x_1, y_1, z_1)$ and $B\,(x_2, y_2, z_2)$, then the force vector $\vec{P}$ in terms of force multiplier is given as:

$$\vec{P} = P_m.\,\overrightarrow{AB} \quad \text{...(34a)}$$

$$= P_m.[(x_2 - x_1)\hat{i} + (y_2 - y_1)\hat{j} + (z_2 - z_1)\hat{k}] \quad \text{...(34b)}$$

7.3. Angle of Inclination of Force Vector with Coordinate Axes

From relation (26), we have

$$d_x = d\cos\alpha \quad \text{or} \quad \cos\alpha = \frac{d_x}{d}$$

or $$\alpha = \cos^{-1}\left(\frac{d_x}{d}\right) = \cos^{-1}\left(\frac{x_2 - x_1}{d}\right) \quad ...(35a)$$

Similarly, we may write

$$\beta = \cos^{-1}\left(\frac{d_y}{d}\right) = \cos^{-1}\left(\frac{y_2 - y_1}{d}\right) \quad ...(35b)$$

and $$\gamma = \cos^{-1}\left(\frac{d_z}{d}\right) = \cos^{-1}\left(\frac{z_2 - z_1}{d}\right) \quad ...(35c)$$

Example 5. *Find a force vector $\vec{P}$ of magnitude 1578 N which is directed from a point A(12, 4, 10) metres towards a point B (– 16, 8, – 4) metres.*

Solution. Given: Magnitude of force (P) = 1578 N; Coordinates of point A= (12, 4, 10); Position vector of point A $(\overrightarrow{OA}) = 12\hat{i} + 4\hat{j} + 10\hat{k}$; Coordinates of point B = (– 16, 8, – 4) and Position vector of point B $(\overrightarrow{OB}) = -16\hat{i} + 8\hat{j} - 4\hat{k}$.

We know that, vector AB in the direction of force $\vec{P}$,

$$\begin{aligned}\overrightarrow{AB} &= \overrightarrow{OB} - \overrightarrow{OA} \\ &= (-16\hat{i} + 8\hat{j} - 4\hat{k}) - (12\hat{i} + 4\hat{j} + 10\hat{k}) \\ &= (-16 - 12)\hat{i} + (8 - 4)\hat{j} + (-4 - 10)\hat{k} \\ &= -28\hat{i} + 4\hat{j} - 14\hat{k}\end{aligned}$$

∴ Unit vector in the direction of force $\vec{P}$,

$$\begin{aligned}\hat{\lambda}_{AB} &= \frac{\overrightarrow{AB}}{|\overrightarrow{AB}|} = \frac{-28\hat{i} + 4\hat{j} - 14\hat{k}}{\sqrt{(-28)^2 + (4)^2 + (-14)^2}} \\ &= \frac{-28\hat{i} + 4\hat{j} - 14\hat{k}}{\sqrt{996}} = \frac{-28\hat{i} + 4\hat{j} - 14\hat{k}}{31.56}\end{aligned}$$

We also know that, Force vector $\vec{P}$ in space,

$$\begin{aligned}&= P \times \hat{\lambda}_{AB} \\ &= 1578 \times \frac{-28\hat{i} + 4\hat{j} - 14\hat{k}}{31.56} = 50\,[-28\hat{i} + 4\hat{j} - 14\hat{k}] \\ &= -1400\hat{i} + 200\hat{j} - 700\hat{k} \quad \textbf{Ans.}\end{aligned}$$

8. MOMENT OF A FORCE

The moment of a force $\vec{P}$ about a point O is a vector which is equal to the product of the force and perpendicular distance of O from the line of action of the force. In vector form,

Moment of force $\vec{P}$ about a point O is given by,

$$\vec{M} = \vec{r} \times \vec{P} \qquad \text{...(36)}$$

where $\vec{r}$ is the position vector of any point on the line of action of the force w.r.t O.

$\therefore$ Magnitude of the moment $\vec{M}$,

$$|\vec{M}| = |\vec{r}||\vec{P}| \sin \theta$$

or

$$M = r P \sin \theta \qquad \text{...(37)}$$

The moment of a force is a *vector*. If a force is acting on the body, it generates the turning moment on the body and hence the body rotates.

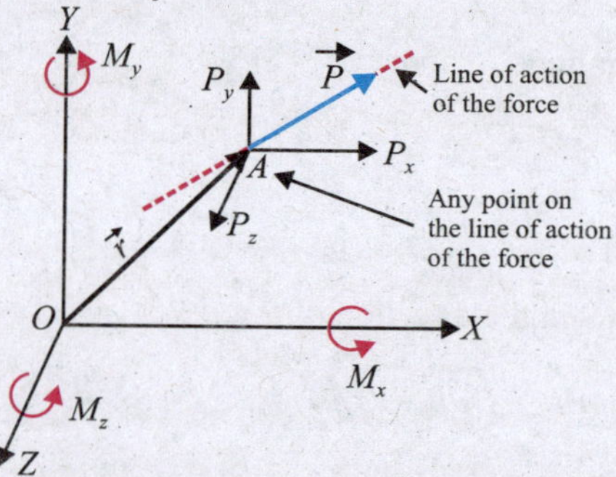

Fig. 6. Moment of a force

Note: The moment of a force $\vec{P}$ about a *directed line* is given by,

$$\vec{M} = (\overrightarrow{AB} \times \vec{P}) \cdot \hat{\lambda} \qquad \text{...(38)}$$

where A is any point on the line of action of force and B is any point on the directed line with unit vector $\hat{\lambda}$ in its direction.

8.1. Moment of a Force in Terms of Scalar Components

Let

$$\vec{P} = P_x\hat{i} + P_y\hat{j} + P_z\hat{k} \text{ and}$$

$$\vec{r} = x\hat{i} + y\hat{j} + z\hat{k}$$

where $\vec{r}$ is the position vector of any point on the line of action of the force w.r.t O.

$\therefore$ Moment of a force $\vec{P}$ about O is given by,

$$\vec{M} = \vec{r} \times \vec{P} = \begin{vmatrix} \hat{i} & \hat{j} & \hat{k} \\ x & y & z \\ P_x & P_y & P_z \end{vmatrix} \qquad \text{...(39a)}$$

$$= (y P_z - z P_y)\hat{i} + (z P_x - x P_z)\hat{j} + (x P_y - y P_x)\hat{k} \qquad \text{...(39b)}$$

where M_x = Moment of $\vec{P}$ about x-axis $= y P_z - z P_y$

$$M_y = \text{Moment of } \vec{P} \text{ about } y\text{-axis} = z\,P_x - x\,P_z$$

$$M_z = \text{Moment of } \vec{P} \text{ about } z\text{-axis} = x\,P_y - y\,P_x$$

9. RESULTANT FORCE OF CONCURRENT FORCES

Consider a number of concurrent forces $\vec{P_1}, \vec{P_2}, \vec{P_3}, ...,$ acts on a particle in space.

Let $\vec{R} = R_x\hat{i} + R_y\hat{j} + R_z\hat{k}$ be the resultant force in terms of scalar components.

The resultant force of concurrent forces is given by,

$$\vec{R} = \vec{P_1} + \vec{P_2} + \vec{P_3} + ... \qquad ...(40)$$

where

$$\vec{P_1} = P_{1x}\hat{i} + P_{1y}\hat{j} + P_{1z}\hat{k},$$

$$\vec{P_2} = P_{2x}\hat{i} + P_{2y}\hat{j} + P_{2z}\hat{k} \text{ and}$$

$$\vec{P_3} = P_{3x}\hat{i} + P_{3y}\hat{j} + P_{3z}\hat{k} \text{ and so on.}$$

Substituting the above values in relation (40), we get

$$\vec{R} = (P_{1x}\hat{i} + P_{1y}\hat{j} + P_{1z}\hat{k}) + (P_{2x}\hat{i} + P_{2y}\hat{j} + P_{2z}\hat{k}) + (P_{3x}\hat{i} + P_{3y}\hat{j} + P_{3z}\hat{k}) + ...$$

$$= (P_{1x} + P_{2x} + P_{3x} + ...)\hat{i} + (P_{1y} + P_{2y} + P_{3y} + ...)\hat{j} + (P_{1z} + P_{2z} + P_{3z} + ...)\hat{k}$$

or

$$\vec{R} = (\Sigma P_X)\hat{i} + (\Sigma P_Y)\hat{j} + (\Sigma P_Z)\hat{k} \qquad ...(41)$$

This is the required relation to find the resultant force on a particle in terms of rectangular components of given concurrent forces.

9.1. Magnitude of Resultant Force

But

$$\vec{R} = R_x\hat{i} + R_y\hat{j} + R_z\hat{k} \qquad ...(42)$$

Equating relation (41) and (42), we get

$$R_x = \Sigma P_X \qquad ...(43a)$$

$$R_y = \Sigma P_Y \qquad ...(43b)$$

$$R_z = \Sigma P_Z \qquad ...(43c)$$

∴ Magnitude of resultant force,

$$R = \sqrt{R_x^2 + R_y^2 + R_z^2} \qquad ...(44a)$$

$$= \sqrt{(\Sigma P_X)^2 + (\Sigma P_Y)^2 + (\Sigma P_Z)^2} \qquad ...(44b)$$

9.2. Inclination of Resultant Force with Coordinate Axes

Let α, β and γ be the angle made by the resultant force $\vec{R}$ with the X, Y and Z axes respectively. The components of resultant force along X, Y and Z axes are:

$$R_x = R\cos\alpha,\ R_y = R\cos\beta,\ R_z = R\cos\gamma \qquad ...(i)$$

where R is the magnitude of resultant force.

From above relation (i), we have

$$R_x = R\cos\alpha$$

or $$\cos\alpha = \frac{R_x}{R}$$

or $$\alpha = \cos^{-1}\left(\frac{R_x}{R}\right) = \cos^{-1}\left(\frac{\Sigma P_X}{R}\right) \quad \text{...(45a)}$$

Similarly, $$\beta = \cos^{-1}\left(\frac{R_y}{R}\right) = \cos^{-1}\left(\frac{\Sigma P_Y}{R}\right) \quad \text{...(45b)}$$

and $$\gamma = \cos^{-1}\left(\frac{R_z}{R}\right) = \cos^{-1}\left(\frac{\Sigma P_Z}{R}\right) \quad \text{...(45c)}$$

10. EQUILIBRIUM OF A PARTICLE IN SPACE

A particle subjected to a number of concurrent forces in space is said to be in equilibrium when the resultant force on it zero.

∴ $$\vec{R} = 0 \quad \text{...(46a)}$$

or $$R_x\hat{i} + R_y\hat{j} + R_z\hat{k} = 0$$

or $$R_x = 0; \quad R_y = 0; \quad R_z = 0 \quad \text{...(46b)}$$

or $$\Sigma P_X = 0; \quad \Sigma P_Y = 0; \quad \Sigma P_Z = 0 \quad \text{...(46c)}$$

These equations are the required equations for the equilibrium of a particle subjected to a number of concurrent forces in space.

SOLVED EXAMPLES

Forces—Composition of Forces and Resolution of a Force

Example 6. *A boom AC lying in vertical plane XY, is supported by a ball and socket joint at C and by cables AE and BD as shown in the Fig. 7. For this system, it is found that the force multiplier of force P_1 acting from B to D is P_{m1} = 250 N/m and that of force P_2 acting from A to E is P_{m2} = 150 N/m. Determine:*

(i) the component of each force along AC.

(ii) the angle made by each force with AC. **(Nagpur University)**

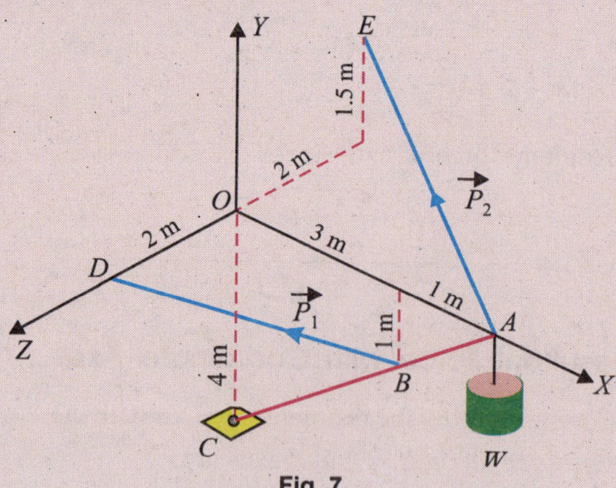

Fig. 7

Solution. Given: Force multiplier of first force P_1 (P_{m1}) = 250 N/m acting from B to D; Force multiplier of second force P_2 (P_{m2}) = 150 N/m acting from A to E; Coordinates of point O = (0, 0, 0); Coordinates of point A = (4, 0, 0); Coordinates of point B = (3, – 1, 0); Coordinates of point C = (0, –4, 0); Coordinates of point D = (0, 0, 2) and Coordinates of point E = (0, 1.5, – 2).

We know that, Force in the direction of BD,

$$\vec{P_1} = P_{m1}\ \overrightarrow{BD}$$

$$= 250\,[(0-3)\,\hat{i} + (0-(-1))\,\hat{j} + (2-0)\,\hat{k}\,]$$

$$= 250\,(-3\,\hat{i} + \hat{j} + 2\,\hat{k}\,) \quad \text{...}(i)$$

Similarly, Force in the direction of AE,

$$\vec{P_2} = P_{m2}\ \overrightarrow{AE}$$

$$= 150\,[(0-4)\,\hat{i} + (1.5-0)\,\hat{j} + (-2-0)\,\hat{k}\,]$$

$$= 150\,(-4\,\hat{i} + 1.5\,\hat{j} - 2\,\hat{k}\,) \quad \text{...}(ii)$$

And vector AC, $\overrightarrow{AC} = [(0-4)\,\hat{i} + (-4-0)\,\hat{j} + (0-0)\,\hat{k}\,] = -4\,\hat{i} - 4\,\hat{j} + 0\,\hat{k}$

∴ Unit vector along AC,

$$\hat{\lambda}_{AC} = \frac{\overrightarrow{AC}}{\left|\overrightarrow{AC}\right|} = \frac{-4\hat{i} - 4\hat{j} + 0\hat{k}}{\sqrt{(-4)^2 + (-4)^2 + 0^2}} = \frac{-4\hat{i} - 4\hat{j} + 0\hat{k}}{5.66} \quad \text{...}(iii)$$

(i) **Component of each force along AC**

We also know that, Component of $\vec{P_1}$ along $\overrightarrow{AC}$,

$$= \vec{P_1} \cdot \hat{\lambda}_{AC}$$

$$= 250\,(-3\,\hat{i} + \hat{j} + 2\,\hat{k}\,) \cdot \left(\frac{-4\hat{i} - 4\hat{j} + 0\hat{k}}{5.66}\right)$$

$$= 44.17\,(-3\,\hat{i} + \hat{j} + 2\,\hat{k}\,) \,.\, (-4\,\hat{i} - 4\,\hat{j} + 0\,\hat{k}\,)$$

$$= 44.17 \times (12 - 4 + 0) = \mathbf{353.36\ N} \qquad \textbf{Ans.}$$

Similarly, Component of $\vec{P_2}$ along $\overrightarrow{AC}$,

$$= \vec{P_2} \cdot \hat{\lambda}_{AC}$$

$$= 150\,(-4\,\hat{i} + 1.5\,\hat{j} - 2\,\hat{k}\,) \cdot \left(\frac{-4\hat{i} - 4\hat{j} + 0\hat{k}}{5.66}\right)$$

$$= 26.5 \times (-4\hat{i} + 1.5\hat{j} - 2\hat{k}).(-4\hat{i} - 4\hat{j} + 0\hat{k})$$

$$= 26.5\,(16 - 6 + 0) = \mathbf{265\ N} \qquad \textbf{Ans.}$$

(ii) **Angle made by each force with AC**

Let α = Angle made by force P_1 with AC and

β = Angle made by force P_2 with AC

We know that, component of P_1 with AC;

$$P_1 \cos\alpha = \vec{P_1}\cdot\hat{\lambda}_{AC}$$

or $$\left|250(-3\hat{i}+\hat{j}+2\hat{k})\right|\cos\alpha = 353.36$$

or $$250\sqrt{(-3)^2+1^2+2^2}\cos\alpha = 353.36$$

or $$250\times 3.742\cos\alpha = 353.36$$

or $$935.5\cos\alpha = 353.36$$

or $$\cos\alpha = 0.3777$$

or $$\alpha = \cos^{-1}(0.3777) = 67.8° \quad \textbf{Ans.}$$

Similarly, component of P_2 with AC;

$$P_2 \cos\beta = \vec{P_2}\cdot\hat{\lambda}_{AC}$$

or $$\left|150(-4\hat{i}+1.5\hat{j}-2\hat{k})\right|\cos\beta = 265$$

or $$150\sqrt{(-4)^2+1.5^2+(-2)^2}\cos\beta = 265$$

or $$150\times 4.717\cos\beta = 265$$

or $$707.55\cos\beta = 265$$

or $$\cos\beta = 0.3745$$

or $$\beta = \cos^{-1}(0.3745) = 68° \quad \textbf{Ans.}$$

Resultant Force of Concurrent Forces

Example 7. *A vertical boom AE is supported by guy wires AB, AC and AD as shown in the Fig. 8. If the tension in wire AD is 750 N, find the forces in wires AB and AC so that the resultant force on A will be vertical.*

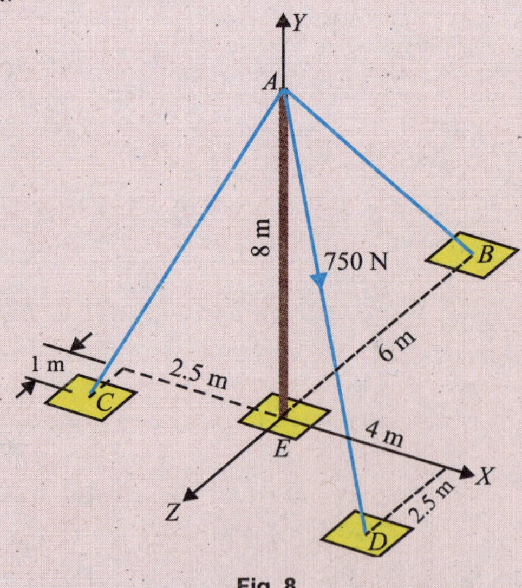

Fig. 8

Solution. Given: Coordinates of points with respect to E as origin and Y and Z axes along A and B respectively are: Coordinates of point E = (0, 0, 0); Coordinates of point A = (0, 8, 0);

Coordinates of point B = (0, 0, –6); Coordinates of point C = (–2.5, 0, 1); Coordinates of point D = (4, 0, 2.5) and Tension in AD (T) = 750 N.

∴ Vector AB, AC, and AD are given as:

$$\overrightarrow{AB} = (0-0)\,\hat{i} + (0-8)\,\hat{j} + (-6-0)\,\hat{k} = -8\,\hat{j} - 6\,\hat{k}$$

$$\overrightarrow{AC} = (-2.5-0)\,\hat{i} + (0-8)\,\hat{j} + (1-0)\,\hat{k} = -2.5\,\hat{i} - 8\,\hat{j} + \hat{k}$$

and $$\overrightarrow{AD} = (4-0)\,\hat{i} + (0-8)\,\hat{j} + (2.5-0)\,\hat{k} = 4\,\hat{i} - 8\,\hat{j} + 2.5\,\hat{k}$$

∴ Tension in wire AD,

$$\vec{T} = T\hat{\lambda}_{AD} = T\frac{\overrightarrow{AD}}{|\overrightarrow{AD}|} = 750 \times \frac{4\hat{i} - 8\hat{j} + 2.5\hat{k}}{\sqrt{4^2 + (-8)^2 + 2.5^2}}$$

$$= 750 \times \frac{4\hat{i} - 8\hat{j} + 2.5\hat{k}}{9.29} = 80.73\,(4\hat{i} - 8\hat{j} + 2.5\hat{k})$$

$$= 322.92\,\hat{i} - 645.84\,\hat{j} + 201.82\,\hat{k}$$

We know that, Force in the direction of AB,

$$\vec{P_1} = P_{m1}\,\overrightarrow{AB} = P_{m1}(-8\,\hat{j} - 6\,\hat{k}) \quad \text{...(where } P_{m1} = \text{Force multiplier)}$$

And Force in the direction of AC,

$$\vec{P_2} = P_{m2}\,\overrightarrow{AC} = P_{m2}(-2.5\,\hat{i} - 8\,\hat{j} + \hat{k})$$

...(where P_{m2} = Force multiplier)

Since the resultant force R is along AE *i.e.*, Y-axis, therefore sum of all forces along EX and EZ must be zero *i.e.*, $\Sigma P_X = 0$ and $\Sigma P_Z = 0$.

∴ $\Sigma P_X = 0$

or $322.92 + 0 - 2.5\,P_{m2} = 0$

or $P_{m2} = 129.17$ N ...(*i*)

And $\Sigma P_Z = 0$

or $201.82 - 6P_{m1} + P_{m2} = 0$

or $201.82 - 6\,P_{m1} + 129.17 = 0$ ...[Using (*i*)]

or $6P_{m1} = 330.99$

or $P_{m1} = 55.16$ N ...(*ii*)

Hence, magnitude of force $\vec{P_1}$,

$$P_1 = |\vec{P_1}| = |P_{m1}(-8\,\hat{j} - 6\,\hat{k})| = P_{m1} \times |(-8\,\hat{j} - 6\,\hat{k})|$$

$$55.16 \times \sqrt{(-8)^2 + (-6)^2}$$

$= 55.16 \times 10 =$ **551.6 N** **Ans.**

And magnitude of force $\vec{P_2}$,

$$P_2\,|\vec{P_2}| = |P_{m2}(-2.5\,\hat{i} - 8\,\hat{j} + \hat{k})| = P_{m2} \times |(-2.5\,\hat{i} - 8\,\hat{j} + \hat{k})|$$

$$= 129.17 \times \sqrt{(-2.5)^2 + (-8)^2 + (1)^2}$$

$$= 129.17 \times 8.44 = \mathbf{1090.2\ N} \qquad \textbf{Ans.}$$

Example 8. *Three concurrent forces $P_1 = 810$ N, $P_2 = 1200$ N and $P_3 = 1600$ N act at a point A as shown in the Fig. 9. The forces P_1 and P_2 are acting along the lines AB and AC respectively. Find the direction of force P_3, so that the resultant of the three forces is parallel to Y-axis. Also, find the magnitude of the resultant force.* **(Nagpur University)**

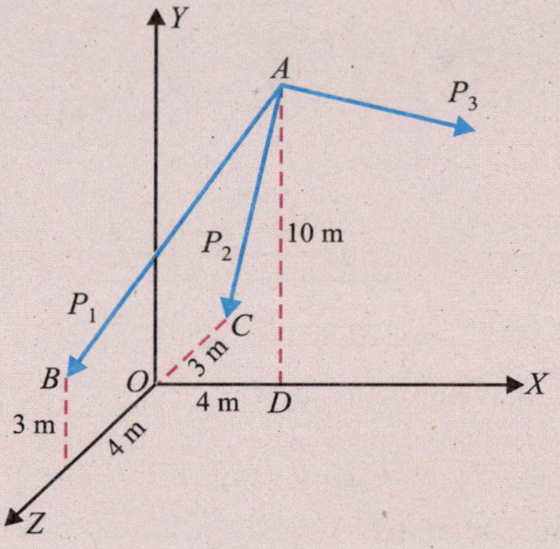

Fig. 9

Solution. Given: First force (P_1) = 810 N; Second force (P_2) = 1200 N; Third force (P_3) = 1600 N; Coordinates of point A = (4, 10, 0); Coordinates of point B = (0, 3, 4) and coordinates of point C = (0, 0, –3).

Direction of Force P_3

We know that, Vectors AB, and AC are given as:

$$\overrightarrow{AB} = (0-4)\hat{i} + (3-10)\hat{j} + (4-0)\hat{k} = -4\hat{i} - 7\hat{j} + 4\hat{k}$$

and

$$\overrightarrow{AC} = (0-4)\hat{i} + (0-10)\hat{j} + (-3-0)\hat{k} = -4\hat{i} - 10\hat{j} - 3\hat{k}$$

We know that, Force in the direction of AB,

$$\vec{P_1} = P_1\,\hat{\lambda}_{AB} = P_1 \frac{\overrightarrow{AB}}{|\overrightarrow{AB}|} = 810 \times \frac{-4\hat{i} - 7\hat{j} + 4\hat{k}}{\sqrt{(-4)^2 + (-7)^2 + 4^2}}$$

$$= 810 \times \frac{-4\hat{i} - 7\hat{j} + 4\hat{k}}{9} = 90\,(-4\hat{i} - 7\hat{j} + 4\hat{k})$$

$$= -360\hat{i} - 630\hat{j} + 360\hat{k} \qquad \text{...}(i)$$

Similarly, Force in the direction of AC,

$$\vec{P_2} = P_2 \hat{\lambda}_{AC} = P_2 \frac{\overrightarrow{AC}}{|\overrightarrow{AC}|} = 1200 \times \frac{-4\hat{i} - 10\hat{j} - 3\hat{k}}{\sqrt{(-4)^2 + (-10)^2 + (-3)^2}}$$

$$= 1200 \times \frac{-4\hat{i} - 10\hat{j} - 3\hat{k}}{11.18} = 107.3(-4\hat{i} - 10\hat{j} - 3\hat{k})$$

$$= -429.2\hat{i} - 1073\hat{j} - 321.9\hat{k} \qquad ...(ii)$$

And Force P_3 in terms of unit vector can be written as,

$$\vec{P_3} = P_3 (l\hat{i} + m\hat{j} + n\hat{k})$$

$$= 1600 (l\hat{i} + m\hat{j} + n\hat{k}) = 1600l\,\hat{i} + 1600m\,\hat{j} + 1600n\,\hat{k} \qquad ...(iii)$$

where l, m and n = Direction cosines of the direction in which force P_3 is acting.

Since the resultant of given system of forces is parallel to Y-axis, therefore sum of all forces along OX and OZ directions must be zero *i.e.*, $\Sigma P_X = 0$ and $\Sigma P_Z = 0$.

$\therefore \quad \Sigma P_X = 0$

or $\quad -360 - 429.2 + 1600\, l = 0$ ...[Taking X components of three forces only]

or $\quad 1600\, l = 789.2$

or $\quad l = 0.493 \qquad ...(iv)$

And $\quad \Sigma P_Z = 0$

or $\quad 360 - 321.9 + 1600\, n = 0$ ...[Taking Z components of three forces only]

or $\quad 1600\, n = -38.1$

or $\quad n = -0.024 \qquad ...(v)$

Since l, m, n are the direction cosines, therefore

$$l^2 + m^2 + n^2 = 1$$

or $\quad (0.493)^2 + m^2 + (-0.024)^2 = 1$

or $\quad m^2 = 0.7564$

or $\quad m = 0.8697$ or -0.8697

or $\quad = -0.8697$ (Since P_3 will be acting downward) $\qquad ...(vi)$

Substituting the values of l, m and n in relation (*iii*), we get

$$\vec{S} = (1600 \times 0.493)\,\hat{i} - 1600\,(-0.8697)\,\hat{j} + (1600 \times -0.024)\,\hat{k}$$

$$= 788.8\,\hat{i} - 1391.5\,\hat{j} - 38.4\,\hat{k} \quad \textbf{Ans.}$$

Magnitude of the Resultant Force *R*

Since the resultant force R is acting along Y-axis, therefore Magnitude of the resultant force R,

$$R = \Sigma P_Z = -630 - 1073 - 1391.5$$

$$= -3094.5 \text{ N}$$

or $\quad R =$ **3094.5 N (downwards)** **Ans.**

Equilibrium of Forces

Example 9. *A load of 2000 N is supported by three cables AB, AC and AD and cable AD lies in YZ plane as shown in the Fig. 10. Find the forces in the cables AB, AC and AD.*

(Aurungabad University)

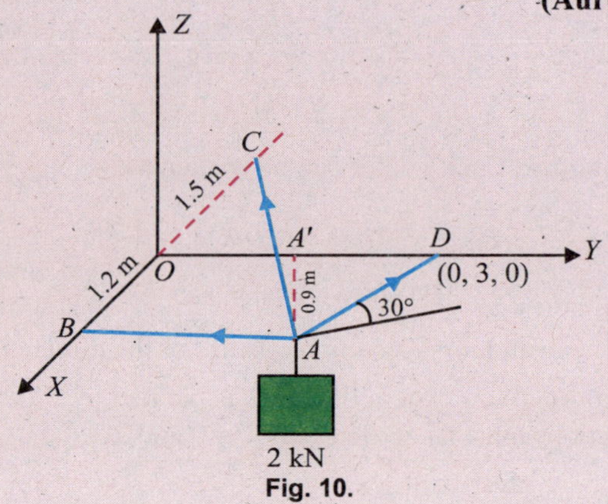

Fig. 10.

Solution. Given: Coordinates of point $B = (1.2, 0, 0)$; Coordinates of point $C = (-1.5, 0, 0)$; Coordinates of point $D = (0, 3, 0)$ and Load supported by cables $(W) = 2$ kN = 2000 N.

Let

P_1 = Force in the cable AB,

P_2 = Force in the cable AC and

P_3 = Force in the cable AD.

From the geometry of the figure, we find that,

Y-coordinates of $A = OA' = OD - A'D = 3 - 0.9 \cot 30° = 1.441$ m

$\therefore$ Co-ordinates of point $A = (0, 1.441 - 0.9)$.

We know that, vector $\overrightarrow{AB}$,

$$\overrightarrow{AB} = (1.2 - 0)\ \hat{i} + (0 - 1.441)\ \hat{j} + [0 - (-0.9)]\ \hat{k}$$

$$= 1.2\hat{i} - 1.441\ \hat{j} + 0.9\hat{k}$$

$\therefore$ Length of $\overrightarrow{AB}$, $\quad AB = \sqrt{(1.2)^2 + (-1.441)^2 + (0.9)^2} = 2.08$ m

Similarly, vector $\overrightarrow{AC}$,

$$\overrightarrow{AC} = (-1.5 - 0)\ \hat{i} + (0 - 1.441)\ \hat{j} + [0 - (-0.9)]\ \hat{k}$$

$$= -1.5\hat{i} - 1.441\ \hat{j} + 0.9\hat{k}$$

$\therefore$ Length of $\overrightarrow{AC}$, $\quad AC = \sqrt{(-1.5)^2 + (-1.441)^2 + (0.9)^2} = 2.266$ m

And Position vector of $\overrightarrow{AD}$,

$$\overrightarrow{AD} = (0 - 0)\hat{i} + (3 - 1.441)\ \hat{j} + [0 - (-0.9)]\ \hat{k}$$

$$= 0\hat{i} + 1.559\ \hat{j} + 0.9\hat{k}$$

$\therefore$ Length of $\overrightarrow{AD}$, $AD = \sqrt{0^2 + (1.559)^2 + (0.9)^2} = 1.8$ m

We also know that, force vector along AB,

$$\vec{P_1} = P_1 \hat{\lambda}_{AB} = P_1 \times \frac{\overrightarrow{AB}}{AB} = P_1 \times \frac{(1.2\hat{i} - 1.441\hat{j} + 0.9\hat{k})}{2.08}$$

$$= P_1 (0.577\hat{i} - 0.693\hat{j} + 0.433\hat{k}) \qquad ...(i)$$

Similarly, force vector along AC,

$$\vec{P_2} = P_2 \hat{\lambda}_{AC} = P_2 \times \frac{\overrightarrow{AC}}{AC} = P_2 \times \frac{(-1.5\hat{i} - 1.441\hat{j} + 0.9\hat{k})}{2.266}$$

$$= P_2 (-0.662\hat{i} - 0.636\hat{j} + 0.397\hat{k}) \qquad ...(ii)$$

And force vector along AD,

$$\vec{P_3} = P_3 \hat{\lambda}_{AD} = P_3 \times \frac{\overrightarrow{AD}}{AD} = P_3 \times \frac{(0\hat{i} + 1.559\hat{j} + 0.9\hat{k})}{1.8}$$

$$= P_3 (0\hat{i} + 0.866\hat{j} + 0.5\hat{k}) \qquad ...(iii)$$

And load $\vec{P_4} = -2000\hat{k}$ N ...(Given) ...(iv)

Since the system is in **equilibrium**, therefore sum of forces along OX, OY and OZ directions must be zero *i.e.*, $\Sigma P_X = 0$, $\Sigma P_Y = 0$ and $\Sigma P_z = 0$.

For equilibrium, equating X components of all the forces equals to zero.

$\therefore$ $\Sigma P_X = 0$

or $0.577 P_1 - 0.662 P_2 = 0$

or $P_1 = 1.147 P_2$ ...(v)

Again, for equilibrium, equating Y components of all the forces equals to zero.

$\therefore$ $\Sigma P_Y = 0$

or $-0.693 P_1 - 0.636 P_2 + 0.866 P_3 = 0$

or $0.693 \times 1.147 P_2 + 0.636 P_2 = 0.866 P_3$ ...(Substituting $P_1 = 1.147 P_2$)

or $0.795 P_2 + 0.636 P_2 = 0.866 P_3$

or $1.431 P_2 = 0.866 P_3$

or $P_2 = 0.605 P_3$ ...(vi)

Similarly, for equilibrium, equating Z components of all the forces equals to zero.

$\therefore$ $\Sigma P_Z = 0$

or $0.433 P_1 + 0.397 P_2 + 0.5 P_3 - 2000 = 0$

Substituting the values of P_1 and P_2 in the above equation, we get

$(0.433 \times 1.147 \times 0.605 P_3) + (0.397 \times 0.605 P_3) + 0.5 P_3 = 2000$

or $0.3 P_3 + 0.24 P_3 + 0.5 P_3 = 2000$

or $1.04 P_3 = 2000$

or $P_3 =$ **1923.08 N** **Ans.**

And $P_2 = 0.605 P_3 = 0.605 \times 1923.08$

$=$ **1163.46 N** **Ans.**

and $P_1 = 1.147 P_2 = 1.147 \times 1163.46$

$=$ **1334.49 N** **Ans.**

Example 10. *For the space frame, there are three members AB, AC and AD as shown in the Fig. 11. If a force of W = 20 kN acting at A in a vertically upward direction, then determine the tensions in AB, AC and AD.* **(Nagpur University)**

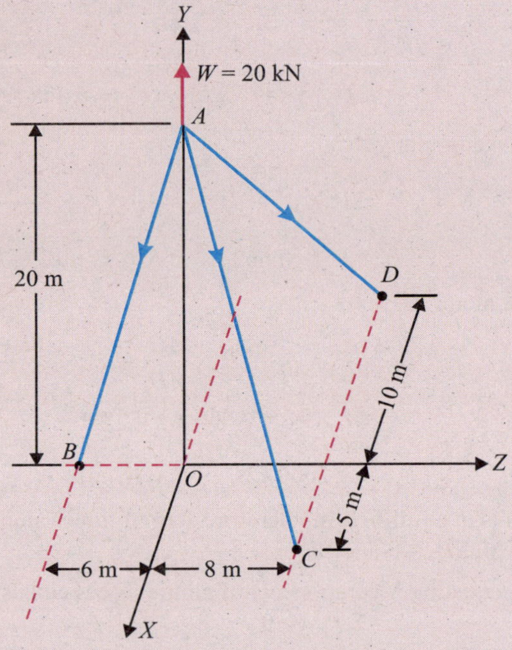

Fig. 11

Solution. Given: Coordinates of point $A = (0, 20, 0)$; Coordinates of point $B = (0, 0, -6)$; Coordinates of point $C = (5, 0, 8)$; Coordinates of point $D = (-10, 0, 8)$ and Force acting at A (W) = 20 kN (vertically upward).

Let P_1 = Tension or force in member AB,

P_2 = Tension or force in member AC, and

P_3 = Tension or force in member AD.

We know that, Force in the member of AB,

$$\vec{P_1} = P_m \overrightarrow{AB} \quad \text{...(where } P_{m1} = \text{Force multiplier)}$$

$$= P_{m1}\,[(0-0)\,\hat{i} + (0-20)\,\hat{j} + (-6-0)\,\hat{k}\,]$$

$$= P_{m1}\,(0\,\hat{i} - 20\,\hat{j} - 6\,\hat{k}\,)$$

Similarly, Force in the member AC,

$$\vec{P_2} = P_{m_2} \overrightarrow{AC} \quad \text{...(where } P_{m2} = \text{Force multiplier)}$$

$$= P_{m2}\,[(5-0)\,\hat{i} + (0-20)\,\hat{j} + (8-0)\,\hat{k}\,]$$

$$= P_{m2}\,(5\,\hat{i} - 20\,\hat{j} + 8\,\hat{k}\,)$$

And Force in the member of AD,

$$\vec{P_3} = P_m \overrightarrow{AD} \quad \text{...(where } P_{m3} = \text{Force multiplier)}$$

$$= P_{m3}\,[(-10-0)\,\hat{i} + (0-20)\,\hat{j} + (8-0)\,\hat{k}\,]$$

$$= P_{m3}\,(-10\,\hat{i} - 20\,\hat{j} + 8\,\hat{k}\,)$$

And vertical force $\vec{P_4} = 20\hat{j}$ kN ...(Given)

Now, forces acting at A are : 20 kN, P_1, P_2 and P_3.

Since the system is in **equilibrium**, therefore sum of forces along OX, OY and OZ directions must be zero *i.e.*, $\Sigma P_X = 0$, $\Sigma P_Y = 0$ and $\Sigma P_Z = 0$.

For equilibrium, equating X components of all the forces equals to zero.

$$\therefore \quad \Sigma P_X = 0$$

$$\text{or} \quad 5P_{m2} - 10P_{m3} = 0$$

$$\text{or} \quad P_{m2} = 2P_{m3} \quad ...(i)$$

Again, for equilibrium, equating Y components of all the forces equals to zero.

$$\therefore \quad \Sigma P_Y = 0$$

$$-20P_{m1} - 20P_{m2} - 20P_{m3} + 20 = 0$$

$$\text{or} \quad 20P_{m1} + 40P_{m3} + 20P_{m3} = 20 \quad (\because P_{m2} = 2P_{m3})$$

$$\text{or} \quad 20P_{m1} + 60P_{m3} = 20$$

$$\text{or} \quad P_{m1} + 3P_{m3} = 1 \quad ...(ii)$$

Similarly, for equilibrium, equating Z components of all the forces equals to zero.

$$\therefore \quad \Sigma P_Z = 0$$

$$\text{or} \quad -6P_{m1} + 8P_{m2} + 8P_{m3} = 0$$

$$\text{or} \quad -6P_{m1} + 8(2P_{m3}) + 8P_{m3} = 0 \quad (\because P_{m2} = 2P_{m3})$$

$$\text{or} \quad -6P_{m1} + 24P_{m3} = 0$$

$$\text{or} \quad P_{m1} - 4P_{m3} = 0 \quad ...(iii)$$

Solving equations (*ii*) and (*iii*), we get

$$P_{m1} = 0.572 \text{ and } P_{m3} = 0.143$$

$$\text{and} \quad P_{m2} = 2P_{m3} = 2 \times 0.143 = 0.286$$

$\therefore$ Tension in member AB,

$$P_1 = |\vec{P_1}| = |P_{m1}(-20\hat{j} - 6\hat{k})| = P_{m1} \times |(-20\hat{j} - 6\hat{k})|$$

$$= 0.572 \times \sqrt{(-20)^2 + (-6)^2} = \mathbf{11.944\ kN} \quad \textbf{Ans.}$$

Similarly, Tension in member AC,

$$P_2 = |\vec{P_2}| = |P_{m2}(5\hat{i} - 20\hat{j} + 8\hat{k})| = P_{m2} \times |(5\hat{i} - 20\hat{j} + 8\hat{k})|$$

$$= 0.286 \times \sqrt{5^2 + (-20)^2 + 8^2} = \mathbf{6.324\ kN} \quad \textbf{Ans.}$$

And Tension in member AD,

$$P_3 = |\vec{P_3}| = |P_{m3}(-10\hat{i} - 20\hat{j} + 8\hat{k})| = P_{m3} \times |(-10\hat{i} - 20\hat{j} + \hat{k})|$$

$$= 0.143 \times \sqrt{(-10)^2 + (-20)^2 + 8^2} = \mathbf{3.396\ kN} \quad \textbf{Ans.}$$

EXERCISE

1. A force of 100 N makes angles of 65°, 75° and 125° respectively with *X*, *Y* and *Z* axes. Express the force in vector form. [**Ans.** $42.26\hat{i} + 25.88\hat{j} - 57.36\hat{k}$]
2. Find the force in vector form, of magnitude 12 N and which is perpendicular to both the forces $\vec{P_1} = -2\hat{i} + \hat{j} - 2\hat{k}$ and $\vec{P_2} = 4\hat{i} - \hat{j} + 3\hat{k}$. [**Ans.** $4\hat{i} - 8\hat{j} - 8\hat{k}$ N]
3. A force acts at origin and makes angles of 65° and 40° with *Y* and *Z* axes respectively. If the *X* component of force is –50 kN, then determine:

   (*i*) the magnitude of force, and

   (*ii*) the angle made by the force with *X* axis. [**Ans.** (*i*) 103.31 kN (*ii*) 118.9°]
4. Find a force vector $\vec{P}$ of magnitude 1822 N which is directed from a point *A* (–6, 8, –2) metres towards a point *B* (8, 2, 8) metres. [**Ans.** $1400\hat{i} - 600\hat{j} + 100\hat{k}$]
5. A vertical beam *AE* is supported by three guy wires *AB*, *AC* and *D* as shown in the Fig. 12. If the tension in wire *AD* is 560 N, then find the forces in wires *AB* and *AC* so that the resultant force on *A* will be vertical. [**Ans.** 568.5 N and 799.7 N]

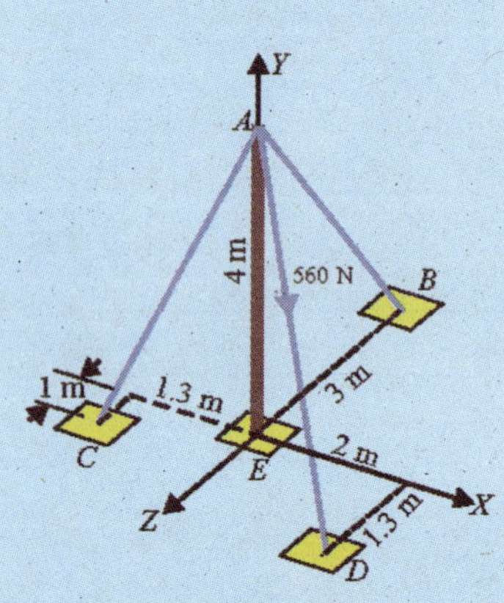

Fig. 12

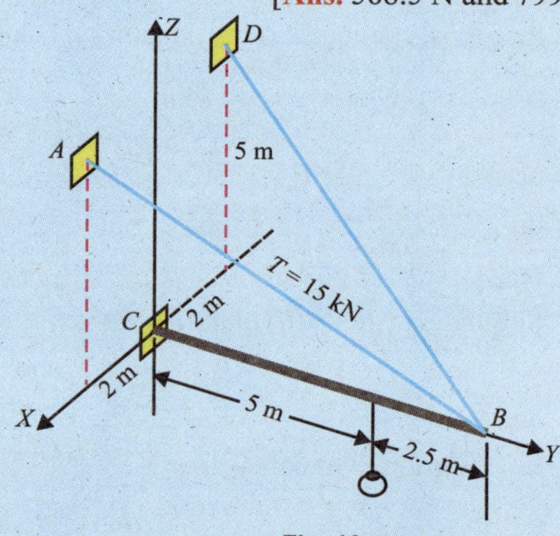

Fig. 13

6. The tension in supporting wire *AB* is 15 kN as shown in Fig. 13. Find the force which the wire exerts on the beam *BC* as a vector $\vec{T}$. Also, determine the angles of $\vec{T}$ with the positive *X*, *Y* and *Z* axes. [**Ans** $\vec{T} = 3.249\hat{i} - 12.184\hat{j} + 8.123\hat{k}$; 77.49°, 144.32° and 57.21°]
7. A boom *AC* lying in vertical plane *XY*, is supported by a ball and socket joint at *C* and by cables *AE* and *BD* as shown in the Fig. 14. For this system, it is found that the force multiplier of force P_1 acting from *B* to *D* is P_{m1} = 225 N/m and that of force P_2 acting from *A* to *E* is P_{m2} = 150 N/m. Determine:

   (*i*) the component of each force along *AC*.

   (*ii*) the angle made by each force with *AC*.

   [**Ans.** (*i*) Component of P_1 along *AC* = 326.3 N; Component of P_2 along *AC* = 344.7 N; (*ii*) Angle made by P_1 with *AC* = 64.9°; Angle made by P_2 with *AC* = 59.8°]

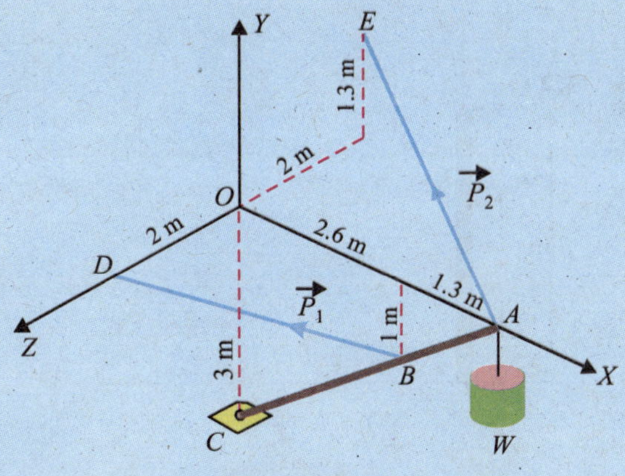

Fig. 14

8. A cantilever frame work is shown in the Fig. 15. Find the shortest distance from point D to line AB. [**Ans.** 2.34 m]

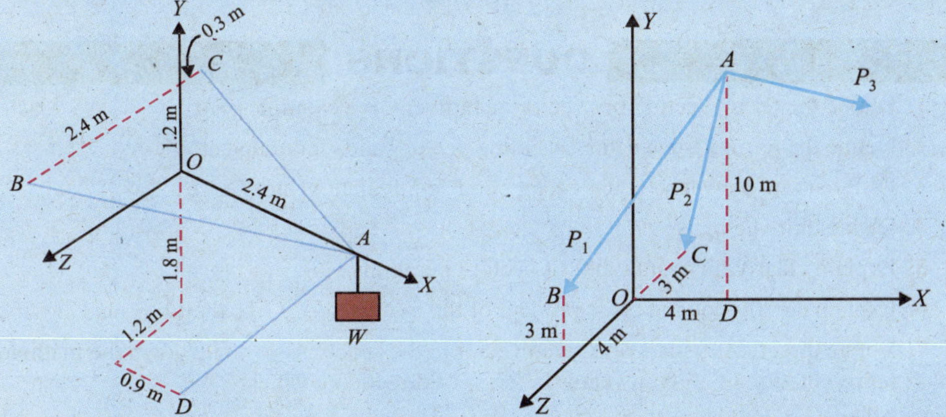

Fig. 15 Fig. 16

9. Three concurrent forces $P_1 = 800$ N, $P_2 = 1200$ N and $P_3 = 1500$ N act at a point A as shown in the Fig. 16. The forces P_1 and P_2 are acting along the lines AB and AC respectively. Find the direction of force P_3, so that the resultant of the three forces is parallel to Y-axis. Also, find the magnitude of the resultant force. [**Ans.** $829.5\,\hat{i} - 1247.4\,\hat{j} - 78\,\hat{k}$; – 3020.4 N]

***10.** A pole AB of length 7 m is acted upon by a 8.4 kN force as shown in the Fig. 17. It is held by a ball and socket point at A and by two cables, CD and CE of equal length. Neglecting the weight of the pole, find the tension in each cable and the reaction at point A.

(Nagpur University)

[**Ans.** 9.65 kN, 9.65 kN, 10.36 kN]

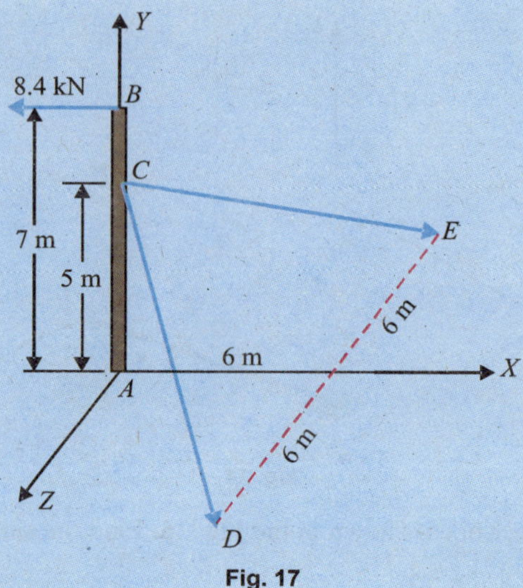

Fig. 17

QUESTIONS

1. Define the term "vector" or "vector quantity". Give examples.
2. Define the terms "free vector", "sliding vector" and "bound vector".
3. Differentiate between "equal vector" and "equivalent vector".
4. Define unit vector.
5. Explain clearly different types of vector operations.
6. Explain clearly dot and cross product of the two vectors. Give their physical significance.
7. Define direction cosines of a given vector in the space. Also, write down the mathematical representation of a given vector in terms of direction cosines.
8. What is meant by moment of a force.
9. What is the effect of a force, when it is acting on a body ?

OBJECTIVE TYPE QUESTIONS

1. Two vectors are said to be equal if
   (*a*) direction is same
   (*b*) their magnitudes are same
   (*c*) originate from the same point
   (*d*) they have same magnitudes and same sense of direction
2. If two vector $\vec{A}$ and $\vec{B}$ are parallel and have equal magnitudes, then
   (*a*) they are equal
   (*b*) they are not equal
   (*c*) they may or may not equal
   (*d*) they have same sense of direction
3. If m is non-zero scalar an $\vec{A}$ is non-zero vector of modulus A, then vector $\vec{A}$ multiplied by scalar m is equal to

(*a*) $\vec{A}$ (*b*) $m\,A$ (*c*) $\vec{A}\,m$ (*d*) none of these

4. If m is non-zero scalar and $\vec{A}$ is non-zero vector of modulus A, then vector $m\,\vec{A}$ is a unit vector if

(*a*) $m = \pm 1$ (*b*) $A = |m|$ (*c*) $A = 1/|m|$ (*d*) $A = 1/|m|^2$

5. If θ is the angle between two vector $\vec{A}$ and $\vec{B}$, the $\vec{A} \cdot \vec{B} = 0$ only if

(*a*) $\theta = 0$ (*b*) $\theta = \pi/4$ (*c*) $\theta = \pi/2$ (*d*) $\theta = \pi$

6. If θ is the angle between two vector $\vec{A}$ and $\vec{B}$, the $\vec{A} \times \vec{B} = 0$ only if

(*a*) $\theta = 0$ (*b*) $\theta = \pi/4$ (*c*) $\theta = \pi/2$ (*d*) $\theta = \pi$

7. If $\hat{i}$, $\hat{j}$ and $\hat{k}$ be the set of orthogonal unit vectors, the $(\hat{i} \cdot \hat{i}) + (\hat{j} \cdot \hat{j}) + (\hat{k} \cdot \hat{k})$ equals to

(*a*) 0 (*b*) 1 (*c*) 2 (*d*) 3

8. If $\hat{i}$, $\hat{j}$ and $\hat{k}$ be the set of orthogonal unit vectors, the $(\hat{i} \times \hat{i}) + (\hat{j} \times \hat{j}) + (\hat{k} \times \hat{k})$ equals to

(*a*) 0 (*b*) 1 (*c*) 2 (*d*) 3

9. The angle between the vector $2\hat{i} - 3\hat{j} + \hat{k}$ an $2\hat{i} + \hat{j} - \hat{k}$ is

(*a*) $\theta = 0$ (*b*) $\theta = \pi/3$ (*c*) $\theta = \pi/4$ (*d*) $\theta = \pi/2$

10. If $\vec{A}$ and $\vec{B}$ are two non-zero vectors, the $(\vec{A} + \vec{B}) \cdot (\vec{A} - \vec{B})$ is equal to

(*a*) $A - B$ (*b*) $A + B$ (*c*) $A^2 - B^2$ (*d*) $A^2 + B^2$

ANSWERS

1. (*d*)	**2.** (*c*)	**3.** (*c*)	**4.** (*c*)	**5.** (*c*)
6. (*a*)	**7.** (*d*)	**8.** (*a*)	**9.** (*d*)	**10.** (*c*)

CHAPTER

34 Pappus–Guldinus Theorems

Contents

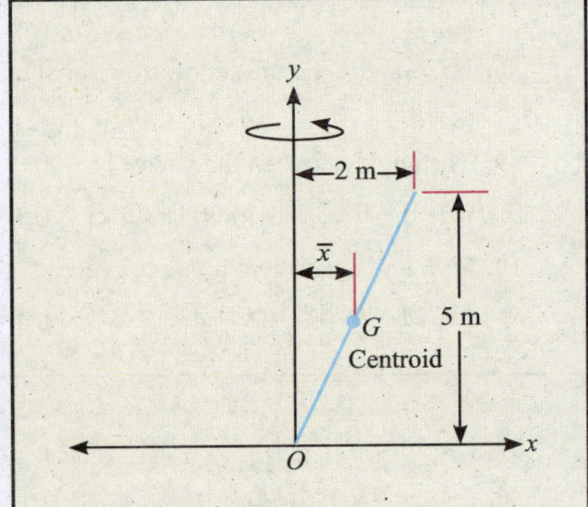

1. INTRODUCTION

When a plane curve rotates about a fixed axis (called *axis of revolution*), it generates a surface, called the **surface of revolution**. Similarly, when a plane area is rotated about a fixed axis, it generated a body, called the **body of revolution.**

Greek scientist **Pappus** and Swiss mathematician **Guldinus** proposed two important theorems which deal with surfaces and bodies of revolution. *Pappus-Guldinus theorems* are also known as **Pappus-Guldinus centroid theorems.**

First theorem deals with the computation of area of surface of revolution generated by rotating a plane curve about a fixed axis which does not intersect the curve. See the examples as shown in Fig. 1.

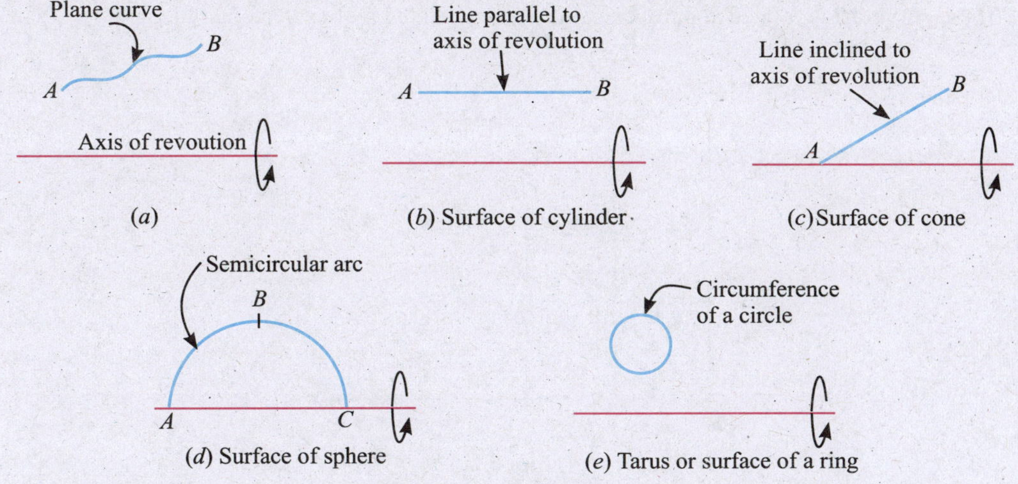

Fig. 1. Generation of surface of revolution

Second theorem deals with the computation of volume of solid /body of revolution generated by rotating a plane area about a fixed axis which does not intersect the area. See the example as shown in Fig. 2.

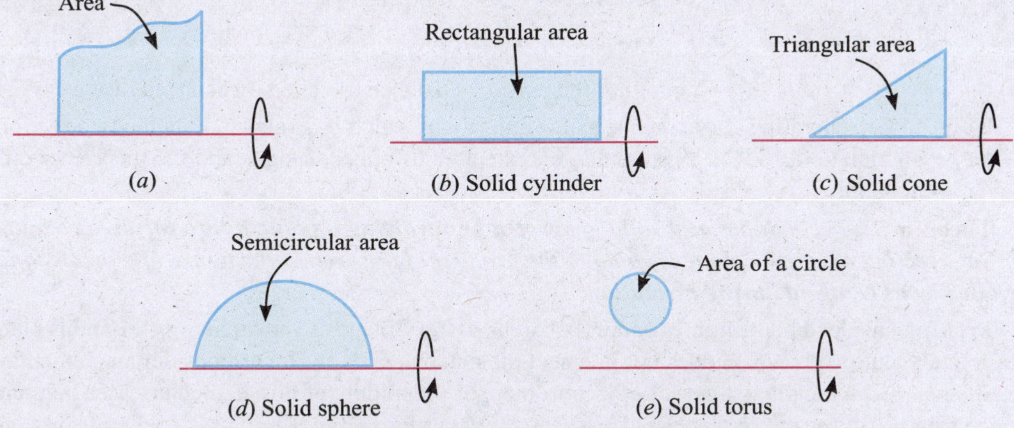

Fig. 2. Generation of solid volume of revolution

2. PAPPUS-GULDINUS THEOREMS

Theorem 1: ***The area of surface generated by revolving a plane curve about a non-intersecting axis in its plane is equal to the product of the length of the curve and the distance travelled by its centroid during revolution.***

Proof: Consider a plane curve '*AB*' of length '*l*' in *xy*-plane as shown in Fig. 3. Let this curve rotate about *x*-axis which does not intersect the curve *AB*, Now, let us consider an elementary length dl_i on the curve so small that it may be thought to be an infinitesimal small portion of a straight line. Let *y*-coordinate of its mid point be y_i and the centroid '*G*' of the entire curve be $\overline{y}$ as shown in Fig. 3.

If the curve AB is rotated about x-axis, the area generated by the elementary length,

$$dA_i = 2\pi y_i dl_i$$

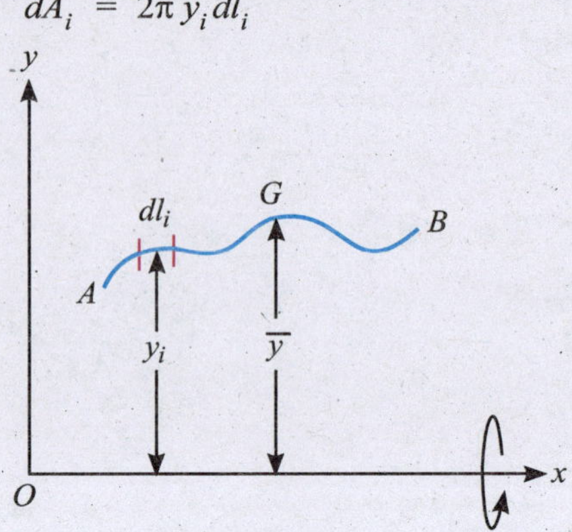

Fig. 3.

∴ Surface area generated by the entire line,

$$A = \int dA_i$$

$$= \int 2\pi y_i \, dl_i$$

$$= 2\pi \int y_i dl_i = 2\pi \times \bar{y}l \qquad ...(\because \int y_i dl_i = \bar{y}l)$$

$$= l \times 2\pi\bar{y}$$

= Length of the curve × Distance travelled by centroid of the curve

Note: If the generating curve intersects the axis about which it is rotated, then the two sections on either side of the axis would generate area of opposite signs and the theorem would not apply.

Theorem 2: ***The volume of a solid generated by revolving a plane figure (or area) about a non-intersecting axis in its plane is equal to the product of the area of the figure and the distance travelled by its centroid during revolution.***

Proof: Consider a plane figure (or area) of area 'A' in xy-plane as shown in Fig. 4. Let this plane area rotate about x-axis which does not intersect the plane area. Now, let us consider an elementary area of area dA_i on the plane figure. Let y-coordinate of the middle of this elementary area be y_i and the centroid 'G' of the entire plane area be $\bar{y}$ as shown in Fig. 4.

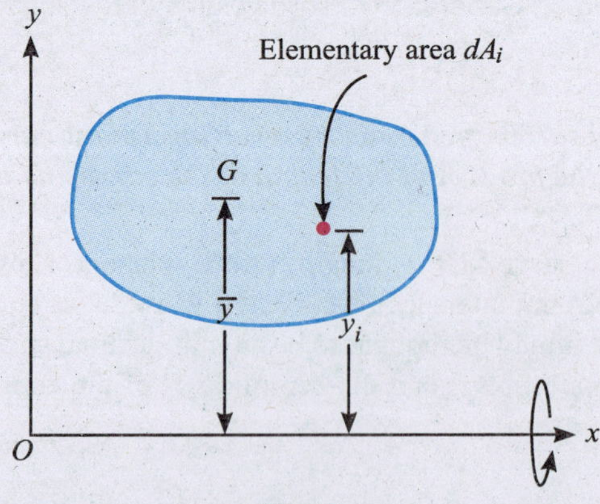

Fig. 4.

When the plane area is rotated about x-axis, the volume generated by the elementary area,

$$dV_i = 2\pi\, y_i\, dA_i$$

∴ Volume generated by the entire plane area,

$$V = \int dV_i$$

$$= \int 2\pi y_i\, dA_i$$

$$= 2\pi \int y_i dA_i = 2\pi \times \bar{y}\, A \qquad ...(\because \int y_i dA_i = \bar{y}A)$$

$$= A \times 2\pi\, \bar{y}$$

= Area of the figure × Distance travelled by centroid of the figure

Application: Pappus-Guldinus Theorems provide a simple way to *determine areas of surfaces of revolution and the volumes of the bodies of revolution.* Conversely, these theorems are used to *determine the centroid of the plane curve* when the area of the surface generated by the plane curve is known or to *determine the centroid of the plane area* when the volume of the body generated by the plane area is known.

Example 1. *Use Pappus-Guldinus theorems to calculate the*

(*a*) *area of a surface of revolution generated by revolving a quarter circular arc of radius 2 m about x-axis as shown in Fig. 5.*

(*b*) *volume of a solid generated by revolving a semicircular circular area of radius 3 m about x-axis as shown in Fig. 6.*

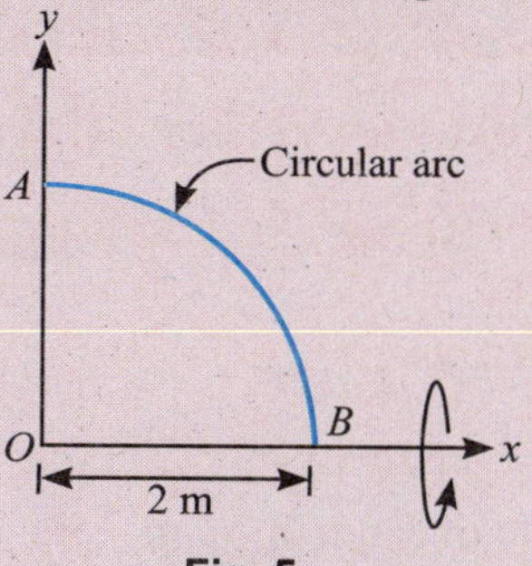

Fig. 5.

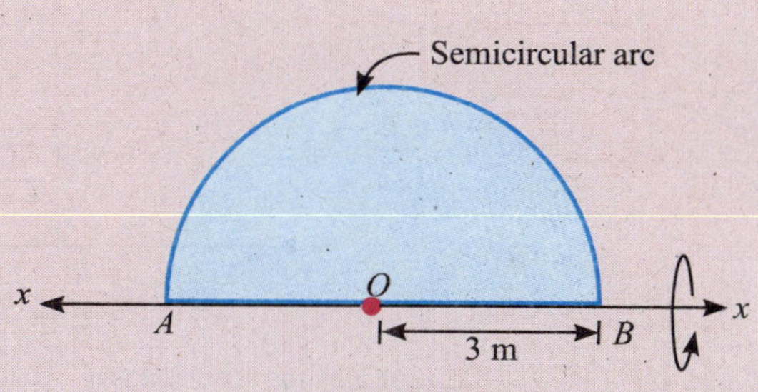

Fig. 6.

Solution. (*a*) Given: Radius of quarter circular arc (r) = 2 m.

We know that distance of centroid of quarter arc from x axis $(\bar{y}) = \dfrac{2r}{\pi}$

Using **1st theorem of Pappus-Guldinus,** area generated by revolving quarter circular arc,

$$A = \text{Length of quarter circular arc} \times \text{Distance travelled by its centroid}$$

$$= \frac{\pi r}{2} \times 2\pi\, \bar{y}$$

$$= \frac{\pi r}{2} \times \left(2\,\pi \times \frac{2r}{\pi}\right)$$

$$= 2\pi r^2 = 2\pi \times 2^2 = 25.13 \text{ m}^2 \quad \textbf{Ans.}$$

(*b*) Given: Radius of semicircular arc (r) = 3 m.

We know that a solid sphere is generated by revolving a semicircular area about x-axis.

We know that distance of centroid of semicircular area from x axis $(\bar{y}) = \dfrac{4r}{3\pi}$

Using **2nd theorem of Pappus-Guldinus,** volume of solid sphere generated by revolving semicircular area,

$$V = \text{Area of semicircle} \times \text{Distance travelled by its centroid}$$

$$= \frac{\pi r^2}{2} \times 2\pi \bar{y}$$

$$= \frac{\pi r^2}{2} \times \left(2\pi \times \frac{4r}{3\pi}\right)$$

$$= \frac{4}{3}\pi r^3 = \frac{4}{3}\pi \times 3^3 = 113.1 \text{ m}^3 \quad \textbf{Ans.}$$

Example 2. *A cone of radius of 2 m and height 5 m is shown in Fig. 7. Using Pappus-Guldinus theorems, determine*

(a) the amount of paint required to paint the inside and outside surfaces of it, if 1 litre of paint covers 15 m².

(b) the volume of the cone.

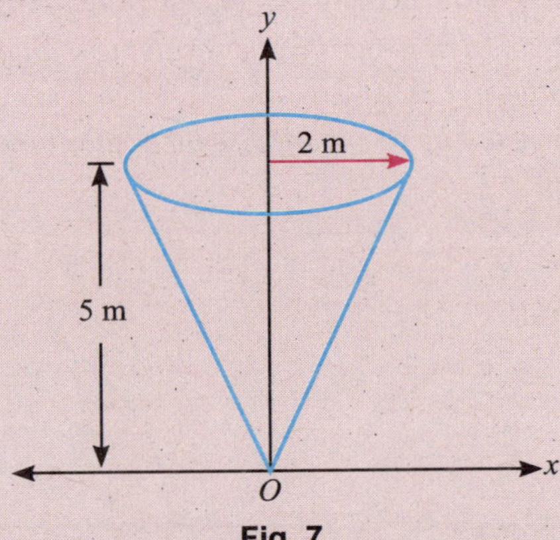

Fig. 7.

Solution. (*a*) Given: Radius of cone (r) = 2 m and height of cone (h) = 5 m.

We know that a cone is generated by revolving a straight line about an axis at which line is inclined.

Here, y axis is the axis of revolution and generating curve is a straight line through the origin as shown in Fig. 8(*a*).

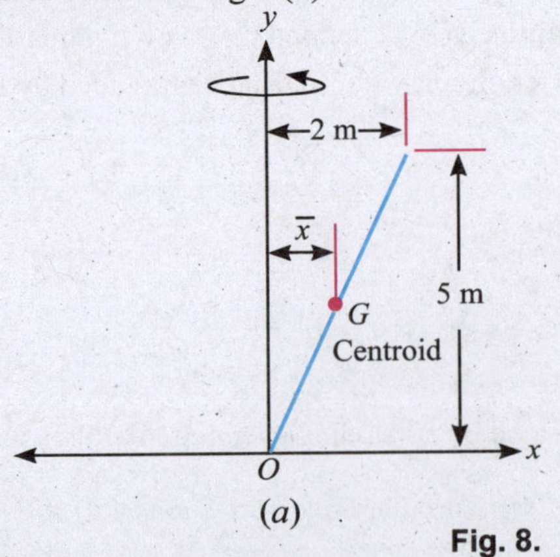

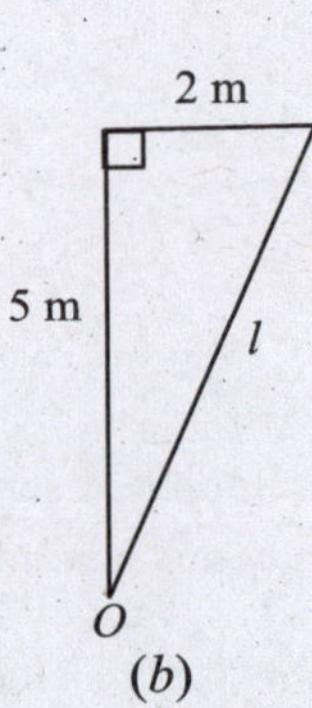

Fig. 8.

We know that distance of centroid of straight line from y axis $(\overline{x}) = \frac{r}{2}$

And length of the curve (straight line) can be found from the Pythagorean theorem:

$$l = \sqrt{5^2 + 2^2} = 5.3852 \text{ m}$$

Using **1st theorem of Pappus-Guldinus,** area of surface generated by revolving the line,

$$A = \text{Length of the line} \times \text{Distance travelled by its centroid}$$

$$= l \times 2\pi\overline{x}$$

$$= l \times \left(2\pi \times \frac{r}{2}\right)$$

$$= \pi\, r\, l = \pi \times 2 \times 5.3852 = 33.8362 \text{ m}^2$$

∴ Amount of paint required to paint the inside and outside surfaces of cone,

$$\text{Volume of paint required} = 2 \times A \times \frac{1 \text{ L}}{15 \text{ m}^2}$$

$$= 2 \times 33.8362 \times \frac{1}{15}$$

$$= 4.5115 \text{ L} \quad \textbf{Ans.}$$

(*b*) Here, the generating area is a triangle as shown in adjacent Fig. 9.

We know that distance of centroid of triangular area from y axis,

$$\overline{x} = \frac{r}{3} = \frac{2}{3} \text{ m}$$

And area of the triangle,

$$A = \frac{1}{2} \times \text{Base} \times \text{Height}$$

$$= \frac{1}{2} \times 5 \times 2 = 5 \text{ m}^2$$

Using **2nd theorem of Pappus-Guldinus,** volume of cone generated by revolving triangular area,

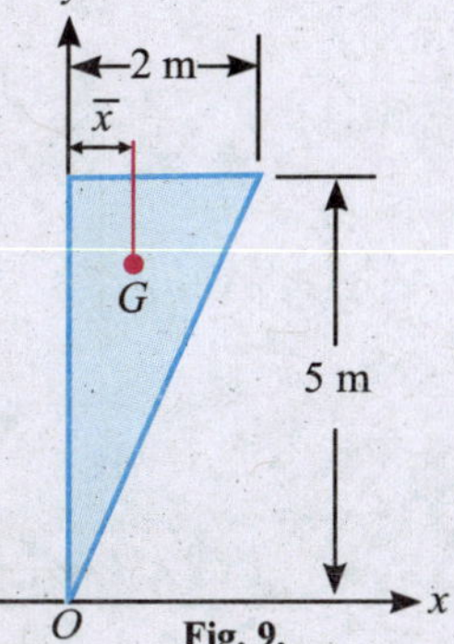

Fig. 9.

$$V = \text{Area of triangle} \times \text{Distance travelled by its centroid}$$

$$= A \times 2\pi\overline{x}$$

$$= 5 \times 2\pi \times \frac{2}{3}$$

$$= 20.94 \text{ m}^3 \quad \textbf{Ans.}$$

EXERCISE

1. Use Pappus-Guldinus theorems to calculate the
   (*a*) area of a surface of revolution generated by revolving a quarter circular arc of radius 10 cm about *y*-axis as shown in Fig. 10.
   (*b*) volume of a solid generated by revolving a semicircular circular area of radius 50 cm about *x*-axis as shown in Fig. 11.

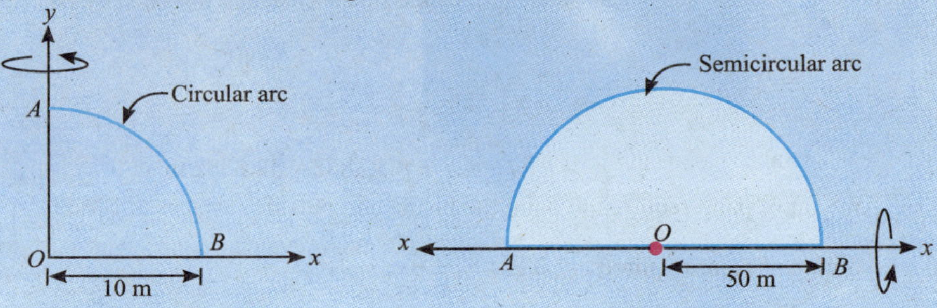

Fig. 10. **Fig. 11.**

(**Ans.** (*a*) 0.0628 m^2; (*b*) 0.524 m^3)

2. A line of length 8 m is rotated about *y*-axis as shown in Fig. 12. Using Pappus-Guldinus theorem to find area and volume of generating cylinder.

(**Ans.** (*a*) 251.33 m^2; (*b*) 502.65 m^3)

Fig. 12. **Fig. 13.**

3. A cone of radius of 3 m and height 10 m is shown in Fig. 13. Using Pappus-Guldinus theorems, determine
   (*a*) the amount of paint required to paint the inside and outside surfaces of cone, if 1 L of paint covers 20 m^2.
   (*b*) the volume of the cone.

(**Ans.** (*a*) 9.84 L; (*b*) 94.25 m^3)

CHAPTER

35 Product of Inertia

Contents

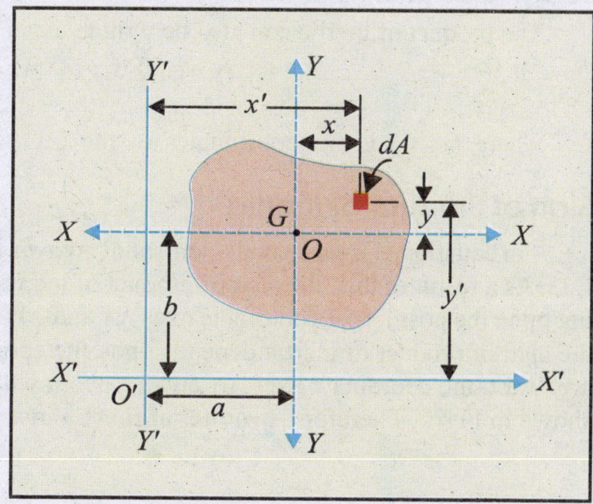

1. PRODUCT OF INERTIA

[DTU, RTU, AUT]

The product of inertia of a plane area is obtained with respect to a set of rectangular axes which completely lying in the plane of the area.

Consider a plane area or section of area A with a set of rectangular axes XX and YY *as* shown in Fig.1. Again, consider a small element of area dA at a distance x and y from the YY-axis and XX-axis, respectively. Then, $\Sigma\, xy\, dA$ is defined as the **product of inertia of the plane area with respect to XX and YY axes.**

Mathematically, product of inertia w.r.t. XX and YY axes is given by,

$$I_{XY} = \Sigma\, xy\, dA = \int_A xy\, dA \quad ...(1)$$

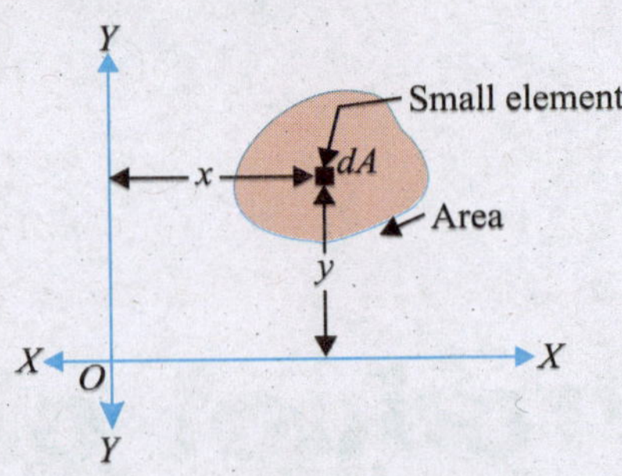

Fig. 1.

Thus, the sum of the products formed by multiplying the area (or sometimes, the mass) of each element of a plane area by the product of the its coordinates of C.G. corresponding to two rectangular axes, is called **product of Inertia.**

The product of inertia can also be written as:

$$I_{XY} = \Sigma x_i y_i A_i = x_1 y_1 A_1 + x_2 y_2 A_2 + x_2 y_2 A_2 + \ldots \quad \ldots(1a)$$

where (x_i, y_i) is the coordinates of the C.G. of area A_i.

Sign of Product of Inertia

In equation (1), each small element of area *dA* is multiplied by the product of its coordinates of C.G. As a result of this, the sign of product of inertia I_{XY} may be *positive*, *negative* or *zero*, depending upon the position of coordinate axes *XX* and *YY* w.r.t. the given area under consideration. Thus, the sign of product of inertia depends upon the coordinates of the various small areas of the plane area w.r.t. the coordinate axes *XX* and *YY* about which the product of inertia is to be found out, as shown in Fig. 2. Therefore, **product of inertia of a plane area may be *positive*, *negative* or *zero*.**

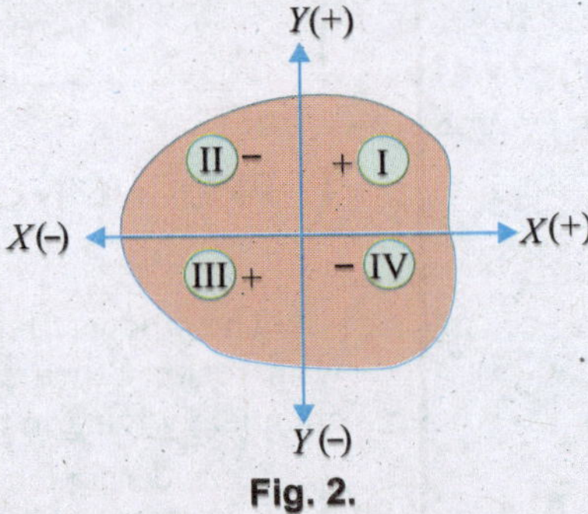

Fig. 2.

Example. For any area in the IInd quadrant, the sign of its *X*-coordinate is *negative* and that of *Y*-coordinate is *positive*, therefore the sign for the product of inertia for the area is *negative* as [(–ve) × (+ve) = (–ve)].

2. *PARALLEL AXES THEOREM FOR PRODUCT OF INERTIA

The parallel theorem for product of inertia is related to find out the product of inertia of a plane area w.r.t. a set of axes which is parallel to the centroidal axes. If the product of inertia of a plane area (or figure) is known about its centroidal axes *XX* and *YY*, it is possible to transform the product of inertia about any other two axes *X′X′* and *Y′Y′* parallel to the centroidal axes *XX* and *YY*.

Consider a plane area A with centroidal axes XX and YY as shown in Fig. 3. Further, consider a small element of area dA at a distance x and y from the YY-axis and XX-axis, respectively.

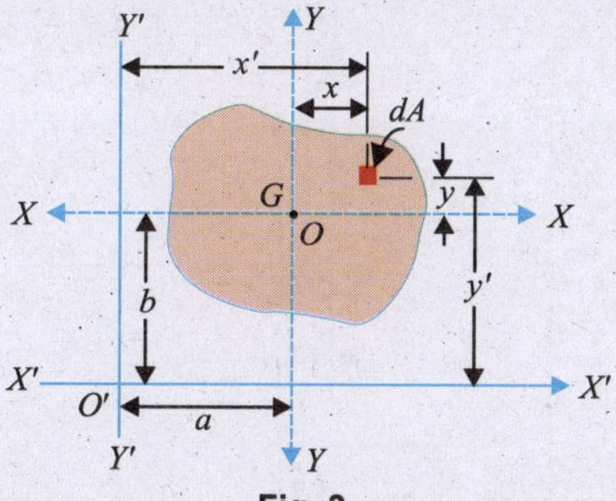

Fig. 3.

We know that the product of inertia of a plane area about its centroidal axes XX and YY,

$$I_{XY} = \int_A x\,y\,dA \qquad \text{...}(i)$$

...(where x and y are the coordinates of C.G. of the small area dA)

And, moment of area of a plane area about its centroidal axis YY,

$$\int_A x\,dA = 0 \qquad \text{...}(ii)$$

Similarly, moment of area of a plane area about its centroidal axis XX,

$$\int_A y\,dA = 0 \qquad \text{...}(iii)$$

Now, the product of inertia of a small element of area dA w.r.t. $X'X'$ and $Y'Y'$ axes,

$$= x'y'\,dA$$

∴ The product of inertia of a plane area w.r.t. $X'X'$ and $Y'Y'$ axes,

$$I_{X'Y'} = \int_A x'y'\,dA$$

$$= \int_A (x+a)\,(y+b)\,dA$$

$$= \int_A (xy + xb + ay + ab)\,dA$$

$$= \int_A xy\,dA + b\int_A x\,dA + a\int_A y\,dA + ab\int_A dA$$

$$= I_{XY} + 0 + 0 + abA \qquad \text{...[Using eqns. (i), (ii) and (iii)]}$$

or

$$I_{X'Y'} = I_{XY} + Aab \qquad \text{...(2)}$$

This is the required **transfer formula for the product of inertia**. This equation (2) is known as **Parallel axes theorem for product of inertia**.

Example 1. *Find the product of inertia of a Z-section about its centroidal axes X and Y as shown in Fig. 4. All dimensions are in mm.*

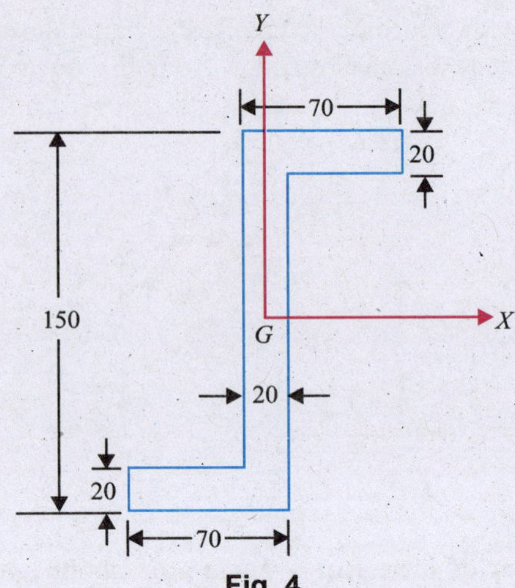

Fig. 4.

Solution. Split up the whole Z-section into three rectangles 1, 2 and 3 separated by a dotted line as shown in Fig. 5. The reference axes *X* and *Y* pass through the centroid *G* of the Z-section as shown in the Fig. 5.

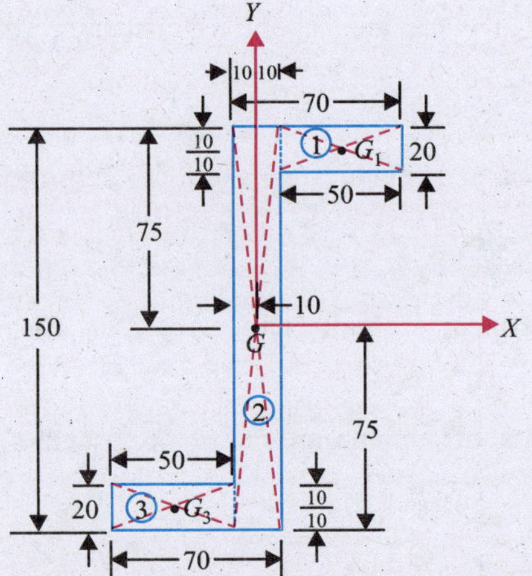

Fig. 5.

The areas and location of their C.G. are as follow:

Rectangle (1):

Area, $A_1 = 50 \times 20 = 1000 \text{ mm}^2$

Coordinates of C.G. (G_1) w.r.t. *XY* axes are:

$$x_1 = \frac{20}{2} + \frac{50}{2} = +35 \text{ mm}; \; y_1 = 75 - 10 = +65 \text{ mm}$$

Rectangle (2):

Area, $A_2 = 150 \times 20 = 3000 \text{ mm}^2$

Coordinates of C.G. (G) w.r.t. *XY* axes are:

$x_2 = 0$ mm; $y_2 = 0$ mm

Rectangle (3):

Area, $A_3 = 50 \times 20 = 1000 \text{ mm}^2$

Coordinates of C.G. (G_3) w.r.t. XY axes are:

$$x_3 = -\frac{20}{2}+\frac{50}{2} = -35 \text{ mm}; \quad y_3 = -(75-10) = -65 \text{ mm}$$

Now, we have to calculate the product of inertia of each area about given X and Y axes.

Let $\quad I_{\overline{XY}}$ = Product of inertia for each area about its centroidal axes parallel to the given X and Y axes

We know that, Product of inertia of rectangle (1) about X and Y axes by using *Parallel axes theorem,*

$$(I_{XY})_1 = I_{\overline{XY}} + A_1 ab$$
$$= 0 + [1000 \times (+35) \times (+65)] \quad ...(\because \; a = +35 \text{ and } b = +65)$$
$$= +2.275 \times 10^6 \text{ mm}^4$$

Similarly, Product of inertia of rectangle (2) about X and Y axes by using *Parallel axes theorem,*

$$(I_{XY})_2 = I_{\overline{XY}} + A_2 ab = 0 + [3000 \times 0 \times 0] \quad ...(\because \; a = 0 \text{ and } b = 0)$$
$$= 0 \text{ mm}^4$$

Similarly, Product of inertia of rectangle (3) about X and Y axes by using *Parallel axes theorem,*

$$(I_{XY})_3 = I_{\overline{XY}} + A_3 ab$$
$$= 0 + [1000 \times (-35) \times (-65)] \quad ...(\because \; a = -35 \text{ and } b = -65)$$
$$= +2.275 \times 10^6 \text{ mm}^4$$

∴ Total product of inertia of the given Z-section w.r.t. X and Y axes,

$$I_{XY} = (I_{XY})_1 + (I_{XY})_2 + (I_{XY})_3$$
$$= (2.275 \times 10^6) + 0 + (2.275 \times 10^6)$$
$$= \mathbf{+4.55 \times 10^6 \text{ mm}^4} \quad \textbf{Ans.}$$

Example 2. *Find the product of inertia for the angle section about the axes XX and YY as shown in Fig. 6. Also, find the product of inertia about parallel axes X'X' and Y'Y' through centroid. All dimensions are in mm.* **(UPTU-UP)**

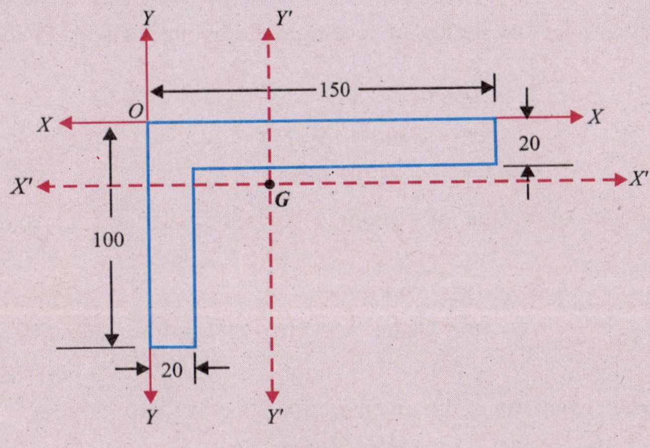

Fig. 6.

Solution.

Spilt up the entire area into two rectangles (1) and (2) separated by a dotted line as shown in Fig. 7.

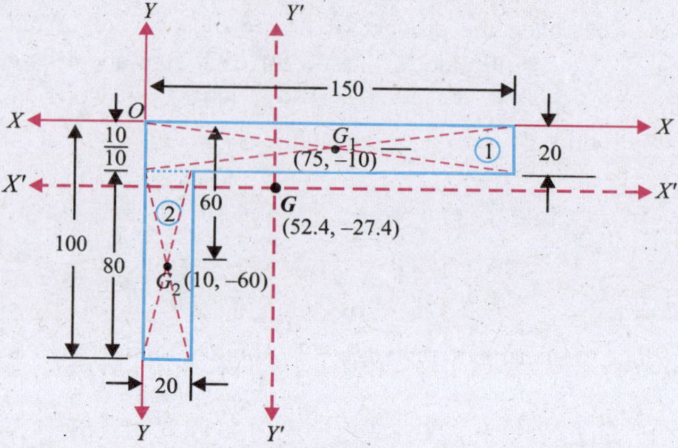

Fig. 7.

The areas and location of their C.G. are as follow:

Rectangle (1):

Area, $A_1 = 150 \times 20 = 3000 \text{ mm}^2$

Coordinates of C.G. (G_1) w.r.t. XY axes are:

$x_1 = +75$ mm; $y_1 = -10$ mm

Rectangle (2):

Area, $A_2 = 80 \times 20 = 1600 \text{ mm}^2$

Coordinates of C.G. (G_2) w.r.t. XY axes are:

$$x_2 = +10 \text{ mm}; \quad y_2 = -\left(20 + \frac{80}{2}\right) = -60 \text{ mm}$$

Product of inertia of the angle section about the axes *XX* and *YY*

First of all, let us find the product of inertia of each area about XX and YY axes.

Let $I_{\overline{XY}}$ = Product of inertia for each area about its centroidal axes parallel to XX and YY axes.

We know that, Product of inertia of rectangle (1) about XX and YY axes by using *Parallel axes theorem*,

$$(I_{XY})_1 = I_{\overline{XY}} + A_1 ab$$
$$= 0 + [3000 \times (+75) \times (-10)] \quad ...(\because a = +75 \text{ and } b = -10)$$
$$= -2.25 \times 10^6 \text{ mm}^4$$

Similarly, Product of inertia of rectangle (2) about XX and YY axes by using *Parallel axes theorem*,

$$(I_{XY})_2 = I_{\overline{XY}} + A_2 ab$$
$$= 0 + [1600 \times (+10) \times (-60)] \quad ...(\because a = +10 \text{ and } b = -60)$$
$$= -0.96 \times 10^6 \text{ mm}^4$$

$\therefore$ Total product of inertia of the given section about XX and YY axes,

$$I_{XY} = (I_{XY})_1 + (I_{XY})_2$$
$$= (-2.25 \times 10^6) + (-0.96 \times 10^6)$$
$$= \mathbf{-3.21 \times 10^6 \text{ mm}^4} \quad \textbf{Ans.}$$

Product of inertia of the section about the axes $X'X'$ and $Y'Y'$ through centroid G

First of all, let us find the centre of gravity of the given section w.r.t. XX and YY axes. We know that coordinates of the centroid G of the section w.r.t. XX and YY axes,

$$\bar{x} = \frac{A_1x_1 + A_2x_2}{A_1 + A_2}$$

$$= \frac{3000 \times 75 + 1600 \times 10}{3000 + 1600} = \frac{241000}{4600} = 52.4 \text{ mm}$$

and

$$\bar{y} = \frac{A_1y_1 + A_2y_2}{A_1 + A_2}$$

$$= \frac{3000 \times (-10) + 1600 \times (-60)}{3000 + 1600}$$

$$= -\frac{126000}{4600} = -27.4 \text{ mm}$$

We know form *Parallel axes theorem* that, Product of inertia of the given section about axes $X'X'$ and $Y'Y'$ through centroid G,

$$I_{X'Y'} = I_{XY} - Aab \qquad ...[\because I_{XY} = I_{X'Y'} + Aab]$$

$$= I_{XY} - (A_1 + A_2)ab$$

$$= -3.21 \times 10^6 - (3000 + 1600) \times (+52.4) \times (-27.4)$$

...[Here $a = 52.4$, $b = -27.4$]

$$= -3.21 \times 10^6 + 6.6 \times 10^6 = \mathbf{+3.39 \times 10^6 \text{ mm}^4} \quad \textbf{Ans.}$$

Example 3. *A plane hatched area is shown in the Fig. 8. Determine the product of inertia of the given area about the axes XX and YY. All dimensions are in mm.*

(AUT-TN, JNTU-AP)

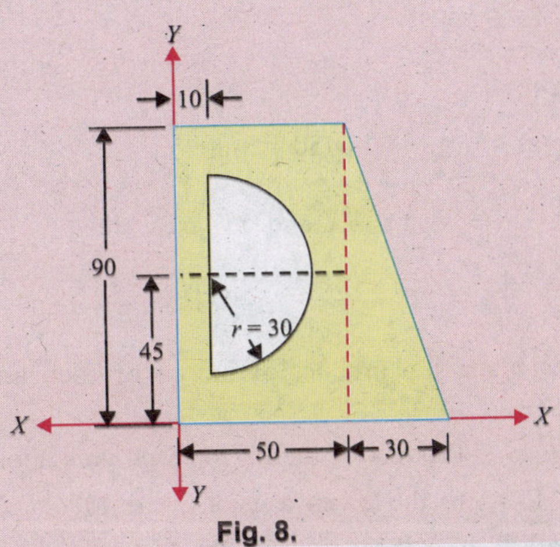

Fig. 8.

Solution. Spilt up the whole plane hatched area into three areas:

(*i*) Rectangle (1) (*ii*) Triangle (2) (*iii*) Semicircle (3)

The position of C.G of each area is shown in Fig. 9.

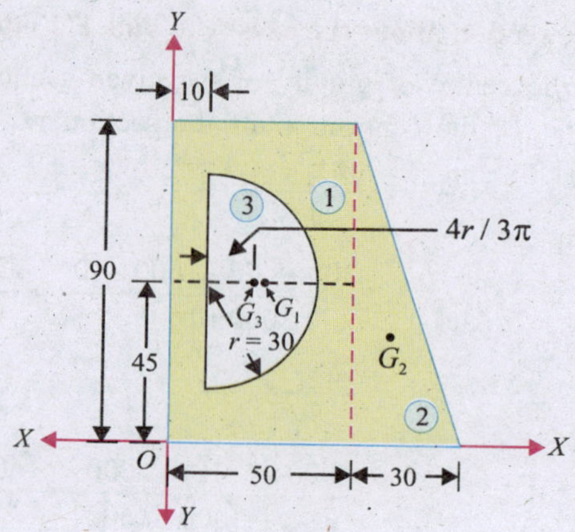

Fig. 9.

The areas and location of their C.G. are as follow:

Rectangle (1):

Area, $A_1 = 90 \times 50 = 4500 \text{ mm}^2$

Coordinates of C.G. (G_1) w.r.t. *XX* and *YY* axes are:

$$x_1 = \frac{50}{2} = +25 \text{ mm}; \quad y_1 = \frac{90}{2} = +45 \text{ mm}$$

Triangle (2):

Area, $A_2 = \frac{1}{2} \times 90 \times 30 = 1350 \text{ mm}^2$

Coordinates of C.G. (G_2) w.r.t. *XX* and *YY* axes are:

$$x_2 = 50 + \frac{30}{3} = +60 \text{ mm}; \quad y_2 = \frac{90}{3} = +30 \text{ mm}$$

Semi-circle (3):

Area, $A_3 = \frac{\pi r^2}{2} = \frac{\pi \times 30^2}{2} = 450\,\pi \text{ mm}^2$

Coordinates of C.G. (G_3) w.r.t. *XX* and *YY* axes are:

$$x_3 = 10 + \frac{4r}{3\pi} = 10 + \frac{4 \times 30}{3\pi} = \left(10 + \frac{40}{\pi}\right) \text{mm}; \quad y_3 = \frac{90}{2} = +45 \text{ mm}$$

Now, we have to calculate the product of inertia of each area about its centroidal axes parallel to the given axes.

Let $I_{\overline{XY}}$ = Product of inertia for each area about its centroidal axes parallel to the given axes *XX* and *YY*

We know form *Parallel axes theorem* that, Product of inertia of rectangle (1) about *XX* and *YY*,

$$(I_{XY})_1 = I_{\overline{XY}} + A_1 ab$$

$$= 0 + [4500 \times (+25) \times (+45)] \quad \ldots(\because a = +25 \text{ and } b = +45)$$

$$= +506.25 \times 10^4 \text{ mm}^4 \ (+)$$

Similarly, Product of inertia of triangle (2) about XX and YY,

$$(I_{XY})_2 = I_{\overline{XY}} + A_2 ab$$

$$= -\frac{(30)^2(90)^2}{72} + 1350 \times (+60) \times (+30)$$

$$\ldots\left[\because \text{For a triangle } I_{\overline{XY}} = -\frac{b^2h^2}{72}\right]$$

$$= -101250 + 2430000 = 2328750 \text{ mm}$$

$$= 232.875 \times 10^4 \text{ mm}^4 \ (+)$$

And, Product of inertia of semi-circle (3) about XX and YY,

$$(I_{XY})_3 = I_{\overline{XY}} + A_3 ab$$

$$= 0 + 450\pi \times \left(10 + \frac{40}{\pi}\right) \times (+45)$$

$$\ldots\left[\because a = \left(10 + \frac{40}{\pi}\right) \text{ and } b = +45\right]$$

$$= 450\,\pi\left(450 + \frac{1800}{\pi}\right)$$

$$= 144.617 \times 10^4 \text{ mm}^4 \ (-)$$

$\therefore$ Total product of inertia of whole plane hatched area about XX and YY,

$$I_{XY\,(\text{total})} = (I_{XY})_1 + (I_{XY})_2 - (I_{XY})_3$$

$$= (506.25 \times 10^4) + (232.875 \times 10^4) - (144.617 \times 10^4)$$

$$= \mathbf{594.508 \times 10^4 \ mm^4} \quad \textbf{Ans.}$$

3. PRINCIPAL AXES AND PRINCIPAL MOMENT OF INERTIA

The axes about which the product of inertia of a plane area is zero, are called the **principal axes**. The moment of inertia about a principal axis is called the **principal moment of inertia**.

For a plane area, moment of inertia is always positive but the product of inertia may be positive, negative or zero.

Notes:

1. The product of inertia is zero about principal axes.
2. The product of inertia about symmetrical axis is zero. Therefore, symmetrical axis is the principal axes of the inertia for the plane figure.

4. *DETERMINATION OF PRINCIPAL MOMENTS OF INERTIA AND DIRECTIONS OF PRINCIPAL AXES **[PTU]**

Consider a plane section of area A with respect to original axes XX and YY passing through 'O'. Also, consider new axes $X'X'$ and $Y'Y'$ inclined at angle θ to the axes XX and YY as shown in Fig. 10.

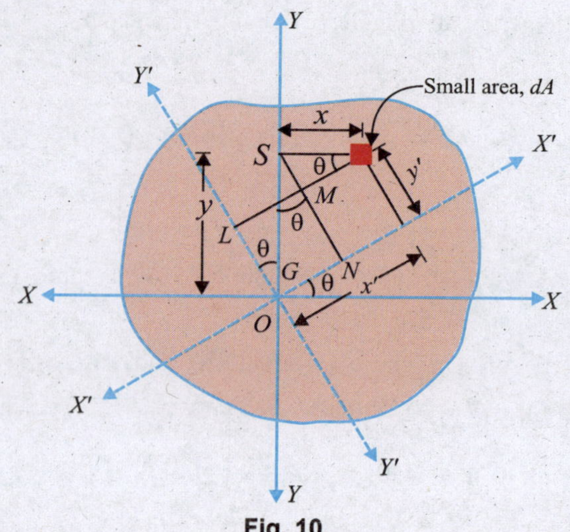

Fig. 10.

Product of Inertia about New Axis $X'X'$ and $Y'Y'$

Let dA = Small area of the plane section,

(x, y) = Coordinates of C.G. of the small area w.r.t. axes XX and YY,

(x', y') = Coordinates of C.G. of the small area w.r.t. axes $X'X'$ and $Y'Y'$ and,

$\angle XOX' = \theta$.

Let us find the relation between coordinates of small area dA in two different systems of axes.

From the geometry of the figure, we find that,

$$x' = LM + x \cos\theta = ON + x\cos\theta = y\sin\theta + x\cos\theta$$

and

$$y' = OL = MN = SN - SM = y\cos\theta - x\sin\theta$$

We know that moments of inertia and product of inertia w.r.t. XX and YY axes are:

$$I_{XX} = \int y^2 dA \qquad \text{...}(ia)$$

$$I_{YY} = \int x^2 dA \qquad \text{...}(ib)$$

and

$$I_{XY} = \int xy dA \qquad \text{...}(ic)$$

∴ Product of inertia of plane area w.r.t. $X'X'$ and $Y'Y'$ axes,

$$I_{X'Y'} = \int x'y' dA$$

$$= \int (y\sin\theta + x\cos\theta)(y\cos\theta - x\sin\theta)\, dA$$

$$= \int (y^2 \sin\theta\cos\theta - xy\sin^2\theta + xy\cos^2\theta - x^2\sin\theta\cos\theta)\, dA$$

$$= \int y^2 \sin\theta\cos\theta\, dA - \int x^2\sin\theta\cos\theta\, dA + \int xy(\cos^2\theta - \sin^2\theta)\, dA$$

$$= \int y^2\left(\frac{2\sin\theta\cos\theta}{2}\right) dA - \int x^2\left(\frac{2\sin\theta\cos\theta}{2}\right) dA + \int xy(\cos^2\theta - \sin^2\theta)\, dA$$

$$= \int y^2 \frac{\sin 2\theta}{2} dA - \int x^2 \frac{\sin 2\theta}{2} dA + \int xy \cos 2\theta \, dA$$

$$= \frac{\sin 2\theta}{2}\left[\int y^2 \, dA - \int x^2 \, dA\right] + \cos 2\theta \int xy \, dA$$

$$= \frac{\sin 2\theta}{2}[I_{XX} - I_{YY}] + \cos 2\theta \, I_{XY} \quad \text{...[Using eqns. } (ia)\text{–}(ic)]$$

or
$$I_{X'Y'} = \frac{I_{XX} - I_{YY}}{2} \sin 2\theta + I_{XY} \cos 2\theta \quad \text{...(3)}$$

This is the *general equation* for finding the product of inertia about any two perpendicular axes if the moments of inertia and product of inertia about two other perpendicular axes are known.

Directions of Principal Axes

We know that *principal axes* for a plane area are the axes about which the product of inertia is zero.

∴ The condition for the axes $X'X'$ and $Y'Y'$ to become the **principal axes** is given by:

$$I_{X'Y'} = 0$$

or
$$\frac{I_{XX} - I_{YY}}{2} \sin 2\theta + I_{XY} \cos 2\theta = 0$$

or
$$\frac{\sin 2\theta}{\cos 2\theta} = -\frac{2I_{XY}}{I_{XX} - I_{YY}}$$

or
$$\tan 2\theta = \frac{2I_{XY}}{I_{YY} - I_{XX}} \quad \text{...(4)}$$

The equation (4) locates the directions of the principal axes by giving two values of θ differing by 90°.

Principal Moments of Inertia

The moment of inertia of plane area w.r.t. $X'X'$ axis,

$$I_{X'X'} = \int y'^2 \, dA$$

$$= \int (y \cos \theta - x \sin \theta)^2 \, dA$$

$$= \int (y^2 \cos^2 \theta + x^2 \sin^2 \theta - 2xy \sin \theta \cos \theta) \, dA$$

$$= \cos^2 \theta \int y^2 dA + \sin^2 \theta \int x^2 dA - (2 \sin \theta \cos \theta) \int xy \, dA$$

$$= I_{XX} \cos^2 \theta + I_{YY} \sin^2 \theta - I_{XY} \sin 2\theta \quad \text{...[Using eqns. } (ia)\text{–}(ic)]$$

$$= I_{XX}\left(\frac{\cos 2\theta + 1}{2}\right) + I_{YY}\left(\frac{1 - \cos 2\theta}{2}\right) - I_{XY} \sin 2\theta$$

or
$$I_{X'X'} = \frac{I_{XX} + I_{YY}}{2} + \frac{I_{XX} - I_{YY}}{2} \cos 2\theta - I_{XY} \sin 2\theta \quad \text{...(5)}$$

The moment of inertia of plane area w.r.t. $Y'Y'$ axis is obtained by changing θ to (θ + π/2) in equation (5). Changing θ to (θ /π/2) in equation (5), we get

$$I_{Y'Y'} = \frac{I_{XX} + I_{YY}}{2} - \frac{I_{XX} - I_{YY}}{2}\cos 2\theta + I_{XY}\sin 2\theta \qquad ...(6)$$

Differentiating equation (5) or equation (6) w.r.t. θ and putting $\frac{d}{d\theta}(I_{X'X'}) = 0$, we get the condition for maximum or minimum moments of inertia.

For maximum or minimum moments of inertia, we have

$$\frac{d}{d\theta}(I_{X'X'}) = 0$$

or $$\frac{d}{d\theta}\left(\frac{I_{XX} + I_{YY}}{2} + \frac{I_{XX} - I_{YY}}{2}\cos 2\theta - I_{XY}\sin 2\theta\right) = 0$$

or $$0 + \frac{I_{XX} - I_{YY}}{2}(-2\sin 2\theta) - I_{XY}(2\cos 2\theta) = 0$$

or $$(I_{YY} - I_{XX})\sin 2\theta - 2\,I_{XY}\cos 2\theta = 0$$

or $$\tan 2\theta = \frac{2I_{XY}}{I_{YY} - I_{XX}} \qquad ...(7)$$

The equation (7) is same as equation (4). This implies that the **axes of minimum and maximum moments of inertia are also the principal axes**. Thus, axes $X'X'$ and $Y'Y'$ will become principal axes if θ is given by equation (4) or (7). This equation will give two values of θ differing by 90°.

Let $UU = X'X'$ = Major principal axis, and
$VV = Y'Y'$ = Minor principal axis.

The value of maximum moment of inertia I_{UU} is given by,

$$I_{UU} = I_{X'X'} = \left(\frac{I_{XX} + I_{YY}}{2}\right) + \sqrt{\left(\frac{I_{XX} - I_{YY}}{2}\right)^2 + (I_{XY})^2} \qquad ...(8)$$

Similarly, value of minimum moment of inertia I_{VV} is given by,

$$I_{VV} = I_{Y'Y'} = \left(\frac{I_{XX} + I_{YY}}{2}\right) - \sqrt{\left(\frac{I_{XX} - I_{YY}}{2}\right)^2 + (I_{XY})^2} \qquad ...(9)$$

Equations (8) and (9) can be used to find the values of principal moment of inertia (I_{UU} and I_{VV}) and while equation (4) will give the inclination of the principal axes.

Note:

Adding equations (8) and (9), we get

$$I_{UU} + I_{VV} = I_{XX} + I_{YY} \qquad ...(10)$$

Adding equations (5) and (6), we get

$$I_{X'X'} + I_{Y'Y'} = I_{XX} + I_{YY} \qquad ...(11)$$

The equations (10) and (11) show that the **sum of moments of inertia about any two perpendicular axes remains constant as the axes are rotated through any angle.** Hence, **sum of moments of inertia of a plane area is independent of the orientation of the axes.**

5. *MOHR'S CIRCLE FOR PRINCIPAL MOMENTS OF INERTIA

Mohr's circle for principal moments of inertia can be drawn by using the following procedure:

1. Take origin O and from it measure $OC = I_{YY}$ and $OD = I_{XX}$ along the horizontal I-axis as shown in Fig. 11.
2. At point C, draw upward perpendicular CF equal to I_{XY} if I_{XY} is positive (and a downward perpendicular if I_{XY} is negative).
3. Bisect CD at point E.
4. Taking point E as the centre and EF as radius, draw a circle cutting OC and OD extended at A and B respectively.

Then, Mohr's circle shown in Fig. 11 gives

OA = Minor principal moment of inertia I_{VV} ,

OB = Major principal moment of inertia I_{UU}, and

Direction AF = Direction of minor principal axis VV and a direction perpendicular to it, is the direction of major principal axis UU.

NOTE: I_{XX} and I_{YY} *are always positive.*

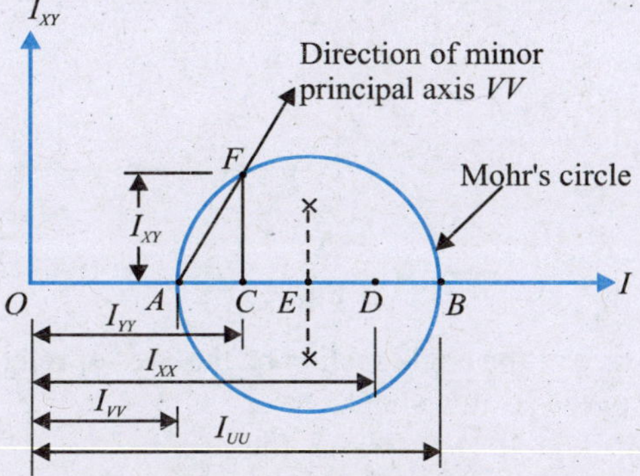

Fig. 11. Mohr's circle

Example 4. *An angle section is shown in Fig. 12. Determine:*

(a) the moments of inertia about its centroidal axes,

(b) the product of inertia about its centroidal axes,

(c) the principal moments of inertia and directions of principal axes by (i) Calculations and (ii) Mohr's circle construction. **(RGPV-MP)**

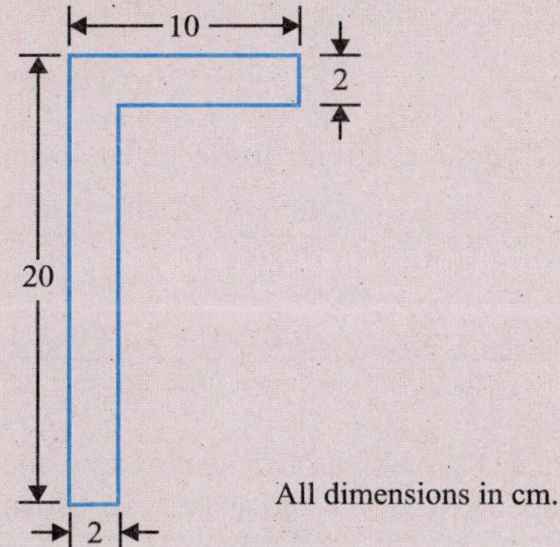

Fig. 12.

Solution.

First of all, let us find the centroid (i.e., centre of gravity) of the given angle section. As the section is not symmetrical about any section, therefore we have to find out the values of $\bar{x}$ and $\bar{y}$ for the angle section. Split up the section into two rectangles (1) and (2) separated by a dotted line as shown in Fig. 13.

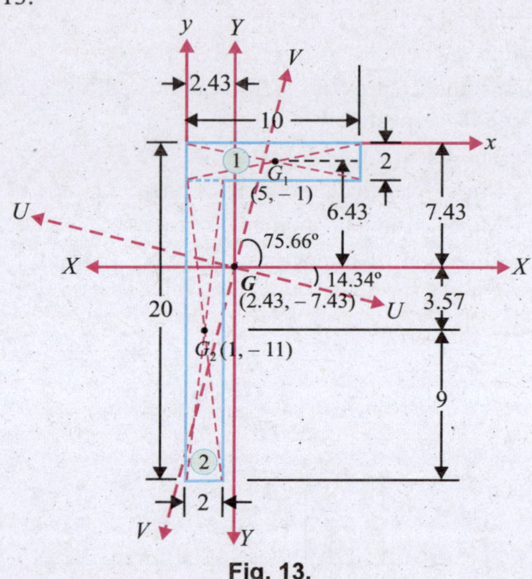

Fig. 13.

Let left face and upper face of the angle section be the axes of reference x and y axes. Now, let us find the centre of gravity of the section about xy axes.

The areas and location of their C.G. are as follow:

***Rectangle* (1):**

Area, $A_1 = 10 \times 2 = 20 \text{ cm}^2$

Coordinates of C.G. (G_1) w.r.t. xy axes are:

$$x_1 = +5 \text{ cm}; \quad y_1 = -1 \text{ cm}$$

***Rectangle* (2):**

Area, $A_2 = 18 \times 2 = 36 \text{ cm}^2$

Coordinates of C.G. (G_2) w.r.t. xy axes are:

$$x_2 = \frac{2}{2} = +1 \text{ cm}; \quad y_2 = -\left(2 + \frac{18}{2}\right) = -11 \text{ cm}$$

We know that, coordinates of the centroid of the given section w.r.t. xy axes,

$$\bar{x} = \frac{A_1 x_1 + A_2 x_2}{A_1 + A_2} = \frac{20 \times 5 + 36 \times 1}{20 + 36} = \frac{136}{56} = 2.43 \text{ cm}$$

and
$$\bar{y} = \frac{A_1 y_1 + A_2 y_2}{A_1 + A_2} = \frac{20 \times (-1) + 36 \times (-11)}{20 + 36} = -\frac{416}{56} = -7.43 \text{ cm}$$

The centroidal axes XX and YY pass through centroid of the section at G (2.43, –7.43) and parallel to the reference axes x-axis and y-axis respectively, as shown in Fig. 13.

(a) Moments of inertia about its centroidal axes *XX* and *YY*

We know that, Moment of inertia of the given section about centroidal axis *XX*,

$$I_{XX} = \left[\frac{10 \times 2^3}{12} + 10 \times 2 \times (7.43 - 1)^2\right] + \left[\frac{2 \times 18^3}{12} + 2 \times 18 \times (11 - 7.43)^2\right]$$

$$= 833.56 + 1430.82 = \mathbf{2264.38\ cm^4} \quad \textbf{Ans.}$$

Similarly, Moment of inertia of the given section about centroidal axis *YY*,

$$I_{YY} = \left[\frac{2 \times 10^3}{12} + 2 \times 10 \times (5 - 2.43)^2\right] + \left[\frac{18 \times 2^3}{12} + 18 \times 2 \times (2.43 - 1)^2\right]$$

$$= 298.76 + 85.61 = \mathbf{384.37\ cm^4} \quad \textbf{Ans.}$$

(b) Product of inertia (I_{XY}) about centroidal axes *XX* and *YY*

Rectangle (1):

$A_1 = 10 \times 2 = 20\ \text{cm}^2$

a_1 = Distance of C.G. (G_1) from *YY* axis $= 5 - 2.43 = +\,2.57$ cm

b_1 = Distance of C.G. (G_1) from *XX* axis $= 7.43 - 1 = +\,6.43$ cm

$I_{\overline{XY}}$ = Product of inertia of rectangle (1) about its centroidal axis passing through G_1

= 0

We know that, Product of inertia of rectangle (1) about *XX* and *YY* axes,

$$(I_{XY})_1 = I_{\overline{XY}} + A_1 a_1 b_1$$

$$= 0 + (10 \times 2) \times (+\,2.57) \times (+\,6.43) = 330.502\ \text{cm}^4$$

Rectangle (2):

$A_2 = 18 \times 2 = 36\ \text{cm}^2$

a_2 = Distance of C.G. (G_2) from *YY* axis $= -\,(2.43 - 1) = -\,1.43$ cm

b_2 = Distance of C.G. (G_2) from *XX* axis $= -\,(11 - 7.43) = -\,3.57$ cm

$I_{\overline{XY}}$ = Product of inertia rectangle (2) about its centroidal axis passing through G_2

= 0

Similarly, Product of inertia of rectangle (2) about *XX* and *YY* axes,

$$(I_{XY})_2 = I_{\overline{XY}} + A_2 a_2 b_2$$

$$= 0 + (18 \times 2)\,(-\,1.43) \times (-\,3.57) = 183.784\ \text{cm}^4$$

∴ Total Product of inertia of angle section about centroidal axes *XX* and *YY*,

$$I_{XY} = (I_{XY})_1 + (I_{XY})_2$$

$$= 330.502 + 183.784 = \mathbf{514.286\ cm^4} \quad \textbf{Ans.}$$

(c) Directions of principal axes and Principal moments of inertia

(i) By Calculations

Directions of principal axes

We know that, the directions/inclination angle (θ) of principal axes with the given axes can be calculated by the relation:

$$\tan 2\theta = \frac{2\,I_{XY}}{I_{YY} - I_{XX}}$$

$$= \frac{2 \times 514.286}{384.37 - 2264.38} = -\,0.5471$$

or $\quad 2\,\theta = \tan^{-1}(-\,0.5471) = -\,28.64^\circ$ or $(180^\circ - 28.64)$

or $\quad \theta = \mathbf{-14.34^\circ}$ **or** $\mathbf{75.66^\circ}$ **Ans.**

The +ve angle taken anti-clockwise and –ve angle is taken clockwise to the given axes *XX* and *YY*. Thus, the principal axes *UU* axis is inclined at – 14.34° to *XX* axis and *VV axis is inclined at* 75.66° *to X-axis* as shown in Fig. 13.

Principal moments of inertia

The moments of inertia about the principal axes UU and VV are the principal moments of inertia.

We know that, Major principal moment of inertia,

$$I_{UU} = \left(\frac{I_{XX} + I_{YY}}{2}\right) + \sqrt{\left(\frac{I_{XX} - I_{YY}}{2}\right)^2 + (I_{XY})^2}$$

$$= \left(\frac{2264.38 + 384.37}{2}\right) + \sqrt{\left(\frac{2264.38 - 384.37}{2}\right)^2 + (514.28)^2}$$

$$= 1324.38 + 1071.49 = \mathbf{2395.87\ cm^4} \quad \textbf{Ans.}$$

Similarly, Minor moment of inertia I_{VV},

$$I_{VV} = \left(\frac{I_{XX} + I_{YY}}{2}\right) - \sqrt{\left(\frac{I_{XX} - I_{YY}}{2}\right)^2 + (I_{XY})^2}$$

$$= \left(\frac{2264.38 + 384.37}{2}\right) - \sqrt{\left(\frac{2264.38 - 384.37}{2}\right)^2 + (514.28)^2}$$

$$= 1324.38 - 1071.49 = \mathbf{252.89\ cm^4} \quad \textbf{Ans.}$$

Alternative Method:

As we know the inclination of *UU* or *VV* axis, therefore by substituting $\theta = -\,14.34^\circ$ in the general equation for product of inertia.

The maximum moment of inertia i.e., major moment of inertia is given as:

$$I_{UU} = I_{\max} = \frac{I_{XX} + I_{YY}}{2} + \frac{I_{XX} - I_{YY}}{2}\cos 2\theta - I_{XY}\sin 2\theta$$

$$= \left(\frac{2264.38 + 384.37}{2}\right) + \left(\frac{2264.38 - 384.37}{2}\right)\cos\,[2\times(-14.34^\circ)]$$

$$-514.28\sin[2\times(-14.34^\circ)]$$

$$= 1324.38 + 824.68 + 246.81 = \mathbf{2395.87\ cm^4}$$

= **Major principal moment of inertia**

Similarly, the minimum moment of inertia i.e., minor moment of inertia is given as:

$$I_{VV} = I_{\min} = \frac{I_{XX} + I_{YY}}{2} - \frac{I_{XX} - I_{YY}}{2}\cos 2\theta + I_{XY}\sin 2\theta$$

$$= \left(\frac{2264.38 + 384.37}{2}\right) - \left(\frac{2264.38 - 384.37}{2}\right)\cos\,[2\times(-14.34^\circ)]$$

$$+\,514.28\,\sin\,[2\times(-14.34^\circ)]$$

$$= 1324.38 - 824.68 - 246.81 = \mathbf{252.89\ cm^4}$$

= **Minor principal moment of inertia**

(ii) By Mohr's Circle Construction

The steps for constructing Mohr's circle are given below:

1. Take origin O.
   From O, measure $OC = I_{YY} = 384.37\ \text{cm}^4$ and $OD = I_{XX} = 2264.38\ \text{cm}^4$ along the horizontal I-axis as shown in Fig. 14.
2. At point C, draw upward perpendicular CF equal to $I_{XY} = 514.28\ \text{cm}^4$ (as I_{XY} is +ve).
3. Bisect CD at point E.
4. Taking point E as the centre and EF as radius, draw a circle cutting OC and OD extended at A and B respectively.

Then,

$$I_{UU} = OB = 2395.87\ \text{cm}^4$$

and, $$I_{VV} = OA = 252.89\ \text{cm}^4$$

Also, minor principal axis is inclined at 75.5°.

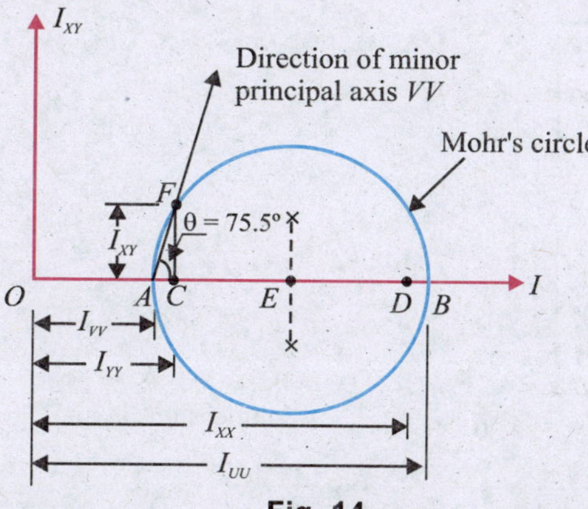

Fig. 14.

EXERCISE

1. Find the product of inertia of a plane area about the axes XX and YY shown in Fig. 15. All dimensions are in mm. [**Ans.** $229.5 \times 10^4\ \text{mm}^4$]

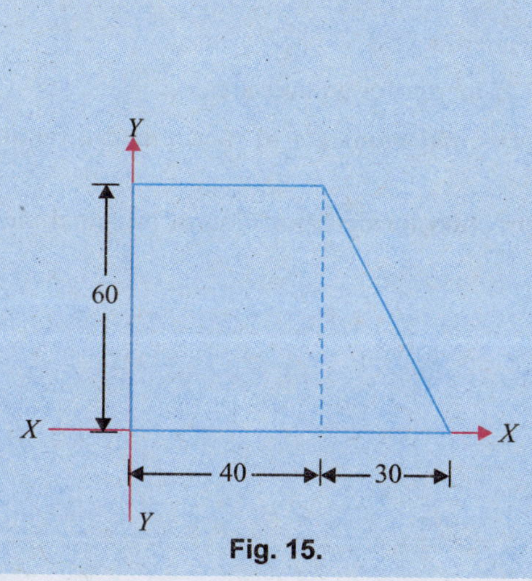

Fig. 15.

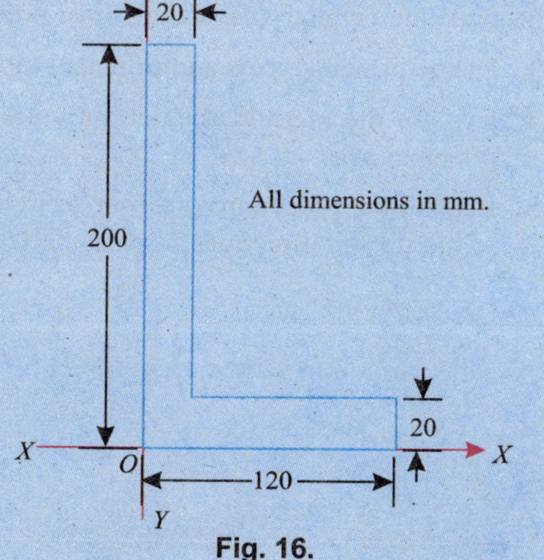

Fig. 16.

2. Find the product of inertia for the angle section about the axes XX and YY shown in Fig. 16. Also, find the product of inertia about parallel axes through its centroid. All dimensions are in mm.

[**Ans.** 540×10^4 mm^4, -720×10^4 mm^4]

3. An angle section is shown in Fig. 17. Determine:
   (*a*) the moments of inertia about its centroidal axes,
   (*b*) the product of inertia about its centroidal axes,
   (*c*) the principal moments of inertia and directions of principal axes by
   (i) Calculations and (ii) Mohr's circle construction.

Note: All dimensions are in mm.

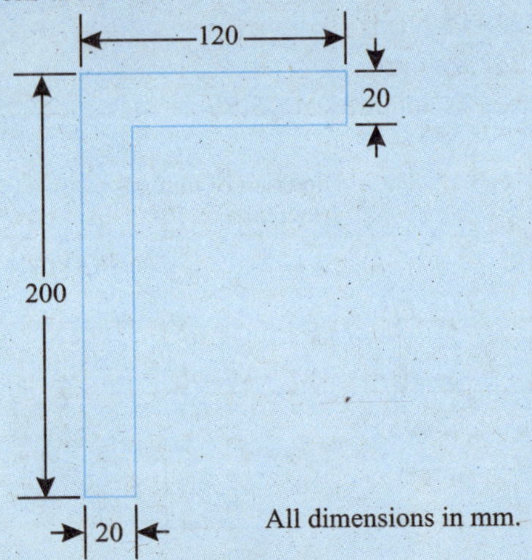

Fig. 17.

[**Ans.** (*a*) 2420×10^4 mm^4, 660×10^4 mm^4 (*b*) $+720 \times 10^4$ mm^4
(*c*) 2267×10^4 mm^4, 403×10^4 mm^4, $-19.7°$ or $70.3°$]

QUESTIONS

1. Define the term "Product of Inertia"?
2. State the parallel axes theorem for product of inertia.
3. Define principal axes and principal moments of inertia for a plane area.
4. Explain the method for determination of principal moments of inertia and directions of principal axes.
5. Explain the procedure of Mohr's circle construction for determination of principal moments of inertia and directions of principal axes.

CHAPTER

36 Rectangular and Cylindrical Coordinate System in Dynamics

Contents

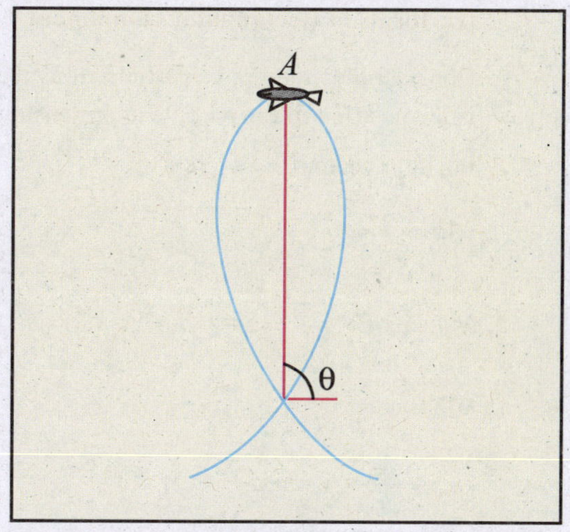

1. INTRODUCTION

A particle is said to be in **motion**, if it changes its position with respect to its surroundings.

A path traced by a particle during its motion is known as **trajectory**. It may be a straight line or a curve. The motion of a particle along a straight line is known as **rectilinear motion**, whereas the motion of a particle along a curved path other than a straight line, is known as **curvilinear motion.**

1.1. Types of Curvilinear Motions

There are two types of curvilinear motions:

1. **Plane curvilinear motion:** If a curved path lies in a plane (two-dimensional), it is called *plane curvilinear motion.*

2. **Space curvilinear motion:** If a curved path lies in a space (three-dimensional), it is called *space curvilinear motion.*

The plane curvilinear motion can be described by using *rectangular coordinate system, normal and tangential coordinate system* and *radial and transverse coordinate system* while space curvilinear motion can be described by using *rectangular coordinate system, cylindrical coordinate system,* and spherical coordinate system.

2. PLANE CURVILINEAR MOTION

The plane curvilinear motion can be described by using following coordinate systems:

1. In Rectangular Coordinates (x, y)
2. In Polar Coordinates (r, θ)
3. In Tangential and Normal coordinates (t, n)

A. In Rectangular Coordinates (x, y):
[In Terms of Rectangular components]

Consider the motion of a particle along a curved path in xy-plane as shown in Fig. 1. If $\vec{r}$ is the position of the particle at any instant, then

1. Position vector $(\vec{r}) = x\hat{i} + y\hat{j}$,

where, $r = |\vec{r}| = \sqrt{x^2 + y^2}$

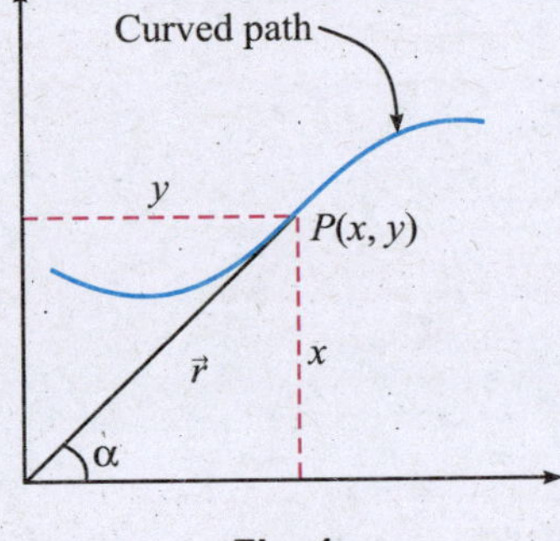

Fig. 1.

2. Velocity $(\vec{v}) = \dfrac{d\vec{r}}{dt} = v_x\hat{i} + v_y\hat{j}$;

where $v_x = \dfrac{dx}{dt} = \dot{x}$,

$v_y = \dfrac{dy}{dt} = \dot{y}$ and

$v = \sqrt{v_x^2 + v_y^2}$

2a. If α is the angle made by the resultant velocity (v) with x-axis, then

$$\tan\alpha = \frac{v_y}{v_x} \quad \text{or} \quad \alpha = \tan^{-1}\left(\frac{v_y}{v_x}\right)$$

3. Acceleration $(\vec{a}) = \dfrac{d\vec{v}}{dt} = a_x\hat{i} + a_y\hat{j}$;

where, $a_x = \dfrac{dv_x}{dt} = \dfrac{d^2x}{dt^2} = \ddot{x}$, $a_y = \dfrac{dv_y}{dt} = \dfrac{d^2y}{dt^2} = \ddot{y}$ and $a = \sqrt{a_x^2 + a_y^2}$

3a. If β is the angle made by the resultant acceleration (α) with x-axis, then

$$\tan\beta = \frac{a_y}{a_x} \quad \text{or} \quad \beta = \tan^{-1}\left(\frac{a_y}{a_x}\right)$$

Example 1. *The motion of a particle is defined by the equations :*

$$x = t^3 + t + 2 \text{ and } y = 3t^2 - 3t + 1;$$

where x and y are in metres and t is in seconds.

(*a*) *Write the expressions for position vector, velocity vector and acceleration vector.*
(*b*) *Calculate the velocity and acceleration of the particle at t = 2 seconds.*

Solution. (*a*) Given : $x = t^3 + t + 2, \quad y = 3t^2 - 3t + 1$

$$v_x = \frac{dx}{dt} = 3t^2 + 1 \qquad v_y = \frac{dy}{dt} = 6t - 3$$

$$a_x = \frac{dv_x}{dt} = 6t \qquad a_y = \frac{dv_y}{dt} = 6$$

(*a*) *Expressions for position vector, velocity vector and acceleration vector*

At any time t, the motion of the particle in vectorial form can be written as:

$$\text{Position vector } (\vec{r}) = x\hat{i} + y\hat{j} = (t^3 + t + 2)\hat{i} + (3t^2 - 3t + 1)\hat{j}$$

$$\text{Velocity vector } (\vec{v}) = v_x\hat{i} + v_y\hat{j} = (3t^2 + 1)\hat{i} + (6t - 3)\hat{j}$$

$$\text{Acceleration } (\vec{a}) = a_x\hat{i} + a_y\hat{j} = 6t\hat{j} + 6\hat{j}$$

(*b*) *Velocity and acceleration of the particle at t = 2 seconds*

Velocity of particle at $t = 2$ seconds

$$\vec{v} = (3 \times 2^2 + 1)\,\hat{i} + (6 \times 2 - 3)\,\hat{j} = 13\hat{i} + 9\hat{j} = v_x\hat{i} + v_y\hat{j}$$

$\therefore \qquad v_x = 13$ m/s and $v_y = 9$ m/s

$$\text{Resultant velocity } (v) = \sqrt{v_x^2 + v_y^2} = \sqrt{13^2 + 9^2} = 15.81 \text{ m/s}$$

If α is the angle made by the resultant velocity (v) with x-axis, then

$$\tan\alpha = \frac{v_y}{v_x} = \frac{9}{13} = 0.692 \text{ or } \alpha = \tan^{-1} 0.692 = 34.68°$$

Since, v_x is positive and v_y is also positive, therefore the resultant lies in the first quadrant. The components of velocity vector have been shown in Fig. 2(a).

Further, acceleration of particle at t = 2 seconds,

$$\vec{a} = (6 \times 2)\hat{i} + 6\hat{j} = 12\hat{i} + 6\hat{j} = a_x\hat{i} + a_y\hat{j}$$

$\therefore \qquad a_x = 12 \text{ m/s}^2$ and $a_y = 6 \text{ m/s}^2$

$$\text{Resultant acceleration } (a) = \sqrt{a_x^2 + a_y^2} = \sqrt{12^2 + 6^2} = 13.42 \text{ m/s}^2$$

If β is the angle made by the resultant acceleration (a) with x-axis, then

$$\tan\beta = \frac{a_y}{a_x} = \frac{6}{12} = 0.5 \quad \text{or} \quad \beta = \tan^{-1} 0.5 = 26.57°$$

Since a_x is positive and a_y is also positive, the resultant acceleration lies in the first quadrant. The components of acceleration vector have been shown in Fig. 2(*b*).

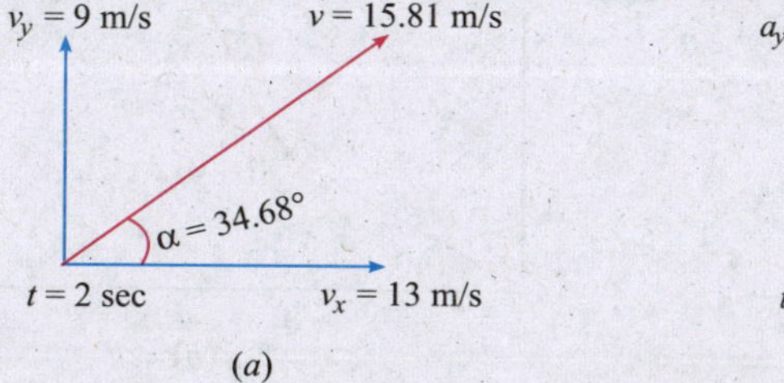

(*a*)

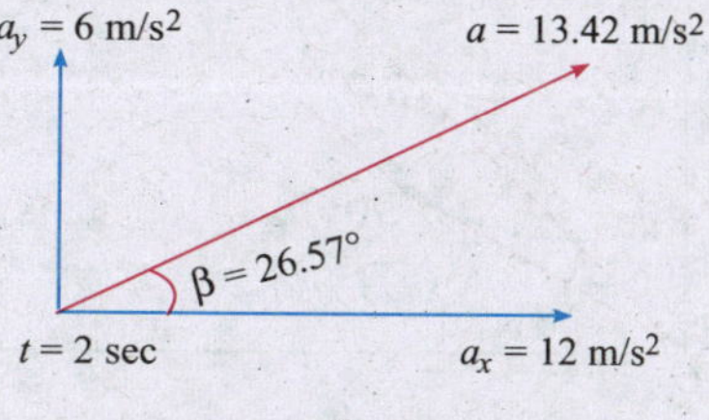

(*b*)

Fig. 2.

B. In Polar Coordinates (r, θ) or Radial and Transverse Coordinates:
[In Terms of Radial and Transverse components]

In this coordinate system,

(*a*) The position, velocity and acceleration of the particle are expressed in terms of polar coordinates *i.e.*, r and θ.

(*b*) The velocity and acceleration of the particle are expressed in the radial and transverse directions.

Consider the motion of a particle along a curved path in *xy*-plane as shown in Fig. 3(*a*). If $\vec{r}$ is the position of the particle at any instant, then

1. Differential position vector $(d\vec{r}) = d\vec{r}_r + d\vec{r}_\theta$

Where, $dr_r = dr$ and $dr_\theta = rd\theta$ as shown in Fig. 3(*b*).

2. Velocity $(\vec{v}) = \vec{v}_r + \vec{v}_\theta$;

where,

Radial component of velocity, $v_r = \dfrac{dr}{dt} = \dot{r}$,

Transverse component of velocity, $v_\theta = r\left(\dfrac{d\theta}{dt}\right) = r\dot{\theta}$, and

Resultant velocity, $v = \sqrt{v_r^2 + v_\theta^2}$

3. Acceleration $(\vec{a}) = \vec{a}_r + \vec{a}_\theta$;

where,

Radial component of acceleration, $a_r = \dfrac{d^2r}{dt^2} - r\left(\dfrac{d\theta}{dt}\right)^2 = \ddot{r} - r\dot{\theta}^2$,

Transverse component of acceleration, $a_\theta = r\dfrac{d^2\theta}{dt^2} + 2\dfrac{dr}{dt}\dfrac{d\theta}{dt} = r\ddot{\theta} + 2\dot{r}\dot{\theta}$, and

Resultant acceleration, $a = \sqrt{a_r^2 + a_\theta^2}$

4. Rectangular coordinates in terms of polar coordinates:

$$x = r\cos\theta,\ y = r\sin\theta$$

5. Polar coordinates in terms of rectangular coordinates:

$$r = \sqrt{x^2 + y^2},\ \theta = \tan^{-1}\left(\frac{y}{x}\right)$$

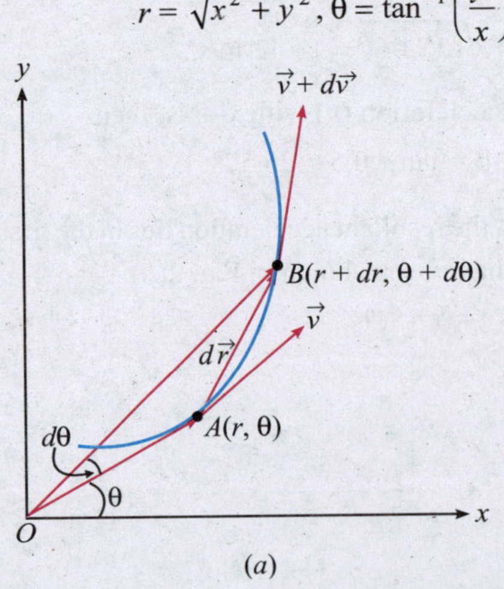

(*a*)

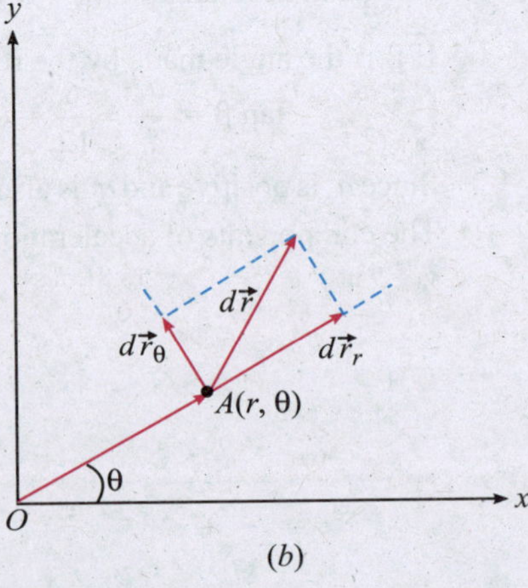

(*b*)

Fig. 3. Polar coordinates (r, θ)

Example 2. *The coordinates of a particle moving in xy-plane are defined by the relations :*

$$r = 4\theta \text{ and } \theta = \frac{t^2}{4} - 1,$$ *where r is in metres, θ is in radians and t is in seconds.*

Calculate the velocity and acceleration of the particle at θ = π/4.

Solution. Given : $\theta = \frac{t^2}{4} - 1$ and $r = 4\theta = 4\left(\frac{t^2}{4} - 1\right) = t^2 - 4.$

Differentiating w.r.t. t, we get

$$\frac{d\theta}{dt} = \frac{t}{2}; \qquad \frac{dr}{dt} = 2t$$

$$\frac{d^2\theta}{dt^2} = \frac{1}{2}; \qquad \frac{d^2r}{dt^2} = 2$$

When $\theta = \frac{\pi}{4}$, then $\frac{\pi}{4} = \frac{t^2}{4} - 1$ or $t = \sqrt{\pi + 4} = 2.67$ sec

Velocity of particle at t = 2.67 sec

Radial velocity, $v_r = \frac{dr}{dt} = 2t = 2 \times 2.67 = 5.34$ m/s

Transverse velocity, $v_\theta = r\frac{d\theta}{dt} = (t^2 - 4)\left(\frac{t}{2}\right) = \frac{t^3}{2} - 2t$

$$= \frac{2.67^3}{2} - 2(2.67) = 9.52 - 5.34 = 4.18 \text{ m/s}$$

Resultant velocity, $v = \sqrt{v_r^2 + v_\theta^2} = \sqrt{(5.34)^2 + (4.18)^2} = 6.75$ m/s

If α is the angle made by the resultant velocity with the radial direction, then

$$\tan\alpha = \frac{4.18}{5.34} = 0.783 \quad \text{or} \quad \alpha = \tan^{-1}0.783 = 38.06°$$

Acceleration of particle at t = 2.67 sec

Radial acceleration, $a_r = \frac{d^2r}{dt^2} - r\left(\frac{d\theta}{dt}\right)^2 = 2 - (t^2 - 4)\left(\frac{t}{2}\right)^2 = 2 - (t^2 - 4)\left(\frac{t^2}{4}\right)$

$$= 2 - \frac{t^4}{4} + t^2 = 2 - \frac{(2.67)^4}{4} + (2.67)^2$$

$$= 2 - 12.71 + 7.13 = -3.58 \text{ m/s}^2$$

Transverse acceleration, $a_\theta = r\frac{d^2\theta}{dt^2} + 2\frac{dr}{dt}\frac{d\theta}{dt} = \left[(t^2 - 4) \times \frac{1}{2}\right] + \left[2 \times 2t \times \frac{t}{2}\right]$

$$= \frac{t^2}{2} - 2 + 2t^2 = \frac{5t^2}{2} - 2 = \frac{5}{2} \times (2.67)^2 - 2$$

$$= 17.82 - 2 = 15.82 \text{ m/s}^2$$

Resultant acceleration, $a = \sqrt{a_r^2 + a_\theta^2} = \sqrt{(-3.58)^2 + (15.82)^2} = 16.22$ m/s²

If β is the angle made by the resultant acceleration with the radial direction, then

$$\therefore \quad \tan\beta = \frac{15.82}{3.58} = 4.419 \quad \text{or} \quad \beta = \tan^{-1}4.419 = 77.25°$$

C. In Tangential and Normal coordinates (*t*, *n*): [In Terms of Tangential and Normal components]

In this system, velocity and acceleration of a moving particle are expressed in tangential and normal directions.

Consider the motion of a particle along a curved path in *xy*-plane as showm in Fig. 4(a). If δs is the small arc length traverse by particle between two points along the path in a small interval of time δt as shown in Fig. 4(b), then

1. Velocity ($\vec{v}$) = $\vec{v}_t$;

where, Tangential component of velocity, $v_t = \dfrac{ds}{dt} = \dot{s}$

Resultant velocity, $v = v_t$

Notes:

Normal component of velocity is zero *i.e.*, $v_n = 0$.

2. Acceleration ($\vec{a}$) = $\vec{a}_t + \vec{a}_n$;

where,

Tangential component, $a_t = \dfrac{dv}{dt}$ ...(Represents rate change of speed)

Normal component, $a_n = \dfrac{v^2}{\rho}$ ...(Represents change in direction)

Resultant acceleration, $a = \sqrt{a_t^2 + a_n^2}$

The tangential acceleration is due to change in magnitute of velocity while *normal acceleration* is due to change in direction of velocity.

2a. If α is the angle made by the resultant with the normal acceleration, then

$$\tan\alpha = \frac{a_t}{a_n} \quad \text{or} \quad \alpha = \tan^{-1}\left(\frac{a_t}{a_n}\right)$$

2b. If β is the angle made by the resultant with the tangential acceleration, then

$$\tan\beta = \frac{a_n}{a_t} \quad \text{or} \quad \beta = \tan^{-1}\left(\frac{a_n}{a_t}\right)$$

Notes:

(*a*) If the magnitude of velocity between two points is constant, then $a_t = 0$.

(*b*) If the direction of motion of particle does not change (in case of rectilinear motion), then radius of curvature $\rho \to \infty$. Therefore $a_n = 0$.

(*c*) If the radius of curvature is constant *i.e.*, particle is moving along a circle, then radius of curvature (ρ) is equal to radius of circle (r). Hence,

Normal acceleration, $a_n = \dfrac{v^2}{r}$.

This normal acceleration is known as **centripetal acceleration** and directed towards the center of the circle.

(*d*) If the equation of a curve path is given by: $y = f(x)$, then radius of curvature (ρ) is obtained by the following relation:

$$\rho = \frac{\left[1+\left(\frac{dy}{dx}\right)^2\right]^{3/2}}{\frac{d^2y}{dx^2}}$$

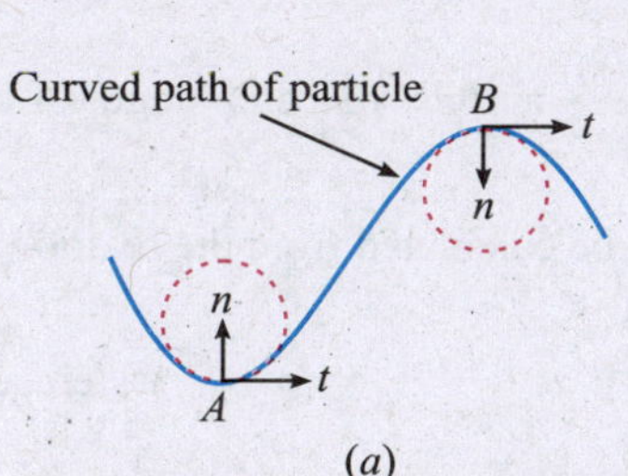

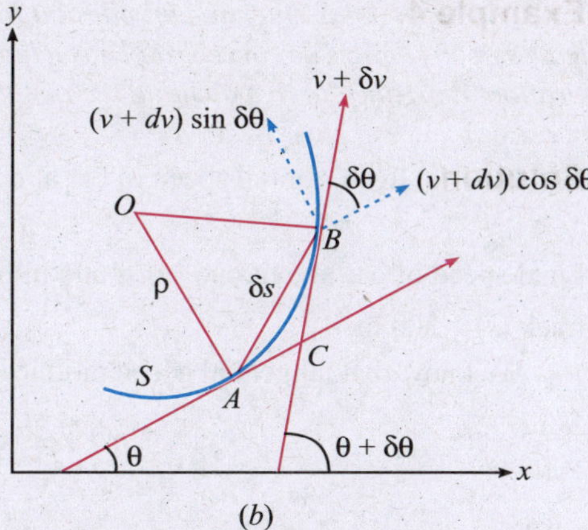

Fig. 4. Tangential and Normal coordinates (*t*, *n*)

Example 3. *An automobile is moving at 90 km/hr on a curved track of radius 200 m and suddenly applies the brakes that causes the speed to decrease to 63 km/hr at a constant rate in 6 seconds. Determine the tangential and normal components of acceleration immediately after the application of brakes and 5 seconds later.*

Solution. Given : Initial speed (u) = 90 km/hr = $\frac{90 \times 1000}{3600}$ = 25 m/s; Radius of curved track (r) = 200 m and final speed (v) = 63 km/hr = $\frac{63 \times 1000}{3600}$ = 17.5 m/s.

Tangential and normal components of acceleration immediately after the application of brakes

We know that tangential acceleration of automobile (a_t),

$$a_t = \frac{dv}{dt} = \text{Rate change of speed}$$

$$= \frac{v-u}{dt} = \frac{17.5-25}{6} = -1.25 \text{ m/s}^2 \quad \text{...(Minus sign shows retardation)}$$

At the time of application of brakes, u = 25 m/s, therefore normal acceleration of automobile,

$$a_n = \frac{u^2}{r} = \frac{25^2}{200} = 3.125 \text{ m/s}^2$$

$$\therefore \quad \text{Total acceleration } (a) = \sqrt{a_n^2 + a_t^2} = \sqrt{(3.125)^2 + (-1.25)^2} = 3.366 \text{ m/s}^2$$

Tangential and normal components of acceleration 5 seconds after the application of brakes

We know that tangential acceleration 5 seconds after the application of brakes,

$$a_t = -1.25 \text{ m/s}^2$$

And velocity of automobile 5 seconds after the application of brakes,

$$v' = u + at = 25 + (-1.25 \times 5) = 18.75 \text{ m/s} \quad \text{...}(\because a = a_t = -1.25 \text{ m/s}^2)$$

Therefore normal acceleration of automobile (a_n),

$$a_n = \frac{(v')^2}{r} = \frac{(18.75)^2}{200} = 1.758 \text{ m/s}^2$$

$\therefore$ Total acceleration $(a) = \sqrt{a_n^2 + a_t^2} = \sqrt{(1.758)^2 + (-1.25)^2} = 2.157 \text{ m/s}^2$

Example 4. *A car enters a curved road (in the form of quarter of a circle) at 40 km/hr and then leaves at 60 km/hr. The curved road has a length of 300 m. If the car is travelling with constant acceleration along the curve, calculate the resultant acceleration at both ends of the curve.*

Solution. Given : Initial speed of car at entry end (u) = 40 km/hr = $\frac{40 \times 1000}{3600}$ = 11.11 m/s;

Final speed of car at exit end (v) = 60 km/hr = $\frac{60 \times 1000}{3600}$ = 16.67 m/s and length of curved track (s) = 300 m.

We know that tangential acceleration(a_t) of car can be calculated from the relation:

$$v^2 - u^2 = 2as$$

or $$16.67^2 - 11.11^2 = 2a_t \times 300 \qquad \text{...(Here, } a = a_t)$$

or $$a_t = \frac{16.67^2 - 11.11^2}{2 \times 300} = 0.257 \text{ m/s}^2$$

Let r be the radius of the circular track, then

$$\frac{1}{4} \times (2\pi r) = 300$$

or $$r = \frac{300 \times 2}{\pi} = 190.99 \text{ m}$$

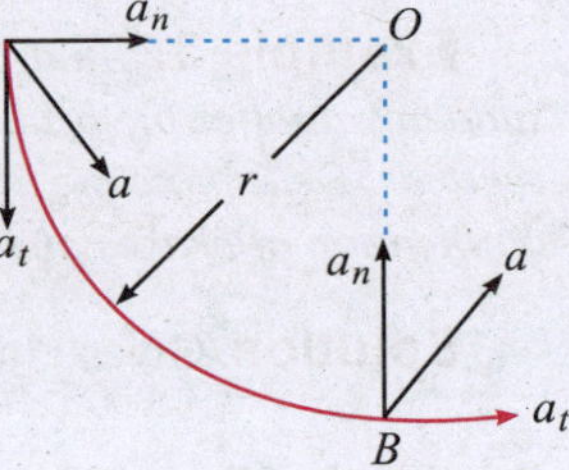

Fig. 5.

At entry end A:

Normal acceleration $(a_n) = \frac{u^2}{r} = \frac{11.11^2}{190.99} = 0.646 \text{ m/s}^2$

Tangential acceleration $(a_t) = 0.257 \text{ m/s}^2$

$\therefore$ Resultant acceleration $(a) = \sqrt{a_n^2 + a_t^2} = \sqrt{(0.646)^2 + (0.257)^2} = 0.695 \text{ m/s}^2$

If α is the angle made by the resultant with tangential acceleration, then

$$\tan \alpha = \frac{a_n}{a_t} = \frac{0.646}{0.257} = 2.514$$

or $$\alpha = \tan^{-1} 2.514 = 68.31^\circ$$

At exit end B:

Normal acceleration $(a_n) = \frac{v^2}{r} = \frac{16.67^2}{190.99} = 1.455 \text{ m/s}^2$

Tangential acceleration $(a_t) = 0.257 \text{ m/s}^2$

$\therefore$ Resultant acceleration $(a) = \sqrt{a_n^2 + a_t^2} = \sqrt{(1.455)^2 + (0.257)^2} = 1.478 \text{ m/s}^2$

If α is the angle made by the resultant with tangential acceleration, then

$$\tan \alpha = \frac{a_n}{a_t} = \frac{1.455}{0.257} = 5.661$$

or $$\alpha = \tan^{-1} 5.661 = 79.98^\circ$$

Example 5. *A car is moving with a uniform speed of 54 km/hr on the circular portion of a road of 250 m radius. Determine the velocity and acceleration component of the car along*

(a) x and y directions, and

(b) normal and tangential directions, at 10 seconds after it has entered the circular portion of the road.

Solution. Given : Uniform speed of car (v) = 54 km/hr = $\dfrac{54 \times 1000}{3600}$ = 15 m/s and radius of circular portion of road (r) = 250 m.

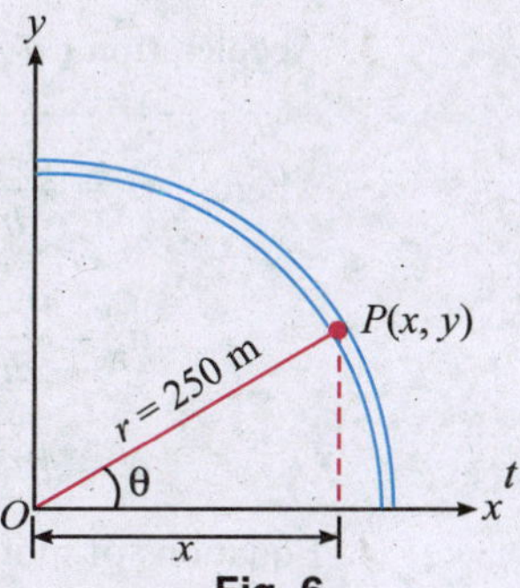

Fig. 6.

We know that angular velocity of car.

$$\omega = \frac{v}{r} = \frac{15}{250} = 0.06 \text{ rad/s}$$

At any instant, the car has the following coordinates:

$$x = r \cos \theta = r \cos \omega t.$$

$$y = r \sin \theta = r \sin \omega t.$$

(*a*) *Components of velocity and acceleration in x and y-directions at t = 10 seconds*

$$v_x = \frac{dx}{dt} = \frac{d}{dt}(r \cos \omega t) = -\,\omega r \sin \omega t$$

$$v_y = \frac{dy}{dt} = \frac{d}{dt}(r \sin \omega t) = \omega r \cos \omega t$$

$$a_x = \frac{dv_x}{dt} = \frac{d}{dt}(-\omega r \sin \omega t) = -\omega^2 r \cos \omega t$$

$$a_y = \frac{dv_y}{dt} = \frac{d}{dt}(\omega r \cos \omega t) = -\omega^2 r \sin \omega t$$

Substituting $t = 10$ s, $r = 250$ m and $\omega = 0.06$ rad/s $= 0.06 \times \dfrac{180}{\pi} = 3.438$ degree/s in above relations, we get

$$v_x = -\,0.06 \times 250 \times \sin(3.438 \times 10) = -8.470 \text{ m/s } (\leftarrow)$$

$$v_y = 0.06 \times 250 \times \cos(3.438 \times 10) = 12.380 \text{ m/s } (\uparrow)$$

$$a_x = -\,(0.06)^2 \times 250 \times \cos(3.438 \times 10) = -\,0.743 \text{ m/s } (\leftarrow)$$

$$a_y = -\,(0.06)^2 \times 250 \times \sin(3.438 \times 10) = -\,0.508 \text{ m/s } (\downarrow)$$

(*b*) *Velocity and acceleration in normal land tangential directions at t = 10 seconds*

$$v_n = 0 \text{ and } v_t = v = 15 \text{ m/s}$$

$$a_n = \frac{v^2}{r} = \frac{15^2}{250} = 0.9 \text{ m/s}^2$$

$$a_t = \frac{dv}{dt} = 0 \qquad \text{...}(\because\ v = \text{constant})$$

2.1 Plane Curvilinear Motion Under Constant Acceleration

Consider the motion of a particle along a curved path in *xy*-plane with constant acceleration. If $\vec{r}$ is the position of the particle at any instant, then

1. Position vector ($\vec{r}$) $= x\hat{i} + y\hat{j}$, where, $r = |\vec{r}| = \sqrt{x^2 + y^2}$

2. Velocity ($\vec{v}$) $= \dfrac{d\vec{r}}{dt} = v_x\hat{i} + v_y\hat{j}$;

   where, $v_x = \dfrac{dx}{dt} = \dot{x},\ v_y = \dfrac{dy}{dt} = \dot{y}$ and $v = \sqrt{v_x^2 + v_y^2}$

3. Acceleration ($\vec{a}$) $= \dfrac{d\vec{v}}{dt} = a_x\hat{i} + a_y\hat{j}$;

   where, $a_x = \dfrac{dv_x}{dt} = \dfrac{d^2x}{dt^2} = \ddot{x}$ = constant,

   $a_y = \dfrac{dv_y}{dt} = \dfrac{d^2y}{dt^2} = \ddot{y}$ = constant, and

   $a = \sqrt{a_x^2 + a_y^2}$

4. Equations of motions for plane curvilinear motion with constant acceleration are:

   (*i*) $\vec{v} = \vec{v}_0 + \vec{a}t$:

   where, $v_x = v_{x0} + a_x t,\ v_y = v_{y0} + a_y t$

   (*ii*) $\vec{r} = \vec{r}_0 + \vec{v}_0 + \frac{1}{2}\vec{a}t^2$;

   where. $x = x_0 + v_{x0}t + \frac{1}{2}a_x t^2,\ y = y_0 + v_{y0}t + \frac{1}{2}a_y t^2$

2.2. Projectile Motion

A particle moving under the effect of vertical and horizontal forces, is called *projectile*. The combined effect of both the forces (components) is to move the particle along a *parabolic path*. The path traced by a projectile in the space is called *trajectory* and the motion is called *projectile motion*.

A. Equations for a projectile motion on horizontal plane:

1. Components of velocity of projection (u),

   Vertical component, $V = u \sin \alpha$

   Horizontal component, $H = u \cos \alpha$

2. Equation of trajectory (or path of a projectile),

$$y = x \tan \alpha - \frac{gx^2}{2u^2 \cos^2 \alpha}$$

3. Time of flight, $t = \dfrac{2u \sin\alpha}{g}$

4. Horizontal range, $R = \dfrac{u^2 \sin 2\alpha}{g}$

5. Maximum height, $H = \dfrac{u^2 \sin^2 \alpha}{2g}$

where. u = Initial velocity or Velocity of projection,
α = Angle of projection with the horizontal, and
g = Acceleration due to gravity.

B. Equations for a projectile motion on an inclined plane:

1. Time of flight, $t = \dfrac{2u\sin(\alpha-\beta)}{g\cos\beta}$

2. Horizontal range, $R = \dfrac{u^2}{g\cos^2\beta}[\sin(2\alpha-\beta)-\sin\beta]$

   (*i*) α for maximum range (R_{max}), $\alpha = \dfrac{\pi}{4}+\dfrac{\beta}{2}$

   (*ii*) Maximum range, $R_{max} = \dfrac{u^2}{g(1+\sin\beta)}$

   where u = Initial velocity of projection,
   α = Angle of projection with the horizontal,
   β = Inclination of inclined plane with the horizontal, and
   g = Acceleration due to gravity.

2.3. Uniform Circular Motion

The motion of a particle moving along a circular path with constant speed, is called *uniform circular motion*. When a particle moves in a circular path with constant speed, then

Normal or Centripetal acceleration, $a_n = \dfrac{v^2}{r}$

It is directed towards the centre, along the radial direction and normal to the path. This acceleration is due to change in the direction of velocity.

3. SPACE CURVILINEAR MOTION

The space curvilinear motion can be described by using following coordinate systems:

A. In Rectangular Coordinates (*x*, *y*, *z*):

Consider the motion of a particle in space. If $\vec{r}$ is the position of the particle at any instant on the curved path, then

1. Position vector $(\vec{r}) = x\hat{i}+y\hat{j}+z\hat{k}$; where, $r = |\vec{r}| = \sqrt{x^2+y^2+z^2}$

2. Velocity $(\vec{v}) = v_x\hat{i}+v_y\hat{j}+y_z\hat{k}$;

   where, $v_x = \dfrac{dx}{dt} = \dot{x},\ v_y = \dfrac{dy}{dt} = \dot{y},\ v_z = \dfrac{dz}{dt} = \dot{z}$ and $v = \sqrt{v_x^2+v_y^2+v_z^2}$

3. Acceleration $(\vec{a}) = a_x\hat{i}+a_y\hat{j}+a_z\hat{k}$;

   where, $a_x = \dfrac{dv_x}{dt} = \dfrac{d^2x}{dt^2} = \ddot{x},\ a_y = \dfrac{dv_y}{dt} = \dfrac{d^2y}{dt^2} = \ddot{y},\ a_z = \dfrac{dv_z}{dt} = \dfrac{d^2z}{dt^2} = \ddot{z}$ and

   Resultant acceleration, $a = \sqrt{a_x^2+a_y^2+a_z^2}$

B. In cylindrical Coordinates (r, θ, z):

The plane curvilinear motion described in two dimensional polar coordinates can be extended to three dimensional space curvilinear motion by using cylindrical coordinates.

1. Position vector $(\vec{R}) = r\vec{\lambda}_r + z\hat{k}$; where $\vec{\lambda}_r$ is unit vector along $\vec{r}$.

2. Velocity $(\vec{v}) = \vec{v}_r + \vec{v}_\theta + \vec{v}_z$;

where, $v_r = \dfrac{dr}{dt} = \dot{r}, v_\theta = r\dfrac{d\theta}{dt} = r\dot{\theta}, v_z = \dfrac{dz}{dt} = \dot{z}$

Resultant acceleration, $v = \sqrt{v_r^2 + v_\theta^2 + v_z^2}$

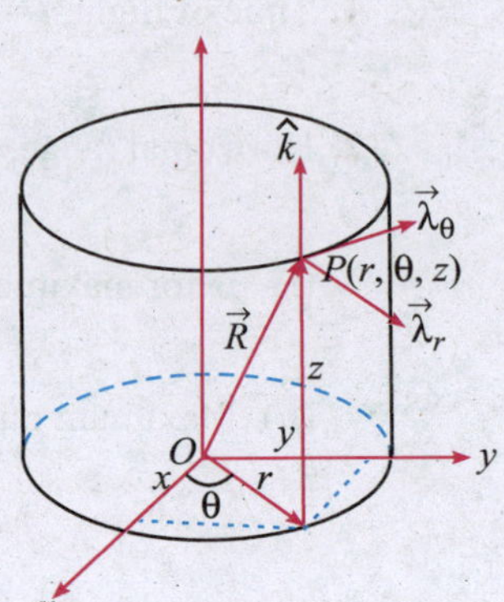

Fig. 7.

3. Acceleration $(\vec{a}) = \vec{a}_r + \vec{a}_\theta + \vec{a}_z$;

where $$a_r = \frac{d^2r}{dt^2} - r\left(\frac{d\theta}{dt}\right)^2 = \ddot{r} - r\dot{\theta}^2,$$

$$a_\theta = r\frac{d^2\theta}{dt^2} + 2\frac{dr}{dt}\cdot\frac{d\theta}{dt} = r\ddot{\theta} + 2\dot{r}\dot{\theta}, \text{ and}$$

$$a_z = \frac{d^2z}{dt^2} = \ddot{z}$$

Resultant acceleration, $a = \sqrt{a_r^2 + a_\theta^2 + a_z^2}$

4. Rectangular coordinates in terms of cylindrical coordinates:

$x = r\cos\theta, y = r\sin\theta, z = z$

5. Cylindrical coordinates in terms of rectangular coordinates:

$$r = \sqrt{x^2 + y^2}, \theta = \tan^{-1}\left(\frac{y}{x}\right), z = z$$

C. In Spherical Coordinates (r, ϕ, θ):

1. Velocity $(\vec{v}) = \vec{v}_r + \vec{v}_\phi + \vec{v}_\theta$;

where, $$v = \frac{dr}{dt} = \dot{r},$$

$$v_\phi = r\frac{d\phi}{dt} = r\dot{\phi}, \text{ and}$$

$$v_\theta = r\frac{d\theta}{dt}\sin\phi = r\dot{\theta}\sin\phi$$

Resultant velocity, $v = \sqrt{v_r^2 + v_\phi^2 + v_\theta^2}$

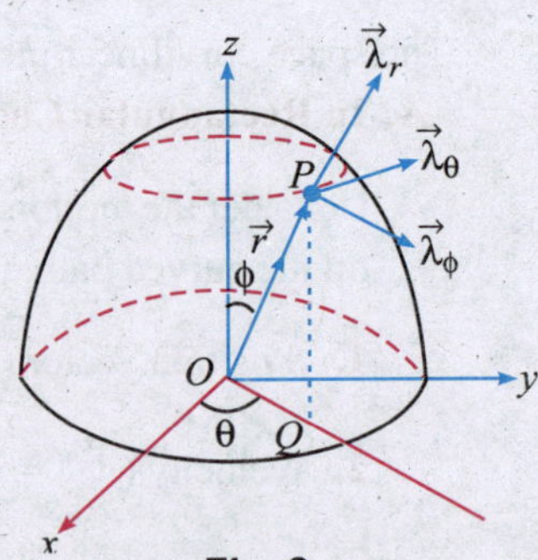

Fig. 8.

2. Acceleration $(\vec{a}) = \vec{a}_r + \vec{a}_\phi + \vec{a}_\theta$;

where, $a_r = \ddot{r} - r\dot{\phi}^2 - r\phi^2\sin^2\phi$

$a_\phi = r\ddot{\phi} + 2\dot{r}\dot{\phi} - r\dot{\theta}^2\sin\phi\cos\phi$, and

$a_\theta = r\ddot{\theta}\sin\phi + 2\dot{r}\dot{\theta}\sin\phi + 2r\dot{\phi}\dot{\theta}\cos\phi$

Resultant acceleration, $a = \sqrt{a_r^2 + a_\theta^2 + a_\theta^2}$

EXERCISE

1. The motion of a particle is given by the relations:

$$x = t^2 + 3t \text{ and } y = t^3 - 8t^2 + 3,$$

where x and y are in metres and l is in seconds. Calculate the velocity and acceleration of the particle at time $t = 2$ seconds. **(Ans.** $v = 21.19$ m/s; $a = 4.47$ m/s^2)

2. A particle moves along a curved path in xy-plane and its coordinates are defined by the equations:

$$x = -4 \sin t \text{ and } y = 3 - 4 \cos t$$

where x and y are in metres and t is in seconds.

(*a*) Show that equation of the curved path is a circle of radius 4 and center (0, 3).

(*b*) Determine the velocity and acceleration of the particle at the instant when it starts motion.

(Ans. (*b*) $v = -4$ m/s, $a = 4$ m/s^2)

3. A particle moves along a curved path in xy-plane defined by the equations:

$$r = \frac{5t^2}{3} \text{ and } \theta = \frac{t^2}{3}$$

where r is in metres, θ is in radians and t is in seconds.

Find the velocity and acceleration of the particle when $\theta = \pi/2$.

(Ans. $v = 13.46$ m/s; $a = 29.25$ m/s^2)

CHAPTER

37

Vibrations

Contents

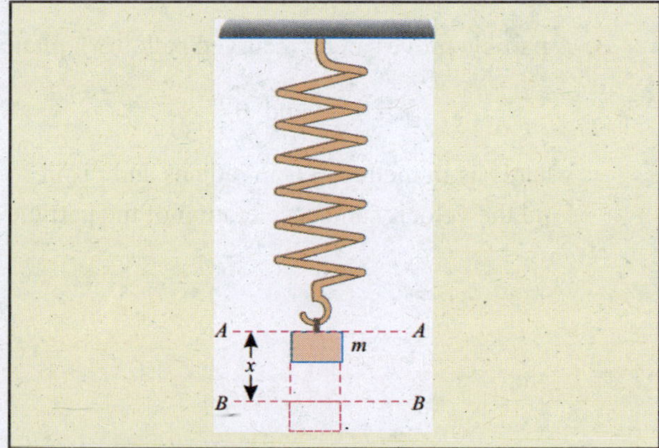

1. INTRODUCTION

Whenever an elastic body (such as a spring, a beam, a shaft etc.) is displaced from its equilibrium position by applying external forces, and then released, it executes a **vibratory motion**. This is due to the reason that, when a body is displaced from its equilibrium position, the work is done on the elastic constraints of the body and is stored as internal forces in the form of elastic or strain energy. Now, if the body is released, these internal forces cause the body to move towards its equilibrium position. If the motion is frictionless, the whole strain energy is converted into kinetic energy during the period when body reaches the equilibrium position, at which it has maximum kinetic energy and due to which the body continues to move in opposite direction. The body passes through the equilibrium position and moves in opposite direction, the whole kinetic energy is utilized to overcome the elastic constraints (forces) and is stored in the form of strain energy due to which body again return to the equilibrium position. In this way, vibratory motion is repeated continuously. Thus, **the vibrations are caused by the elastic forces**.

In mechanical engineering, for design and operation, the analysis of vibration in machine parts is becoming very important, as the trend is towards higher linear speeds which can only result from higher rotational speeds of prime-movers. Gas and steam turbines running at 10000 r.p.m. or more are very common these days. Any machine part should be capable of vibrating because it is not absolutely rigid but elastic. To check vibration, rotational speed should not coincide or be very close to natural frequency of vibration is order that the amplitude of vibration is not so large as to cause permanent failure of the shaft on which the motors are keyed.

2. IMPORTANT TERMS

The following terms, which are commonly used in connection with the vibratory motions, should be clearly understood at this stage:

1. **Amplitude.** The maximum displcement of a body from its mean position is called **amplitude** of vibration. The amplitude of vibration is usually expressed in metres.
2. **Time period or Period of vibration.** The interval of time taken by the vibratory motion to repeat itself is called a **time period**. The period of vibration is usually expressed in seconds.
3. **Cycle.** The motion completed during one time period is called **cycle**.
4. **Frequency.** The number of cycles of vibratory motion completed in a one second (or unit interval of time) is called **frequency**. The frequency is expressed in hertz (Hz) which is equal to one cycle per second.
5. **Degree of freedom.** The independent coordinate with which we define displaced shape of the structure is called **degree of freedom**.
   A system with a single coordinate function is said to **one degree of freedom system**. Similarly a system with two coordinates functions is said to be **two degree of freedom system** and so on.

3. TYPES OF VIBRATIONS

Vibrations may be classified as follows:

1. **Free or natural vibrations.** When a body (system) vibrates under the action of forces inherent in the system without assistance of any impressed forces, vibrations are said to be **free or natural vibrations**. The frequency of free or natural vibration of the system are called **natural frequency**.
   Note: According to the actuating force on the body, the free vibrations may be **damped** or **undamped**:
2. **Forced vibrations.** When a body (system) vibrates under the influence of external force, the vibrations are said to be **forced vibrations**. The external force applied to the body is a periodic disturbing force created by unbalance. The vibrations have the same *frequency asthe applied force* (and are independent of the natural frequency of vibration).
   When the frequency of the external force is same as that of the natural vibrations, resonance takes place. This results in amplitudes of vibrations which may be dangerously large.
   Note: According to the actuating force on the body, the forced vibrations may be **damped** or **undamped**:
3. **Undamped Vibrations.** When the amplitude of vibration is constant over every cycle of vibration due to absence of dissipative forces, the motion is said to be **undamped vibration**.
4. **Damped vibrations.** When there is a reduction in amplitude over every cycle of vibration due to frictional resistance to the motion, the motion is said to be **damped vibration**. This is owing to the fact that a certain amount of energy possessed by the vibrating system is *always dissipated in overcoming frictional resistances to the motion.*
   Note: Resistance to oscillation is known as **damping**.

In this chapter, we shall study only about free or natural vibrations.

4. TYPES OF FREE VIBRATIONS

There are three types of free or natural vibrations which are important from the the subject point of view :

1. Longitudinal vibrations
2. Transverse vibrations
3. Torsional vibrations

Consider a weightless constraint (spring or shaft) whose one end is fixed and other end having a torsional end (i.e., carrying a heavy disc) as shown in Fig. 1. This system may execute any one of three types of vibrations i.e., the longitudinal, transverse or torsional vibrations.

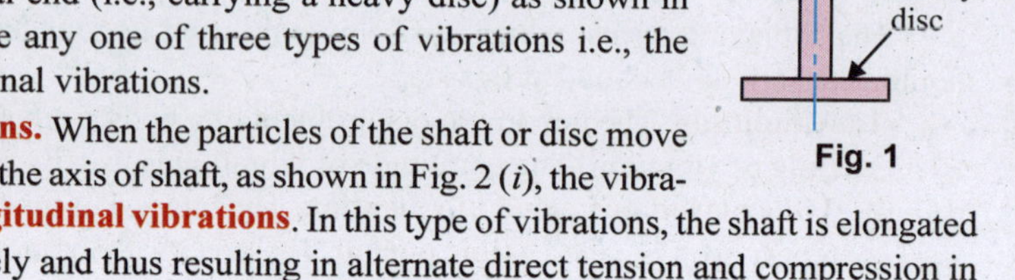

Fig. 1

1. **Longitudinal vibrations.** When the particles of the shaft or disc move in parallel planes along the axis of shaft, as shown in Fig. 2 (*i*), the vibrations are known as **longitudinal vibrations**. In this type of vibrations, the shaft is elongated and shortened alternately and thus resulting in alternate direct tension and compression in the shaft.

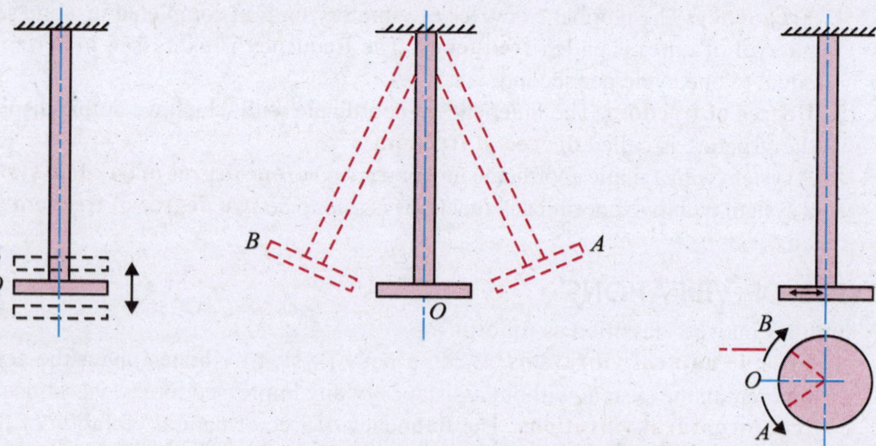

(*i*) Longitudinal vibrations (*ii*) Transverse vibrations (*iii*) Torsional vibrations

Fig. 2. Types of free vibrations.

2. **Transverse Vibrations.** When the particles of shaft or disc moves approximately perpendicular to the axis of the shaft, as shown in Fig. 2 (*ii*), the vibrations are known as **transverse vibrations**. In this type of vibrations, the shaft is straight and bent alternately and thus subjected to tension and compression as a result of bending. Hence, bending stresses are induced in the shaft.

3. **Torsional vibrations.** When the particles of a shaft or disc move in a circle about the axis of shaft or rotation, then the vibrations are called **torsional vibrations**. In this type of vibrations, the shaft is twisted and untwisted alternately on account of vibratory motion of disc and torsional shear stresses are induced in it.

In all the above three cases, if the limit of proportionality is not exceeded, obviously the *restoring forces* in cases (*i*) and (*ii*) and the *restoring couple* in case (*iii*), offered by the elastic constraint, will be *directly proportional to displacement* from the original equilibrium (mean) position.

Hence, the acceleration towards the equilibrium position is directly proportional to the displacement from the equilibrium position and therefore the vibration is **simple harmonic**.

5. FREE LONGITUDINAL VIBRATIONS

5.1. Natural Frequency of Free Longitudinal Vibrations

The natural frequency of the free longitudinal vibrations may be determined by any of the following three methods:

1. Equilibrium method
2. Energy method
3. Rayleigh's method

1. Equilibrium Method

Consider a body which is suspended from a constraint (*i.e.*, spring) of negligible mass as shown in Fig. 3 (*a*).

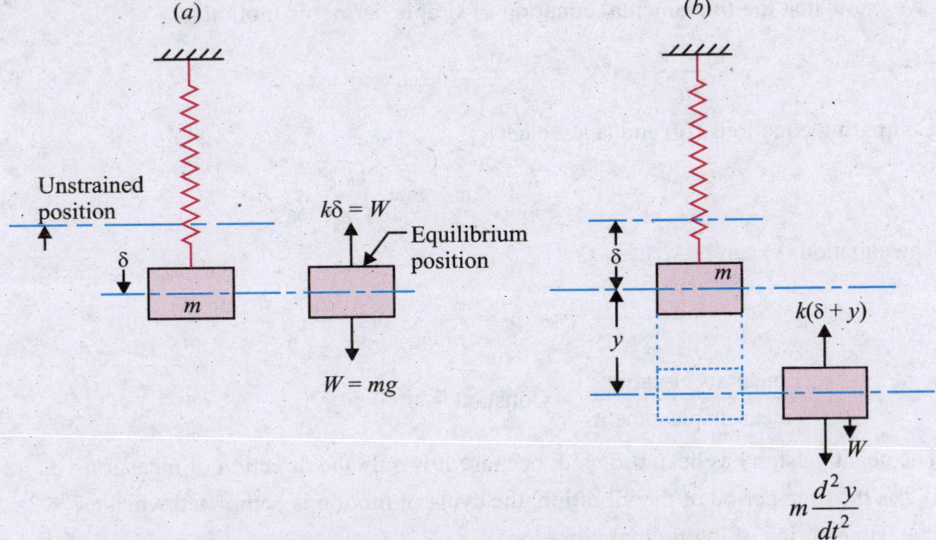

Fig. 3. Natural frequency of free longitudinal vibrations.

Let m = Mass of the body suspended from the spring, in kg,

W = Weight of the body = mg, in newtons,

k = Stiffness of the spring; usually expressed in N/m,

δ = Static deflection of the spring in metres due to weight W newtons, and

y = Displacement given to the body vertically by the external force, in metres.

In the equilibrium position as shown in Fig. 3(*a*), the gravitational pull W (= mg) acting on the body is balanced by a force of the spring.

$$\therefore \qquad W = k.\delta \quad \text{or} \quad mg = k.\delta \qquad \ldots(i)$$

The mass is now displaced from its equilibrium position by a distance y and is then released. After time t,

Restoring force developed in the spring,

$$= W - k(\delta + y) = k\delta - k\delta - ky = -ky \qquad \ldots(ii)$$

... (Taking upward force as negative)

And the unbalanced forces and acceleration, imparted to the body is given by Newton's second law of motions as:

$$\text{Accelerating Force} = \text{Mass} \times \text{Acceleration}$$

$$= m\frac{d^2y}{dt^2} \quad \text{...(iii)}$$

... (Taking downward force as positive)

Equating equations (*ii*) and (*iii*), the equation of motion of the body after time *t* is given by,

$$m\frac{d^2y}{dt^2} = -ky$$

or
$$m\frac{d^2y}{dt^2} + ky = 0 \quad \text{or} \quad \frac{dy^2}{dt^2} + \frac{k}{m} \times x = 0 \quad \text{...(iv)}$$

We know that the fundamental equation of simple harmonic motion is:

$$\frac{d^2y}{dt^2} + \omega^2 y = 0 \quad \text{...(v)}$$

Comparing equations (*iv*) and (*v*), we get

$$\omega^2 = \frac{k}{m} \quad \text{or} \quad \omega = \sqrt{\frac{k}{m}} \quad \text{...(vi)}$$

The equation (*v*) can be written as:

$$\frac{\left(d^2y/dt^2\right)}{y} = -\omega^2 = \frac{k}{m} - \quad \text{or} \quad \frac{dy^2/dt^2}{y} \quad \frac{k}{m} = \omega^2$$

$$\frac{\text{Linear acceleration}}{\text{Linear displacement}} = \text{Constant} = \omega^2 \quad \text{...(vii)}$$

The negative sign has been dropped, because it is only the direction of measuring y.

If T is the time period of the vibration, the cycle of motion is completed when $\omega T = 2\pi$

∴ Time period of logitudinal vibration,

$$T = \frac{2\pi}{\omega} = 2\pi\sqrt{\frac{m}{k}} \quad \text{...(viii)}$$

and, Natural frequency of logitudinal vibration,

$$f_n = \frac{1}{T} = \frac{1}{2\pi}\sqrt{\frac{k}{m}} = \frac{1}{2\pi}\sqrt{\frac{g}{\delta}} \quad \text{...(ix)}$$

∵ $(mg = k\delta)$

Taking the value of g as 9.81 m/s^2 and δ in metres, the equation (*ix*) can be written as:

$$f_n = \frac{1}{2\pi}\sqrt{\frac{9.81}{\delta}} = \frac{0.4985}{\sqrt{\delta}}\ \text{Hz} \quad \text{...(x)}$$

Notes:

1. In the equations (*viii*) and (*ix*), the mass of spring m_s was not taken into consideration.
2. If mass of spring is to be taken into accoun $\frac{m_s}{3}$, should be added to mass m and the equa-

tions (*viii*) and (*ix*) will be

$$T = 2\pi\sqrt{\frac{m+(m_s/3)}{k}} \quad \ldots(xi)$$

$$\text{and } f_n = \frac{1}{2\pi}\sqrt{\frac{k}{m+(m_s/3)}} \quad \ldots(xii)$$

(Refer Article 5.2. also)

2. Energy Method

We know that the kinetic energy of a body is due to its motion and potential energy of that body is w.r.t. to a certain datum position which is equal to the amount of work required to move it from the datum position.

In case of a conservative system, the equations of motion may be determined from energy considerations. Whenever a conservative system is set in motion, the mechanical energy in the system is partially kinetic (due to motion of the body) and partially potential (with respect to certain datum position).

In the case of vibrations, the datum position is the mean or equilibrium position at which the potential energy of the body (or the system) in zero.

In the free vibrations (a case of a conservative system), no energy is transferred to the system or from the system. Therefore, sum of kinetic energy (K.E.) and potential energy (P.E.) must be a constant quantity which is same at all the times during the vibration. It implies that

$$\frac{d}{dt}(\text{K.E.} + \text{P.E.}) = 0 \quad \ldots(i)$$

We know that kinetic energy,

$$\text{K.E.} = \frac{1}{2}m\left(\frac{dy}{dt}\right)^2 = \left(\because \ \text{K.E.} = \frac{1}{2}mv^2\right)$$

and Potential energy,

$$\text{P.E.} = \left(\frac{0+ky}{2}\right)y = \frac{1}{2}ky^2 \quad \ldots(\because \text{P.E.} = \text{Mean force} \times \text{Displacement})$$

Substituting the above values in equation (*i*), we get

$$\frac{d}{dt}\left[\frac{1}{2}m\left(\frac{dy}{dt}\right)^2 + \frac{1}{2}ky^2\right] = 0$$

$$\text{or } \frac{1}{2}\times m\times 2\times\frac{dy}{dt}\times\frac{d^2y}{dt^2} + \frac{1}{2}k\times 2y\times\frac{dy}{dt} = 0$$

$$\text{or} \quad m\times\frac{d^2y}{dt^2} + k\,.\,y = 0 \quad \text{or} \quad \frac{d^2y}{dt^2} + \frac{k}{m}\times x = 0 \quad \text{...(Same as before)}$$

The time period (T) and natural frequency (f_n) of the free vibration may be obtained as discussed in the "Equilibrium method".

3. Rayleigh's Method

According to this method, *the maximum kinetic energy at the mean position is equal to the maximum potential energy (or strain energy) at the extreme position.*

In case of free vibrations, the motion executed by the vibration is assumed to be simple harmonic.

$\therefore$ $$y = Y \sin(\omega \, . \, t) \quad \ldots(i)$$

where, y = Displacement of the body from the mean position after time t seconds, and

Y = Maximum displacement of the body from mean position to extreme position.

Now, differentiating equation (*i*) w.r.t. time t, we get $\frac{dy}{dt} = \omega \times Y \cos(\omega . t)$

The maximum kinetic energy and maximum potential energy are determined by using *end conditions* at mean position and at extreme position.

At mean position: $t = 0$ and velocity is maximum and hence K.E. maximum.

$\therefore$ Maximum velocity at mean position,

$$v \left.\frac{dy}{dt}\right|_{t=0} = \omega . Y \quad (\because \cos(\omega \times 0) = 1)$$

Since velocity is maximum at mean position, therefore K.E. is maximum.

$\therefore$ Maximum K.E. at the mean position.

$$= \frac{1}{2} m v^2 = \frac{1}{2} = m\,\omega^2 Y^2 \quad \ldots(ii)$$

At extreme position: $y = Y$ and Potential energy is maximum.

$\therefore$ Maximum potential energy at the extreme position,

$$\left(\frac{0 + k.Y}{2}\right) = \frac{1}{2}\,kY^2 \quad \ldots(iii)$$

According to this method, maximum kinetic energy at mean position is equal to the maximum potential energy at the extreme point *i.e.*, equating equations (*ii*) and (*iii*), we get

$$\frac{1}{2} m\,\omega^2 Y^2 = \frac{1}{2} kY^2$$

or $$\omega^2 \frac{k}{m} = \text{ or } \omega \sqrt{\frac{k}{m}}$$

We know that time period of vibration,

$$T = \frac{2\pi}{\omega} = 2\pi \sqrt{\frac{m}{k}} \quad \ldots\text{(Same as before)}$$

and natural frequency of vibration,

$$f_n = \frac{1}{T} = \frac{1}{2\pi}\sqrt{\frac{k}{m}} \quad \ldots\text{(Same as before)}$$

Note: In all the above expressions, ω is known as **natural circular frequency** and is generally denoted by ω_n.

5.2. Effect of Inertia of the Constraint in Longitudinal Vibrations

In deriving the mathematical formula for natural frequency of longitudinal vibrations in previous article, the inertia of constraint (*i.e.*, shaft) is neglected. Here, we consider the inertia of constraint.

Consider a constraint (*i.e.*, shaft) whose one end is fixed and other end is free as shown in Fig. 4.

Let m_c = Mass of constraint per unit length,

l = Length of the constraint,

m_C = Total mass of constraint = $m_c \cdot l$, and

v = Longitudinal velocity of the free en.

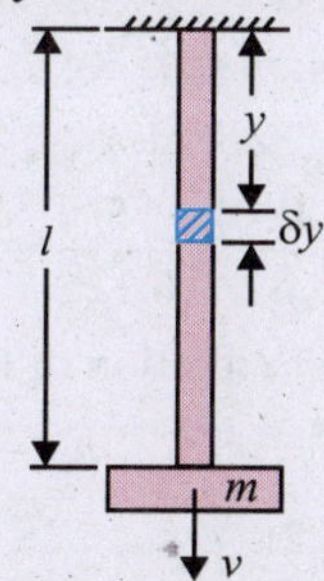

Fig. 4. Effect of inertia of the constraint in longitudinal vibrations.

Consider a small element of the constraint of length δy and at a distance y from the fixed end.

∴ Velocity of the small element,

$$= \frac{y}{l} \times v$$

and, kinetic energy possessed by the small element

$$= \frac{1}{2} \text{ Mass} \times (\text{Velocity})^2$$

$$= \frac{1}{2} \times (m_c \cdot \delta y) \times \left(\frac{y}{l} \times v\right)^2 = \frac{m_c v^2 y^2}{2l^2} \times \delta y$$

∴ Total kinetic energy possessed by the constraint of length l,

$$= \int_0^l \frac{m_c v^2 y^2}{2l^2} \times dy = \frac{m_c v^2}{2l^2} \int_0^l y^2 dy = \frac{m_c v^2}{2l^2} \left[\frac{y^3}{3}\right]_0^l$$

$$= \frac{m_c \, v^2}{2l^2} \times \frac{l^3}{3} = \frac{1}{2} m_c v^2 \times \frac{l}{3}$$

$$= \frac{1}{2}\left(\frac{m_c \, l}{3}\right) v^2 = \frac{1}{2}\left(\frac{m_C}{3}\right) v^2 \quad \ldots(i)$$

...(Substituting $m_c l = m_C$)

If the constraint is assumed to be of negligible mass and a mass $\frac{m_c}{3}$ is placed at the free end, then total energy possessed by the constraint

$$= \frac{1}{2}\left(\frac{m_C}{3}\right) v^2 \quad \ldots(ii) \text{ [Same as eqn. } (i)]$$

Thus, the two systems are dynamically same. Therefore, **inertia of the constraint may be allowed for by adding one-third of its mass to the disc at the free end.**

Hence, when the mass of the constraint m_C and the mass of the disc at the free end m are given, then natural frequency of logitudinal vibration,

$$f_n = \frac{1}{2\pi} \sqrt{\frac{k}{m + \frac{m_C}{3}}} \quad \ldots(ii)$$

Example 1. *A weightless helical spring which extends 0.125 mm under a weight of 5 N, is made to support a mass of 45 kg. The spring-mass system is displaced vertically through 10 mm and released. Find the natural frequency of longitudinal vibration of the system.*

Solution. Given: $\delta = 0.125$; $W = 5$ N; $m = 45$ kg

We know that stiffness of the spring,

$$k = \frac{W}{\delta} = \frac{5}{0.125} = 40 \text{ N/mm} = 40 \times 10^3 \text{ N/m}$$

∴ Natural frequency of longitudinal vibration of the system,

$$f_n = \frac{1}{2\pi}\sqrt{\frac{k}{m}} = \frac{1}{2\pi}\sqrt{\frac{40 \times 10^3}{45}} = \mathbf{4.75 \text{ Hz}} \qquad \textbf{Ans.}$$

Example 2. *A cantilever shaft 100 mm and 1 m long has a disc of 200 kg at its free end. Find the natural frequency of logitudinal vibration. Take G = 200 GN/m².*

Solution. Given: $d = 100$ mm $= 0.1$ m; $m = 200$ kg; $l = 1$ m; $G = 200$ GN/m^2 $= 200 \times 10^9$ N/m^2

We know that cross-sectional area of the shaft,

$$A = \frac{\pi}{4} \times d^2 = \frac{\pi}{4} \times (0.1)^2 = 7.854 \times 10^{-3} \text{ m}^2$$

and static deflection for a cantilever shaft carrying a point load at its free end,

$$\delta = \frac{Wl}{AE} = \frac{(m.g)l}{AE}$$

$$= \frac{200 \times 9.81 \times 1}{7.854 \times 10^{-3} \times 200 \times 10^9} = 1.249 \times 10^{-6} \text{ m}$$

∴ Natural frequency of longitudinal vibration,

$$f_n = \frac{0.4985}{\sqrt{\delta}} = \frac{0.4985}{\sqrt{1.249 \times 10^{-6}}} = \mathbf{446.05 \text{ Hz}} \qquad \textbf{Ans.}$$

Example 3. *A flywheel of mass 500 kg is mounted on a vertical shaft, whose both ends are fixed and its diameter is 50 mm as shown in Fig. 5. Find the natural frequency of longitudinal vibrations. Take Young's modulus for the shaft material is 200 GN/m².*

Solution. Given: $m = 500$ kg; $d = 50$ mm $= 50 \times 10^{-3}$ m, $E = 200$ GN/m^2 $= 200 \times 10^9$ N/m

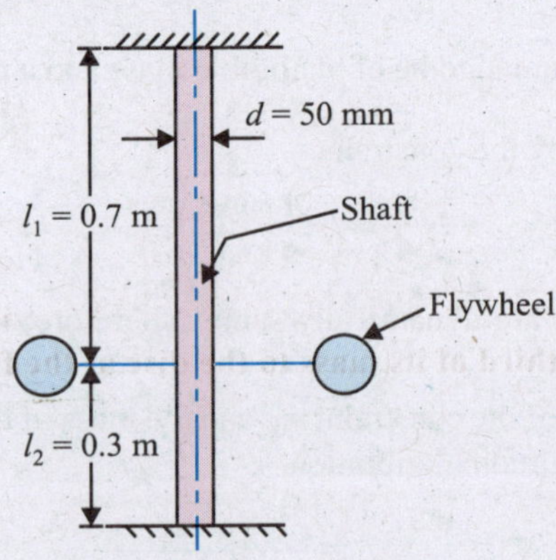

Fig. 5.

We know that cross-sectional area of the vertical shaft,

$$A = \frac{\pi}{4}\times d^2 = \frac{\pi}{4}\times(50\times10^{-3})^2 = 1.96 \times 10^{-3}\ \text{m}^2$$

and moment of inertia of the shaft,

$$I = \frac{\pi}{64}\times d^4 = \frac{\pi}{64}\times(50\times10^{-3})^4 = 0.307 \times 10^{-6}\ \text{m}^4$$

Let mass of flywheel carried by length $l_1 = m_1$

Then, mass of flywheel carried by length $l_2 = m - m_1$

We know that extension of length l_1,

$$= \frac{W_1 l_1}{AE} = \frac{m_1.g\,.l_1}{AE} \qquad \ldots(i)$$

Similarly, compression of length of l_2,

$$\frac{W_2 l_2}{AE} = \frac{(m-m_1)g\,.l_2}{AE} \qquad \ldots(ii)$$

Since the system in equilibrium, therefore extension of length l_1 must be equal to compression of length l_2. Equating equations (*i*) and (*ii*), we get

$$\frac{m_1 g\, l_1}{AE} = \frac{(m-m_1)g l_2}{AE}$$

or $\quad m_1 l_1 = (m - m_1) l_2$

or $\quad m_1 \times 0.7 = (500 - m_1) \times 0.3 = 150 - 0.3\, m_1$

or $\quad m_1 = 150$ kg

∴ Extension of length l_1,

$$\delta = \frac{m_1 g\, l_1}{AE} = \frac{150\times9.81\times0.7}{(1.96\times10^{-3})\times(200\times10^9)} = 2.628 \times 10^{-6}\ \text{m}$$

We also know that natural frequency of longitudinal vibration,

$$f_n = \frac{0.4985}{\sqrt{\delta}} = \frac{0.4985}{\sqrt{2.628\times10^{-6}}} = 307.5\ \text{Hz}$$ **Ans.**

6. FREE TRANSVERSE VIBRATIONS

6.1. Natural Frequency of Free Transverse Vibrations

Consider a constraint (*i.e.*, shaft) of negligible mass whose one end is fixed and other end carries a body of weight W as shown in Fig. 6

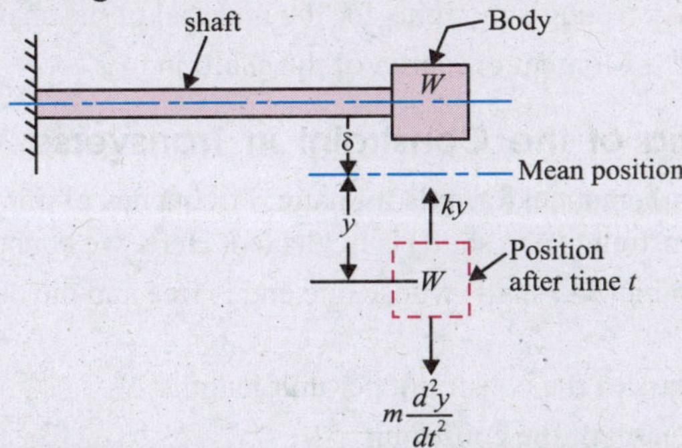

Fig. 6. Natural frequency of free transverse vibration.

Let, k = Stiffness of the shaft,

δ = Static deflection of shaft due to weight of the body W,

y = Displacement of the body from the mean position after time t, and

m = Mass of the body $\frac{W}{g}$.

Since the body is now displaced from its mean position by a distance 'y' as shown in figure and is then released. Therefore after time t,

$$\text{Restoring force} = -ky \qquad \ldots(i)$$

and

$$\text{Accelerating force} = \text{Mass} \times \text{Acceleration}$$

$$= m \times \frac{d^2y}{dt^2} \qquad \ldots(ii)$$

Equating equations (i) and (ii), the equation of motion of body after time t is:

$$m \times \frac{d^2y}{dt^2} = -ky$$

or

$$m\frac{d^2y}{dt^2} + ky = 0$$

or

$$\frac{d^2y}{dt^2} + \frac{k}{m} \times y = 0 \qquad \ldots\text{(Same as before)}$$

Hence, the time period (T) and natural frequency of transverse vibrations (f_n) are same as that of longitudinal vibrations.

$$T = 2\pi\sqrt{\frac{m}{k}}$$

and

$$f_n = \frac{1}{T} = \frac{1}{2\pi}\sqrt{\frac{k}{m}} = \frac{1}{2\pi}\sqrt{\frac{g}{\delta}}$$

Note: The value of static deflection δ in this case is identical with the static deflection of a cantilever beam loaded at free end and it is given as :

$$\delta = \frac{Wl^3}{3EI} \text{ metres}$$

where, W = Load at the free end, in newtons,

l = Length of the shaft, in metres,

E = Young's modulus for the material of the shaft, in N/m^2, and

I = Moment of inertia of the shaft, in m^4.

6.2 Effect of Inertia of the Constraint in Transverse Vibration

In deriving the mathematical formula for natural frequency of transverse vibrations in previous article, the inertia of constraint (*i.e.*, shaft) is neglected. Here, we consider the inertia of constraint.

Consider a constraint (*i.e.*, shaft) whose one end is free and the other end is fixed, as shown in Fig. 7.

Let, m_c = Mass of the constraint per unit length,

l = Length of the constraint,

m_C = Total mass of the constraint = $m_c \times l$, and

v = Transverse velocity of the free end.

Consider a small element of the constraint of length δx and at a distance x from the fixed end

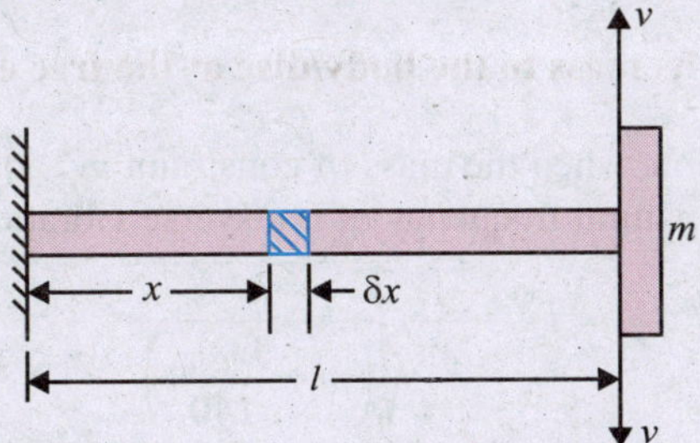

Fig. 7. Effect of inertia of the constraint in transverse vibrations.

We know that velocity of this element is given by,

$$= \left[\frac{3lx^2 - x^3}{2l^3} \times v\right]$$

$\therefore$ K.E. of the element $= \frac{1}{2} \times (\text{Mass of small element}) \times (\text{Velocity})^2$

$$= \frac{1}{2}(m_c \times \delta x)\left[\frac{3lx^2 - x^3}{2l^3} \times v\right]^2$$

and Total kinetic energy (K.E.) of the constraint (or shaft),

$$= \int_0^l \frac{1}{2} \times m_c \left[\frac{3lx^2 - x^3}{2l^3} \times v\right]^2 dx$$

$$= \frac{m_c\, v^2}{8l^6} \int_0^l \left(9l^2 x^4 - 6l\, x^5 + x^6\right) dx$$

$$= \frac{m_c v^2}{8l^6}\left[\frac{9l^2 x^5}{5} - \frac{6lx^6}{6} + \frac{x^7}{7}\right]_0^l$$

$$= \frac{m_c\, v^2}{8l^6}\left[\frac{9l^7}{5} - \frac{6l^7}{6} + \frac{l^7}{7}\right]$$

$$= \frac{m_c\, v^2}{8l^6}\left(\frac{33l^7}{35}\right)$$

$$= \frac{33}{280} \times m_c\, lv^2 = \frac{1}{2}\left(\frac{33}{140} \times m_c \times l\right)v^2$$

$$= \frac{1}{2}\left(\frac{33}{140} \times m_C\right)v^2 \qquad \text{(Substituting } m_c \times l = m_C\text{)} \quad \ldots(i)$$

If the constraint is assumed to be of negligible mass and a mass $\frac{33m_C}{140}$ is placed at the free end, then, total K.E. possessed by the constraint,

$$= \frac{1}{2}\left(\frac{33m_C}{140}\right)v^2 \qquad \text{[Same as } (i)\text{]} \qquad \ldots(ii)$$

Thus, the two systems are dynamically same. Therefore, **inertia of the constraint may be allowed for by addi $\frac{33}{140}$ g of its mass to the body/disc at the free end.**

From the above discussion, when the mass of constraint m_C and the mass of the body/disc m at the free end are given, then natural frequency of transverse vibration,

$$f_n = \frac{1}{2\pi}\sqrt{\frac{k}{\left(m + \frac{33m_C}{140}\right)}} \qquad \ldots(iii)$$

Notes:

(*i*) When both the ends of the constraint are *fixed* and the body/disc is situated in the middle of it, then it can be proved that inertia of constraint may be allowed for by adding $\frac{13}{35}$ of its mass to the body/disc. In this case, natural frequency of transverse vibration,

$$f_n = \frac{1}{2\pi}\sqrt{\frac{k}{m + \frac{13m_C}{35}}} \qquad \ldots(iv)$$

(*ii*) When the constraint is like a *simply supported beam*, then it can be proved that inertia of constraint may be allowed for by adding $\frac{17}{35}$ of its mass to the body/disc.

In this case, natural frequency of transverse vibration,

$$f_n = \frac{1}{2\pi}\sqrt{\frac{k}{m + \frac{17m_C}{35}}} \qquad \ldots(v)$$

Example 4. *A cantilever shaft 100 mm and 1 m long has a disc of 200 kg at its free end as shown in Fig 8. Find the natural frequency of transverse vibration. Take G = 200 GN/m.*

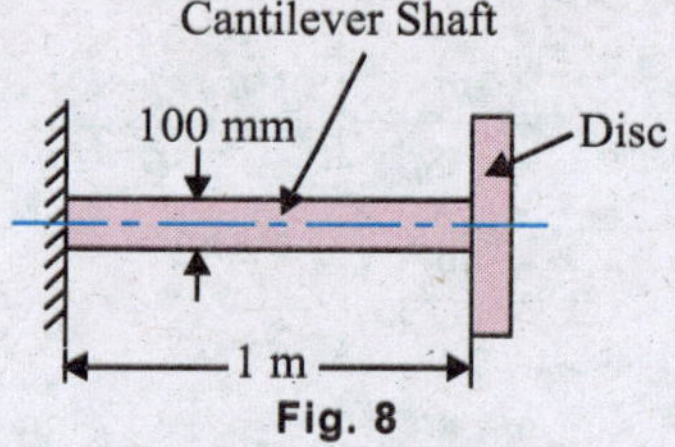

Fig. 8

Solution. Given: $d = 100$ mm $= 0.1$ m; $m = 200$ kg; $l = 1$ m; $G = 200$ GN/m^2 $= 200 \times 10^9$ N/m^2

We know that moment of inertia of the shaft,

$$I = \frac{\pi}{64} \times d^4 = \frac{\pi}{64} \times (0.1)^4 = 4.909 \times 10^{-6} \text{ m}^4$$

and the static deflection for a cantilever shaft carrying a point load at its free end,

$$\delta = \frac{Wl^3}{3EI} = \frac{(m.g)l^3}{3EI}$$

$$= \frac{(200 \times 9.81) \times 1^3}{3 \times (200 \times 10^9) \times (4.909 \times 10^{-6})} = 0.666 \times 10^{-3} \text{ m}$$

∴ Natural frequency of transverse vibration,

$$f_n = \frac{0.4985}{\sqrt{\delta}} = \frac{0.4985}{\sqrt{0.666 \times 10^{-3}}} = \textbf{19.05 Hz} \quad \textbf{Ans.}$$

Example 5. *A flywheel of mass 500 kg is mounted on a horizontal shaft, whose both ends are fixed and its diameter is 50 mm, as shown in Fig. 9. Find the natural frequency of transverse vibrations for the system. Take G = 200 GN/m.*

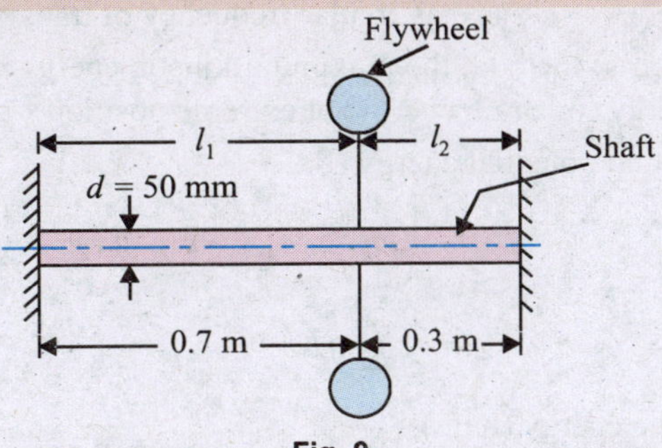

Fig. 9

Solution. Given: $d = 50$ mm $= 0.05$ m; $m = 500$ kg; $l_1 = 0.7$ m; $l_2 = 0.3$, $l = 0.7 + 0.3 = 1.0$ m;

$$G = 200 \text{ GN/m}^2 = 200 \times 10^9 \text{ N/m}^2$$

We know that moment of inertia of the shaft,

$$I = \frac{\pi}{64} \times d^4 = \frac{\pi}{64} \times (0.05)^4 = 0.307 \times 10^{-6} \text{ m}^4$$

We also know that the static deflection for a shaft when it is fixed at both ends and carries an eccentric point load,

$$\delta = \frac{W l_1^3 l_2^3}{3EIl^3} = \frac{(m.g) l_1^3 l_2^3}{3EIl^3}$$

$$= \frac{(500 \times 9.81) \times (0.7)^3 \times (0.3)^3}{3 \times (200 \times 10^9) \times (0.307 \times 10^{-6}) \times (1)^3}$$

$$= 2.466 \times 10^{-4} \text{ m}$$

∴ Natural frequency of transverse vibration,

$$f_n = \frac{1}{2\pi} \sqrt{\frac{g}{\delta}} = \frac{1}{2\pi} \sqrt{\frac{9.81}{2.466 \times 10^{-4}}} = \textbf{31.74 Hz} \quad \textbf{Ans.}$$

6.3 Natural Frequency of Free Transverse Vibrations for a Shaft Subjected to a Number of Point Loads

Consider a shaft *AB* of negligible mass loaded with a number of point loads W_1, W_2, W_3 etc. in newtons, as shown in Fig. 10. Let m_1, m_2, m_3 etc. be the corresponding masses in kg. The natural frequency of such a shaft may be found out by the following two methods:

1. Energy (or Rayleigh's) method

2. Dunkerley's method.

1. Energy (or Rayleigh's) method:

Let the total deflections under the points loads W_1, W_2, W_3 etc. be y_1, y_2, y_3 etc. as shown in Fig. 10.

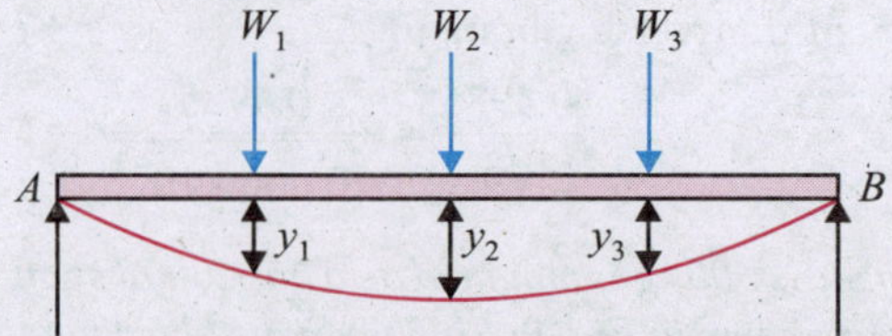

Fig. 10. Shaft loaded with a number of point loads.

Let ω = Circular frequency of transverse vibration.

In energy (or Rayleigh's) method, the maximum kinetic energy at mean position is equal to the maximum potential energy (or strain energy) at extreme position.

We know that maximum potential energy (P.E.),

$$= \frac{1}{2}m_1 . g . y_1 + \frac{1}{2}m_2 . g . y_2 + \frac{1}{2}m_3 . g . y_3 + \ldots$$

$$= \frac{1}{2}\sum m . g . y$$

And, maximum kinetic energy (K.E.),

$$= \frac{1}{2}m_1(\omega . y_1)^2 + \frac{1}{2}m_2(\omega . y_2)^2 + \frac{1}{2}m_3(\omega . y_3)^2 + \ldots$$

$$= \frac{1}{2} \times \omega^2 \Sigma\, m.\, y^2$$

Equating the maximum K.E. to the maximum P.E., we get

$$\frac{1}{2}\Sigma\, m.g.y = \frac{1}{2} \times \omega^2 \Sigma\, m.\, y^2$$

or

$$\omega^2 = \frac{g\Sigma\, m.y}{\Sigma\, m.y^2} = \text{ or } \omega = \sqrt{\frac{g\,\Sigma\, m . y}{\Sigma\, m.\, y^2}} \qquad \ldots(i)$$

∴ Natural frequency of transverse vibration,

$$f_n = \frac{\omega}{2\pi} = \frac{1}{2\pi}\sqrt{\frac{g\,\Sigma\, m.y}{\Sigma\, m . y^2}} \qquad \ldots(ii)$$

2. Dunkerley's method:

When a shaft carries a number of point loads and the uniformly distributed load, the natural frequency of transverse vibration of the shaft can be obtained from *Dunkerley's empirical formula*.

According to this empirical formula:

$$\frac{1}{(f_n)^2} = \frac{1}{(f_{n1})^2} + \frac{1}{(f_{n2})^2} + \frac{1}{(f_{n3})^2} + \ldots + \frac{1}{(f_{ns})^2} \qquad \ldots(i)$$

where f_n = Natural frequency of transverse vibration of the shaft carrying point loads and the uniformly distributed load,

f_{n1}, f_{n2}, f_{n3} etc. = Natural frequency of transverse vibration of each point load, and

f_{ns} = Natural frequency of transverse vibration of the uniformly distributed load (or due to the mass of the shaft).

Now, consider a shaft AB carrying a number of point loads and a uniformly distributed load as shown in Fig. 11.

Let $\delta_1, \delta_2, \delta_3$ etc. = Static deflections due to the point loads W_1, W_2, W_3 etc. when considered separately, and

δ_s = Static deflection due to the uniformly distributed load or due to the mass of the shaft.

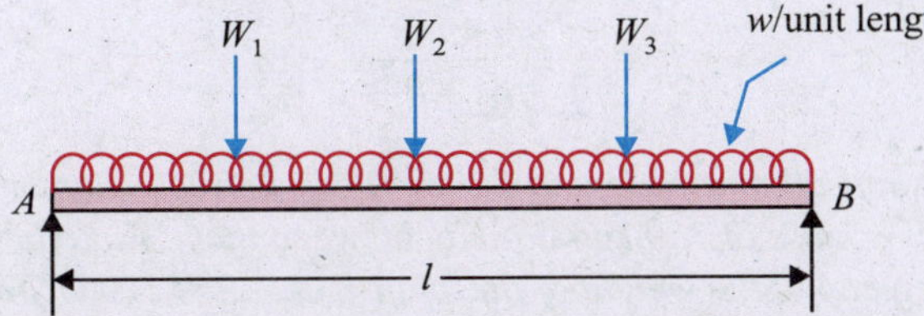

Fig. 11. Shaft carrying a number of point loads and a uniformly distributed load.

We know that natural frequency of transverse vibration due to point load W_1,

$$f_{n1} = \frac{0.4985}{\sqrt{\delta_1}} \text{Hz}$$

Similarly, natural frequencies of transverse vibrations due to point loads W_2, W_3 etc.,

$$f_{n2} = \frac{0.4985}{\sqrt{\delta_2}} \text{Hz}$$

$$f_{n3} = \frac{0.4985}{\sqrt{\delta_3}} \text{Hz etc.}$$

We also know that natural frequency of transverse vibration due to a uniformly distributed load or weight of the shaft,

$$f_{ns} = \frac{0.5615}{\sqrt{\delta_s}} \text{Hz}$$

According to *Dunkerley's empirical formula*, the natural frequency of the whole system can be calculated by the relation as:

$$\frac{1}{(f_n)^2} = \frac{1}{(f_{n1})^2} + \frac{1}{(f_{n2})^2} + \frac{1}{(f_{n3})^2} + \ldots + \frac{1}{(f_{ns})^2}$$

$$= \frac{\delta_1}{(0.4985)^2} + \frac{\delta_2}{(0.4985)^2} + \frac{\delta_3}{(0.4985)^2} + \ldots + \frac{\delta_s}{(0.5615)^2}$$

$$= \frac{1}{(0.4985)^2}\left[\delta_1 + \delta_2 + \delta_3 + \ldots + \frac{\delta_s}{1.27}\right]$$

or

$$f_n = \frac{0.4985}{\sqrt{\delta_1 + \delta_2 + \delta_3 + \ldots + \frac{\delta_s}{1.27}}} \text{Hz} \quad \ldots(ii)$$

Notes:

1. The values of $\delta_1, \delta_2, \delta_3$ etc. for *a simply supported shaft* may be obtained from the relation:

$$\delta = \frac{Wa^2b^2}{3EIl} \quad \ldots(iii)$$

where δ = Static deflection due to a point load W,
a and b = Distances of point load W from the ends,
E = Young's modulus of shaft's material,
I = Moment of inertia of the shaft, and
l = Total length of the shaft.

2. When there is no uniformly distributed load or mass of the shaft is negligible, then $\delta_s = 0$.

$$\therefore \quad f_n = \frac{0.4985}{\delta_1 + \delta_2 + \delta_3 + ...} \text{Hz} \qquad ...(iv)$$

Example 6. *A steel shaft 30 mm wide and 50 mm deep is freely supported at two points 1.2 metre apart, and carries a mass of 150 kg mid-way between them. Find the frequency of natural transverse vibration, neglecting the weight of the shaft. Take, E = 270 GN/m².*

What will be the frequency of transverse vibration, if an additional mass of 100 kg is distributed uniformly along the length of the shaft?

Solution. Given: b = 30 mm = 0.03 m; d = 50 mm = 0.05 m; l = 1.2 m; m = 150 kg; E = 270 GN/m²; = 270×10^9 N/m²; Additional mass = 100 kg (distributed uniformly along the length).

We know that moment of inertia of rectangular steel shaft,

$$I = \frac{bd^3}{12} = \frac{0.03 \times (0.05)^3}{12} = 31.25 \times 10^{-8} \text{ m}^4$$

Natural frequency of transverse vibration due to only point load of 150 kg

We know that static deflection at centre due to point load at mid-point for the simply supported steel shaft,

$$\delta_1 = \frac{Wl^3}{48EI} = \frac{(m.g)l^3}{48\,EI}$$

$$= \frac{(150 \times 9.81) \times 1.2^3}{48 \times (270 \times 10^9) \times 31.25 \times 10^{-8}} = 6.278 \times 10^{-4} \text{ m}$$

$\therefore$ Natural frequency of transverse vibration due to point load of 150 kg,

$$f_{n1} = \frac{0.4985}{\sqrt{\delta_1}} = \frac{0.4985}{\sqrt{6.278 \times 10^{-4}}} = \mathbf{19.89\ Hz} \quad \textbf{Ans.} \qquad ...(i)$$

Natural frequency of transverse vibration after adding additional distributed load

First of all, we find natural frequency of transverse vibration due to only distributed load. We know the deflection due to a uniformly distributed load of 100 kg at centre,

$$\delta_s = \frac{5wl^4}{384EI} = \frac{5Wl^3}{384EI} \qquad (\because W = w \times l)$$

$$= \frac{5 \times (100 \times 9.81) \times (1.2)^3}{384 \times (270 \times 10^9) \times (31.25 \times 10^{-8})} = 2.616 \times 10^{-4} \text{ m}$$

$\therefore$ Natural frequency of transverse vibration due to uniformly distributed load of 100 kg,

$$f_{ns} = \frac{0.5615}{\sqrt{\delta_s}} = \frac{0.5615}{\sqrt{2.616 \times 10^{-4}}} = 34.72 \text{ Hz} \qquad ...(ii)$$

Now, we find the natural frequency of transverse vibration when a shaft carries a point load and a uniformly distributed load.

According to **Dunkerley's emperical formula**, the natural frequency of transverse vibration (f_n) can be calculated as:

$$\frac{1}{f_n^2} = \frac{1}{f_{n1}^2} + \frac{1}{f_{ns}^2}$$

Substituting the values of f_{n1} and f_{ns}, from equations (*i*) and (*ii*), we have

$$\frac{1}{f_n^2} = \frac{1}{(19.89)^2} + \frac{1}{(34.72)^2}$$

$$= (2.5277 \times 10^{-3}) + (0.8295 \times 10^{-3}) = 3.3572 \times 10^{-3}$$

$\therefore$ $$f_n = \sqrt{\frac{1}{3.3572 \times 10^{-3}}} = \textbf{17.26 Hz} \quad \textbf{Ans.}$$

Alternative method:

$$f_n = \frac{0.4985}{\sqrt{\delta_1 + \frac{\delta_s}{1.27}}} = \frac{0.4985}{\sqrt{6.278 \times 10^{-4} + \frac{2.616 \times 10^{-4}}{1.27}}}$$

$$= \frac{0.4985}{0.0289} = \textbf{17.26 Hz} \quad \textbf{Ans.}$$

Example 7. *A shaft having 200 mm diameter and 1.5 m long is simply supported at its ends. It is subjected to three isolated loads of 1000 N, 1750 N and 2000 N at distances of 0.4 m, 0.6 m and 0.8 m, from the left hand support respectively. If the density of shaft material is 8000 kg/m³, find the natural frequency of free transverse vibrations of the shaft. Take E = 200 GN/m².*

Solution. Given: $d = 200$ mm $= 0.2$ m; $l = 1.5$ m; $W_1 = 1000$ N ; $W_2 = 1750$ N; $W_3 = 2000$ N; $\rho = 8000$ kg/m^3; $E = 200$ GN/m$^2 = 200 \times 10^9$ N/m^2

The loaded shaft is shown in Fig. 12.

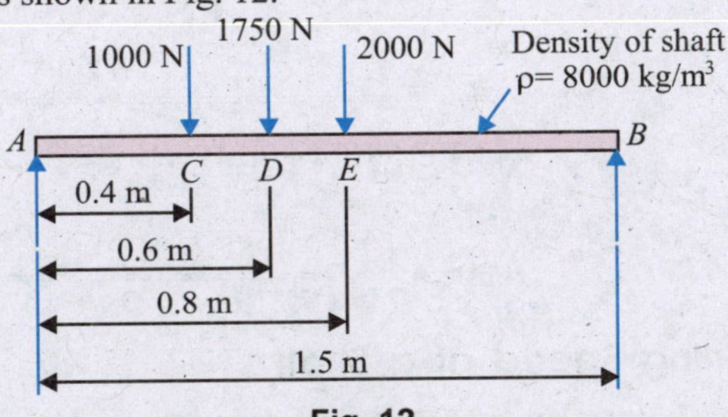

Fig. 12.

We know that moment of inertia of the shaft,

$$I = \frac{\pi}{64} \times d^4 = \frac{\pi}{64} \times (0.2)^4 = 7.854 \times 10^{-5} \text{ m}$$

and static deflection due to an isolated/point load W,

$$\delta = \frac{Wa^2b^2}{3EIl}$$

$\therefore$ Static deflection due to a point load of 1000 N only,

$$\delta_1 = \frac{1000 \times (0.4)^2 (1.1)^2}{3 \times (200 \times 10^9) \times (7.854 \times 10^{-5}) \times 1.5} = 2.739 \times 10^{-6} \text{ m}$$

...(In this case $a = 0.4$ m, $b = 1.1$ m)

Similarly, static deflection due to a point load of 1750 N only,

$$\delta_2 = \frac{1750 \times (0.6)^2 (0.9)^2}{3 \times (200 \times 10^9) \times (7.854 \times 10^{-5}) \times 1.5} = 7.219 \times 10^{-6} \text{ m}$$

...(In this case $a = 0.6$ m, $b = 0.9$ m)

and, static deflection due to a point of load 2000 N only,

$$\delta_3 = \frac{2000 \times (0.8)^2 (0.7)^2}{3 \times (200 \times 10^9) \times (7.854 \times 10^{-5}) \times 1.5} = 8.873 \times 10^{-6} \text{ m}$$

...(In this case $a = 0.8$ m, $b = 0.7$ m)

Since the density of shaft material is 8000 kg/m^2, therefore mass of the shaft,

$$m = \text{Volume} \times \text{Density} \left[\pi \times \left(\frac{d}{2}\right)^2 \times l\right] \times \rho = \frac{\pi}{4} \times d^2 \times l \times \rho$$

$$= \frac{\pi}{4} \times 0.2^2 \times 1.5 \times 8000 = \simeq 377 \text{ kg.}$$

∴ Weight of the shaft, $W = mg = 377 \times 9.81 = 3698.37$ N

We know that static deflection of the shaft due to its own weight at centre,

$$\delta_s = \frac{5}{384} \times \frac{Wl^3}{EI} = \frac{5}{384} \times \frac{3698.37 \times (1.5)^3}{200 \times 10^9 \times 7.854 \times 10^{-5}}$$

$$= 10.347 \times 10^{-6} \text{ m}$$

∴ Natural frequency of the free transverse vibration of the shaft,

$$f_n = \frac{0.4985}{\sqrt{\delta_1 + \delta_2 + \delta_3 + \dfrac{\delta_s}{1.27}}}$$

$$= \frac{0.4985}{\sqrt{2.739 \times 10^{-6} + 7.219 \times 10^{-6} + 8.873 \times 10^{-6} + \dfrac{10.347 \times 10^{-6}}{1.27}}}$$

$$= \frac{0.4985 \times 10^3}{\sqrt{2.739 + 7.219 + 8.873 + 8.147}} = \mathbf{95.98\ Hz} \quad \textbf{Ans.}$$

6.4. Critical or Whirling Speed of a Shaft

In actual practice, the centre of gravity (c.g.) of a rotating shaft which carries different mounting and accessaries, is displaced from the axis of its rotation (though the amount of displacement may be small). This happens due to *one or more* of the following *reasons*:

- The flywheel, disc, gear, pulley or rotor may be eccentrically mounted.
- The shaft may not be straight, due to rough handling or poor workmanship.
- The shaft bends due to its own weight when it is horizontal.
- The difference of pulls on two sides of the belt, taking drive from the shaft.
- The unbalanced magnetic pull may exist in case of electric machines.

Due to any one or more of the reasons mentioned above, when the different mountings and accesseries are put on the shaft, the centre of gravity of the rotating system does not coincide with the axis of the shaft or centre line of supporting bearings, when shaft is stationery. It implies that centre of gravity of the rotating system is at a certain distance from axis of rotation. Due to this, it is subjected to centrifugal force. This force will bend the shaft which will further increase the distance of centre of gravity from the axis of rotation and hence this correspondingly increases the centrifugal force (which further increases the distance of centre of gravity from the axis of rotation). In this way, the effect is cumulating and ultimately the shaft *may even fail*. The bending of the shaft depends on the eccentricity of the centre of the mass of the rotating system as well as upon the speed at which the shaft rotates.

The speed of shaft at which it tends to vibrate violently in the transverse direction is called "**Critical speed**" or "**Whirling speed**" or "**Whipping speed**" of the shaft.

Or, the speed of shaft at which it runs so that the additional deflection of it from the axis of rotation becomes infinite, is called **Critical speed**.

Consider a shaft of negligible mass (as compared to rotor) fitted with a rotor as shown in Fig. 13 (*a*). The point *O* is on the shaft axis and *G* is the centre of gravity (c.g.) of the rotor. The centre line of the bearing and the axis of the shaft coincide when the shaft is stationray. Fig. 13 (*b*) shows the shaft when rotating about the axis of rotation with a uniform angular speed ω rad/s.

Let, m = Mass of the rotor,

e = Initial eccentricity *i.e.*, initial distance of the centre of gravity (or mass) of the rotor from the centre line of the bearing or shaft axis, when the shaft is stationary,

y = Additional displacement of centre of gravity of the rotor from the axis of rotation due to centrifugal force,

s = Stiffness of the shaft *i.e.,* the load required per unit deflection of the shaft, and

ω_c = Critical speed or whirling speed of the shaf.

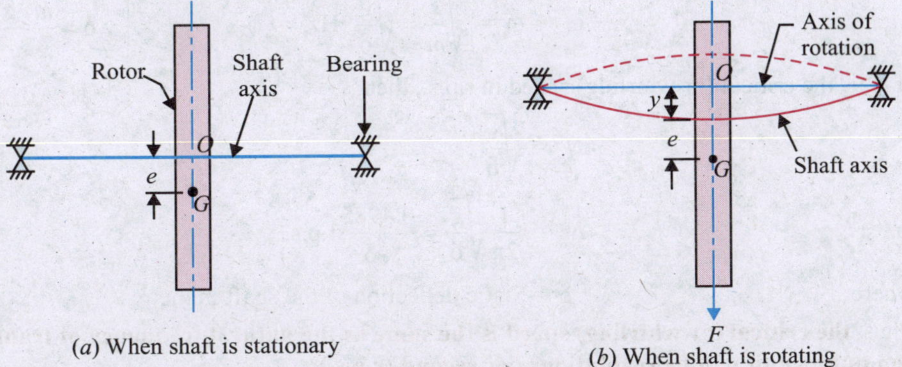

(*a*) When shaft is stationary

(*b*) When shaft is rotating

Fig. 13. Critical or whirling speed of a shaft.

Since the shaft is rotating at a uniform speed of ω rad/s, therefore centrifugal force acting radially outwards through *G* causing deflection of shaft, is given as:

$$F = m\,\omega^2(y + e)$$

Also, the shaft behaves like a spring. Therefore force resisting the deflection y,

$$= s \,.\, y$$

For the equilibrium position,

$$m\,\omega^2\,(y + e) = s \,.\, y$$

or $$m \,.\, \omega^2 \,.\, y + m \,.\, \omega^2 \,.\, e = s \,.\, y$$

or $$y(s - m \,.\, \omega^2) = m \,.\, \omega^2 \,.\, e$$

$$\therefore \qquad y = \frac{m \cdot \omega^2 \cdot e}{s - m \cdot \omega^2} = \frac{\omega^2 \cdot e}{(s/m) - \omega^2} \qquad \ldots(i)$$

We know that circular frequency (or natural angular vibrations),

$$\omega_n = \sqrt{\frac{s}{m}} \quad \text{or} \quad \frac{s}{m} = (\omega_n)^2 \qquad \ldots(ii)$$

From equations (*i*) and (*ii*),

$$y = \frac{\omega^2 \cdot e}{(\omega_n)^2 - \omega^2} \qquad \ldots(iii)$$

From the above equation, when $\omega > \omega_n$, the value of y will be negative and the shaft deflects in the opposite direction as shown dotted in Fig. 13. (*b*).

To get the values of y *always positive*, both plus and minus signs are taken.

$$\therefore \qquad y = \pm \frac{\omega^2 e}{(\omega_n)^2 - \omega^2} = \frac{\pm e}{\left(\frac{\omega_n}{\omega}\right)^2 - 1} = \frac{\pm e}{\left(\frac{\omega_c}{\omega}\right)^2 - 1} \qquad \ldots(iv)$$

...(Substituting $\omega_n = \omega_c$)

It is obvious from the above expression that when $\omega_n = \omega_c$, the value of y becomes *infinite*. Therefore ω_c is the *critical* or *whirling speed*.

∴ Critical (or whirling) speed of the shaft,

$$\omega_c = \omega_n = \sqrt{\frac{s}{m}} = \sqrt{\frac{g}{\delta}} \text{ Hz} \qquad \left[\because \delta = \frac{m \cdot g}{s}\right] \qquad \ldots(v)$$

If N_c is the critical (or whirling) speed in r.p.s., then

$$2\pi N_c = \sqrt{\frac{g}{\delta}}$$

or

$$N_c = \frac{1}{2\pi}\sqrt{\frac{g}{\delta}} = \frac{0.4985}{\sqrt{\delta}} \text{ r.p.s.} \qquad \ldots(vi)$$

where δ = Static deflection of the shaft in metres.

Thus, **the critical (or whirling) speed is the same as the natural frequency of transverse vibration but its unit will be revolutions per second (r.p.s.).**

Notes:

1. The value of e is taken as *positive* if the c.g. of the rotor *does not lie between* the centre line of the shaft and centre line of the bearings (as in the above article). On the other hand, the valure of e is taken as *negative* if the centre of gravity of the rotor *lies between* the centre line of the shaft and centre line of the bearings.
2. A shaft supported *in short bearings* or *ball bearings* is assumed to be a *simply supported shaft* while the shaft supported in *long bearings* or *journal bearings* is assumed to have *both ends fixed.*
3. Whenever a shaft may be subjected to a number of point loads, uniformly distributed load or combination of both, the critical speed of a shaft is equal to the frequency of transverse vibration in r.p.s. The Dunkeley's method may be used for determining the frequency.

Example 8. *A horizontal shaft of 15 mm diameter is supported at the ends by two short bearings and a rotor of 10 kg mass is mounted mid-way on it. The bearing span is 500 mm. Owing to some manufacturing defect, the centre of gravity of the disc is 0.1 mm away from the geometric centre of the rotor. If the system rotates at 3000 r.p.m., determine:*

(*i*) *The amplitude of the steady state vibration;*

(*ii*) *The dynamic force transmitted to each bearing.*

Take E = 200 GN/m².

Solution. Given: $d = 15$ mm $= 0.015$ m; $m = 10$ kg; $l = 500$ mm $= 0.5$ m; $e = 0.1$ mm $= 0.0001$ m; $N = 3000$ r.p.m.; $E = 200$ GN/m^2 $= 200 \times 10^9$ N/m^2

(*i*) Amplitude of the steady state vibration (*y*)

We know that moment of inertia of the shaft,

$$I = \frac{\pi}{64} \times d^4 = \frac{\pi}{64} \times (0.015)^4 = 2.485 \times 10^{-9} \text{ m}^4$$

Since the shaft is supported in short bearings, therefore it is assumed to be simply supported shaft at its ends.

We know that static deflection at centre of the shaft due to mass of rotor,

$$\delta = \frac{Wl^3}{48EI} \qquad \text{...}(i)$$

and stiffness of the shaft,

$$s = \frac{W}{\delta} = \frac{48EI}{l^3} \qquad \text{...[Using eqn. }(i)]$$

$$= \frac{48 \times 200 \times 10^9 \times 2.485 \times 10^{-9}}{(0.5)^3} = 190848 \text{ N/m}$$

We also know that critical frequency,

$$\omega_n = \sqrt{\frac{s}{m}} = \sqrt{\frac{190848}{10}} = 138.15 \text{ rad/s}$$

and angular frequency of shaft,

$$\omega = \frac{2\pi N}{60} = \frac{2\pi \times 3000}{60} = 314.16 \text{ rad/s}$$

∴ Amplitude of the steady state vibration,

$$y = \frac{\omega^2 e}{(\omega_n)^2 - \omega^2} = \frac{(314.16)^2 \times 0.0001}{(138.15)^2 - (314.16)^2}$$

$$= -1.24 \times 10^{-4} \text{ m} = \mathbf{-0.124 \text{ mm}} \qquad \textbf{Ans.}$$

The *negative sign shows that the displacement is out of phase with centrifugal force.*

(*ii*) The dynamic force transmitted to each bearing

The load on *each* bearing,

$$= \frac{\text{Load due to weight}}{2} + \frac{\text{Dynamic load}}{2}$$

$$= \frac{mg}{2} + \frac{s \times y}{2} = \frac{10 \times 9.81}{2} + \frac{190848 \times 1.24 \times 10^{-4}}{2}$$

$$= 49.05 + 11.83 = \mathbf{60.88 \text{ N}} \qquad \textbf{Ans.}$$

Example 9. *A hollow shaft 1.5 m long, having external and internal diameters 75 mm and 40 mm respectively, is supported in flexible bearings at the ends. It carries two wheels (each of mass 60 kg) one at the centre of the shaft and the other at a distance of 0.375 m from the centre towards left. If the density of the shaft material and modulus of elasticity of the shaft material are 7700 kg/m³ and 200 GN/m² respectively, find the lowest whirling speed of the shaft, considering the mass of the shaft.*

Solution. Given: $d_o = 75$ mm $= 0.075$ m; $d_i = 40$ mm $= 0.04$ m; $l = 1.5$ m; $m_1 = m_2 = 60$ kg; $\rho = 7700$ kg/m³; $E = 200$ GN/m² $= 200 \times 10^9$ N/m²

The loaded shaft is shown in Fig. 14.

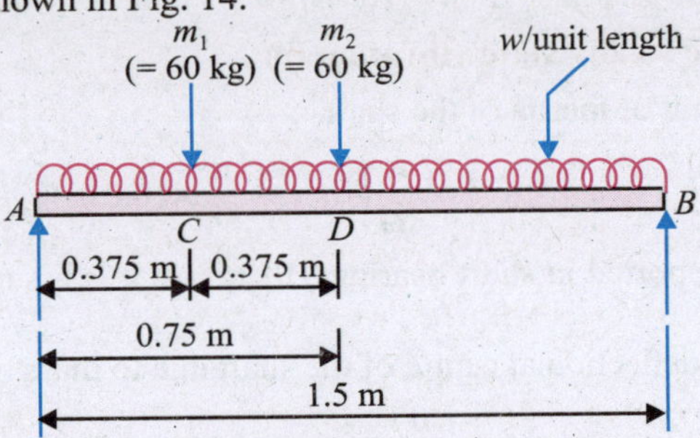

Fig. 14.

We know that moment of inertia of the hollow shaft,

$$I = \frac{\pi}{64}\left[(d_o)^4 - (d_i)^4\right] = \frac{\pi}{64}\left[(0.075)^4 - (0.04)^4\right]$$

$$= 1.427 \times 10^{-6} \text{ m}^4$$

and mass of the shaft per metre length,

$$m_s = \text{Area} \times \text{Length} \times \text{Density} = \frac{\pi}{4}\left[(d_o)^2 - (d_i)^2\right] \times l \times \rho$$

$$= \frac{\pi}{4}\left[(0.075)^2 - (0.04)^2\right] \times 1 \times 7700 = 24.34 \text{ kg/m}$$

We know that the static deflection due to a point load W,

$$\delta = \frac{Wa^2b^2}{3EIl} = \frac{(m.g)\, a^2b^2}{3EIl}$$

∴ Static deflection due to a mass of 60 kg at C,

$$\delta_1 = \frac{(m_1 g)a^2b^2}{3EIl} = \frac{(60 \times 9.81) \times (0.375)^2 \times (1.125)^2}{3 \times 200 \times 10^9 \times 1.427 \times 10^{-6} \times 1.5}$$

$$= 81.57 \times 10^{-6} \text{ m} \quad \text{...(Here } a = 0.375 \text{ m, } b = 1.125 \text{ m)}$$

Similarly, static deflection due to a mass of 60 kg at D,

$$\delta_s = \frac{(m_2 g)a^2b^2}{3EIl} = \frac{(60 \times 9.81) \times (0.75)^2 \times (0.75)^2}{3 \times 200 \times 10^9 \times 1.427 \times 10^{-6} \times 1.5}$$

$$= 145.01 \times 10^{-6} \text{ m} \quad \text{...(Here } a = b = 0.75 \text{ m)}$$

Also, static deflection due to uniformly distributed load or mass/ weight of the shaft w,

$$\delta = \frac{5wl^4}{384EI} = \frac{5\,(m_s \times g)\, l^4}{384\, EI} \qquad (\Theta\ w = m_s \times g)$$

$$= \frac{5 \times (24.34 \times 9.81) \times (1.5)^4}{384 \times 200 \times 10^9 \times 1.427 \times 10^{-6}}$$

$$= 55.15 \times 10^{-6}\,\text{m}$$

We know that frequency of transverse vibrations,

$$f_n = \frac{0.4985}{\sqrt{\delta_1 + \delta_2 + \dfrac{\delta_s}{1.27}}}$$

$$= \frac{0.4985}{\sqrt{81.57 \times 10^{-6} + 145.01 \times 10^{-6} + \dfrac{55.15 \times 10^{-6}}{1.27}}}$$

$$= \frac{0.4985 \times 10^3}{\sqrt{81.57 + 145.01 + 43.425}}$$

$$= 30.34\ \text{Hz}$$

Since the whirling speed of the shaft (N_c) in r.p.s. is equal to the frequency of transverse vibration in Hz, therefore,

$$N_c = 30.34\ \text{r.p.s} = 30.34 \times 60 = \textbf{1820.4 r.p.m.} \qquad \textbf{Ans.}$$

7. FREE TORSIONAL VIBRATIONS

When the particles of a shaft or disc move in a circle about the axis of shaft or rotation, then the vibrations are called **torsional vibrations**. In this case, the shaft is twisted and untwisted alternately and torsional shear stresses are induced in it.

In this section, we shall discuss the frequency of torsional vibrations for different rotating systems.

7.1 Natural Frequency of Free Torsional Vibrations

Consider a constraint (or shaft) of negligible mass whose one end is fixed and the other end carrying a heavy disc as shown in Fig. 15.

Let m = Mass of the disc, in kg,

I = Mass moment of inertia of disc = mk^2 in kg-m^2,

k = Radius of gyration, in metres,

θ = Angular displacement of the shaft from mean position after time t, in radians, and

q = Torsional stiffness of the shaft, in Nm.

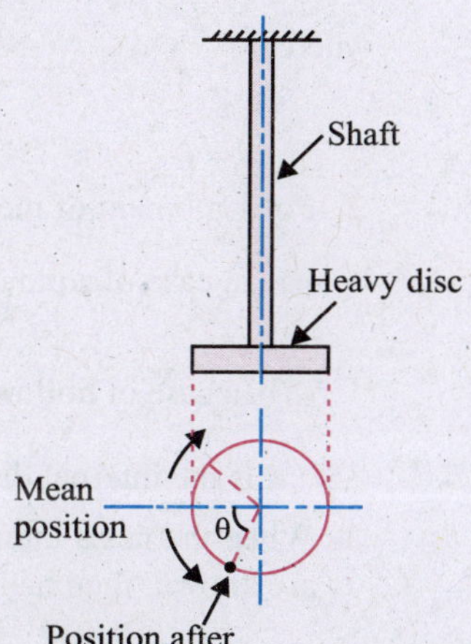

Fig. 15. Natural frequency of free torsional vibrations.

The disc or shaft is now displaced from it mean position by an angular displacement θ as shown in Fig. 15 and is then released. Therefore after time t,

$$\text{Accelerating force} = I \times \frac{d^2\theta}{dt^2} \qquad \ldots(i)$$

and restoring force developed in the shaft, $= -\,q\,.\,\theta$ ...(*ii*)

Equating equations (*i*) and (*ii*), the equation of motion after time t is given as:

$$I \times \frac{d^2\theta}{dt^2} = -q \cdot \theta$$

or
$$I \times \frac{d^2\theta}{dt^2} + q\,\theta = 0$$

or
$$\frac{d^2\theta}{dt^2} + \frac{q}{I} \times \theta = 0 \quad \ldots(iii)$$

We know that the fundamental equation of simple harmonic motion is given as

$$\frac{d^2\theta}{dt^2} + \omega^2 x = 0 \quad \ldots(iv)$$

Comparing equations (*iii*) and (*iv*), we have

$$\omega^2 = \frac{q}{I} \quad \text{or} \quad \omega = \sqrt{\frac{q}{I}} \quad \ldots(v)$$

∴ Time period of torsional vibration,

$$T = \frac{2\pi}{\omega} = 2\pi\sqrt{\frac{I}{q}} \quad \ldots(vi)$$

and natural frequency of torsional vibration,

$$f_n = \frac{1}{T} = \frac{1}{2\pi}\sqrt{\frac{q}{I}} \quad \ldots(vii)$$

Notes:

1. In the above relation, the value of the torsional stiffness q may be obtained from the **torsion equation** as follows

$$\frac{T}{J} = \frac{C\theta}{l} \quad \text{or} \quad \frac{T}{\theta} = \frac{C \cdot J}{l}$$

$$\therefore \quad = \frac{C \cdot J}{l} \qquad \left(\because \frac{T}{\theta} = q\right)$$

where C = Modulus of rigidity of the shaft material,
J = Polar moment of inertia of the shaft cross-section and
l = Length of the shaft.

2. Polar moment of inertia in different cases:

   (*i*) In case of solid circular shaft, $J = \frac{\pi}{32} d^4$, where d is the diameter of shaft.

   (*ii*) In case of hollow circular shaft, $J = \frac{\pi}{32}\left(d_o^4 - d_i^4\right)$, where d_o is external diameter and d_i is the internal diameter of the shaft.

3. When the mass moment of inertia of the constraint I_c and the mass moment of the disc I are known, then natural frequency of torsion vibration,

$$f_n = \frac{1}{2\pi}\sqrt{\frac{q}{I + (I_c/3)}} \quad \ldots(viii)$$

Example 10. *A vertical shaft of 100 mm diameter is fixed at both the ends and a flywheel is mounted on it as shown in Fig. 16. The flywheel has a mass of 1200 kg and its radius of gyration is 0.65 m. Find the natural frequency of torsional vibrations, if the modulus of rigidity for the shaft material is 84 GN/m².*

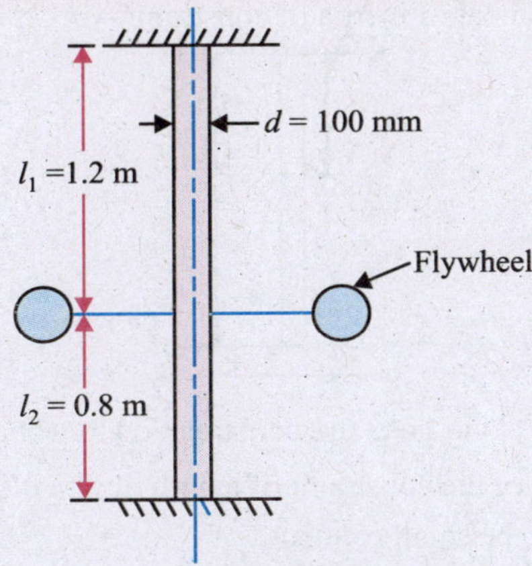

Fig. 16

Solution. Given: $d = 100$ mm $= 0.1$ m; $l_1 = 1.2$ m; $l_2 = 0.8$ m; $m = 1200$ kg; $k = 0.65$ m; $C = 84$ GN/m^2 $= 84 \times 10^9$ N/m^2

We know that polar moment of inertia of the shaft,

$$J = \frac{\pi}{32} d^4 = \frac{\pi}{32} \times (0.1)^4 = 9.817 \times 10^{-6} \text{ m}^4$$

∴ Torsional stiffness of the shaft for length l_1,

$$q_1 = \frac{C.J}{l_1} = \frac{84 \times 10^9 \times 9.817 \times 10^{-6}}{1.2} = 687.19 \times 10^3 \text{ N-m}$$

Similarly, torsional stiffness of the shaft for length l_2,

$$q_2 = \frac{C.J}{l_2} = \frac{84 \times 10^9 \times 9.817 \times 10^{-6}}{0.8} = 1030.785 \times 10^3 \text{ N-m}$$

∴ Total torsional stiffness of the shaft,

$$q = q_1 + q_2 = 687.19 \times 10^3 + 1030.785 \times 10^3$$
$$= 1717.975 \times 10^3 \text{ N-m}$$

We also know that moment of inertia of the flywheel,

$$I = mk^2 = 1200 \times (0.65)^2 = 507 \text{ kg-m}^2$$

∴ Natural frequency of torsional vibrations,

$$f_n = \frac{1}{2\pi}\sqrt{\frac{q}{I}} = \frac{1}{2\pi}\sqrt{\frac{1717.975 \times 10^3}{507}} = 9.26 \text{ Hz}$$ **Ans.**

7.2 EFFECT OF INERTIA OF THE CONSTRAINT ON TORSIONAL VIBRATIONS

Consider a constraint (*i.e.*, shaft) whose one end is fixed and other end is free as shown in Fig. 17.

Let ω = Angular velocity of free end of constraint,

l = Length of constraint,

m = Mass of constraint per unit length,

m_c = Total mass of constraint = ml,

k = Radius of gyration of constraint, and

I_c = Total moment of inertia of constraint = $m_c.k^2 = m.l.^2$.

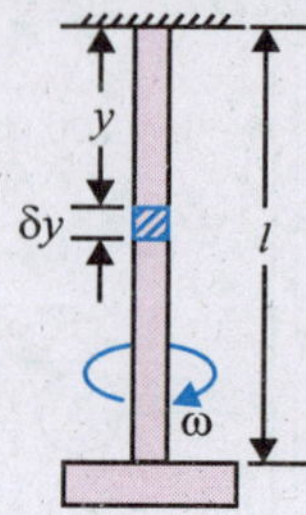

Fig. 17. Effect of inertia of the constraint on torsional vibrations.

Consider a small element of the constraint of length δy at a distance y from the fixed end.

∴ Moment of inertia of the small element,

$$\delta I = (m.\delta y)k = \frac{\delta y}{l} \times m.k^2 \times l \qquad \text{...(Dividing and multiplying by } l\text{)}$$

$$= \frac{\delta y}{l} \times I_c$$

and angular velocity of the small element,

$$\delta = \frac{\omega}{l} \times y$$

Also, Kinetic energy possessed by the small element δy,

$$= \frac{1}{2} \times \delta I \times (\delta\omega)^2 = \frac{1}{2} \times \left(\frac{\delta y}{l} \times I_c\right) \times \left(\frac{\omega}{l} \times y\right)^2$$

$$= \frac{1}{2} \times \frac{I_c \omega^2}{l^3} \times y^2 \delta y$$

∴ Total kinetic energy of the constraint is given as:

$$\text{K.E} = \int_0^l \frac{1}{2} \times \frac{I_c \omega^2}{l^3} y^2 dx = \frac{1}{2} \times \frac{I_c \omega^2}{l^3} \left[\frac{y^3}{3}\right]_o^l = \frac{1}{2} \times \frac{I_c \omega^2}{l^3} \times \frac{l^3}{3}$$

$$= \frac{1}{2} \times \left(\frac{I_c}{3}\right) \omega^2 \qquad \text{...}(i)$$

If the constraint is assumed to be of negligible mass and a mass whose moment of inertia $\frac{I_c}{3}$ is placed at the free end, then total kinetic energy of the constraint,

$$= \frac{1}{2} \times \left(\frac{I_c}{3}\right) \omega^2 \qquad \text{...(Same as equ. } (i)\text{)}$$

Thus, the two systems are dynamically same. Therefore, inertia of the constraint may be allowed for by adding $\frac{I_c}{3}$ to its mass moment of inertia I of the disc/ body at the free end.

From the above discussion, when the mass moment of inertia of constraint I_c and mass moment of inertia of the body/disc I are given, then natural frequency of torsional vibration,

$$f_n = \frac{1}{2\pi}\sqrt{\frac{q}{I + \frac{I_c}{3}}} \qquad \text{...(ii)}$$

7.3 Free Torsional Vibrations of a Single Rotor System

Consider a single rotor system as shown in Fig. 18.

We have already discussed the natural frequency of free torsional vibration for a shaft fixed at one end and carrying a rotor at the free end is given by,

$$f_n = \frac{1}{2\pi}\sqrt{\frac{q}{I}} = \frac{1}{2\pi}\sqrt{\frac{CJ}{l.I}} \qquad \left(\because q = \frac{CJ}{l}\right)$$

where C = Modulus of rigidity of shaft material,

J = Polar moment of inertia of shaft,

l = Length of the shaft,

m = Mass of rotor,

k = Radius of gyration of rotor, and

I = Mass moment of inertia of rotor = mk^2.

It may be observed that the amplitude of torsional vibration is zero at A (known as node) and maximum at B, as shown in Fig. 18.

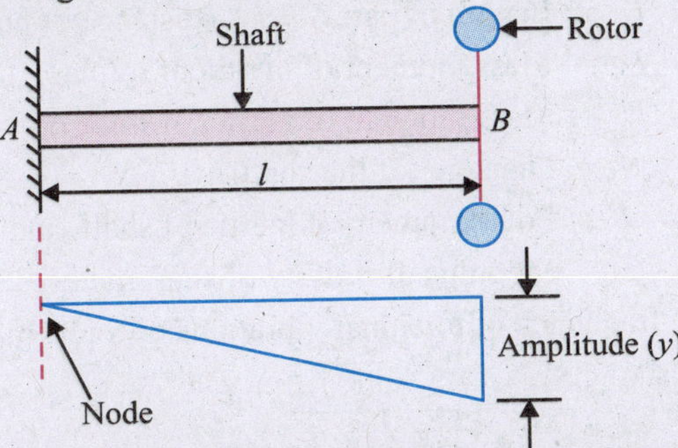

Fig. 18. Free torsional vibrations of a single rotor system.

Thus, **node** is the point or section of the shaft whose amplitude of torsional vibration is zero. It implies that the shaft remains uneffected by the vibration at the **node**.

7.4 Free Torsional Vibrations of a Two-Rotor System

Consider a two-rotor system which consists of a shaft with two rotors A and B at its ends as shown in Fig. 19. In this system, the torsional vibrations occur only when the two rotors move in *opposite directions i.e.,* if rotor A moves in clockwise direction, while rotor B moves in anti-clockwise direction and *vice-versa.* Thus, the mode of vibration will be such that the two rotors will *always be* 180° *out of phase but have the same frequency.*

In that case, a particular section of the shaft would neither get twisted nor untwisted as if it were *damped* at that section along the length of the shaft. The point at which the shaft behaves as if it were damped is known as **node point** (N). The UNV is known as **elastic** *line for the shaft.*

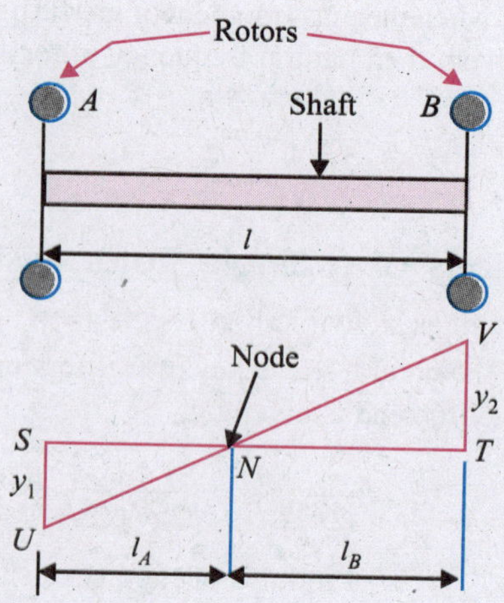

Fig. 19. Free torsional vibrations of a two-rotor system.

It may be observed, that the node lies at N where the amplitude of vibration is zero as shown in Fig. 19.

The node point N can be assumed as a fixed end and the shaft can assumed as two seperate shafts NS and NT each fixed to one of its ends at N and carrying rotors at other free ends.

Let l = Total length of the shaft $= l_A + l_B$

l_A = Length of part NS *i.e.*, distance of node from rotor A,

l_B = Length of part NT, *i.e.*, distance of node from rotor B,

I_A = Mass moment of inertia of rotor A,

I_B = Mass moment of inertia of rotor B,

d = Diameter of the shaft,

J = Polar moment of inertia of shaft, and

C = Modulus of rigidity of shaft material.

We know that natural frequency of torsional vibrations of rotor A,

$$f_n = \frac{1}{2\pi}\sqrt{\frac{C.J}{l_A . I_A}} \quad ...(i)$$

Similarly, natural frequency of torsional vibrations of rotor B,

$$f_{nB} = \frac{1}{2\pi}\sqrt{\frac{C.J}{l_B . I_B}} \quad ...(ii)$$

In this system, two rotors have same frequency.

$$\therefore \quad f_{nA} = f_{nB}$$

or

$$\frac{1}{2\pi}\sqrt{\frac{C.J}{l_A . I_A}} = \frac{1}{2\pi}\sqrt{\frac{C.J}{l_B . I_B}}$$

or

$$l_A . I_A = l_B . I_B \quad \text{or} \quad l_A = \frac{l_B . I_B}{I_A} \quad ...(iii)$$

We also know that,

$$l = l_A + l_B \quad ...(iv)$$

To determine the position of node and natural frequency of torsional vibration for a two rotor system:

(*i*) Solving equations (*iii*) and (*iv*), we may find the *values* of l_A and l_B and hence the *position of node.*

(*ii*) Substituting the values of l_A or l_B in equations (*i*) or (*ii*), the natural frequency of torsional vibrations for a two-rotor system may be evaluated.

7.5 Free Torsional Vibrations of a Three-Rotor System

Consider a three-rotor system, which consists of a shaft and three rotors *A*, *B* and *C* as shown in Fig. 20(*a*). The rotors *A* and *C* are attached to the shaft at its ends and the rotor *B* is attached in between *A* and *C*.

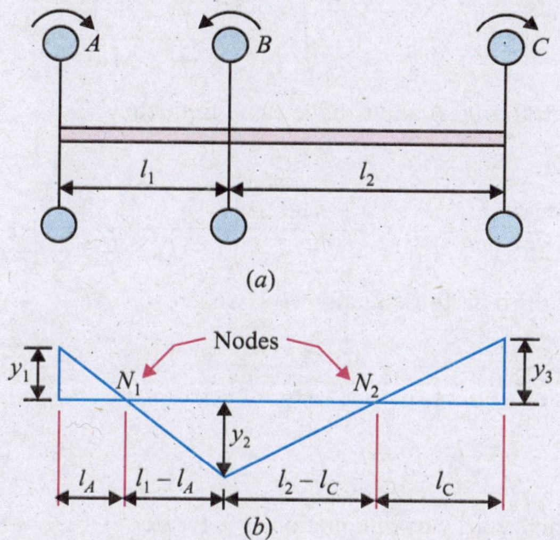

Fig. 20. Free torsional vibrations of a three-rotor system.

In this system, the torsional vibrations may occur in two ways, that is, either one node or two nodes. In each case, the two rotors rotate in one direction and the third rotor rotates in opposite direction with the same frequency.

Let the rotors *A* and *C* rotate in the same direction (clockwise direction) and the rotor *B* in opposite direction (anti-clockwise direction) as shown in Fig. 20 (*a*). Let N_1 and N_2 be the node points of such a system as depicted in Fig. 20 (*b*).

As already discussed in previous article that the shaft may be considered as fixed end at the nodes.

For Rotors A and C

Natural frequency of torsional vibration for rotor *A*,

$$f_A = \frac{1}{2\pi}\sqrt{\frac{C.J}{l_A.I_A}} \qquad ...(i)$$

Similarly, natural frequency of torsional vibration for rotor *C*,

$$f_{nC} = \frac{1}{2\pi}\sqrt{\frac{C.J}{l_C.I_C}} \qquad ...(ii)$$

For Rotors B

In rotor B, the resisting torque is supplied by two lengths $(l_1 - l_A)$ and $(l_2 - l_C)$ between the nodes N_1 and N_2. Since, each length is twisted through the same angular displacement, therefore total torsional stiffness of the rotor B is equal to the sum of the separate stiffness *i.e.*,

Total torsional stiffness of rotor B

$$= \text{Torsional stiffness due to } (l_1 - l_A) + \text{Torsional stiffness due to } (l_2 - l_C)$$

$$= \frac{C.J}{l_1 - l_A} + \frac{C.J}{l_2 - l_C} = C.J\left(\frac{1}{l_1 - l_A} + \frac{1}{l_2 - l_C}\right)$$

∴ Natural frequency of torsional vibration for rotor B,

$$f_{nB} = \frac{1}{2\pi}\sqrt{\frac{C.J}{I_B}\left[\frac{1}{l_1 - l_A} + \frac{1}{l_2 - l_C}\right]} \qquad \text{...(iii)}$$

In this system, three rotors A, B and C have same frequency.

$$\therefore \qquad f_{nA} = f_{nB} = f_{nC}$$

or

$$\frac{1}{2\pi}\sqrt{\frac{C.J}{l_A.I_A}} = \frac{1}{2\pi}\sqrt{\frac{C.J}{I_B}\left[\frac{1}{l_1 - l_A} + \frac{1}{l_2 - l_C}\right]} = \frac{1}{2\pi}\sqrt{\frac{C.J}{l_C.I_C}} \qquad \text{...(iv)}$$

Taking first part and third part of equation (*iv*), we get

$$\frac{1}{2\pi}\sqrt{\frac{C.J}{l_A.I_A}} = \frac{1}{2\pi}\sqrt{\frac{C.J}{l_C.I_C}}$$

or

$$l_A.I_A = l_C.I_C \quad \text{or} \quad l_A = \frac{l_C.I_A}{I_A} \qquad \text{...(v)}$$

Taking second part and third part of equation (*iv*), we get

$$= \frac{1}{2\pi}\sqrt{\frac{C.J}{I_B}\left[\frac{1}{l_1 - l_A} + \frac{1}{l_2 - l_C}\right]} = \frac{1}{2\pi}\sqrt{\frac{C.J}{l_C.l_C}}$$

or

$$= \frac{1}{I_B}\left[\frac{1}{l_1 - l_A} + \frac{1}{l_2 - l_C}\right] = \frac{1}{l_C.I_C} \qquad \text{...(vi)}$$

Solving equations (*v*) and (*vi*) by substituting the value of l_A from equations (*v*) in equations (*vi*), a quadratic equation in l_C is obtained. Thus, there are two values of l_C and correspondingly two values of l_A. One value of l_A and the corresponding value of l_C gives the *position of two nodes*. The frequency obtained by substituting the value of l_A or l_C in equations (*i*) or (*ii*) is known as **two node frequency.** But in the other pair of values, one gives the position of *single node* and the other is beyond the physical limits of the equation. In this case, the frequency obtained is known as **fundamental frequency** or **single node frequency.**

Notes:

1. When the rotors A and B rotate in the same direction and rotor C in opposite direction, then the torsional vibrations occur with a single node as shown in Fig. 21 (*b*). In this case $l_A > l_1$ *i.e.*, the node lies between the rotor B and C, but it *does not give the actual value of the note.*

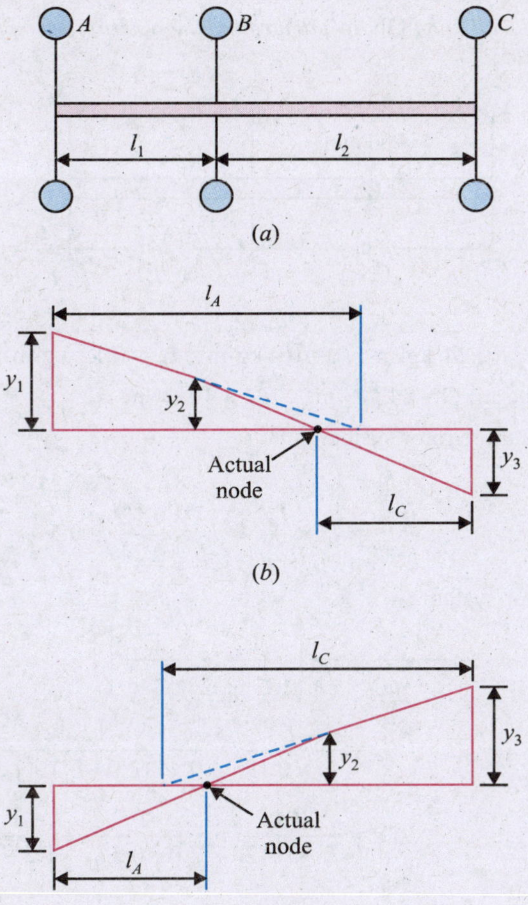

Fig. 21

2. When the rotors B and C rotate in the same direction and the rotor A in opposite direction, then the torsional vibrations also occur with a single mode as shown in Fig. 21 (*c*). In this case $l_C > l_2$ *i.e.,* node lies between the rotors A and B, but it *does not give the actual value of the node.*

3. When the amplitude of the rotor A (y_1) is known, then the amplitude of the rotor B,

$$y_2 = \frac{l_A - l_1}{l_A} \times y_1$$

and amplitude of the rotor C,

$$y_3 = \frac{l_C}{l_C - l_2} \times y_2$$

Since there are two values of l_A and l_C, therefore there will be *two values of amplitude for one node and two node vibrations.*

Example 11. *A centrifugal pump is driven directly by a single cylinder oil engine. The rotating mass of the engine, flywheel and pump with shaft is equivalent to a three rotor system as shown in Fig. 22. The mass moment of inertia of rotors A, B and C are 0.24, 0.4 and 0.12 kg-m^2. If modulus* *of rigidity of the shaft material is 84* GN/m^2, *determine the position of the nodes and frequencies of torsional vibration.*

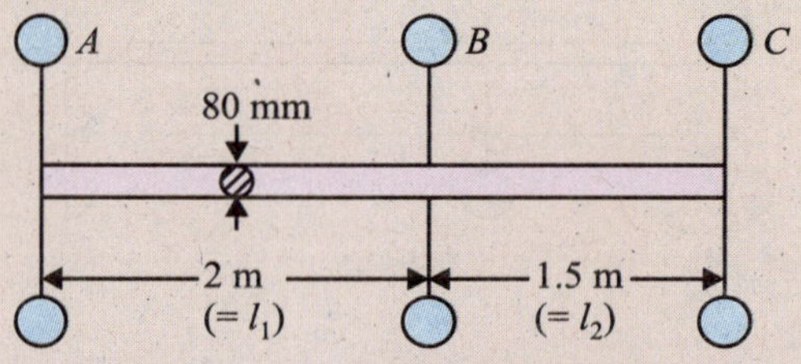

Fig. 22.

Solution. Given: $I_A = 0.24$ kg-m^2; $I_B = 0.4$ kg-m^2; $I_C = 0.12$ kg-m^2; d = 80 mm = 0.08 m, l_1 = 2.0 m; l_2 = 1.5 m; C = 84 GN/m^2 = 84 × 10^9 N/m^2.

We know that for a three rotor system,

$$l_A \cdot I_A = l_C \cdot I_C$$

or
$$l_A = l_C \times \frac{I_C}{I_A} = l_C \times \frac{0.12}{0.24} = 0.5\, l_C$$

Also, for a three rotor system

$$\frac{1}{l_C \cdot I_C} = \frac{1}{I_B}\left[\frac{1}{l_1 - l_A} + \frac{1}{l_2 - l_C}\right]$$

or
$$= \frac{1}{l_C \times 0.12} = \frac{1}{0.4}\left[\frac{1}{(2.0 - l_A)} + \frac{1}{1.5 - l_C}\right]$$

or
$$= \frac{0.4}{l_C \times 0.12} = \frac{1}{2.0 - 0.5\, l_C} + \frac{1}{1.5 - l_C}$$

...(Substituting $l_A = 0.5\, l_C$)

or
$$= \frac{1}{0.3 l_C} = \frac{1.5 - l_C + 2.0 - 0.5\, l_C}{(2.0 - 0.5\, l_C)\,(1.5 - l_C)}$$

or
$$= \frac{1}{0.3\, l_C} = \frac{3.5 - 1.5\, l_C}{3 - 2.0\, l_C - 0.75\, l_C + 0.5\, l_C^2}$$

or
$$= \frac{1}{0.3\, l_C} = \frac{3.5 - 1.5\, l_C}{3 - 2.75\, l_C + 0.5\, l_C^2}$$

or $\quad 3 - 2.75\, l_C + 0.5\, l_C^{\,2} = 1.05\, l_C - 0.45\, l_C^{\,2}$

or $\quad 0.95\, l_C^{\,2} - 3.8\, l_C + 3 = 0$

Solving quadratic equation, the values of l_C are given as:

$$l_C = \frac{3.8 \pm \sqrt{(3.8)^2 - 4 \times 0.95 \times 3}}{2 \times 0.95} = \frac{3.8 \pm 1.74}{1.9}$$

$$= \frac{3.8 + 1.74}{1.9} \quad \text{or} \quad \frac{3.5 - 1.74}{1.9}$$

$$= 2.916 \text{ m} \quad \text{or} \quad 1.084 \text{ m}$$

and
$$l_A = 0.5\, l_C = 1.458 \text{ m} \quad \text{or} \quad 0.542 \text{ m}$$

Here, we see that the value of $l_C = 2.916$ m, and the corresponding value of $l_A = 1.458$ m gives the position of *single node* as shown in Fig. 23 (*b*). The value of $l_C = 1.084$ m and the corresponding value of $l_A = 0.542$ m gives the position of *two nodes,* as shown in Fig. 23 (*c*).

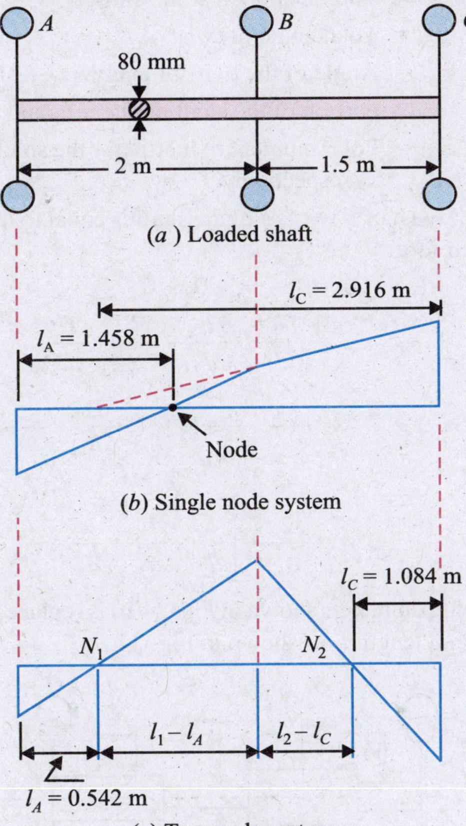

Fig. 23

We know that the polar moment inertia of the shaft,

$$J = \frac{\pi}{32} \times d^4 = \frac{\pi}{32} \times (0.08)^4 = 4.02 \times 10^{-6}\ \text{m}^4$$

∴ Natural frequency of torsional vibration for a *single node system,*

$$f_{n1} = \frac{1}{2\pi}\sqrt{\frac{C.J}{l_A.I_A}} = \frac{1}{2\pi}\sqrt{\frac{84 \times 10^9 \times 4.02 \times 10^{-6}}{1.458 \times 0.24}} = \textbf{156.35 Hz} \quad \textbf{Ans.}$$

Similarly, natural frequency of torsional vibration for a *two-node system,*

$$f_n = \frac{1}{2\pi}\sqrt{\frac{C.J}{l_A.I_A}} = \frac{1}{2\pi}\sqrt{\frac{84 \times 10^9 \times 4.02 \times 10^{-6}}{0.542 \times 0.24}} = \textbf{256.43 Hz} \quad \textbf{Ans.}$$

7.6 TORSIONALLY EQUIVALENT SHAFT

In actual practice, the shaft may have different diameters for different lengths. But theoretically such a shaft may be replaced by a torsionally equivalent shaft of uniform diameter. Two shafts are said to be *torsionally equivalent* to one another, if these *twist through the same.*

Consider a shaft of varying diameters d_1, d_2, d_3 and d_4 over lengths l_1, l_2, l_3 and l_4 respectively, carrying two rotors A and B at its ends as shown in Fig. 24(a).

Let
T = Torque applied at the ends,
θ = Total angle of twist of B w.r.t. A,
$\theta_1, \theta_2, \theta_3, \theta_4$ = Angle of the twist of lengths l_1, l_2, l_3 and l_4 w.r.t. to adjoining portion of the shaft on the left, and
J_1, J_2, J_3, J_4 = Polar moment of inertia for the shafts of diameters d_1, d_2, d_3 and d_4 respectively.

Since the total angle of twist of B w.r.t. A of the shaft is equal to the sum of the angle of twists of lengths l_1, l_2, l_3 and l_4, therefore

$$\theta = \theta_1 + \theta_2 + \theta_3 + \theta_4$$

$$= \frac{Tl_1}{CJ_1} + \frac{Tl_2}{CJ_2} + \frac{Tl_3}{CJ_3} + \frac{Tl_4}{CJ_4}$$

$$= \frac{Tl_1}{C \times \frac{\pi}{32} \times d_1^4} + \frac{Tl_2}{C \times \frac{\pi}{32} \times d_2^4} + \frac{Tl_3}{C \times \frac{\pi}{32} \times d_3^4} + \frac{Tl_4}{C \times \frac{\pi}{32} \times d_4^4}$$

$$= \frac{32}{\pi} \cdot \frac{T}{C} \left[\frac{l_1}{d_1^4} + \frac{l_2}{d_2^4} + \frac{l_3}{d_3^4} + \frac{l_4}{d_4^4} \right] \quad ...(i)$$

Now, the shaft of varying diameters shown in Fig. 24(a) is replaced by a torsionally equivalent shaft of uniform diameter d and length l_e as shown in Fig. 24(b).

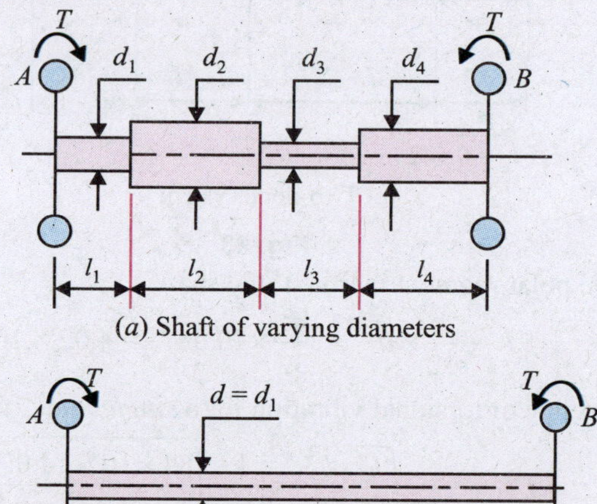

(a) Shaft of varying diameters

(b) Torsionally equivalent shaft

Fig. 24.

∴ Angle of twist of torsionally equivalent shaft,

$$\theta = \frac{Tl_e}{CJ} = \frac{Tl_e}{C \times \frac{\pi}{32} d^4} = \frac{32}{\pi} \cdot \frac{T}{C} \cdot \frac{l_e}{d^4} \quad ...(ii)$$

Since two shafts are torsionally equivalent to one another, therefore these two shafts must have the same total angle of twist when equal opposing torques T are applied at their opposite ends.

Equating equations (i) and (ii), we ge

$$\frac{32}{\pi}\cdot\frac{T}{C}\cdot\frac{l_e}{d^4} = \frac{32}{\pi}\cdot\frac{T}{C}\left[\frac{l_1}{d_1^4}+\frac{l_2}{d_2^4}+\frac{l_3}{d_3^4}+\frac{l_4}{d_4^4}\right]$$

or
$$l_e = l_1\left(\frac{d}{d_1}\right)^4 + l_2\left(\frac{d}{d_2}\right)^4 + l_3\left(\frac{d}{d_3}\right)^4 + \left(\frac{d}{d_4}\right)^4 \quad \text{...(iii)}$$

For simplicity of arithmetical calculation, the diameter d of the torsionally equivalent shaft is assumed equal to one of the diameters existing on the actual shaft. Assuming $d = d_1$, the above equation can be written as:

$$l_e = l_1 + l_2\left(\frac{d_1}{d_2}\right)^4 + l_3\left(\frac{d_1}{d_3}\right)^4 + l_4\left(\frac{d_1}{d_4}\right)^4 \quad \text{...(iv)}$$

This expression gives the length of torsionally equivalent shaft.

Example 12. *Two rotors A and B are attached to the ends of a shaft 2 m long. The mass of rotor A is 900 kg and its radius of gyration is 0.85 m; the corresponding values of rotor B are 700 kg and 0.55 m respectively. The shaft is 95 mm in diameter for first 0.9 m; 60 mm for next 0.6 m; and 50 mm for the remainder of its length. If the modulus of rigidity of shaft material is 80 GN/m^2, determine:*

(i) *The location of the node, and*

(ii) *The natural frequency of free torsional vibrations of the system.*

Solution. Given: $l = 2$ m; $l_1 = 0.9$ m; $d_1 = 95$ mm $= 0.095$ m; $l_2 = 0.06$ m; $d_2 = 60$ mm $= 0.06$ m; $l_3 = 0.05$ m; $d_3 = 50$ mm $= 0.05$ m; $m_A = 900$ kg, $k_A = 0.85$ m; $m_B = 700$ kg, $k_B = 0.55$ m; $C = 80$ GN/m^2 $= 80 \times 10^9$ N/m^2.

The actual shaft is shown in Fig. 25(a). First of all, we find the length of torsionally equivalent shaft assuming its diameter as $d_1 = 95$ mm as shown in Fig. 25(b).

We know that length of equivalent shaft assuming of 95 mm diameter,

$$le = l_1 + l_2\left(\frac{d_1}{d_2}\right)^4 + l_3\left(\frac{d_1}{d_3}\right)^4$$

$$= 0.90 + 0.6\left(\frac{0.095}{0.06}\right)^4 + 0.5\left(\frac{0.095}{0.05}\right)^4$$

$$= 0.90 + 3.77 + 6.52 = 11.19 \text{ m} \quad \text{...(i)}$$

(i) Location of the node

Suppose the node of the equivalent shaft lies at N as shown in Fig. 25(c).

Let l_A = Distance of the node from rotor A, and

l_B = Distance of the node from rotor B.

We know that mass moment of inertia of rotor A,

$$I_A = m_A k_A^2 = 900 \times (0.85)^2 = 650.25 \text{ kg-m}^2,$$

Similarly, mass moment of inertia of rotor B,

$$I_B = m_B k_B^2 = 700 \times (0.55)^2 = 211.75 \text{ kg-m}^2$$

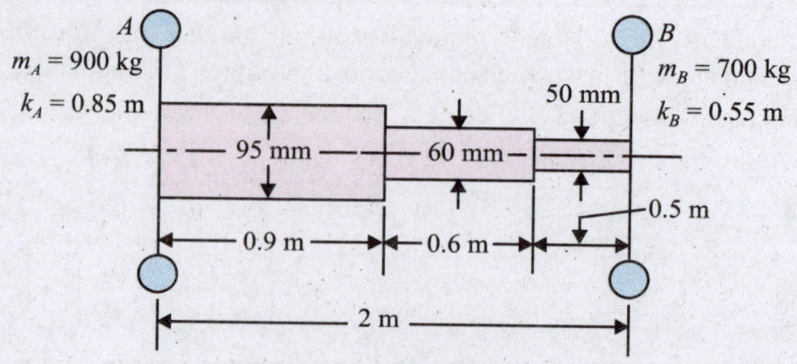

(*a*) Shaft of varying diameters

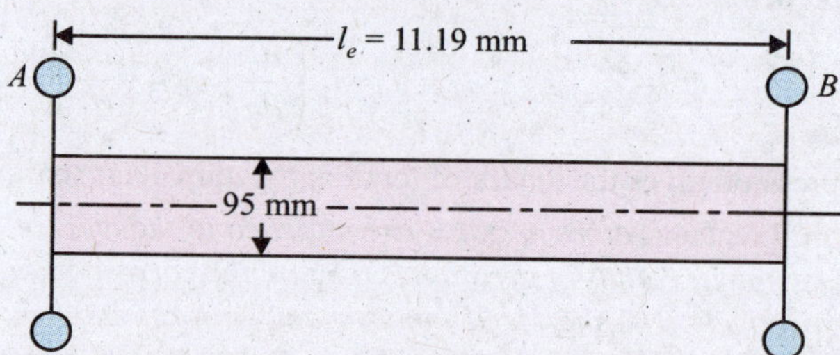

(*b*) Torsionally equivalent shaft

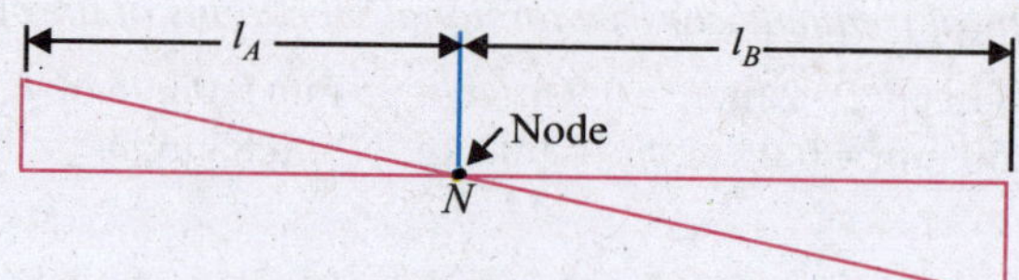

(*c*) Position of node on equivalent shaft

Fig. 25.

We know that for two rotor system,

$$l_A \cdot I_A = l_B \cdot I_B$$

or

$$l_A = \frac{l_B \, I_B}{I_A} = \frac{l_B \times 211.75}{650.25} = 0.326 \, l_B \qquad \text{...}(ii)$$

Also,

$$l_A + l_B = l_e \qquad \text{...}(iii)$$

Substituting the value of l_e and l_A from equations (*i*) and (*ii*), we get

$$0.326 \, l_B + l_B = 11.19 \quad \text{or} \quad l_B = 8.44 \text{ m}$$

and

$$l_A = l_e - l_B = 11.19 - 8.44 = 2.75 \text{ m}$$

Hence, the node lies at 2.75 m from rotor *A* or 8.44 m from rotor *B* on the *equivalent shaft.*

∴ Position of node on the *original shaft* from rotor *A*,

$$= l_1 + (l_A - {}_1) \left(\frac{d_2}{d_1} \right)^4$$

$$= 0.9 + (2.75 - 09) \left(\frac{0.06}{0.095} \right)^4 = \mathbf{1.194 \text{ m}} \quad \textbf{Ans.}$$

(*ii*) **Natural frequency of free torsional vibrations**

We know that polar moment of inertia of the equivalent shaft,

$$J = \frac{\pi}{32} \times d_1^4 = \frac{\pi}{32} \times (0.095)^4 = 7.996 \times 10^{-6} \text{ m}^4$$

∴ Natural frequency of free torsional vibrations,

$$f_n = f_{nA} \text{ (or } f_{nB})$$

$$= \frac{1}{2\pi}\sqrt{\frac{C\,.\,J}{l_A\,.\,I_A}} = \frac{1}{2\pi}\sqrt{\frac{80 \times 10^9 \times 7.996 \times 10^{-6}}{2.75 \times 650.25}}$$

$$= \mathbf{3.01 \approx 3 \text{ Hz}} \quad \textbf{Ans.}$$

8. CAUSES OF VIBRATIONS IN MACHINES, THEIR HARMFUL EFFECTS AND REMEDIES

(A) Causes of Vibrations:

In the machines, some of the important causes of vibrations are given below:

1. Unbalancing of rotating machine components
2. Unbalancing of reciprocating machine components
3. Misalignment of rotating components such as couplings etc.
4. Using simple spur gears for power transmission
5. Using bent shaft
6. Worn out of the moving components such as teeth of the gears, ball or rolling bearings slides, drive belts etc. for power transmission
7. Looseness in transmission components such as belts, chains etc.
8. Looseness in fastenings of the moving components
9. Loose or Non-rigid machine foundations, causing settlement of foundation which may, in turn cause misalignment of machine components
10. Improper isolation of vibrations from other vibrating sources and machines installed nearby
11. Impact taking components of the machines
11. Impact of working fluids in prime movers
13. More material contact between two components such as base plates on the foundations for the pedestal bearings

(B) Harmful Effects of Vibrations:

In the machines, some of the *harmful effects of vibrations* are given below:

1. Reduction in functional life of machine components due to their excessive wearing
2. Excessive stresses and strains are induced in the machine components
3. Failure of machine components due to fatigue failure
4. Reduction in useful life of buildings and structures around the machines
5. Human beings around the machines are subjected to discomfort in the form of physical and mental strains when machines produce heavy noise due to vibration

(C) Remedies

It is impossible to eliminate the vibrations completely, but these can be reduced by adopting various remedies. Some of the remedies are given below:

1. Balancing of unbalanced rotating components
2. Partial balancing of reciprocating components
3. Correcting the misalignment of rotating components and checking it periodically
4. Using helical gears instead of spur gears for power transmission
5. Using straight shaft
6. Timely replacement of worn out moving parts with excessive clearance
7. Proper tightening and locking of fastenings and checking it periodically
8. Isolating vibrations from other sources and machines by installing vibration insulation pads in the machine foundations
9. Making sufficiently strong machine foundations on compact ground so that they do not yield or settle under the load of the machine.

EXERCISE

1. A weightless helical spring elogates 0.15 mm under a load of 12 N and is made to support a mass of 75 kg. The spring-mass system is displaced vertically through 25 mm and released. Determine the natural frequency of longitudinal vibration of the system. **[Ans.** 5.198 Hz]
2. A vertical shaft of 50 mm diameter fixed at both ends, carries a flywheel as shown in Fig. 26. The flywheel has a mass of 500 kg and its radius of gyration is 0.5 m. Find the natural frequency of longitudinal vibrations. Take $E = 200\ \text{GN/m}^2$ for the shaft material. **[Ans.** 235 Hz]

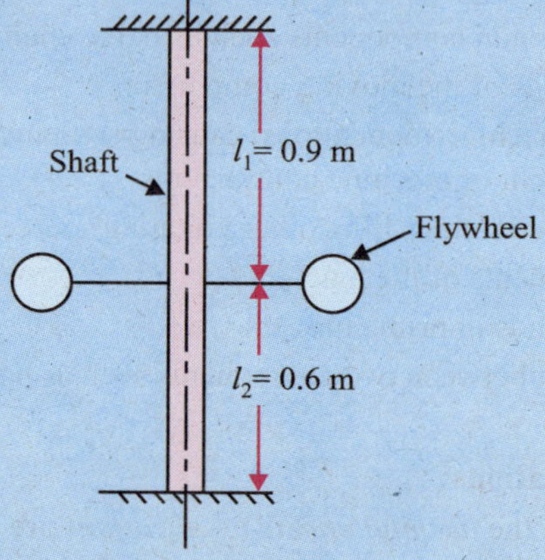

Fig. 26

3. A shaft having 50 mm diameter and 3 metres long is simply supported at its ends. It carries three point loads of 500 N, 750 N, and 375 N at 1 m, 2 m, and 2.5 m from the left support. Find the frequency of transverse vibrations, neglecting the weight of the shaft. Take: $E = 200\ \text{GN/m}^2$ for the shaft material. **[Ans.** 4.96 Hz]
4. A steel shaft 25 mm wide and 50 mm deep is freely supported at two points 1 metre apart. It carries a mass of 100 kg mid-way between them. If $E = 270\ \text{GN/m}^2$ for the shaft material, then find the frequency of natural transverse vibration, neglecting the weight of the bar.

   What will be the frequency of vibration, if an additional mass of 100 kg is distributed uniformly along the length of the shaft ? **[Ans.** 29.24 Hz; 23.93 Hz]

5. A shaft of 50 mm diameter fixed at both ends, carries a flywheel as shown in Fig. 27. The flywheel has a mass of 500 kg and its radius of gyration is 0.5 m. Find the natural frequency of transverse vibrations, neglecting the weight of the shaft. Take $E = 200$ GN/m^2 for the shaft material. **[Ans.** 14.17 Hz]

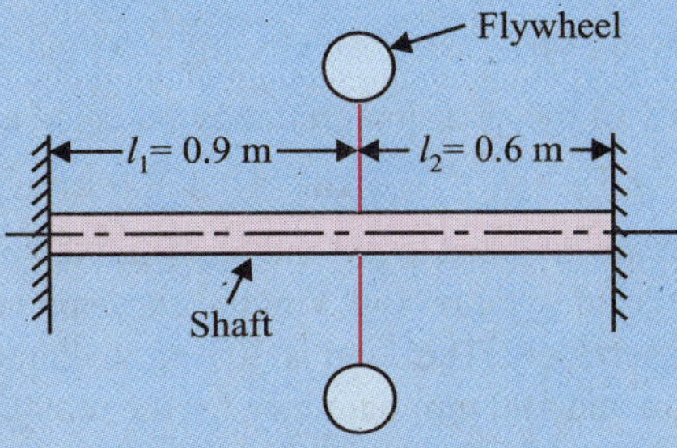

Fig. 27

6. A horizontal shaft of 20 mm diameter is supported at the ends by two bearings and a rotor of mass 12 kg is mounted mid-way on it. The bearing span is 800 mm. Owing to some manufacturing defect, the centre of gravity of the disc is 0.1 mm away from the geometric centre of the rotor. If the system rotates at 3000 r.p.m. If $E = 200$ GN/m^2 for shaft material, then determine

   (*i*) The amplitude of the steady state vibration

   (*ii*) The dynamic force transmitted to the bearing

   [Ans. (*i*) 0.1142 mm (*ii*) 67.27 N]

7. A vertical shaft of 50 mm diameter fixed at both ends, carries a flywheel as shown in Fig. 28. The flywheel has a mass of 500 kg and its radius of gyration is 0.5 m. Find the natural frequency of torsional vibrations. Take $E = 200$ GN/m^2 for the shaft material.

   [Ans. 5.32 Hz]

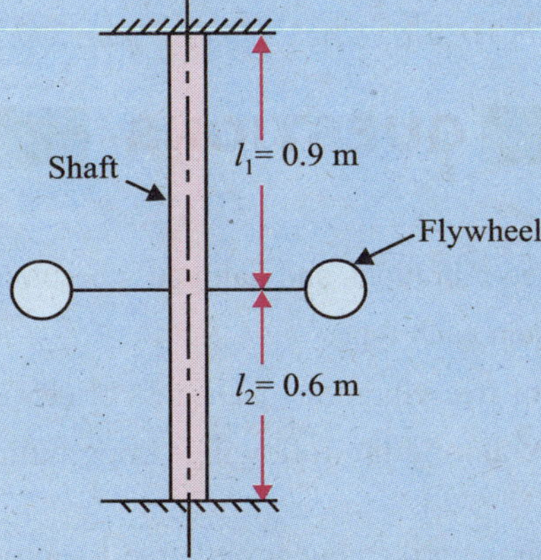

Fig. 28

8. A shaft of 150 mm diameter and 1.5 metre long has one of its ends fixed and the other end carries a disc of mass 750 kg at a radius of gyration of 675 mm. Determine the natural frequency of torsional vibrations. Take $E = 80$ GN/m^2 for the shaft material.

   [Ans. 14 Hz]

9. A steel shaft 2.25 m long is 95 mm in diameter for first 0.9 m of its length, 60 mm in diameter for next 0.75 m of its length and 50 mm in diameter for the remainder of its

length. The shaft carries two flywheels (or rotors) at two ends, the first having a mass of 900 kg and 0.85 m radius of gyration located at 95 mm diameter end and the second having a mass of 700 kg and 0.55 m radius of gyration located at the other end. Determine

(*i*) the location of the node, and

(*ii*) the natural frequency of free torsional vibrations of the system.

The modulus of rigidity of the shaft material may be taken as 80 GN/m^2.

[**Ans.** (*i*) 3.3 m from first flywheel ; (*ii*) 2.75 Hz]

10. The two rotors *A* and *B* are attached to the end of a shaft 0.5 m long. The mass of the rotor *A* is 300 kg and its radius of gyration is 0.3 m; the corresponding values of the rotor *B* are 500 kg and 0.45 m respectively. The shaft is 70 mm in diameter for the first 0.25 m; 120 mm for the next 0.07 m and 100 mm diameter for the remaining length. The modulus of rigidity for the shaft material is 80 GN/m^2. Find

(*i*) The position of the node, and

(*ii*) The natural frequency of free torsional vibrations.

[**Ans.** (*i*) 0.225 m from rotor *A* ; (*ii*) 27.3 Hz]

11. Three rotors *A, B* and *C* having moment of inertia of 2000, 6000 and 3500 kg.m^2 respectively are carried on a uniform shaft of 350 mm diameter. The length of the shaft between the rotors *A* and *B* is 6 m and between *B* and *C* is 32 m. If modulus of rigidity for the shaft material is 80 GN/m^2, then find the natural frequency of the torsional vibrations.

[**Ans.** 6.16 Hz for a *single node system*; 18.27 Hz for a *double node system*]

12. Three rotors *A, B* and *C* having moment of inertia of 0.15, 0.3 and 0.09 kg.m^2 respectively are carried on a uniform shaft of 70 mm diameter. The length of the shaft between the rotors *A* and *B* is 1.5 m and between *B* and *C* is 1 m. If modulus of rigidity for the shaft material is 84 GN/m^2, then find the natural frequency of the torsional vibrations.

[**Ans.** 171 Hz for a *single node system*; 277 Hz for a *double node system*]

QUESTIONS

1. What is vibratory motion ?
2. Define: free vibrations, forced vibrations and damped vibrations.
3. Classify vibrations and explain each type.
4. What are the general types of free vibrations ? Discuss briefly each type with neat sketches.
5. Explain the main difference among the following: Longitudinal, transverse and torsional vibrations.
6. Derive an expression for the natural frequency of free longitudinal vibration for a vertical spring of negligible mass fixed at one end and carrying a mass at other end by

   (*i*) Equilibrium method

   (*ii*) Energy method

   (*iii*) Rayleigh's method
7. Discuss the effect of inertia of the shaft in longitudinal vibrations.

8. Derive an expression for the natural frequency of free transverse vibration for a cantilever shaft of negligible mass fixed at one end and carrying a body (*i.e.*, disc) at other end by

   (*i*) Equilibrium method

   (*ii*) Energy method

   (*iii*) Rayleigh's method

9. Discuss the effect of inertia of the shaft in transverse vibrations.

10. Derive an expression for the natural frequency of free transverse vibration for a simply supported shaft subjected to a number of point loads, by

    (*i*) Energy (or Rayleigh's) method

    (*ii*) Dunkerley's method

11. Define 'whirling speed' or 'critical speed' of a shaft. Prove that the whirling speed for a rotating shaft is the same as the natural frequency of transverse vibration in revolutions per second (r.p.s).

12. Derive an expression for the natural frequency of free torsional vibrations for a shaft of negligible mass fixed at one end and carrying a load on the free end.

13. Assuming the effect of inertia of a shaft, derive an expression for the natural frequency of free torsional vibrations for that shaft.

14. Derive an expression for the natural frequency of free torsional vibrations for a two rotar system.

15. Derive an expression for the natural frequency of free torsional vibrations for a three rotar system.

16. What is meant by tortionally equivanent shaft ? Derive an expression for the equivalent length for a shaft of varying diameters (or stepped shaft).

17. Explain the causes of vibrations in machines, their harmful effects and their remedies.

OBJECTIVE TYPE QUESTIONS

1. The particles of the disc mounted at the free end of a shaft move perpendicular to the its axis of rotation. Such an effect is caused by

   (*a*) Transverse vibrations (*b*) Torsional vibrations

   (*c*) Longitudinal vibrations (*d*) None of these

2. Torsional vibrations are said to occur when the particles of the disc mounted at the free end of a shaft, moves

   (*a*) parallel to its axis (*b*) perpendicular to its axis

   (*c*) in a circle about its axis (*d*) None of these

3. When there is a reduction in amplitude over every cycle of vibration, then the motion of body is said to have

   (*a*) force vibration (*b*) full vibration

   (*c*) damped vibration (*d*) None of these

4. Which of the following statements is *incorrect?*

   (*a*) Free vibrations have external force supplied at the ends only.

(*b*) Vibrations in machines may be due to loose fittings and lack of balance.

(*c*) Forced vibrations are independent of the natural frequency of vibrations.

(*d*) A vibrating system is said to be stable if the amplitude of vibration decreases with time.

5. When a body is subjected to transverse vibrations, the stress induced in a body will be

(*a*) shear stress
(*b*) compressive stress
(*c*) tensile stress
(*d*) None of these

6. Which of the following statements is *incorrect* in the context of transverse vibrations?

(*a*) Shaft is bent alternately.

(*b*) Bending stresses are induced in the shaft.

(*c*) Particles of the shaft move approximately perpendicular to the axis of the shaft.

(*d*) The restoring force need not be directly proportional to the displacement of the particles from the mean position.

7. For the longitudinal and transverse vibrations:

(*a*) Both the time period and the frequency are same

(*b*) Both the time period and frequency are different

(*c*) The time period is same but the natural frequency is different

(*d*) The natural frequency is same but the time period is different

8. In a spring-mass system, the mass of the system is made doubled and the spring stiffness is half. The natural frequency of longitudinal vibrations

(*a*) is doubled
(*b*) is halved
(*c*) is quadrupled
(*d*) remains unaffected.

9. Which type of vibration is induced when the stretched string of a guitar is operated ?

(*a*) Free longitudinal
(*b*) Forced transverse
(*c*) Torsional
(*d*) Non-linear vibration.

10. The whipping speed of a rotating shaft carrying a mass at the centre of its span is

(*a*) equal to the natural frequency of transverse vibrations

(*b*) less than the natural frequency of transverse vibrations

(*c*) more than the natural frequency of transverse vibrations

(*d*) None of these

11. Which of the following statements is *incorrect?*

(*a*) The whirling speed of shaft can be reduced by decreasing the shaft stiffness.

(*b*) The secondary critical speed of shaft occurs at twice the primary critical speed.

(*c*) At critical speed the shaft becomes dynamically unstable with large lateral amplitude.

(*d*) The critical speed of a shaft is affected by eccentricity, span of shaft and the disc diameter.

12. The factor(s) which affect(s) the critical speed of the shaft is/are

(*a*) eccenticity
(*b*) diameter of the disc
(*c*) span of the shaft
(*d*) All of these

13. If a shaft is rotating at a speed less than the critical speed, the phase difference between displacement and centrifugal force would be:

(*a*) 0° (*b*) 45°
(*c*) 90° (*d*) 180°

14. If a shaft is rotating at a speed greater than the critical speed, the phase difference between displacement and centrifugal force would be:

(*a*) 0° (*b*) 45°
(*c*) 90° (*d*) 180°

15. A circular disc of moment of inertia I is attached to the free end of a vertical shaft. If torsional stiffness of shaft is q and mass of shaft is neglected, the natural frequency of torsional vibrations will be:

(*a*) $2\pi\sqrt{q \cdot I}$ (*b*) $2\pi\sqrt{\frac{q}{I}}$

(*c*) $\frac{1}{2\pi}\sqrt{\frac{I}{q}}$ (*d*) $\frac{1}{2\pi}\sqrt{\frac{q}{I}}$

16. The amplitude of torsional vibration at a nodal point in a shaft is:

(*a*) zero (*b*) maximum
(*c*) minimum (*d*) None of these

17. In three-rotor system, the amplitude of torsional vibration is maximum

(*a*) at ends of the shaft (*b*) at nodal point(s) of the shaft
(*c*) in between the shaft (*d*) None of these

18. A shaft with two rotors at its ends will have :

(*a*) no node (*b*) one node
(*c*) two nodes (*d*) three nodes

19. A shaft with three rotors will have :

(*a*) zero node (*b*) one node
(*c*) two nodes (*d*) either 'a' or 'b'

20. A shaft of varying diameters d_1, d_2 and d_3 over lengths l_1, l_2 and l_3 respectively, has two rotors at the ends. If this shaft is replaced by a torsionally equivalent shaft of length l_e and of uniform diameter d, then length of the equivalent shaft assuming $d = d_1$ is

(*a*) $l_e = l_1 + l_2 + l_3$ (*b*) $l_e = d_1 + d_2\left(\frac{l_1}{l_2}\right)^4 + d_3\left(\frac{l_1}{l_3}\right)^4$

(*c*) $l_e = l_1 + l_2\left(\frac{d_1}{d_2}\right)^4 + l_3\left(\frac{d_1}{d_3}\right)^4$ (*d*) $l_1 + \frac{l_1}{l_2}\left(\frac{d_1}{d_2}\right)^4 + \frac{l_1}{l_3}\left(\frac{d_1}{d_3}\right)^4$

ANSWERS

1. (*a*)	2. (*c*)	3. (*c*)	4. (*a*)	5. (*c*)	6. (*d*)
7. (*a*)	8. (*b*)	9. (*b*)	10. (*c*)	11. (*c*)	12. (*d*)
13. (*a*)	14. (*d*)	15. (*d*)	16. (*a*)	17. (*c*)	18. (*b*)
19. (*d*)	20. (*c*)				

APPENDIX

List of Appendix

Topics

1. Units and Conversions.

2. Properties of Plane Areas.

Units and Conversions

Fundamental Units

The measurement of physical quantities is one of the most important operations in engineering. Every quantity is measured in terms of some arbitrary, but internationally accepted units, called *fundamental units.*

All the physical quantities, met with in Engineering Mechanics, are expressed in terms of three fundamental quantities, *i.e.*

1. Length **2. Mass** **3. Time.**

Derived Units

Sometimes, the units are also expressed in other units (which are derived from fundamental units) known as *derived units e.g.* units of area, velocity, acceleration, pressure etc.

Systems of Units

There are only four systems of units, which are commonly used and universally recognised. These are known as :

1. C.G.S. units **2. F.P.S. units** **3. M.K.S. units** **4. S.I. units.**

In this book, we shall use only the S.I. system of units, as the future courses of studies are conduced in this system of units only.

S.I. Units (International System of Units)

The eleventh *General Conference of Weights and Measures has recommended a unified and systematically constituted system of fundamental and derived units for international use. This system of units is now being used in many countries.

In India, the Standards of Weights and Measures Act of 1956 (vide which we switched over to M.K.S. units) has been revised to recognise all the S.I. units in industry and commerce.

S.I. Units

S.I. Units are divided into three classes:

1. Base Units.
2. Derived Units.
3. Supplementary Units.

Table 1. S.I. Base Units

Quantity		S.I. Unit	
Name	Symbol	Name	Symbol
length		metre	m
mass		kilogram	kg
time		second	s
electric current		ampere	A
thermodynamic temperature		kelvin	K
luminious intensity		candela	cd
amount of substance		mole	mol

Table 2. S.I. Derived Units Expressed in Terms of Base Units

Quantity		S.I. Unit	
Name	Symbol	Name	Symbol
area		square metre	m^2
volume		cubic metre	m^3
		litre	L
speed, velocity		metre per second	m/s
acceleration		metre per second squared	m/s^2
wave number		1 per metre	m^{-1}
density, mass density		kilogram per cubic metre	kg/m^3
concentration (of amount of substance)		mole per cubic metre	mol/m^3
activity (radioactive)		1 per second	s^{-1}
specific volume		cubic metre per kilogram	m^3/kg
luminance		candela per square metre	cd/m^2

Table 3. S.I. Derived Units with Special Names

Quantity		S.I. Units			
Name	Symbol	Name	Symbol	In S.I. other units	In S.I. base units
frequency	v	hertz	Hz	—	s^{-1}
force	F	newton	N	—	m-kg-s^{-2}
pressure	P	pascal	Pa	N/m^2	m^{-1}-kg-s^{-2}
energy	E	joule	J	N-m	m^2-kg-s^{-2}
work	W				
quantity of heat power	P				
power	P	watt	W	J/S	m^2-kg-s^{-3}
electric charge	e	coulomb	C	A-s	s-A
electric tension	V	volt	V	W/A	m^2-kg-s^{-3}-A^{-1}
capacitance	C	farad	F	C/V	m^{-2}-kg^{-1}-s^4
electric resistance	R	ohm	Ω	V/A	m^2-kg-s^{-3}-A^{-2}
conductance		siemens	S	A/V	m^{-2}-kg^{-1}-s^3-A^2
magnetic flux		weber	Wb	V.S.	m^2-kg-s^{-2}-A^{-1}
magnetic flux density			tesla	T	Wb/m^2 kg-s^{-2}-A^{-}
1					
magnetic induction					
inductance	I	henry	H	Wb/A	m^2-kg.s^{-2}-A^{-2}
luminous flux		lumen	lm	—	cd-sr
illuminance		lux	lx	—	m^{-2}-cd-sr
atomic mass		atomic mass unit	u or amu		
molarity	M	molar	M		
normality	N	normal	N		

Table 4. **Examples of S.I. Derived Units Expressed in Terms of Special Names**

Quantity		S.I. Units		
Name	Symbol	Name	Symbol	In S.I. base units
dynamic viscosity		pascal second	Pa-s	m^{-1}-kg-s^{-1}
moment of force		metre newton	N-m	m^2-kg-s^{-2}
surface tension		newton per metre	N/m	kg-s^{-2}
heat flux density, irradiance		watt per square metre	W/m^2	kg-s^{-2}
heat capacity, entropy		joule per kelvin	J/K	m^2-kg-s^{-2}-K^{-1}
specific heat capacity, specific entropy		joule per kilogram kelvin	J/(kg-K)	m^2-s^{-2}-K^{-1}
specific energy		joule per kilogram	J/kg	m^2-s^{-2}
thermal conductivity		watt per metre kelvin	W/(m-K)	m-kg-s^{-3}-K^{-1}
energy density		joule per cubic metre	J/m^3	m^{-1}-kg-s^{-2}
electric field strength		volt per metre	V/m	m-kg-s^{-3}-A^{-1}
electric charge density		coulomb per cubic metre	C/m^3	m^{-3}-s-A
electric flux density		coulomb per square metre	C/m^2	m^{-2}-s-A
permittivity		farad per metre	F/m	m^{-3}-kg^{-4}-s^4-A^4
current density		ampere per square metre	A/m^2	—
magnetic field strength		ampere per metre	A/m	—
permeability		henry per metre	H/m	m-kg-s^{-2}- A^{-2}
molar energy		joule per mole	J/mol	m^2-kg-s^{-2}- mol^{-1}
molar heat capacity		joule per mole kelvin	J/(mol-K)	m^2-kg-s^{-2}-K^{-1}-mol^{-1}

The S.I. units assigned to third class called "Supplementary Units" may be regarded either as base units or as derived units. Refer Tables 5 and 6.

Table 5. **S.I. Supplementary Units**

Quantity		S.I. Units	
Name	Symbol	Name	Symbol
plane angle	$\alpha, \beta, \gamma, \theta$	radian	rad
solid angle		steradian	sr

Table 6. **S.I. Derived Units Formed by Using Supplementary Units**

Quantity		S.I. Units	
Name	Symbol	Name	Symbol
angular velocity		radian per second	rad/s
angular acceleration		radian per second squared	rad/s^2
radiant intensity		watt per steradian	W/sr
radiance		watt per square metre steradian	W/(m^2- sr)

Table 7. S.I. Prefixes

Factor	Prefix	Symbol	Factor	Prefix	Symbol
10^{12}	tera	T	10^{-1}	deci	d
10^{9}	giga	G	10^{-2}	centi	c
10^{6}	mega	M	10^{-3}	milli	m
10^{3}	kilo	k	10^{-6}	micro	μ
10^{2}	hecto	h	10^{-9}	nano	n
10^{1}	deca	da	10^{-12}	pico	p
			10^{-15}	fasnto	f
			10^{-18}	atto	a

Conversion Factors

1. Area

$$1\ \text{m}^2 = 10^4\ \text{cm}^2 = 10^6\ \text{mm}^2$$
$$1\ \text{cm}^2 = 10^{-4}\ \text{m}^2 = 10^2\ \text{mm}^2$$
$$1\ \text{mm}^2 = 10^{-2}\ \text{cm}^2 = 10^{-6}\ \text{m}^2$$

2. Volume

$$1\ \text{m}^3 = 10^6\ \text{cm}^3 = 1000\ \text{L}$$
$$1\ \text{cm}^3 = 10^{-6}\ \text{m}^3 = 10^{-3}\ \text{L}$$
$$1\ \text{L} = 10^{-3}\ \text{m}^3 = 10^3\ \text{cm}^3$$

3. Mass

$$1\ \text{kg} = 1000\ \text{g} = 6.02 \times 10^{26}\ \text{u} = 6.85 \times 10^{-2}\ \text{slug}$$
$$1\ \text{g} = 10^{-3}\ \text{kg} = 6.02 \times 10^{23}\ \text{u} = 6.85 \times 10^{-5}\ \text{slug}$$
$$1\ \text{u} = 1.66 \times 10^{-27}\ \text{kg} = 1.66 \times 10^{-24}\ \text{g} = 1.138 \times 10^{-28}\ \text{slug}$$
$$1\ \text{slug} = 14.59\ \text{kg} = 14.59 \times 10^3\ \text{g} = 8.78 \times 10^{27}\ \text{u}$$

4. Density

$$1\ \text{kg/m}^3 = 10^{-3}\ \text{g/cm}^3$$
$$1\ \text{g/cm}^3 = 10^3\ \text{kg/m}^3$$

5. Force (in Absolute Units)

$$1\ \text{N} = 1\ \text{kg-m/s}^2 = 10^3\ \text{g-m/s}^2 = 10^5\ \text{g-cm/s}^2 = 10^5\ \text{dyne}$$
$$1\ \text{dyne} = 1\ \text{g-cm/s}^2 = 10^{-5}\ \text{N} = 10^{-5}\ \text{kg-m/s}^2$$

Force (in Gravitational Units)

$$1\ \text{N} = 10^5\ \text{dyne} = 0.102\ \text{kg(f)} = 0.102 \times 10^{-3}\ \text{t(f)}$$
$$1\ \text{dyne} = 10^{-5}\ \text{N} = 0.102 \times 10^{-5}\ \text{kg(f)} = 0.102 \times 10^{-8}\ \text{t(f)}$$
$$1\ \text{kg(f)} = 9.81\ \text{N} = 9.81 \times 10^5\ \text{dyne} = 10^{-3}\ \text{t(f)}$$
$$1\ \text{t(f)} = 10^3\ \text{kg(f)} = 9.81 \times 10^3\ \text{N} = 9.81\ \text{kN} = 9.81 \times 10^8\ \text{dyne}$$

Engineering Constants and Expressions in S.I. Units

Engineering Constants & Expressions	M.K.S. System	S.I. Units
1. Value of g	9.81 kg-m/kg(f)-s^2	1 kg-m/N-s^2
2. Universal gas constant (R_0)	848 kg(f)-m/kg-mole-K	$848 \times 9.81 = 8314$ J/kg-mole-K
		(1 kg(f)-m = 9.81 J)
3. Gas constant (R)	29.27 kg(f)-m/kg-K	$\frac{8314}{29} = 287$ J/kg-K
	for air	for air
4. Specific heat (for air)	$c_v = 0.17$ kcal/kg-K	$c_v = 0.17 \times 4.184$
		$= 0.71128$ kJ/kg-K
	$c_p = 0.24$ kcal/kg-K	$c_p = 0.24 \times 4.184$
		$= 1$ kJ/kg-K
5. Flow through nozzle-exit velocity (C_2)	$91.5\sqrt{U}$ where U is in kcal	$44.7\sqrt{U}$ where U is in kJ
6. Refrigeration 1 ton	= 50 kcal/min	= 210 kJ/min
7. **Heat transfer**		
The Stefan Boltzman	$Q = \sigma T^4$ kcal/m^2-h	$Q = \sigma T^4$ W/m^2-h
Law is given by:	when $\sigma = 4.9 \times 10^{-8}$ kcal/h-m^2-K^4	when $\sigma = 5.67 \times 10^{-8}$ W/m^2-K^4

Properties of Plane Areas

Notatión:

A = Area

$\bar{x}, \bar{y}$ = Position of Centroid C of the area

= Distances of Centroid C of the area from edges of the area

I_x, I_y = Moments of Inertia with respect to the x and y axes, respectively

I_{xy} = Product of Inertia with respect to the x and y axes

$I_P = I_x + I_y$ = Polar Moment of Inertia with respect to the origin of the x and y axes

I_{BB} = Moment of Inertia with respect to axis B-B

For Students - Simple Areas

☐

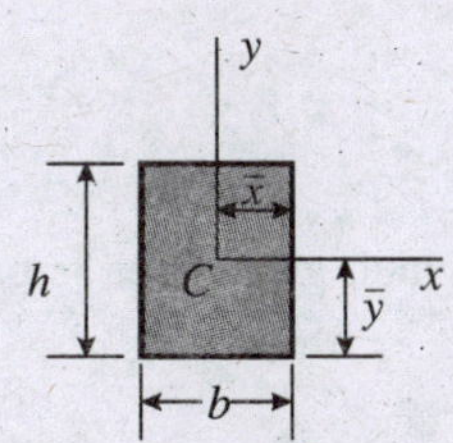

Rectangle (Origin of axes at centroid C of given area)

$$A = bh \qquad \bar{x} = \frac{b}{2} \qquad \bar{y} = \frac{h}{2}$$

$$I_x = \frac{bh^3}{12} \qquad I_y = \frac{hb^3}{12} \qquad I_{xy} = 0 \qquad I_P = \frac{bh}{12}\left(h^2 + b^2\right)$$

☐

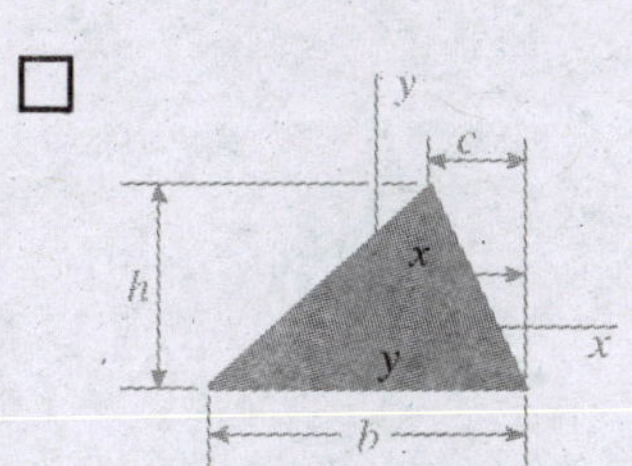

Triangle (Origin of axes at centroid C of given area)

$$A = \frac{bh}{2} \qquad \bar{x} = \frac{b+c}{3} \qquad \bar{y} = \frac{h}{3}$$

$$I_x = \frac{bh^3}{36} \qquad I_y = \frac{bh}{36}\left(b^2 - bc + c^2\right)$$

$$I_{xy} = \frac{bh^2}{72}(b - 2c) \qquad I_P = \frac{bh}{36}\left(h^2 + b^2 - bc + c^2\right)$$

☐

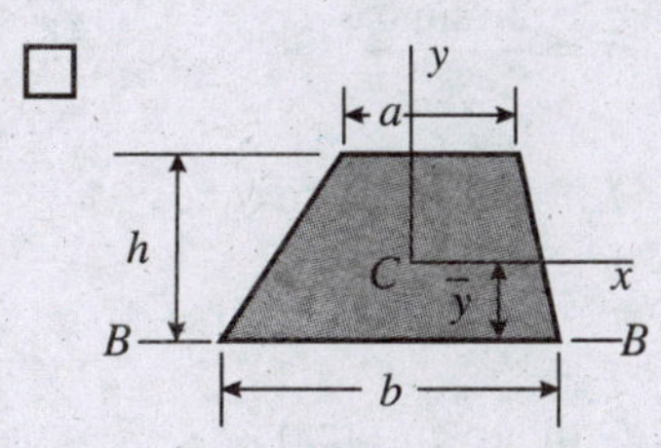

Trapezoid (Origin of axes at centroid C of given area)

$$A = \frac{h(a+b)}{2} \qquad \bar{y} = \frac{h(2a+b)}{3(a+b)}$$

$$I_x = \frac{h^3\left(a^2 + 4ab + b^2\right)}{36(a+b)} \qquad I_{BB} = \frac{h^3(3a+b)}{12}$$

☐

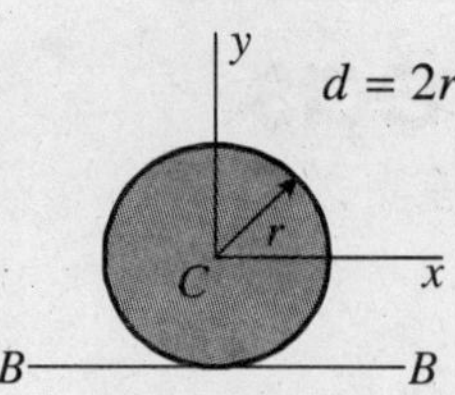

Circle (Origin of axes at centroid C of given area)

$$A = \pi r^2 = \frac{\pi d^2}{4} \qquad \bar{x} = r \qquad \bar{y} = r$$

$$I_x = I_y = \frac{\pi r^4}{4} = \frac{\pi d^4}{64}$$

$$I_{xy} = 0 \qquad I_P = \frac{\pi r^4}{2} = \frac{\pi d^4}{32} \qquad I_{BB} = \frac{5\pi r^4}{4} = \frac{5\pi d^4}{64}$$

☐

Semicircle (Origin of axes at centroid C of given area)

$$A = \frac{\pi r^2}{2} \qquad \bar{y} = \frac{4r}{3\pi}$$

$$I_x = \frac{(9\pi^2 - 64) r^4}{72\pi} \approx 0.11\, r^4 \qquad I_y = \frac{\pi r^4}{8}$$

$$I_{xy} = 0 \qquad I_{BB} = \frac{\pi r^4}{8}$$

☐

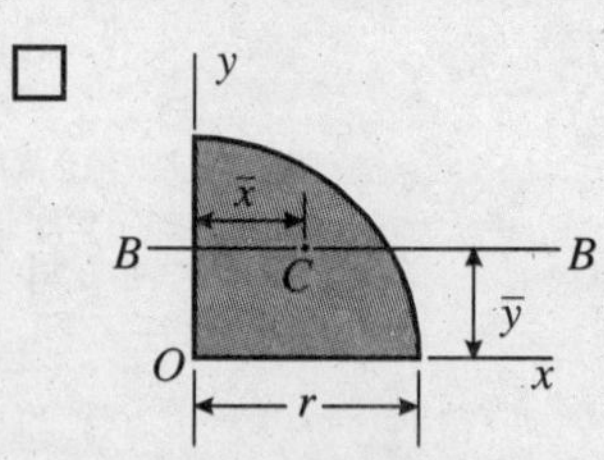

Quarter Circle

(Origin of axes at center O of circle of which given area is part)

$$A = \frac{\pi r^2}{4} \qquad \bar{x} = \bar{y} = \frac{4r}{3\pi}$$

$$I_x = I_y = \frac{\pi r^4}{16} \qquad I_{xy} = \frac{r^4}{8}$$

$$I_{BB} = \frac{(9\pi^2 - 64) r^4}{144\,\pi} \approx 0.055\, r^4$$

☐

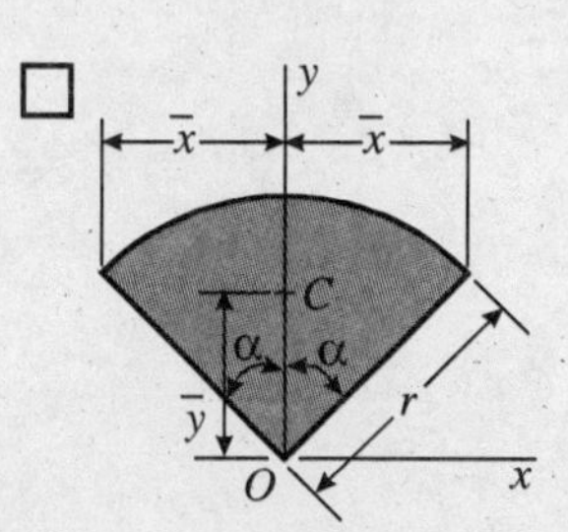

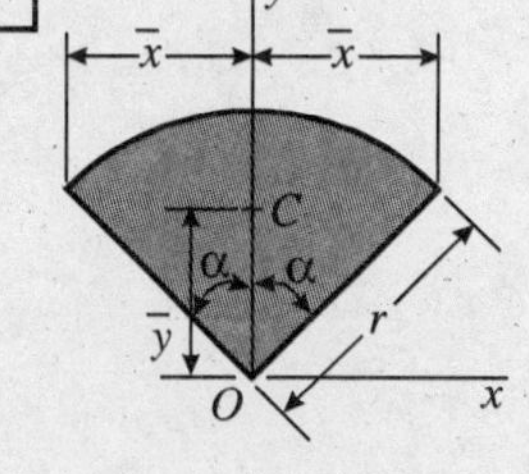

Circular sector

(Origin of axes at center O of circle of which given area is part)

α = angle in radians ...(for $\alpha \le \pi/2$)

$$A = \alpha r^2 \qquad \bar{x} = r \sin\alpha \qquad \bar{y} = \frac{2r \sin\alpha}{3\alpha}$$

$$I_x = \frac{r^4}{4}(\alpha + \sin\alpha\cos\alpha) \qquad I_y = \frac{r^4}{4}(\alpha - \sin\alpha\cos\alpha)$$

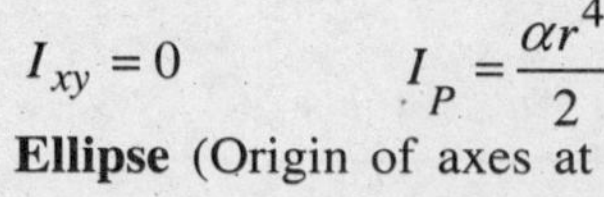

$$I_{xy} = 0 \qquad I_P = \frac{\alpha r^4}{2}$$

☐

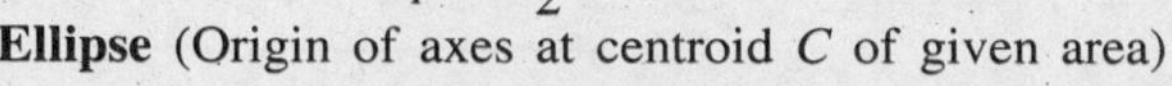

Ellipse (Origin of axes at centroid C of given area)

$$A = \pi ab \qquad \bar{x} = a \qquad \bar{y} = b$$

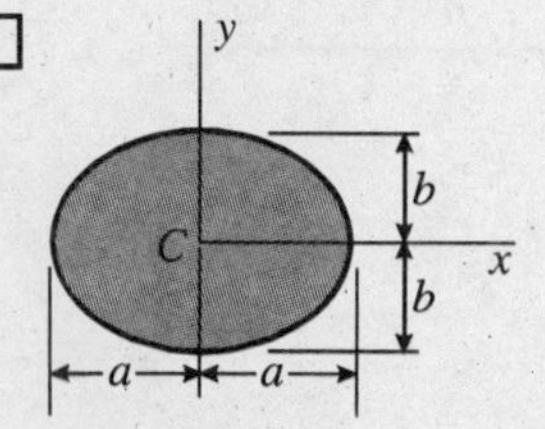

$$I_x = \frac{\pi ab^3}{4} \qquad I_y = \frac{\pi ba^3}{4}$$

$$I_{xy} = 0 \qquad I_P = \frac{\pi ab}{4}(b^2 + a^2)$$

Circumference $\approx \pi\left[1.5(a+b) - \sqrt{ab}\right]$ ...(for $a/3 \le b \le a$)

$$\approx 4.17\, b^2 / a + 4a$$

For Engineering Professionals

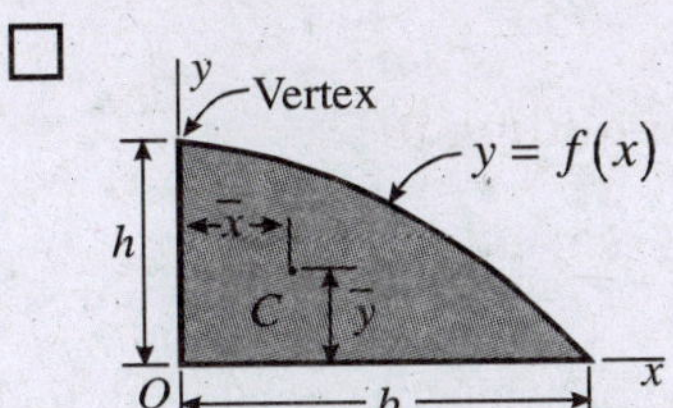

Parabolic Semi-segment or Semi Parabola

(Origin of axes at corner O)

$$y = f(x) = h\left(1 - \frac{x^2}{b^2}\right)$$

$$A = \frac{2bh}{3} \qquad \bar{x} = \frac{3b}{8} \qquad \bar{y} = \frac{2h}{5}$$

$$I_x = \frac{16bh^3}{105} \qquad I_y = \frac{2hb^3}{15} \qquad I_{xy} = \frac{b^2h^2}{12}$$

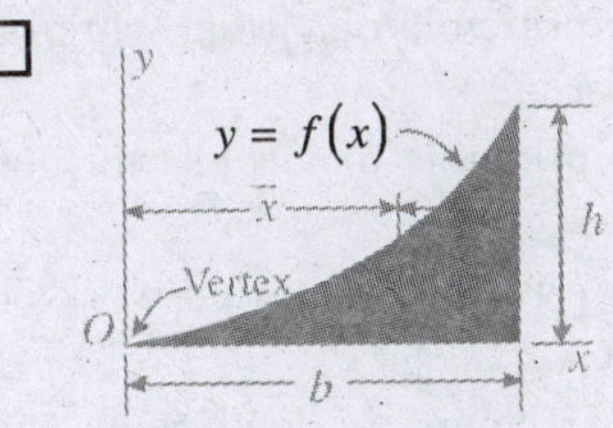

Parabolic Spandrel

(Origin of axes at corner or vertex O or pont of tangency)

$$y = f(x) = \frac{hx^2}{b^2}$$

$$A = \frac{bh}{3} \qquad \bar{x} = \frac{3b}{4} \qquad \bar{y} = \frac{3h}{10} \qquad \bar{x}' = \frac{b}{4}$$

$$I_x = \frac{bh^3}{21} \qquad I_y = \frac{hb^3}{5} \qquad I_{xy} = \frac{b^2h^2}{12}$$

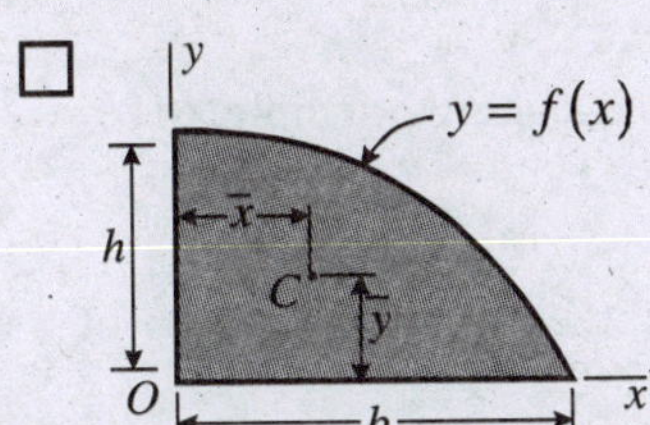

Semi-Segment of *n* th degree (Origin of axes at corner O)

$$y = f(x) = h\left(1 - \frac{x^n}{b^n}\right) \qquad \ldots(\text{for } n > 0)$$

$$A = bh\left(\frac{n}{n+1}\right) \qquad \bar{x} = \frac{b(n+1)}{2(n+2)} \qquad \bar{y} = \frac{hn}{2n+1}$$

$$I_x = \frac{2bh^3n^3}{(n+1)(2n+1)(3n+1)} \qquad I_y = \frac{hb^3n}{3(n+3)}$$

$$I_{xy} = \frac{b^2h^2n^2}{4(n+1)(n+2)}$$

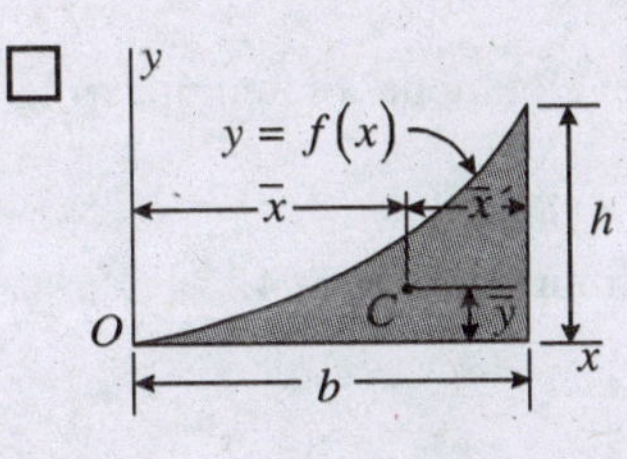

Spandrel of *n* th degree (Origin of axes at point of tangency O)

$$y = f(x) = \frac{hx^n}{b^n} \qquad (n > 0)$$

$$A = \frac{bh}{n+1} \qquad \bar{x} = \frac{b(n+1)}{n+2} \qquad \bar{y} = \frac{h(n+1)}{2(2n+1)} \qquad \bar{x}' = \frac{b}{n+2}$$

$$I_x = \frac{bh^3}{3(3n+1)} \qquad I_y = \frac{hb^3}{n+3} \qquad I_{xy} = \frac{b^2h^2}{4(n+1)}$$

INDEX

Engineering Mechanics

(Alphabetical Sequencing of Articles corresponding to Page Numbers)

A

B

C

D

K

L

M

N

O

P

R

S

T

U

V

W